Theory of Waveguides and Transmission Lines

Theory of Waveguides and Transmission Lines

Edward F. Kuester

CRC Press
Taylor & Francis Group
Boca Raton London New York

CRC Press is an imprint of the
Taylor & Francis Group, an **informa** business

First edition published 2021
by CRC Press
6000 Broken Sound Parkway NW, Suite 300, Boca Raton, FL 33487-2742

and by CRC Press
2 Park Square, Milton Park, Abingdon, Oxon, OX14 4RN

© 2021 Taylor & Francis Group, LLC

CRC Press is an imprint of Taylor & Francis Group, LLC

ISBN: 978-1-4987-3087-7 (hbk)
ISBN: 978-0-367-54044-9 (pbk)
ISBN: 978-1-315-37004-0 (ebk)

Typeset in CMR
by Nova Techset Private Limited, Bengaluru & Chennai, India

**Visit the Taylor & Francis Web site at
http://www.taylorandfrancis.com**

**and the CRC Press Web site at
http://www.crcpress.com**

Dedication

To Nancy, for whom reciprocity always holds

Contents

Preface

The following story is true. There was a little boy, and his father said, "Do try to be like other people. Don't frown." And he tried and tried, but could not. So his father beat him with a strap; and then he was eaten up by lions. Reader, if young, take warning by his sad life and death. . . . if you are different, you had better hide it, and pretend to be solemn and wooden-headed. Until you make your fortune. . . . In particular, if you are going to write a book, . . . be rigorous, that will cover a multitude of sins. And do not frown.

—Oliver Heaviside,
Electromagnetic Theory, volume III.
London: "The Electrician" Printing
and Publishing Company, 1912, p. 1.

This book is a treatment of the concepts underlying the functioning of waveguides and transmission lines, both on their own and as parts of a larger system. Although many specific structures are covered, the aim is to give an understanding of basic principles rather than an encyclopedic listing of all such transmission channels. Likewise, several topics, such as anisotropic media, while undeniably important, have been omitted in the interest of focusing on fundamental ideas in as clear a manner as possible. The reader interested in probing the subject at greater breadth or depth (or in examining its history) will find extensive references to the technical literature at the end of each chapter. Even these are not exhaustive, but represent sources that the author has found particularly enlightening on a variety of topics. They can be thought of as a sort of tourist guide, indicating the sites that the author has particularly enjoyed visiting.

The first eight chapters of this book deal with various types of electromagnetic waveguides and transmission lines—particularly with their mode properties. Chapters 1–4 deal with so-called "classical transmission lines", which are distributed networks intermediate between ordinary lumped-element circuits and full electromagnetic field descriptions of waveguides. The basic properties of classical transmission lines are the same as those of any waveguide, so we will study them, their connections to lumped-element networks and the distortion of pulses by them at some length before going on to the full field analysis of waveguide modes. Chapters 5–8 examine the modes of various specific kinds of waveguide from the point of view of their electromagnetic fields, including traditional hollow waveguides, dielectric (including optical) waveguides, printed transmission lines such as microstrip, and more. As the treatment progresses, it is hoped that the features common to all varieties of waveguide will begin to be apparent. With this background, the stage is set for the last part of the book (Chapters 9–13), in which the problems of excitation and scattering of waveguide modes, their coupling and resonant properties are addressed. The organizing principle here is the Lorentz reciprocity theorem and its variants, which provide a unifying theme for the analysis of waveguides as interconnecting portions of real systems, and shows how they perform various functions in these systems.

Throughout the book, the emphasis is on analytical description of the physical phenomena, rather than extensive presentation of numerically computed data. Of course, modern electromagnetic modeling software makes possible highly accurate simulation of a great many waveguiding structures, but the author believes that the analytical formulas make possible an insight that is more difficult to arrive at through numerical modeling, and the latter is always available to the engineer who wishes to perform more precise design.

The reader is assumed to have had an initial exposure to electromagnetic fields (typically one year at the junior level in a US undergraduate curriculum) as well as a background in electric circuit analysis and a basic introduction to transmission line analysis. These topics are reviewed briefly in this book as needed for our purposes, but the reader is urged to refer to one of several texts on electromagnetic fields recommended in the bibliographical notes at the end of Chapter 1, and to one of the texts in circuit analysis cited at the end of Appendix A for suitable foundation material in these subjects. Many other books to which reference is made at the end of each chapter are listed in the Bibliography following the Appendices. References to journal articles are made only at the end of chapters, where full citations to specific papers are given.

The book has benefited greatly from comments by Dr. Christopher Holloway, Profs. David Chang, Zoya Popović, John Dunn, the late Carl Johnk and the late Branko Popović. The author would also like to express his gratitude to the Electromagnetics Laboratory at the Technical University of Delft, The Netherlands, and to the Laboratoire d'Électromagnetisme et d'Acoustique at the École Polytechnique Fédérale de Lausanne, Switzerland, for their hospitality during his temporary visits with them. Some of the material in this book was developed in the generous atmospheres provided by these groups. Finally, the author is grateful to the hundreds of students in the graduate course on waveguides and transmission lines at the University of Colorado Boulder who have endured since 1980 the evolution of the course notes that have become this book. Their questions and feedback have helped the quality of presentation immeasurably, and as a result the book became better every year. The author would welcome receiving any corrections or improvements that future readers may suggest. But at some point the work must be declared complete, whatever improvements may still be possible. Kirsten finished her book; it is high time that I finished mine.

Author

Edward F. Kuester received a BS from Michigan State University, East Lansing in 1971, and MS and PhD degrees from the University of Colorado at Boulder, in 1974 and 1976, respectively, all in electrical engineering.

Since 1976, he has been with the Department of Electrical, Computer and Energy Engineering at the University of Colorado at Boulder, where he is currently a Professor Emeritus. In 1979, he was a Summer Faculty Fellow at the Jet Propulsion Laboratory, Pasadena, CA. From 1981–1982, he was a Visiting Professor at the Technische Hogeschool, Delft, The Netherlands. In 1992 and 1993, Dr. Kuester was an Invited Professor at the École Polytechnique Fédérale de Lausanne, Switzerland. He has held the position of Visiting Scientist at the National Institute of Standards and Technology (NIST), Boulder, CO in 2002, 2004 and 2006. His research interests include the modeling of electromagnetic phenomena of guiding and radiating structures, metamaterials, applied mathematics and applied physics. He is the coauthor of two books, author or co-author of chapters in six others, and has translated two books from Russian. He is co-holder of two US patents, and author or coauthor of more than 100 papers in refereed technical journals.

Dr. Kuester is a Life Fellow of IEEE, and a member of URSI Commission B.

1 MODES OF A CLASSICAL TRANSMISSION LINE

1.1 INTRODUCTION

In the broadest sense, all of electrical engineering deals with some sort of guided-wave system. Maxwell's equations and Poynting's theorem indicate to us that, even in an elementary low-frequency circuit, energy is stored and power is transferred, not by wires or circuit elements, but by the electromagnetic fields surrounding them. The configurations themselves serve merely to "arrange" or "guide" the fields in an advantageous manner. As operating frequency increases, this physical picture remains the same, but the quasi-static approximation of circuit theory which was useful at lower frequencies breaks down. Moreover, the electrical parameters of materials generally change at higher frequencies (conductors become more lossy, atomic and molecular vibrations begin to influence the dielectric properties, and so on). Circuits designed on quasi-static principles no longer operate efficiently, and new structures, along with new methods to analyze them, must be found.

In this same broad context, of course, antennas could also be thought of as guided-wave structures, especially so if designed with a highly directional pattern. The advantage of an antenna in a communication system is that no structure need be erected over a long distance. However, even in the best antenna systems it is difficult to achieve a really good degree of "guidance", and tens or hundreds of dB of signal may be lost as a result. This is in part because no actual antenna guides a signal in a single direction exclusively, and some sort of intermediate structure must be provided along the way if most or all of the signal is to reach its destination. If such a structure had no material losses, it could in principle guide all the energy from a source at one end to a load at the other, in sharp contrast to even a lossless antenna system. It is such structures as this which we have in mind when we talk about waveguides, and which is the subject of this book.

Of course, the distinction is not a precise one. If an open waveguide (such as a two-wire transmission line) changes direction or encounters a discontinuity, energy can be radiated as with an antenna. Therefore, for our purposes a *waveguide* will be defined to be any arrangement of matter in space whose electrical properties (permittivity ϵ, conductivity σ, permeability μ) and geometrical structure vary neither with the time t, nor with some given direction in space, say the z-direction. This *axial* or *longitudinal* direction may be part of a Cartesian (x, y, z) or cylindrical (ρ, ϕ, z) system of coordinates. More general coordinate systems (u_1, u_2, z) can also be used, and are sometimes invoked to deal with special situations. We have in mind *artificial* structures made for the purpose of guiding waves, but there are naturally occurring cases of waveguides (such as the so-called earth-ionosphere waveguide) as well. We will not explicitly deal with natural waveguides in this book.

You may already be familiar with some types of waveguides: the rectangular metallic waveguide (Figure 1.1(a)), the circular coaxial waveguide (Figure 1.1(b)), and the two-wire transmission line (Figure 1.1(c)). Perhaps less familiar are planar guiding structures and those involving dissimilar dielectrics, such as optical fibers, dielectric channel waveguides, microstrip transmission lines and fin-lines (Figures 1.2(a)-(d)). Each of these was developed in response to a particular need to provide wave guidance for a certain frequency range or under other special conditions. Additionally, each is also tailored in response to various technological constraints or requirements (such as the development of integrated microwave circuits).

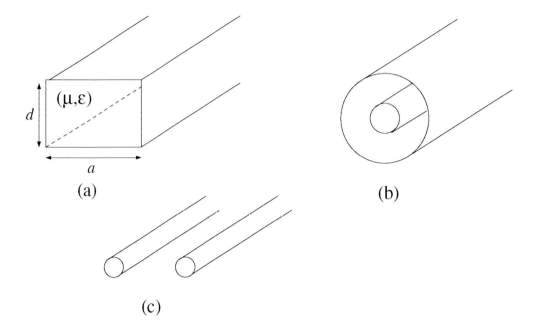

Figure 1.1   (a) Rectangular metallic waveguide; (b) circular coaxial waveguide; (c) two-wire transmission line.

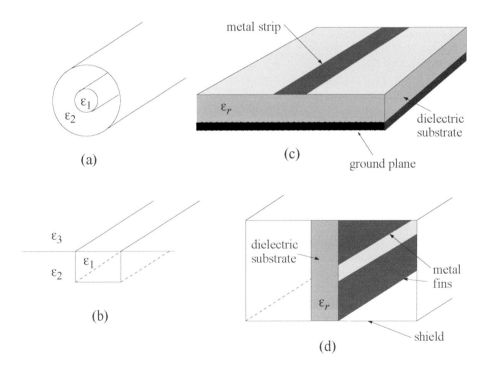

Figure 1.2   (a) Optical fiber waveguide; (b) dielectric channel waveguide; (c) microstrip; (d) finline.

According to our definition, a waveguide is infinitely long and has a constant cross-section. In practice, of course, only finite lengths of waveguides are actually used; we will refer to these as *sections* or *segments* of waveguide. Before we can treat segments of waveguide, however, we must first study the properties of infinitely long waveguides. In this chapter, we will do this for a simple but important structure—the so-called *classical transmission line*. In Chapters 5-8, we will look at more elaborate types of waveguiding structures.

1.2   CLASSICAL TRANSMISSION LINE (DISTRIBUTED CONSTANT MODEL)

The most readily understood waveguide is perhaps the classical transmission line. This line is not really associated with any particular physical structure, but rather serves as an approximate model for a wide variety of waveguides met with in practice. The classical line will also be an important concept in the exact representation of fields in *any* waveguide, as will be seen in Chapter 10, and serves as an analogy for virtually *any* form of one-dimensional wave propagation.

The classical transmission line can be described by a distributed circuit model as shown in Figure 1.3. Consider an incremental length of line (which for definiteness can be thought of as a pair of wires) lying between the locations z and $z + \Delta z$ along the z-axis. We assume that the wires present a series resistance of r (Ω/m) and a series inductance of l (H/m) per unit length along z, and a shunt capacitance c (F/m) and shunt conductance g (S/m) per unit length as well. We will further assume that all connections of this transmission line to external circuit elements (such as generators and loads) are such that at any point z on the line, the current on the "positive" (upper) conductor is equal but opposite to the current in the "reference" or "ground" (lower) conductor, as indicated in Figure 1.3. This restricts the type of external circuit connection that can be made between different points z on the line (especially of the "feedback" type), and forces the pair of terminals at any point of a transmission line to behave as a *port* (see Section 2.1).

The equivalent circuit of Figure 1.3 that results implies the relations

$$0 = (\text{r}\Delta z)i(z,t) + (\text{l}\Delta z)\frac{\partial i(z,t)}{\partial t} + v(z + \Delta z, t) - v(z,t) \tag{1.1}$$

$$0 = (\text{g}\Delta z)v(z + \Delta z, t) + (\text{c}\Delta z)\frac{\partial v(z + \Delta z, t)}{\partial t} + i(z + \Delta z, t) - i(z,t)$$

by Kirchhoff's circuit laws. Dividing these equations by Δz and letting $\Delta z \to 0$, we obtain the equations

$$\frac{\partial v(z,t)}{\partial z} = -\text{r}i(z,t) - \text{l}\frac{\partial i(z,t)}{\partial t}$$

$$\frac{\partial i(z,t)}{\partial z} = -\text{g}v(z,t) - \text{c}\frac{\partial v(z,t)}{\partial t} \tag{1.2}$$

which are known as the telegrapher's equations, or transmission-line equations. They provide a relationship between the voltage v and current i at any point along the line. The parameters r, l, g and c are called the distributed parameters of the line.

It should be noted that other distributed circuit models for the transmission line will also result in equations (1.2). If we can associate the resistance per unit length r_A and inductance per unit length l_A with the upper conductor, and r_B and l_B with the lower conductor as shown in Figure 1.4, the same telegrapher's equations (1.2) result as do from the circuit of Figure 1.3, if we put $\text{r} = \text{r}_A + \text{r}_B$ and $\text{l} = \text{l}_A + \text{l}_B$. The difference is that in the model of Figure 1.3, the "ground" conductor is at the same voltage for all values of z, while in Figure 1.4, this is not necessarily the case. This means that the telegrapher's

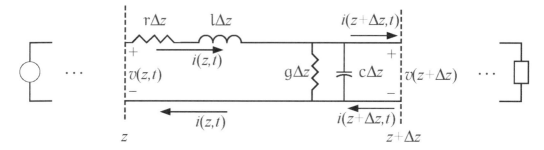

Figure 1.3 Distributed network model of the classical transmission line.

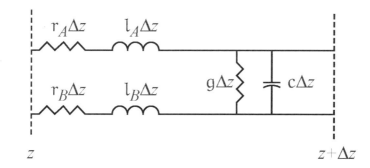

Figure 1.4 Alternative distributed network model of the classical transmission line.

equations by themselves are incapable of giving any information about the voltage difference between points at different values of z on a transmission line. Either this information will be unobtainable (and may, in fact, be irrelevant under certain conditions), or it will follow only from information about external circuit connections.

Except for Chapter 4, where pulse phenomena will be discussed, we will be concerned in this book with time-harmonic phenomena at an angular frequency $\omega = 2\pi f$. Voltages, currents, fields or sources of this type can be obtained from complex (phasor) functions which are independent of t. For example,

$$v(z,t) = \mathrm{Re}\left\{e^{j\omega t}V(z)\right\}$$
$$i(z,t) = \mathrm{Re}\left\{e^{j\omega t}I(z)\right\} \tag{1.3}$$

and so on. We will adopt the convention that, except where noted, a quantity denoted by a capital letter is the phasor counterpart of the function of time denoted by the corresponding small letter. We follow electrical engineering practice and use $j = \sqrt{-1}$ to represent the imaginary unit.

For time-harmonic voltage and current, equations (1.2) become ordinary differential equations:

$$\frac{dV(z)}{dz} = -zI(z)$$
$$\frac{dI(z)}{dz} = -yV(z) \tag{1.4}$$

where z and y represent respectively the distributed series impedance and shunt admittance per unit length of the line:

$$z = r + j\omega l; \qquad y = g + j\omega c \tag{1.5}$$

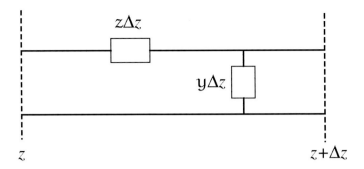

Figure 1.5 General distributed network model of the classical transmission line.

It is possible to model transmission lines with more complicated distributed networks using generalized versions of z and y:

$$z = r(\omega) + jx(\omega); \qquad y = g(\omega) + jb(\omega) \tag{1.6}$$

where x is the series reactance per unit length, and b is the shunt susceptance per unit length (Figure 1.5). All these line parameters may depend on frequency, possibly in complicated ways. Examples of such behavior are illustrated in examples 1.2.2 and 1.2.3, and in problems p1-3 through p1-7.

When z and y are independent of z (the line is then said to be uniform), these equations have a well-known solution:

$$
\begin{aligned}
V(z) &= V_0^+ e^{-\gamma z} + V_0^- e^{\gamma z} \\
I(z) &= \frac{V_0^+}{Z_c} e^{-\gamma z} - \frac{V_0^-}{Z_c} e^{\gamma z}
\end{aligned}
\tag{1.7}
$$

where V_0^+ and V_0^- are complex constants which will be determined by the strengths of the sources (generators) connected elsewhere on the line. The quantity

$$Z_c = \frac{z}{\gamma} = \frac{\gamma}{y} = |Z_c| e^{j\phi z} \quad (\Omega) \tag{1.8}$$

is called the characteristic impedance of the line. We sometimes also use the characteristic admittance $Y_c = 1/Z_c$. The quantity

$$\gamma = \alpha + j\beta = \sqrt{zy} \quad (\text{m}^{-1}) \tag{1.9}$$

is called the complex propagation coefficient of the line, with its real part α denoted as the attenuation coefficient, and β the phase coefficient (or sometimes, casually, the propagation constant). The terms proportional to V_0^+ in (1.7) are called the *forward wave*, while those proportional to V_0^- are called the *reverse* wave, for reasons that we will clarify below.

For a transmission line (and in later chapters, for any waveguide mode) whose complex propagation coefficient γ has an imaginary part β not equal to zero, it is customary to define several auxiliary quantities in terms of β:

1. Guide wavelength:

$$\lambda_g \equiv \frac{2\pi}{|\beta|} \tag{1.10}$$

2. Phase velocity:

$$v_p = \frac{\omega}{\beta} \tag{1.11}$$

3. Group velocity:

$$v_g = \frac{1}{d\beta/d\omega} \tag{1.12}$$

The group velocity will be seen in Chapter 4 to describe the velocity with which the envelope of a modulated signal travels along the line. Under many conditions, it is also the velocity with which energy travels along the line. The phase velocity on the other hand, is generally not the velocity of any physically observable quantity.

The sign of the square root taken when calculating γ (which in turn will affect the value of Z_c) must be chosen carefully. All transmission lines considered in this book are assumed to be *passive*, meaning that the real parts r and g of z and y are nonnegative. We then choose the square root in (1.9) such that

$$\mathrm{Re}(\gamma) = \alpha \geq 0 \tag{1.13}$$

The solution for α which obeys this constraint can be obtained by squaring (1.9) and separating into real and imaginary parts:

$$\alpha^2 - \beta^2 = \mathrm{Re}(zy) = rg - xb$$

$$2\alpha\beta = \mathrm{Im}(zy) = rb + gx$$

Eliminating β between these two equations gives a quadratic equation for α^2, whose nonnegative solution gives α as:

$$\alpha = \sqrt{\frac{|zy| + \mathrm{Re}(zy)}{2}} \tag{1.14}$$

where

$$|zy| = \sqrt{(r^2 + x^2)(g^2 + b^2)}$$

and all square roots are taken to be nonnegative.

If α as computed from (1.14) is positive, then the value of β is uniquely determined as:

$$\beta = \frac{rb + gx}{2\alpha} \tag{1.15}$$

and may be either positive or negative (a wave for which $\beta > 0$ is called a *direct wave*, while one for which $\beta < 0$ is called a *backward wave*). From (1.15), it can be shown that the real part of the characteristic impedance given by (1.8) is:

$$\mathrm{Re}(Z_c) = \frac{g|z| + r|y|}{2\alpha|y|} \tag{1.16}$$

and thus is always nonnegative. The characteristic impedance thus lies in the right half of the complex plane: $-\pi/2 \leq \phi_Z \leq \pi/2$. We will see below that this implies that a direct wave carries power in the same direction as phase moves, while a backward wave carries power in the opposite direction to the phase. The ordinary classical transmission line depicted in Figure 1.3 supports only direct waves if l and c are positive, but other types of transmission line can sometimes support backward waves.

If $\alpha = 0$, then (1.13) does not specify the value of β uniquely. This case can only arise if two conditions are both satisfied:

i) the line is lossless ($r = g = 0$), and
ii) $xb > 0$.

We then have $\beta^2 = xb$. In this case, we will determine the sign of β by requiring that the characteristic impedance (which by (1.8) must be real) be positive, in agreement with the fact that $\text{Re}(Z_c) \geq 0$ in the case when $\alpha > 0$. The result of this requirement is

$$\beta = +\sqrt{xb} \qquad \text{(lossless line, $x > 0$ and $b > 0$)} \qquad (1.17)$$

or

$$\beta = -\sqrt{xb} \qquad \text{(lossless line, $x < 0$ and $b < 0$)} \qquad (1.18)$$

Again, β may be either positive or negative, and the wave is denoted as a direct or backward wave accordingly.

Our choice for the sign of the square root determining γ is based on a desire to attach a physical meaning to each of the terms in equations (1.7) representing waves on the transmission line. If $\alpha > 0$, the forward wave of (1.7) is one that decays exponentially in the positive z direction. Now, the complex power flow in the z-direction at a point z on the transmission line (Figure 1.6) is:

$$\hat{P}(z) = \frac{1}{2} V(z) I^*(z) \qquad (1.19)$$

where * denotes the complex conjugate. The time-average power flow P_{av} is the real part of this complex power. Thus the time-average power in the forward wave (in the absence of the reverse wave) is

$$P_{av}(z) = \frac{|V_0^+|^2 e^{-2\alpha z}}{2} \text{Re}\left(\frac{1}{Z_c^*}\right) \qquad (1.20)$$

and therefore if power flow is nonzero, this wave is also the one that carries energy in the positive z-direction, because $\text{Re}(Z_c) \geq 0$. If $\alpha = 0$, the choice of (1.17) or (1.18) for the phase constant of the forward wave also results in energy flow in the positive z-direction. Likewise, the reverse wave in (1.7) transports energy (if at all) in the negative z-direction, and if it decays, does so in this direction as well. In any event, the forward waves should move toward larger values of z from a source located at smaller values of z, while the reverse wave moves toward smaller values of z from a source located at larger values of z. This allows

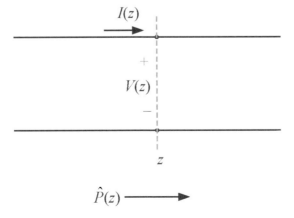

Figure 1.6 Power flow on a classical transmission line.

us to decide which wave should be present on a section of line which extends to $+\infty$ or to $-\infty$ in the z-direction.

Let us next consider the flow of power along the line when both forward and reverse waves are present. By (1.7) the complex power at z is

$$\hat{P}(z) = \frac{1}{2Z_c^*} \left\{ |V_0^+|^2 e^{-2\alpha z} - |V_0^-|^2 e^{2\alpha z} + 2j \mathrm{Im}\left[V_0^- (V_0^+)^* e^{2j\beta z}\right] \right\} \tag{1.21}$$

When Z_c is real,

$$P_{\mathrm{av}} = \frac{|V_0^+|^2 e^{-2\alpha z} - |V_0^-|^2 e^{2\alpha z}}{2Z_c} \tag{1.22}$$

This states that the powers carried individually by the forward and reverse waves superpose to give the total power flow with no interaction between them. On the other hand, if Z_c is imaginary ($\phi_Z = \pm\pi/2$), there can be time average power flow only if both waves are present simultaneously:

$$P_{\mathrm{av}} = \mp \mathrm{Im}\left[V_0^- (V_0^+)^* e^{2j\beta z}\right] \frac{1}{|Z_c|} \tag{1.23}$$

and no such power superposition principle holds in this case. On a lossy line intermediate between these two cases, some contribution from both terms will be present. The complex oscillatory power on the other hand is simply

$$\hat{P}_{\mathrm{osc}} = VI/2 = \left[(V_0^+)^2 e^{-2\gamma z} - (V_0^-)^2 e^{2\gamma z}\right] \frac{1}{2Z_c} \tag{1.24}$$

regardless of the values of Z_c and γ, and contains no cross terms.

It is convenient to express the amplitudes $V_0^\pm$ in another form, so as to make the power flow on the line more explicit. Let us introduce new variables a_0 and b_0, called respectively the forward and reverse *wave amplitudes* at $z = 0$, through the equations

$$V_0^+ = N_V a_0 \qquad V_0^- = N_V b_0 \tag{1.25}$$

where N_V is a suitably chosen *normalizing factor*. The question now is: how do we best choose this factor? Sometimes, an attempt is made to choose it so as to make $|a_0|^2$ and $|b_0|^2$ represent the time-average power carried by the forward and reverse waves respectively. From the preceding discussion, it is seen that this cannot be done in general: for example, if Z_c is imaginary the time-average power flow may be zero even if $a_0 \neq 0$. Moreover, attempting to specify the time-average power leaves the phase angle of N_V completely undetermined, and we would have to resort to arbitrary criteria in order to render it unique. On the other hand, the physically less important complex oscillatory power is expressed in a mathematically simpler form via the forward and reverse waves.

We will thus choose to normalize the forward and reverse modes by requiring that each carries a complex *oscillatory* power of $1e^{j0}$ watt in its direction of energy transport (or attenuation) at the position $z = 0$ when $a_0 = 1$ or $b_0 = 1$ as appropriate. We will find that in important special cases, this normalization will still lead to a simple form for P_{av}. Thus, if $a_0 = 1$ and $b_0 = 0$, we have the normalized forward mode. Its voltage has phase angle $\phi_Z/2$ and its current has phase angle $-\phi_Z/2$ at $z = 0$. The normalized reverse mode ($a_0 = 0$, $b_0 = 1$) has the same voltage as the forward mode, but the opposite current at $z = 0$. Setting $\hat{P}_{\mathrm{osc}} = 1$ for the normalized forward mode, or $\hat{P}_{\mathrm{osc}} = -1$ for the normalized reverse mode gives

$$N_V = \sqrt{2Z_c} = \sqrt{2|Z_c|} e^{j\phi_Z/2} \tag{1.26}$$

where ϕ_Z lies in the interval $[-\pi/2, +\pi/2]$, as noted above.

From (1.24), then, a_0^2 and $-b_0^2$ represent respectively the complex oscillatory powers carried through the position $z = 0$ in the $+z$-direction by the forward and reverse waves on the transmission line. With this choice of normalization condition, equations (1.21) and (1.24) for the complex and oscillatory powers at any z become

$$\hat{P}(z) = e^{j\phi z} \left\{ |a_0|^2 e^{-2\alpha z} - |b_0|^2 e^{2\alpha z} + 2j \, \text{Im} \, (b_0 a_0^* e^{2j\beta z}) \right\} \tag{1.27}$$

and

$$\hat{P}_{\text{osc}} = a_0^2 e^{-2\gamma z} - b_0^2 e^{2\gamma z} \tag{1.28}$$

respectively. When Z_c is real, the time-average power from (1.27) reduces to

$$P_{\text{av}}(z) = |a_0|^2 e^{-2\alpha z} - |b_0|^2 e^{2\alpha z} \tag{1.29}$$

which is analogous to (1.28).

We can see in (1.7) the basic concept of a *mode*. A mode in the present context is a combination of a voltage and current (later, it will be a combination of electric and magnetic fields) which propagate along z according to the common propagation factor $\exp(j\omega t - \gamma z)$ and which maintain a constant relationship between V and I. Thus in (1.7) we identify the forward mode (whose amplitude decreases with, or whose energy moves toward, increasing z) as

$$V^+(z) = \sqrt{2Z_c} a_0 e^{-\gamma z}; \qquad I^+(z) = \sqrt{\frac{2}{Z_c}} a_0 e^{-\gamma z}$$

and the reverse mode (whose amplitude decreases with or energy moves toward decreasing z) as

$$V^-(z) = \sqrt{2Z_c} b_0 e^{\gamma z}; \qquad I^-(z) = -\sqrt{\frac{2}{Z_c}} b_0 e^{\gamma z}$$

In each case, the voltage and current associated with each mode vary with z according to the same law, $\exp(\mp\gamma z)$, and the ratio $V(z)/I(z)$ is a constant, $\pm Z_c$, independent of the location z. Moreover, the overall amplitude factor (a_0 or b_0) common to V and I is arbitrary, and is only determined by the conditions of excitation of the line.

The wave amplitude of a mode at an arbitrary position along a uniform transmission line is denoted by $a(z)$ for the forward mode, and by $b(z)$ for the reverse mode, where

$$a(z) = a_0 e^{-\gamma z} \qquad b(z) = b_0 e^{\gamma z} \tag{1.30}$$

It is also sometimes convenient to define normalized *total* voltages and currents, defined by

$$\tilde{V}(z) = a(z) + b(z) \qquad \tilde{I}(z) = a(z) - b(z) \tag{1.31}$$

when we wish not to clutter our formulas with factors like $\sqrt{2Z_c}$. Evidently, the actual voltage and current will be given by

$$V(z) = \sqrt{2Z_c} \tilde{V}(z) \qquad I(z) = \sqrt{\frac{2}{Z_c}} \tilde{I}(z) \tag{1.32}$$

The wave amplitudes $a(z)$ and $b(z)$ are given in terms of the voltage and current by:

$$a(z) = \frac{V(z) + Z_c I(z)}{\sqrt{8Z_c}}; \qquad b(z) = \frac{V(z) - Z_c I(z)}{\sqrt{8Z_c}} \tag{1.33}$$

1.2.1   EXAMPLE: THE CONVENTIONAL TRANSMISSION LINE

The simplest lossless classical transmission line is one in which $x = \omega l$ and $b = \omega c$, with l and c both positive and independent of frequency. Then we have a propagating mode with $\alpha = 0$,

$$\beta = \omega\sqrt{lc} \tag{1.34}$$

and

$$Z_c = \sqrt{\frac{l}{c}} \tag{1.35}$$

These formulas are familiar from introductory texts on transmission lines. From (1.11) and (1.12), we see that

$$v_p = v_g = \frac{1}{\sqrt{lc}} \tag{1.36}$$

1.2.2   EXAMPLE: AN UNCONVENTIONAL TRANSMISSION LINE

For a lossless line in which one (but not both) of x or b is negative[1],

$$Z_c = \frac{jx}{\sqrt{-xb}} = \frac{\sqrt{-xb}}{jb} \tag{1.37}$$

where $\sqrt{-xb}$ is taken to be positive. Now Z_c is seen to be positive imaginary if $x > 0$ and $b < 0$, and negative imaginary if $b > 0$ and $x < 0$.

An example of this behavior is furnished by the coaxial cable which has its inner conductor interrupted at frequent periodic intervals as shown in Figure 1.7. The effect of this upon the distributed constant model of this transmission line is the insertion of an additional distributed capacitance into the line, in the form of an elastance[2] $s\Delta z$ in *series* with the series inductance $l\Delta z$ for a small length Δz of the transmission line. Using the distributed constant model for this line, we see that the shunt admittance per unit length remains

$$y = jb = j\omega c$$

while the series impedance in a length Δz of the line is

$$z\Delta z = j\omega l\Delta z + \frac{1}{j\omega\left(\frac{1}{s\Delta z}\right)}$$

or

$$z = jx = j\omega l + \frac{s}{j\omega}$$

Observe that the series capacitance in a length Δz of line is taken to be proportional to $1/\Delta z$ rather than to Δz. This is because as Δz increases, the effect is of combining the gap capacitances in series, which results in an overall capacitance inversely proportional to the number of gaps in Δz, and therefore to Δz itself.

If we define a *cutoff frequency* as

$$\omega_c = \sqrt{\frac{s}{l}} \tag{1.38}$$

---

[1] As we will see below, this can happen when using the transmission line equations to model plane wave problems, for sufficiently large incidence angles, or when modeling unconventional transmission lines.

[2] Elastance S, sometimes called the potential coefficient and denoted p, is the inverse of capacitance C and has units of inverse farads, sometimes called darafs.

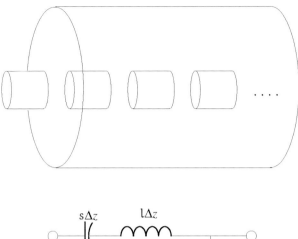

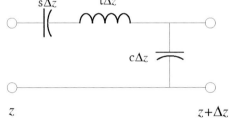

Figure 1.7 Transmission line with periodically interrupted conductor, and its differential equivalent network.

(for reasons to be clarified below), then if $\omega > \omega_c$, by the same procedure employed for a conventional transmission line, we compute the propagation constant to be

$$\gamma = \sqrt{zy} = \pm j\frac{\sqrt{\omega^2 - \omega_c^2}}{v_{p\infty}} \tag{1.39}$$

and the corresponding characteristic impedance to be

$$Z_c = \frac{\gamma}{y} = \pm Z_{c\infty}\sqrt{1 - \frac{\omega_c^2}{\omega^2}} \tag{1.40}$$

where

$$v_{p\infty} = \frac{1}{\sqrt{lc}} \tag{1.41}$$

is the phase velocity of the line in the limit as $\omega \to \infty$ (i. e., what we would get if the gaps were not present), and

$$Z_{c\infty} = \sqrt{\frac{l}{c}} \tag{1.42}$$

is the characteristic impedance we have without gaps (or as $\omega \to \infty$). The choice of sign in (1.39) must be made so that $Z_c > 0$ in (1.40), so the upper $(+)$ sign is chosen.

For $\omega < \omega_c$, the line does not support a propagating mode, but instead one which decays exponentially, because $\gamma = \alpha$ is real for $\omega < \omega_c$:

$$\alpha = \frac{\sqrt{\omega_c^2 - \omega^2}}{v_{p\infty}}, \tag{1.43}$$

the plus sign being chosen to obtain $\alpha > 0$. In this frequency range where the mode is cutoff, the characteristic impedance is purely imaginary; in fact, by (1.37) it is negative imaginary:

$$Z_c = \frac{\alpha}{y} = -jZ_{c\infty}\sqrt{\frac{\omega_c^2}{\omega^2} - 1} \tag{1.44}$$

According to (1.23), this means that both a forward and reverse mode must be present in order for time-average power to flow along the line. A section of such a line functions as a high-pass filter.

Another feature of note is that, even in the propagating regime $\omega > \omega_c$, the phase and group velocities are functions of frequency, with

$$v_p = \frac{v_{p\infty}}{\sqrt{1 - \omega_c^2/\omega^2}} \tag{1.45}$$

and

$$v_g = v_{p\infty}\sqrt{1 - \omega_c^2/\omega^2} \tag{1.46}$$

We see that the phase velocity is always greater than that of the gapless line, while the group velocity is always smaller. The phenomenon of v_p depending on ω (so that $v_p \neq v_g$) is called *dispersion*, and means that transient signals launched on a dispersive line have each of their frequency components phase shifted by a different amount due to propagation along the line. Re-assembling the frequencies at the load end of the line results in a distorted pulse. This distortion is studied in more detail in Chapter 4.

1.2.3 EXAMPLE: A BACKWARD-WAVE TRANSMISSION LINE

Consider next a transmission line for which the distributed series impedance is that of an elastance as in the previous example, while the shunt admittance is that of inductance, as shown in Figure 1.8. This could be realized in practice by a kind of "mushroom" structure, using gap-separated upper sections as in the previous example, each connected to ground by a wire possessing a small inductance. Because the inductance when cascaded is in a parallel connection, it is better to characterize it as a *reluctance* per unit length r_m.[3] From the

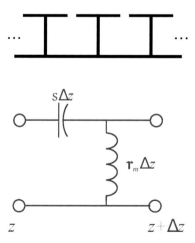

Figure 1.8 Differential section of a backward-wave transmission line.

---

[3] Reluctance R_m is the inverse of inductance, and has units of inverse henries, also known as sturgeons (!).

distributed network model for this line, we see that the shunt admittance per unit length is

$$y = jb = \frac{r_m}{j\omega}$$

while the series impedance in a length Δz of the line is

$$z = jx = \frac{s}{j\omega}$$

Then

$$\gamma = \sqrt{zy} = \pm j\frac{\sqrt{sr_m}}{\omega} \tag{1.47}$$

and the corresponding characteristic impedance is

$$Z_c = \frac{\gamma}{y} = \mp\sqrt{\frac{s}{r_m}} \tag{1.48}$$

Observe that we must now choose the *lower* sign to ensure $Z_c > 0$, but this means that

$$\gamma = j\beta = -j\frac{\sqrt{sr_m}}{\omega} \tag{1.49}$$

which means that even though power flow is in the $+z$-direction, phase travels in the negative z-direction. In fact, the phase velocity is negative:

$$v_p = \frac{\omega}{\beta} = -\frac{\omega^2}{\sqrt{sr_m}}, \tag{1.50}$$

while the group velocity is positive:

$$v_g = \frac{1}{d\beta/d\omega} = +\frac{\omega^2}{\sqrt{sr_m}}, \tag{1.51}$$

both being strongly frequency-dependent. For this reason, this line is termed a *backward-wave* structure because phase travels in the opposite direction from the power. Structures like this have application in distributed amplifiers, and in three-dimensional versions, as negative refraction elements.

1.3   REFLECTED WAVES, THE SMITH CHART AND CHAIN PARAMETERS

In analysis of transmission-line problems, we define the total impedance (or line impedance) at any point z along the line as

$$Z(z) = \frac{V(z)}{I(z)} = Z_c\frac{1 + \rho(z)}{1 - \rho(z)} \tag{1.52}$$

where for a uniform line,

$$\rho(z) = \frac{V_0^-}{V_0^+}e^{2\gamma z} \tag{1.53}$$

is the position-dependent reflection coefficient on the line, if V_0^+ and V_0^- are the constants appearing in (1.7). The reflection coefficient can also be expressed in terms of the wave amplitude description: $\rho(z) = b(z)/a(z)$.

The inverse relation to (1.52) allows the reflection coefficient to be computed from the line impedance:

$$\rho(z) = \frac{Z(z) - Z_c}{Z(z) + Z_c} \tag{1.54}$$

The normalized line impedance, defined as

$$\bar{Z}(z) = Z(z)/Z_c \tag{1.55}$$

is free from explicit dependence on the characteristic impedance, and can easily be related to $\rho(z)$:

$$\bar{Z}(z) = \frac{1 + \rho(z)}{1 - \rho(z)}; \qquad \rho(z) = \frac{\bar{Z}(z) - 1}{\bar{Z}(z) + 1} \tag{1.56}$$

Another way of expressing the amount of reflected wave at a given point on a transmission line is in terms of the *(voltage) standing wave ratio* [(V)SWR], defined as

$$\text{VSWR} = \frac{1 + |\rho(z)|}{1 - |\rho(z)|} \tag{1.57}$$

Since $|\rho(z)| \leq 1$ for a transmission line with real characteristic impedance (if no gain or sources are present), it is clear that the VSWR is real and $1 \leq \text{VSWR} \leq \infty$ in this case. In the absence of attenuation on the line ($\alpha = 0$), the VSWR is the same everywhere on a uniform, sourcefree section of line.

A common way of displaying graphically the relative amount of reverse to forward mode content on a line is by means of the Smith chart (Figure 1.9). The Smith chart represents the unit circle of the complex ρ-plane, $|\rho| \leq 1$. What makes it distinctive, however, is that it is calibrated in the real and imaginary parts of the normalized impedance $\bar{Z} = \bar{R} + j\bar{X}$ instead of real and imaginary parts of ρ. As the location of the observation point z along the line moves, the locus of $\rho(z)$ moves in a circle or a spiral on the Smith chart.

Although individual determination of the amplitude constants V_0^+ and V_0^- in (1.7) requires knowledge of the sources on the line, and will be deferred until Chapter 3, we can determine the ratio V_0^-/V_0^+ which results when a lumped impedance element is connected to the line. Consider, for example, the situation of Figure 1.10, where a load impedance Z_L is connected at $z = d$ on a transmission line whose characteristic impedance is Z_c and whose propagation coefficient is γ. At $z = d$, we must have, by Kirchhoff's circuit laws and Ohm's law:

$$Z_L = \frac{V(d)}{I(d)} = Z(d)$$

or,

$$\bar{Z}(d) = Z_L/Z_c$$

By (1.56), we see that

$$\rho(d) = \frac{Z_L/Z_c - 1}{Z_L/Z_c + 1} = \frac{Z_L - Z_c}{Z_L + Z_c} \tag{1.58}$$

On the other hand, by (1.53),

$$\rho(0) = \frac{V_0^-}{V_0^+} = e^{-2\gamma d}\rho(d) = e^{-2\gamma d}\left(\frac{Z_L - Z_c}{Z_L + Z_c}\right) \tag{1.59}$$

and therefore the input impedance presented at the terminals at $z = 0$ is, by (1.56),

$$Z(0) \;\; = \;\; Z_c\bar{Z}(0) = Z_c\frac{1 + \rho(0)}{1 - \rho(0)}$$

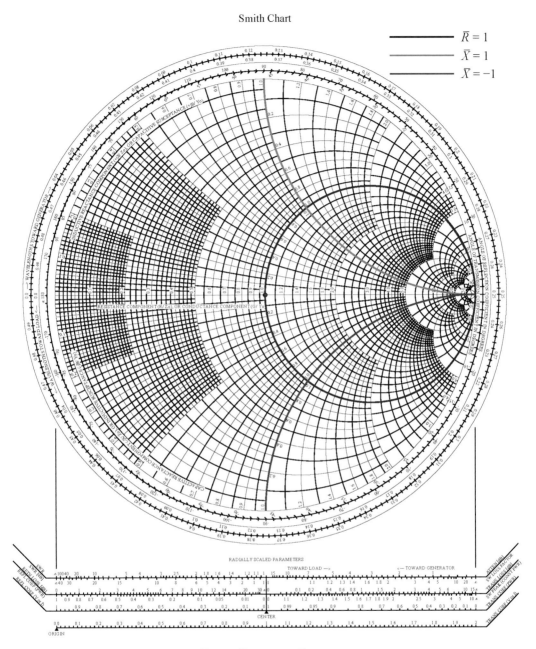

Figure 1.9 The Smith Chart, with $\bar{R} = 1$, $\bar{X} = 1$ and $\bar{X} = -1$ circles emphasized. (© 2015 IEEE. Used with permission.)

$$= Z_c \frac{Z_L \cosh \gamma d + Z_c \sinh \gamma d}{Z_L \sinh \gamma d + Z_c \cosh \gamma d} \tag{1.60}$$

To use this formula in the special case when $\gamma = j\beta$ is imaginary, make the replacements $\cosh \gamma d = \cos \beta d$ and $\sinh \gamma d = j \sin \beta d$.

A more powerful means of analyzing sections of uniform transmission line is to use a relationship between voltage and current at one end and those at the other end. Suppose a section of transmission line lies in the interval $z_1 < z < z_2$. Regardless of what is connected

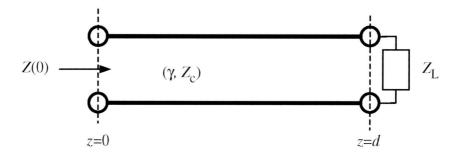

Figure 1.10 Section of transmission line terminated with a load impedance.

to either end of the line, the relations (1.7) can be used to express $V(z_1)$ and $I(z_1)$ in terms of $V(z_2)$ and $I(z_2)$. We can in fact write

$$
\begin{aligned}
V(z_1) &= \mathcal{A}(z_1, z_2)V(z_2) + \mathcal{B}(z_1, z_2)I(z_2) \\
I(z_1) &= \mathcal{C}(z_1, z_2)V(z_2) + \mathcal{D}(z_1, z_2)I(z_2)
\end{aligned}
\tag{1.61}
$$

which are true regardless of the values of the two amplitudes $V_0^\pm$ in (1.7) if

$$
\begin{aligned}
\mathcal{A}(z_1, z_2) &= \cosh \gamma(z_2 - z_1) \\
\mathcal{B}(z_1, z_2) &= Z_c \sinh \gamma(z_2 - z_1) \\
\mathcal{C}(z_1, z_2) &= \frac{1}{Z_c} \sinh \gamma(z_2 - z_1) \\
\mathcal{D}(z_1, z_2) &= \cosh \gamma(z_2 - z_1)
\end{aligned}
\tag{1.62}
$$

The quantities $\mathcal{A}(z_1, z_2)$, $\mathcal{B}(z_1, z_2)$, $\mathcal{C}(z_1, z_2)$ and $\mathcal{D}(z_1, z_2)$ are examples of elements of a chain matrix, which is discussed more generally in Section 2.1.2. In particular, we note here the properties

$$
\mathcal{A}(z_1, z_2)\mathcal{D}(z_1, z_2) - \mathcal{B}(z_1, z_2)\mathcal{C}(z_1, z_2) = 1
\tag{1.63}
$$

and

$$
\begin{aligned}
\mathcal{A}(z_1, z_2) &= \mathcal{D}(z_2, z_1) \\
\mathcal{B}(z_1, z_2) &= -\mathcal{B}(z_2, z_1) \\
\mathcal{C}(z_1, z_2) &= -\mathcal{C}(z_2, z_1) \\
\mathcal{D}(z_1, z_2) &= \mathcal{A}(z_2, z_1)
\end{aligned}
\tag{1.64}
$$

which turn out to be true in general, even for cascaded sections of line, or nonuniform lines as considered in Section 1.5.

To see how (1.61) can be applied, let the load impedance $Z_L = Z(d)$ be connected at $z = d$ as in Figure 1.10. Then put $z_1 = 0$ and $z_2 = d$ in (1.61), divide the two equations and obtain:

$$
Z(0) = \frac{\mathcal{A}(0, d)V(d) + \mathcal{B}(0, d)I(d)}{\mathcal{C}(0, d)V(d) + \mathcal{D}(0, d)I(d)} = \frac{\mathcal{A}(0, d)Z(d) + \mathcal{B}(0, d)}{\mathcal{C}(0, d)Z(d) + \mathcal{D}(0, d)}
\tag{1.65}
$$

which we readily see reduces to (1.60). Another useful result is the expression of the voltage across the load impedance in terms of the voltage at $z = 0$. From (1.61), we have

$$
\frac{V(d)}{V(0)} = \frac{V(d)}{\mathcal{A}(0, d)V(d) + \mathcal{B}(0, d)I(d)} = \frac{Z(d)}{\mathcal{A}(0, d)Z(d) + \mathcal{B}(0, d)} = \frac{Z_L}{Z_L \cosh \gamma d + Z_c \sinh \gamma d}
\tag{1.66}
$$

Other such relations can be quickly obtained in a similar way using the chain relations (1.61) for a uniform classical transmission line.

1.3.1 EXAMPLE: REFLECTION AND TRANSMISSION AT A SECTION OF MISMATCHED LINE

To illustrate the foregoing ideas, consider the example of a wave incident from the left on a transmission line of characteristic impedance Z_{c1} and propagation constant γ_1 to a section of different transmission line of length d, whose characteristic impedance is Z_{c2} and propagation constant is γ_2, as shown in Figure 1.11. From (1.65) or (1.60), the input impedance seen by an incident voltage wave $V_1^+ e^{-\gamma_1 z}$ in $z < 0$ is

$$Z(0) = Z_{c2} \frac{Z_{c1} + Z_{c2} \tanh \gamma_2 d}{Z_{c1} \tanh \gamma_2 d + Z_{c2}} \tag{1.67}$$

so that the reflection coefficient at $z = 0$ is

$$\rho = \rho(0) = \frac{(Z_{c2}^2 - Z_{c1}^2) \sinh \gamma_2 d}{(Z_{c2}^2 + Z_{c1}^2) \sinh \gamma_2 d + 2 Z_{c1} Z_{c2} \cosh \gamma_2 d} \tag{1.68}$$

We can similarly define a (voltage) *transmission coefficient* τ at $z = d$ as

$$\tau \equiv \frac{V(d)}{V_1^+} = \frac{V(d)}{V(0)}(1 + \rho) = \frac{2 Z_{c1} Z_{c2}}{(Z_{c2}^2 + Z_{c1}^2) \sinh \gamma_2 d + 2 Z_{c1} Z_{c2} \cosh \gamma_2 d} \tag{1.69}$$

Note that while ρ is uniquely defined as the ratio of reflected to incident waves at the same position on the same transmission line, the transmission coefficient may have many definitions, because the transmitted wave in question may be at a different position on the line, or on a different transmission line altogether. It is therefore defined specifically for any particular application.

Let us consider two special cases in detail. In both cases, the characteristic impedance Z_{c1} will be assumed to be real. In Case A, we will also assume that Z_{c2} is real, while $\gamma_2 = j\beta_2$ is imaginary. The central transmission line is thus lossless, and supports a propagating mode. We then have

$$\rho = \frac{j(Z_{c2}^2 - Z_{c1}^2) \sin \beta_2 d}{j(Z_{c2}^2 + Z_{c1}^2) \sin \beta_2 d + 2 Z_{c1} Z_{c2} \cos \beta_2 d} \tag{1.70}$$

and

$$\tau = \frac{2 Z_{c1} Z_{c2}}{j(Z_{c2}^2 + Z_{c1}^2) \sin \beta_2 d + 2 Z_{c1} Z_{c2} \cos \beta_2 d} \tag{1.71}$$

As d varies, $|\rho|$ varies between 0 (when d is an integer multiple of λ_{g2} on the central section of line) and a maximum of

$$\frac{Z_{c2}^2 - Z_{c1}^2}{Z_{c2}^2 + Z_{c1}^2}$$

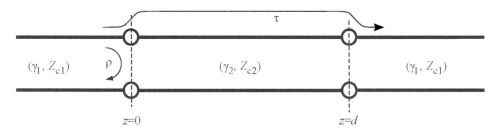

Figure 1.11 Section of transmission line inserted in a different line.

when d is an odd multiple of a quarter wavelength. The value of $|\tau|$ likewise varies between a maximum of 1 and a minimum of

$$\frac{2Z_{c2}Z_{c1}}{Z_{c2}^2 + Z_{c1}^2}$$

at these same values of d.

For Case B, we suppose that the central section of line is lossless but cutoff, so that $Z_{c2} = jX_{c2}$ is imaginary, while $\gamma_2 = \alpha_2$ is real, as is the case for $\omega < \omega_c$ in Example 1.2.2. From (1.68) and (1.69) we now have

$$\rho = -\frac{(X_{c2}^2 + Z_{c1}^2)\sinh\alpha_2 d}{(Z_{c1}^2 - X_{c2}^2)\sinh\alpha_2 d + 2jZ_{c1}X_{c2}\cosh\alpha_2 d} \tag{1.72}$$

and

$$\tau = \frac{2jZ_{c1}X_{c2}}{(Z_{c1}^2 - X_{c2}^2)\sinh\alpha_2 d + 2jZ_{c1}X_{c2}\cosh\alpha_2 d} \tag{1.73}$$

In contrast with Case A, the behavior of $|\rho|$ and $|\tau|$ is now monotonic with d. Indeed, if $\alpha_2 d \gg 1$, then $|\rho| \simeq 1$, while

$$|\tau| \simeq \frac{4Z_{c1}|X_{c2}|}{Z_{c1}^2 + X_{c2}^2}e^{-\alpha_2 d}$$

which, although small, is not exactly zero. This nonzero transmitted wave is the result of what is called "tunneling," analogous to the similar phenomenon in quantum mechanics.

1.4 TRANSMISSION LINES AS MODELS FOR PLANE WAVE PROPAGATION

Although we derived the telegrapher's equations from a distributed network model for the behavior of two parallel conductors of appreciable length, they are also capable of modeling a wide variety of other physical processes involving wave propagation. We will see in later chapters that the propagation of any waveguide mode can be described in this way. In this section we will illustrate the analogy between the classical transmission line and the propagation of plane waves in layered media.

Consider the case of a time-harmonic plane wave propagating in a lossless homogeneous medium whose permittivity is ϵ and whose permeability is μ (in $z < 0$). In $z > 0$ is a lossless one-dimensionally layered medium in $z > 0$ as shown in Figure 1.12. For now, we will think of this region as having piecewise constant permeability $\mu(z)$ and permittivity $\epsilon(z)$, but in

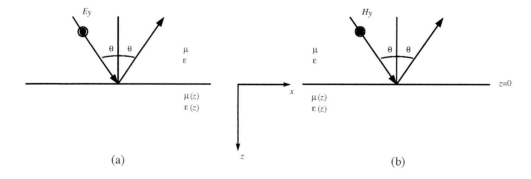

Figure 1.12 Obliquely incident plane waves onto a plane inhomogeneous medium: (a) perpendicular (TE) polarization; (b) parallel (TM) polarization.

the end, what we do below will turn out to be valid for arbitrary variations of the material parameters with z (even in $z < 0$).

The fields of the incident wave in $z < 0$ contain a factor $e^{j(\omega t - k_x x)}$, where $k_x = k \sin \theta$, θ is the angle of incidence to the z-axis, and $k = \omega \sqrt{\mu \epsilon}$ is the wavenumber in the region $z < 0$. We further assume that the incident fields have no dependence on y. The nature of the geometry implies that the total field anywhere will possess both these properties as well. The exponential factor $e^{-jk_x x}$ is present in the fields for all values of z, *with the same value of* k_x (otherwise continuity of tangential **E** and **H** could not hold). We use this fact to define an angle of propagation $\theta(z)$ at each point in $z > 0$ from the relation

$$k_x = k \sin \theta = k(z) \sin \theta(z) = \text{constant} \tag{1.74}$$

where the local wavenumber of the medium is defined as $k(z) = \omega \sqrt{\mu(z)\epsilon(z)}$. Eqn. (1.74) is a generalized form of *Snell's Law* for a layered medium.

We thus put $\partial/\partial y \equiv 0$ and suppress the common exponential factor above from all the fields. With these assumptions, the Maxwell equations (D.3) in a source-free region ($\mathbf{J}_{\text{ext}} = 0$) reduce to:

$$-\frac{\partial E_y}{\partial z} = -j\omega\mu H_x; \qquad -\frac{\partial H_y}{\partial z} = j\omega\epsilon E_x$$

$$\frac{\partial E_x}{\partial z} + jk_x E_z = -j\omega\mu H_y; \qquad \frac{\partial H_x}{\partial z} + jk_x H_z = j\omega\epsilon E_y$$

$$-jk_x E_y = -j\omega\mu H_z; \qquad -jk_x H_y = j\omega\epsilon E_z \tag{1.75}$$

It will be observed that this set of six equations in six scalar unknowns is *decoupled*: the only nonzero field components might be E_y, H_x and H_z; or H_y, E_x and E_z. The first of these situations is called perpendicular polarization (that is, the E-field is perpendicular to the plane of incidence determined by the direction of propagation and the normal to the surface of the medium; this polarization is also sometimes called TE, meaning the field is *Transverse Electric* to these directions, or s, from *senkrecht*—the German word for perpendicular), while the second is called parallel (or TM or p) polarization. A general plane wave independent of y would be made up of a superposition of the two polarizations.

If H_z or E_z respectively is eliminated from the set of three scalar equations that result from Maxwell's equations in either of these polarizations, we have

$$-\frac{\partial E_y}{\partial z} = -j\omega\mu H_x; \qquad \frac{\partial H_x}{\partial z} = j\omega\left(\epsilon - \frac{k_x^2}{\omega^2\mu}\right)E_y \tag{1.76}$$

$$-\frac{\partial H_y}{\partial z} = j\omega\epsilon E_x; \qquad \frac{\partial E_x}{\partial z} = -j\omega\left(\mu - \frac{k_x^2}{\omega^2\epsilon}\right)H_y \tag{1.77}$$

Eqns. (1.76) and (1.77) are identical with the telegrapher's equations:

$$\frac{dV}{dz} = -j\mathsf{x}(z)I(z); \qquad \frac{dI}{dz} = -j\mathsf{b}(z)V(z) \tag{1.78}$$

if the correspondences for V, I, b and x given in Table 1.1 are used. More general problems involving losses will result in the yet more general form of the telegrapher's equations:

$$\frac{dV}{dz} = -\mathsf{z}(z)I(z); \qquad \frac{dI}{dz} = -\mathsf{y}(z)V(z) \tag{1.79}$$

where $\mathsf{z}(z)$ and $\mathsf{y}(z)$ can be fairly arbitrary functions of z. This form is more general than (1.4) in that it is true for *nonuniform* transmission lines, i. e., lines whose parameters may vary along the z-direction.

The implication of this equivalence is that we may solve reflection and transmission problems for plane waves using only the telegrapher's equations and any methods developed to solve transmission line problems. For example, if $k_x = 0$ in a homogeneous region, we have that the most general solution is a pair of plane waves traveling in the $\pm z$-directions:

$$V(z) = E^+ e^{-jkz} + E^- e^{jkz}; \qquad I(z) = \frac{E^+}{\zeta} e^{-jkz} - \frac{E^-}{\zeta} e^{jkz}$$

analogous to the transmission-line solutions of (1.7). Here $E^\pm$ are complex constants and $\zeta = \sqrt{\mu/\epsilon}$ is the wave impedance of the medium. For arbitrary values of k_x in either polarization, the general solution is a pair of obliquely traveling plane waves:

$$V(z) = E^+ e^{-jk_z z} + E^- e^{jk_z z}; \qquad I(z) = \frac{E^+}{Z_c} e^{-jk_z z} - \frac{E^-}{Z_c} e^{jk_z z} \qquad (1.80)$$

where the phase constant in the z-direction is

$$k_z = \sqrt{\mathsf{x}\mathsf{b}} = \sqrt{k^2 - k_x^2} = k \cos\theta \qquad (1.81)$$

and the equivalent characteristic impedance

$$Z_c = \frac{\mathsf{x}}{k_z} = \frac{k_z}{\mathsf{b}}$$

is given in Table 1.1 for the two polarizations.

When using the transmission line equations to model plane wave problems (or when modeling transmission lines with unusual distributed parameters—see problem p1-3 and example 1.2.2), $\mathsf{x}(z)$ or $\mathsf{b}(z)$ may become negative for incidence angles sufficiently close to $\pi/2$ as a result of $\cos\theta(z)$ becoming imaginary. The signs of the square roots taken when calculating $\gamma = jk_z$ and Z_c must then be chosen carefully. The situation is similar to that of example 1.2.2. We choose the square root for γ as discussed in connection with (1.9), so that β is positive if $\gamma = j\beta$ is pure imaginary, and $\gamma = \alpha > 0$ if it is real (in which case $\cos\theta$ becomes negative imaginary). The characteristic impedance should be calculated from (1.37). This point is illustrated by the example in the next subsection.

Table 1.1

Equivalent variables for transmission line and plane wave problems.

Trans. line	Plane wave (TE)	Plane Wave (TM)
$V(z)$	$E_y(z)$	$E_x(z)$
$I(z)$	$-H_x(z)$	$H_y(z)$
$\mathsf{x}(z)$	$\omega\mu(z)$	$\omega\mu(z) - \frac{k_x^2}{\omega\epsilon(z)} = \omega\mu(z)\cos^2\theta(z)$
$\mathsf{b}(z)$	$\omega\epsilon(z) - \frac{k_x^2}{\omega\mu(z)} = \omega\epsilon(z)\cos^2\theta(z)$	$\omega\epsilon(z)$
$Z_c(z)$	$\frac{\zeta(z)}{\cos\theta(z)}$	$\zeta(z)\cos\theta(z)$
$\gamma(z)$	$j\omega\sqrt{\mu(z)\epsilon(z)}\cos\theta(z)$	$j\omega\sqrt{\mu(z)\epsilon(z)}\cos\theta(z)$

1.4.1 EXAMPLE: REFLECTION AND TRANSMISSION AT AN INTERFACE

From what has been said above, the problem of plane wave reflection at the plane interface between two media can be reduced to that of two transmission lines connected at the point $z = 0$. If the line occupying $z < 0$ has characteristic impedance Z_{c1}, while that of the line in $z > 0$ is Z_{c2}, the reflection coefficient of a wave incident on line 1 is easily obtained by recognizing that the semi-infinite line 2, on which no incoming wave is present, is equivalent to a lumped load impedance $Z_L = Z_{c2}$ connected to line 1 at $z = 0$. From (1.58), the (voltage) reflection coefficient for the incident wave on line 1 is:

$$\rho = \frac{Z_{c2} - Z_{c1}}{Z_{c2} + Z_{c1}} \tag{1.82}$$

The transmitted wave on line 2 must have a voltage at $z = 0$ equal to the total voltage on line 1 at this same point, by Kirchhoff's law. Thus,

$$V_2(0) = V_1(0) = (1 + \rho)V_1^{\text{inc}}$$

so that the (voltage) transmission coefficient τ is

$$\tau = \frac{V_2(0)}{V_1^{\text{inc}}} = 1 + \rho = \frac{2Z_{c2}}{Z_{c2} + Z_{c1}} \tag{1.83}$$

It is evident that the reflection and transmission coefficients are functions only of the relative characteristic impedances of the two lines, and will involve their propagation coefficients only indirectly (through the angles of incidence and transmission).

Now let us use this model to treat TE-polarized plane waves. Let μ_1 and ϵ_1 be the parameters of the region $z < 0$ from which the incident wave comes, and μ_2 and ϵ_2 be the parameters of the region $z > 0$. Since k_x is the same in both regions to ensure continuity of the fields for all x, we have Snell's law:

$$k_x = k_2 \sin \theta_2 = k_1 \sin \theta_1 \tag{1.84}$$

where θ_1 is the angle of incidence in region 1, and θ_2 is the angle of propagation of the transmitted wave in region 2. Then from (1.82), (1.83) and Table 1.1, we obtain for ρ and τ the *electric-field* reflection and transmission coefficients given by the classical Fresnel formulas:

$$\rho_{\text{TE}} = \frac{\zeta_2 \cos \theta_1 - \zeta_1 \cos \theta_2}{\zeta_2 \cos \theta_1 + \zeta_1 \cos \theta_2} \tag{1.85}$$

$$\tau_{\text{TE}} = \frac{2\zeta_2 \cos \theta_1}{\zeta_2 \cos \theta_1 + \zeta_1 \cos \theta_2} \tag{1.86}$$

Note that because there is only one component of electric field in this polarization, the voltage reflection and transmission coefficients on the transmission-line analog problem are identical to the Fresnel coefficients ρ_{TE} and τ_{TE} which are ratios of the electric field.

In the TM polarization, V is the analog of only one component (E_x) of the total electric field, whereas the Fresnel coefficients ρ_{TM} and τ_{TM} are ratios of *total* E-fields. In general, therefore, ρ_{TM} and τ_{TM} will not be the same as the ρ and τ obtained from the equivalent transmission line problem, except when the incidence angle $\theta_1 = 0$. The reader is asked to fill in the details of the determination of the TM Fresnel coefficients in problem p1-15.

If transmission line 2 has, say, $x_2 > 0$ but $b_2 < 0$, then its characteristic impedance is positive imaginary, and the reflected wave on line 1 acquires a phase shift only, with no reduction in amplitude:

$$\rho = e^{j\chi} \tag{1.87}$$

where

$$\chi = 2\tan^{-1}\sqrt{-\frac{\mathsf{x}_1\mathsf{b}_2}{\mathsf{x}_2\mathsf{b}_1}} = \pi - 2\tan^{-1}\sqrt{-\frac{\mathsf{x}_2\mathsf{b}_1}{\mathsf{x}_1\mathsf{b}_2}} \qquad (1.88)$$

In particular, when modeling TE plane wave reflection when $k_x = k_2\sin\theta_2 = k_1\sin\theta_1 > k_2$, we have from Snell's law that $\sin\theta_2 > 1$, and hence $\cos\theta_2$ must be negative imaginary (θ_2 is a complex angle). Then

$$\rho_{\mathrm{TE}} = e^{j\chi_{\mathrm{TE}}(\theta)} \qquad (1.89)$$

where (since $\cos\theta_2 = -j\sqrt{\sin^2\theta_2 - 1} = -j\sqrt{k_1^2\sin^2\theta_1 - k_2^2}/k_2$)

$$\chi_{\mathrm{TE}}(\theta) = \pi - 2\tan^{-1}\left(\frac{\zeta_2}{\zeta_1}\frac{k_2\cos\theta_1}{\sqrt{k_1^2\sin^2\theta_1 - k_2^2}}\right) \qquad (1.90)$$

Thus, the magnitude $|\rho_{\mathrm{TE}}|$ of the reflection coefficient becomes equal to 1, and we have the phenomenon of total reflection, which serves as the basis surface wave guidance in dielectric waveguides.

1.5 NONUNIFORM TRANSMISSION LINES AND THE WKB APPROXIMATION

We often encounter *nonuniform* transmission lines, either as actual transmission lines or as models of mathematically analogous wave propagation problems. When the variation of parameters is discontinuous, we simply connect terminals of sections of different lines together, and enforce continuity of current and voltage as boundary conditions. When the line parameters vary continuously with z, however, we can no longer employ the simple exponential solutions for forward and reverse going waves that are obtained for the uniform line.

In most cases, the transmission-line equations cannot be solved exactly for arbitrary continuous variations of z and y. One can take the approach of breaking the line into a cascaded connection of a large number of line sections whose line parameters are constant within each section. Although the problem could now be handled by ordinary techniques of transmission-line analysis, it leads to equations which are unwieldy to solve. If the changes in line parameters with z are gradual enough, it is possible to obtain an explicit approximate solution to the nonuniform transmission line problem that is considerably simpler. This approximation is known as the WKB (Wentzel-Kramers-Brillouin) approximation.

1.5.1 LOWEST-ORDER WKB APPROXIMATION

To derive this approximation to lowest order, let us consider a lossless nonuniform transmission line lying between $0 < z < d$ and terminated with a load impedance Z_L at the point $z = d$ (it is not difficult to extend the analysis to lossy lines). Let us temporarily replace this line by an approximate one having piecewise uniform values of x and b as shown in Figure 1.13. In each uniform section (section i will be designated as the interval $z_{i-1} < z < z_i$), the characteristic impedance and propagation coefficient have the constant values Z_{ci} and γ_i respectively. If the characteristic impedance Z_c of this line varies sufficiently slowly from one section (i, say) to the next ($i+1$), then a $+z$-traveling voltage wave in section i:

$$V^+(z) = V_i^+ e^{-\gamma_i z} \qquad (1.91)$$

upon encountering the junction point z_i to the next section $i+1$ will undergo nearly zero reflection, and be transmitted with a voltage transmission coefficient that is nearly 1. More

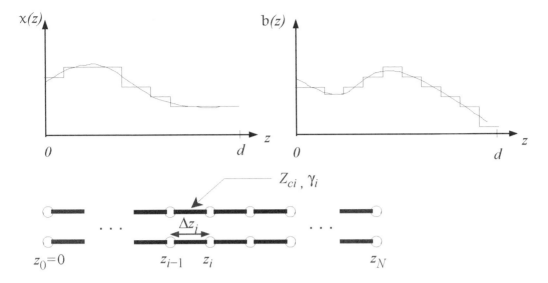

Figure 1.13 Piecewise constant approximation to a nonuniform transmission line.

precisely, by (1.82) and (1.83), we have

$$\tau = 1 + \rho = \sqrt{1 - \rho^2}\sqrt{\frac{1 + \rho}{1 - \rho}} = \sqrt{1 - \rho^2}\sqrt{\frac{Z_{c,i+1}}{Z_{ci}}} \simeq \sqrt{\frac{Z_{c,i+1}}{Z_{ci}}} \qquad (1.92)$$

where we have neglected terms of the order of the very small quantity ρ^2. Hence the forward voltage wave at the beginning of section $i + 1$ will be approximately

$$V^+(z_i^+) \simeq \sqrt{\frac{Z_{c,i+1}}{Z_{ci}}}V^+(z_i^-) \qquad (1.93)$$

In terms of wave amplitudes, this is simply

$$a(z_i^+) \simeq a(z_i^-) \qquad (1.94)$$

In other words, a is continuous in this approximation.

Applying this result to each of the junctions between transmission line sections, we get the wave amplitude at a junction z_N in terms of that at $z_0 = 0$:

$$a(z_N^-) \simeq a(0^+)\exp\left(-\sum_{i=1}^{N}\gamma_i\Delta z_i\right) \qquad (1.95)$$

In terms of the voltage, we have

$$V^+(z_N^-) \simeq V^+(0^+)\sqrt{\frac{Z_{cN}}{Z_{c1}}}\exp\left(-\sum_{i=1}^{N}\gamma_i\Delta z_i\right) \qquad (1.96)$$

where $\Delta z_i = d_i = z_i - z_{i-1}$ is the length of section i.

If we pass to the limit as all the Δz_i's go to zero and the number of sections N goes to infinity in such a way that the position $z_N = z$ on the nonuniform line segment remains constant, then the sum in the exponent (which is a Riemann sum) passes over into an integral, and we get

$$a(z) \simeq a(0)e^{-Q(z)} \qquad (1.97)$$

or

$$V^+(z) \simeq V^+(0)\sqrt{\frac{Z_c(z)}{Z_c(0)}}\,e^{-Q(z)} \tag{1.98}$$

where

$$Q(z) = \int_0^z \gamma(z')\,dz' \tag{1.99}$$

$\gamma(z)$ is the position-dependent propagation coefficient corresponding to the z-dependent values of the line parameters x and b:

$$\gamma(z) = j\beta(z) = j\sqrt{\mathsf{x}(z)\mathsf{b}(z)}$$

and $Z_c(z)$ is the position-dependent characteristic impedance

$$Z_c(z) = \sqrt{\frac{\mathsf{x}(z)}{\mathsf{b}(z)}}$$

The formulas to be derived can easily be extended to lossy lines or nonpropagating modes by using the formulas

$$\gamma(z) = \sqrt{\mathsf{z}(z)\mathsf{y}(z)}$$

and

$$Z_c = \sqrt{\frac{\mathsf{z}(z)}{\mathsf{y}(z)}}$$

for the more general case. As was done for the uniform line, we choose the square roots so that Z_c is positive real if γ is pure imaginary, and otherwise $\mathrm{Re}(\gamma) > 0$. We write

$$Q(z) = Q_r(z) + jQ_i(z) \tag{1.100}$$

where

$$Q_r(z) = \int_0^z \alpha(z')\,dz' \tag{1.101}$$

and

$$Q_i(z) = \int_0^z \beta(z')\,dz' \tag{1.102}$$

The term $Q_i(z)$ is called a phase integral, because it represents the accumulated phase reduction of the voltage wave on the nonuniform line. We can similarly call $Q_r(z)$ the attenuation integral, because it represents the accumulated attenuation along the nonuniform line. Finally, since the reflection at each incremental section has been assumed to be small, it must be true that the line impedance is tracking with the characteristic impedance,

$$Z(z) = \frac{V^+(z)}{I^+(z)} \simeq Z_c(z)$$

Therefore, the current of this $+z$-traveling wave is given by

$$I^+(z) \simeq \frac{V^+(0)}{\sqrt{Z_c(0)Z_c(z)}}\,e^{-Q(z)} \tag{1.103}$$

Analogous equations can be obtained for the voltage and current of a $-z$-traveling wave:

$$b(z) \simeq b(0)e^{+Q(z)} \tag{1.104}$$

$$V^-(z) \simeq V^-(0)\sqrt{\frac{Z_c(z)}{Z_c(0)}}e^{+Q(z)} \tag{1.105}$$

$$I^-(z) \simeq -\frac{V^-(0)}{\sqrt{Z_c(0)Z_c(z)}}e^{+Q(z)} \tag{1.106}$$

We then see by examining the resulting equations that the input impedance at the point $z = 0$ of an inhomogeneous section of transmission line between $z = 0$ and $z = d$ loaded by an impedance Z_L is given by

$$Z(0) \simeq Z_c(0)\frac{Z_L \cosh Q(d) + Z_c(d)\sinh Q(d)}{Z_c(d)\cosh Q(d) + Z_L \sinh Q(d)} \tag{1.107}$$

which corresponds to a reflection coefficient of

$$\rho(0) = \frac{Z(0) - Z_c(0)}{Z(0) + Z_c(0)} \simeq \rho_{\mathrm{WKB}}(0) \equiv \rho_L e^{-2Q(d)} \tag{1.108}$$

at $z = 0$, where

$$\rho_L = \frac{Z_L - Z_c(d)}{Z_L + Z_c(d)} \tag{1.109}$$

is the reflection coefficient of Z_L at the load point $z = d$. These results can be used in place of (1.60) or (1.59) when the section of line is nonuniform. Note that in this approximation, no reflection is introduced by the nonuniformity of the line; if the load is matched at the far end $z = d$ of the line [$Z_L = Z_c(d)$], then it will be matched at the near end $z = 0$ as well ($\rho(0) \simeq 0$).

More generally, a section of nonuniform line can be represented in the lowest WKB approximation by the chain parameter description (1.61), where now the chain parameters are given by the expressions

$$\begin{aligned}
\mathcal{A}(z_1, z_2) &= \sqrt{\frac{Z_c(z_1)}{Z_c(z_2)}}\cosh[Q(z_2) - Q(z_1)] \\
\mathcal{B}(z_1, z_2) &= \sqrt{Z_c(z_1)Z_c(z_2)}\sinh[Q(z_2) - Q(z_1)] \\
\mathcal{C}(z_1, z_2) &= \frac{1}{\sqrt{Z_c(z_1)Z_c(z_2)}}\sinh[Q(z_2) - Q(z_1)] \\
\mathcal{D}(z_1, z_2) &= \sqrt{\frac{Z_c(z_2)}{Z_c(z_1)}}\cosh[Q(z_2) - Q(z_1)]
\end{aligned} \tag{1.110}$$

which reduces to (1.62) for the case of a section of uniform line. The proof is left to the reader as an exercise. Equation (1.110) can be used to prove (1.107), for example.

1.5.2   FIRST-ORDER (RAYLEIGH-BREMMER) WKB APPROXIMATION

If the load impedance Z_L in (1.107) is matched to the local characteristic impedance $Z_c(d)$ at the end of the nonuniform line section, we get $Z(0) = Z_c(0)$ independent of the electrical length Δ of the section, so to this order of approximation this section would be ideally matched to a uniform line of characteristic impedance $Z_c = Z(0)$ connected to the left of this section in $z < 0$. However, this result is not exactly true, as we have assumed that the small incremental reflections at each of the junctions between piecewise constant line sections in Figure 1.13 can be neglected. A better approximation is obtained if we account for this reflection as follows.

We temporarily assume that the load reflection coefficient $\rho_L = 0$. At the point z_i, the forward wave is reflected with an incremental reflection coefficient

$$\Delta\rho|_{z_i} = \frac{Z_{c,i+1} - Z_{ci}}{Z_{c,i+1} + Z_{ci}} \simeq \Delta z_i \frac{Z_c'(z_i)}{2Z_c(z_i)}$$

Seen from the input point $z = 0$, we must include a round-trip factor of phase or attenuation for the wave, so the reflection coefficient appears there as

$$\Delta\rho|_{z=0} \simeq e^{-2Q(z_i)} \Delta z_i \frac{Z_c'(z_i)}{2Z_c(z_i)}$$

Summing up all the contributions from these incremental reflections, and passing from a sum to an integral as we let the number of piecewise constant sections of line become infinite as before, we have

$$\rho(0) \simeq \rho_B^+(0) \equiv \int_0^d e^{-2Q(z)} \frac{Z_c'(z)}{2Z_c(z)}\, dz \tag{1.111}$$

where $\rho_B^+(0)$ is the Rayleigh-Bremmer reflection coefficient, named for the scientists who first devised this approximation. The notation $\rho_B^+(0)$ indicates that this is the cumulative reflection coefficient of an initially forward wave, evaluated at $z = 0$, due to the incremental reflections from nonuniformity of the characteristic impedance (and not reflection from the load impedance at the end of the line). It can be shown by more detailed study of this approximation that it is valid whenever

$$\left| \frac{Z_c'(z)}{4\gamma(z)Z_c(z)} \right| \ll 1 \tag{1.112}$$

which in turn implies that the reflection coefficient resulting from the accumulation of these incremental reflections is small.

If the load reflection coefficient is not zero, there will be an additional contribution to the reverse wave at $z = 0$ coming from the reverse wave that starts at $z = d$ and passes through the length of nonuniform line according to the lowest order WKB approximation of the previous subsection. The total reverse wave amplitude at $z = 0$ is therefore:

$$b(0) \simeq \rho_B^+(0)a(0) + e^{-Q(d)}b(d) \tag{1.113}$$

where the wave amplitudes $a(z)$ and $b(z)$ are defined in (1.33). In a similar manner, the wave emerging from the $z = d$ end of the line is given by

$$a(d) \simeq \rho_B^-(d)b(d) + e^{-Q(d)}a(0) \tag{1.114}$$

where

$$\rho_B^-(d) \equiv -e^{-2Q(d)} \int_0^d e^{2Q(z)} \frac{Z_c'(z)}{2Z_c(z)}\, dz \tag{1.115}$$

is the Rayleigh-Bremmer reflection coefficient of a reverse ($-z$ traveling) wave seen at $z = d$, which is derived in a similar way to (1.111). Expressing the wave amplitudes in terms of voltages and currents, and rearranging the resulting equations in the form of the chain relations (1.61), we find that the chain parameters of a nonuniform line in the Rayleigh-Bremmer (first-order) approximation are:

$$\mathcal{A}(0,d) = \frac{1}{2}\sqrt{\frac{Z_c(0)}{Z_c(d)}}\left\{ e^{Q(d)}\left[1+\rho_B^+(0)\right]\left[1-\rho_B^-(d)\right]+e^{-Q(d)}\right\}$$

$$\mathcal{B}(0,d) = \frac{1}{2}\sqrt{Z_c(0)Z_c(d)}\left\{ e^{Q(d)}\left[1+\rho_B^+(0)\right]\left[1+\rho_B^-(d)\right]-e^{-Q(d)}\right\}$$

$$\mathcal{C}(0,d) = \frac{1}{2}\frac{1}{\sqrt{Z_c(0)Z_c(d)}}\left\{ e^{Q(d)}\left[1-\rho_B^+(0)\right]\left[1-\rho_B^-(d)\right]-e^{-Q(d)}\right\}$$

$$\mathcal{D}(0,d) = \frac{1}{2}\sqrt{\frac{Z_c(d)}{Z_c(0)}}\left\{ e^{Q(d)}\left[1-\rho_B^+(0)\right]\left[1+\rho_B^-(d)\right]+e^{-Q(d)}\right\} \tag{1.116}$$

Once again, the algebraic details of this derivation are left to the reader as an exercise.

Example: Exponential Line

As an example, suppose the propagation coefficient on the nonuniform section $\gamma = j\beta$ is constant, while the characteristic impedance

$$Z_c = Z_{c0}e^{(z/d)\ln(Z_L/Z_{c0})}$$

varies exponentially between the value Z_{c0} (equal to the characteristic impedance of a uniform line connected to the left of the terminals $z = 0$) and the load impedance Z_L. Substitution into (1.111) gives

$$\rho(0) \simeq \rho_B^+(0) = \frac{\ln(Z_L/Z_{c0})}{2d}\int_0^d e^{-2j\beta z}\,dz = e^{-j\beta d}\frac{\sin\beta d}{2\beta d}\ln\left(\frac{Z_L}{Z_{c0}}\right) \tag{1.117}$$

According to the general criterion for validity of the WKB approximation, for slow enough transitions that $\beta d > 4\ln(Z_L/Z_{c0})$ (say), this result should be accurate. Comparing (1.117) with the exact result (F.13) in section F.1, we find that this result is accurate to better than 2.5% for $\beta d > 1.3$. The largest magnitude attained by $\rho(0)$ for $\beta d > 2.7$ is no greater than 0.15.

1.5.3   TURNING POINTS

When the rate of change of the characteristic impedance of a nonuniform transmission line in z becomes too rapid, the approximation that reflections at successive interfaces between piecewise uniform line sections are nearly zero breaks down. This will occur, for example, at points for which $c(z)$ or $l(z)$ has a simple zero—values of z at which this occurs are said to be *turning points*. The analysis must be modified near turning points, and a rigorous derivation is provided in Appendix F. Here, a heuristic analysis will be made to account for the reflection from a turning point (it should be emphasized that this heuristic method fails to give the proper value for the transmission coefficient at a turning point, and this quantity must be obtained from the method of Appendix F).

Suppose for definiteness that we have a capacitive turning point at $z = z_t$. That is, $x(z_t) \neq 0$, $b(z_t) = 0$ and $b'(z_t) \neq 0$. We will further assume that b is a decreasing function of z, so that in fact $b'(z_t) < 0$. If b is otherwise still a slowly varying function of z, then the lowest order WKB approximation will break down only in a relatively small neighborhood of the turning point: $z_t - \delta z < z < z_t + \delta z$. We now replace $b(z)$ in this neighborhood by the piecewise constant expression

$$\tilde{b}(z) = b(z_t - \delta z) \simeq -\delta z b'(z_t)$$

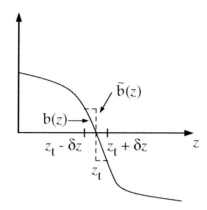

Figure 1.14 Modified value of $b(z)$ near a turning point z_t.

for $z_t - \delta z < z < z_t$, and

$$\tilde{b}(z) = b(z_t + \delta z) \simeq \delta z\, b'(z_t)$$

for $z_t < z < z_t + \delta z$. Thus, $\tilde{b}$ has a small discontinuity at the turning point, but joins continuously with $b(z)$ outside this neighborhood (see Figure 1.14).

The ordinary WKB method can be applied to the entire range of z now, except at the turning point, where we can obtain the reflection coefficient at the discontinuity by usual methods (and to lowest order, this is the only contribution to the overall reflection coefficient). A wave incident from the left ($z < 0$) will thus propagate from $z = 0$ to $z = z_t$ and back, accumulating a round-trip phase factor of $\exp[-2Q(z_t)]$, and will additionally encounter total reflection at z_t, with a reflection coefficient of

$$\rho \simeq \frac{j\sqrt{\dfrac{x(z_t)}{-\delta z\, b'(z_t)}} - \sqrt{\dfrac{x(z_t)}{-\delta z\, b'(z_t)}}}{j\sqrt{\dfrac{x(z_t)}{-\delta z\, b'(z_t)}} + \sqrt{\dfrac{x(z_t)}{-\delta z\, b'(z_t)}}} = e^{j\pi/2} \qquad \text{(capacitive turning point)} \qquad (1.118)$$

The voltage exponentially decays beyond the turning point, and if a second turning point is not located too close to the first, we may ignore any penetration of the wave beyond the first one, and regard it as if it were isolated.

Thus, in this case, a capacitive turning point located at z_t in an inhomogeneous section of line beginning at $z = 0$ produces an input impedance seen at $z = 0 < z_t$ of:

$$Z(0) = Z_c(0)\frac{1 + e^{-2Q(z_t)+j\pi/2}}{1 - e^{-2Q(z_t)+j\pi/2}} = Z_c(0)\frac{j\cosh Q(z_t) + \sinh Q(z_t)}{\cosh Q(z_t) + j\sinh Q(z_t)} \qquad (1.119)$$

It is as if the effect of the portion of transmission line beyond the turning point were that of a lumped load reactance of normalized impedance[4] $Z_L/Z_c = j$ placed at the turning point, with the line section in $z > z_t$ removed. This case corresponds to the TE polarization of plane waves in Table 1.1 if the permittivity ϵ is a function of position, while μ remains constant (the usual circumstance).

A similar derivation for the case of an inductive turning point [$c(z_t) \neq 0$, $l(z_t) = 0$ and $l'(z_t) < 0$] gives a voltage reflection coefficient of

$$\rho = \exp(-j\pi/2) \qquad \text{(inductive turning point)} \qquad (1.120)$$

---

[4]Of course, $Z_c(z_t) = \infty$, so we must understand this idea in a limiting sense.

at the turning point, and thus an input impedance seen at $z = 0 < z_t$ of:

$$Z(0) = Z_c(0)\frac{1 + e^{-2Q(z_t) - j\pi/2}}{1 - e^{-2Q(z_t) - j\pi/2}} = Z_c(0)\frac{-j\cosh Q(z_t) + \sinh Q(z_t)}{\cosh Q(z_t) - j\sinh Q(z_t)} \tag{1.121}$$

because the phase of the characteristic impedance just beyond the turning point is now $-\pi/2$ rather than $+\pi/2$ as for the capacitive turning point. This case corresponds to the TM polarization of plane waves in Table 1.1 if the permittivity ϵ is a function of position, while μ remains constant.

The amplitude of the transmitted evanescent wave in $z > z_t$ is not obtained correctly using this trick; the transmission coefficient must be found by the method of Appendix F, and we merely quote the result here [see (F.76), (F.81) and (F.82)]. If $z_0 < z_t$ and $z_1 > z_t$,

$$\tau(z_1, z_0) = \frac{V^{\mathrm{tr}}(z_1)}{V^{\mathrm{inc}}(z_0)} \simeq e^{-j[Q_i(z_t) - Q_i(z_0)] - [Q_r(z_1) - Q_r(z_t)]}\sqrt{\frac{Z_c(z_1)}{Z_c(z_0)}} \tag{1.122}$$

where the phase of the square root of the characteristic impedance $\sqrt{Z_c(z_1)}$ for $z > z_t$ is understood to be $+\frac{\pi}{4}$ if $\mathrm{Im}\,[Z_c(z_1)] > 0$ (a capacitive turning point), or $-\frac{\pi}{4}$ if $\mathrm{Im}\,[Z_c(z_1)] < 0$ (an inductive turning point).

Example: Reflection of TE Plane Wave from a Turning Point

As an application, we consider the reflection of an obliquely incident TE-polarized plane wave from a half-space with constant material parameters ϵ_1 and μ_0 in $z < 0$ onto a region in $z > 0$ with linearly decreasing permittivity $\epsilon = \epsilon_1(1 - z/d)$, where d is a length scale which indicates the rate of variation of ϵ with z.[5] With $k_c = k_1$, this problem is equivalent to a nonuniform transmission line problem with $\mathsf{l}(z) = \mu_0$ and $\mathsf{c}(z) = \epsilon_1(\cos^2\theta - z/d)$. There is a turning point at $z_t = d\cos^2\theta$, and so we have

$$Q(z_t) = j\omega\int_0^{z_t}\sqrt{\mathsf{l}(z)\mathsf{c}(z)}\,dz = \frac{2}{3}jk_1\frac{z_t^{3/2}}{d^{1/2}} = \frac{2}{3}jk_1 d\cos^3\theta$$

Thus, the impedance seen at $z = 0$ is

$$Z(0) = j\zeta_1\frac{1 + \sin 2\vartheta}{\cos 2\vartheta}$$

where $\zeta_1 = \sqrt{\mu_0/\epsilon_1}$ and $\vartheta = \frac{2}{3}k_1 d\cos^3\theta$, and the resulting reflection coefficient there is

$$\rho(0) = \exp\left[2j\tan^{-1}\left(\frac{\cos\vartheta}{1 + \sin\vartheta}\right)\right]$$

Perhaps the most noticeable difference between the behavior of this reflection coefficient and that of a wave totally reflected from an abrupt dielectric interface is that as $\theta \to \pi/2$, the phase here approaches $\pi/2$, whereas that of the abrupt interface approaches π. This has implications for the use of this effect in dielectric waveguide operation, as will be seen in Section 7.2.

---

[5] Of course, this permittivity becomes negative for $z > d$, but we need to assume only that this linear variation holds up to the turning point $z_t < d$, after which the permittivity could, for example, stabilize to a constant without affecting the result we are about to derive.

1.6 NOTES AND REFERENCES

The reader will find it useful from time to time to refer to a good introductory text in electromagnetics for such topics as Maxwell's equations, elements of vector analysis, elementary transmission-line theory, and so on. The following are recommended:

Javid and Brown (1963);
Johnk (1975);
Magid (1972);
Portis (1978).

but many others are also suitable for one or more of these topics.

Examples of natural waveguides in terrestrial environments are discussed in

Bremmer (1949);
Budden (1961);
Delogne (1982).

The transmission-line equations (with negligible series inductance) have also been found to provide a good model for the description of electrical signals on neurons; see:

Tuckwell (1988).

Acoustic phenomena are also frequently modeled by the classical transmission-line equations. See, for example,

Morse and Ingard (1968), Chapter 9;
Fletcher and Rossing (1991).

The latter book shows how a wide variety of musical instruments can be described in terms of transmission lines. The reader familiar with electric circuits and transmission lines will find the terminology used in these books quite familiar. Interesting examples in the area of active and nonlinear transmission lines (topics not covered in this book) are given in

Scott (1970);
Wong (1993);
Remoissenet (1996).

What is perhaps the earliest published account of an experimental demonstration of electrical transmission line phenomena is given in:

W. Watson, "An account of the Experiments made by several Gentlemen of the Royal Society, in order to discover whether or no the electrical Power, when the Conductors thereof were not supported by Electrics *per se* would be sensible at great Distances: With an Inquiry concerning the respective Velocities of Electricity and Sound: To which is added an appendix, containing some further Inquiries into the Nature and Properties of Electricity," *Phil. Trans. Roy. Soc. London*, vol. 45, pp. 49-91 (1748).

W. Watson, "An account of the Experiments made by some Gentlemen of the Royal Society, in order to measure the absolute Velocity of Electricity," *Phil. Trans. Roy. Soc. London*, vol. 45, pp. 491-496 (1748).

The description of the sometimes reluctant participants in these experiments is quite amusing in places, especially given that no one seems to have been permanently harmed in the process. The telegraphers' equations were first derived by Lord Kelvin more than a century later:

W. Thomson (Lord Kelvin), "On the theory of the electric telegraph," *Proc. Roy. Soc. London*, vol. 7, pp. 382-399 (1855) [also in his *Mathematical and Physical Papers*, vol. 2. Cambridge, UK: University Press, 1884, pp. 61-76].

although in this original work, the series inductance of the line was neglected. A few years later, Kirchhoff would include the inductance for a special case, but Heaviside appears to have been the first to present the complete equations for the general case:

O. Heaviside, "On the extra current," *Phil. Mag.*, ser. 5, vol. 2, pp. 135-145 (1876) [also in his *Electrical Papers*, vol. 1. New York: Chelsea, 1970, pp. 53-61].

The classical transmission line with unusual series or shunt parameters can be used as a model for TE or TM modes in hollow metallic waveguides:

Milton and Schwinger (2006), Chapter 13.

(see also Chapter 8). It also serves as a model for voltage and current in distributed transformer windings that have significant capacitance between adjacent turns:

Bewley (1963), Chapters 15-16.
Rüdenberg (1968), Chapters 19-21.

One-, two- and three-dimensional versions of the backward-wave transmission line are described in

Caloz and Itoh (2006).

wherein the circuits are also used as models for metamaterials.

The representation of plane wave reflection and transmission problems using equivalent transmission-line networks is discussed in

Solimeno, *et al.* (1986), Chapter 3, Section 7.

We have used the common designation "WKB" for the approximation presented in Section 1.5.1, although properly speaking this refers only to the researchers who introduced the technique into the field of quantum mechanics. It can be found in the mathematical literature a century before that, and is also referred to as the Liouville-Green (LG) approximation. The history of the method is discussed in

Olver (1974), Chapter 6.

More detailed treatments of the WKB method and propagation in inhomogeneous media are to be found in many sources, for example:

Brekhovskikh (1960), Section 16;
Sodha and Ghatak (1977);
Marcuse (1982), Chapter 11;
Snyder and Love (1983), Chapter 35;
Solimeno, *et al.* (1986), Chapter 3.

The approach used here was first used by Rayleigh

Lord Rayleigh, "On the propagation of waves through a stratified medium, with special reference to the question of reflection," *Proc. Roy. Soc. London A*, vol. 86, pp. 207-226 (1912).

in a special case, but is usually attributed to Bremmer:

> H. Bremmer, "The W.K.B. approximation as the first term of a geometric-optical series," *Commun. Pure Appl. Math.*, vol. 4, pp. 105-115 (1951).

where higher order approximations are also derived. The phase shift due to reflection at a turning point is usually treated by recourse to matched asymptotic expansions and Airy functions, as for instance in:

> Brekhovskikh, *loc. cit.*;
> Marcuse (1982), Section 11.5.

The (less rigorous) approach we present in this chapter is due to

> E. Persico, "Dimostrazione elementare del metodo di Wentzel e Brillouin," *Nuov. Cim.*, vol. 15, pp. 133-138 (1938).
>
> S. Kaliski, "The WKB method in the 'classical' approach to quantum mechanics," *Bull. Acad. Polonaise Sci., ser. Sci. Tech.*, vol. 18, pp. 553-559 (1970).
>
> G. B. Hocker and W. K. Burns, "Modes in diffused optical waveguides of arbitrary index profile," *IEEE J. Quantum Electron.*, vol. 11, pp. 270-276 (1975).
>
> D. Marcuse, "Elementary derivation of the phase shift at a caustic," *Appl. Opt.*, vol. 15, pp. 2949-2950 (1976).
>
> F. S. Crawford, "Heuristic derivation of the 'WKB penetration' phase constant for bound states," *Amer. J. Phys.*, vol. 56, pp. 374-375 (1988).

In any case, knowledge of the phase shift by itself does not repair the defect of the WKB solution that the approximate voltage and current can become singular at the turning point. There have been some modifications of the WKB method which do not suffer from this drawback, and yet avoid the use of Airy functions (or Bessel functions of fractional order as in Appendix F); see in particular:

> S. N. Stolyarov and Yu. A. Filatov, "Reflection of electromagnetic waves from inhomogeneous layers," *Radio Eng. Electron. Phys.*, vol. 28, no. 12, pp. 24-28 (1983).

The modified formulas in this paper also remain valid in other situations where ordinary WKB fails, notably when $Z_c(z)$ has a step function behavior. The exact solution of the exponentially tapered transmission line problem can be found in Section F.1 of the Appendices for comparison with the WKB approximation given in this chapter. The WKB solution can be used to design optimal low-reflection transitions between widely differing impedances. The optimal (equal-ripple) high-pass taper was given in

> R. W. Klopfenstein, "A transmission line taper of improved design," *Proc. IRE*, vol. 44, pp. 31-35 (1956).
>
> D. Kajfez and J. O. Prewitt, "Correction to 'A transmission line taper of improved design'," *IEEE Trans. Micr. Theory Tech.*, vol. 21, p. 364 (1973).

Least-square optimized tapers over more restricted frequency bands can be designed using shorter lengths of nonuniform line:

> Kuznetsov and Stratonovich (1964), Chap. 4.

Yao-Wen Hsu and E. F. Kuester, "Direct synthesis of passband impedance matching with nonuniform transmission lines," *IEEE Trans. Micr. Theory Tech.*, vol. 58, pp. 1012-1021 (2010).

1.7   PROBLEMS

p1-1 Derive equation (1.16).

p1-2 Let the values $r(> 0)$, $g(> 0)$ and b of a classical transmission line be given. For what value of x does the attenuation constant α attain its minimum possible value? What is this minimum value, and what are the corresponding values of β and Z_c?

p1-3 A wire helix located parallel to a ground plane as shown in part (a) of the figure can serve as a transmission line. If the wire and ground plane are lossless, and losses in the intervening medium can be neglected, then we can put $r = g = 0$. However, in addition to the series inductance $l\Delta z$ and shunt capacitance $c\Delta z$ shown in the distributed circuit model of Figure 1.3, there are additional effects which should be taken into account. There is a capacitive effect between adjacent windings of the helix, which can be modeled as an *elastance* $s\Delta z$ in parallel with $l\Delta z$ as shown in part (b) of the figure.

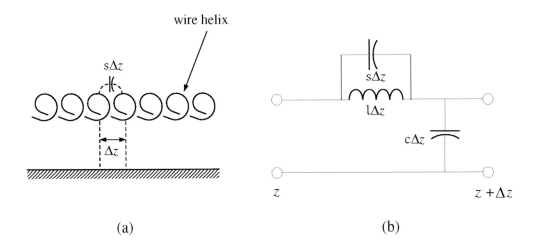

(a)   (b)

(a) Give a physical argument for why the additional capacitance term should have the elastance form $s\Delta z$ rather than a capacitance form $c_w\Delta z$ (as is the case for the shunt capacitance). In other words, as $\Delta z \to 0$, why should this additional capacitance go to infinity rather than to zero?

(b) Using the distributed circuit model for this line, calculate the characteristic impedance Z_c and complex propagation coefficient γ.

(c) Explain the difference in the behavior of Z_c and γ as functions of ω for this line from that of the conventional line described by (1.8) and (1.9).

p1-4 Consider a classical transmission line whose distributed-parameter model is as shown in the figure below.

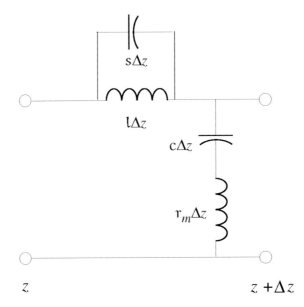

This is like the line of problem p1-3, except that there is an additional term in the shunt impedance—a reluctance per unit length r_m. Obtain expressions for the propagation constant and characteristic impedance of this line. If we assume that $sc > r_m l$, determine whether γ and Z_c are real or imaginary in various ranges of the frequency, and specify their algebraic signs in each case.

p1-5 Consider a classical transmission line whose distributed-parameter model is as shown in the figure below.

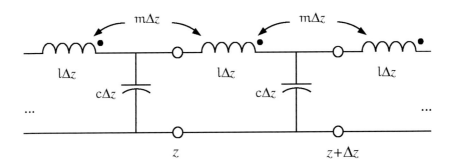

This differs from the classical distributed-network model of a transmission line (Figure 1.3) by the inclusion of a mutual inductance m per unit length between nearest-neighbor series inductances of the circuit model. Derive the telegrapher's equations that apply to this transmission line (be sure to observe the "dot convention" for the mutual inductances). Show that it is equivalent to a line of Figure 1.3 if the series inductance per unit length l is replaced by an equivalent value l_{eq}, and give an expression for l_{eq}.

p1-6 Consider a classical transmission line whose distributed-parameter model is as shown in the figure below.

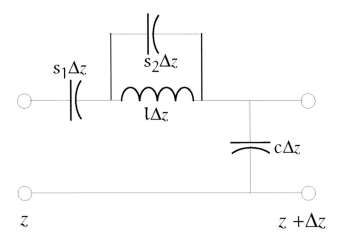

This is like the line of example 1.2.2, except that there is an additional term parallel to the series inductance: an elastance per unit length s_2. Define the frequencies

$$\omega_{c1} \equiv \sqrt{\frac{s_1 s_2}{l(s_1 + s_2)}} \qquad \omega_{c2} \equiv \sqrt{\frac{s_2}{l}}$$

Make a qualitative plot of the series reactance per unit length x versus frequency, noting in particular what happens when ω is near 0, ω_{c1}, ω_{c2} and ∞. Obtain expressions for the propagation constant and characteristic impedance of this line. Determine whether γ and Z_c are real or imaginary in the various ranges of the frequency, and specify their algebraic signs in each case. What kind of filter does this line resemble?

p1-7 Consider a classical transmission line whose distributed-parameter model is as shown in the figure below.

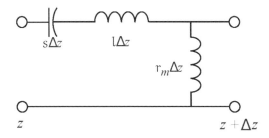

(a) Draw a sketch of a physical transmission line that would have this as its distributed-network model.

(b) Obtain expressions for the propagation constant and characteristic impedance of this line, using the notation

$$\omega_c = \sqrt{\frac{s}{l}}$$

and distinguishing between the ranges $\omega < \omega_c$ and $\omega > \omega_c$. Determine whether γ and Z_c are real or imaginary in each frequency range, and specify their algebraic signs in each case.

p1-8 Use equation (A.32) to prove that, on a passive transmission line which is terminated in a load impedance Z_L with real part greater than or equal to zero, the line impedance $Z(z)$ will have nonnegative real part for all z. Assume that the line parameters y and z appearing in (1.4) can be nonconstant functions of z whose real parts are greater than or equal to zero everywhere.

p1-9 Consider a lossy transmission line whose characteristic impedance has a phase angle ϕ_Z with $-\pi/2 < \phi_Z < \pi/2$.

(a) Using the result of problem p1-8, prove that the set of possible values of the reflection coefficient $\rho(z)$ lies in the interior of the circle

$$|\rho(z) + j\tan\phi_Z| \leq \sec\phi_Z$$

(b) Show that the time-average power flow obtained from (1.21) at the point $z = 0$ is given by

$$P_{\text{av}} = P_{\text{av,inc}}\left[1 - |\rho|^2 - 2\text{Im}(\rho)\tan\phi_Z\right]$$

where $\rho = \rho(0)$, and if the mode is normalized,

$$P_{\text{av,inc}} = |a_0|^2\cos\phi_Z$$

Show that $P_{\text{av}} \geq 0$, and thus that the condition $|\rho| > 1$ shown to be possible in part (a) does not violate conservation of energy.

p1-10 A length d of lossless transmission line (Z_{c1}, β_1) is connected between a real load impedance Z_L and a second lossless line (Z_{c2}, β_2) as shown. Show that the most general conditions needed in order for no reflected wave to exist on the second line are that either

$$Z_{c1} = \sqrt{Z_L Z_{c2}} \quad \text{and} \quad \beta_1 d = \frac{(2n+1)\pi}{2}, \quad n = 0, 1, 2, \ldots$$

or $Z_{c2} = Z_L$, with

$$\beta_1 d = \frac{(2n)\pi}{2}, \quad n = 0, 1, 2, \ldots \quad \text{if } Z_{c1} \neq Z_{c2}$$

$$\beta_1 d = \text{ arbitrary if } Z_{c1} = Z_{c2}$$

In the first case, we have the so-called quarter-wave transformer.

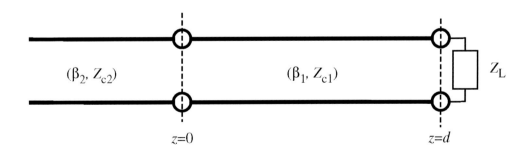

(β_2, Z_{c2}) (β_1, Z_{c1}) Z_L

$z=0$ $z=d$

p1-11 Show how to use a single short-circuited transmission-line stub of length d_2 to provide a matched $(Z_{in} = Z_c)$ input impedance to a transmission line section loaded with an arbitrary impedance Z_L. Specifically, given an arbitrary complex value of Z_L, and real values of β and Z_c, give formulas (not a Smith-chart algorithm) for d_1 and d_2 for the matching arrangement in the figure, which is known as a single-stub tuner. Assume that all transmission lines are lossless ($\alpha = 0$ and Z_c is real).

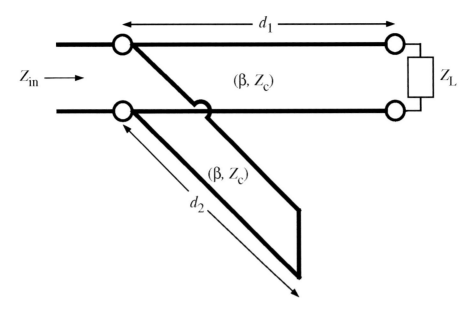

p1-12 A lossless semi-infinite transmission line of a given real characteristic impedance Z_{c1} and propagation constant β_1 is to be matched to a second lossless line of given real characteristic impedance $Z_{c2} \neq Z_{c1}$ and propagation constant β_2, using two lengths l_1 and l_2 of these same types of line, connected as shown below.

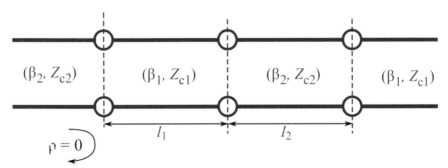

Show that the most general conditions needed in order for no reflected wave to exist on the second line are that

$$\tan \beta_1 l_1 = \tan \beta_2 l_2 = \sqrt{\frac{Z_{c1} Z_{c2}}{Z_{c1}^2 + Z_{c1} Z_{c2} + Z_{c2}^2}}$$

In other words, no lines of characteristic impedance other than Z_{c1} and Z_{c2} are needed to achieve this impedance match. Show also that the total electrical length $\beta_1 l_1 + \beta_2 l_2$ can always be made less than the electrical length $\beta d = \pi/2$ of a quarter-wave transformer.

p1-13 Derive equations (1.61) and (1.62).

p1-14 A length d of the transmission line described in Section 1.2.2 is terminated with a real load impedance $Z_L = R_L$. At the other end of the line, a voltage generator of strength V_G is connected. The parameters V_G, R_L, s, l and c are all assumed to be real and independent of frequency. Make a rough sketch of the ratio $|V_L/V_G|$ vs.

ω, where V_L is the voltage across R_L. You don't have to make precise calculations, just indicate what the ratio does near $\omega = 0$, $\omega = \omega_c$ and as $\omega \to \infty$.

p1-15 By the method used to obtain (1.85) and (1.86), find the (electric field) Fresnel reflection and transmission coefficients ρ_{TM} and τ_{TM} for an obliquely incident, parallel-polarized plane wave at the interface between two media. Note: As normally defined, these coefficients are the ratios of *total* electric fields (which in this polarization have both x and z components). The transmission-line analogy of Table 1.1 produces only ratios of E_x in this polarization—be sure to convert your answer to conform to the usual definitions, and label your figure to indicate directions of positive electric field. For the case of total reflection (when $\mu_{r1}\epsilon_{r1} > (k_x/k_0)^2 > \mu_{r2}\epsilon_{r2}$ or in other words $k_1^2 > k_1^2 \sin^2\theta_1 > k_2^2$), express the reflection coefficient in the form

$$\rho_{\text{TM}} = e^{j\chi_{\text{TM}}}$$

and give an expression for $\chi_{\text{TM}}(\theta)$.

p1-16 A layer of material with permittivity ϵ and permeability μ lies between the planes $z = 0$ and $z = d$. Outside this layer (in $z < 0$ and $z > d$) is free space. Using the transmission line equivalence derived in Section 1.4, find the total reflection coefficient of a TE-polarized plane wave incident at an angle θ to the z-axis from the surface $z = 0$ of this layer.

p1-17 A layer of material with permittivity ϵ and permeability μ lies between the planes $z = 0$ and $z = d$. Outside this layer (in $z < 0$ and $z > d$) is free space. Using the transmission line equivalence derived in Section 1.4, find the total reflection coefficient of a TM-polarized plane wave incident at an angle θ to the z-axis from the surface $z = 0$ of this layer.

p1-18 Consider a lossless nonuniform transmission line with constant $\gamma = j\beta$ but characteristic impedance that varies as

$$Z_c(z) = Z_{c0} \exp\left[\frac{z}{d}\ln\left(\frac{Z_L}{Z_{c0}}\right) + A\sin\left(\frac{2\pi z}{d}\right)\right]$$

where A is a real, dimensionless constant. This impedance, like the example of the exponential line treated in the notes, varies continuously between the values of Z_{c0} at $z = 0$ and some real load impedance Z_L to be connected at $z = d$. The additional term allows the slope of $Z_c(z)$ within the interval $0 < z < d$ to be adjusted by choosing the value of A appropriately. We will investigate whether this can achieve further reduction of the reflection coefficient at the input of the nonuniform line section.

(a) Use equation (1.111) to obtain an expression for the Rayleigh-Bremmer reflection coefficient $\rho_B^+(0)$ at the input when the line is loaded with the impedance Z_L at $z = d$.

(b) Choose a small positive value of A, and using this choice of A plot $\left|\rho_B^+(0)\right|$ vs. βd for $0 < \beta d < 10$. What is the effect of the modified variation of Z_c in this case?

(c) Choose a small negative value of A, and using this choice of A plot $\left|\rho_B^+(0)\right|$ vs. βd for $0 < \beta d < 10$. What is the effect of the modified variation of Z_c in this case?

(d) By varying the values of A used in parts (b) and (c) above, can you in some sense optimize the reflection coefficient? Note that if your choice of A causes $\left|\rho_B^+(0)\right| > 1$ for some frequencies, then the Rayleigh-Bremmer approximation is no longer valid, and this choice of A is not permitted.

p1-19 Consider a lossless nonuniform transmission line with constant β but characteristic impedance which varies as

$$Z_c(z) = Z_{c0} \exp\left\{ \frac{z}{d} \ln\left(\frac{Z_L}{Z_{c0}}\right) + \frac{A}{d^3} z(z-d)\left(z - \frac{d}{2}\right) \right\}$$

where A is a real, dimensionless constant. This impedance, like the example of the exponential line treated in the notes, varies continuously between the values of Z_{c0} at $z = 0$ and some real load impedance Z_L to be connected at $z = d$. The additional term allows the slope of $Z_c(z)$ to be adjusted by choosing the value of A appropriately. We will investigate whether this can achieve further reduction of the reflection coefficient at the input of the nonuniform line section.

(a) Use equation (1.111) to obtain an expression for the reflection coefficient $\rho(0)$ at the input when the line is loaded with the impedance Z_L at $z = d$.

(b) Use whatever means (analytical calculation or numerical computation on a computer) you wish to determine the value of A which minimizes the magnitude of the second highest maximum of $|\rho(0)|$ as a function of βd (the largest maximum occurs at $\beta d = 0$, and cannot be reduced in size). Use the value $Z_L = 4Z_{c0}$ for the load impedance. Plot your result for $|\rho(0)|$ vs. βd, along with comparisons to the result obtained in the notes for the exponential line, and to the value for the case when Z_L directly terminates the line at $z = 0$, with no nonuniform line present.

p1-20 Prove (1.110), using (1.98), (1.103), (1.105) and (1.106). Then use it to obtain (in the lowest-order WKB approximation) an expression for the ratio $V(d)/V(0)$ of output to input voltage for a section of nonuniform transmission line terminated at $z = d$ with a load impedance Z_L.

p1-21 Consider a nonuniform transmission line on which both the characteristic impedance and the propagation constant vary with position in $0 < z < d$. Suppose that $\gamma = j\beta$, with

$$\beta(z) = \beta_0 \left(\frac{1}{2} + \frac{z}{d}\right)$$

and

$$Z_c(z) = Z_{c0} \exp\left[\frac{1}{2}\left(\frac{z}{d} + \frac{z^2}{d^2}\right) \ln\frac{Z_L}{Z_{c0}}\right]$$

where β_0, Z_{c0} and Z_L are positive constants. Plot these two functions vs. z/d, using values such that $Z_L/Z_{c0} = 4$, and compare them to the corresponding functions from the example of Section 1.5.2 (taking $\beta = \beta_0$ for that case). Obtain an expression for the Rayleigh-Bremmer reflection coefficient $\rho_B^+(0)$ of this section of line. How does it compare to that of the exponentially tapered line from the example of Section 1.5.2?

p1-22 Prove (1.116).

p1-23 If a load impedance Z_L is connected to the end $z = d$ of a nonuniform transmission line, such that the reflection coefficient at this end is

$$\rho_L = \frac{Z_L - Z_c(d)}{Z_L + Z_c(d)} \neq 0$$

show that the overall reflection coefficient from the input end $z = 0$ is given in the Rayleigh-Bremmer approximation by

$$\rho(0) \simeq \rho_B^+(0) + \frac{\rho_{\text{WKB}}(0)}{1 - \rho_L \rho_B^-(d)}$$

Give a physical interpretation of this result.

2 MULTIPORT NETWORK THEORY: MATRIX DESCRIPTIONS

Transmission lines are used to interconnect circuit elements in far more complicated ways than those indicated in Chapter 1. Analysis and design of such networks is made more systematic by the introduction of the concepts of multiport networks as used in lumped-element circuit theory. There are, however, certain aspects of this formalism that are special to the inclusion of transmission lines in the network.

2.1 SOURCELESS LINEAR TWO-PORT NETWORKS

Networks composed of lumped elements and sections of transmission line are often viewed in terms of their connection points to the outside world. These *terminals* are nodes of the network, at which connections to external circuits may be made. Not all nodes of a circuit are used as terminals, only those to which external connections are to be made. We will consider networks with $N + 1$ terminals, where $N \geq 1$. At each terminal there is a current flow into the network, and the sum of all terminal currents for a finite network must be zero by Kirchhoff's current law.

Often it is convenient to consider networks whose terminals appear in pairs called *ports*, such that the total current into the network at each port is zero (that is, any current entering the network through one terminal of a port must leave through the other terminal). One terminal of each port is identified as positive for voltage reference. Voltage across the port terminals with this convention is called the port voltage, and the current going into the network through the positive port terminal is called the port current. Unless provided for by external circuit connections or other considerations, it is not permissible to measure a voltage between terminals belonging to two different ports (see the discussion of the classical transmission line in Section 1.2). One such possible external connection arises in the identification of an $(N + 1)$-terminal network as an N-port network. To do this, we choose one of the network terminals and designate it as "ground" or "reference". Each of the other terminals then forms a port together with a terminal connected to the ground terminal. Because each port shares a common terminal connection, it is possible in this case to talk of voltage between any of the port terminals.

2.1.1 IMPEDANCE AND ADMITTANCE MATRICES

Consider a two-port network containing only linear lumped elements and/or segments of linear classical transmission line, but no impressed sources (Figure 2.1). A linear relationship must therefore exist among the terminal variables (the port voltages and currents):

$$V_1 = Z_{11}I_1 + Z_{12}I_2 \tag{2.1}$$
$$V_2 = Z_{21}I_1 + Z_{22}I_2 \tag{2.2}$$

or, in more compact matrix form:

$$[V] = [Z][I] \tag{2.3}$$

Figure 2.1   The linear, sourceless two-port network.

where

$$[V] = \begin{bmatrix} V_1 \\ V_2 \end{bmatrix}$$

$$[I] = \begin{bmatrix} I_1 \\ I_2 \end{bmatrix}$$

$$[Z] = \begin{bmatrix} Z_{11} & Z_{12} \\ Z_{21} & Z_{22} \end{bmatrix}$$

A network is said to be *reciprocal* (see Section A.2.1 of Appendix A) if for any two *states* a and b of the network (that is, for the port voltages and currents resulting from any two sets of external sources connected to the ports of the network), we have

$$(V_1^a I_1^b - V_1^b I_1^a) = (-V_2^a I_2^b + V_2^b I_2^a) \tag{2.4}$$

Eqn. (2.4) can be expressed in compact matrix-vector notation as:

$$[I^b]^T [V^a] = [I^a]^T [V^b] \tag{2.5}$$

where the superscript ^{T} denotes the transpose of a matrix or vector, e. g.,

$$[I]^T = [I_1, I_2] \quad ; \quad [Z]^T = \begin{bmatrix} Z_{11} & Z_{21} \\ Z_{12} & Z_{22} \end{bmatrix}$$

But since (2.3) holds for any state of this junction, (2.5) implies

$$\begin{aligned}
[I^b]^T [Z][I^a] &= [I^a]^T [Z][I^b] \\
&= \left\{ [I^a]^T [Z][I^b] \right\}^T \\
&= [I^b]^T [Z]^T [I^a]
\end{aligned} \tag{2.6}$$

since a scalar quantity is its own transpose. Since (2.6) must hold for *any* $[I^a]$ and $[I^b]$, we conclude that

$$[Z] = [Z]^T \tag{2.7}$$

In other words, $[Z]$ is a *symmetric* matrix for a reciprocal network, which means that

$$Z_{12} = Z_{21} \tag{2.8}$$

This reciprocity law holds only for certain kinds of two-port networks. Later, we will see how to characterize the property of reciprocity in terms of the electromagnetic material parameters of the network. In Appendix A, however, we show how the reciprocity property follows from the properties of the lumped elements and transmission lines which constitute the network.

We might also characterize the junction in terms of an admittance matrix $[Y]$:

$$[I] = [Y][V] \tag{2.9}$$

Clearly $[Y] = [Z]^{-1}$ when the inverse exists, and it is evident that $[Y]$ is also a symmetric matrix when the network is reciprocal. In detail, we have

$$Z_{11} = \frac{Y_{22}}{D_Y}; \qquad Z_{12} = -\frac{Y_{12}}{D_Y} \quad \text{and} \quad Y_{11} = \frac{Z_{22}}{D_Z}; \qquad Y_{12} = -\frac{Z_{12}}{D_Z}$$

$$Z_{21} = -\frac{Y_{21}}{D_Y}; \qquad Z_{22} = \frac{Y_{11}}{D_Y} \qquad\qquad Y_{21} = -\frac{Z_{21}}{D_Z}; \qquad Y_{22} = \frac{Z_{11}}{D_Z} \tag{2.10}$$

where

$$D_Y = Y_{11}Y_{22} - Y_{12}Y_{21}; \qquad D_Z = Z_{11}Z_{22} - Z_{12}Z_{21} = \frac{1}{D_Y}$$

are the determinants of $[Y]$ and $[Z]$ respectively.

As is well known from circuit theory, many networks with two ports can produce the same port responses as a network with a given impedance matrix $[Z]$. The equivalent T-network for a reciprocal two-port network which arises from a 3-terminal network is shown in Figure 2.2. Also commonly used in this case is the equivalent Π-network, which is most easily displayed in terms of the elements of the admittance matrix (Figure 2.3). Note that the impedances and admittances appearing in these equivalent circuits may not each be physically realizable; some, for example, could have negative real parts. Nevertheless, the

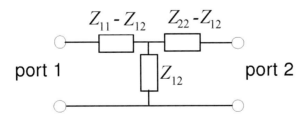

Figure 2.2 Equivalent T-network for a reciprocal two-port.

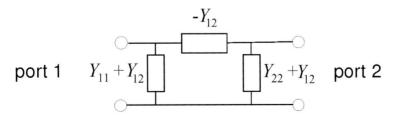

Figure 2.3 Equivalent Π-network for a reciprocal two-port.

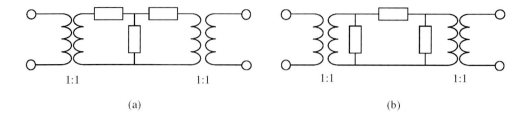

(a)   (b)

Figure 2.4 Generalized T- and Π-network equivalent circuits for reciprocal sourceless two-ports.

equivalent circuit still serves a useful purpose in the analysis of transmission-line/lumped-element circuits.

The T- and Π-networks given above are suitable if only legitimate port connections are allowed to be made. That is, if external connections made to ports 1 and 2 are such that the current into one terminal of each port is always equal to the current leaving the other terminal of the port, these equivalent circuits will suffice. However, if the external connection could result in other current conditions (for example, an impedance is connected between the positive terminals of ports 1 and 2, or there is a possibility that the "ground" terminals of ports 1 and 2 might support a voltage difference), we must insert 1:1 ideal transformers at each port to allow for this. Thus, we have the more general equivalent circuits shown in Figure 2.4. If the network is not reciprocal, we cannot use the T or Π equivalent networks. In such cases, it may be useful to use an equivalent network containing dependent voltage or current sources as shown in Figure 2.5. Such networks can be applied, for instance, in modeling a circuit using software such as SPICE.

2.1.2 CHAIN MATRICES

Some idealized circuit elements do not have impedance or admittance matrices because one or more of their elements is infinite. An example of this is the ideal transformer (Figure 2.6). This device is characterized by the relation

$$\left[\begin{array}{c} V_1 \\ I_1 \end{array} \right] = \left[\begin{array}{cc} 1/n & 0 \\ 0 & n \end{array} \right] \left[\begin{array}{c} V_2 \\ -I_2 \end{array} \right] \tag{2.11}$$

where n is the turns ratio of the transformer. Relation (2.11) is a special case of the use of the so-called *chain matrix* or $\mathcal{ABCD}$-*matrix* of which the general form is

$$\left[\begin{array}{c} V_1 \\ I_1 \end{array} \right] = \left[\begin{array}{cc} \mathcal{A} & \mathcal{B} \\ \mathcal{C} & \mathcal{D} \end{array} \right] \left[\begin{array}{c} V_2 \\ -I_2 \end{array} \right] \tag{2.12}$$

We note two things about this representation. First, the sign of I_2 is reversed in (2.12) because the usual application of the chain matrix is in cascaded two-port networks as shown in Figure 2.7. Here $-I_{a2}$ of an output side becomes $+I_{b1}$ of the next input side, and the effect of the cascaded circuit is simply to multiply the corresponding chain matrices:

$$\left[\begin{array}{cc} \mathcal{A} & \mathcal{B} \\ \mathcal{C} & \mathcal{D} \end{array} \right] = \left[\begin{array}{cc} \mathcal{A}_a & \mathcal{B}_a \\ \mathcal{C}_a & \mathcal{D}_a \end{array} \right] \left[\begin{array}{cc} \mathcal{A}_b & \mathcal{B}_b \\ \mathcal{C}_b & \mathcal{D}_b \end{array} \right] \tag{2.13}$$

Second, we observe that (2.12) gives the voltage and current at port 1 in terms of those at port 2; in other words, it gives what are usually considered to be "input" quantities as a

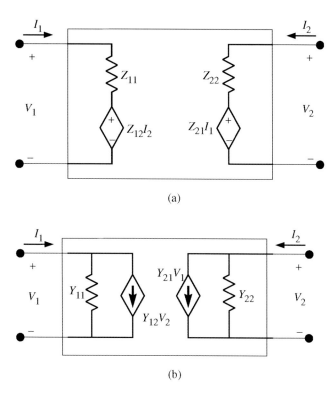

(a)

(b)

Figure 2.5 Equivalent networks with dependent sources for nonreciprocal two-ports: (a) in terms of Z-parameters, (b) in terms of Y-parameters.

function of the "output". If we wish to utilize the chain matrix in the "reverse" direction, we have

$$\left[\begin{array}{c} V_2 \\ -I_2 \end{array} \right] = \left[\begin{array}{cc} \mathcal{A} & \mathcal{B} \\ \mathcal{C} & \mathcal{D} \end{array} \right]^{-1} \left[\begin{array}{c} V_1 \\ I_1 \end{array} \right] \tag{2.14}$$

where

$$\left[\begin{array}{cc} \mathcal{A} & \mathcal{B} \\ \mathcal{C} & \mathcal{D} \end{array} \right]^{-1} = \frac{1}{\mathcal{AD} - \mathcal{BC}} \left[\begin{array}{cc} \mathcal{D} & -\mathcal{B} \\ -\mathcal{C} & \mathcal{A} \end{array} \right] \tag{2.15}$$

We have already encountered a special case of the chain matrix in our description of the terminal behavior of transmission line sections: equation (1.62).

Comparing (2.12) with (2.2), we can obtain the chain parameters in terms of the impedance parameters and vice versa:

$$\begin{aligned} \mathcal{A} &= \frac{Z_{11}}{Z_{21}} \quad \text{and} \quad Z_{11} = \frac{\mathcal{A}}{\mathcal{C}} \\ \mathcal{B} &= \frac{Z_{22}Z_{11}}{Z_{21}} - Z_{12} \qquad Z_{12} = \frac{\mathcal{AD} - \mathcal{BC}}{\mathcal{C}} \\ \mathcal{C} &= \frac{1}{Z_{21}} \qquad Z_{21} = \frac{1}{\mathcal{C}} \\ \mathcal{D} &= \frac{Z_{22}}{Z_{21}} \qquad Z_{22} = \frac{\mathcal{D}}{\mathcal{C}} \end{aligned} \tag{2.16}$$

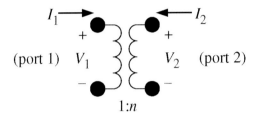

Figure 2.6   The ideal transformer.

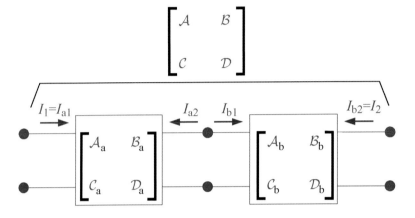

Figure 2.7   Cascaded networks in chain matrix description.

From (2.16) we can verify that the determinant of the chain matrix is

$$\mathcal{AD} - \mathcal{BC} = \frac{Z_{12}}{Z_{21}} \tag{2.17}$$

which is equal to 1 if the network is reciprocal. For the special case of the ideal transformer, it can be seen from (2.11)-(2.16) that the impedance matrix would need to have $Z_{21} \to \infty$, in such a way that $Z_{22}/Z_{21} \to n$, $Z_{11}/Z_{21} \to 1/n$ and $Z_{12}/Z_{21} \to 1$.

For reciprocal networks, one of the most natural equivalent networks expressed in terms of the chain parameters is shown in Figure 2.8. The network is a cascade connection of two

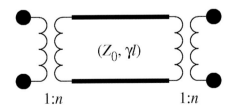

Figure 2.8   Equivalent circuit for reciprocal two-port based on chain parameters.

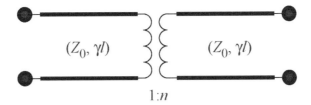

Figure 2.9 Alternative equivalent circuit for reciprocal two-port based on chain parameters.

ideal transformers and a length of transmission line. The transmission line parameters are:

$$Z_0 = \sqrt{\frac{\mathcal{B}}{\mathcal{C}}}; \qquad \gamma l = \cosh^{-1} \sqrt{\mathcal{AD}}$$

while the transformer turns ratios are given by:

$$n = \left(\frac{\mathcal{D}}{\mathcal{A}}\right)^{1/4}$$

An alternative form of this circuit that contains only one transformer but two sections of transmission line is shown in Figure 2.9. For this network the parameters are given by

$$Z_0 = \sqrt{\frac{\mathcal{AB}}{\mathcal{CD}}}; \qquad \gamma l = \frac{1}{2}\cosh^{-1}\sqrt{\mathcal{AD}}; \qquad n = \sqrt{\frac{\mathcal{D}}{\mathcal{A}}}$$

2.1.3 HYBRID AND IMAGE PARAMETERS

For active two-ports like transistors or other devices that are essentially current amplifiers, we can conveniently use the hybrid parameters (H-parameters). These are defined by the relationships

$$V_1 = H_{11}I_1 + H_{12}V_2 \qquad (2.18)$$
$$I_2 = H_{21}I_1 + H_{22}V_2 \qquad (2.19)$$

An equivalent circuit corresponding to the H-parameters is shown in Figure 2.10.

Finally, occasional use is made of the *image parameters* of a two-port network. Image parameters are used in the classical design of filters, and in the design of long transmission circuits in which an impedance match must be maintained. The image impedances Z_{I1} and

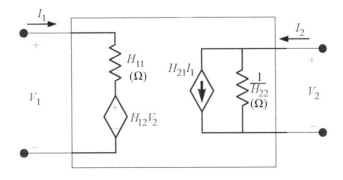

Figure 2.10 Equivalent circuit for a two-port based on its hybrid parameters.

Z_{I2} are defined by the two following requirements. First, if Z_{I2} is connected to port 2, the impedance seen looking into port 1 is Z_{I1}; conversely, if Z_{I1} is connected to port 1, the impedance seen looking into port 2 is Z_{I2}. The remaining image parameters for a two-port are the image transfer constants Γ_{12} and Γ_{21}. If port 2 is terminated with Z_{I2}, then we put

$$e^{-2\Gamma_{12}} = -\frac{V_2 I_2}{V_1 I_1}$$

and analogously for Γ_{21} (which equals Γ_{12} for a reciprocal network). The equivalent circuit of Figure 2.8 can also be related to the image parameters as well, with

$$Z_0 = \sqrt{Z_{I1} Z_{I2}}; \qquad \gamma l = \Gamma_{12}; \qquad \text{and} \qquad n = \left(\frac{Z_{I2}}{Z_{I1}}\right)^{1/4}$$

2.1.4   THE SCATTERING MATRIX AND RELATED REPRESENTATIONS

For two-port networks, we will adopt the following conventions:

(i) A transmission line is assumed to be connected to each of the ports, with separate z-axes for each line defined such that the $+z$-direction points *into* the two-port network.

(ii) The characteristic impedance of the transmission line connected to port i is denoted $Z_{c,i}$.

(iii) We define a *forward* (incident) wave amplitude a_i and a *reverse* (reflected) wave amplitude b_i at each port as [cf. (1.30)-(1.33)]:

$$
\begin{aligned}
a_i &= \frac{V_i + Z_{c,i} I_i}{\sqrt{8 Z_{c,i}}} \\
b_i &= \frac{V_i - Z_{c,i} I_i}{\sqrt{8 Z_{c,i}}}
\end{aligned}
\tag{2.20}
$$

if $Z_{c,i} \neq 0$. The factor involving $Z_{c,i}^{-1/2}$ is introduced into (2.20) so that whenever $Z_{c,i}$ is real, the time-average power $\frac{1}{2}\mathrm{Re}(V_i I_i^*)$ entering port i is equal to $(|a_i|^2 - |b_i|^2)$.

(iv) We define a *characteristic impedance matrix* $[Z_c]$ as the diagonal matrix whose nonzero elements are the characteristic impedances of the transmission lines associated with ports 1 and 2:

$$[Z_c] = \begin{bmatrix} Z_{c,1} & 0 \\ 0 & Z_{c,2} \end{bmatrix} \tag{2.21}$$

It is particularly simple to calculate arbitrary powers of such a matrix, viz.:

$$[Z_c]^q = \begin{bmatrix} Z_{c,1}^q & 0 \\ 0 & Z_{c,2}^q \end{bmatrix} \tag{2.22}$$

for any real number q.

In situations where we are concerned with standing wave ratios in the system, a sort of generalization of the reflection coefficient known as the scattering matrix $[S]$ is the most convenient way to describe the junction. Introducing the incident and reflected wave amplitudes (a_1, a_2) and (b_1, b_2) from V_1, V_2, I_1 and I_2 according to (2.20), we define the $[S]$-matrix so that

$$[b] = [S][a] \tag{2.23}$$

for any possible $[a]$ and $[b]$, where

$$[a] = \begin{bmatrix} a_1 \\ a_2 \end{bmatrix} \quad ; \quad [b] = \begin{bmatrix} b_1 \\ b_2 \end{bmatrix} \quad ; \quad [S] = \begin{bmatrix} S_{11} & S_{12} \\ S_{21} & S_{22} \end{bmatrix} \tag{2.24}$$

From (2.20), we have that

$$
\begin{aligned}
[a] &= \frac{1}{2}[2Z_c]^{-1/2} \{[V] + [Z_c][I]\} \\
[b] &= \frac{1}{2}[2Z_c]^{-1/2} \{[V] - [Z_c][I]\}
\end{aligned}
\tag{2.25}
$$

Since relations (2.23), (2.25) and (2.3) must hold for *all* possible $[I]$, we find that

$$[S] = [Z_c]^{-1/2} \{[Z] - [Z_c]\} \{[Z] + [Z_c]\}^{-1} [Z_c]^{1/2} \tag{2.26}$$

Note that if the characteristic impedance at both ports is the same ($Z_{c1} = Z_{c2}$), the matrix $[Z_c]$ is proportional to the identity matrix

$$[\mathbf{1}] = \begin{bmatrix} 1 & 0 \\ 0 & 1 \end{bmatrix} \tag{2.27}$$

and in this case (2.26) simplifies to

$$[S] = \{[Z] - [Z_c]\} \{[Z] + [Z_c]\}^{-1} \tag{2.28}$$

It is sometimes convenient to use a normalized impedance matrix $[\bar{Z}]$ analogous to the scalar normalized impedance $\bar{Z}$ encountered in classical transmission-line theory. There are many ways in which we could define such a matrix ($[Z_c]^{-1}[Z]$, for example), but only one simple definition yields a matrix which is symmetric if $[Z]$ is symmetric:

$$[\bar{Z}] = [Z_c]^{-1/2}[Z][Z_c]^{-1/2} \tag{2.29}$$

and

$$[Z] = [Z_c]^{1/2}[\bar{Z}][Z_c]^{1/2} \tag{2.30}$$

(which reduce to $[Z] = Z_c[\bar{Z}]$ if $Z_{c1} = Z_{c2}$) so that for a reciprocal two-port:

$$[\bar{Z}]^T = [\bar{Z}] \tag{2.31}$$

Likewise, we can define the normalized admittance matrix

$$[\bar{Y}] = [\bar{Z}]^{-1} = [Z_c]^{1/2}[Y][Z_c]^{1/2} \tag{2.32}$$

which is also symmetric for reciprocal networks.

Using the normalized impedance matrix, we have from (2.26):

$$[S] = \{[\bar{Z}] - [\mathbf{1}]\} \{[\bar{Z}] + [\mathbf{1}]\}^{-1} \tag{2.33}$$

Since $[\bar{Z}]$ is symmetric for a reciprocal network, it is readily seen that $[S]$ is also symmetric in this case:

$$
\begin{aligned}
[S]^T &= \{[\bar{Z}] + [\mathbf{1}]\}^{-1} \{[\bar{Z}] - [\mathbf{1}]\} \\
&= [\mathbf{1}] - 2\{[\bar{Z}] + [\mathbf{1}]\}^{-1} \\
&= \{[\bar{Z}] - [\mathbf{1}]\} \{[\bar{Z}] + [\mathbf{1}]\}^{-1} \\
&= [S] \tag{2.34}
\end{aligned}
$$

i. e., $S_{12} = S_{21}$. Note the similarity of (2.33) to the relationship between the reflection coefficient and the impedance on a classical transmission line (1.56). Rearranging (2.33) to solve for $[\bar{Z}]$, we have

$$[\bar{Z}] = \{[\mathbf{1}] - [S]\}^{-1} \{[\mathbf{1}] + [S]\} \tag{2.35}$$

This formula, too, is analogous to the corresponding expression for impedance in terms of reflection coefficient on a transmission line. The relationship between the elements of $[S]$ and $[Z]$ can be given in detail as:

$$
\begin{aligned}
S_{11} &= \frac{(Z_{11} - Z_{c1})(Z_{22} + Z_{c2}) - Z_{12}Z_{21}}{(Z_{11} + Z_{c1})(Z_{22} + Z_{c2}) - Z_{12}Z_{21}} \\
S_{12} &= \frac{2Z_{12}\sqrt{Z_{c1}Z_{c2}}}{(Z_{11} + Z_{c1})(Z_{22} + Z_{c2}) - Z_{12}Z_{21}} \\
S_{21} &= \frac{2Z_{21}\sqrt{Z_{c1}Z_{c2}}}{(Z_{11} + Z_{c1})(Z_{22} + Z_{c2}) - Z_{12}Z_{21}} \\
S_{22} &= \frac{(Z_{11} + Z_{c1})(Z_{22} - Z_{c2}) - Z_{12}Z_{21}}{(Z_{11} + Z_{c1})(Z_{22} + Z_{c2}) - Z_{12}Z_{21}}
\end{aligned} \tag{2.36}
$$

and

$$
\begin{aligned}
Z_{11} &= Z_{c1}\frac{(1 + S_{11})(1 - S_{22}) + S_{12}S_{21}}{(1 - S_{11})(1 - S_{22}) - S_{12}S_{21}} \\
Z_{12} &= \sqrt{Z_{c1}Z_{c2}}\frac{2S_{12}}{(1 - S_{11})(1 - S_{22}) - S_{12}S_{21}} \\
Z_{21} &= \sqrt{Z_{c1}Z_{c2}}\frac{2S_{21}}{(1 - S_{11})(1 - S_{22}) - S_{12}S_{21}} \\
Z_{22} &= Z_{c2}\frac{(1 - S_{11})(1 + S_{22}) + S_{12}S_{21}}{(1 - S_{11})(1 - S_{22}) - S_{12}S_{21}}
\end{aligned} \tag{2.37}
$$

Another matrix which is useful for representing two-ports in cascaded connection is the (wave-amplitude) transmission matrix (sometimes called the transfer matrix or $[T]$-matrix). For this,

$$
\begin{bmatrix} b_2 \\ a_2 \end{bmatrix} = [T] \begin{bmatrix} a_1 \\ b_1 \end{bmatrix} \tag{2.38}
$$

$$[T] = [T_{II}][T_I]$$

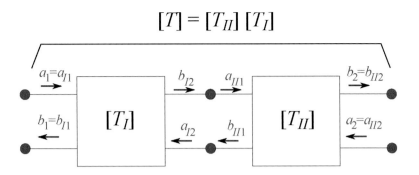

Figure 2.11 Cascaded networks in T-matrix description.

In contrast with the chain matrix, the "output" variables appear on the left side of this equation, and the input variables on the right. Because the reflected wave from one junction becomes the incident wave at the next, an overall $[T]$-matrix for cascaded networks (Figure 2.11) is obtained by matrix multiplication, this time with the second network's matrix on the left:

$$\begin{bmatrix} b_2 \\ a_2 \end{bmatrix} = [T_{II}][T_I] \begin{bmatrix} a_1 \\ b_1 \end{bmatrix} \tag{2.39}$$

The elements of $[T]$ are most easily related to those of $[S]$, by comparing (2.38) to (2.23):

$$T_{11} = S_{21} - \frac{S_{11}S_{22}}{S_{12}} \tag{2.40}$$

$$T_{12} = \frac{S_{22}}{S_{12}} \tag{2.41}$$

$$T_{21} = -\frac{S_{11}}{S_{12}} \tag{2.42}$$

$$T_{22} = \frac{1}{S_{12}} \tag{2.43}$$

For the determinant of $[T]$, we have

$$\det[T] \equiv \Delta_T = \frac{S_{21}}{S_{12}} \tag{2.44}$$

which is equal to 1 if reciprocity holds.

A matrix related to $[T]$ which is sometimes encountered is the $[R]$-matrix, for which

$$\begin{bmatrix} b_1 \\ a_1 \end{bmatrix} = [R] \begin{bmatrix} a_2 \\ b_2 \end{bmatrix} \tag{2.45}$$

A comparison of (2.38) and (2.45), taking into account (2.44), shows that

$$R_{11} = \frac{T_{11}}{\Delta_T} \tag{2.46}$$

$$R_{12} = -\frac{T_{21}}{\Delta_T} \qquad (2.47)$$

$$R_{21} = -\frac{T_{12}}{\Delta_T} \qquad (2.48)$$

$$R_{22} = \frac{T_{22}}{\Delta_T} \qquad (2.49)$$

This matrix, too, finds application in cascading problems.

2.1.5 EXAMPLE: EQUIVALENT NETWORK FOR A LENGTH OF CLASSICAL TRANSMISSION LINE

Consider the two-port consisting of a length l of classical transmission line whose characteristic impedance is Z_0 and propagation coefficient $\gamma = j\beta$, as shown in Figure 2.12(a). The characteristic impedance associated with both ports is $Z_{c1} = Z_{c2} = Z_0$. The scattering matrix for this circuit is readily obtained, since waves pass through it without reflection:

$$b_2 = e^{-j\beta l} a_1 \qquad (2.50)$$
$$b_1 = e^{-j\beta l} a_2 \qquad (2.51)$$

and therefore

$$[S] = \begin{bmatrix} 0 & e^{-j\beta l} \\ e^{-j\beta l} & 0 \end{bmatrix} \qquad (2.52)$$

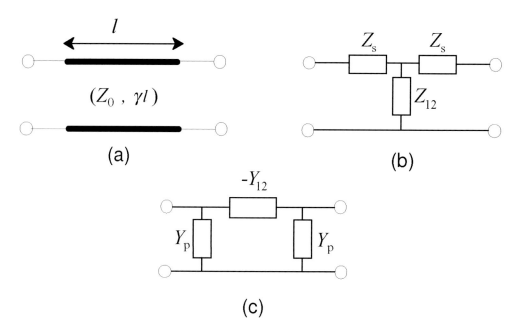

Figure 2.12 (a) A finite section of classical transmission line; (b) equivalent T-network; (c) equivalent Π-network.

The normalized impedance matrix $[\bar{Z}]$ is then obtained from (2.35) as:

$$
\begin{aligned}
[\bar{Z}] &= \{[\mathbf{1}] - [S]\}^{-1}\{[\mathbf{1}] + [S]\} \\
&= \frac{1}{1 - e^{-2j\beta l}}
\begin{bmatrix}
1 & e^{-j\beta l} \\
e^{-j\beta l} & 1
\end{bmatrix}
\begin{bmatrix}
1 & e^{-j\beta l} \\
e^{-j\beta l} & 1
\end{bmatrix} \\
&=
\begin{bmatrix}
-j\cot\beta l & -j\csc\beta l \\
-j\csc\beta l & -j\cot\beta l
\end{bmatrix}
\end{aligned}
\tag{2.53}
$$

Using

$$
[Z_c] =
\begin{bmatrix}
Z_0 & 0 \\
0 & Z_0
\end{bmatrix}
$$

to denormalize $[\bar{Z}]$, we have

$$
[Z] =
\begin{bmatrix}
-jZ_0\cot\beta l & -jZ_0\csc\beta l \\
-jZ_0\csc\beta l & -jZ_0\cot\beta l
\end{bmatrix}
\tag{2.54}
$$

Based upon the equivalent T-network of Figure 2.2, we can assert that the length of transmission line can be represented by the equivalent lumped circuit of Figure 2.12(b), where

$$
\begin{aligned}
Z_s &= jZ_0\tan\frac{\beta l}{2} \\
Z_{12} &= -jZ_0\csc\beta l
\end{aligned}
\tag{2.55}
$$

In a similar way, we find that the Π-equivalent network of Figure 2.12(c) has parameter values:

$$
\begin{aligned}
Y_p &= \frac{j}{Z_0}\tan\frac{\beta l}{2} \\
-Y_{12} &= -\frac{j}{Z_0}\csc\beta l
\end{aligned}
\tag{2.56}
$$

Finally, use of (2.16) yields the chain parameters

$$
\begin{aligned}
\mathcal{A} &= \cos\beta l \\
\mathcal{B} &= jZ_0\sin\beta l \\
\mathcal{C} &= \frac{j}{Z_0}\sin\beta l \\
\mathcal{D} &= \cos\beta l
\end{aligned}
\tag{2.57}
$$

which is a special case of (1.62) for lossless transmission lines.

2.1.6 EXAMPLE: USE OF CHAIN MATRICES

In Figure 2.13, we show two simple two-ports consisting of a single impedance Z connected in series or a single admittance Y connected in shunt between the ports. By inspection, the

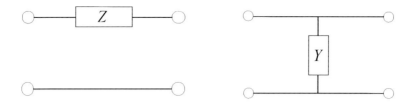

Figure 2.13 Series and shunt connections of impedance as two-port networks.

chain matrix of the series impedance is seen to be

$$
\begin{bmatrix} \mathcal{A} & \mathcal{B} \\ \mathcal{C} & \mathcal{D} \end{bmatrix} = \begin{bmatrix} 1 & Z \\ 0 & 1 \end{bmatrix}
$$

(2.58)

and that of the shunt admittance is

$$
\begin{bmatrix} \mathcal{A} & \mathcal{B} \\ \mathcal{C} & \mathcal{D} \end{bmatrix} = \begin{bmatrix} 1 & 0 \\ Y & 1 \end{bmatrix}
$$

(2.59)

If a load impedance Z_L is connected to port 2 of a sourceless two-port, it is readily shown that the input impedance looking into port 1 is

$$
Z_{\text{in},1} = \frac{\mathcal{A} Z_L + \mathcal{B}}{\mathcal{C} Z_L + \mathcal{D}}
$$

(2.60)

Likewise, the voltage at port 2 is expressed in terms of the voltage at port 1 as:

$$
V_2 = V_1 \frac{Z_L}{\mathcal{A} Z_L + \mathcal{B}}
$$

(2.61)

Note the similarity to (1.60) and (1.66), which are the special cases of (2.60) and (2.61) when the two-port is a segment of transmission line.

2.1.7 EXAMPLE: REALIZATION OF A 1:2 IDEAL TRANSFORMER

Ideal transformers are often useful circuit elements, providing as they do impedance transformation over a wide frequency range. Unfortunately, an actual transformer will fail to produce the desired behavior, increasingly so as frequency increases. This is due to losses in the transformer's magnetic core, and to leakage and parasitic (inductive, capacitive or resistive) effects which cannot be neglected as is the case at lower frequencies. Thus, it is important to be able to accomplish this circuit function in another way that is free of these drawbacks, even if wide bandwidth must be sacrificed.

One way of doing this is to use a length l ($= \lambda/2$ at the design frequency) of transmission line, one terminal from each end connected together at terminal 0 as shown in Figure 2.14. We will view this network as a two-port, at first taking terminals 1 and 0 to form port 1, and terminals 2 and 0 to form port 2, with terminals 1 and 2 taken to be the positive reference nodes and the corresponding port voltages denoted as V_{10} and V_{20} respectively. From (2.57), we have $\mathcal{A} = \mathcal{D} = -1$ and $\mathcal{B} = \mathcal{C} = 0$, independent of the characteristic impedance of the transmission line. From (2.61), we find that $V_{10} = -V_{20}$: since $l = \lambda/2$, the voltage at one end of this loop will be the negative of that at the other end.

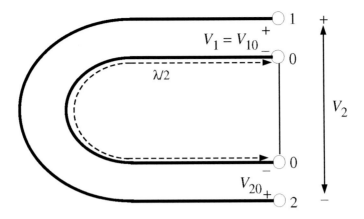

Figure 2.14 Half-wave transmission-line loop as 1:2 transformer.

Now let us change our point of view so that the second port is now taken to be the terminal pair 1-2, with terminal 1 being the positive reference as shown. Port 1 remains defined as before, so that $V_1 = V_{10}$. On the other hand, at the new port 2 we have $V_2 = V_{10} - V_{20} = 2V_1$, and a simple calculation shows that the current at port 1 is twice that at port 2, meaning that this device behaves as a 1:2 ideal transformer, but only at the design frequency f_0. It deviates from ideal performance away from that frequency, but the lack of the drawbacks of low-frequency transformer implementations makes it a popular method to accomplish this circuit function at RF frequencies and above.

2.1.8 EXAMPLE: A TWO-TERMINAL ANTENNA AS A TWO-PORT

Wire antennas are used from very low frequencies up to UHF and beyond for transmission as well as reception. They typically operate in a *balanced* mode with respect to the "ground" (i. e., the environment) as shown in Figure 2.15(a). In this mode, the current at each of the antenna terminals is the same, though oppositely directed. In contrast to the *monopole* mode of operation shown in Figure 2.15(b), the balanced mode of operation is relatively insensitive to the presence of other conductors (especially the earth) in the vicinity of the antenna. An equivalent circuit for this antenna incorporating both modes of operation is

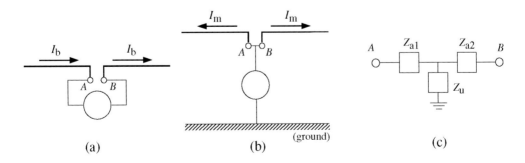

Figure 2.15 Center-fed dipole antenna: (a) balanced mode; (b) monopole mode; (c) equivalent circuit.

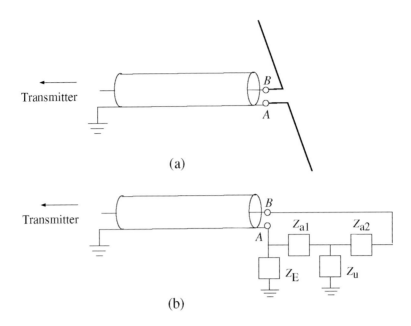

Figure 2.16 (a) Antenna driven by coaxial line; (b) equivalent circuit seen by TEM mode of coax.

shown in Figure 2.15(c). What is normally referred to as the antenna impedance is given by $Z_a = Z_{a1} + Z_{a2}$, while the "unbalanced" antenna impedance Z_u represents the impedance of interaction between the antenna and ground. In other words, the antenna must generally be regarded as a two-port rather than a one-port network, and the parameters of this two-port depend on the relationship of the antenna to its environment in a way that is not easily modeled or predicted, especially if the monopole mode of operation is produced.

If a shielded transmission line such as a circular coaxial cable is used to feed the antenna (in order to avoid interference from stray fields), there is a possibility that the antenna will be driven partially in the undesirable monopole mode. This is because currents flowing on the exterior of the coaxial shield interact with the external environment in a complicated way, and present an equivalent impedance Z_E connected from the shield side of the coax to the ground (there is also a generally different impedance Z_{EB} seen between the inner conductor of the coax and ground, but in practice this is so large as to be negligible by comparison). The driven antenna of Figure 2.16(a) thus presents the equivalent circuit of Figure 2.16(b). Unless the partial antenna impedances Z_{a1} and Z_{a2} are small compared to Z_u and Z_E, we readily see that the excitation of the antenna is significantly unbalanced: current will flow through the impedance Z_u representing the monopole mode of the antenna. A means of compensating for this effect must be found, and a popular one is the balun, which is described in the next section.

2.2 MULTIPORT NETWORKS

Networks with three or more ports (Figure 2.17) can also be characterized by the $[Z]$, $[Y]$ or $[S]$ matrices. Their definitions are direct generalizations from the two-port case: eqns. (2.3), (2.9) and (2.23), where now $[V]$, $[I]$, $[b]$ and $[a]$ are N-element column vectors in the case of an N-port network. Again, $[Z]$, $[Y]$ and $[S]$ are symmetric if the network is reciprocal. In addition, eqns. (2.26)-(2.35) relating $[Z]$ to $[S]$ are still valid. A generalization of the

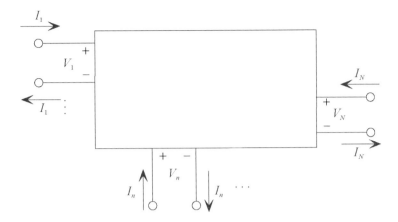

Figure 2.17 An N-port network.

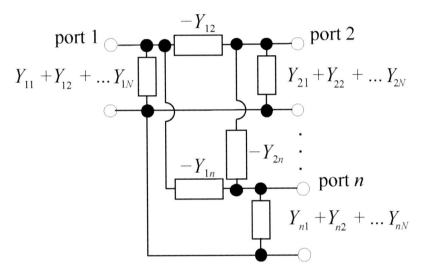

Figure 2.18 Equivalent circuit for a sourceless reciprocal N-port network in terms of its Y-parameters (not all elements shown).

equivalent Π-network of Figure 2.3 holds for sourceless reciprocal N-ports, and is shown in Figure 2.18. In general, the number N^2 of independent parameters needed to describe an arbitrary N-port network grows rapidly with the number of ports, but properties such as reciprocity and symmetry allow the number of required parameters to be greatly reduced, as some of the examples in this section will show.

2.2.1   EXAMPLE: ANALYSIS OF BALUNS

Many applications in RF and microwave circuit design call for a balanced current flow or voltage balance at two terminals of a network. Driving a two-terminal antenna in a balanced manner as described in Section 2.1.8 above is one such example; as a second example, mixers often employ balanced circuit topologies in order to minimize noise and spurious signal generation. Due to the physical environment, however, the two terminals in question cannot usually be considered as a single port. There is a third "terminal" present (the "ground") which together with the other two terminals forms a two-port, rather than

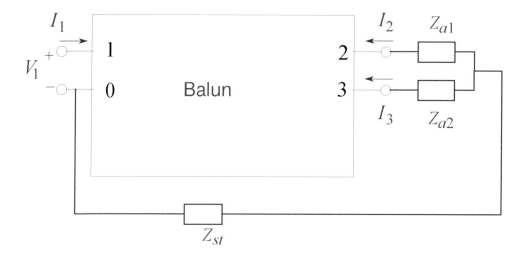

Figure 2.19 Two-port load driven through a balun network.

a one-port network from the original pair of terminals. A *balun* (*bal*anced to *un*balanced converter) is a network designed to accomplish the task of guaranteeing the balance of voltage or current at a pair of terminals in a circuit. More specifically, a balun is a *three-port* network designed so that a source connected to port 1 always results in balanced currents or voltages at ports 2 and 3 (that is, $I_2 = -I_3$ or $V_2 = -V_3$), *regardless of the properties of the two-port connected to ports 2 and 3, or of any other external connections.* It is usually desirable for the balun to be lossless, and convenient (though not necessary) for it to be reciprocal.

We will examine first the example of a *current balun*, used primarily to ensure the balance of currents in antennas. Consider the case of a two-port antenna as considered before, driven through a balun as shown in Figure 2.19. Port 1 of the balun consists of terminals 1 and 0, port 2 of terminals 2 and 0, and port 3 of terminals 3 and 0. We demand that no current flow through the "stray" impedance $Z_{\rm st} = Z_E + Z_u$, regardless of its value, and those of Z_{a1} and Z_{a2}. We assume for simplicity that the balun is reciprocal. If the balun has an admittance matrix $[Y]$, then from the current balance condition $I_2 + I_3 = 0$ we find

$$0 = I_2 + I_3 = (Y_{12} + Y_{13})V_1 + (Y_{22} + Y_{23})V_2 + (Y_{33} + Y_{23})V_3 \qquad (2.62)$$

For this to happen regardless of the values of the antenna impedances (and therefore of the values of V_1, V_2 and V_3), we must have

$$Y_{13} = -Y_{12} \quad \text{and} \quad Y_{22} = Y_{33} = -Y_{23} \qquad (2.63)$$

A lumped element realization of a current balun can be obtained by applying conditions (2.63) to the equivalent circuit of Figure 2.18. The resulting circuit can be drawn as shown in Figure 2.20. The characteristic bridge structure comprising two elements of Y_{12} and two of $-Y_{12}$ is called a Boucherot bridge. Of course, if the balun is a passive structure, we must have $Y_{12} = jB$. In a lossless structure, $\pm B$ would be, for example, a pair of inductors and a pair of capacitors. They could also be realized by short-circuited or open-circuited transmission-line stubs. A wide variety of lumped-element or transmission-line baluns can be realized in this way.

The elements Y_{11} and Y_{22} do not affect the balun operation, and can be used to achieve impedance matching if desired. Because a lossless, reciprocal three-port network cannot be

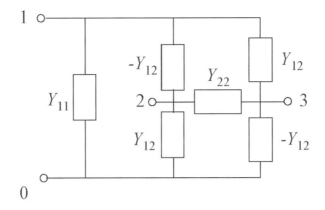

Figure 2.20 The Boucherot bridge (or lattice) balun.

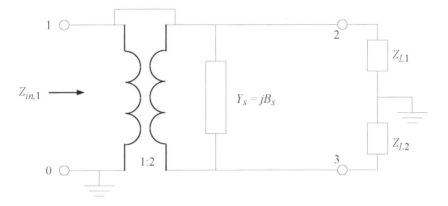

Figure 2.21 Balun using ideal transformer.

matched at all ports (i.e., cannot have $S_{11} = S_{22} = S_{33} = 0$; see Section 2.3), we must expect Y_{11} and Y_{22} to have nonzero real parts if both the input and output ports are to be matched, similar to a Wilkinson power divider. If we choose $Y_{11} = Y_{22} = 0$, we can still match port 1 (achieve $S_{11} = 0$ without loss) by an appropriate choice of B. If the characteristic impedance at port 1 is Z_{c1}, while those at ports 2 and 3 are both equal to Z_{c2}, then it can be shown that port 1 will be matched if we require

$$B = \pm \frac{1}{\sqrt{Z_{c1}(2Z_{c2})}} \tag{2.64}$$

The proof is left as an exercise.

Another type of balun is the *voltage* balun, often used in, for example, balanced mixer circuits. One realization of the voltage balun uses a 1:2 ideal transformer as shown in Figure 2.21. The two load impedances Z_{L1} and Z_{L2} require to have identical voltages appear across them, even if $Z_{L1} \neq Z_{L2}$. An additional shunt susceptance B_s connected across the secondary transformer winding does not affect the balancing action of the circuit nor introduce losses. The multiport circuit formed by terminals 1-0, 2-0 and 3-0 does not possess an impedance matrix, but does enforce $V_2 = -V_3 = V_1$: the balun doubles the voltage from port 1 to that appearing across terminals 2 and 3 while maintaining the balanced voltage condition.

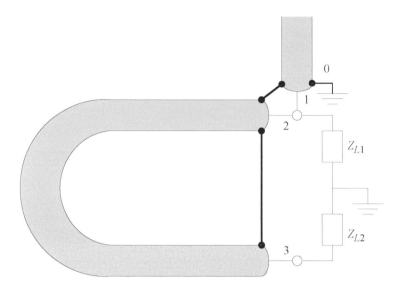

Figure 2.22   Coaxial loop balun.

The total balanced-mode impedance $Z_L = Z_{L1} + Z_{L2}$ of the load is transformed by the voltage balun into an impedance of

$$Z_{\text{in},1} = \frac{Z_L/4}{1 + jB_s Z_L}$$

seen looking into port 1. If $Z_{\text{in},1}$ is to be real ($= R_{\text{in},1}$), then B_s must be chosen to impedance-match the balanced-mode load impedance $Z_L = R_L + jX_L$:

$$B_s = \frac{X_L}{|Z_L|^2}$$

whence

$$R_{\text{in},1} = \frac{|Z_L|^2}{4R_L}$$

The balun thus performs an impedance transformation in addition to its primary function. If $X_L = 0$ and thus $B_s = 0$, the transformation ratio is 4:1. A 300 Ω load impedance, for example, can be matched to a 75 Ω coaxial transmission line using this type of balun. These are widely used in television antenna applications.

At frequencies above a few megahertz, the performance of a traditional low-frequency transformer will begin to degrade, and in practice we must use instead a loop of transmission line (such as coaxial cable) to achieve this purpose as described in Section 2.1.7. The resulting circuit is shown in Figure 2.22. Naturally, field fringing effects at the coax junctions which have been neglected here will modify these results to some extent. In Chapter 11 we will see how such effects are included by the use of appropriate equivalent circuits. We may then use them to refine this simplified circuit model of the balun.

2.2.2   ANALYSIS OF FOUR-PORTS BY EVEN AND ODD EXCITATION

Many networks exhibit internal symmetries that constrain the possible values of their network parameters. A common example of this is the four-port network shown in Figure 2.23. Only positive terminals of each port are shown; the other terminal is understood to be

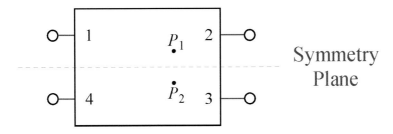

Figure 2.23 Symmetrical four-port network.

ground. We assume that this network possesses a plane of symmetry from top to bottom. More precisely, if we exchange the wave inputs to ports 1 and 4, as well as to ports 2 and 3 ($a_1 \leftrightarrow a_4$ and $a_2 \leftrightarrow a_3$), then the output wave variables will be exchanged in the same way ($b_1 \leftrightarrow b_4$ and $b_2 \leftrightarrow b_3$). We can then easily show that the following relations must exist between elements of the scattering matrix:

$$S_{44} = S_{11}; \quad S_{33} = S_{22}; \quad S_{13} = S_{42}; \quad S_{31} = S_{24}; \quad S_{12} = S_{43}; \quad S_{21} = S_{34} \tag{2.65}$$

and also

$$S_{14} = S_{41}; \quad S_{23} = S_{32} \tag{2.66}$$

(these latter being true even if the network is not reciprocal).

The analysis of such a network can be greatly simplified by the technique of even and odd excitations. An even excitation is one where the incident wave amplitudes (marked with an additional subscript e) above the plane of symmetry are equal to the corresponding ones below the plane:

$$a_{4e} = a_{1e}; \quad a_{3e} = a_{2e} \tag{2.67}$$

In this excitation, every voltage (with respect to ground) at a point P_2 inside the network below the symmetry plane is identical with the corresponding voltage at the symmetrically located point P_1 above the symmetry plane, and as a consequence, no currents can flow across it inside the network. On the other hand, an odd excitation (marked with an additional subscript o) is one where

$$a_{4o} = -a_{1o}; \quad a_{3o} = -a_{2o} \tag{2.68}$$

and in this case every voltage at the point P_2 is equal to the negative of the corresponding voltage at P_1, and as a consequence, the voltage must be everywhere equal to zero at the symmetry plane inside the network.

These observations allow us to reduce the analysis of the original four-port to that of two separate two-port networks corresponding to even and odd excitations and containing only ports 1 and 2. In the case of even excitation, ports 3 and 4 (along with everything in the network below the symmetry plane) are eliminated, and any internal network connections are left open-circuited at the location of that plane. For odd excitation, everything at the symmetry plane is connected to ground. The resulting simplified networks are described by even and odd scattering matrices respectively:

$$
\begin{bmatrix} b_{1e} \\ b_{2e} \end{bmatrix} = [S_e] \begin{bmatrix} a_{1e} \\ a_{2e} \end{bmatrix} ; \quad \begin{bmatrix} b_{1o} \\ b_{2o} \end{bmatrix} = [S_o] \begin{bmatrix} a_{1o} \\ a_{2o} \end{bmatrix} \tag{2.69}
$$

where

$$
[S_e] = \begin{bmatrix} \rho_{1e} & \tau_{12e} \\ \tau_{21e} & \rho_{2e} \end{bmatrix} ; \quad [S_o] = \begin{bmatrix} \rho_{1o} & \tau_{12o} \\ \tau_{21o} & \rho_{2o} \end{bmatrix} \tag{2.70}
$$

with $\rho_{1e}, \rho_{2e}, \tau_{12e}$ and τ_{21e} being the even-excitation reflection and transmission coefficients, and $\rho_{1o}, \rho_{2o}, \tau_{12o}$ and τ_{21o} the odd-excitation reflection and transmission coefficients, respectively, which can be found by analysis of the reduced even- and odd-excitation networks.

The scattering matrix of the original four-port is now found as follows. Suppose first that we choose $a_{1o} = a_{1e}$, while all excitations incident at ports 2 and 3 are taken to be zero. By superposition we then have $a_1 = a_{1e} + a_{1o} = 2a_{1e}$, $a_4 = a_{1e} - a_{1o} = 0$ and $a_2 = a_3 = 0$. we can now compute, for example,

$$
S_{11} = \left. \frac{b_1}{a_1} \right|_{a_2=a_3=a_4=0} = \frac{b_{1e} + b_{1o}}{2a_{1e}} = \frac{\rho_{1e} + \rho_{1o}}{2} \tag{2.71}
$$

In similar fashion we can compute all the other scattering parameters of the four-port, and we obtain

$$
S_{11} = \frac{\rho_{1e} + \rho_{1o}}{2}; \quad S_{21} = \frac{\tau_{21e} + \tau_{21o}}{2}; \quad S_{31} = \frac{\tau_{21e} - \tau_{21o}}{2}; \quad S_{41} = \frac{\rho_{1e} - \rho_{1o}}{2}
$$
$$
S_{12} = \frac{\tau_{12e} + \tau_{12o}}{2}; \quad S_{22} = \frac{\rho_{2e} + \rho_{2o}}{2}; \quad S_{32} = \frac{\rho_{2e} - \rho_{2o}}{2}; \quad S_{42} = \frac{\tau_{12e} - \tau_{12o}}{2} \tag{2.72}
$$

while the third and fourth columns of the scattering matrix can be found using relations (2.65)-(2.66).

2.2.3 EXAMPLE: RING AND HYBRID NETWORKS

A hybrid network is a four-port network in which, when matched loads are placed at each port, an incident wave at one port is split in some ratio and delivered as outgoing waves at two of the remaining ports, while there are no outgoing waves at the incident port or the last remaining ("isolated") port. Hybrid networks can be used in conjunction with transmit-receive switches to form duplexers: three-port networks which transfer all power from one port to a second port, and all power incident at the second port to a third (either at the same frequency or at different frequencies, permitting a receiver and transmitter to use the same antenna, for example).[1] To achieve duplexer functionality in a lossless three-port network requires a nonreciprocal network (called a circulator); however, a reciprocal four-port hybrid can be used for this purpose if the fourth port is terminated in a matched load. Hybrid networks are also used in power combiners and repeater amplifier circuits to avoid

---

[1] A *duplexer* is different than a *diplexer*, which is a three-port network that takes a signal incident at one port, and transfers that signal to a second port over a certain frequency range while delivering the signal to a third port at all other frequencies. A VHF-UHF splitter for television cables is an example of a diplexer. A device that splits a signal into more than two frequency ranges is called a *multiplexer*.

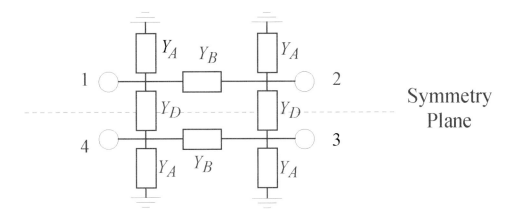

Figure 2.24 Lumped-element hybrid network.

feedback which results in oscillation. The most general analysis of a hybrid network is quite complicated. We will simplify matters by considering a special case.

One way of realizing a lossless hybrid network is a ring network as shown in Figure 2.24. The admittances Y_A, Y_B and Y_D are imaginary. The analysis of this network can be simplified by the use of even and odd excitations, using the admittance matrix at first rather than the scattering matrix. We start by assuming an even excitation: $V_1 = V_4$ and $V_2 = V_3$. The bottom half of the network can be removed and open circuits left at the symmetry plane positions. The result is a cascade of three two-port networks consisting of: (i) Y_A in shunt, (ii) Y_B in series and (iii) Y_A in shunt again. The chain matrices of these three two-ports (from eqns. (2.58) and (2.59)) are multiplied together to get an overall chain matrix:

$$\begin{bmatrix} \mathcal{A} & \mathcal{B} \\ \mathcal{C} & \mathcal{D} \end{bmatrix} = \begin{bmatrix} 1 & 0 \\ Y_A & 1 \end{bmatrix} \begin{bmatrix} 1 & \frac{1}{Y_B} \\ 0 & 1 \end{bmatrix} \begin{bmatrix} 1 & 0 \\ Y_A & 1 \end{bmatrix} = \begin{bmatrix} 1 + \frac{Y_A}{Y_B} & \frac{1}{Y_B} \\ Y_A\left(2 + \frac{Y_A}{Y_B}\right) & 1 + \frac{Y_A}{Y_B} \end{bmatrix} \tag{2.73}$$

Terminating one port of this network with the characteristic impedance $Z_c = 1/Y_c$ and using (2.73) and (2.60), we find the input impedance at the other port to be

$$Z_{\text{in},1e} = \frac{Y_A + Y_B + Y_c}{Y_A(Y_A + Y_B + Y_c) + Y_B(Y_A + Y_c)} \tag{2.74}$$

from which we can obtain the even-mode voltage reflection and transmission coefficients ρ_{1e} and τ_{21e}. In particular, we have

$$\rho_{1e} = \frac{Z_{\text{in},1e} - Z_c}{Z_{\text{in},1e} + Z_c} = \frac{Y_c^2 - Y_A^2 - 2Y_A Y_B}{Y_A^2 + Y_c^2 + 2Y_A Y_B + 2Y_B Y_c + 2Y_A Y_c} \tag{2.75}$$

In a similar manner, we can use odd excitation, $V_1 = -V_4$ and $V_2 = -V_3$, to arrive at odd mode reflection and transmission coefficients ρ_{1o} and τ_{21o}. The equivalent circuit is the same as that of even mode excitation, except that the shunt admittances Y_A are replaced by $Y_A + 2Y_D$, since the impedances $1/Y_D$ must be cut in half and grounded in the middle. Note that because of the left-right symmetry of this network, we have $\rho_{2e,o} = \rho_{1e,o}$ and $\tau_{12e,o} = \tau_{21e,o}$ (the latter also being true by reciprocity).

We demand that the scattering parameters of this reciprocal network obey $S_{11} = S_{22} = S_{33} = S_{44} = 0$ and $S_{23} = S_{14} = 0$ in order to provide the desired hybrid behavior described above. This requires that $\rho_{1e} = \rho_{1o} = 0$ if the same characteristic impedance reference Z_c is used at all four ports. If $\rho_{1e} = 0$, we must have

$$Y_c^2 = Y_A(Y_A + 2Y_B) \tag{2.76}$$

while if $\rho_{1o} = 0$, we need

$$Y_c^2 = (Y_A + 2Y_D)(Y_A + 2Y_B + 2Y_D) \tag{2.77}$$

Equation (2.77) is identical with (2.76) if $Y_D = 0$, but in this case the circuits on opposite sides of the symmetry plane would be completely decoupled, and we would arrive at $S_{31} = S_{42} = 0$, a special case without practical utility, so from here on we assume that $Y_D \neq 0$. We can then obtain from (2.76) and (2.77) the constraints

$$Y_A + Y_B + Y_D = 0; \qquad Y_B = \pm j\sqrt{Y_c^2 - Y_D^2} \tag{2.78}$$

To find S_{21} and S_{31}, we first compute the even and odd transmission coefficients. Since $\rho_{1e} = 0$, we can use voltage division to get

$$\tau_{21e} = (1 + \rho_{1e})\frac{\frac{1}{Y_A + Y_c}}{\frac{1}{Y_A + Y_c} + \frac{1}{Y_B}} = \frac{Y_B}{Y_A + Y_B + Y_c} \tag{2.79}$$

In a similar manner we find

$$\tau_{21o} = \frac{Y_B}{-Y_A - Y_B + Y_c} \tag{2.80}$$

From (2.72) we then obtain

$$S_{21} = \frac{Y_B Y_c}{Y_c^2 - Y_D^2} = \pm j\frac{Y_c}{\sqrt{Y_c^2 - Y_D^2}} \tag{2.81}$$

and

$$S_{31} = -\frac{Y_B(Y_A + Y_B)}{Y_c^2 - Y_D^2} = \mp j\frac{Y_D}{\sqrt{Y_c^2 - Y_D^2}} \tag{2.82}$$

the upper or lower signs being taken according to the one chosen for Y_B in (2.78). The fraction of the power incident at port 1 that is delivered to port 2 (sometimes called the coupling factor) is thus

$$|S_{21}|^2 = \frac{Y_c^2}{Y_c^2 - Y_D^2} \tag{2.83}$$

(which is < 1 for a lossless network since Y_c is real while Y_D is imaginary). Thus, by suitable choices of element admittances in the network, a hybrid network can be designed to provide any desired degree of power division between the coupled ports. Combining (2.78) and (2.83), we obtain the following expressions for the admittances required to realize a specified coupling factor:

$$Y_B = \pm\frac{jY_c}{|S_{21}|}; \qquad Y_D = \pm jY_c\sqrt{\frac{1}{|S_{21}|^2} - 1}; \qquad Y_A = jY_c\left(\mp\frac{1}{|S_{21}|} \mp \sqrt{\frac{1}{|S_{21}|^2} - 1}\right) \tag{2.84}$$

The phase angle of S_{21} is $\pm 90°$, while that of S_{31} is $0°$ or $180°$, depending on the signs of Y_B and Y_D.

It is not necessary to use lumped elements to realize this hybrid network. A set of transmission lines can be connected into a ring network, and replacing them with their equivalent Π-networks reduces the ring to the lumped element form of Figure 2.24. At higher frequencies this is almost always preferable to the use of lumped elements. Details will be left as an exercise.

2.3 POWER AND ENERGY RELATIONS

An impedance Z, connected for example as a load on a classical transmission line, can be considered as a passive one-port network with possibly many lumped circuit elements or even sections of transmission line internal to it. Suppose that such a network consisting of lumped elements only (all capacitances and conductances are located in tree branches, while all resistances and inductances are in the links) is connected to the end of a transmission line at the point $z = z_1$. From the general properties of such networks as outlined in Appendix A, we can deduce the following relation between the complex power delivered to the port and the internal stored energies and dissipated powers in the network (see (A.20)):

$$\frac{V_1 I_1^*}{2} = p_{L,\mathrm{av}} + 2j\omega \left(w_{M,\mathrm{av}} - w_{E,\mathrm{av}} \right)$$

where $V_1 = V(z_1)$ and $I_1 = I(z_1)$ are the port voltage and current, while

$$p_{L,\mathrm{av}} = \frac{1}{2} \left\{ [I_{\mathrm{int}}]^\dagger [R][I_{\mathrm{int}}] + [V_{\mathrm{int}}]^\dagger [G][V_{\mathrm{int}}] \right\}$$

(here $[I_{\mathrm{int}}]^\dagger = [I_{\mathrm{int}}^*]^T$ denotes the *Hermitian conjugate* of $[I_{\mathrm{int}}]$) is the time-average power lost to the circuit conductances and resistances, and

$$w_{E,\mathrm{av}} = \frac{1}{4} [V_{\mathrm{int}}]^\dagger [C][V_{\mathrm{int}}]; \qquad w_{M,\mathrm{av}} = \frac{1}{4} [I_{\mathrm{int}}]^\dagger [L][I_{\mathrm{int}}]$$

are the time-average electric and magnetic stored energies in the circuit capacitances and inductances respectively. Here the column vector $[I_{\mathrm{int}}]$ denotes the internal link currents and the column vector $[V_{\mathrm{int}}]$ denotes the internal tree voltages of the one-port network to distinguish them from the port voltages and currents. The matrices $[R]$, $[L]$, $[G]$ and $[C]$ are constituents of the branch admittance matrix $[G] + j\omega[C]$ and link impedance matrix $[R] + j\omega[L]$ of Appendix A.

If $Z = Z_{11} = V_1/I_1$ is the impedance seen looking into the one-port, then

$$Z_{11} = \frac{2}{|I_1|^2} \left(p_{L,\mathrm{av}} + 2j\omega W_{\mathrm{excess}} \right) \tag{2.85}$$

where

$$W_{\mathrm{excess}} = w_{M,\mathrm{av}} - w_{E,\mathrm{av}} \tag{2.86}$$

is the excess of the time-average stored magnetic energy over the time-average stored electric energy. Note that it is purely a matter of whether $w_{M,\mathrm{av}} > w_{E,\mathrm{av}}$ or $w_{M,\mathrm{av}} < w_{E,\mathrm{av}}$ that determines whether the imaginary part X_{11} of Z_{11} (i. e., the reactance) of the one-port is positive (inductive) or negative (capacitive). Thus, any equivalent circuit that we construct for the one-port, while it may result in different total electric and magnetic stored energies, must exhibit the same difference between them for a given port voltage or current.

As with one-port circuits, the capacitive or inductive nature of two-port sourceless networks is dependent upon whether predominantly electric or magnetic energy is stored in the junction. It is more convenient to express the general relations for this in terms of the scattering matrix, rather than $[Z]$ or $[Y]$. We start by applying the expression (A.20) for complex power flow to a network as in the steps leading up to (2.85). In terms of port voltages and currents, the complex power delivered to the network through its two ports is now

$$\frac{1}{2}(V_1 I_1^* + V_2 I_2^*) = \frac{1}{2}[I]^\dagger[V]$$

In place of (2.85) we now have

$$\frac{1}{2}[I]^\dagger[V] = P + 2j\omega W_{\text{excess}} \tag{2.87}$$

where P is the sum of the powers dissipated in resistances or conductances of the network, while W_{excess} is the excess of time-average stored magnetic energy over stored electric energy in the volume V. The network is said to be *passive* if $P \geq 0$. Using the wave amplitudes $[a]$ and $[b]$ defined in (2.25), we have

$$\{[a] - [b]\}^\dagger\{[a] + [b]\} = P + 2j\omega W_{\text{excess}} \tag{2.88}$$

Introducing the scattering matrix $[S]$ from (2.23), we have from (2.88):

$$[a]^\dagger\left\{[\mathbf{1}] + [S] - [S]^\dagger - [S]^\dagger[S]\right\}[a] = P + 2j\omega W_{\text{excess}} \tag{2.89}$$

Separating real and imaginary parts, we get

$$[a]^\dagger\{[\mathbf{1}] - [S]^\dagger[S]\}[a] = P \tag{2.90}$$

and

$$[a]^\dagger\{[S] - [S]^\dagger\}[a] = 2j\omega W_{\text{excess}} \tag{2.91}$$

For a passive network, (2.90) implies

$$[a]^\dagger[S]^\dagger[S][a] \leq [a]^\dagger[a] \tag{2.92}$$

for any possible value of $[a]$; we write this condition more succinctly as

$$[S]^\dagger[S] \leq [\mathbf{1}] \tag{2.93}$$

Note that this does *not* mean that the corresponding elements of the matrices on the left and right sides of (2.93) satisfy the inequality, but merely that (2.92) is true for any $[a]$ (see Appendix B).

If a two-port is lossless, then $P = 0$ in (2.90), and we must have

$$[S]^\dagger[S] = [\mathbf{1}] \tag{2.94}$$

A matrix which obeys (2.94) is said to be *unitary*. A consequence of (2.94) and relation (2.33) is that

$$[\bar{Z}] = -[\bar{Z}]^\dagger \tag{2.95}$$

For a reciprocal network, the matrix $[\bar{Z}]$ is symmetric, and eqns. (2.31) and (2.95) imply that all the elements of the normalized impedance matrix are imaginary (reactive). If the characteristic impedance matrix is purely real, this conclusion holds for the elements of $[Z]$ as well. This is only natural for a network (and transmission lines) with no losses present. On the other hand, putting $a_i = 0$ for all $i \neq n$, but $a_n \neq 0$ in (2.91), we get

$$\text{Im}(S_{nn}) = \frac{\omega}{|a_n|^2} W_{\text{excess}} \tag{2.96}$$

We see that $\text{Im}(S_{nn})$ has the same sign as W_{excess}.

Relations (2.90) and (2.91) are true for any choice of $[a]$, but for general lossy two-port networks they do not yield simple relations between elements of $[S]$ and the excess energy W_{excess}. An exception to this is the case of scattering from a single shunt impedance element Z_{12} connected across a transmission line, in which case the elements of $[Z]$ are all equal: $Z_{11} = Z_{12} = Z_{22}$. The scattering matrix for such a shunt impedance can be shown from (2.33) to have the form

$$[S] = \frac{1}{2Z_{12} + Z_0} \begin{bmatrix} -Z_0 & 2Z_{12} \\ 2Z_{12} & -Z_0 \end{bmatrix} \tag{2.97}$$

if both ports are referred to the same characteristic impedance Z_0. It is clear from (2.97) that $\text{Im}(S_{11})$ is negative or positive accordingly as $\text{Im}(Z_{12})$ is negative or positive (so long as Z_0 is real). From (2.96), we see that for this simple shunt element, $\text{Im}(Z_{12})$ is positive (inductive) if W_{excess} is positive and negative (capacitive) if W_{excess} is negative. In a similar way, for the case of scattering from a single admittance element $-Y_{12}$ connected in series in a transmission line (cf. Figure 2.3), the scattering matrix is

$$[S] = \frac{1}{1 - 2Z_0 Y_{12}} \begin{bmatrix} 1 & -2Z_0 Y_{12} \\ -2Z_0 Y_{12} & 1 \end{bmatrix} \tag{2.98}$$

Similarly to the previous example, we see that $\text{Im}(Y_{12})$ has the same sign as W_{excess}.

For a lossless N-port network, relations (2.94), (2.95) and (2.96) are still true. The symmetry of a $[Z]$, $[Y]$ or $[S]$ matrix for a reciprocal N-port network means that only $N(N + 1)/2$ of the N^2 complex matrix elements are independent of each other. For a lossless network, eqn. (2.94) yields a further $N(N + 1)/2$ *real*, independent equations which the elements of $[S]$ must satisfy. In this way, the number of parameters we must determine in order to describe completely the behavior of a multiport network can be considerably reduced. Further reduction is possible if physical symmetries of the network can be exploited.

2.3.1 EXAMPLE: LOSSLESS, RECIPROCAL TWO-PORT NETWORK

For a lossless, reciprocal two-port, the diagonal elements of (2.94) give

$$\left. \begin{array}{c} |S_{11}|^2 + |S_{12}|^2 = 1 \\ |S_{12}|^2 + |S_{22}|^2 = 1 \end{array} \right\} \quad \Rightarrow \quad |S_{11}| = |S_{22}| = \sqrt{1 - |S_{12}|^2} \tag{2.99}$$

The physical interpretation of (2.99) is straightforward: all power incident from this lossless network at either port must either be reflected or transmitted. The off-diagonal elements of (2.94) give

$$S_{11}^* S_{12} + S_{12}^* S_{22} = 0 \tag{2.100}$$

or

$$S_{11}^* S_{12} = -S_{12}^* S_{22} \tag{2.101}$$

The physical interpretation of (2.101) is not so obvious. Let us first dispense with two special cases that are relatively trivial as far as two-ports are concerned. If $|S_{12}| = 0$, then each port would consist of a simple reactance to ground, uncoupled to the other port, such that $|S_{11}| = |S_{22}| = 1$. This case is really that of two separate lossless one-ports. If on the other hand $|S_{12}| = 1$, so that $|S_{11}| = |S_{22}| = 0$, no constraint exists on the phase of S_{12}; any length of uniform transmission line with $Z_c = Z_{c1} = Z_{c2}$ could serve as the two-port in this case. Assume therefore that $|S_{12}|$ equals neither 1 nor 0. Then the magnitudes of both sides of (2.101) yield no new conditions. However, expressing the S-parameters in terms of magnitudes and phases as $S_{11} = |S_{11}| e^{j\phi_{11}}$, etc., then from the phases of (2.101) we get

$$2\phi_{12} = \pi + \phi_{11} + \phi_{22} \pmod{2\pi} \tag{2.102}$$

Thus, for a lossless, reciprocal two-port network, only three real parameters are needed to (essentially) completely specify the scattering matrix; for example, $|S_{12}|$, ϕ_{11} and ϕ_{22}. The remaining information is then obtained from (2.99) and (2.101) (although ϕ_{12} could always be replaced by $\phi_{12} + \pi$, so that S_{12} is only determined to within a $\pm$ sign).

2.3.2   EXAMPLE: LOSSLESS, RECIPROCAL THREE-PORT NETWORK

As an example of the power/energy relations for multiports with $N > 2$, consider the restrictions placed upon a three-port network that is lossless and reciprocal. We ask whether such a network can also be matched at all three ports. The corresponding conditions are (2.94), (2.34) and $S_{11} = S_{22} = S_{33} = 0$ respectively. The latter two conditions result in

$$[S] = \begin{bmatrix} 0 & S_{12} & S_{13} \\ S_{12} & 0 & S_{23} \\ S_{13} & S_{23} & 0 \end{bmatrix} \tag{2.103}$$

The diagonal terms of (2.94) yield the equations

$$1 = |S_{12}|^2 + |S_{13}|^2 = |S_{12}|^2 + |S_{23}|^2 = |S_{13}|^2 + |S_{23}|^2 \tag{2.104}$$

so we have

$$|S_{12}| = |S_{13}| = |S_{23}| = \frac{1}{\sqrt{2}} \tag{2.105}$$

But from the off-diagonal terms of (2.94), we have

$$S_{13}S_{23}^* = S_{12}S_{23}^* = S_{12}S_{13}^* = 0 \tag{2.106}$$

which is clearly contradictory to (2.105). Thus, a three-port network that is lossless, reciprocal and matched is not possible.

2.4   NOTES AND REFERENCES

Basic network theory and applications for RF and microwave circuits are treated in

Montgomery, Dicke and Purcell (1965);
Collin (2001);
Kerns and Beatty (1967);
Helszajn (1994);
Maas (1998);
Pozar (2005).

Both Kerns and Beatty, as well as Pozar, give tables presenting conversions among the various matrix representations for multiport networks, of which we have only given a few examples here. The reader is warned that Pozar assumes that the reference (characteristic) impedances at the ports are identical, while there are minor typographical errors in Kerns and Beatty, as well as different definitions of the wave amplitudes assumed. More advanced aspects of network theory are given in

Carlin and Giordano (1964);

Newcomb (1966);

T. Nishide and A. Matsumoto, "Multiport image parameter theory," *Recent Developments in Microwave Multiport Networks* [Monograph Series, Research Institute of Applied Electricity no. 18]. Sapporo, Japan: Research Institute of Applied Electricity, 1970, pp. 103-183.

Various specific types of multiport networks involving transmission lines (including filters) are discussed in

Matsumoto (1970);

Karakash (1950);

Matthaei, Young and Jones (1964).

For more information on baluns, consult

King (1965), pp. 220-224;

Dworsky (1979), pp. 215-221;

Reich, Ordnung, Krauss and Skalnik (1953), pp. 204-212;

Weeks (1968);

Balanis (1982), pp. 365-368;

Uchida (1967), Section 5.3;

Matsumoto (1970), Section X.2;

Sevick (1990);

and

S. Frankel, "Reactance networks for coupling between unbalanced and balanced circuits," *Proc. IRE*, vol. 29, pp. 486-493 (1941).

N. Nagai and A. Matsumoto, "Application of distributed-constant network theory to balun transformers," *Electron. Commun. Japan* vol. 50, no. 5, pp. 114-121 (1967).

N. Nagai, "Theory and experiments on multiwire multiports," *Recent Developments in Microwave Multiport Networks* [Monograph Series, Research Institute of Applied Electricity no. 18]. Sapporo, Japan: Research Institute of Applied Electricity, 1970, pp. 1-68.

T. Nishide, N. Nagai and A. Matsumoto, "Balance-to-unbalance transformers," *Recent Developments in Microwave Multiport Networks* [Monograph Series, Research Institute of Applied Electricity no. 18]. Sapporo, Japan: Research Institute of Applied Electricity, 1970, pp. 69-85.

N. Nagai, T. Nishide and A. Matsumoto, "Multiwire-line baluns and unbaluns," *Bull. Res. Inst. Appl. Electricity*, vol. 22, pp. 24-51 (1970).

Hu Shuhao, "The balun family," *Microwave J.*, vol. 30, no. 9, pp. 227-229 (1987).

R. Sturdivant, "Balun designs for wireless, ... mixers, amplifiers and antennas," *Appl. Microwave*, vol. 5, no. 3, pp. 34-44 (1993).

V. Trifunović and B. Jokanović, "Review of printed Marchand and double Y baluns: Characteristics and applications," *IEEE Trans. Micr. Theory Tech.*, vol. 42, pp. 1454-1462 (1994).

Hybrid networks are treated in

> Matthaei, Young and Jones (1964), Chap. 13.
> Sazonov, Gridin and Mishustin (1982), Section 2.10.

and

- A. Alford, "High frequency bridge circuits and high frequency repeaters," *U. S. Patent no. 2,147,809*, Feb. 21, 1939.
- L. Young, "Branch guide directional couplers," *Proceedings of the National Electronics Conference*, vol. 12, 1-3 October 1956, Chicago, pp. 723-732.
- J. Reed and G. J. Wheeler, "A method of analysis of symmetrical four-port networks," *IRE Trans. Micr. Theory Tech.*, vol. 4, pp. 246-252 (1956).
- C. Y. Ho, "Some results on the design of N (2, 3, 4)-way branch line type power splitters," *IEEE 1976 Region V Conference*, 14-16 April 1976, Austin, TX, pp. 45-50.
- R. G. Manton, "Hybrid networks and their uses in radio-frequency circuits," *Radio Electron. Eng.*, vol. 54, pp. 473-489 (1984).
- V. F. Fusco and S. B. D. O'Caireallain, "Lumped element hybrid networks for GaAs MMICs," *Micr. Opt. Technol. Lett.*, vol. 2, pp. 19-23 (1989).

The paper by Reed and Wheeler and the book by Sazonov *et al.* describe the even and odd excitation technique for the analysis of symmetrical multiport networks.

2.5   PROBLEMS

p2-1 Two classical transmission lines of different characteristic impedances and propagation coefficients are connected at $z = 0$ as shown. Choosing the terminal planes A_1 at $z = 0^-$ and A_2 at $z = 0^+$, obtain the scattering matrix $[S]$ and the impedance matrix $[Z]$ for this junction, associating the characteristic impedance Z_{c1} with port 1 and Z_{c2} with port 2.

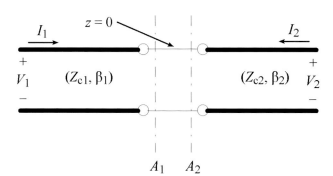

p2-2 In the circuit of problem p2-1, a shunt capacitor is added to the circuit at $z = 0$. Find $[S]$ and $[Z]$ for this modified circuit.

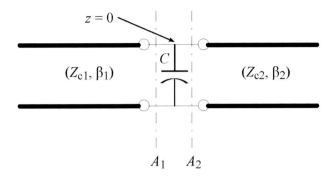

p2-3 Re-examine the quarter-wave transformer of problem p1-10 if we now account for the effects of fringing fields at the junctions between line sections of different characteristic impedance. Suppose that the equivalent circuit is as given below, assume that $\omega C_{1,2} Z_{c1,2,3} \ll 1$, but is not zero, and find what new values of Z_{c2} and d must be used in order to preserve the impedance match condition of the ideal matching circuit.

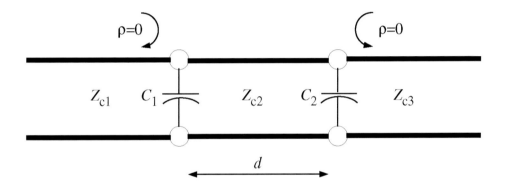

p2-4 In the circuit of problem p2-1, a series inductor is added to the circuit at $z = 0$. Find the $[S]$-matrix for this circuit.

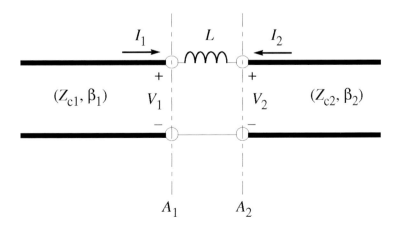

p2-5 From (2.26) or (2.33), find the conditions which must be satisfied by the elements of the $[Z]$ matrix of a source-free two-port network in order for $S_{11} = 0$ to be true. Based on this condition, and the equivalent circuit of Figure 2.2, construct a lossless matching network based on lossless lumped elements (i.e., inductors and capacitors) that will give zero reflection on a transmission line of characteristic impedance Z_{c1} connected to port 1, if a matched load of $Z_L = Z_{c2}$ is connected to port 2. Is it possible to simultaneously match in the other direction (that is, using the same network, obtain zero reflection on a line of characteristic impedance Z_{c2} connected to port 2 with a load impedance of Z_{c1} connected to port 1)? Assume that Z_{c1} and Z_{c2} are given real quantities.

p2-6 (a) Consider a nonuniform classical transmission line in the lowest order WKB approximation, as described by the chain matrix parameters (1.110) with $z_1 = 0$ and $z_2 = d$. Show that the scattering matrix for this line in this approximation is given by

$$[S] = \begin{bmatrix} 0 & e^{-Q(d)} \\ e^{-Q(d)} & 0 \end{bmatrix}$$

where the port characteristic impedances are chosen as $Z_{c1} = Z_c(0^+)$ and $Z_{c2} = Z_c(d^-)$. Determine the $[Z]$ matrix in the lowest-order WKB approximation as well.

(b) Consider this same line in the first-order WKB approximation described by the chain matrix parameters (1.116). Obtain the $[S]$ matrix and the $[Z]$ matrix of the nonuniform line to this higher accuracy.

p2-7 Show that a sourceless linear reciprocal two-port network possessing an impedance matrix representation can be represented by the equivalent network shown below, that is an alternative to the equivalent T-network of Figure 2.2.

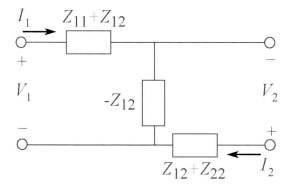

Note that, unlike the T-network, in this equivalent circuit there is not a common ground node between the two ports.

p2-8 Find the impedance matrix elements Z_{11}, Z_{12}, Z_{21} and Z_{22} of the lattice network shown below (note that Z_b and Z_c do not directly connect to each other; the "wire" of Z_b passes over that of Z_c without contact).

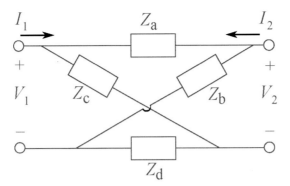

From these expressions, obtain expressions for the elements of the chain matrix of this network.

p2-9 Consider the two-port networks shown in the figure below.

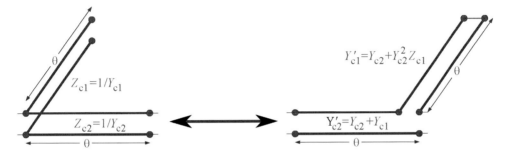

The electrical length $\theta = \beta l$ of each transmission line section is the same, although the characteristic impedance of each section is different, as shown. Use the results of Section 2.1.5 to prove that these two networks are equivalent to each other at all frequencies. This equivalence is one of *Kuroda's identities*.

p2-10 Obtain expressions for the hybrid (H-) parameters of a two-port network in terms of

    (a) the $\mathcal{ABCD}$ parameters,
    (b) the Z-parameters, and
    (c) the Y-parameters.

Do *not* assume that the network is reciprocal. In each case, state what conditions (if any) must be satisfied for your formulas to be valid.

p2-11 Show that the chain parameters of a two-port network can be expressed in terms of the scattering parameters as:

$$\mathcal{A} = \sqrt{\frac{Z_{c1}}{Z_{c2}}} \frac{(1 + S_{11})(1 - S_{22}) + S_{12}S_{21}}{2S_{21}}$$

$$\mathcal{B} = \sqrt{Z_{c1}Z_{c2}} \frac{(1 + S_{11})(1 + S_{22}) - S_{12}S_{21}}{2S_{21}}$$

$$\mathcal{C} = \frac{1}{\sqrt{Z_{c1}Z_{c2}}} \frac{(1 - S_{11})(1 - S_{22}) - S_{12}S_{21}}{2S_{21}}$$

$$\mathcal{D} = \sqrt{\frac{Z_{c2}}{Z_{c1}}} \frac{(1 - S_{11})(1 + S_{22}) + S_{12}S_{21}}{2S_{21}}$$

Do not assume that the network is reciprocal, or that $Z_{c1} = Z_{c2}$. Also, obtain expressions for the S-parameters in terms of the chain parameters.

p2-12 A voltage balun is a three-port network for which $V_3 = -V_2$, regardless of external connections to the network (that is, regardless of the values of I_1, I_2 and I_3). Derive the conditions which must be satisfied by the $[Z]$ matrix of such a network. Find one example of a circuit that obeys these conditions (use the result of problem p2-15).

p2-13 In certain situations, there may arise the need for an "unbalun"—a three-port network for which $I_2 = I_3$ regardless of any external circuits connected to it. If the unbalun is to be lossless and reciprocal, derive the additional constraints on the Y-parameters necessary to achieve the desired behavior. Obtain a lumped element circuit that performs as an unbalun.

p2-14 Design a reciprocal three-port network that will force the relation $b_3 = Ab_2$ for a given complex constant A, regardless of any external circuit connections. That is, find a matrix $[Y]$, $[Z]$ or $[S]$ that ensures this behavior, and find a lumped element equivalent circuit that corresponds to this matrix. Determine whether there are any restrictions on the value of A if the network is to be passive, or if it is to be lossless.

p2-15 Show that a sourceless linear reciprocal three-port network possessing an impedance matrix representation can be represented by the equivalent network shown below.

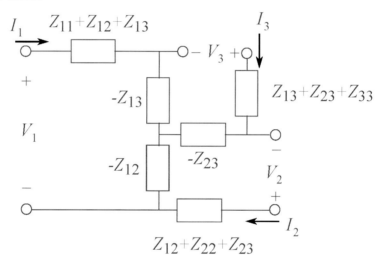

Note that this equivalent circuit does not have a common ground node for all three ports.

p2-16 The branch-line hybrid is a four-port device with lengths of transmission line connected between adjacent ports in the ring arrangement shown in the figure below.

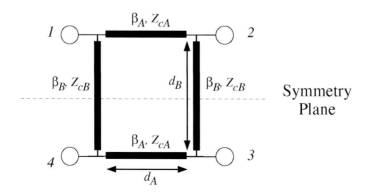

(a) Use the equivalent Π-network for each transmission-line section given in Figure 2.12(c) to reduce this network to the lumped-element hybrid shown in Figure 2.24. Express Y_A, Y_B and Y_D in terms of the transmission-line parameters Z_{cA}, Z_{cB}, $\beta_A d_A$ and $\beta_B d_B$.

(b) Write the constraints (2.78) in terms of the transmission-line parameters. If a 3 dB hybrid is desired ($|S_{21}|^2 = 0.5$), express the condition resulting from (2.83) in terms of the transmission-line parameters.

(c) Obtain one possible design for a 3 dB branch-line hybrid satisfying the constraints found in part (b). [Hint: $d_A = \lambda_A/4$ and $d_B = \lambda_B/4$ results in a well-known version of this device.]

p2-17 Repeat problem p2-16, but do parts (b) and (c) for the case of an arbitrary value of $|S_{21}|$. Assume that $d_A = \lambda_A/4$ for part (c).

p2-18 Repeat problem p2-16, but assume that d_A is shorter than $\lambda_A/4$ for part (c). What is the smallest length possible for d_A that still permits operation as a hybrid network? Comment in a general way on what advantages or disadvantages this design might have compared to choosing $d_A = \lambda_A/4$.

p2-19 Prove that condition (2.64) results in $S_{11} = 0$ for the Boucherot bridge balun.

p2-20 Prove the relations (2.65) and (2.66).

p2-21 Provide details of the derivations of (2.72).

p2-22 A complex load impedance, when connected to a lossless transmission line of some given characteristic impedance Z_c, produces a nonzero, complex reflection coefficient ρ_L. Find the most general conditions which must be satisfied by a passive, lossless, reciprocal two-port network so that when it is connected in cascade with the load, the transmission line now has a reflection coefficient of zero. Your solution will indicate that some degree of freedom is possible in the choice of matching network. Give a particular example of such a matching network, either by specifying its equivalent circuit or giving its scattering matrix in terms of ρ_L.

p2-23 As seen from the example of Section 2.3.2, a lossless three-port network cannot simultaneously be both matched at each port ($S_{11} = S_{22} = S_{33} = 0$) and reciprocal. If we loosen these restrictions somewhat, let us see what may become possible. Suppose we want to have a lossless reciprocal network that will be matched at port 1 ($S_{11} = 0$), and split the power $|a_1|^2$ incident at port 1 into some given fractions $|S_{21}|^2$ and $|S_{31}|^2$ emerging from ports 2 and 3 respectively.

(a) Show that we must have $|S_{21}|^2 + |S_{31}|^2 = 1$.

(b) Show that $|S_{22}| + |S_{33}| = 1$. From this, show that if equal powers $|a_2|^2 = |a_3|^2$ are incident at ports 2 and 3, then at least 50% of that combined power must be reflected back from those ports.

(c) What is the largest fractional crosstalk power $|S_{23}|^2$ that can pass between ports 2 and 3? For what values of $|S_{22}|$ and $|S_{33}|$ does this maximum occur?

(d) Suppose the reference planes (positions) of ports 2 and 3 are chosen so that the phase angles $\angle S_{22} = \angle S_{33} = 0$. Show that if an incident wave is applied to port 1 only, the scattered waves emerging from ports 2 and 3 are either in phase, or out of phase by $180°$.

p2-24 Consider a reciprocal two-port network, matched at port 1 ($S_{11} = 0$), but possibly lossy. Show that

$$|S_{12}|^2 \leq 1 - |S_{22}|$$

If the network is lossless, show that $S_{22} = 0$ and $|S_{12}| = 1$.

3 CLASSICAL TRANSMISSION LINES: EXCITATION AND COUPLING

3.1 INDUCED AND IMPRESSED CURRENTS AND VOLTAGES; EQUIVALENT CIRCUITS

It is well to be clear about terminology when we talk about sources, either in analysis of circuits or of classical transmission lines. Let us state what we mean by the term "externally generated" sources. In circuit theory, we make use of ideal voltage and current sources to simplify analysis and conceptual understanding of the subject. These sources are regarded as capable of sustaining voltages or currents we can "control", whereas other voltages and currents at different places in the circuit (or even the current through a voltage generator, for example) are subject to conditions that prevail in the rest of the circuit. Of course, no such ideal sources can be achieved in practice because the circuit into which a real source is connected will always affect it (load it down) to some degree. No *real* voltage source produces constant voltage regardless of what load is connected to it (especially if that load is a short circuit!). Nevertheless, we find the concept of an ideal generator a very useful one in the analysis of networks. We say that such a generator is an *impressed* source. Currents or voltages which disappear when the impressed sources are turned off are referred to as *induced* currents or voltages.

3.1.1 THEVENIN AND NORTON EQUIVALENT CIRCUITS

It is often the case in circuit analysis that a network can be partitioned into two sections, one of which can be replaced by a different subnetwork without affecting the voltages or currents in the other. The two alternative subnetworks are said to be equivalent to each other as far as their effect on the second subnetwork is concerned. Examples of such equivalence are Thévenin and Norton equivalent circuits. Let us pursue this idea a little further than is usually done in introductory expositions of circuit theory, at the same time providing a connection with the ideas of impressed and induced sources.

Consider an *active* multiport network (that is, one which may contain internal impressed sources in addition to passive components such as resistors, capacitors, etc.). If the network is linear and the impedance or admittance matrix representation exists when the internal sources are absent, it is possible to relate the port voltages and currents in at least one of the following forms:

$$[V] = [Z][I] + [V^{\text{ext}}] \tag{3.1}$$

or

$$[I] = [Y][V] - [I^{\text{ext}}] \tag{3.2}$$

where the impressed sources are related by $[V^{\text{ext}}] = -[Z][I^{\text{ext}}]$. We observe that if the internal sources are deactivated (i. e., current sources replaced by open circuits and voltage sources by short circuits), then the terms $[V^{\text{ext}}]$ and $[I^{\text{ext}}]$ must disappear from these expressions, and we recover the impedance and admittance matrix descriptions of the resulting sourceless multiport. On the other hand, $[V^{\text{ext}}]$ represents the voltages appearing at the ports when the ports are open-circuited ($[I] = 0$) and $-[I^{\text{ext}}]$ the currents that flow when

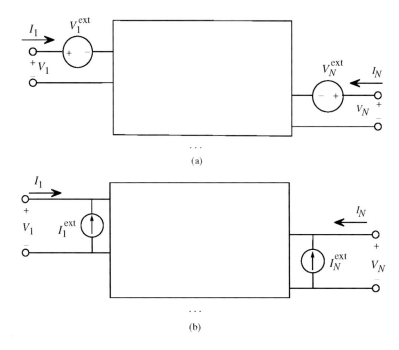

Figure 3.1 Thévenin (a) and Norton (b) equivalent circuits for an active multiport.

the ports are short-circuited. Thus, (3.1) and (3.2) are, in essence, statements of generalized Thévenin and Norton equivalence theorems, respectively. That is, any such network can be replaced by an equivalent circuit consisting of a sourceless multiport (characterized by the matrix $[Z]$ or $[Y]$), with additional ideal impressed voltage sources in series with each port, or ideal impressed current sources in parallel, as shown in Figure 3.1. The voltages and currents induced in any circuitry connected external to the multiport will remain the same (though of course internal voltages and currents will generally be different).

3.1.2 HUYGENS (WAVE) SOURCES

Other kinds of equivalence principles can be stated as well. Suppose (Figure 3.2(a)) that some source network drives a passive load Z_L as shown. A current I and a voltage V are induced at the terminals of the load by this generator, although they are not independent (since the load forces the relation $V/I = Z_L$ to hold). Now let the sources within the source network be turned off, and *impressed* generators for V and I be inserted as shown in Figure 3.2(b). Elementary circuit theory shows that the same voltage and current (V, I) appear at Z_L, while the generator impedance Z_s to the left has neither voltage across it nor current through it, and therefore may have any value at all (*not* necessarily the Thévenin equivalent impedance of the original source network). The presence of the element Z_s is thus now irrelevant to the voltage and current being produced at Z_L. In fact, suppose that we replace both these impedances by semi-infinite transmission lines of characteristic impedance $Z_c = Z_L$ as shown in Figure 3.2(c). Then no wave is launched on the transmission line to the left of the sources, and only a wave traveling in the $+z$-direction to the right of the sources is produced. This combination of the sources V and I is called a *Huygens* or *wave* equivalent source, its characteristic property being that it produces a response only to one side of it.

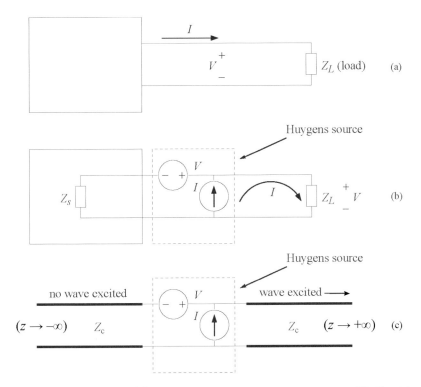

Figure 3.2 (a) Original network; (b) equivalent impressed Huygens source (V, I) with arbitrary passive impedance Z_s; (c) load and generator impedances replaced by semi-infinite transmission lines.

Once V and I have been made impressed sources, however, we can adjust Z_s to have any convenient value without affecting the current or voltage at the load. If we set $Z_s = 0$, the current generator is now shorted out and can be removed (that is, its removal makes no difference to the current induced in Z_L, which is now assured by the voltage generator). As far as Z_L is concerned, then, the circuit of Figure 3.2(a) is completely equivalent to that of Figure 3.3(a). In this case, the current flow I is an induced one, and passes through the voltage source rather than the current source. The voltage V remains an impressed one, and the load continues to "see" the same conditions it has throughout all the modifications of the source network on the left. Conversely, we could make Z_s equal to ∞ (an open circuit) and remove the voltage generator, leaving the situation of Figure 3.3(b). In this case the voltage is induced, the current is impressed, and the load conditions continue to be the same.

This equivalence principle is an application of what is known in circuit theory as the substitution theorem. Unlike the Thévenin and Norton theorems, the equivalent sources used now depend on values of induced currents or voltages in the original circuit. Since these are dependent on the nature of the external circuit (here the load impedance), the resulting equivalent circuit is not the same if the external circuit is changed. The implications of this in circuit theory are perhaps not as profound as are those of the analogous equivalence principles in field theory, to be examined in Chapter 10. Nevertheless, this discussion shows that there are many equivalent impressed source arrangements capable of inducing the same voltages and currents in a given portion of a circuit—including one that contains sections of classical transmission line.

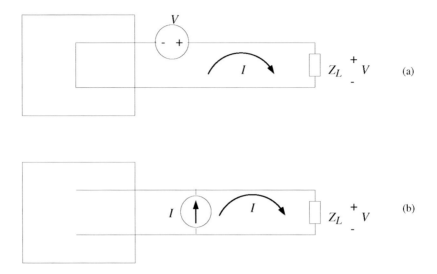

Figure 3.3 (a) Z_s made equal to zero, impressed V, induced I; (b) Z_s made equal to infinity, impressed I, induced V.

One such important application is the excitation of a wave in a single direction only on a transmission line. Consider the Huygens source connected in an infinite transmission line as shown in Figure 3.4. If the voltage source and current source are chosen such that the relation $V_0 = Z_c I_0$ holds, then there will be a forward wave produced to the right of the source (since the semi-infinite section of line presents an impedance of Z_c), but nothing is produced to the left of the source. By (1.33), the forward wave amplitude at a point just to the right of the wave source will be

$$a = a^{\text{ext}} = \frac{V_0}{\sqrt{2Z_c}}$$

and we regard a^{ext} as the strength of this wave source. Accordingly, we will represent such a source schematically as in Figure 3.5. If we had chosen $V_0 = -Z_c I_0$ instead, the wave source would have produced nothing to the right, but a reverse wave amplitude

$$b = b^{\text{ext}} = -\frac{V_0}{\sqrt{2Z_c}}$$

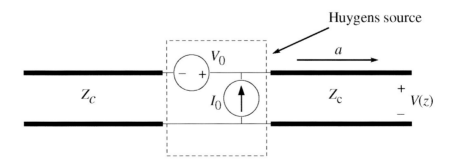

Figure 3.4 Huygens source in a transmission line.

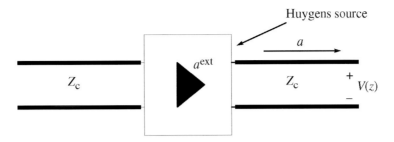

Figure 3.5 Schematic representation of a Huygens source in a transmission line.

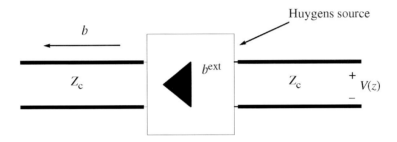

Figure 3.6 Schematic representation of a reverse Huygens source in a transmission line.

to its left. We denote this source schematically as shown in Figure 3.6.

Now let us place both $[V^{\text{ext}}]$ and $[I^{\text{ext}}]$ at the ports of a network. Instead of (3.1) or (3.2), we now have

$$[V] + [V^{\text{ext}}] = [Z]\left([I] + [I^{\text{ext}}]\right) \tag{3.3}$$

Defining the impressed wave sources $[a^{\text{ext}}]$ and $[b^{\text{ext}}]$ to be [compare (2.25)]:

$$[a^{\text{ext}}] = \frac{1}{2}[2Z_c]^{-1/2}\left\{[V^{\text{ext}}] + [Z_c][I^{\text{ext}}]\right\}$$

$$[b^{\text{ext}}] = -\frac{1}{2}[2Z_c]^{-1/2}\left\{[V^{\text{ext}}] - [Z_c][I^{\text{ext}}]\right\} \tag{3.4}$$

let us also suppose that $[V^{\text{ext}}] = -[Z_c][I^{\text{ext}}]$ so that $[a^{\text{ext}}] = 0$. Then $[V^{\text{ext}}] = [2Z_c]^{1/2}[b^{\text{ext}}]$ and $[I^{\text{ext}}] = -\sqrt{2}[Z_c]^{-1/2}[b^{\text{ext}}]$ combine to form a wave source which launches waves only *away* from the multiport as shown in Figure 3.7, and represent the equivalent effect of sources inside the original multiport without introducing spurious incident waves traveling toward it. The resulting relation between $[a]$ and $[b]$ is then:

$$[b] = [S][a] + [b^{\text{ext}}] \tag{3.5}$$

3.2   EXCITATION OF A CLASSICAL TRANSMISSION LINE

Consider the excitation of a uniform classical transmission line in terms of the distributed constant model of Section 1.2. In place of the purely sourceless network model for an incremental length Δz of the line as shown in Figure 1.3, we now suppose there are impressed series voltage generators and shunt current generators distributed in a continuous manner along the line. Our network model is as shown in Figure 3.8 (note that shunt voltage sources

Figure 3.7 Wave sources for an active multiport.

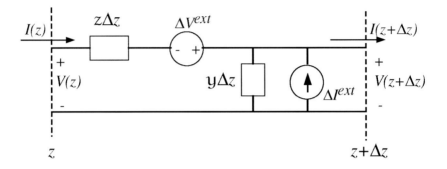

Figure 3.8 Distributed source excitation of a classical transmission line.

and series current sources are not used, as they would force the line voltage and current to specified values, rendering the problem trivial). Following the derivation of Section 1.2, we can apply Kirchhoff's laws to Figure 3.8, and take the limit as $\Delta z \to 0$. We obtain the telegrapher's equations with distributed sources:

$$\frac{dV(z)}{dz} = -zI(z) + E_{\text{ext}}(z)$$

$$\frac{dI(z)}{dz} = -yV(z) + H_{\text{ext}}(z) \tag{3.6}$$

where

$$E_{\text{ext}}(z) = \lim_{\Delta z \to 0} \frac{\Delta V^{\text{ext}}}{\Delta z}$$

$$H_{\text{ext}}(z) = \lim_{\Delta z \to 0} \frac{\Delta I^{\text{ext}}}{\Delta z} \tag{3.7}$$

are the series voltage and shunt current sources per unit length impressed at each point along the line. We give them a notation suggestive of field quantities because their units (volt/m and amp/m) are those of electric and magnetic fields respectively. However, no physical interpretation as actual electric or magnetic fields is necessarily implied here.

Now, from here on, we assume that the transmission line is uniform (z and y are constant). Further, the impressed sources E_{ext} and H_{ext} are assumed to be zero outside some interval $z_1 < z < z_2$. Therefore, for $z \geq z_2$, we expect only a forward wave to be present:

$$V(z) = Z_c I(z) = \sqrt{2Z_c} a_{02} e^{-\gamma z} \qquad (z \geq z_2) \tag{3.8}$$

To the left of the sources, we have only a reverse wave:

$$V(z) = -Z_c I(z) = \sqrt{2Z_c} b_{01} e^{\gamma z} \qquad (z \leq z_1) \tag{3.9}$$

In other words, V and I are totally produced by the localized sources, and there are no incoming waves from infinity.

We can solve (3.6) by the method of variation of parameters. We introduce the variable wave amplitudes $a_0(z)$ and $b_0(z)$ by generalization from the constant values defined in Chapter 1 in terms of $V(z)$ and $I(z)$:

$$\begin{aligned}
V(z) &= [a_0(z)e^{-\gamma z} + b_0(z)e^{\gamma z}]\sqrt{2Z_c} \\
I(z) &= [a_0(z)e^{-\gamma z} - b_0(z)e^{\gamma z}]\sqrt{\frac{2}{Z_c}}
\end{aligned} \tag{3.10}$$

or,

$$\begin{aligned}
a_0(z) &= \frac{1}{2}e^{\gamma z}\frac{V(z) + Z_c I(z)}{\sqrt{2Z_c}} \\
b_0(z) &= \frac{1}{2}e^{-\gamma z}\frac{V(z) - Z_c I(z)}{\sqrt{2Z_c}}
\end{aligned} \tag{3.11}$$

Differential equations for a_0 and b_0 are obtained by substituting (3.10) into (3.6), and using (1.8) and (1.9). We obtain

$$a_0'(z) = e^{\gamma z} A_{\text{ext}}(z); \qquad b_0'(z) = -e^{-\gamma z} B_{\text{ext}}(z) \tag{3.12}$$

where

$$A_{\text{ext}}(z) = \frac{E_{\text{ext}}(z) + Z_c H_{\text{ext}}(z)}{2\sqrt{2Z_c}}; \qquad B_{\text{ext}}(z) = -\frac{E_{\text{ext}}(z) - Z_c H_{\text{ext}}(z)}{2\sqrt{2Z_c}} \tag{3.13}$$

are per-unit-length wave sources—continuously distributed versions of the Huygens sources in (3.4). In (3.12), these sources are multiplied by the factors $e^{\pm\gamma z}$ that account for the position of each elemental source on the line. Equations (3.12) can be readily integrated to obtain a_0 and b_0, if appropriate initial or boundary conditions are available.

The boundary conditions are found from (3.8) and (3.9). We should have

$$\left.\begin{aligned}
a_0(z) &\equiv 0 \qquad \text{for } z \leq z_1 \\
b_0(z) &\equiv 0 \qquad \text{for } z \geq z_2
\end{aligned}\right\} \tag{3.14}$$

Then, integrating (3.12) yields

$$a_0(z) = \int_{z_1}^{z} A_{\text{ext}}(z')e^{\gamma z'}\, dz' \qquad (z > z_1) \tag{3.15}$$

and

$$b_0(z) = \int_{z}^{z_2} B_{\text{ext}}(z')e^{-\gamma z'}\, dz' \qquad (z < z_2) \tag{3.16}$$

Note that since E_{ext} and H_{ext} are zero for $z \leq z_1$ and $z \geq z_2$, $a_0(z)$ becomes a constant a_{02} for $z \geq z_2$ and $b_0(z)$ the constant b_{01} for $z \leq z_1$, as stated earlier in (3.8) and (3.9). In fact,

$$a_{02} = \int_{z_1}^{z_2} A_{\text{ext}}(z')e^{\gamma z'}\, dz'$$

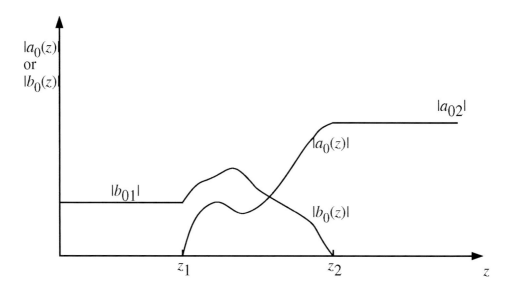

Figure 3.9 Forward and reverse mode amplitudes as functions of z.

$$b_{01} = \int_{z_1}^{z_2} B_{\text{ext}}(z')e^{-\gamma z'}\,dz' \tag{3.17}$$

Thus, the forward-mode amplitude gradually changes from zero (at z_1) and becomes constant after passing z_2. The opposite occurs for a reverse-mode amplitude, as shown in Figure 3.9. The constancy of $a_0(z)$ and $b_0(z)$ in source-free regions is also immediately obvious from (3.12).

For a line with no attenuation ($\gamma = j\beta$), the amplitude of the forward wave a_{02} can be maximized relative to that of the reverse wave b_{01} in a variety of ways. The most obvious is to use both distributed voltage and current sources such that $E_{\text{ext}}(z) = Z_c H_{\text{ext}}(z)$. If only one type of source ($E_{\text{ext}}(z)$, say) is available, a_{02} can still be enhanced by matching the source's phase progression in z with that of the desired wave, $e^{-j\beta z}$:

$$E_{\text{ext}}(z) = (\text{const})e^{-j\beta z}$$

The integral for a_{02} in (3.17) increases linearly as the source region is increased, since the integrand is constant. The integral for b_{01} on the other hand, has a highly oscillatory integrand, which upon integration tends to cancel itself resulting in a much lower magnitude than for a_{02}. The reverse choice of source phasing can be used to enhance the excitation of the reverse wave.

3.2.1 MODELING OF LUMPED SOURCES

Ordinary lumped voltage and current sources can be modeled in the language of the distributed sources E_{ext} and H_{ext}. However, we need to employ a Dirac delta-function to do so. Suppose a series voltage source V_0 and shunt current source I_0 are connected to the line at a point $z = z_0$ as shown in Figure 3.10. From Kirchhoff's laws at z_0, the currents and voltages on either side of z_0 will be related by

$$V(z_0^+) - V(z_0^-) = V_0 \tag{3.18}$$

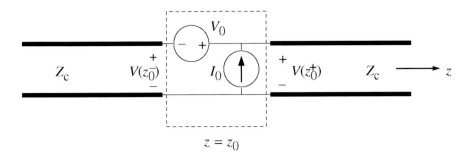

Figure 3.10 Lumped sources at $z = z_0$.

$$I(z_0^+) - I(z_0^-) = I_0 \qquad (3.19)$$

On the other hand, if we integrate both sides of (3.6) from z_0^- to z_0^+, we have

$$V(z_0^+) - V(z_0^-) = \int_{z_0^-}^{z_0^+} E_{\text{ext}}(z)\,dz$$

$$I(z_0^+) - I(z_0^-) = \int_{z_0^-}^{z_0^+} H_{\text{ext}}(z)\,dz \qquad (3.20)$$

since I and V suffer (at worst) step discontinuities across z_0. Comparing (3.20) with (3.18) and (3.19), we find that the lumped generators can be expressed as

$$E_{\text{ext}}(z) = V_0\delta(z - z_0)$$
$$H_{\text{ext}}(z) = I_0\delta(z - z_0) \qquad (3.21)$$

Thus, the description of eqn. (3.6) is capable of describing lumped excitations as well as distributed ones.

3.2.2 EQUIVALENT CIRCUITS FOR DISTRIBUTED SOURCES

As discussed in the previous section, we sometimes find it convenient to use the concept of equivalent sources in circuit or transmission line problems. If we are not interested in the behavior of $V(z)$ and $I(z)$ within the source region $z_1 < z < z_2$, then it is possible to replace the actual sources with other, equivalent ones inside the source region which produce the same V and I exterior to it. One possibility would be the use of a Thévenin or Norton equivalent circuit for a section of line excited by distributed sources. Another is to replace the sources by lumped sources V_0 and I_0 located at z_0 within the source region as shown in Figure 3.10.

If z_0 is a point in the interval $[z_1, z_2]$, let us require that $V(z)$ and $I(z)$, or what is the same thing, $a_0(z)$ and $b_0(z)$, be identical outside of $[z_1, z_2]$ whether the original or the equivalent sources are used. Now for the original sources, $a_0(z)$ and $b_0(z)$ are given by (3.14) or by the constants (3.17) as appropriate. With the equivalent sources (3.21), eqn. (3.14) remains true, while (3.17) becomes

$$a_{02} = \frac{1}{2}e^{\gamma z_0}\frac{V_0 + Z_c I_0}{\sqrt{2Z_c}}$$

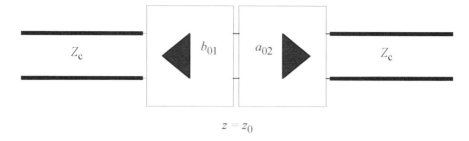

Figure 3.11 Lumped wave sources at $z = z_0$.

$$b_{01} = -\frac{1}{2} e^{-\gamma z_0} \frac{V_0 - Z_c I_0}{\sqrt{2 Z_c}} \tag{3.22}$$

Thus, an equivalent circuit for Figure 3.10 can be constructed using wave sources, as shown in Figure 3.11.

For (3.22) to be the same as (3.17), the equivalent sources must be given by

$$V_0 = \int_{z_1}^{z_2} \left[E_{ext}(z) \cosh \gamma(z - z_0) + Z_c H_{ext}(z) \sinh \gamma(z - z_0) \right] dz$$

$$I_0 = \int_{z_1}^{z_2} \left[H_{ext}(z) \cosh \gamma(z - z_0) + \frac{E_{ext}(z)}{Z_c} \sinh \gamma(z - z_0) \right] dz \tag{3.23}$$

If the source region is electrically small, $|\gamma(z_2 - z_1)| \ll 1$, then $\cosh \simeq 1$ and $\sinh \simeq 0$, so that

$$V_0 \simeq \int_{z_1}^{z_2} E_{ext}(z)\, dz$$

$$I_0 \simeq \int_{z_1}^{z_2} H_{ext}(z)\, dz \tag{3.24}$$

which is consistent with the delta-function (lumped-source) idealization (3.21) pictured in Figure 3.10. This concept of equivalent source representation turns out to have important application in the treatment of coupled lines as well as for the excitation itself.

The wave sources can also be placed at both $z = z_1$ and $z = z_2$ to form an equivalent circuit like that of Figure 3.7. Leaving the details to the reader, we can obtain the equivalent circuit shown in Figure 3.12, wherein the segment of transmission line between z_1 and z_2 is now sourceless, and the values of the voltage and current sources are given by

$$V_{01} = Z_c I_{01} = -\frac{1}{2} \int_{z_1}^{z_2} \left[E_{ext}(z') - Z_c H_{ext}(z') \right] e^{-\gamma z'}\, dz'$$

$$V_{02} = Z_c I_{02} = \frac{1}{2} \int_{z_1}^{z_2} \left[E_{ext}(z') + Z_c H_{ext}(z') \right] e^{\gamma z'}\, dz' \tag{3.25}$$

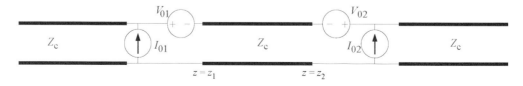

Figure 3.12 Lumped wave sources placed outside the source region.

Further variations can be made by replacing the transmission line in $z_1 < z < z_2$ with either of the equivalent circuits shown in Figure 2.12, or by using Thévenin or Norton equivalent sources instead of wave sources at $z = z_1$ and $z = z_2$.

3.3 MULTICONDUCTOR TRANSMISSION LINES

Suppose that E_{ext} and H_{ext} are due to the fields of another transmission line adjacent to the one being excited. In other words, they arise because of mutual capacitance and mutual inductance with the second line, and can therefore be considered as dependent sources. Then we say that the two transmission lines are coupled, and indeed, not only does the second line serve as a source of excitation for the first, but also *vice versa*. In general, when two or more parallel transmission lines are placed in proximity to each other (such as adjacent printed circuit transmission lines on a common substrate, or the conductors in a "ribbon cable" interconnecting circuit boards in a computer), coupling will occur between the lines. This may either be a desired effect, which can be used to achieve actions such as directional coupling, or undesired, in which case the phenomenon of crosstalk between the transmission line circuits arises. In this section, we give an overview of the behavior of coupled transmission lines, leaving specific configurations to a later chapter.

3.3.1 MATRIX TELEGRAPHER'S EQUATIONS

Consider two coupled lines as shown in Figure 3.13. One of the conductors is regarded as the "ground" conductor, and the voltages $V^{(1)}$ and $V^{(2)}$ at any point z along the system are measured from each of the other conductors to this reference. Positive currents $I^{(1)}$ and $I^{(2)}$ are defined on each conductor in the axial direction z. Note that the conductor number is denoted here by a superscript, rather than a subscript. We will at first limit ourselves to the lossless case in order to focus on the essential concepts.

If conductors 1 and 2, in addition to their shunt capacitances (c_{11} or c_{22}) and series inductances (l_{11} or l_{22}) per unit length, also exhibit a mutual inductance l_{12} and mutual capacitance c_{12} per unit length as coupling between them (see Figure 3.14), then by a similar procedure as in Section 1.2, we can obtain the telegraphers' equations for this multiconductor transmission line:

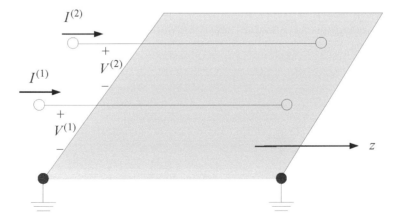

Figure 3.13 Two coupled transmission lines.

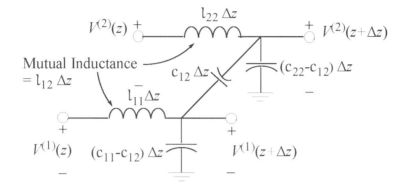

Figure 3.14 Distributed network model of coupled transmission lines.

$$\frac{dV^{(1)}(z)}{dz} = -j\omega l_{11}I^{(1)}(z) - j\omega l_{12}I^{(2)}(z)$$

$$\frac{dV^{(2)}(z)}{dz} = -j\omega l_{12}I^{(1)}(z) - j\omega l_{22}I^{(2)}(z)$$

$$\frac{dI^{(1)}(z)}{dz} = -j\omega c_{11}V^{(1)}(z) + j\omega c_{12}V^{(2)}(z)$$

$$\frac{dI^{(2)}(z)}{dz} = j\omega c_{12}V^{(1)}(z) - j\omega c_{22}V^{(2)}(z) \qquad (3.26)$$

Notice that we have assumed reciprocity in putting $c_{21} = c_{12}$ and $l_{21} = l_{12}$. Note too that the equivalent circuit for the capacitive part of a length Δz of this multiconductor line consists of a capacitance $(c_{11} - c_{12})\Delta z$ from conductor 1 to ground, a capacitance $(c_{22} - c_{12})\Delta z$ from conductor 2 to ground, and a capacitance $c_{12}\Delta z$ from conductor 1 to conductor 2.

Comparing (3.26) with (3.6), we recognize that the coupling term $-j\omega l_{12}I^{(2)}(z)$ can be interpreted as a distributed external voltage source $E_{ext}^{(1)}(z)$ exciting conductor 1, although it is a source dependent on the current in wire 2. Likewise, $j\omega c_{12}V^{(2)}(z)$ is a distributed dependent current source $H_{ext}^{(1)}(z)$ for conductor 1. Each conductor continuously excites the other by this mechanism, providing a coupling between the transmission lines.

Equations (3.26) can be written in the matrix-vector form:

$$\frac{d[V(z)]}{dz} = -j\omega[l][I(z)]$$

$$\frac{d[I(z)]}{dz} = -j\omega[c][V(z)] \qquad (3.27)$$

where $[l]$ and $[c]$ are given by

$$[l] = \begin{bmatrix} l_{11} & l_{12} \\ l_{12} & l_{22} \end{bmatrix} \qquad (3.28)$$

$$[c] = \begin{bmatrix} c_{11} & -c_{12} \\ -c_{12} & c_{22} \end{bmatrix} \qquad (3.29)$$

and the voltage and current column vectors are:

$$[V(z)] = \begin{bmatrix} V^{(1)}(z) \\ V^{(2)}(z) \end{bmatrix}$$

$$[I(z)] = \begin{bmatrix} I^{(1)}(z) \\ I^{(2)}(z) \end{bmatrix}$$

In case there are losses or unconventional aspects to the multiconductor line, we replace $j\omega[\mathsf{l}]$ by $[\mathsf{z}]$ and $j\omega[\mathsf{c}]$ by $[\mathsf{y}]$ in (3.27):

$$\begin{aligned} \frac{d[V(z)]}{dz} &= -[\mathsf{z}][I(z)] \\ \frac{d[I(z)]}{dz} &= -[\mathsf{y}][V(z)] \end{aligned} \qquad (3.30)$$

where

$$[\mathsf{z}] = \begin{bmatrix} \mathsf{z}_{11} & \mathsf{z}_{12} \\ \mathsf{z}_{12} & \mathsf{z}_{22} \end{bmatrix}; \qquad [\mathsf{y}] = \begin{bmatrix} \mathsf{y}_{11} & \mathsf{y}_{12} \\ \mathsf{y}_{12} & \mathsf{y}_{22} \end{bmatrix} \qquad (3.31)$$

(note that $\mathsf{y}_{12} = -j\omega\mathsf{c}_{12}$ for conventional lossless coupled lines). From here on, we will treat the general case unless otherwise stated.

The flow of energy along a multiconductor line is described by a superposition of the powers on each of the conductor/ground pairs. Thus the complex power $\hat{P}(z)$ passing the point z is

$$\hat{P}(z) = \frac{1}{2}\left[V^{(1)}(z)I^{(1)*}(z) + V^{(2)}(z)I^{(2)*}(z)\right] = \frac{1}{2}[I(z)]^{\dagger}[V(z)] \qquad (3.32)$$

and the complex oscillatory power is

$$\hat{P}_{\mathrm{osc}}(z) = \frac{1}{2}\left[V^{(1)}(z)I^{(1)}(z) + V^{(2)}(z)I^{(2)}(z)\right] = \frac{1}{2}[I(z)]^{T}[V(z)] \qquad (3.33)$$

Although we have explicitly considered only the case of two conductors with ground, all of what follows can be applied to the case of a line with N conductors plus ground, simply by considering the vectors to be $N \times 1$ column vectors and the matrices to be $N \times N$ square matrices.

3.3.2 MODES OF MULTICONDUCTOR LINES

Suppose we now search for mode solutions of (3.30):

$$[V] = [V_m]e^{-\gamma_m z}; \qquad [I] = [I_m]e^{-\gamma_m z} \qquad (3.34)$$

where a subscript is used to indicate the mode index m. Then the system of equations

$$\gamma_m[V_m] = [\mathsf{z}][I_m]; \qquad \gamma_m[I_m] = [\mathsf{y}][V_m] \qquad (3.35)$$

must have a nontrivial solution. If we eliminate $[I_m]$ from (3.35), we obtain

$$\gamma_m^2 [V_m] = [G][V_m] \tag{3.36}$$

where

$$[G] = [z][y] = \begin{bmatrix} z_{11}y_{11} + z_{12}y_{12} & z_{11}y_{12} + z_{12}y_{22} \\ z_{12}y_{11} + z_{22}y_{12} & z_{12}y_{12} + z_{22}y_{22} \end{bmatrix} = \begin{bmatrix} G_{11} & G_{12} \\ G_{21} & G_{22} \end{bmatrix} \tag{3.37}$$

These expressions can be simplified somewhat if we define impedance and admittance coupling coefficients as:

$$K_Z = \frac{z_{12}}{\sqrt{z_{11}z_{22}}}; \qquad K_Y = \frac{y_{12}}{\sqrt{y_{11}y_{22}}} \tag{3.38}$$

Consider hypothetical uncoupled transmission lines whose modes have the characteristic impedances $\hat{Z}_{ci}$ and propagation constants $\hat{\gamma}_i$ given by

$$\hat{Z}_{ci} = \sqrt{\frac{z_{ii}}{y_{ii}}}; \qquad \hat{\gamma}_i = \sqrt{z_{ii}y_{ii}} \tag{3.39}$$

for $i = 1, 2$. The signs of all square roots are taken according to the usual rules for the real parts of the characteristic impedance and propagation constant. In terms of the quantities defined above, the elements of the matrix $[G]$ defined in (3.37) can be written as

$$
\begin{aligned}
G_{11} &= \hat{\gamma}_1^2 + K_Z K_Y \hat{\gamma}_1 \hat{\gamma}_2 \\
G_{22} &= \hat{\gamma}_2^2 + K_Z K_Y \hat{\gamma}_1 \hat{\gamma}_2 \\
G_{12} &= \sqrt{\frac{\hat{Z}_{c1}}{\hat{Z}_{c2}}} \sqrt{\hat{\gamma}_1 \hat{\gamma}_2} \left(\hat{\gamma}_1 K_Y + \hat{\gamma}_2 K_Z \right) \\
G_{21} &= \sqrt{\frac{\hat{Z}_{c2}}{\hat{Z}_{c1}}} \sqrt{\hat{\gamma}_1 \hat{\gamma}_2} \left(\hat{\gamma}_2 K_Y + \hat{\gamma}_1 K_Z \right)
\end{aligned} \tag{3.40}
$$

We see that γ_m^2 must be an eigenvalue of the matrix $[G]$. If we had eliminated $[V_m]$, we would have found that γ_m^2 must also be an eigenvalue of $[y][z] = [G]^T$, because $[z]^T = [z]$ and $[y]^T = [y]$ (note that $[G]$ is not a symmetric matrix). Thus, it is natural that $[z][y]$ and $[y][z]$ have the same eigenvalues, although their eigenvectors (either $[V_m]$ or $[I_m]$ respectively) will be different.

To determine γ_m^2, we must obtain the solution of the determinantal equation

$$\det\left\{ [G] - \gamma_m^2 [\mathbf{1}] \right\} = 0 \quad \Rightarrow \quad (G_{11} - \gamma_m^2)(G_{22} - \gamma_m^2) - G_{12}G_{21} = 0 \tag{3.41}$$

which is

$$\gamma_m^2 = \frac{G_{11} + G_{22}}{2} \pm \sqrt{\left(\frac{G_{11} - G_{22}}{2} \right)^2 + G_{12}G_{21}} \tag{3.42}$$

If $(G_{11} - G_{22})^2 + 4G_{12}G_{21} \neq 0$, there will be two distinct values of γ_m^2, which we denote γ_1^2 and γ_2^2 corresponding to the choice of plus or minus sign in (3.42) (the modes are said to be nondegenerate in this case). Since only γ_m^2 is determined in this way, both $+\gamma_m$ and $-\gamma_m$ are possible solutions: a forward mode and a reverse mode for each value of m. The sign of γ_m will be determined by the requirement that time-average power is carried in the $+z$-direction $[\text{Re}(\hat{P}_m) \geq 0$, where the complex power $\hat{P}_m$ of a mode is given by (3.59) below], or that we have decay in the $+z$-direction $[\text{Re}(\gamma_m) > 0]$, as was done for the single transmission line in Chapter 1.

The voltage eigenvector $[V_m]$ can be obtained from (3.36) once γ_m has been found. Put

$$[V_m] = \begin{bmatrix} V_m^{(1)} \\ \\ V_m^{(2)} \end{bmatrix} \tag{3.43}$$

and examine the first row of (3.36):

$$G_{11}V_m^{(1)} + G_{12}V_m^{(2)} = \gamma_m^2 V_m^{(1)} \tag{3.44}$$

which means that the ratio of wire voltages for mode m is

$$R_m \equiv \frac{V_m^{(2)}}{V_m^{(1)}} = \frac{\gamma_m^2 - G_{11}}{G_{12}} \tag{3.45}$$

This ratio is different for the two modes if they are nondegenerate. If $\det[G] = (\det[z])(\det[y]) = 0$, then one of the $\gamma_m = 0$ and we have a special case that must be handled in a different way and will not be considered here. Otherwise, $\gamma_m \neq 0$ and we may from (3.35) obtain $[I_m]$ if $[V_m]$ is known, or vice versa:

$$[I_m] = \frac{1}{\gamma_m}[y][V_m]; \qquad [V_m] = \frac{1}{\gamma_m}[z][I_m] \tag{3.46}$$

For every mode with eigenvalue γ_m that satisfies (3.35), there will thus also be a corresponding reverse mode with $\gamma_{-m} = -\gamma_m$ for which we can choose

$$[V_{-m}] = [V_m]; \qquad [I_{-m}] = -[I_m] \tag{3.47}$$

We adopt the convention that modes denoted by positive values of m are forward modes (as defined at the end of this subsection), while those with negative m are reverse modes.

The eigenvectors can be normalized as we have done for the single transmission line by requiring that the complex oscillatory power in a forward mode equals 1:

$$\frac{1}{2}[V_m]^T[I_m] = 1 \qquad (m > 0) \tag{3.48}$$

Using (3.46), this normalization condition can also be expressed in either of the forms:

$$[I_m]^T[z][I_m] = [V_m]^T[y][V_m] = 2\gamma_m \qquad (m > 0) \tag{3.49}$$

Since the ratio of the wire voltages is known from (3.45), we can solve for both $V_m^{(1)}$ and $V_m^{(2)}$ from (3.49).

The eigenvectors can also be shown to obey so-called *orthogonality* properties. An orthogonality property can be obtained by premultiplying the first of (3.35) by $[I_n]^T$, the second by $[V_n]^T$, generating two further equations by interchanging m and n, and combining so as to cancel the terms on the right sides. The result is

$$(\gamma_m - \gamma_n)\left\{[V_m]^T[I_n] + [V_n]^T[I_m]\right\} = 0$$

If $\gamma_m \neq \gamma_n$, then we have the orthogonality property

$$[V_m]^T[I_n] + [V_n]^T[I_m] = 0 \tag{3.50}$$

If $\gamma_m = \gamma_n$ even though $m \neq n$, it is possible to use Gram-Schmidt orthogonalization procedures (see Section B.2) to ensure that (3.50) holds in that case as well. We will assume from here on that this has been done. Putting $m \to -m$ in (3.50), adding the result to (3.50) itself and using (3.47), we obtain the somewhat simpler version of orthogonality

$$[V_m]^T[I_n] = 0 \qquad (m \neq \pm n) \tag{3.51}$$

The form of eqn. (3.51) suggests the reason that this property is called orthogonality. If two vectors $\mathbf{A}$ and $\mathbf{B}$ in space obey the relation $\mathbf{A} \cdot \mathbf{B} = 0$, they are perpendicular (or orthogonal) to each other. For column vectors, the analog of the dot product is the multiplication of the transpose of one by the other as on the left side of (3.51). Additional orthogonality-type properties can be obtained by using (3.46) in conjunction with (3.51). If $\gamma_m \neq 0$ and $\gamma_n \neq 0$, we get

$$[I_m]^T[z][I_n] = 0 \qquad (m \neq \pm n) \tag{3.52}$$

and

$$[V_m]^T[y][V_n] = 0 \qquad (m \neq \pm n) \tag{3.53}$$

The orthogonality and normalization properties (together known as *orthonormalization*) can be conveniently expressed as a single equation if we introduce the matrices

$$[M_V] = \left[[V_1][V_2]\right] = \begin{bmatrix} V_1^{(1)} & V_2^{(1)} \\ V_1^{(2)} & V_2^{(2)} \end{bmatrix}; \qquad [M_I] = \left[[I_1][I_2]\right] = \begin{bmatrix} I_1^{(1)} & I_2^{(1)} \\ I_1^{(2)} & I_2^{(2)} \end{bmatrix} \tag{3.54}$$

The normalization property (3.48) and the orthogonality property (3.51) imply that

$$[M_V]^T[M_I] = 2[\mathbf{1}] \tag{3.55}$$

or in other words,

$$[M_V]^{-1} = \frac{1}{2}[M_I]^T; \qquad [M_I]^{-1} = \frac{1}{2}[M_V]^T \tag{3.56}$$

Using these matrices, equations (3.35) for both values of m can be combined into the single forms

$$[\gamma][M_V] = [z][M_I]; \qquad [\gamma][M_I] = [y][M_V] \tag{3.57}$$

where $[\gamma]$ denotes the matrix

$$[\gamma] = \begin{bmatrix} \gamma_1 & 0 \\ 0 & \gamma_2 \end{bmatrix} \tag{3.58}$$

Complex power flow in a single mode (m) is described in a similar fashion. At $z = 0$,

$$\hat{P}_m = \frac{1}{2}[I_m]^\dagger[V_m] = \frac{1}{2\gamma_m}[I_m]^\dagger[z][I_m] = \frac{1}{2\gamma_m^*}[V_m]^\dagger[y]^\dagger[V_m] \tag{3.59}$$

and thus the time-average power flow in this mode is

$$\hat{P}_{\mathrm{av},m} = \frac{1}{2}\mathrm{Re}\left([I_m]^\dagger[V_m]\right) = \frac{1}{2}\mathrm{Re}\left([I_m]^\dagger\frac{[z]}{\gamma_m}[I_m]\right) = \frac{1}{2}\mathrm{Re}\left([V_m]^\dagger\frac{[y]^\dagger}{\gamma_m^*}[V_m]\right) \tag{3.60}$$

which must be nonnegative for a forward mode. If the attenuation constant α_m differs from zero, we can show that the sign of the propagation constant γ_m of the forward wave must be chosen such that $\alpha_m > 0$ in order to achieve $\hat{P}_{\mathrm{av},m} \geq 0$ (see problem p3-14).

Let us emphasize that the modes of a multiconductor transmission line are features of the entire coupled system, and not of one or the other "positive" conductor; in general currents and voltages must be present on both conductors simultaneously. For this reason, these modes are sometimes called *system modes* or *normal modes* to distinguish them from modes of either conductor (plus ground) in isolation.

3.3.3 PROPAGATING MODES ON LOSSLESS MULTICONDUCTOR LINES

A propagating mode ($\gamma_m = j\beta_m$, with β_m real) can occur only in the special case when the multiconductor line is lossless ($[z] = j[x]$ and $[y] = j[b]$, with $[x]$ and $[b]$ real). In this situation, (3.35) becomes

$$\beta_m[V_m] = [x][I_m]; \qquad \beta_m[I_m] = [b][V_m] \tag{3.61}$$

We see from this that it must then be possible to choose both $[V_m]$ and $[I_m]$ as real vectors in this case, and therefore the matrices $[M_V]$ and $[M_I]$ will be real as well. When considering whether a propagating mode with no attenuation on lossless coupled transmission lines is a forward wave, we choose the sign of β_m so that $\mathrm{Re}(\hat{P}_m) > 0$. The complex power (3.59) now becomes

$$\hat{P}_m = \frac{1}{2}[I_m]^T[V_m] = \frac{1}{2\beta_m}[I_m]^T[x][I_m] = \frac{1}{2\beta_m}[V_m]^T[b][V_m] \tag{3.62}$$

which is real. The sign of β_m is thus chosen so as to make

$$\frac{1}{2\beta_m}[I_m]^T[x][I_m] = \frac{1}{2\beta_m}[V_m]^T[b][V_m] > 0 \tag{3.63}$$

for a forward mode.

In the case of a lossless conventional multiconductor line described by (3.26), we have $[x] = \omega[l]$, $[b] = \omega[c]$, and the characteristic impedances and propagation constants of the modes of the hypothetical uncoupled transmission lines are

$$\hat{Z}_{ci} = \sqrt{\frac{l_{ii}}{c_{ii}}}; \qquad \hat{\beta}_i = \omega\sqrt{l_{ii}c_{ii}} \tag{3.64}$$

while $K_Y = -K_C$ and $K_Z = K_L$, where

$$K_L = \frac{l_{12}}{\sqrt{l_{11}l_{22}}}; \qquad K_C = \frac{c_{12}}{\sqrt{c_{11}c_{22}}} \tag{3.65}$$

Thus in this case,

$$\begin{aligned}
G_{11} &= -\hat{\beta}_1^2 + K_L K_C \hat{\beta}_1 \hat{\beta}_2 \\
G_{22} &= -\hat{\beta}_2^2 + K_L K_C \hat{\beta}_1 \hat{\beta}_2 \\
G_{12} &= \sqrt{\frac{\hat{Z}_{c1}}{\hat{Z}_{c2}}} \sqrt{\hat{\beta}_1 \hat{\beta}_2} \left(\hat{\beta}_1 K_C - \hat{\beta}_2 K_L \right) \\
G_{21} &= \sqrt{\frac{\hat{Z}_{c2}}{\hat{Z}_{c1}}} \sqrt{\hat{\beta}_1 \hat{\beta}_2} \left(\hat{\beta}_2 K_C - \hat{\beta}_1 K_L \right)
\end{aligned} \tag{3.66}$$

3.3.4 TOTAL VOLTAGES AND CURRENTS ON A MULTICONDUCTOR LINE

Because of the orthogonality and linear independence of the eigenvectors of the modes, we can represent arbitrary voltages and currents on the multiconductor line as superpositions of mode eigenvectors. For example, if only forward modes are known to be present, then we write

$$[V^+(z)] = a_1[V_1]e^{-\gamma_1 z} + a_2[V_2]e^{-\gamma_2 z}; \qquad [I^+(z)] = a_1[I_1]e^{-\gamma_1 z} + a_2[I_2]e^{-\gamma_2 z} \qquad (3.67)$$

where a_1 and a_2 are complex constants (actually modal wave amplitudes) which are to be determined. To make the notation more compact, we create the column vector

$$[a] = \begin{bmatrix} a_1 \\ a_2 \end{bmatrix} \qquad (3.68)$$

and the diagonal matrix

$$[E(z)] = \begin{bmatrix} e^{-\gamma_1 z} & 0 \\ 0 & e^{-\gamma_2 z} \end{bmatrix} \qquad (3.69)$$

Now (3.67) can be written more compactly as

$$[V^+(z)] = [M_V][E(z)][a]; \qquad [I^+(z)] = [M_I][E(z)][a] \qquad (3.70)$$

where $[M_V]$ and $[M_I]$ are the matrices defined in (3.54). By (3.70) we see that for any $[V^+]$ and $[I^+]$, we have

$$[V^+(z)] = [Z_c][I^+(z)] \qquad (3.71)$$

where the *matrix* characteristic impedance $[Z_c]$ is defined by

$$[Z_c] = [M_V][M_I]^{-1} = \frac{1}{2}[M_V][M_V]^T \qquad (3.72)$$

using (3.56). This matrix represents the $[Z]$-matrix of the two-port termination which must be used at the end of a multiconductor line section if no reflected waves are to exist. Using (3.57), it can be expressed in various alternative forms:

$$[Z_c] = [\mathbf{z}][M_I][\gamma]^{-1}[M_I]^{-1} = [\mathbf{y}]^{-1}[M_I][\gamma][M_I]^{-1} = [M_V][\gamma][M_V]^{-1}[\mathbf{y}]^{-1} = [M_V][\gamma]^{-1}[M_V]^{-1}[\mathbf{z}] \qquad (3.73)$$

Setting $z = 0$ in (3.70) allows us to determine $[a]$:

$$[a] = [M_V]^{-1}[V^+(0)] = [M_I]^{-1}[I^+(0)] \qquad (3.74)$$

so we can rewrite the forward-going voltage (3.70) as

$$[V^+(z)] = [M_V][E(z)][M_V]^{-1}[V^+(0)]; \qquad [I^+(z)] = [M_I][E(z)][M_I]^{-1}[I^+(0)] \qquad (3.75)$$

We can similarly express the reverse voltage waves as

$$[V^-(z)] = [M_V][E(-z)][M_V]^{-1}[V^-(0)]; \qquad [I^-(z)] = [M_I][E(-z)][M_I]^{-1}[I^-(0)] \qquad (3.76)$$

These reverse waves obey

$$[V^-(z)] = [M_V][E(-z)][b]; \qquad [I^-(z)] = -[M_I][E(-z)][b] \qquad (3.77)$$

(where $[b]$ is the vector of reverse wave amplitudes) and

$$[V^-(z)] = -[Z_c][I^-(z)] \tag{3.78}$$

Combining the forward and reverse waves and using (3.71) and (3.78) gives the total voltage and current on the coupled lines:

$$[V(z)] = [V^+(z)] + [V^-(z)] = [Z_c]\left([I^+(z)] - [I^-(z)]\right) \tag{3.79}$$

and

$$[I(z)] = [I^+(z)] + [I^-(z)] \tag{3.80}$$

If only forward modes (3.67) are present, the complex power carried along the multiconductor line is found from (3.32) to be

$$
\begin{aligned}
\hat{P}(z) &= \frac{1}{2}[I^+(z)]^\dagger[V^+(z)] = \frac{1}{2}\left\{a_1^*[I_1]^\dagger e^{-\gamma_1^* z} + a_2^*[I_2]^\dagger e^{-\gamma_2^* z}\right\}\left\{a_1[V_1]e^{-\gamma_1 z} + a_2[V_2]e^{-\gamma_2 z}\right\} \\
&= \frac{1}{2}\left\{|a_1|^2[I_1]^\dagger[V_1]e^{-2\alpha_1 z} + |a_2|^2[I_2]^\dagger[V_2]e^{-2\alpha_2 z}\right. \\
&\quad \left. + a_1 a_2^*[I_2]^\dagger[V_1]e^{-(\alpha_1+\alpha_2)z-j(\beta_1-\beta_2)z} + a_2 a_1^*[I_1]^\dagger[V_2]e^{-(\alpha_1+\alpha_2)z-j(\beta_2-\beta_1)z}\right\}
\end{aligned}
\tag{3.81}
$$

which is significantly more complicated than the expression for power carried by a single mode. In addition to the first two terms, which represent power carried by each mode in the absence of the other, there are "cross-powers" represented by the third and fourth terms of (3.81), which indicate that the powers carried by individual modes do not obey superposition in general.

However, in the special case of a propagating mode on a lossless line, we saw in the previous section that the eigenvectors $[I_m]$ are real, so that $[I_m]^\dagger = [I_m]^T$. Using the normalization and orthogonality properties (3.48) and (3.51), we can then reduce (3.81) to the simpler form

$$\hat{P}(z) = |a_1|^2 + |a_2|^2 \tag{3.82}$$

Thus, one consequence of orthogonality of propagating modes on lossless multiconductor lines is that the total complex power flow in the forward modes is equal to the sum of the powers carried by the individual modes. This is one reason that so much importance is attached to the study of modes on transmission lines and waveguides.

3.3.5   REFLECTION ON A MULTICONDUCTOR LINE

Reflection on a multiconductor line is treated in a manner analogous to that on a single transmission line. The matrix voltage reflection coefficient $[\rho(z)]$ at an arbitrary position is defined by

$$[V^-(z)] = [\rho(z)][V^+(z)] \tag{3.83}$$

which must hold for any combination of incident waves $[V^+(z)]$. Using (3.71) and (3.78) we also find that

$$[I^-(z)] = -[Z_c]^{-1}[\rho(z)][Z_c][I^+(z)] \tag{3.84}$$

If the matrix line impedance $[Z(z)]$ is defined (again, by analogy with the case of a single transmission line) from:

$$[V(z)] = [Z(z)][I(z)] \tag{3.85}$$

for any possible $[V(z)]$ and $[I(z)]$, then

$$[\rho(z)] = [Z_c]\left([Z(z)] + [Z_c]\right)^{-1}\left([Z(z)] - [Z_c]\right)[Z_c]^{-1} \tag{3.86}$$

in analogy with (1.54).

If an arbitrary two-port termination characterized by the load impedance matrix $[Z_L]$ is connected at $z = d$, the voltage and current at the load are related by $[V(d)] = [Z_L][I(d)]$, and we have $[Z(d)] = [Z_L]$, so that

$$[\rho(d)] = [Z_c] ([Z_L] + [Z_c])^{-1} ([Z_L] - [Z_c]) [Z_c]^{-1} \tag{3.87}$$

generalizing the result (1.58) for a single transmission line. Note that merely connecting a shunt load impedance from each conductor to ground at the end of the line is *not* in general sufficient to suppress all reflected waves: a full two-port network whose impedance matrix representation is $[Z_L] = [Z_c]$ must be connected to the end of the coupled lines. This point is explored further in problem p3-11.

3.3.6 EXCITATION OF A MULTICONDUCTOR LINE

Suppose that a pair of ideal voltage sources $V_g^{(1)}$ and $V_g^{(2)}$ are connected between conductors 1 and 2 respectively and ground at the end $z = 0$ of a semi-infinite two-conductor line located in $z > 0$. Only forward modes are excited on the coupled lines, since there is no reflection. Defining the column vector

$$[V_g] = \begin{bmatrix} V_g^{(1)} \\ V_g^{(2)} \end{bmatrix}$$

and observing that $[V^+(0)] = [V_g]$, we have from (3.75) that

$$[V^+(z)] = [M_V][E(z)][M_V]^{-1}[V_g] = \frac{1}{2}[M_V][E(z)][M_I]^T[V_g] \tag{3.88}$$

or

$$
\begin{aligned}
V^{(1)+}(z) &= \frac{1}{2}\left\{ V_1^{(1)} e^{-\gamma_1 z}[I_1]^T[V_g] + V_2^{(1)} e^{-\gamma_2 z}[I_2]^T[V_g] \right\} \\
V^{(2)+}(z) &= \frac{1}{2}\left\{ V_1^{(2)} e^{-\gamma_1 z}[I_1]^T[V_g] + V_2^{(2)} e^{-\gamma_2 z}[I_2]^T[V_g] \right\}
\end{aligned}
\tag{3.89}
$$

Note that even if a voltage source is connected to only one of the conductors ($V_g^{(2)} = 0$, say), in general nonzero voltages will appear on both conductors due to the coupling. This "cross-excitation" is known as crosstalk, and is often undesirable, especially in high-speed digital circuits. An example of this behavior is considered in problem p3-16.

3.3.7 EXAMPLE: THE SYMMETRIC THREE-CONDUCTOR LINE

Consider the case when the two ungrounded conductors are situated identically relative to the ground conductor. Then $c_{11} = c_{22}$ and $l_{11} = l_{22}$. Here, the solution is considerably simpler than in the general case, and can be written out explicitly as follows. The capacitive and inductive coupling coefficients (3.65) reduce to

$$K_C = \frac{c_{12}}{c_{11}}; \qquad K_L = \frac{l_{12}}{l_{11}} \tag{3.90}$$

respectively, in this case. Because of the symmetry of this configuration, its modes will be either even or odd with respect to conductors 1 and 2. For the even mode,

$$[V_e] = \begin{bmatrix} V_e^{(1)} \\ V_e^{(1)} \end{bmatrix}; \qquad [I_e] = \begin{bmatrix} I_e^{(1)} \\ I_e^{(1)} \end{bmatrix} \tag{3.91}$$

while for the odd mode

$$[V_o] = \begin{bmatrix} V_o^{(1)} \\ -V_o^{(1)} \end{bmatrix}; \qquad [I^{(o)}] = \begin{bmatrix} I_o^{(1)} \\ -I_o^{(1)} \end{bmatrix} \tag{3.92}$$

From (3.46) we have

$$I_e^{(1)} = V_e^{(1)} \frac{\omega(c_{11} - c_{12})}{\beta_e} \tag{3.93}$$

and

$$I_o^{(1)} = V_o^{(1)} \frac{\omega(c_{11} + c_{12})}{\beta_o} \tag{3.94}$$

The propagation coefficients $\beta_{e,o}$ for these modes are obtained by substitution of (3.91)-(3.94) into (3.35):

$$\beta_e^2 = \hat\beta_1^2(1 - K_C)(1 + K_L); \qquad \beta_o^2 = \hat\beta_1^2(1 + K_C)(1 - K_L) \tag{3.95}$$

where $\hat\beta_1 = \hat\beta_2$ is the propagation coefficient in the absence of coupling ($K_C = K_L = 0$), given in (3.64). Enforcing the condition that $\hat{P}_{av,m} \geq 0$ in (3.60), we find that

$$\begin{aligned} \beta_e &= +\hat\beta_1\sqrt{(1 - K_C)(1 + K_L)} \\ \beta_o &= +\hat\beta_1\sqrt{(1 + K_C)(1 - K_L)} \end{aligned} \tag{3.96}$$

For both of these to be propagating modes (β_e and β_o both real), we must have $-1 \leq K_C \leq +1$ and $-1 \leq K_L \leq +1$.

We speak of the even and odd mode scalar characteristic impedances defined by

$$Z_{ce} = \frac{V_e^{(1)}}{I_e^{(1)}} = \hat{Z}_{c1}\sqrt{\frac{1 + K_L}{1 - K_C}}; \qquad Z_{co} = \frac{V_o^{(1)}}{I_o^{(1)}} = \hat{Z}_{c1}\sqrt{\frac{1 - K_L}{1 + K_C}} \tag{3.97}$$

where $\hat{Z}_{c1} = \hat{Z}_{c2}$ is the characteristic impedance of one of the uncoupled lines, given in (3.64). To get the matrix characteristic impedance for this case, we first use the normalization condition (3.48) or (3.49) to obtain the values of $V_e^{(1)}$ and $V_o^{(1)}$:

$$V_{e,o}^{(1)} = \sqrt{Z_{c(e,o)}} \tag{3.98}$$

We can now construct

$$[M_V] = \begin{bmatrix} \sqrt{Z_{ce}} & \sqrt{Z_{co}} \\ \sqrt{Z_{ce}} & -\sqrt{Z_{co}} \end{bmatrix} \tag{3.99}$$

and using this in (3.72) we get

$$[Z_c] = \frac{1}{2} \begin{bmatrix} Z_{ce} + Z_{co} & Z_{ce} - Z_{co} \\ Z_{ce} - Z_{co} & Z_{ce} + Z_{co} \end{bmatrix} \tag{3.100}$$

The two-port load impedance network corresponding to this matrix characteristic impedance is shown in Figure 3.15. Termination of each conductor at the end of this line by a separate impedance is not sufficient to prevent the appearance of reflected waves unless $Z_{ce} = Z_{co}$, but this cannot happen in the ordinary case when $K_L > 0$ and $K_C > 0$.

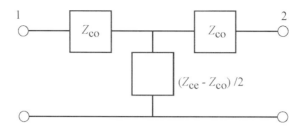

Figure 3.15 Matched load impedance T-network for symmetric three-conductor line.

Now suppose that this coupled-line system is terminated at the far end ($z = d$) by a two-port network whose impedance matrix is equal to the characteristic impedance matrix of the coupled lines. Suppose also that $K_C \neq K_L$, and that the initial voltages at the near end ($z = 0$) are set to $V^{(1)}(0) = V_G$ by a generator, and $V^{(2)}(0) = 0$ by a short circuit to ground. We wish to determine the voltages $V^{(1)}(d)$ and $V^{(2)}(d)$ at the far end of the coupled lines. Because the lines are terminated by a matched network ($[Z_L] = [Z_c]$), there is no reflection from the load ($[\rho] = 0$) and only forward-traveling waves need be considered [as in (3.67) and (3.75)]. By (3.75),

$$[V(z)] = [M_V][E(z)][M_V]^{-1}[V(0)] \tag{3.101}$$

where from (3.69)

$$[E(z)] = \begin{bmatrix} e^{-j\beta_e z} & 0 \\ 0 & e^{-j\beta_o z} \end{bmatrix}, \tag{3.102}$$

and $\beta_1 = \beta_e \neq \beta_o = \beta_2$, where

$$\beta_e = \hat{\beta}_1\sqrt{(1 - K_C)(1 + K_L)}; \qquad \beta_o = \hat{\beta}_1\sqrt{(1 + K_C)(1 - K_L)}$$

We have from (3.99) that

$$[M_V] = \begin{bmatrix} \sqrt{Z_{ce}} & \sqrt{Z_{co}} \\ \sqrt{Z_{ce}} & -\sqrt{Z_{co}} \end{bmatrix} \tag{3.103}$$

and thus

$$[M_V]^{-1} = \frac{1}{2} \begin{bmatrix} \frac{1}{\sqrt{Z_{ce}}} & \frac{1}{\sqrt{Z_{ce}}} \\ \frac{1}{\sqrt{Z_{co}}} & -\frac{1}{\sqrt{Z_{co}}} \end{bmatrix} \tag{3.104}$$

and so from (3.101) at $z = d$, after carrying out the matrix multiplications we get:

$$[V(d)] = \begin{bmatrix} V_G e^{-j\beta_{av}d}\cos(\Delta\beta d) \\ -jV_G e^{-j\beta_{av}d}\sin(\Delta\beta d) \end{bmatrix} \tag{3.105}$$

Here $\beta_{av} = (\beta_e + \beta_o)/2$ and $\Delta\beta = (\beta_e - \beta_o)/2$. Note that we can make $V^{(1)}(d) = 0$ if $\cos(\Delta\beta d) = 0$, or $d = (2m + 1)l_c$ for some integer m, where l_c is the coupling length:

$$l_c = \frac{\pi}{|\beta_e - \beta_o|} = \frac{\pi\beta_{av}}{\hat{\beta}_1^2|K_L - K_C|} = \frac{\hat{\lambda}_1^2}{2\lambda_{av}|K_L - K_C|} \tag{3.106}$$

Here, $\hat{\lambda}_1 = 2\pi/\hat{\beta}_1$ and $\lambda_{av} = 2\pi/\beta_{av}$. What has happened is that the even and odd modes have become $180°$ out of phase relative to their in-phase condition at $z = 0$. This behavior forms the basis of operation of so-called interference couplers, most widely used in integrated optics, but applicable in a variety of other contexts as well. If $K_L = K_C$, as happens for TEM coupled lines, this kind of coupling will not happen at all, and the only coupling mechanism will be due to whatever mismatch happens at the load end when $[Z_L] \neq [Z_c]$.

3.4 NOTES AND REFERENCES

The most familiar circuit equivalence theorem is undoubtedly Thévenin's theorem. A few texts, for example:

Carlin and Giordano (1964), Sects. 3.7 and 3.8.

present generalizations of Thévenin's theorem to multiport networks. The generalization is in fact contained in a paper by Helmholtz which predates Thévenin's paper by 30 years:

H. L. F. von Helmholtz, "Ueber einige Gesetze der Vertheilung elektrischer Ströme in körperlichen Leitern mit Anwendung auf die thierisch-elektrischen Versuche," *Ann. Phys. Chem.*, vol. 89, pp. 211-233, 353-377 (1853) [also in H. L. F. von Helmholtz, *Wissenschaftliche Abhandlungen*, vol. 1. Leipzig: Johann Ambrosius Barth, 1882, pp. 475-519].

A summary of Helmholtz' result can be found in:

Koenigsberger (1965), pp. 99-103.

Other sources for the generalized version of this theorem are:

H. Ataka, "On an extension of Thévenin's theorem," *Phil. Mag.*, ser. 7, vol. 25, pp. 663-666 (1938).

L. Sartre, "Sur une généralisation du théorème de Thévenin et son application au calcul des courants, dans le cas d'une charge dyssymétrique de réseaux polyphasés," *Rev. Gén. Élec.*, vol. 61, pp. 238-240 (1952).

L. A. Zadeh, "On passive and active networks and generalized Norton's and Thévenin's theorems," *Proc. IRE*, vol. 44, p. 378 (1956).

L. A. Zadeh, "Multipole analysis of active networks," *IRE Trans. Circ. Theory*, vol. 4, pp. 97-105 (1957).

M. A. Shakirov, "A multidimensional equivalent source theorem," *Radioelectron. Commun. Syst.*, vol. 21, no. 5, pp. 24-28 (1978).

Related to the multiport Thévenin theorem is a theorem on maximum power transfer from a multiport network containing sources. See:

C. A. Desoer, "The maximum power transfer theorem for *n*-ports," *IEEE Trans. Circ. Theory*, vol. 20, pp. 328-330 (1973).

A. T. de Hoop, "The *N*-port receiving antenna and its equivalent electrical network," *Philips Res. Repts.*, vol. 30 (suppl.), pp. 302*-315* (1975).

The relation of the Thévenin theorem to the *S*-matrix representation is discussed in

D. Kajfez and D. R. Wilton, "Network representation of receiving multiport antennas," *Arch. Elek. Übertragungstech.*, vol. 30, pp. 450-454 (1976).

Huygens or wave equivalent sources for circuits and transmission lines are discussed in

Haus and Adler (1959), Chapter 5.

and also in

H.-J. Butterweck, "Die Ersatzwellenquelle," *Arch. Elek. Übertrag.*, vol. 14, pp. 367-372 (1960).

K. Hartmann, "Noise characterization of linear circuits," *IEEE Trans. Circ. Syst.*, vol. 23, pp. 581-590 (1976).

V. N. Dikyj and E. F. Zaytsev, "Calculation of noise in multi-terminal networks with specified wave parameters," *Radio Eng. Electron. Phys.*, vol. 22, no. 12, pp. 80-85 (1977).

R. Pauli and W. Konig, "Wave sources (linear network design and analysis)," *Arch. Elek. Übertragungstech.*, vol. 41, pp. 365-368 (1987).

The substitution theorem (which is the analog of Love's equivalence principle and Schelkunoff's induction theorem for electromagnetic fields; see Chapter 9) is given in:

Desoer and Kuh (1969), Chapter 16.

The relationship to their electromagnetic counterparts is given in

S. A. Schelkunoff, "On diffraction and radiation of electromagnetic waves," *Phys. Rev.*, vol. 56, pp. 308-316 (1939).

R. F. Harrington, "Field equivalence theorems and their circuit analogues," *Elec. Eng.*, vol. 73, pp. 923-927 (1954).

I. V. Lindell, "Huygens' principle in electromagnetics," *IEE Proc. pt. A*, vol. 143, pp. 103-105 (1996).

A discussion of the telegrapher's equations for the source region can be found in

King (1965), Chapter 1.
Johnson (1965), Sect. 4.19;
Mashkovtsev, Tsibizov and Emelin (1966), Chapters 1 and 4;
Felsen and Marcuvitz (1973), Chapter 2;
Frankel (1977), Chapters 4 and 10.

and also in

C. S. Lindquist, "Transmission line response to independent distributed sources," *Proc. IEEE*, vol. 56, pp. 1353-1354 (1968).

The related question of nonradiating sources in a transmission line was considered by

A. Sihvola, G. Kristensson and I. V. Lindell, "Nonradiating sources in time-domain transmission-line theory," *IEEE Trans. Micr. Theory Tech.*, vol. 45, pp. 2155-2159 (1997).

The classic paper on multiconductor transmission lines is

J. R. Carson and R. S. Hoyt, "Propagation of periodic currents over a system of parallel wires," *Bell Syst. Tech. J.*, vol. 6, pp. 495-545 (1927).

This paper still rewards the reader who is willing to seek it out. More recent treatments of the subject are found in

R. Levy, "Directional couplers," in *Advances in Microwaves*, vol. 1 (L. Young, ed.). New York: Academic Press, 1966, pp. 115-209.

Uchida (1967);

Matick (1969), Chapters 7 and 8;

Gupta and Singh (1974), Chapter 5;

Garg, Bahl and Bozzi (2013), Chapter 8;

Frankel (1977), Chapters 3 and 7;

Edwards (1981), Chapter 6 and Appendix A.

Paul (1992), Chapter 10.

Paul (1994).

A good collection of important papers on this topic is to be found in

Young (1972).

Other relevant papers are

I. A. D. Lewis, "Analysis of a transmission-line type of thermionic-amplifier valve," *Proc. IEE (London), part IV*, vol. 100, pp. 16-24 (1953).

B. M. Oliver, "Directional electromagnetic couplers," *Proc. IRE*, vol. 42, pp. 1686-1693 (1954).

J. Brown, "Propagation in coupled transmission line systems," *Quart. J. Mech. Appl. Math.*, vol. 11, pp. 235-243 (1958).

E. G. Vlostovskii, "Theory of coupled transmission lines," *Telecommun. Radio Eng.*, pt. 2, vol. 22, no. 4, pp. 87-93 (1967).

M. M. Krage and G. I. Haddad, "Characteristics of coupled microstrip transmission lines—I: Coupled-mode formulation of inhomogeneous lines," *IEEE Trans. Micr. Theory Tech.*, vol. 18, pp. 217-222 (1970).

J. C. Isaacs and N. A. Strakhov, "Crosstalk in uniformly coupled transmission lines," *Bell Syst. Tech. J.*, vol. 52, pp. 101-115 (1973).

K. D. Marx, "Propagation modes, equivalent circuits, and characteristic terminations for multiconductor transmission lines with inhomogeneous dielectrics," *IEEE Trans. Micr. Theory Tech.*, vol. 21, pp. 450-457 (1973).

C. R. Paul, "On uniform multimode transmission lines," *IEEE Trans. Micr. Theory Tech.*, vol. 21, pp. 556-558 (1973).

M. Kh. Zakhar-Itkin, "The reciprocity theorem and the matrix telegraph equations for multimode transmission lines," *Radio Eng. Electron. Phys.*, vol. 19, no. 11, pp. 76-85, 1974.

V. K. Tripathi, "Asymmetric coupled transmission lines in an inhomogeneous medium," *IEEE Trans. Micr. Theory Tech.*, vol. 23, pp. 734-739 (1975).

R. A. Speciale, "Even- and odd-mode waves for nonsymmetrical coupled lines in non-homogeneous media," *IEEE Trans. Micr. Theory Tech.*, vol. 23, pp. 897-908 (1975).

C. R. Paul, "Useful matrix chain parameter identities for the analysis of multiconductor transmission lines," *IEEE Trans. Micr. Theory Tech.*, vol. 23, pp. 756-760 (1975).

M. Kh. Zakhar-Itkin, "Energy conservation law and nonself-adjoint operators in waveguide theory," *Radio Eng. Electron. Phys.*, vol. 21, no. 10, pp. 7-15, 1976.

L. N. Deryugin, O. A. Kurdyumov and V. F. Terichev, "Operating mechanism of directional couplers on two lines with distributed electromagnetic coupling," *Radiophys. Quantum Electron.*, vol. 21, pp. 208-209 (1978).

J. O. Scanlon, "Theory of microwave coupled-line networks," *Proc. IEEE*, vol. 68, pp. 209-231 (1980).

D. E. Bockelman and W. R. Eisenstadt, "Combined differential and common-mode scattering parameters: Theory and simulation," *IEEE Trans. Micr. Theory Tech.*, vol. 43, pp. 1530-1539 (1995).

C. R. Paul, "Decoupling the multiconductor transmission line equations," *IEEE Trans. Micr. Theory Tech.*, vol. 44, pp. 1429-1440 (1996).

R. Schwindt and C. Nguyen, "Generalized scattering parameters for two coupled transmission-line structures in an inhomogeneous medium," *Micr. Opt. Technol. Lett.*, vol. 12, pp. 335-342 (1996).

Applications of multiconductor transmission lines are numerous. In addition to directional couplers, they can be used to realize baluns, hybrid networks, power combiners, splitters and filters. A sample of the vast literature on the subject is

A. Matsumoto and N. Nagai, "Synthesis of multiports with multiwire sections," *Developments in Multiwire Networks* [Monograph Series, Research Institute of Applied Electricity no. 12]. Sapporo, Japan: Research Institute of Applied Electricity, 1964, pp. 1-13.

N. Nagai, "Theory and experiments on multiwire multiports," *Recent Developments in Microwave Multiport Networks* [Monograph Series, Research Institute of Applied Electricity no. 18]. Sapporo, Japan: Research Institute of Applied Electricity, 1970, pp. 1-68.

K. Watanabe, N. Nagai and K. Hatori, "Study on power dividers," *Recent Developments in Microwave Multiport Networks* [Monograph Series, Research Institute of Applied Electricity no. 18]. Sapporo, Japan: Research Institute of Applied Electricity, 1970, pp. 87-101.

Active and nonlinear coupled transmission lines can be used as models of distributed amplifiers; see

C. S. Lindquist, "Uniform transmission line response to independent distributed sources," *Proc. IEEE*, vol. 56, pp. 1740-1741 (1968).

C. S. Lindquist, "Active transmission lines—Characteristic impedances and propagation functions," *Proc. IEEE*, vol. 57, pp. 1422-1423 (1969).

and

Wong (1993).

3.5 PROBLEMS

p3-1 Show that a uniform semi-infinite section of transmission line to the left of the point $z = 0$ on which an incident forward voltage wave $V^i(z)$ is present can be replaced by the Thévenin equivalent circuit shown below, where Z_c is the characteristic impedance of the line.

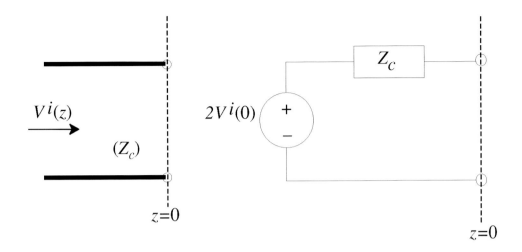

p3-2 A section of uniform transmission line occupies the interval $[z_1, z_2]$, and is excited by the impressed distributed sources $E_{\text{ext}}(z)$ and $H_{\text{ext}}(z)$ within this region. Find the *two-port* Thévenin and Norton equivalent circuits for this transmission line section, with the ports chosen at the ends $z = z_1$ and $z = z_2$.

p3-3 Let the distributed sources for a uniform lossless classical transmission line be $H_{\text{ext}}(z) \equiv 0$ and

$$E_{\text{ext}}(z) = e^{-j\beta_s z}$$

for $z > 0$, and $E_{\text{ext}}(z) = 0$ for $z < 0$, where β_s is some real constant. Determine expressions for the wave amplitudes $a_0(z)$ and $b_0(z)$ produced by this source. Graph $|a_0(z)|$ and $|b_0(z)|$ vs. z for the cases $\beta_s = 0$, $\beta_s = 0.99\beta$ and $\beta_s = \beta$, where β is the propagation coefficient of the transmission line. For purposes of these graphs, choose β and Z_c numerically equal to 1.

p3-4 Repeat problem p3-3, but with the distributed source

$$E_{\text{ext}}(z) = e^{-j\beta z + jBz^2}$$

where B is some real constant. Your formulas for the wave amplitudes will not be expressible in terms of elementary functions. Using suitable numerical software, graph $|a_0(z)|$ and $|b_0(z)|$ vs. z for the cases $B = 3$ and $B = -3$, and comment on the results. Assume that $Z_c = 1$ and $\beta = 4$.

p3-5 Show that the equivalent sources (3.23) when used in (3.22) produce the same amplitudes a_0 and b_0 as do the actual sources given in (3.17).

p3-6 Let the distributed sources for a uniform lossless classical transmission line be $H_{\text{ext}}(z) \equiv 0$ and

$$E_{\text{ext}}(z) = e^{-j\beta_s z}$$

for $0 < z < z_0$, and $E_{\text{ext}}(z) = 0$ elsewhere, where β_s is some real constant. Determine expressions for the wave amplitudes $a_0(z)$ and $b_0(z)$ produced by this source. Graph $|a_0(z)|$ and $|b_0(z)|$ vs. z for the cases $\beta_s = 0$, $\beta_s = 0.9\beta$ and $\beta_s = \beta$, where β is the propagation coefficient of the transmission line. For purposes of these graphs, choose β and Z_c numerically equal to 1, and $z_0 = 10\pi$.

p3-7 On a uniform classical transmission line, some distributed external voltage $E_{\text{ext}}(z)$ and current $H_{\text{ext}}(z)$ sources are located between z_1 and z_2. Suppose that

$$E_{\text{ext}}(z) = \frac{dF}{dz} + zG$$

and

$$H_{\text{ext}}(z) = \frac{dG}{dz} + \mathsf{y}F$$

for some functions $F(z)$ and $G(z)$ which vanish outside the interval $z_1 < z < z_2$. Show that these sources are nonradiating in the sense that $V(z)$ and $I(z)$ on the transmission line are identically equal to zero for $z \leq z_1$ and for $z \geq z_2$. For the case where

$$F = F_0 \sin \beta z; \qquad G = G_0 \sin \beta z \qquad \text{for } 0 < z < \frac{\pi}{\beta}$$

while

$$F = G = 0 \qquad \text{elsewhere}$$

with F_0 and G_0 being constants, plot the functions $|V(z)|$ and $|I(z)|$ versus z between $z_1 = 0$ and $z_2 = \pi/\beta$ (you may assume any convenient values for the parameters to make your plots).

p3-8   For a lossless transmission line with $\gamma = j\beta$, distributed sources are present for $z > 0$ in the form

$$F(z) = F_1 \qquad \text{for } 0 < z < \frac{\lambda_g}{2}$$

$$F(z) = -F_1 \qquad \text{for } \frac{\lambda_g}{2} < z < \lambda_g$$

$$F(z) = F_1 \qquad \text{for } \lambda_g < z < \frac{3\lambda_g}{2}$$

and so on, where we abbreviate

$$F(z) = \frac{E_{\text{ext}}(z) + Z_c H_{\text{ext}}(z)}{2\sqrt{2Z_c}},$$

$\lambda_g = \frac{2\pi}{\beta}$, and F_1 is a constant. Find an expression for the wave amplitude $a_0(z)$ of the forward wave that is excited by this stepwise constant distributed source. Plot the magnitude of $a_0(z)$ vs. z for $0 < z < 3\lambda_g$, using the values $\lambda_g = 1$ and $F_1 = 1$.

p3-9   In the interval $0 < z < d$ of a lossless transmission line with $\gamma = j\beta$, there are distributed sources

$$E_{\text{ext}}(z) = Z_c H_{\text{ext}}(z) = \frac{V_0 z}{d^2}$$

where V_0 is a constant with units of voltage. Obtain expressions for the wave amplitudes $a_0(z)$ and $b_0(z)$ excited by these sources. Assuming that $V_0^2/8Z_c = 1$ watt, plot the following:

    (a) $|a_0(z)|^2$ as a function of z/d for $0 < z/d < 1$ and $\beta d = 6$; and

    (b) $|a_{02}|^2$ as a function of βd for $0 < \beta d < 10$.

Explain physically why $|a_{02}|^2$ decreases as βd increases.

p3-10   Use equations (3.6) to derive the relation

$$\frac{d}{dz}[V(z)I^*(z)] = -\mathsf{z}|I(z)|^2 - \mathsf{y}^*|V(z)|^2 + I^*(z)E_{\text{ext}}(z) + V(z)H_{\text{ext}}^*(z)$$

where $\mathsf{z} = \mathsf{r} + j\mathsf{x}$ and $\mathsf{y} = \mathsf{g} + j\mathsf{b}$ are the per-unit-length line parameters. If the time-average power carried past the position z on the line by the waves is

$$P_{\text{av}}(z) = \text{Re}\left[\frac{V(z)I^*(z)}{2}\right]$$

then obtain an expression for the difference $P_{\text{av}}(z_2) - P_{\text{av}}(z_1)$ between the time-average powers carried at the two ends z_1 and z_2 of an interval containing externally impressed sources. Your expression should contain the functions $V(z)$, $I(z)$, $E_{\text{ext}}(z)$ and $H_{\text{ext}}(z)$ but not their explicit forms, such as (3.10) for example. Give a physical interpretation of your result. [Hint: Compare it to the Poynting theorem of electromagnetics.]

p3-11 Consider a pair of lossless coupled transmission lines as in Section 3.3.7. The inductive and capacitive coupling coefficients are equal: $K_C = K_L \equiv K$. A length l of these coupled lines forms a four-port network: port 1 is on line 1 at $z = 0$, port 2 on line 2 at $z = 0$, port 3 on line 1 at $z = l$, and port 4 on line 2 at $z = l$. Assume that *uncoupled* transmission lines of characteristic impedance $Z_c^{(0)} = \sqrt{Z_{ce} Z_{co}}$ are connected to each port.

(a) Find the reflection coefficient matrix $[\rho]$ at the load end $z = l$.

(b) Use the result of part (a) to find the scattering matrix of this four-port network. [Hint: Make use of the symmetries of this structure to simplify calculations.]

(c) By examining the dependence of S_{11}, S_{21}, S_{31} and S_{41} on βl, comment on the possible practical function of this four-port network.

p3-12 Consider the transmission line of problem p1-3, coupled to a straight wire coaxial to and inside the helix as shown.

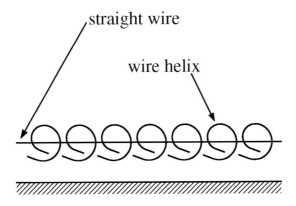

straight wire

wire helix

In (3.28), we should replace $\omega[l]$ by the matrix $[x]$, where

$$[x] = \begin{bmatrix} \dfrac{\omega l_{11}}{1 - \omega^2/\omega_c^2} & 0 \\[2ex] 0 & \omega l_{22} \end{bmatrix}$$

while $[b] = \omega[c]$ and $[c]$ is given as (3.29) as usual. Find the values of the complex propagation constants $\gamma_{1,2}$ for this system of coupled transmission lines. Note that in this example, we are not guaranteed that γ will be pure imaginary, or even real. Plot the values of $\gamma_{1,2}$ (both real and imaginary parts) versus ω/ω_c for $0 < \omega/\omega_c < 4$ if $\omega_c = 2\pi \times 10^6 \ \sec^{-1}$,

$$\frac{1}{\sqrt{l_{22} c_{22}}} = 3 \times 10^8 \text{m/sec}$$

$$\frac{l_{11} c_{11}}{l_{22} c_{22}} = 3$$

and

$$\frac{l_{11} c_{12}^2}{l_{22} c_{22}^2} = \frac{l_{11} c_{11}}{l_{22} c_{22}} \frac{c_{12}^2}{c_{11} c_{22}^2} = 0.2$$

p3-13 Consider the transmission line of problem p1-3, but containing two coupled helical conductors as well as the ground conductor. In (3.30), use $[z] = j[x]$, where

$$[x] = \begin{bmatrix} \frac{\omega l_{11}}{1 - \omega^2/\omega_c^2} & 0 \\ 0 & \frac{\omega l_{22}}{1 - \omega^2/\omega_c^2} \end{bmatrix}$$

while $[y] = j\omega[c]$ and $[c]$ is given as in (3.29) as usual.

Find the values of the complex propagation constants $\gamma_{1,2}$ for this system of coupled transmission lines. Note that in this example, we are not guaranteed that γ will be pure imaginary, or even real. Plot the values of $\gamma_{1,2}$ (both real and imaginary parts) versus ω/ω_c for $0 < \omega/\omega_c < 4$, assuming that: $\omega_c = 2\pi \times 10^6 \text{ sec}^{-1}$, $l_{11} = l_{22}$,

$$\frac{1}{\sqrt{l_{11}c_{11}}} = 3 \times 10^8 \text{m/sec}$$

$$\frac{c_{11}}{c_{22}} = 1.2$$

and

$$\frac{c_{12}}{c_{22}} = 0.2$$

p3-14 Consider a multiconductor transmission line that is lossy in general:

$$\text{Re}\left([I]^\dagger [z][I]\right) \geq 0 \qquad \text{for any vector } [I]$$

$$\text{Re}\left([V]^\dagger [y][V]\right) \geq 0 \qquad \text{for any vector } [V]$$

If a mode $[V_m]$, $[I_m]$, γ_m of this line obeys (3.35) with $\text{Re}(\gamma_m) = \alpha_m \neq 0$, show that the time-average power

$$\hat{P}_{\text{av},m} = \frac{1}{2}\text{Re}\left([I_m]^\dagger [V_m]\right) = \frac{1}{2}\text{Re}\left([I_m]^\dagger \frac{[z]}{\gamma_m}[I_m]\right) = \frac{1}{2}\text{Re}\left([V_m]^\dagger \frac{[y]^\dagger}{\gamma_m^*}[V_m]\right)$$

carried by this mode is nonnegative if $\alpha_m > 0$ [this is the analog of the nonnegative value of (1.16) for a single transmission line].

p3-15 A conventional transmission line and a backward-wave transmission line are oriented parallel to each other and are coupled along their length. The lines are assumed lossless, so that

$$[z] = j\,[x]\,; \qquad [y] = j\,[b]$$

The line parameters are given by

$$x_{11} = \omega l_{11}; \quad x_{22} = -\frac{s_{22}}{\omega};$$

$$[b] = \begin{bmatrix} \omega c_{11} & -\omega c_{12} \\ -\omega c_{12} & -\frac{r_{m22}}{\omega} \end{bmatrix}$$

Here s_{22} is the series elastance per unit length and r_{m22} is the shunt reluctance per unit length of the second line (cf. problem p1-4). Moreover, we assume a zero series mutual reactance between the lines: $x_{12} = 0$. Define a reference frequency ω_r by

$$\omega_r^2 = \sqrt{\frac{r_{m22}s_{22}}{l_{11}c_{11}}},$$

a capacitive coupling coefficient

$$K_C = \frac{c_{12}}{c_{11}},$$

a phase velocity

$$v_{11} = \frac{1}{\sqrt{l_{11}c_{11}}},$$

and the characteristic impedances

$$Z_{c11} = \sqrt{\frac{l_{11}}{c_{11}}}; \quad Z_{c22} = \sqrt{\frac{s_{22}}{r_{m22}}}$$

The significance of ω_r is that at this frequency, we have $\beta_{11} = -\beta_{22}$, where

$$\beta_{11} = \frac{\omega}{v_{11}}; \qquad \beta_{22} = -\frac{\sqrt{r_{m22}s_{22}}}{\omega}$$

(a) Express $[x]$ and $[b]$ in terms of only the parameters ω_r, K_C, v_{11}, Z_{c11} and Z_{c22} defined above, as well as the operating frequency ω.

(b) Obtain expressions for the two eigenvalues γ_1^2 and γ_2^2 of this coupled transmission line system in terms of the parameters ω_r, K_C, v_{11}, Z_{c11} and Z_{c22} and the operating frequency ω only. Under what conditions might $\gamma_{1,2}^2$ have both real and imaginary parts different from zero?

(c) Use the following numerical values for the parameters of these coupled transmission lines:

$$v_{11} = 2 \times 10^8 \text{ m/sec}; \quad \omega_r = 2\pi \times 10^9 \text{ rad/sec};$$

$$K_C = 0.6; \quad Z_{c11} = 50 \ \Omega; \quad Z_{c22} = 200 \ \Omega$$

Plot the real parts α_1 and α_2 and the imaginary parts β_1 and β_2 of γ_1 and γ_2 versus ω/ω_r for $0 < \omega/\omega_r < 4$, being careful to use the correct signs for β_1 and β_2. You should observe some unusual, and perhaps unexpected, behavior for ω between approximately $0.7\,\omega_r$ and $2.4\,\omega_r$. Such modes as these, for which γ can be complex even in the absence of losses, have been called *complex modes*.

p3-16 Consider the coupled-line example of Section 3.3.7, but now assume that $c_{11} \neq c_{22}$ and $l_{11} \neq l_{22}$, and define

$$K_C = \frac{c_{12}}{\sqrt{c_{11}c_{22}}}; \qquad K_L = \frac{l_{12}}{\sqrt{l_{11}l_{22}}}$$

and let $\hat{\beta}_{1,2} > 0$ be the propagation coefficients in the absence of coupling:

$$\hat{\beta}_1^2 = \omega^2 l_{11}c_{11}; \qquad \hat{\beta}_2^2 = \omega^2 l_{22}c_{22}$$

such that $0 < K_{L,C} < 1$, and l_{11}, l_{22}, c_{11} and c_{22} are all positive. The normal modes of this structure will no longer be even and odd modes in general. Show that $\beta_m = \hat{\beta}_0\sqrt{1 \pm \Delta}$, where

$$\hat{\beta}_0 = \sqrt{\frac{1}{2}\left(\hat{\beta}_1^2 + \hat{\beta}_2^2\right) - \hat{\beta}_1\hat{\beta}_2 K_L K_C}$$

and

$$\Delta = \frac{1}{\hat{\beta}_0^2} \sqrt{\left(\frac{\hat{\beta}_1^2 - \hat{\beta}_2^2}{2} \right)^2 + \hat{\beta}_1 \hat{\beta}_2 \left(\hat{\beta}_1^2 + \hat{\beta}_2^2 \right) (K_L^2 + K_C^2) \left(\frac{\hat{\beta}_1 \hat{\beta}_2}{\hat{\beta}_1^2 + \hat{\beta}_2^2} - \frac{K_L K_C}{K_L^2 + K_C^2} \right)}$$

are both real quantities, chosen to be greater than or equal to zero.

p3-17 Starting from the result of problem p3-16, obtain expressions for the voltage ratios

$$R_m \equiv \frac{V_m^{(2)}}{V_m^{(1)}}$$

of the normal modes.

(a) If $\hat{\beta}_1$ is not too different from $\hat{\beta}_2$, but $K_L \neq K_C$, we will have

$$\frac{\hat{\beta}_1 \hat{\beta}_2}{\hat{\beta}_1^2 + \hat{\beta}_2^2} > \frac{K_L K_C}{K_L^2 + K_C^2}$$

In this case, show that the two values of R_m have opposite signs, similar to the even and odd modes of the symmetric coupled lines. In this case, the mode with $R_m = R_c > 0$ is called the c-mode, while the mode with and $R_m = R_\pi < 0$ is called the π-mode. Show that we might have either $\beta_c > \beta_\pi$ or $\beta_c < \beta_\pi$, depending on the values of $\hat{\beta}_1$, $\hat{\beta}_2$, K_L and K_C.

(b) If

$$\frac{\hat{\beta}_1 \hat{\beta}_2}{\hat{\beta}_1^2 + \hat{\beta}_2^2} < \frac{K_L K_C}{K_L^2 + K_C^2}$$

show that both voltage ratios have the same sign. Is the ratio for the mode with $\beta_m = \hat{\beta}_0 \sqrt{1 + \Delta}$ larger or smaller in magnitude than that with $\beta_m = \hat{\beta}_0 \sqrt{1 - \Delta}$?

p3-18 Suppose that a section of coupled lines has been chosen to have the coupling length $l = l_c$ from (3.106) at a certain design frequency ω_0. If the actual operating frequency ω is different from ω_0, only a part of the power initially launched in wire 1 will have been transferred to wire 2. Assume that the coupling coefficients K_C and K_L, as well as the uncoupled phase velocities $\omega/\hat{\beta}_1$ and $\omega/\hat{\beta}_2$, are independent of frequency. Determine the range of frequencies (as fractions of ω_0) over which at least 90% of the power from wire 1 will be transferred to wire 2, and from this the fractional operating bandwidth of the coupler. Show that this bandwidth is independent of the phase velocities and coupling coefficients.

4 PULSE PROPAGATION AND DISTORTION

4.1 INTRODUCTION

There is, of course, no such thing as a continuous wave (CW) time-harmonic signal which lasts from time infinitely past to infinitely into the future. What we deal with in practice (whether in a circuit, on a transmission line or in a waveguide) are signals which are of finite duration in time. That is, the signal has a beginning and an end. It is of interest to determine how such real signals propagate and distort when passing through such systems.

4.2 THE FOURIER TRANSFORM AND THE ANALYTIC SIGNAL

A wide class of functions $f(t)$ of time t can be described in terms of a frequency spectrum $F(\omega)$ via the Fourier transform (see Appendix G):

$$f(t) = \frac{1}{2\pi} \int_{-\infty}^{\infty} F(\omega) e^{j\omega t} \, d\omega \tag{4.1}$$

$$F(\omega) = \int_{-\infty}^{\infty} f(t) e^{-j\omega t} \, dt \tag{4.2}$$

A *signal* $f(t)$ (it can be a voltage, a current, an electric or magnetic field, etc.) must be a *real* function of time t. Because it is real, a constraint is imposed on its Fourier transform $F(\omega)$ by virtue of (4.2):

$$F(-\omega) = F^*(\omega) \tag{4.3}$$

for real ω, where * denotes the complex conjugate. We sometimes represent $F(\omega)$ in a polar form

$$F(\omega) = A_F(\omega) e^{-j\phi_F(\omega)} \tag{4.4}$$

where $A_F(\omega)$ is real but not necessarily positive, defined in such a way that the phase lag $\phi_F(\omega)$ is a continuous function of ω. From (4.3), we observe that A_F is an even function of ω, while ϕ_F is odd. Note that this representation is not necessarily the usual polar decomposition of the complex function $F(\omega)$, because in that expression the amplitude is always nonnegative. In the version used here, $A_F(\omega)$ may be negative, so that if it passes through a zero $\phi_F(\omega)$ may remain continuous and *both* ϕ_F and A_F can be differentiated at all real nonnegative values of ω. The difference is illustrated in Figure 4.1. The DC part $F(0)$ of the spectrum must be real, and for definiteness we define $\phi_F(0) = 0$, so that $A_F(0) = F(0)$, and define the functions $A_F(\omega)$ and $\phi_F(\omega)$ at other frequencies by analytic continuation from $\omega = 0$.

Because of these properties, the information about $F(\omega)$ at negative frequencies is redundant, and (4.1) can be written as an integral over positive frequencies only:

$$f(t) = \text{Re}[f_A(t)] \tag{4.5}$$

where

$$f_A(t) \equiv \frac{1}{\pi} \int_0^{\infty} F(\omega) e^{j\omega t} \, d\omega = |f_A(t)| e^{j\phi_A(t)} \tag{4.6}$$

109

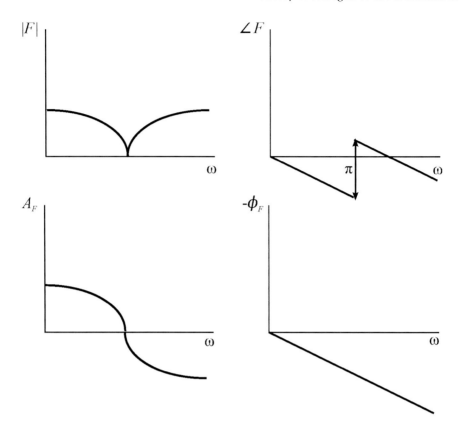

Figure 4.1 The differences between $|F|$ and A_F, and between $\angle F$ and $-\phi_F$.

is known as the *analytic signal* corresponding to $f(t)$, a complex function whose magnitude is $|f_A(t)|$ and whose phase is $\phi_A(t)$. Comparing its definition (4.6) with (4.1), we see that the analytic signal has a Fourier transform $F_A(\omega)$ which is zero for negative frequencies and equal to $2F(\omega)$ for positive frequencies.

The reason for the name analytic signal comes from the behavior of f_A for *complex* values of t. Indeed, let $\text{Im}(t) > 0$. Then

$$f_A(t) = \frac{1}{\pi} \int_0^\infty F(\omega) e^{j\omega t}\, d\omega = \frac{1}{\pi} \int_0^\infty \left[\int_{-\infty}^\infty f(t') e^{-j\omega t'}\, dt' \right] e^{j\omega t}\, d\omega \qquad (4.7)$$

$$= \frac{1}{\pi} \int_{-\infty}^\infty f(t') \left[\int_0^\infty e^{j\omega(t-t')}\, d\omega \right] dt'$$

But

$$\int_0^\infty e^{j\omega(t-t')}\, d\omega = \frac{j}{t-t'} \qquad \text{if } \text{Im}(t) > 0 \qquad (4.8)$$

so we get a formula for directly computing f_A for complex values of its argument in terms of the values of f for real values of t only:

$$f_A(t) = \frac{j}{\pi} \int_{-\infty}^\infty \frac{f(t')\, dt'}{t-t'} \qquad \text{if } \text{Im}(t) > 0 \qquad (4.9)$$

This formula shows that f_A is an analytic function of the complex variable t in the upper half of the complex plane. By taking the limit as $\text{Im}(t) \to 0$ from above, it can be shown that the expression for the analytic signal for real values of t is:

$$f_A(t) = f(t) + jH[f(t)] \tag{4.10}$$

where the Hilbert transform of f is given by

$$H[f(t)] = \frac{1}{\pi} \lim_{\delta \to 0} \left(\int_{-\infty}^{t-\delta} + \int_{t+\delta}^{\infty} \right) \frac{f(t')\, dt'}{t - t'} \tag{4.11}$$

The energy or power associated with a signal $f(t)$ is proportional to the square of the signal, $f^2(t)$. In fact, applying the signal to a square-law detector such as a diode produces an output signal proportional to this energy or power. In terms of the analytic signal, some algebraic manipulation applied to the square of (4.5) gives

$$f^2(t) = \frac{1}{2}|f_A(t)|^2 + \frac{1}{2}\mathrm{Re}[f_A^2(t)] = \frac{1}{2}|f_A(t)|^2 \left[1 + \cos 2\phi_A(t)\right] \tag{4.12}$$

If $|f_A(t)|$ is slowly varying (i. e., has primarily low-frequency spectral components) while $\phi_A(t)$ changes rapidly by comparison (so that $f(t)$ is a narrowband signal as considered in Section 4.2.2 below), the signal $f^2(t)$ can be passed through a filter which will eliminate the second term of (4.12):[1]

$$\left[f^2(t)\right]_{\text{filtered}} = \frac{1}{2}|f_A(t)|^2 \tag{4.13}$$

We see that in such a case the magnitude of the analytic signal is closely related to the output of a filtered detector (a so-called *envelope detector*). Such a detector is used to demodulate amplitude-modulated (AM) signals.

If the signal is differentiated with respect to t (this can be accomplished by applying the signal voltage to a series capacitor C followed by a shunt resistor R, provided that the RC time constant is small compared to time durations characteristic of the variation of the signal) before being detected and filtered, the output will be

$$\left\{[f'(t)]^2\right\}_{\text{filtered}} = \frac{1}{2}|f_A'(t)|^2 = \frac{1}{2}\left[\frac{d}{dt}|f_A(t)|\right]^2 + \frac{1}{2}|\phi_A'(t)|^2|f_A(t)|^2 \tag{4.14}$$

Thus, an output dependent on the phase (or at least its time derivative) of the analytic signal can also be readily produced by such a *slope detector* circuit. If $|f_A(t)|$ is constant, then the output of this slope detector is proportional to $|\phi_A'(t)|^2$ alone. This is the principle of demodulation of frequency-modulated (FM) signals.

The analytic signal is a useful tool for handling modulated signals. For example, let $r(t)$ be a given *complex* function of time (not necessarily an analytic signal itself), with Fourier transform $R(\omega)$. Consider the signal

$$f(t) = \mathrm{Re}\left[r(t)e^{j\omega_0 t}\right] \tag{4.15}$$

where ω_0 is a nonnegative frequency called the *carrier frequency*, and $r(t)$ is called the *baseband signal* or envelope. Some examples of this kind of signal are shown in Figure 4.2. Figure 4.2(b) shows an amplitude-modulated signal ($r(t)$ is real), while Figure 4.2(c) depicts a frequency-modulated signal ($r(t)$ has constant amplitude but varying phase). The unmodulated carrier is shown in Figure 4.2(a). The Fourier spectrum of $f(t)$ is readily computed:

---

[1] This is analogous to taking the time-average value of the power or energy of a time-harmonic signal in a circuit (see Appendix A.3).

(a)

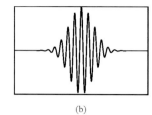
(b)

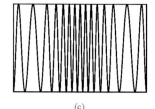
(c)

Figure 4.2 (a) Unmodulated carrier signal (i. e., $r(t) = 1$); (b) AM modulated signal with $r(t) = e^{-t^2/T^2}$; (c) FM modulated signal with $r(t) = e^{j\omega_0 t \cos \omega_m t}$.

$$F(\omega) = \frac{1}{2} \left[R(\omega - \omega_0) + R^*(-\omega - \omega_0) \right] \tag{4.16}$$

(note that (4.16) obeys (4.3) as it must). Then using the Fourier frequency integral representation for $r(t)$ in terms of $R(\omega)$, we find from (4.6) that the analytic signal corresponding to $f(t)$ is

$$f_A(t) = \frac{1}{2\pi} \int_0^\infty \left[R(\omega - \omega_0) + R^*(-\omega - \omega_0) \right] e^{j\omega t} \, d\omega \tag{4.17}$$

We next ask whether it is always possible to find a baseband envelope function $r(t)$ for any given arbitrary signal $f(t)$. The answer is yes, but the resultant $r(t)$ is not unique. A natural choice for it would be

$$r(t) = e^{-j\omega_0 t} f_A(t) \tag{4.18}$$

but we are left to determine what ω_0 should be. A potential approach would be to choose ω_0 to be where "most" of the spectrum $F(\omega)$ of the signal is concentrated. For example, this might be

$$\omega_0 = \frac{\int_0^\infty \omega |F(\omega)|^2 \, d\omega}{\int_0^\infty |F(\omega)|^2 \, d\omega} \tag{4.19}$$

But this might not be very representative of the actual signal. If, for example, the spectrum $F(\omega)$ were concentrated around two widely separated frequencies ω_{01} and ω_{02} as shown in Figure 4.3, eqn. (4.19) would predict a value of ω_0 somewhere between ω_{01} and ω_{02}, where almost no spectral content of the signal is located. In such a case, it is preferable to split the signal as $f(t) = f_1(t) + f_2(t)$, such that the spectrum of each term is concentrated around

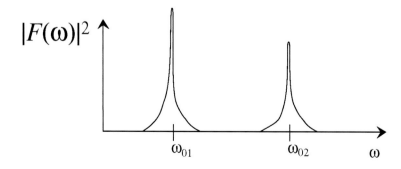

Figure 4.3 Signal spectrum concentrated at two different frequencies.

a single frequency. Thus, a fairly general form for an arbitrary signal might be

$$f(t) = \text{Re}\left[\sum_{i=1}^{N} r_i(t)e^{j\omega_{0i}t}\right] \tag{4.20}$$

for suitably chosen values of ω_{0i}. However the ω_{0i} are selected, we observe that the spectrum of the analytic signal corresponding to each term on the right side of (4.20) will be:

$$F_{Ai}(\omega) = \left\{ \begin{array}{cc} 2F_i(\omega) = R_i(\omega - \omega_{0i}) + R_i^*(-\omega - \omega_{0i}) & \text{for } \omega > 0 \\ 0 & \text{for } \omega < 0 \end{array} \right\} \tag{4.21}$$

4.2.1 ENERGY MOMENTS AND PULSE WIDTHS OF SIGNALS

Equation (4.19) is an example of a "moment" of a pulse, which is a way of characterizing the time or frequency behavior of a signal of general form. Although exact or numerical evaluation of the Fourier transform of a pulse can give us detailed information, it is often sufficient to know only a certain few parameters of the pulse in order to understand how it is distorted after passing through a network or propagation channel.

Two of the most important features of a pulse are the "center" and "duration", which can be defined in terms of its *energy moments*. The definition of the energy moments of a pulse depends on the notion of a weighted average of a function with respect to the energy in the pulse. Thus, if $x(t)$ is a function of time (not necessarily real), we define its average value with respect to the energy in a pulse $f(t)$ to be:

$$\langle x \rangle_f \equiv \frac{\int_{-\infty}^{\infty} x(t)|f(t)|^2 \, dt}{\int_{-\infty}^{\infty} |f(t)|^2 \, dt} \tag{4.22}$$

provided the integrals exist. If $X(\omega)$ is a function of frequency, we define its average value with respect to the energy in the spectrum $F(\omega)$ of the pulse $f(t)$ as

$$\langle X \rangle_F \equiv \frac{\int_0^{\infty} X(\omega)|F(\omega)|^2 \, d\omega}{\int_0^{\infty} |F(\omega)|^2 \, d\omega} \tag{4.23}$$

where we only carry out the average over positive frequencies because of (4.3). It is to be noted that the energies of the time-domain and frequency-domain versions of a pulse are related by taking $f = g$ in Parseval's theorem (G.6):

$$\int_{-\infty}^{\infty} |f(t)|^2 \, dt = \frac{1}{\pi} \int_0^{\infty} |F(\omega)|^2 \, d\omega \tag{4.24}$$

Evidently, ω_0 as given by (4.19) is equal to $\langle \omega \rangle_F$.

The nth-order temporal moments of a pulse $f(t)$ are defined as $\langle t^n \rangle_f$. The first-order temporal moment is called the center $t_c(f)$ of the pulse:

$$t_c(f) \equiv \langle t \rangle_f \tag{4.25}$$

This definition is analogous to that of the expected value of a quantity in quantum mechanics or the mean value of a random variable in probability theory. Now, by entry G.1.4 of Table G.1, we have from (G.6) that the center of a pulse can also be expressed by the ratio:

$$t_c(f) = \frac{j \int_{-\infty}^{\infty} \frac{dF(\omega)}{d\omega} F^*(\omega) \, d\omega}{\int_{-\infty}^{\infty} |F(\omega)|^2 \, d\omega} \tag{4.26}$$

Expressing $F(\omega)$ in the form (4.4) and removing integrals of odd functions of ω, we may cast $t_c(f)$ in terms of a spectral average value:

$$t_c(f) = \langle \phi'_F(\omega) \rangle_F \tag{4.27}$$

where

$$\phi'_F(\omega) = t_{g,F}(\omega) \tag{4.28}$$

represents an intrinsic group delay time of the portion of the spectrum at the frequency ω (see (4.41) below). The pulse center is thus a spectral average of this intrinsic group delay time.

We can by similar means define a pulse duration or pulse width for an arbitrary signal $f(t)$. We do this by using the second-order moment of the signal, namely

$$T_d^2(f) \equiv \frac{2 \int_{-\infty}^{\infty} [t - t_c(f)]^2 |f(t)|^2 \, dt}{\int_{-\infty}^{\infty} |f(t)|^2 \, dt} \tag{4.29}$$

so that $T_d(f)$ itself will be called the pulse duration. It will be seen that the quantity T_d^2 is analogous to the variance of a random variable. The integrals in the numerator of (4.29) can be expressed as integrals over the spectrum $F(\omega)$, as was done for t_c in (4.26). We find:

$$T_d^2(f) = 2 \left\langle [t_{g,F}(\omega)]^2 \right\rangle_F - 2t_c^2(f) \tag{4.30}$$

We should note that in spite of the apparent generality of these results, the moment description may not always be useful in practice. For example, if the "pulse" consists of two separate disturbances widely separated in time (analogous to the separation of frequencies in the case of Figure 4.3), the center of the pulse defined by (4.25) will be located somewhere between the two disturbances, where there is little signal at all. Likewise, the duration (4.29) of this signal will be much wider than that of either of the individual disturbances which constitute the signal. It is best to resolve a general signal into a superposition of pulses (similar to (4.20)), each of which is identifiably a single "bump," and then to apply the method of pulse moments to each term separately.

As the foregoing example illustrates, definitions of pulse center and duration based on the energy moments are not always the most suitable ones in practice. This is particularly true of digital signals, for which the energy distribution of a pulse is not as important as the voltage level relative to the threshold values for certain logic gates. For these signals, concepts such as the rise and fall times of a signal are at least as important as the duration, and more so than the center. They are also best defined using not $|f(t)|^2$ but $f(t)$ itself or perhaps the envelope $r(t)$. We will employ a particular case of this idea later when developing the quasi-Gaussian approximation for evaluating pulse distortion.

4.2.2   BAND-LIMITED SIGNALS

As we have seen, not every complex function $r(t)$ can be multiplied by a single factor $e^{j\omega_0 t}$ to yield an analytic signal. One important class of envelope functions which can is that for which $R(\omega)$ is bandlimited; that is,

$$R(\omega) = 0 \quad \text{for} \quad |\omega| > \omega_b$$

where $\omega_b > 0$ denotes the bandwidth of $r(t)$. Then if $\omega_0 > \omega_b$, we have $R^*(-\omega - \omega_0) = 0$ for $\omega > 0$, and the analytic signal (4.17) simplifies to

$$f_A(t) = r(t)e^{j\omega_0 t} \tag{4.31}$$

This example is thus analogous to the reconstruction of a sinusoidal function of time from a time-independent phasor quantity by multiplying by $e^{j\omega t}$ and taking the real part. The function $r(t)$ serves as a complex *envelope* to modulate the carrier $e^{j\omega_0 t}$. If $r(t) = |r(t)|e^{j\phi_r(t)}$, then the signal (4.15) is said to be an *amplitude-modulated* signal if $\phi_r(t)$ is constant, while if $|r(t)|$ is constant the signal is said to be *frequency-* or *phase-modulated*. In general, both kinds of modulation may be present simultaneously.

The energy (4.12) in a band-limited signal is thus given by

$$f^2(t) = \frac{1}{2}|r(t)|^2 + \frac{1}{2}\text{Re}[r^2(t)e^{2j\omega_0 t}] \tag{4.32}$$

The Fourier spectrum of each term on the right side of (4.32) can be found as follows. For the first term, we write

$$\begin{aligned}
|r(t)|^2 &= \frac{1}{4\pi^2}\int_{-\infty}^{\infty}\int_{-\infty}^{\infty} R(\omega_1)R^*(\omega_2)e^{j(\omega_1-\omega_2)t}\,d\omega_1 d\omega_2 \\
&= \frac{1}{4\pi^2}\int_{-\infty}^{\infty}\int_{-\infty}^{\infty} R(\omega_1)R^*(\omega_1-\omega)e^{j\omega t}\,d\omega_1 d\omega \\
&= \frac{1}{2\pi}\int_{-\infty}^{\infty}\left[\frac{1}{2\pi}\int_{-\infty}^{\infty} R(\omega_1)R^*(\omega_1-\omega)\,d\omega_1\right]e^{j\omega t}\,d\omega
\end{aligned} \tag{4.33}$$

where the change of variable $\omega = \omega_1 - \omega_2$ was made. By comparison of (4.33) to (4.1), we see that the Fourier transform of $|r(t)|^2$ is

$$\frac{1}{2\pi}\int_{-\infty}^{\infty} R(\omega_1)R^*(\omega_1-\omega)\,d\omega_1$$

from which we conclude that if $R(\omega) \equiv 0$ for $|\omega| > \omega_b$, then the spectrum of $|r(t)|^2$ vanishes for $|\omega| > 2\omega_b$. In a similar fashion, we find that the spectrum of the second term on the right side of (4.32) lies in the range $2(\omega_0 - \omega_b) < \omega < 2(\omega_0 + \omega_b)$. Therefore if $2\omega_b < 2(\omega_0 - \omega_b)$ (that is, $\omega_0 > 2\omega_b$), then the spectra of the terms on the right side of (4.32) do not overlap, and the second can be removed in principle by passing $f^2(t)$ through a low-pass filter. A process like this is the basis of most practical demodulation schemes.

4.3 NARROWBAND PULSE DISTORTION BY A LINEAR TIME-INVARIANT NETWORK

Although communication systems normally contain portions which are nonlinear, much of the signal channel can typically be characterized as a linear, time-invariant system. The basic ideas of pulse propagation on a transmission line or waveguide can be found in the analysis of the simpler problem of predicting the distortion of a pulse in such a system.

Consider a network for which we define two signals (each of which may be a voltage, a current, a field component, or any linear combination of such quantities): the input $f(t)$ and the output $w(t)$. Because the network is linear and time-invariant, the Fourier transforms of the input and output are related by some complex transfer function $H(\omega)$:

$$W(\omega) = H(\omega)F(\omega) \tag{4.34}$$

The transfer function depends on the properties of the network, but not on the input or output signal. It will prove convenient in what follows to write the transfer function in an exponential form:

$$H(\omega) = A(\omega)e^{-j\Theta(\omega)} \tag{4.35}$$

where $A(\omega)$ and $\Theta(\omega)$ are real, *continuous* functions of frequency defined as for the function F in Section 4.2. Here $A(\omega)$ is called the amplitude of the transfer function, while $\Theta(\omega)$ is the phase delay of the transfer function. For definiteness we define $\Theta(0) = 0$, so that $A(0) = H(0)$.

The energy in the input and output pulses can be related using (4.24):

$$
\begin{aligned}
\int_{-\infty}^{\infty} |w(t)|^2 \, dt &= \frac{1}{\pi} \int_0^{\infty} |W(\omega)|^2 \, d\omega = \frac{1}{\pi} \int_0^{\infty} |H(\omega)|^2 |F(\omega)|^2 \, d\omega \\
&\leq \frac{H_{\max}^2}{\pi} \int_0^{\infty} |F(\omega)|^2 \, d\omega = H_{\max}^2 \int_{-\infty}^{\infty} |f(t)|^2 \, dt \qquad (4.36)
\end{aligned}
$$

where $H_{\max}$ is the maximum value of $|H(\omega)|$ over all frequencies. Equality in (4.36) occurs only if $|H(\omega)|$ is constant. This relation expresses that the energy contained in the output pulse cannot exceed that of the input pulse multiplied by the maximum gain $H_{\max}^2$ of the system. We have a similar constraint on the analytic signals:

$$
\int_{-\infty}^{\infty} |w_A(t)|^2 \, dt \leq H_{\max}^2 \int_{-\infty}^{\infty} |f_A(t)|^2 \, dt \qquad (4.37)
$$

By the convolution theorem (G.4), the output signal $w(t)$ can be written as

$$
w(t) = \int_{-\infty}^{\infty} f(t')h(t - t') \, dt' \qquad (4.38)
$$

where $h(t)$ is the inverse Fourier transform of $H(\omega)$. If values of or a formula for $h(t)$ is available, (4.38) furnishes a way to compute the output pulse without using the Fourier transform. In fact, virtually all linear systems encountered in practice are *causal*: there can be no output before there is an input. In mathematical terms, this means that $h(t) = 0$ for all $t < 0$, and thus (4.38) can be written as

$$
w(t) = \int_{-\infty}^{t} f(t')h(t - t') \, dt' \qquad (4.39)
$$

However, the integral (4.39) can rarely be carried out analytically, and its accurate numerical integration can often either not be done at all, or requires very long computation times due to the rapid oscillation of $f(t)$ at the carrier frequency. We therefore explore several approximate techniques by which to analyze the distortion of a pulse by a linear system.

4.3.1   QUASI-STATIONARY APPROXIMATION (QSA)

Suppose now that $F(\omega)$ is narrowband, in the sense that it is significantly different from zero only near a nonnegative carrier frequency ω_0 (and of course, near $-\omega_0$ as well, but since we will be using the analytic signal negative frequencies will not need to be considered). More precisely, we need the "bandwidth" of $H(\omega)$ to be large compared to that of the input signal in the following sense. Near ω_0, the phase of the transfer function has a predominantly linear dependence of its phase on frequency, while the attenuation can be regarded as approximately constant. We will emphasize these features of the transfer function by introducing a new function Θ_d (the distortion part of Θ) according to

$$
\Theta(\omega) \equiv \omega_0 t_{p0} + (\omega - \omega_0)t_{g0} + \Theta_d(\omega) \qquad (4.40)
$$

The first and second terms come from the first two terms of the Taylor series expansion of $\Theta(\omega)$ about $\omega = \omega_0$. Here we have abbreviated the values of Θ and its derivatives at ω_0 as

$$
\Theta_0 \equiv \Theta(\omega_0) \equiv \omega_0 t_{p0}
$$

$$\Theta'_0 \equiv \Theta'(\omega_0) \equiv t_{g0} \tag{4.41}$$

and so on. The real quantity t_{p0} is called the *phase delay time* and t_{g0} the *group delay time* of the transfer function at the frequency ω_0. We use a similar notation for A and its derivatives at ω_0. The definition of Θ_d is such that it and its first derivative with respect to ω at ω_0 are equal to zero. This assures that $\Theta_d(\omega)$ is in some sense much more slowly varying than $\Theta(\omega)$ as a function of frequency near ω_0.

The analytic signal of the output is now:

$$w_A(t) = \frac{1}{\pi} \int_0^\infty H(\omega) F(\omega) e^{j\omega t} \, d\omega \tag{4.42}$$

If $F(\omega)$ is given by (4.16) and corresponds to an analytic signal of the form (4.31), then

$$F(\omega) = \frac{1}{2} R(\omega - \omega_0)$$

and we can express (4.42) as

$$w_A(t) = \frac{1}{2\pi} e^{j\omega_0 \tau_p} \int_{-\infty}^\infty H_d(\omega_0 + u) R(u) e^{ju\tau_g} \, du \tag{4.43}$$

where

$$\tau_p \equiv t - t_{p0} \tag{4.44}$$

is the phase-delayed time,

$$\tau_g \equiv t - t_{g0} \tag{4.45}$$

is the group-delayed time, and we have made the change of integration variable $\omega = \omega_0 + u$. The function

$$H_d(\omega) \equiv A(\omega) e^{-j\Theta_d(\omega)} \tag{4.46}$$

is the portion of the transfer function responsible for distorting the envelope: $A(\omega)$ distorts the amplitude, while $\Theta_d(\omega)$ distorts the phase.

Since the most significant contributions to the integral in (4.43) come from frequencies close to ω_0, it is appropriate to expand H_d into a Taylor series about this frequency, and integrate term by term. Using the envelope representation (4.31) for f_A, the result is

$$
\begin{aligned}
w_A(t) &= \frac{1}{2\pi} e^{j\omega_0 \tau_p} \sum_{n=0}^\infty \frac{H_{dn}}{n!} \int_{-\infty}^\infty u^n R(u) e^{ju\tau_g} \, du \\
&= e^{j\omega_0 \tau_p} \sum_{n=0}^\infty \frac{H_{dn}}{n!} \left(-j \frac{d}{d\tau_g} \right)^n r(\tau_g) \\
&= e^{j\omega_0 \tau_p} \sum_{n=0}^\infty \frac{H_{dn}}{n!} (-j)^n \frac{d^n r(\tau_g)}{d\tau_g^n}
\end{aligned}
\tag{4.47}
$$

in which we have abbreviated the derivatives of H_d at ω_0 as:

$$H_{dn} = \left. \frac{d^n H_d(\omega)}{d\omega^n} \right|_{\omega=\omega_0} \tag{4.48}$$

In terms of the quantities A_0, A'_0, A''_0, Θ''_0, etc., the first few coefficients (4.48) are given by:

$$H_{d0} = A_0 \tag{4.49}$$

$$H_{d1} = A_0' \tag{4.50}$$

$$H_{d2} = A_0'' - jA_0\Theta_0'' \tag{4.51}$$

Equation (4.47) is called the *quasi-stationary* expansion for the distortion of a signal by a linear time-invariant system. It may be convergent or not, but retention of the first few terms in the series is usually its most practical application.

Using the foregoing results, we can write out the first three terms of expansion (4.47) as:

$$w_A(t) \simeq e^{j\omega_0\tau_p} \left\{ A_0 r(\tau_g) - jA_0' r'(\tau_g) - \frac{1}{2} \left[A_0'' - jA_0\Theta_0'' \right] r''(\tau_g) \right\} \tag{4.52}$$

We will call equation (4.52) the quasi-stationary approximation (QSA) of the pulse distortion. A far simpler (but also less accurate) approximation is obtained by keeping only the first term in curly brackets on the right side of (4.52):

$$w_A(t) \simeq A_0 e^{j\omega_0\tau_p} r(\tau_g) \tag{4.53}$$

To this order of precision, the output is described by an analytic signal whose phase is delayed by the phase delay time t_{p0} of the network at the carrier frequency, but whose envelope is delayed by the group delay time t_{g0}. An additional change of amplitude by the factor A_0 also takes place, but no distortion of the shape of the signal occurs. Despite the relative shift between phase and envelope, when the output signal is applied to an amplitude detector, the result will be a scaled and time-delayed version of what we get when the same is done to the input. The other terms in the curly brackets of (4.52) cause a distortion of the shape of the input pulse. The kind and amount of distortion depends on the form of the input pulse, as well as on the parameters A_0', A_0'' and Θ_0'' which are characteristics of the linear system.

4.3.2   EXAMPLE: AM GAUSSIAN PULSE (QSA)

Consider an input signal which has a Gaussian amplitude envelope:

$$r(t) = e^{-t^2/2T^2} \tag{4.54}$$

where the positive, real constant T is the characteristic duration of the pulse (for $|t| > 2T$, the amplitude is less than 14% of its maximum value at $t = 0$, and the energy is less than 2% of its maximum value).[2] If $\omega_0 T \gg 1$, then $r(t)e^{j\omega_0 t}$ is to a high degree of accuracy an analytic signal. The Gaussian pulse closely approximates the output of many practical sources, notably lasers, so its study is of importance.

We will suppose for simplicity that the transfer function has constant amplitude, so that $A_0' = A_0'' = 0$, and further assume that $A(\omega) = A_0 = 1$. Then the analytic signal of the output is given by (4.52) as

$$w_A(t) \simeq e^{j\omega_0\tau_p} \left[1 + \frac{j\Delta}{2} \left(\frac{\tau_g^2}{T^2} - 1 \right) \right] e^{-\tau_g^2/2T^2} \tag{4.55}$$

---

[2]Note that a calculation shows that if f is the Gaussian pulse corresponding to (4.54), then the pulse duration as given by (4.29) is $T_d(f) = T$.

where

$$\Delta = \frac{\Theta_0''}{T^2} \tag{4.56}$$

is called the quadratic dispersion parameter of this pulse due to its passage through the linear system. Examination of this expression reveals that two things have happened to the signal:

1. The output phase is $\omega_0 \tau_p + \tan^{-1}\left[\frac{\Delta}{2}\left(\frac{\tau_g^2}{T^2} - 1\right)\right]$. This means that the carrier frequency ω_0 has been "chirped": the time-dependent phase factor $\exp(j\omega_0 t)$ has changed to a nonlinear function of time, a kind of frequency modulation. This will be important if an FM detector is applied to the signal.

2. The shape of the amplitude envelope of the pulse has changed. The output power envelope $|w_A(t)|^2$ (which is obtained as a result of AM detection) is

$$|w_A(t)|^2 = e^{-\tau_g^2/T^2}\left[1 + \left(\frac{\Delta}{2}\right)^2\left(\frac{\tau_g^2}{T^2} - 1\right)^2\right] \tag{4.57}$$

Note that the output maximum $(1 + \Delta^2/4)$ occurring at $t = 0$ is larger than that of the input pulse. Indeed, we can show that (4.57) violates the energy constraint (4.37). This is because w_A itself has only been computed accurate to terms of order Δ, and so can only be expected to obey the energy constraint up to that order of accuracy. More terms of (4.52) would need to be kept in order to alleviate this difficulty.

The major effect of nonlinear dependence of the phase of the transfer function on frequency has been to change the duration of the pulse to a new value T_{out}. But because of the failure of this approximation to obey energy conservation as noted above, it is not easy to characterize succinctly the value of T_{out}. Since information is often sent through networks as a series of pulses, it is clear that adjacent output pulses will have to be separated by at least T_{out} in order to prevent pulse overlap and loss of data. The network thus imposes a maximum data rate of about $1/T_{\text{out}}$ pulses per second for this type of input pulse. We therefore need a more accurate means of evaluating T_{out}.

4.3.3   QUASI-GAUSSIAN APPROXIMATION (QGA)

Because the output of many narrowband pulse sources has an approximately Gaussian envelope, it is worthwhile to investigate the distortion of such pulses by a method which makes use of the particular properties of such pulses. Let us consider again the envelope (4.54), but now allow for both phase and amplitude variations of this envelope by considering that T may be complex. In order to have a realistic pulse with finite energy, this envelope must decay as $t \to \pm\infty$. Thus, we require that $\text{Re}(1/T^2) > 0$, or equivalently $|\text{Re}(T)| > |\text{Im}(T)|$. From the table of Fourier transforms in Appendix G, we find that the Fourier transform of (4.54) is

$$R(\omega) = T\sqrt{2\pi}e^{-\omega^2 T^2/2} \tag{4.58}$$

Now if $r(t)$ is a more general complex envelope, but is still narrowband with a single maximum of its spectral amplitude $|R(\omega)|$ occurring at $\omega = 0$, it might seem reasonable to assume that its Fourier transform could be approximated for small enough $|\omega|$ by a function of the form

$$R(\omega) \simeq R_0 T_0 \sqrt{2\pi}e^{-\omega^2 T_0^2/2}$$

where R_0 and T_0 are some complex constants. However, this function has the property that $R'(0) = 0$, and this will not necessarily be true just because $|R(\omega)|$ has a maximum at $\omega = 0$. A more suitable approximate functional form is

$$R(\omega) \simeq R_0 T_0 \sqrt{2\pi} e^{-\omega^2 T_0^2/2} e^{-j\omega T_c} \tag{4.59}$$

where T_c is the temporal location of the pulse center, also a complex quantity in general. The value of R_0 can be found by matching the value of the spectrum $R(\omega)$ in (4.59) exactly at $\omega = 0$, but we will not need this quantity explicitly in the final formula (4.67) below. Matching the first derivative of R at $\omega = 0$ allows us to determine T_c:

$$T_c = j \frac{R'(0)}{R(0)} = \frac{\int_{-\infty}^{\infty} t r(t)\, dt}{\int_{-\infty}^{\infty} r(t)\, dt} \tag{4.60}$$

By matching the second derivative at $\omega = 0$ we obtain T_0:

$$T_0^2 = -\frac{R''(0)}{R(0)} - T_c^2 = \frac{\int_{-\infty}^{\infty} t^2 r(t)\, dt}{\int_{-\infty}^{\infty} r(t)\, dt} - \left(\frac{\int_{-\infty}^{\infty} t r(t)\, dt}{\int_{-\infty}^{\infty} r(t)\, dt} \right)^2 \tag{4.61}$$

The second equalities in (4.60) and (4.61) follow from differentiating the Fourier integral directly. Observe that (4.61), which involves the complex envelope $r(t)$, is different than (4.29), which uses the energy of the pulse; definition (4.61) is specific to approximately Gaussian pulses, and is in fact identical to (4.29) for an exact Gaussian. In addition to $|\text{Re}(T_0)| > |\text{Im}(T_0)|$, the condition

$$R(0) = \int_{-\infty}^{\infty} r(t)\, dt \neq 0$$

is also required for (4.59) to be a valid approximation. Narrowband pulses whose spectrum vanishes at $\omega = 0$ can be handled by a modification of this procedure, which we leave as an exercise for the reader.

Now let us re-examine the procedure which led to the quasi-stationary approximation. In order to obtain a different approximation which more closely obeys the energy constraint, we will make an approximation to the term $\exp[-j\Theta_d(\omega_0 + u)]$ in (4.46) that has a magnitude of one (in contrast to the Taylor series expansion used in the QSA, which does not):

$$e^{-j\Theta_d(\omega_0 + u)} \simeq e^{-ju^2 \Theta_0''/2} \tag{4.62}$$

On the other hand, the amplitude of the transfer function will be treated in the same way as before:

$$A(\omega_0 + u) \simeq A_0 + u A_0' + \frac{1}{2} u^2 A_0'' \tag{4.63}$$

Then from (4.43) the analytic signal of the output pulse corresponding to the input signal envelope $r(t)$ (as approximated by (4.59)) is approximately

$$w_A(t) \simeq \frac{R_0 T_0 \sqrt{2\pi} e^{j\omega_0 \tau_p}}{2\pi} \int_{-\infty}^{\infty} e^{ju(\tau_g - T_c)} \left[A_0 + u A_0' + \frac{1}{2} u^2 A_0'' \right] e^{-u^2(T_0^2 + j\Theta_0'')/2}\, du \tag{4.64}$$

We now change to a new integration variable

$$v = u \sqrt{\frac{T_0^2 + j\Theta_0''}{T_0^2}} = u \sqrt{1 + j\Delta_0}$$

in (4.64), where

$$\Delta_0 = \frac{\Theta_0''}{T_0^2} \tag{4.65}$$

and re-approximate the resulting terms by means of the original envelope using (4.59):

$$R_0 T_0 \sqrt{2\pi} e^{-v^2 T_0^2/2} \simeq R(v) e^{jvT_c}$$

and thus obtain

$$w_A(t) \simeq \frac{e^{j\omega_0 \tau_p}}{2\pi} \frac{1}{\sqrt{1+j\Delta_0}} \int_{-\infty}^{\infty} e^{jv\left(\frac{\tau_g - T_c}{\sqrt{1+j\Delta_0}} + T_c\right)} \left[A_0 + vA_0' \frac{1}{\sqrt{1+j\Delta_0}} + \frac{1}{2}v^2 A_0'' \frac{1}{1+j\Delta_0} \right] R(v)\, dv \tag{4.66}$$

This can finally be evaluated in terms of a "stretched" version of r and its derivatives:

$$\boxed{w_A(t) \simeq \frac{e^{j\omega_0 \tau_p}}{\sqrt{1+j\Delta_0}} \left[A_0 r\left(\tau_d\right) - jA_0' \frac{d}{d\tau_g} r\left(\tau_d\right) - \frac{A_0''}{2} \frac{d^2}{d\tau_g^2} r\left(\tau_d\right) \right]} \tag{4.67}$$

where

$$\tau_d = \frac{\tau_g - T_c}{\sqrt{1+j\Delta_0}} + T_c \tag{4.68}$$

Equation (4.67) is called the quasi-Gaussian approximation (QGA) of the pulse distortion. Note that the constant R_0 is not used in the QGA; it is implicitly contained in the function $r(t)$. Only the constants T_c and T_0 need be computed. Use of the QGA requires that an analytical expression for $r(t)$ be available so that it can be evaluated at complex values of time. Alternatively, equations (4.9) and (4.31) could be used to evaluate r if its argument has a positive imaginary part.

In some situations, it may not be sufficiently accurate to approximate the Fourier transform of the envelope by that of a single, time-shifted Gaussian pulse. One option is to use the sum of several Gaussian pulses, each time-shifted by a different amount, or with a different pulse width, or both. Determining the various constants in such a representation is more complicated, and would involve matching the values of higher-order derivatives of the exact and approximate envelope spectra at $\omega = 0$. It is usually simpler to compute the pulse distortion directly by a numerical evaluation of the Fourier transform if the envelope is so complicated as to require such a higher-order approximation.

If $A_0' = A_0'' = 0$, then in the QGA the output pulse envelope is stretched in the time coordinate and scaled in amplitude from that of the input, but in some sense still retains the essential features of the original pulse shape. The interpretation of the complex stretching factor in (4.67) is not obvious in general, and we must examine particular pulse shapes to determine its significance. We will do this for the case of the AM Gaussian pulse in the next subsection.

4.3.4   EXAMPLE: AM GAUSSIAN PULSE (QGA)

We return to the example of the pulse envelope (4.54) (with T taken to be real). In this case, $T_0 = T$ and $T_c = 0$, and if we assume again a constant amplitude transfer function with $A_0 = 1$, (4.67) gives

$$w_A(t) \simeq e^{j\omega_0 \tau_p} \sqrt{\frac{T^2}{T^2 + j\Theta_0''}} r\left(\tau_g \sqrt{\frac{T^2}{T^2 + j\Theta_0''}}\right) = e^{j\omega_0 \tau_p} \sqrt{\frac{1}{1+j\Delta_0}} e^{-\tau_g^2/[2T^2(1+j\Delta_0)]} \tag{4.69}$$

where Δ_0 is the quadratic dispersion parameter defined in (4.56). In order to re-examine the phase and amplitude of this signal, we express the "stretched" term in the exponent as:

$$-\frac{\tau_g^2}{2T^2(1+j\Delta_0)} = -\frac{\tau_g^2}{2T^2}\frac{1-j\Delta_0}{1+\Delta_0^2}$$

which allows us to observe that:

1. The output phase is

$$\omega_0\tau_p - \frac{1}{2}\tan^{-1}\Delta_0 + \frac{\tau_g^2}{2T^2}\frac{\Delta_0}{1+\Delta_0^2}$$

   Again we see that a "chirping" of the signal has occurred, in this approximation as a quadratic function of time. This amounts to a frequency modulation that is linear in time.

2. The output power envelope $|w_A(t)|^2$ is now

$$|w_A(t)|^2 = \frac{T}{T_{\text{out}}}e^{-\tau_g^2/T_{\text{out}}^2} \tag{4.70}$$

   where we can now easily identify the width of the output pulse as

$$T_{\text{out}}^2 = T^2(1+\Delta_0^2) = T^2 + \left(\frac{\Theta_0''}{T}\right)^2 \tag{4.71}$$

The zeroth-order and second-order QSA (4.53) and (4.52), as well as the QGA (4.67) are compared with the exact solution in Figure 4.4. The transfer function used in this example is

$$H(\omega) = e^{-jT_f\sqrt{\omega^2-\omega_c^2}}, \tag{4.72}$$

a high-pass filter that models a waveguide mode or the unconventional transmission line described in Section 1.2.2. We use the values $\omega_0 = 2\pi \times 10^7$ rad/s, $\omega_c = 1.9\pi \times 10^7$ rad/s, $T = 1$ μs and $T_f = 3$ μs. These parameters give a group delay time of $t_{g0} = 9.6$ μs and a quadratic dispersion parameter of $\Delta_0 = -1.415$. The "exact" solution is obtained by

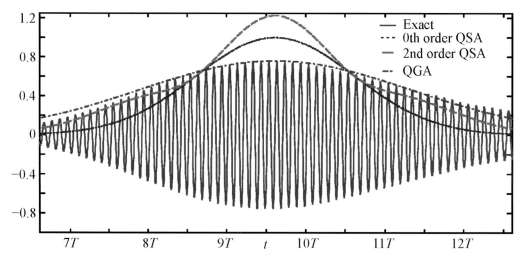

Figure 4.4 Comparison of exact output $w(t)$ with QSA and QGA envelopes $|w_A(t)|$ for an AM Gaussian input pulse to the transfer function (4.72) (see text for parameters).

numerical integration of (4.39) using entry 5 of Table G.2. For clarity, only the envelopes $|w_A(t)|$ are shown for the QSA and QGA. We can see that the 0th-order QSA is both too narrow and has too large a maximum envelope value as compared to the exact result, while the 2nd-order QSA predicts the pulse width more accurately but has an even larger maximum envelope value. The QGA is much more accurate, reproducing not only the correct value of the pulse peak, but also the correct width of the pulse. The only noticeable error in the QGA is a small time shift of the pulse envelope, which is present in the QSA approximations as well.

We note from (4.71) that the duration T_{out} of the output pulse has been increased by comparison with that of the input pulse T. As a function of T, T_{out} becomes large both for small T (that is, very narrow input pulses are severely broadened) and for large T (where, of course, we started out with a broad pulse). For a given phase characteristic Θ_0'' of the transfer function, there is an optimum value of T for which T_{out} will be minimum: $T = T_{\text{opt}} = \sqrt{|\Theta_0''|}$, giving $T_{\text{out,opt}} = \sqrt{2}T_{\text{opt}} = \sqrt{2|\Theta_0''|}$, the narrowest pulse of this form that can emerge from the network.

Since information is often sent through networks as a sequence of such pulses, adjacent pulses have to be separated in time by at least T_{out} in order to prevent overlap and loss of data. The network itself thus imposes a maximum pulse rate of roughly

$$f_p = \frac{1}{T_{\text{out,opt}}} = \frac{1}{\sqrt{2|\Theta_0''|}} \tag{4.73}$$

Gaussian pulses per second. Of course, what the actual data rate is will depend on details such as threshold detection levels, encoding schemes and other implementation specifics, discussion of which is beyond the scope of this text, but it can be expected to have the form

$$f_d = \frac{\text{constant}}{T_{\text{out,opt}}}$$

where the constant depends on the implementation.

4.3.5 CHANGE OF PULSE CENTER AND DURATION BY A LOSSLESS NETWORK

The QSA or QGA is useful in practice if only a few terms of the expansion provide an accurate prediction of the distortion of the pulse. This in turn requires not only that the pulse be narrowband, but also that the pulse shape is sufficiently simple that easy analytical expressions can be computed, as is the case for the Gaussian pulse, for example. To predict the distortion of more general pulses, we must either resort to numerical calculation of the relevant Fourier integral (4.43) or content ourselves with less complete information about the output pulse which will nevertheless be useful in the design of communication systems. In this subsection, we compute the time delay of the pulse center and change in pulse duration of an arbitrary signal due to passage through a lossless network. We do this using the definitions (4.26) and (4.29) based on the energy moments.

If an arbitrary pulse is input to a *lossless* linear time-invariant system whose transfer function is $H(\omega) = \exp[-j\Theta(\omega)]$, then the total phase of the output spectrum $W(\omega)$ will be $-\phi_F(\omega) - \Theta(\omega)$, while $|W(\omega)| = |F(\omega)|$. Thus, the center of the output pulse w can be written as

$$t_c(w) = \langle \phi_F'(\omega) + \Theta'(\omega) \rangle_F = t_c(f) + t_{g,\text{av}}(f) \tag{4.74}$$

where

$$t_{g,\text{av}}(f) \equiv \langle t_g(\omega) \rangle_F \tag{4.75}$$

is the average value of the group delay time $t_g(\omega) = \Theta'(\omega)$, weighted by the spectral energy $|F(\omega)|^2$ at each frequency ω. In other words, the center of the output pulse is delayed from the center of the input pulse by the average group delay time $t_{g,\text{av}}(f)$. The result (4.74) is exact for a pulse of arbitrary shape (narrowband or not), provided the system transfer function is lossless. If the pulse is narrowband about a carrier frequency ω_0, we have $t_{g,\text{av}}(f) \simeq t_{g0}$ as is obtained in the QSA and QGA.

The duration of the output signal is given by

$$T_d^2(w) = T_d^2(f) + 4C(t_{g,F}, t_g) + 2\left\langle [t_g(\omega)]^2 \right\rangle_F - 2\left\langle t_g(\omega) \right\rangle_F^2 \tag{4.76}$$

where $C(t_{g,F}, t_g)$ is a sort of covariance function between the group delay time of the input signal and that of the transfer function:

$$C(t_{g,F}, t_g) \equiv \left\langle t_g(\omega) t_{g,F}(\omega) \right\rangle_F - \left\langle t_g(\omega) \right\rangle_F \left\langle t_{g,F}(\omega) \right\rangle_F \tag{4.77}$$

and may be either positive or negative.

4.4 PROPAGATION AND DISTORTION OF PULSES; GROUP VELOCITY

The results of the previous section can be directly applied to the propagation of pulses on a transmission line. Suppose that a forward voltage wave is launched on a semi-infinite classical transmission line (in $z > 0$) by connection of an ideal voltage source $v_G(t) = v(0, t)$ at the terminals $z = 0$. This is taken to be the input signal. If the output signal is chosen to be $v(z, t)$, the voltage at a point z and time t on the line, then the transfer function will be

$$H(\omega) = e^{-\gamma(\omega)z} = e^{-\alpha(\omega)z - j\beta(\omega)z}$$

as is appropriate for a forward-traveling wave. The results of the previous section apply, with $A_0 = e^{-\alpha_0 z}$, $\Theta_0 = \omega_0 t_{p0} = \omega_0 z/v_{p0}$, $A_0' = -\alpha_0' z A_0$, $\Theta_0' = t_g = \beta_0' z = z/v_{g0}$, etc. The subscripts $_0$ denote quantities to be evaluated at $\omega = \omega_0$, while v_p and v_g denote the phase and group velocities respectively, defined in (1.11) and (1.12). If only the first term in the curly brackets of (4.52) is retained, we have for the analytic signal of the output pulse

$$v_A(z, t) = e^{-\alpha_0 z} e^{j\omega_0(t - z/v_{p0})} r(t - z/v_{g0}) \tag{4.78}$$

We see from this result that the group velocity v_{g0} at the center or carrier frequency of the pulse is the velocity at which the envelope function $r(t)$ travels in this approximation, while the phase of the carrier is traveling at the phase velocity v_{p0}.

Let us emphasize that (4.78), along with the concepts of group velocity and phase velocity, makes sense only if the approximations leading to this result are valid. That is, the neglected terms in (4.52) or (4.67) must be small. As z increases, terms such as $\Theta_0'' = \beta_0'' z$ will become significant, and distortion of the input pulse will begin to be evident. To examine this effect more closely, consider the case when attenuation is absent: $\alpha(\omega) \equiv 0$. In (4.43), we approximate

$$H_d(\omega_0 + u) \simeq e^{-j\frac{1}{2}\beta_0'' z u^2}$$

to obtain

$$v_A(z, t) = \frac{1}{2\pi} e^{j\omega_0 \tau_p} \int_{-\infty}^{\infty} R(u) e^{ju\tau_g} e^{-j\frac{1}{2}\beta_0'' z u^2} \, du \tag{4.79}$$

where $\tau_p = t - z/v_{p0}$ and $\tau_g = t - z/v_{g0}$. But the most important values of u in this integral are those within the bandwidth of the envelope function $r(t)$, i.e., $|u| \le \omega_b$. Therefore, we can neglect the u^2 term in (4.79) (and therefore predict that distortion will be minimal) provided that

$$\left| \frac{\omega_b^2 \beta_0'' z}{2} \right| \ll 1$$

or equivalently

$$z \ll \frac{2a_0}{\omega_b^2} \equiv z_d$$

The quantity $a_0 \equiv 1/|\beta_0''|$, whose units are m/sec^2, is sometimes called the *group acceleration* of the transmission line, and z_d the *dispersion length*. The smaller $|\beta_0''|$ is, the longer the distance a pulse can be transmitted without distortion. On the other hand, a shorter pulse (with larger ω_b) will distort after a shorter distance than will a pulse of longer duration. Of course, the other derivatives of α and β can also influence pulse distortion, and must be accounted for in more accurate predictions of this effect.

When the results (4.69)-(4.71) for a Gaussian input pulse are applied to propagation along a lossless transmission line, we find that the characteristic duration of the pulse changes from T to

$$\sqrt{T^2 + (\beta_0'' z)^2/T^2} \equiv T_{\text{out}}(z) \tag{4.80}$$

Initially, the pulse duration is nearly unchanged, but as z increases, the duration increases. This will put an upper limit on the distance a pulse may travel in a dispersive transmission line before overlap with adjacent pulses occurs. At this point, a repeater capable of reshaping the pulse would have to be inserted into the transmission line if resolvable pulse propagation over longer distances is desired.

4.4.1 PROPAGATION OF ARBITRARY PULSES

The energy moment description of pulse distortion can be applied to the propagation of arbitrary pulses along a transmission line. We replace

$$t_g(\omega) \to \frac{z}{v_g(\omega)}$$

and we find from (4.74) that the center of the pulse (now a function of z) is given by

$$t_c(w) = t_c(f) + \frac{z}{v_{g,\text{av}}(f)} \tag{4.81}$$

where

$$\frac{1}{v_{g,\text{av}}(f)} = \left\langle \frac{1}{v_g(\omega)} \right\rangle_F \tag{4.82}$$

is an average group velocity for the signal on the line, and is the velocity with which the center of the pulse travels. This is true for *any* signal f, but if f is a narrowband signal with spectrum concentrated around ω_0, we see that the average group velocity becomes approximately v_{g0}, in agreement with our earlier result (4.78).

The pulse duration, from (4.76), is obtained from:

$$T_{d,\text{out}}^2(z) = T_d^2(f) + 4zC(t_{g,F}, \frac{1}{v_g}) + 2z^2 \left\{ \left\langle \left[\frac{1}{v_g(\omega)} \right]^2 \right\rangle_F - \left\langle \frac{1}{v_g(\omega)} \right\rangle_F^2 \right\} \tag{4.83}$$

Thus, the square of the pulse duration is a quadratic function of z (a result we obtained approximately for a narrowband pulse in (4.80)). Eventually, the duration increases linearly with z if z is large enough. Note that if $C(t_{g,F}, \frac{1}{v_g}) < 0$, the pulse duration will initially decrease to some minimum value before beginning to increase again. This condition, which involves selecting an appropriate phase characteristic of the input pulse spectrum relative to the frequency dependence of the group velocity, can be used to counteract the natural tendency of pulses to spread and overlap with adjacent ones in a pulse stream. The technique is called *chirping*, because it involves a frequency modulation of the input pulse's carrier frequency.

4.5 MULTIMODE BROADENING OF NARROWBAND PULSES

In recent years, various kinds of multimode waveguides (most notably in the optical frequency band) have been used in practical communication systems. At optical frequencies especially, the deterioration in fractional bandwidth this entails is more than compensated for by the higher carrier frequency as compared to a microwave system. In this section, we will examine the effect of multimode propagation on the temporal distortion of a pulse.

In contrast with single-mode waveguide systems, the signal detector in a multimode waveguide is usually not deeply concerned with the relative phases or other details of the portions of the signal appearing in each mode. The modes are excited in large numbers, and somewhat indiscriminately, so the detection process obtains only information about the sum of the powers contained in the modes. The primary distorting effect is that each mode travels with a different group velocity from the others, and the result seen at the detector is a considerably broadened version of that produced by the source. Pictorially, the situation is as shown in Figure 4.5. Denote by $|g_{Am}(z,t)|^2$ the detected output pulse present in the mth mode of the waveguide. We assume at first that an equal amount of energy is imparted to each mode at $z = 0$ as shown in Figure 4.5(a). As z increases, in the simplest approximation the pulse in each mode will travel undistorted but delayed by the group delay $t_{gm} = z/v_{gm}$ of the mode (Figure 4.5(b)).

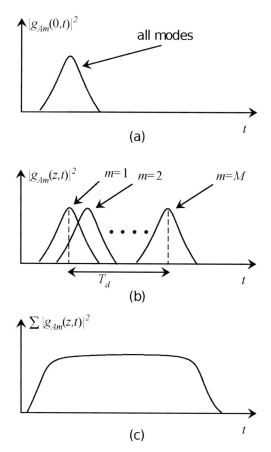

Figure 4.5 (a) Mode intensities at $z = 0$; (b) mode intensities at $z > 0$; (c) total intensity at detector for $z > 0$.

The spread in the pulse centers of the various modes is then the difference between the largest and smallest mode group delays:

$$T_d = (t_{gm})_{\max} - (t_{gm})_{\min} = z \left(\frac{1}{v_{g,\min}} - \frac{1}{v_{g,\max}} \right)$$

This is called the *delay difference* or *delay spread* between the slowest and fastest modes, often expressed as delay difference per unit length,

$$\frac{T_d}{z} = \frac{1}{v_{g,\min}} - \frac{1}{v_{g,\max}} \tag{4.84}$$

In many waveguides, the group velocities of the individual modes are closely and roughly evenly spaced. Natural filtering effects of the detector tend to smooth out the ripple which would otherwise be expected when contributions from all modes are added together. Thus, the response of the detector at the output of a multimode waveguide will appear as in Figure 4.5(c). If the original pulse width at $z = 0$ is small compared to T_d, the detected pulse at the output will have a half-width of about $T_d/2$.

This broadening of the pulse would appear to be much more severe than that which results from dispersion of a single mode. However, there are several factors which alleviate the multimode distortion effects to some extent. For one thing, some modes are usually excited with larger amplitudes than other modes, and the amplitudes often vary slowly with mode number m. Also, the attenuation constant α_m of the mth mode is frequently larger for the modes that are excited poorly, so the amplitudes of these modes are reduced even further when $z > 0$. In this case, we end up with a situation like that of Figure 4.6, with the reduction in some mode amplitudes serving to reduce the width of the detected pulse. Local irregularities that cause inter-coupling between the modes can also reduce the width of the detected pulse. The price one pays for the reduction in pulse width is the power that is lost in mode attenuation.

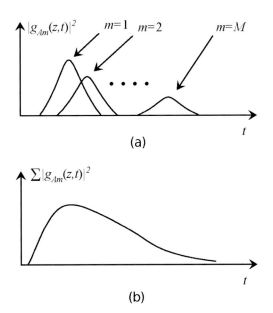

Figure 4.6 Nonuniform mode amplitudes: (a) mode intensities at $z > 0$; (b) total intensity at detector for $z > 0$.

4.6   DISTORTION FACTOR AND FIDELITY FACTOR

As higher data rates have been required of communication systems, the limitations of narrowband signals have made it necessary to utilize so-called ultrawideband (UWB) signals in applications such as radar, location, tracking and sensors. A commonly used descriptor of the distortion of wideband signals is the *distortion factor*, and the related *fidelity factor*.

We define the distortion factor D of the output signal relative to the input signal as

$$D \equiv \min_{A, \tau} M(A, \tau) \tag{4.85}$$

where

$$M(A, \tau) = \frac{\int_{-\infty}^{\infty} [w(t) - Af(t - \tau)]^2 \, dt}{\int_{-\infty}^{\infty} w^2(t) \, dt} \tag{4.86}$$

is a measure of the deviation of the output signal from the input. In other words, we accept as undistorted any output that is an amplitude-scaled replica of the input, possibly delayed in time, and the distortion is the fractional integrated squared error of the best fit to such a replica. For any given τ, we can carry out the minimization with respect to the scale parameter A explicitly by demanding that

$$\frac{\partial M}{\partial A} = 0 \tag{4.87}$$

The unique solution is $A = A_0(\tau)$, where

$$A_0(\tau) = \frac{\int_{-\infty}^{\infty} w(t) f(t - \tau) \, dt}{\int_{-\infty}^{\infty} f^2(t) \, dt} \tag{4.88}$$

and the resulting expression for M for this optimal value of A is

$$M_0(\tau) \equiv M[A_0(\tau), \tau] = 1 - \frac{\left[\int_{-\infty}^{\infty} w(t) f(t - \tau) \, dt\right]^2}{\int_{-\infty}^{\infty} f^2(t) \, dt \int_{-\infty}^{\infty} w^2(t) \, dt} \tag{4.89}$$

The distortion defined in (4.85) can now be expressed as

$$D \equiv \min_{\tau} M_0(\tau) \tag{4.90}$$

A fidelity factor FF can be defined in terms of the correlation between the input and output signals as

$$\text{FF} \equiv \max_{\tau} \frac{\left|\int_{-\infty}^{\infty} w(t) f(t - \tau) \, dt\right|}{\sqrt{\int_{-\infty}^{\infty} f^2(t) \, dt \int_{-\infty}^{\infty} w^2(t) \, dt}} \tag{4.91}$$

so that

$$D = 1 - \text{FF}^2 \tag{4.92}$$

It is not hard to show that $0 \leq D \leq 1$ and $0 \leq \text{FF} \leq 1$.

The fidelity factor can also be expressed in terms of integrals in the frequency domain using Parseval's theorem (G.6). Since $W(\omega) = H(\omega) F(\omega)$, we have

$$\text{FF} = \max_{\tau} \frac{\left|\int_{-\infty}^{\infty} H(\omega) |F(\omega)|^2 e^{j\omega\tau} \, d\omega\right|}{\sqrt{\int_{-\infty}^{\infty} |F(\omega)|^2 \, d\omega \int_{-\infty}^{\infty} |H(\omega) F(\omega)|^2 \, d\omega}} \tag{4.93}$$

Because f, w and h are all real functions of time, their Fourier transforms obey (4.3), meaning that all integrals over frequency can be expressed as integrals over nonnegative frequencies only, and we obtain:

$$\mathrm{FF} = \max_{\tau} C(\tau) \tag{4.94}$$

where

$$C(\tau) \equiv \frac{\left| \mathrm{Re} \left[\int_0^\infty H(\omega)\,|F(\omega)|^2\,e^{j\omega\tau}\,d\omega \right] \right|}{\sqrt{\int_0^\infty |F(\omega)|^2\,d\omega \int_0^\infty |H(\omega)F(\omega)|^2\,d\omega}} \tag{4.95}$$

is a real function of τ. For most applications, a fidelity factor of 0.9 or larger is required to achieve adequate system performance. Unfortunately, the maximization of $C(\tau)$ is not possible analytically in general, and must be carried out by numerical methods.

4.6.1 EXAMPLE: EXPONENTIAL PULSE IN A LOW-PASS FILTER

An example for which the fidelity factor can be evaluated explicitly involves an exponential pulse

$$f(t) = e^{-t/T}\theta(t) \qquad \Rightarrow \qquad F(\omega) = \frac{T}{1 + j\omega T} \tag{4.96}$$

where θ is the *Heaviside unit step function*, defined as

$$\begin{aligned} \theta(t) &= 1 \qquad (t > 0) \\ &= 0 \qquad (t < 0) \end{aligned} \tag{4.97}$$

and T is the time constant of the pulse. Let this pulse be the input to a simple low-pass filter with transfer function

$$H(\omega) = \frac{1}{1 + j\omega/\omega_c} \tag{4.98}$$

where ω_c is the (angular) corner frequency of the filter. From standard tables of integrals, we can find that

$$\int_0^\infty |F(\omega)|^2\,d\omega = \frac{\pi T}{2} \tag{4.99}$$

$$\int_0^\infty |H(\omega)F(\omega)|^2\,d\omega = \frac{\pi T}{2}\frac{\omega_c T}{\omega_c T + 1} \tag{4.100}$$

$$\begin{aligned} \mathrm{Re} \int_0^\infty & H(\omega)\,|F(\omega)|^2\,e^{j\omega\tau}\,d\omega \\ &= \frac{\pi \omega_c T^2}{2}\frac{e^{\tau/T}}{\omega_c T + 1} \qquad (\tau < 0) \\ &= \frac{\pi \omega_c T^2}{2}\left[\frac{e^{-\tau/T}}{\omega_c T - 1} - \frac{2e^{-\omega_c \tau}}{(\omega_c T - 1)(\omega_c T + 1)} \right] \qquad (\tau > 0) \end{aligned} \tag{4.101}$$

Therefore,

$$\begin{aligned} C(\tau) &= \sqrt{\frac{\omega_c T}{\omega_c T + 1}}\,e^{\tau/T} \qquad (\tau < 0) \\ &= \sqrt{\frac{\omega_c T}{\omega_c T + 1}}\,\frac{(\omega_c T + 1)e^{-\tau/T} - 2e^{-\omega_c \tau}}{\omega_c T - 1} \qquad (\tau > 0) \end{aligned} \tag{4.102}$$

It can be shown that the maximum value of $C(\tau)$ in this example will occur for $\tau = \tau_0 > 0$, so requiring $C'(\tau) = 0$ to determine τ_0 gives

$$\tau_0 = \frac{T}{\omega_c T - 1} \ln \frac{2\omega_c T}{\omega_c T + 1} \tag{4.103}$$

and from (4.94), (4.102) and (4.103), we obtain

$$\text{FF} = 2^{\frac{1}{1 - \omega_c T}} \left(\frac{\omega_c T}{1 + \omega_c T} \right)^{\frac{1 + \omega_c T}{2(1 - \omega_c T)}} \tag{4.104}$$

As expected, pulses with shorter duration compared with the time constant of the filter are more severely distorted—in fact, $\text{FF} \simeq 2\sqrt{\omega_c T}$ if $\omega_c T \ll 1$. In the opposite limit, $\text{FF} \to 1$ as $\omega_c T \to \infty$. When $\omega_c T = 1$, we must evaluate (4.103) and (4.104) by taking the limit, and we find that $\tau_0 = 0.5T$ and $\text{FF} = \sqrt{2/e} \simeq 0.858$.

4.7   WAVEFRONT DISTORTION

Modern communication systems make frequent use of information encoded in digital form, a series of rectangular (or nearly so) pulses in the time domain. Digital signals are inherently wideband, as very high frequency constituents are needed to reproduce the fast rise and fall times of such pulses. It is thus important to be able to predict the distortion of step-like functions of time when passed through a linear system.

For this purpose, it is more suitable to use the Laplace transform than the Fourier transform to analyze transient behavior. Let us assume that all functions of time to be considered will be identically zero for times $t < 0$. In other words, all functions will contain the *Heaviside unit step function* (4.97). A rectangular voltage pulse of strength V_0 and duration T, beginning at $t = 0$, can be written as

$$v(t) = V_0 \left[\theta(t) - \theta(t - T) \right] \tag{4.105}$$

Other digital pulses can be expressed in similar fashion.

The Laplace transform $F(s)$ of a function $f(t)$ is defined as

$$F(s) = \int_0^\infty f(t) e^{-st} \, dt \tag{4.106}$$

We see that it is similar to the Fourier transform (4.2) for functions which vanish when $t < 0$, if we make the replacement of the real frequency ω by a complex frequency $s = j\omega$. If s is permitted to have a positive real part, the integral in (4.106) will converge even for functions f which do not decay to zero as $t \to \infty$. Although there is an inversion formula for the Laplace transform analogous to (4.1) to reconstruct $f(t)$ from a knowledge of $F(s)$, it requires integration in the complex s-plane and as such is beyond the scope of this text. We will be content to reconstruct time-domain functions by reference to a table of Laplace transforms, of which a few of the most relevant to our needs are given in Appendix G.

One approximate technique for calculating the distortion of a step function input to a linear system whose transfer function is $H(s)$ can be derived using Heaviside's first expansion theorem, which is given in Section G.2.1 of Appendix G. As a simple example, consider a unit step function $f(t) = \theta(t)$ which is a voltage applied across a series combination of a resistor R and a capacitor C. The output $w(t)$ is to be the voltage across the capacitor. Elementary circuit analysis shows that the transfer function will be

$$H(s) = \frac{1}{1 + sRC} \tag{4.107}$$

Then

$$W(s) = H(s)F(s) = \frac{1}{s}H(s) = \frac{1}{RCs^2} - \frac{1}{(RC)^2 s^3} + \cdots \tag{4.108}$$

Comparing to the expansion (G.9), we can identify the expansion coefficients for W as $W_0 = W_1 = 0$, and

$$W_i = (-1)^i \frac{1}{(RC)^{i-1}} \qquad (i \geq 2) \tag{4.109}$$

Using (G.10), we obtain

$$w(t) = \theta(t) \sum_{i=1}^{\infty} (-1)^{i+1} \left(\frac{1}{RC}\right)^i \frac{t^i}{i!} = \theta(t)\left[1 - e^{-t/RC}\right] \tag{4.110}$$

which is, of course, a well-known result showing how the initially sharp input step pulse is distorted by the network.

The point to be recognized here is that even in more complicated examples where the series (G.10) cannot be identified in simple closed form, we may compute a finite number of the initial terms of the series and use them to approximate the distortion of the initial part of the pulse for small values of t (the so-called *wavefront*). In systems containing transmission lines or other delay elements, the transfer function may have a large s expansion more complicated than (G.9), containing an additional exponential factor:

$$H(s) = e^{-sT_f}\left[H_{f0} + \frac{H_{f1}}{s} + \frac{H_{f2}}{s^2} + \cdots\right] \tag{4.111}$$

where the constant T_f is the limiting value of the phase and group delay times as complex frequency approaches infinity. When the input is again a unit step function, the transform of the output is

$$W(s) = e^{-sT_f}\left[\frac{H_{f0}}{s} + \frac{H_{f1}}{s^2} + \frac{H_{f2}}{s^3} + \cdots\right] \tag{4.112}$$

and from (G.10) and the time-shift rule from Table G.3, we get

$$w(t) = \theta(t - T_f) \sum_{i=0}^{\infty} H_{fi} \frac{(t - T_f)^i}{i!} \tag{4.113}$$

which expresses the distortion of the wavefront after it is delayed by T_f. The initial output response is often quite small relative to its value at later times. This initial part of the response is called a *forerunner* or *precursor* to distinguish it from the main part of the output signal. Under certain conditions the forerunner signal can be comparable in magnitude to the main signal, in which case the differences in time delay can once again lead to undesirable interference in the digital information and bit error conditions. The result is a reduction of attainable bit transmission rates in the system.

4.8   MOMENT MATCHING AND AWE

While the wavefront (early-time) expansion (4.113) is accurate in predicting the initial shape of the output pulse, as time increases it is necessary to include a larger and larger number of terms in the expansion in order to maintain this accuracy. This makes (4.113) an inefficient means of forecasting the distortion of the entire pulse. The trouble is that while the transfer function has been well approximated for large values of s, no provision has been made for accuracy for small s, where information about the pulse shape at later times is contained.

To handle this aspect of pulse behavior, let $h(t)$ be the *impulse response* of the system [that is, the output $w(t)$ when the input $f(t) = \delta(t)$ is a delta function]. Then we write $H(s)$ as the Laplace transform of $h(t)$, and expand the factor e^{-st} as a power series in s to obtain the so-called *moment expansion* of $H(s)$:

$$
\begin{aligned}
H(s) &= \int_0^\infty h(t)e^{-st}\,dt \\
&= \sum_{n=0}^\infty \frac{(-s)^n}{n!} \int_0^\infty h(t)t^n\,dt \\
&= \sum_{n=0}^\infty m_n s^n
\end{aligned}
\tag{4.114}
$$

where

$$
m_n = \frac{(-1)^n}{n!} \int_0^\infty h(t)t^n\,dt
\tag{4.115}
$$

are called the moments of $h(t)$. Equation (4.114) is an expansion of $H(s)$ about $s = 0$, complementing the expansion (4.111) about $s = \infty$.

An approximation for the output of the system valid for late times t is obtained if we choose an approximation $H_{ap}(s)$ to $H(s)$ which matches that of $H(s)$ itself as $s \to 0$, and use it to compute $W(s)$ and thus $w(t)$. To be useful, such an approximation should have an easily computed inverse Laplace transform. We choose the form

$$
H_{ap}(s) = e^{-sT_f} \left[H_{f0} + \sum_{i=1}^q \frac{R_i}{s - p_i} \right]
\tag{4.116}
$$

where T_f and H_{f0} have the same meaning as in (4.111), while the p_i are a set of poles in the left half of the complex s-plane, and R_i are the residues of these poles. It is straightforward to see that the approximate impulse response corresponding to (4.116) will be

$$
h_{ap}(t) = H_{f0}\delta(t - T_f) + \theta(t - T_f) \sum_{i=1}^q R_i e^{p_i(t - T_f)}
\tag{4.117}
$$

The approximate response $w_{ap}(t)$ to a unit step function input $f(t) = \theta(t)$ will similarly be

$$
w_{ap}(t) = \theta(t - T_f) \left[\left(H_{f0} - \sum_{i=1}^q \frac{R_i}{p_i} \right) + \sum_{i=1}^q \frac{R_i}{p_i} e^{p_i(t - T_f)} \right]
\tag{4.118}
$$

We determine the $2q$ unknown constants R_i and p_i in (4.116) by expanding $H_{ap}(s)$ about $s = 0$ and $s = \infty$. We first obtain l equations by matching the terms proportional to s^{-1}, $s^{-2}, \ldots s^{-l}$ as $s \to \infty$. Since the behavior of the s^0 term is already matched, we get no new information from it. If $l = 0$, we do not match any terms in negative powers of s. As $s \to \infty$,

$$
H_{ap}(s) = e^{-sT_f} \left[H_{f0} + \sum_{n=0}^\infty \frac{1}{s^{n+1}} \sum_{i=1}^q R_i p_i^n \right]
\tag{4.119}
$$

so our possible matching conditions there are:

$$
s^{-1}: \qquad H_{f1} = \sum_{i=1}^q R_i
\tag{4.120}
$$

$$s^{-2}: \qquad H_{f2} = \sum_{i=1}^{q} R_i p_i \qquad (4.121)$$

and so on up to the terms in s^{-l}.

We obtain the remaining $(2q - l)$ conditions necessary to determine all the constants by matching the terms proportional to s^0, s^1, s^2, ... s^{2q-l-1} as $s \to 0$. In this limit (remembering that we must also expand the term e^{-sT_f} in a power series in s),

$$H_{ap}(s) = \left(H_{f0} - \sum_{i=1}^{q} \frac{R_i}{p_i} \right) + s \left[-T_f \left(H_{f0} - \sum_{i=1}^{q} \frac{R_i}{p_i} \right) - \sum_{i=1}^{q} \frac{R_i}{p_i^2} \right] + \cdots \qquad (4.122)$$

so our possible matching conditions are:

$$s^0: \qquad m_0 = H_{f0} - \sum_{i=1}^{q} \frac{R_i}{p_i} \qquad (4.123)$$

$$s^1: \qquad m_1 = -T_f m_0 - \sum_{i=1}^{q} \frac{R_i}{p_i^2} \qquad (4.124)$$

and so on up to the terms in s^{2q-l-1}.

This technique (especially when we choose $l = 0$), is called *asymptotic waveform evaluation* (AWE), and has been widely applied to simplify the calculation of time responses of complicated systems. In the simplest situation ($q = 1$), we may solve these equations explicitly (this is left as an exercise). If $q = 2$, we already have to do some clever algebra to get the explicit solution, and for higher orders of approximation these equations must be solved numerically to obtain the approximate response. We will outline here the explicit calculation for the case $q = l = 2$ to show what is involved. Let us write (4.120)-(4.121) in the matrix form

$$\begin{bmatrix} 1 & 1 \\ p_1 & p_2 \end{bmatrix} [R] = [H_f] \qquad (4.125)$$

where

$$[R] = \begin{bmatrix} R_1 \\ R_2 \end{bmatrix}; \qquad [H_f] = \begin{bmatrix} H_{f1} \\ H_{f2} \end{bmatrix}$$

Likewise, we express (4.123)-(4.124) as

$$\begin{bmatrix} \frac{1}{p_1} & \frac{1}{p_2} \\ \frac{1}{p_1^2} & \frac{1}{p_2^2} \end{bmatrix} [R] = [M] \qquad (4.126)$$

where

$$[M] = \begin{bmatrix} H_{f0} - m_0 \\ -m_0 T_f - m_1 \end{bmatrix} \equiv \begin{bmatrix} M_1 \\ M_2 \end{bmatrix}$$

Solving (4.126) for $[R]$ and substituting into (4.125) we get

$$\begin{bmatrix} p_1 + p_2 & -p_1 p_2 \\ p_1^2 + p_1 p_2 + p_2^2 & -p_1 p_2 (p_1 + p_2) \end{bmatrix} [M] = [H_f] \tag{4.127}$$

Introducing the new unknown variables $r_1 = p_1 + p_2$ and $r_2 = p_1 p_2$, we can after a little algebra solve the system of equations (4.127) as

$$r_1 = \frac{M_2 H_{f2} - M_1 H_{f1}}{M_2 H_{f1} - M_1^2}; \qquad r_2 = \frac{M_1 H_{f2} - H_{f1}^2}{M_2 H_{f1} - M_1^2} \tag{4.128}$$

The poles are then obtained as solutions of a quadratic equation:

$$p_{1,2} = \frac{r_1 \pm \sqrt{r_1^2 - 4r_2}}{2} \tag{4.129}$$

and finally the residues are obtained from the solution of (4.125):

$$R_1 = \frac{H_{f2} - p_2 H_{f1}}{p_1 - p_2}; \qquad R_2 = \frac{p_1 H_{f1} - H_{f2}}{p_1 - p_2} \tag{4.130}$$

4.8.1 EXAMPLE

As an example, consider the transfer function

$$H(s) = e^{-T_f \sqrt{s^2 + \omega_c^2}} \tag{4.131}$$

with an input taken to be the rectangular pulse given by (4.105). This transfer function can represent, for example, the effect of a transmission line with a low-frequency cutoff of ω_c and high-frequency time delay T_f. The exact step function response of this system can be found from entries 4 and 8 of Table G.4 in Appendix G:

$$w_{\text{step}}(t) = \theta(t - T_f) \left[1 - \omega_c T_f \int_{T_f}^{t} \frac{J_1 \left(\omega_c \sqrt{u^2 - T_f^2} \right)}{\sqrt{u^2 - T_f^2}} \, du \right] \tag{4.132}$$

where J_1 is a Bessel function (see Appendix C). The response to the rectangular input pulse (4.105) is then

$$w(t) = V_0 \left[w_{\text{step}}(t) - w_{\text{step}}(t - T) \right] \tag{4.133}$$

and this is plotted in Figure 4.7. Note how there is considerable "sag" and some ringing in the output pulse after the initial arrival at $t = T_f$. Such deviations from ideal pulse shapes can lead to false triggering of logic gates, or failure to trigger when expected, increasing the bit error rate of digital systems. The first- ($q = 1$, $l = 1$) and second-order ($q = 2$, $l = 2$) AWE approximations to the output response are also shown in Figure 4.7. We see that the second-order AWE gives a certain reproduction of the ringing, but only for the first minimum, while the first-order approximation does not reproduce the ringing at all. It is therefore important to use an AWE of sufficiently high order if accurate results are desired.

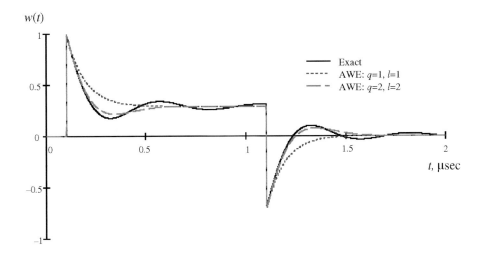

Figure 4.7 Response of the linear system (4.131) to a rectangular step input (4.105): $V_0 = 1$, $\omega_c = 4\pi \times 10^6$ sec^{-1}, $T_f = 10^{-7}$ sec, $T = 10^{-6}$ sec.

4.9 NOTES AND REFERENCES

General treatments of modulation and detection are to be found in

Rowe (1965), Chapters 3 and 4;
Thomas (1969), Chapter 7.
Stremler (1990), Chapters 5 and 6.

Relatively few books treat pulse propagation in much detail. See, for example,

Jackson (1975), Chapter 7;
Javid and Brown (1963), Chapter 13.

A nice overview, and English translations of the classic papers of Sommerfeld and Brillouin on the subject are in:

Brillouin (1960).

This book deals mainly with a detailed study of the precursor problem, as do the following (but for propagation channels with various dispersion laws):

Carslaw and Jaeger (1963), Chapter 9;
Kuznetsov and Stratonovich (1964).

A more detailed and precise treatment of the Sommerfeld-Brillouin problem can be found in

Oughstun and Sherman (1994).

Detailed comparison of precursors to group-delayed main signals is made in

H. Hosono and T. Hosono, "Superluminal group velocities in passive media," *IEICE Trans. Electron.*, vol. E87-C, pp. 1578-1585 (2004).

General treatments of pulse dispersion can be found in:

Vainshtein and Vakman (1983);
Vakman (1998);
Agrawal (2001), Chapter 3.

as well as the papers:

L. A. Vainshtein, "Propagation of pulses," *Sov. Phys. Uspekhi*, vol. 19, pp. 189-205 (1976).

D. E. Vakman and L. A. Vainshtein, "Amplitude, phase, frequency—fundamental concepts of oscillation theory," *Sov. Phys. Uspekhi*, vol. 20, pp. 1002-1016 (1977).

For more information on signal analysis in general and analytic signals in particular, see:

Franks (1969), Section 4.4;
Vainshtein and Vakman (1983);
Cohen (1995), Chapters 1-4;
Vakman (1998).

The quasi-stationary approximation (of which there are many variations) is usually attributed to Carson and Fry in their treatment of FM distortion, but was actually anticipated in a slightly different form over 50 years earlier by Gouy. Many subsequent developments have been made, including to AM signals, and we list only a small sample of the literature here:

L. G. Gouy, "Sur la propagation des ondes lumineuses, eu égard à la dispersion," *J. Math. Pures Appl.*, ser. 3, vol. 8, pp. 335-356 (1882).

J. R. Carson and T. C. Fry, "Variable frequency electric circuit theory with application to the theory of frequency-modulation," *Bell Syst. Tech. J.*, vol. 16, pp. 513-540 (1937).

F. L. H. M. Stumpers, "Distortion of frequency-modulated signals in electrical networks," *Communication News*, vol. 9, pp. 82-92 (1948).

F. Zweig, P. M. Schultheiss and C. A. Wogrin, "On the response of linear systems to signals modulated in both amplitude and frequency," *IRE Trans. Circ. Theory*, vol. 2, pp. 367-369 (1955).

J W. Head and C. G. Mayo, "The response of a network to a frequency-modulated input voltage," *Proc. IEE (London)*, part C, vol. 105, pp. 509-512 (1958).

Baghdady (1961), Chapter 19.

M. L. Liou, "Noise in an FM system due to an imperfect linear transducer," *Bell Syst. Tech. J.*, vol. 45, pp. 1537-1561 (1966).

D. Anderson, J. Askne and M. Lisak, "Wave packets in an absorptive and strongly dispersive medium," *Phys. Rev. A*, vol. 12, pp. 1546-1552 (1975).

A collection of the classic papers on distortion of FM signals is:

Klapper (1970).

Our quasistationary approximation is based on the paper by Zweig *et al.* (1955).

The quasi-Gaussian approximation is an extension of a method of treating the broadening of a Gaussian pulse using a second-order Taylor series expansion of the propagation constant. This method has been treated in many papers, of which a sample follows:

C. G. B. Garrett and D. E. McCumber, "Propagation of a Gaussian light pulse through an anomalous dispersion medium," *Phys. Rev. A*, vol. 1, pp. 305-313 (1970).

F. P. Kapron and D. B. Keck, "Pulse transmission through a dielectric optical waveguide," *Applied Optics*, vol. 10, pp. 1519-1523 (1971).

E. F. Kuester and D. C. Chang, "Single-mode pulse dispersion in optical waveguides," *IEEE Trans. Micr. Theory Tech.*, vol. 23, pp. 882-887 (1975).

D. Marcuse, "Pulse distortion in single-mode fibers," *Applied Optics*, vol. 19, pp. 1653-1660 (1980).

Third-order dispersion terms can be accounted for using Airy functions:

J. R. Wait, "Distortion of pulsed signals when the group delay is a nonlinear function of frequency," *Proc. IEEE*, vol. 58, pp. 1292-1294 (1970).

The analysis of the evolution of arbitrary pulses in terms of their moments is due to:

S. H. Moss, "Pulse distortion," *Proc. IEE (London), part IV*, vol. 98, pp. 37-42 (1951).

L. C. Baird, "Moments of a wave packet," *Amer. J. Phys.*, vol. 40, pp. 327-329 (1972).

D. G. Anderson and J. I. H. Askne, "Wave packets in strongly dispersive media," *Proc. IEEE*, vol. 62, pp. 1518-1523 (1974).

D. Anderson, J. Askne and M. Lisak, " The velocity of wave packets in dispersive and slightly absorptive media," *Proc. IEEE*, vol. 63, pp. 715-717 (1975).

H. M. Bradford, "Propagation and spreading of a pulse or wave packet," *Amer. J. Phys.*, vol. 44, pp. 1058-1063 (1976).

H. M. Bradford, "Propagation of a step in the amplitude or envelope of a pulse or wave packet," *Amer. J. Phys.*, vol. 47, pp. 688-694 (1979).

G. Bonnet, "Au delà d'une vitesse de groupe: vitesse d'onde et vitesse de signal," *Ann. Télécommun.*, vol. 38, pp. 345-366 and 471-487 (1983).

D. E. Vakman, "Evolution of the parameters of a pulse propagating with dispersion and attenuation," *Sov. J. Commun. Technol. Electron.*, vol. 31, no. 7, pp. 100-105 (1986).

D. Anderson and M. Lisak, "Propagation characteristics of frequency-chirped super-Gaussian optical pulses," *Opt. Lett.*, vol. 11, pp. 569-571 (1986).

D. Anderson and M. Lisak, "Analytic study of pulse broadening in dispersive optical fibers," *Phys. Rev. A*, vol. 35, pp. 184-187 (1987).

Lin Guo-Cheng, Sui Cheng-Hua and Lin Quang, "Non-Gaussian pulse propagation and pulse quality factor using intensity moment method," *Chin. Phys. Lett.*, vol. 16, pp. 415-417 (1999).

whose treatment we have followed. Various definitions of pulse width and rise time are considered in

Brown (1963), Chapter 5.

A good introduction to multimode pulse distortion can be found in:

D. Gloge, "Dispersion in weakly guiding fibers," *Applied Optics*, vol. 10, pp. 2442-2445 (1971).

D. Gloge, A. R. Tynes, M. A. Duguay and J. W. Hansen, "Picosecond pulse distortion in optical fibers," *IEEE J. Quantum Electron.*, vol. 8, pp. 217-221 (1972).

Effects of mode coupling and attenuation on multimode pulse propagation are treated in

D. Marcuse, "Pulse propagation in multimode dielectric waveguides," *Bell Syst. Tech. J.*, vol. 51, pp. 1199-1232 (1972).

S. Geckeler, "Pulse broadening in optical fibers with mode mixing," *Appl. Opt.*, vol. 18, pp. 2192-2198 (1979).

An overview of UWB systems is given in:

S. Roy, J. R. Foerster, V. S. Somayazulu and D. G. Leeper, "Ultrawideband radio design: The promise of high-speed, short-range wireless connectivity," *Proc. IEEE*, vol. 92, pp. 295-311 (2004).

The fidelity factor was first introduced in:

D. Lamensdorf and L. Susman, "Baseband-pulse-antenna techniques," *IEEE Ant. Prop. Mag.*, vol. 36, no. 1, pp. 20-30 (1994).

but the distortion factor predates this work by many decades; see

R. H. Direen, E. F. Kuester and M. Elmansouri, "Characterisation of distortion of narrowband signals by linear time-invariant systems," *IET Micr. Ant. Prop.*, vol. 11, pp. 1872-1879 (2017).

and the references therein.

The discussion of wavefronts and precursors is based on:

E. O. Schulz-DuBois, "Sommerfeld pre- and post-cursors in the context of waveguide transients," *IEEE Trans. Micr. Theory Tech.*, vol. 18, pp. 455-460 (1970).

A. Karlsson and S. Rikte, "Time-domain theory of forerunners," *J. Opt. Soc. Amer. A*, vol. 15, pp. 487-502 (1998).

P. M. Jordan and A. Puri, "Digital signal propagation in dispersive media," *J. Appl. Phys.*, vol. 85, pp. 1273-1282 (1999).

For Heaviside's first expansion theorem, see

van der Pol and Bremmer (1987).

The AWE was first introduced in

L. T. Pillage and R. A. Rohrer, "Asymptotic waveform evaluation for timing analysis," *IEEE Trans. Computer-Aided Des.*, vol. 9, pp. 352-366 (1990).

Some of the subsequent developments of the technique can be found in

Chiprout and Nakhla (1994);
Pillage, Rohrer and Visweswariah (1995).

Finally, here are several papers on off-the-beaten-track subjects related to pulse distortion:

A. M. Bel'skii and A. P. Khapalyuk, "Propagation of a space-limited pulse through an isotropic medium," *J. Appl. Spectroscopy*, vol. 17, pp. 947-951 (1972).

I. A. Man'kin, "Waveguide excitation by nonstationary extraneous currents," *Radio Eng. Electron. Phys.*, vol. 24, no. 4, pp. 16-23 (1979).

F. Mainardi, "Energy velocity for hyperbolic dispersive waves," *Wave Motion*, vol. 9, pp. 201-208 (1987).

4.10 PROBLEMS

p4-1 If the AM Gaussian signal described by (4.54) is applied to a system with nonuniform loss or gain [so that $A(\omega)$ is not constant in frequency], find an expression for the distortion of the signal using the first three terms of the quasi-stationary expansion. Compare this result to that of the quasi-Gaussian approximation. How is the distortion of the signal changed by the fact that A_0', $A_0'' \neq 0$?

p4-2 Consider a Gaussian pulse that is frequency-modulated in addition to being amplitude-modulated, with

$$r(t) = \exp\left[-\frac{t^2}{2T^2} + jM(t)\omega_0 t\right]$$

where $M(t) = t/T_m$ is a linear frequency modulation function with a modulation rate $1/T_m$, and T_m is a constant. Using the quasi-Gaussian approximation, calculate the distortion of the amplitude envelope of this pulse due to propagation for a distance z along a lossless transmission line with $\gamma(\omega) = j\beta(\omega)$. Show that for certain values of T_m the pulse amplitude duration (as inferred from the amplitude envelope $|w_A(t)|^2$) may initially get smaller as z increases before eventually increasing as it does for a pure AM Gaussian pulse.

p4-3 A gradual form of the unit step function has the form

$$u(t) = \frac{1}{1 + e^{-2t/T_R}}$$

where the constant T_R is a measure of the rise time of the step. Two such functions can be combined to give an approximately rectangular pulse envelope:

$$r(t) = u\left(t + \frac{T}{2}\right) - u\left(t - \frac{T}{2}\right)$$

where T is the duration of the pulse.

    (a) Under what conditions on T, T_R and the carrier frequency ω_0 will this be a narrowband pulse?

    (b) Use the quasi-stationary expansion (up to the first three terms) to find an expression for the output $w(t)$ of a lossless ($A_0 = 1$, $A_0' = A_0'' = 0$) linear time-invariant network when the input is a narrowband signal with envelope $r(t)$ and carrier frequency ω_0. Plot the square of the magnitude of the analytic signal w_A versus τ_g/T for $\Delta \equiv \Theta_0''/T_R^2 = 0$ (no distortion), 1, 2 and 3. Use the parameter values $\omega_0 T = 20$, $\omega_0 T_R = 10$.

    (c) Repeat part (b) using the quasi-Gaussian approximation (4.67).

p4-4 If an input pulse $f(t)$ is applied to a lossy linear time-invariant network, show that the center of the output pulse $w(t)$ is related to that of the input pulse by

$$t_c(w) = t_c(f) + t_{g,\mathrm{av}}(p)$$

where $p(t)$ is a function whose Fourier transform is $F(\omega)A(\omega)$, F is the Fourier transform of f, and A is the real magnitude of the network transfer function.

p4-5 A capacitor C is connected to the end of a lossless transmission line with characteristic impedance Z_c. Both C and Z_c are real and independent of frequency.

    (a) Obtain an expression for the frequency-dependent reflection coefficient $\rho(\omega)$ of this load.

    (b) Consider the reflection coefficient to be a transfer function (i. e., the input is the incident voltage and the output is the reflected voltage).

Obtain expressions for the phase delay time t_{p0} and group delay time t_{g0} at a carrier frequency of ω_0. If $\omega_0 = 2\pi f_0$, where $f_0 = 100$ MHz, $Z_c = 50\ \Omega$ and $C = 100$ pF, compute these two delay times.

(c) Given the same parameter values as in part (b), what is the smallest possible duration of a reflected Gaussian pulse from this load, based on the quasi-Gaussian approximation?

p4-6 An inductor L in series with a resistor R is connected as a load to the end of a lossless transmission line with characteristic impedance Z_c. Each of L, R and Z_c is real and independent of frequency.

(a) Obtain an expression for the frequency-dependent reflection coefficient $\rho(\omega)$ of this load.

(b) Consider the reflection coefficient to be a transfer function (i. e., the input is the incident voltage and the output is the reflected voltage). Obtain expressions for the phase delay time t_{p0} and group delay time t_{g0} at a carrier frequency of ω_0. If $\omega_0 = 2\pi f_0$, where $f_0 = 1$ GHz, $Z_c = 50\ \Omega$, $R = 10\ \Omega$ and $L = 15$ nH, compute these two delay times.

(c) Given the same parameter values as in part (b), what is the smallest possible duration of a reflected Gaussian pulse from this load, based on the quasi-Gaussian approximation?

p4-7 Use the quasi-Gaussian approximation to study the distortion by a linear network of a "super-Gaussian" pulse whose envelope is

$$e^{-t^4/2T^4}$$

where T is a positive real constant. Assume that $A_0' = A_0'' = 0$. Give expressions for the output signal's phase and power envelope. How does the duration of the output pulse compare to that of the input pulse (take the output pulse duration T_{out} to be that value of t at which $|w_A(t)|^2$ has decreased to $1/e$ of its maximum value)? What is the narrowest pulse of such a form that can emerge from this network?

p4-8 If the spectrum of a narrowband pulse envelope is zero at DC (that is, if $R(0) = 0$), derive a modified version of the quasi-Gaussian approximation that describes the distortion of this pulse in passing through a linear system.

p4-9 An inductor L and a capacitor C connected in parallel are connected as a load to the end of a lossless transmission line with characteristic impedance Z_c. The values of L, C and Z_c are all real and independent of frequency.

(a) Obtain an expression for the frequency-dependent reflection coefficient $\rho(\omega)$ of this load. To simplify the appearance of the formula, abbreviate

$$\omega_r = \frac{1}{\sqrt{LC}}; \qquad Z_r = \sqrt{\frac{L}{C}}$$

(b) Consider the reflection coefficient to be a transfer function (i. e., the input is the incident voltage and the output is the reflected voltage). Obtain expressions for the phase delay time t_{p0} and group delay time t_{g0} at a carrier frequency of ω_0. If $\omega_0 = 2\pi f_0$, where $f_0 = 1$ GHz, $Z_c = 50\ \Omega$, $L = 7.958$ nH and $C = 3.183$ pF, compute these two delay times.

(c) Given the same parameter values as in part (b), what is the smallest possible duration of a reflected AM Gaussian pulse from this load, based on the quasi-Gaussian approximation?

p4-10 Consider a classical transmission line with $g = 0$, and r, l and c all positive and independent of frequency. Assume that $\frac{1}{\sqrt{lc}} = c$ (i. e., that the velocity of propagation on this transmission line in the absence of loss is equal to the speed of light).

    (a) Obtain expressions for $\alpha(\omega)$ and $\beta(\omega)$. Simplify these expressions by using the notation $x = (r/\omega l)^2$.

    (b) Obtain expressions for the phase velocity v_p and the group velocity v_g as functions of x. Show that $v_p < c$ if $x \neq 0$.

    (c) Determine the conditions under which v_g can be greater than c. What is the maximum value that v_g can have?

    (d) Explain why $v_g > c$ as found for certain cases in part (c) does not violate any fundamental physical law (such as relativity). You may wish to compare the values of α and β for the value of x that makes v_g maximum as found in part (c).

p4-11 Consider a classical transmission line whose line parameters l, c, r and g are all independent of frequency. Find the unique condition relating the values of these line parameters which guarantees that both α and v_p will be independent of frequency. Under this condition, find the voltage $v(z, t)$ of a forward wave on the line if the voltage at $z = 0$ is some function $v(0, t) = f(t)$.

p4-12 A pulse transmission system consists of a generator, a destination (load) and a number N of independent transmission lines connected between the generator and the destination. The generator sends out an identical stream of narrowband pulses of center frequency ω_0 onto each transmission line. Each line is of identical length d, but has a different group velocity ($v_{g,i}$ for the ith transmission line) and phase velocity ($v_{p,i}$) at ω_0. The destination is assumed to combine all the arriving voltage pulses by adding them together, without reflection back along the transmission lines. If the pulse stream from the generator is a sequence of rectangular pulses (that is, the envelopes are rectangular) of height V_p, duration ("on time") T_p and time between pulses ("off time") T_o, give an expression for the maximum length d of the transmission lines for which pulses arriving at the destination do not overlap with adjacent ones arriving from any of the connected transmission lines. Assume that the transmission lines are labeled such that line $i = 1$ has the lowest group velocity $v_{g,1}$ of them all, and as i increases, so does the group velocity, until we reach line $i = N$, for which $v_{g,N}$ is the largest of all.

p4-13 The exponential pulse (4.96) is input to a high-pass filter whose transfer function is

$$H(\omega) = \frac{j\omega/\omega_c}{1 + j\omega/\omega_c}$$

Evaluate the function $C(\tau)$ defined by (4.95), and obtain an expression for the fidelity factor valid for any value of $\omega_c T$. Discuss the meaning of your result.

p4-14 For the transfer function (4.131), obtain expressions for the first three terms of the wavefront expansion (4.113) of the output due to an input unit step function. For the parameter values used in Figure 4.7, compute and plot on the same graph the response of the system to the rectangular input pulse (4.105) using

    (a) just the $i = 0$ term,

    (b) the $i = 0$ and $i = 1$ terms, and

    (c) the $i = 0, 1$ and 2 terms of (4.113).

Comment on the accuracy of each of these approximations compared to the exact result, and to the two AWE results shown in Figure 4.7.

p4-15   (a) Obtain explicit expressions for the constants R_1 and p_1 that appear in the first-order ($q = 1$, $l = 1$) AWE approximation (4.118) for the output response of a linear system to a unit step-function input. These expressions should be in terms of the appropriate properties of the transfer function $H(s)$.

(b) Apply the result of part (a) to calculate the response of the linear system whose transfer function is (4.131) to a rectangular input pulse. Use the parameters $V_0 = 1$, $\omega_c = 2\pi \times 10^6$ sec^{-1}, $T_f = 10^{-7}$ sec and $T = 10^{-6}$ sec. Plot your result and compare to the exact and second-order AWE responses as in Figure 4.7.

5 HOLLOW METALLIC WAVEGUIDES

In this chapter, we begin our study of the electromagnetic analysis of the modes of various specific types of waveguide. We start with what were historically the earliest waveguides to find wide application at microwave frequencies: hollow metallic tubes. Hollow rectangular waveguides (cf. Figure 1.1(a)) are still among the most commonly used waveguides for microwave frequencies. Even as high as 60 GHz, they remain the medium of choice for many applications. However, as frequency increases even further, their smaller size can make them hard to fabricate (the accurate machining of the corners and polishing of the metal surfaces being some of the problems). Moreover, other properties of these guides, such as wall losses, etc., are not always ideal for some applications, and hollow waveguides of other cross-sectional shapes are used instead. Although generally used at medium microwave frequencies, hollow metallic waveguides can be employed at frequencies as low as 300 MHz or as high as several Terahertz (1 THz = 10^{12} Hz).

In this chapter, we will consider the general properties of hollow metallic waveguides, with simply-connected (one-piece) boundaries. First we study the simple, though highly idealized, case of a parallel-plate waveguide, which exhibits many of the important properties common to all metallic guides, and then we proceed to consider waveguides of arbitrary cross-section. Two methods for analyzing special cross-sectional shapes will be discussed: separation of variables (which will be used to study the case of the circular cross-section), and the ray method (which will be used to study the properties of rectangular and triangular shapes in this chapter, and extended in Chapter 7 to deal with a wider variety of guiding structures).

5.1 THE PARALLEL-PLATE WAVEGUIDE

Even though a given structure may fit our definition of a waveguide as a structure of infinite length in the z-direction whose material properties and geometry are independent of z, it may not actually guide field energy in this direction unless appropriate physical mechanisms are built into it to confine or localize the waves in the cross-section. This may be done by metallic walls or dielectric boundaries, but in any case, the shape and size of these boundaries will play a critical role in determining the characteristics of the guided modes of the structure. For waveguides of arbitrary structure, we can appeal to Maxwell's equations and associated boundary conditions to answer the question of whether guided waves will exist. Indeed, this is the only rigorous way to address the problem, and the framework for doing so will be outlined in the next section.

On the other hand, there are many important concepts in the theory of guided waves that have their basis in fairly simple ideas from plane wave solutions of Maxwell's equations.[1] Let us illustrate this for the case of perhaps the simplest of all waveguides, the *parallel-plate* waveguide shown in Figure 5.1. This guide consists of two perfectly conducting planes, infinite in the y- and z-directions, located at $x = 0$ and $x = a$, and separated by a homogeneous medium of electrical properties μ and ϵ. Consider the plane wave

$$\mathbf{E} = \mathbf{u}_x a_0 E_0 e^{-jkz}; \quad \mathbf{H} = \mathbf{u}_y a_0 \frac{E_0}{\zeta} e^{-jkz}$$

---

[1] For a brief summary of Maxwell's equations and the properties of their solutions which will be needed in this book, refer to Appendix D.

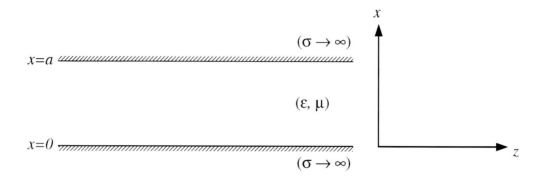

Figure 5.1 The parallel-plate waveguide.

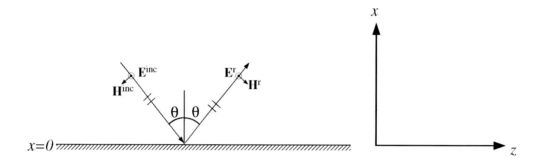

Figure 5.2 Reflection of a plane wave by a perfectly conducting plane (perpendicular polarization).

where a_0 is an arbitrary (constant) wave amplitude and E_0 is a reference field amplitude.[2] This field is not perturbed by the presence of the conducting walls, so long as the surface charge and current densities necessary to terminate the normal $\mathbf{E}$ and tangential $\mathbf{H}$ fields on the walls are present also. Hence, if we identify $\gamma = jk$ and

$$\boldsymbol{\mathcal{E}} = \mathbf{u}_x E_0; \quad \boldsymbol{\mathcal{H}} = \mathbf{u}_y \frac{E_0}{\zeta} \tag{5.1}$$

we have found a source-free field solution for this waveguide of the form:

$$\begin{aligned}
\mathbf{E}(x, y, z) &= a_0 e^{-\gamma z} \boldsymbol{\mathcal{E}}(x, y) \\
\mathbf{H}(x, y, z) &= a_0 e^{-\gamma z} \boldsymbol{\mathcal{H}}(x, y)
\end{aligned} \tag{5.2}$$

A field of such a form is called a *mode* of the waveguide.

Plane waves propagating in directions other than the z-direction will be perturbed by the presence of the walls: what happens, of course, is that they are reflected. For example, consider the effect of an obliquely incident plane wave whose electric field is polarized perpendicular to the plane of incidence as shown in Figure 5.2. For an incidence angle of θ with respect to the normal to the wall, let

$$\mathbf{E}^{inc} = \mathbf{u}_y a^E E_1 e^{-jk(-x \cos \theta + z \sin \theta)} \tag{5.3}$$

---

[2]Chosen, for example, so that when $a_0 = 1$, unit power is carried by the wave.

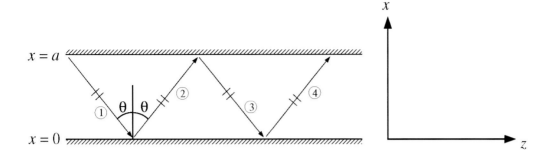

Figure 5.3 Successive ray reflections in a parallel-plate waveguide.

where again E_1 is some constant, eventually to be chosen so that a^E is a wave amplitude. In order to satisfy the boundary condition $E_y = 0$ at the wall, a second (reflected) plane wave must be present, propagating *away* from the wall at the same angle θ to the normal as shown in Figure 5.2. The resulting total electric field is

$$\mathbf{E} = \mathbf{u}_y a^E \left\{ E_1 e^{-jk(-x\cos\theta + z\sin\theta)} + E_2 e^{-jk(x\cos\theta + z\sin\theta)} \right\} \tag{5.4}$$

Of course, we know that the reflection coefficient at the wall is -1, so that $E_2 = -E_1$, but let us not use that knowledge for the time being. A wave incident on the upper wall of the guide similarly produces a reflected wave propagating downward. Accounting for all possible reflections of these plane waves as in Figure 5.3, we see that all of the wave vectors of these plane waves must be in one of two directions, $\mathbf{k} = \mathbf{k}_1 = k(\mathbf{u}_x \cos\theta + \mathbf{u}_z \sin\theta)$ or $\mathbf{k} = \mathbf{k}_2 = k(-\mathbf{u}_x \cos\theta + \mathbf{u}_z \sin\theta)$, corresponding to the waves labeled 1 or 2 in Figure 5.3. Thus, the general form of a field with this polarization must be that of (5.4) above (waves 3 and 4 of Figure 5.3 have the same form as waves 1 and 2 respectively, and do not need to be included since an appropriate adjustment of the values of E_1 and E_2 could be used to account for them, and we assume this has been done). We will, though somewhat inaccurately, refer to the directions of propagation of these plane waves as *rays*. However, it should always be kept in mind that the field is not localized to the schematized arrows representing these rays in the figures, but is rather made up of plane waves which extend to infinity in the y and z directions and are uniform in any direction perpendicular to their directions of propagation.

The allowed values of the angle θ are found by enforcing the boundary conditions $E_y = 0$ at $x = 0$ and $x = a$. From $x = 0$ we get $E_1 = -E_2$, while at $x = a$ we find

$$E_1 e^{jka\cos\theta} = -E_2 e^{-jka\cos\theta}$$

Thus, the two conditions together imply that

$$\cos\theta = \frac{m\pi}{ka}; \quad \sin\theta = \left[1 - \left(\frac{m\pi}{ka}\right)^2 \right]^{1/2} \tag{5.5}$$

and

$$\mathbf{E} = \mathbf{u}_y a_m^E (2jE_1) \sin\frac{m\pi x}{a} e^{-j[k^2 - (\frac{m\pi}{a})^2]^{1/2} z} \tag{5.6}$$

for $m = 1, 2, 3, \ldots$. By identifying

$$\gamma_m = jk\sin\theta = j[k^2 - (\frac{m\pi}{a})^2]^{1/2} \tag{5.7}$$

and

$$\mathcal{E} = \mathbf{u}_y(2jE_1)\sin\frac{m\pi x}{a} \qquad (5.8)$$

we have clearly obtained another set of modes for this waveguide.

So long as the wavenumber k is greater than the "cutoff" wavenumber $k_{c,m} = m\pi/a$, we can interpret the field of such a mode as the result of plane waves bouncing back and forth between the plates at the angle $\theta = \cos^{-1}(m\pi/ka)$. Below cutoff ($k < k_c$), the angle θ becomes complex, and we must interpret (5.6) as two inhomogeneous plane waves, attenuating with the coordinate z as $\exp\left\{-\left[(m\pi/a)^2 - k^2\right]^{1/2}z\right\}$. There is thus a critical or cutoff frequency $\omega_{c,m} = k_{c,m}/\sqrt{\mu\epsilon}$ such that when $\omega > \omega_{c,m}$, the TE$_m$ mode propagates ($\gamma_m = j\beta_m$), while when $\omega < \omega_{c,m}$, the mode attenuates in the z-direction ($\gamma_m = \alpha_m$). The cutoff phenomenon is common to modes of many different waveguide structures.

We can construct another set of modes for this structure by considering parallel-polarized plane waves. It is most convenient in this case to deal with the magnetic field, which for a wave incident at an angle θ is

$$\mathbf{H}^{inc} = \mathbf{u}_y a^H H_1 e^{-jk(-x\cos\theta + z\sin\theta)} \qquad (5.9)$$

where H_1 is the amplitude of the magnetic field and a^H is a wave amplitude. The requirement that the tangential electric field should vanish at the perfectly conducting boundary is equivalent to requiring that $\partial H_y/\partial x$ be zero there. Following basically the same analysis as in the previous case, we find another set of allowable mode fields with

$$\mathbf{H} = \mathbf{u}_y a_m^H(2H_1)\cos\frac{m\pi x}{a}e^{-j[k^2-(\frac{m\pi}{a})^2]^{1/2}z}; \qquad m = 0, 1, 2, ... \qquad (5.10)$$

which are yet a further set of modes of the structure. Note that the $m = 0$ mode of (5.10) is, in fact, none other than the special mode (5.1) obtained above. The expression for γ_m is the same as in (5.7), while the field distribution is different, viz.:

$$\mathcal{H} = \mathbf{u}_y(2H_1)\cos\frac{m\pi x}{a} \qquad (5.11)$$

We have displayed here the forward modes of this waveguide; a similar procedure can be used to obtain the reverse modes. A number of other observations can be made about the properties of parallel-plate waveguide modes:

(i) Whether a mode propagates with no attenuation ($\gamma = j\beta$) or decays exponentially ($\gamma = \alpha$) depends upon the frequency ω, the plate separation a, and the mode number m. For a given frequency (and hence wavenumber k), there will only be a finite number of propagating modes.

(ii) The factor $e^{-\gamma_m z}$ which determines the change in the magnitude or phase of the mode field with z is different for each mode of the waveguide. If a spatial field distribution resulting from the superposition of several modes is present at $z = 0$, this distribution is distorted as it propagates to cross-sections at other values of z.

(iii) For a propagating mode of this structure, since $\beta \leq k$, the phase velocity v_p given by equation (1.11) will always be greater than or equal to the speed of light in the medium filling the guide. The group velocity v_g is always less than or equal to that of a plane wave in the filling medium.

(iv) The phase velocity v_p varies with ω and hence differs from the group velocity v_g for all but the TEM mode. For the other modes, a signal containing more than one frequency component will be dispersed as it propagates along the guide, as discussed in the previous chapter.

The similarities between each of the modes of the parallel-plate waveguide and those of the classical transmission line are striking. In both cases, an amplitude constant a_m is present which cannot be determined without information about the sources. Moreover, the propagation factor of $\exp(-\gamma z)$ is present, and forward and reverse modes exist in both cases. This close parallel between the classical transmission line and the modes of a waveguide will be exploited later on in Chapters 10-11 to describe waveguide excitation and waveguide networks in terms of equivalent transmission-line circuits.

5.2   WAVEGUIDE MODES IN GENERAL

5.2.1   MODES

Outside of a region containing impressed sources, i.e., any place where the current density $\mathbf{J}_{\text{ext}}$ is zero, a general waveguide can be expected to support mode fields, which vary with the axial coordinate z and time t as $\exp(j\omega t - \gamma z)$, and have the form (5.2) analogous to the mode solutions (1.7) for a classical transmission line. For a given value of ω, not all values of γ will correspond to nontrivial field solutions. For example, only the values given by (1.9) yield nonzero mode voltages and currents on a classical transmission line. Moreover, as we have seen, the modes of a parallel-plate waveguide have only the allowed values of propagation coefficient:

$$\gamma = \pm\gamma_m = \pm\sqrt{\left(\frac{m\pi}{a}\right)^2 - \omega^2\mu\epsilon}$$

where m is a nonnegative integer and μ and ϵ are the permeability and permittivity of the medium filling the guide. The values γ_m are known as *characteristic values* or *eigenvalues* of the modes of this waveguide. This case illustrates a common situation which occurs when the waveguide is completely made up of lossless media: the allowed values of $\gamma = \alpha + j\beta$ are either pure imaginary (corresponding to the so-called *propagating* modes of the guide) or pure real numbers (corresponding to *evanescent* modes).

The modal field, while it may differ in magnitude and phase, will not change its characteristic distribution $\mathcal{E}(x,y)$ or $\mathcal{H}(x,y)$ from one cross-section to another. The field distributions $\mathcal{E}$ and $\mathcal{H}$ are *vector* functions of the cross-sectional coordinates (x,y) or (ρ,ϕ). Note that this does *not* necessarily mean that $\mathcal{E}_z = 0$ or $\mathcal{H}_z = 0$; it only means that none of the components of $\mathcal{E}$ or $\mathcal{H}$ depends on z.

It may happen (especially in simple geometries) that for certain modes $\mathcal{E}_z$ or $\mathcal{H}_z$ will vanish identically. It has become customary to use the following terminology to classify modes:

1. $\mathcal{E}_z = 0$, $\mathcal{H}_z = 0$: TEM (transverse electromagnetic) mode.
2. $\mathcal{E}_z = 0$, $\mathcal{H}_z \neq 0$: TE (transverse electric) mode; sometimes called an H-mode.
3. $\mathcal{E}_z \neq 0$, $\mathcal{H}_z = 0$: TM (transverse magnetic) mode; sometimes called an E-mode.
4. $\mathcal{E}_z \neq 0$, $\mathcal{H}_z \neq 0$: Hybrid mode; the terminologies HE-mode and EH-mode are sometimes used; many other terms are also used in specific contexts.

In general, a waveguide mode will be hybrid; TEM, TE and TM modes exist only under certain conditions. We have already seen that the parallel-plate waveguide can support TEM (eqn. (5.1)), TE (5.8) and TM (5.11) modes. We will see in Chapter 5 that any metallic waveguide filled with a homogeneous medium supports only TE and TM modes if its boundary wall is simply-connected (i.e., is all of one piece). TEM modes can exist on multi-conductor structures such as those of Figures 1.1(b) and (c). In Chapters 6, 7 and 8, we will study structures which support hybrid modes, such as the microstrip transmission line and the dielectric optical waveguide.

5.2.2 REDUCED MAXWELL EQUATIONS FOR MODES

Mode fields must satisfy the source-free version of Maxwell's equations (D.3), namely:

$$
\begin{aligned}
\nabla \times \mathbf{E} &= -j\omega\mu\mathbf{H} \\
\nabla \times \mathbf{H} &= j\omega\epsilon\mathbf{E}
\end{aligned}
\tag{5.12}
$$

Substituting the form (5.2) of a modal field into (5.12) gives us the following general equations for $\mathcal{E}$ and $\mathcal{H}$ in Cartesian coordinates:

$$
\begin{vmatrix}
\mathbf{u}_x & \mathbf{u}_y & \mathbf{u}_z \\
\dfrac{\partial}{\partial x} & \dfrac{\partial}{\partial y} & -\gamma \\
\mathcal{E}_x & \mathcal{E}_y & \mathcal{E}_z
\end{vmatrix}
= -j\omega\mu\mathcal{H}
$$

$$
\begin{vmatrix}
\mathbf{u}_x & \mathbf{u}_y & \mathbf{u}_z \\
\dfrac{\partial}{\partial x} & \dfrac{\partial}{\partial y} & -\gamma \\
\mathcal{H}_x & \mathcal{H}_y & \mathcal{H}_z
\end{vmatrix}
= j\omega\epsilon\mathcal{E}
\tag{5.13}
$$

We can split the left side of each of equations (5.13) into two terms, introducing a convenient shorthand in the process:

$$
\begin{vmatrix}
\mathbf{u}_x & \mathbf{u}_y & \mathbf{u}_z \\
\dfrac{\partial}{\partial x} & \dfrac{\partial}{\partial y} & -\gamma \\
\mathcal{E}_x & \mathcal{E}_y & \mathcal{E}_z
\end{vmatrix}
=
\begin{vmatrix}
\mathbf{u}_x & \mathbf{u}_y & \mathbf{u}_z \\
\dfrac{\partial}{\partial x} & \dfrac{\partial}{\partial y} & 0 \\
\mathcal{E}_x & \mathcal{E}_y & \mathcal{E}_z
\end{vmatrix}
+
\begin{vmatrix}
\mathbf{u}_x & \mathbf{u}_y & \mathbf{u}_z \\
0 & 0 & -\gamma \\
\mathcal{E}_x & \mathcal{E}_y & \mathcal{E}_z
\end{vmatrix}
= \nabla_T \times \mathcal{E} - \gamma\mathbf{u}_z \times \mathcal{E}
\tag{5.14}
$$

where we have introduced the subscript "T" to denote the transverse (xy or $\rho\phi$) part of a vector or operator. Thus, for instance,

$$
\begin{aligned}
\mathcal{E}_T &= \mathbf{u}_x\mathcal{E}_x + \mathbf{u}_y\mathcal{E}_y \\
&= \mathbf{u}_\rho\mathcal{E}_\rho + \mathbf{u}_\phi\mathcal{E}_\phi \\
\nabla_T\Phi &= \mathbf{u}_x\frac{\partial\Phi}{\partial x} + \mathbf{u}_y\frac{\partial\Phi}{\partial y} \\
&= \mathbf{u}_\rho\frac{\partial\Phi}{\partial\rho} + \mathbf{u}_\phi\frac{1}{\rho}\frac{\partial\Phi}{\partial\phi} \\
\nabla_T \cdot \mathbf{A} &= \frac{1}{\rho}\frac{\partial}{\partial\rho}(\rho A_\rho) + \frac{1}{\rho}\frac{\partial A_\phi}{\partial\phi} \\
&= \frac{\partial A_x}{\partial x} + \frac{\partial A_y}{\partial y} \\
\nabla_T \times \mathbf{A} &= \mathbf{u}_x\frac{\partial A_z}{\partial y} - \mathbf{u}_y\frac{\partial A_z}{\partial x} + \mathbf{u}_z\left(\frac{\partial A_y}{\partial x} - \frac{\partial A_x}{\partial y}\right) \\
&= \mathbf{u}_\rho\frac{1}{\rho}\frac{\partial A_z}{\partial\phi} - \mathbf{u}_\phi\frac{\partial A_z}{\partial\rho} + \mathbf{u}_z\left[\frac{1}{\rho}\frac{\partial}{\partial\rho}(\rho A_\phi) - \frac{1}{\rho}\frac{\partial A_\rho}{\partial\phi}\right]
\end{aligned}
\tag{5.15}
$$

Hence, ∇_T is the ∇ operator with $\frac{\partial}{\partial z}$ set to zero whenever it appears. Note that $\nabla_T \times \mathbf{A}$ *does* have a z-component and *does* involve A_z as well. Using this notation, (5.13) can be written as

$$\nabla_T \times \boldsymbol{\mathcal{E}} - \gamma \mathbf{u}_z \times \boldsymbol{\mathcal{E}} = -j\omega\mu\boldsymbol{\mathcal{H}}$$
$$\nabla_T \times \boldsymbol{\mathcal{H}} - \gamma \mathbf{u}_z \times \boldsymbol{\mathcal{H}} = j\omega\epsilon\boldsymbol{\mathcal{E}} \tag{5.16}$$

Yet another form for these equations is obtained if we realize that the *transverse* part of $\nabla_T \times \mathbf{A}$ can be written

$$(\nabla_T \times \mathbf{A})_T = \nabla_T \times (\mathbf{u}_z A_z) = -\mathbf{u}_z \times \nabla_T A_z$$

We can then *partially* split each of (5.16) into transverse and longitudinal parts:

$$-\mathbf{u}_z \times \nabla_T \mathcal{E}_z - \gamma \mathbf{u}_z \times \boldsymbol{\mathcal{E}}_T = -j\omega\mu\boldsymbol{\mathcal{H}}_T \tag{5.17}$$

$$\nabla_T \times \boldsymbol{\mathcal{E}}_T = -j\omega\mu\mathbf{u}_z\mathcal{H}_z \tag{5.18}$$

$$-\mathbf{u}_z \times \nabla_T \mathcal{H}_z - \gamma \mathbf{u}_z \times \boldsymbol{\mathcal{H}}_T = j\omega\epsilon\boldsymbol{\mathcal{E}}_T \tag{5.19}$$

$$\nabla_T \times \boldsymbol{\mathcal{H}}_T = j\omega\epsilon\mathbf{u}_z\mathcal{E}_z \tag{5.20}$$

Furthermore, if we take the cross product of $\mathbf{u}_z$ with eqn. (5.17), we get

$$\nabla_T\mathcal{E}_z + \gamma\boldsymbol{\mathcal{E}}_T = -j\omega\mu\mathbf{u}_z \times \boldsymbol{\mathcal{H}}_T \tag{5.21}$$

because $\mathbf{u}_z \times (\mathbf{u}_z \times \mathbf{A}_T) = -\mathbf{A}_T$ for any transverse vector $\mathbf{A}_T$. Using this equation and equation (5.19), we can solve for *all* of the transverse fields in terms of $\mathcal{E}_z$ and $\mathcal{H}_z$:

$$\boldsymbol{\mathcal{E}}_T = -\frac{1}{\omega^2\mu\epsilon + \gamma^2}\left[-j\omega\mu\mathbf{u}_z \times \nabla_T\mathcal{H}_z + \gamma\nabla_T\mathcal{E}_z\right] \tag{5.22}$$

$$\boldsymbol{\mathcal{H}}_T = -\frac{1}{\omega^2\mu\epsilon + \gamma^2}\left[j\omega\epsilon\mathbf{u}_z \times \nabla_T\mathcal{E}_z + \gamma\nabla_T\mathcal{H}_z\right] \tag{5.23}$$

In other words, the components $\mathcal{E}_z$ and $\mathcal{H}_z$ are special in that if we know them we can find all other field components from (5.22) and (5.23). It is the special nature of $\mathcal{E}_z$ and $\mathcal{H}_z$ which led to the TEM/TE/TM/Hybrid mode classification scheme outlined earlier in this section.

5.2.3 FORWARD AND REVERSE MODES

In a waveguide made up of linear isotropic material (the situation to which we limit ourselves in this text), there corresponds to each forward mode that propagates (or attenuates) in the $+z$-direction another (reverse) mode that propagates (or attenuates) in the $-z$-direction. Let us adopt the convention that a forward mode will have a propagation coefficient γ_m with either a positive real part or (if $\text{Re}(\gamma_m) = 0$) a positive real complex power flow in the $+z$ direction. Let $\boldsymbol{\mathcal{E}}_m$, $\boldsymbol{\mathcal{H}}_m$ be the fields of such a forward mode, where $m(> 0)$ is an index designating the mode in question (in certain situations, m may stand for a double index, as will be seen for example in the case of TE_{mn} or TM_{mn} modes of a hollow rectangular waveguide).

To each forward mode there is a corresponding reverse mode (with propagation coefficient $\gamma_{-m} = -\gamma_m$) which we will designate with an index $m < 0$. It is clear upon examination of (5.17)–(5.20) that its fields can be found by reversing the signs of the *longitudinal component of* $\mathcal{E}$ and the *transverse components of* $\mathcal{H}$:

$$\begin{aligned} \mathcal{E}_{-mT} &= \mathcal{E}_{mT} \;\; ; \;\; \mathcal{E}_{-mz} = -\mathcal{E}_{mz} \\ \mathcal{H}_{-mT} &= -\mathcal{H}_{mT} \;\; ; \;\; \mathcal{H}_{-mz} = \mathcal{H}_{mz} \end{aligned} \tag{5.24}$$

Of course, we could equally well have reversed the signs of $\mathcal{E}_T$ and $\mathcal{H}_z$, but (5.24) is the commonly used convention which will be followed here.

5.2.4 NORMALIZED MODE FIELDS, POWER AND STORED ENERGY

The fields of a waveguide mode contain an undetermined amplitude constant which can be related to the complex power (or alternatively to the complex oscillatory power) carried by the mode. As with the case of the classical transmission line, it is convenient (when possible) to deal with mode fields that are *normalized* by having their field amplitudes adjusted so that unit power is carried by the mode through its cross section.[3] In this section, we will consider how to deal with this issue for the modes of a waveguide.

Modes of a Lossless Waveguide

Let us first consider the case when the media from which the waveguide is constructed are lossless (ϵ and μ are real and any boundary walls are perfectly conducting). Denote the cross-sectional area of the waveguide by S_0, and the boundary contour of this cross section by C_0. The results of problem p5-3 show that, for any mode carrying nonzero complex power, it must be possible to choose the arbitrary amplitude of the mode fields such that $\mathcal{E}_T$ is real, so long as γ is either pure real (a cutoff mode, in which case $\mathcal{H}_T$ is imaginary) or pure imaginary (a propagating mode, in which case $\mathcal{H}_T$ is real). Supposing that this has been done, we have

$$\hat{P}_m \equiv \frac{1}{2} \int_{S_0} \mathcal{E}_m \times \mathcal{H}_m^* \cdot \mathbf{u}_z \, dS = \left[\frac{1}{2} \int_{S_0} \mathcal{E}_m \times \mathcal{H}_m \cdot \mathbf{u}_z \, dS \right]^* \equiv \hat{P}_{m,\mathrm{osc}}^* \qquad \text{(when } \mathcal{E}_{mT} \text{ is real)} \tag{5.25}$$

For a propagating mode, the phase of $\hat{P}_m$ is either 0 (in which case, we call the mode a forward mode, in analogy with the treatment for the classical transmission line in Chapter 1) or π (in which case, we call the mode a reverse mode). For a cutoff mode the phase of $\hat{P}_m$ is $\pm\pi/2$ [see eqns. (J.28) and (J.23) of Appendix J]. We have

$$\hat{P}_m = \frac{1}{2} \int_{S_0} (\mathcal{E}_{mT} \times \mathcal{H}_{mT}^*) \cdot \mathbf{u}_z \, dS = \begin{array}{l} \text{real if } m \text{ is a propagating mode} \\[1em] \text{imaginary if } m \text{ is a cutoff mode} \end{array} \tag{5.26}$$

This remains true if $\mathcal{E}_T$ is multiplied by an arbitrary complex constant, although clearly the phase of $\hat{P}_{m,\mathrm{osc}}$ will be altered by such a multiplication.

We can thus see that attempting to normalize a mode field by imposing the condition that its complex power is 1 is only possible for a propagating mode, and even then the phase

---

[3] In certain exceptional cases, the power carried by a mode may be zero, and it thus cannot be normalized in this way. In these cases, there arise so-called *complex modes* (see Appendix J), which can have propagation coefficients that are neither pure real nor pure imaginary even for a lossless waveguide. Although of some practical importance, the study of these modes is beyond the scope of this book, and the reader is referred to the references at the end of Appendix J for further information.

angle of the field amplitude is left arbitrary. We have already observed similar behavior in Chapter 1 for the classical transmission line. Whatever normalization condition is imposed on the mode fields of a lossless waveguide, the complex power of the mode will retain the same phase angle as discussed above, but we could constrain the complex *oscillatory* power to be ± 1, according to whether we have a forward mode or a reverse mode, in analogy with the normalization imposed on the classical transmission line modes in Chapter 1. We thus demand that the complex amplitude of the fields of any mode will hereafter be understood to have been constrained by the following *normalization condition*:

$$
\begin{aligned}
\hat{P}_{m,\text{osc}} &= \frac{1}{2} \int_{S_0} \boldsymbol{\mathcal{E}}_m \times \boldsymbol{\mathcal{H}}_m \cdot \mathbf{u}_z \, dS \\
&= 1 \qquad \text{(forward mode)} \\
&= -1 \qquad \text{(reverse mode)}
\end{aligned}
\tag{5.27}
$$

A forward mode field of arbitrary amplitude will have the form (5.2), and thus a_m^2 represents the complex oscillatory power associated with the mode, while b_m^2 represents the complex oscillatory power carried in the negative z-direction by a reverse mode.

Note that equation (5.25) is no longer necessarily true for a normalized mode field, since $\boldsymbol{\mathcal{E}}_{mT}$ may no longer be real (when the mode is cutoff, for instance). But even after normalization has been imposed, it is still true that $|\hat{P}_m| = |\hat{P}_{m,\text{osc}}|(=1)$, and from (J.23) the normalization condition for a forward mode on a lossless waveguide can be expressed in terms of the complex power as

$$
\hat{P}_m = \frac{1}{2} \int_{S_0} \boldsymbol{\mathcal{E}}_{mT} \times \boldsymbol{\mathcal{H}}_{mT}^* \cdot \mathbf{u}_z \, dS = \left\{ \begin{array}{ll} \pm 1 & (\gamma_m = j\beta_m) \\ \\ \pm j u_m & (\gamma_m = \alpha_m) \end{array} \right\} \qquad (m \gtrless 0)
\tag{5.28}
$$

where $u_m = +1$ if the mode stores more magnetic than electric energy per unit length, and $u_m = -1$ if the mode stores more electric than magnetic energy per unit length.

Modes of a Lossy Waveguide

For a lossy waveguide wherein all modes suffer some attenuation, a forward mode is one whose fields contain the factor $e^{-\gamma_m z}$ with $\alpha_m = \text{Re}(\gamma_m) > 0$, while a reverse mode is one whose fields vary as $e^{\gamma_m z}$ under the same constraint on α_m. As remarked in Appendix D, when a waveguide is not lossless, no simple relationship exists between the complex power and the complex oscillatory power of a mode—not even their magnitudes are necessarily identical. The only case of a lossy waveguiding structure where $|\hat{P}_m| = |\hat{P}_{m,\text{osc}}|$ is the classical transmission line (and a few waveguide equivalents to it), where the power is expressed in terms of a single voltage and a single current. Since it is the complex oscillatory power rather than $\hat{P}_m$ that is closely connected with the orthogonality property to be derived in Chapter 9, it will again prove more convenient to use (5.27) as the normalization condition, which will uniquely determine the fields to within a plus or minus sign.

It is worth mentioning that the oscillatory power of any waveguide mode can also be expressed in terms of the oscillatory stored energy per unit length of the guide. From (J.4) we have that

$$
\hat{P}_{m,\text{osc}} = \frac{j\omega}{\gamma_m} U_{m,\text{osc}}
\tag{5.29}
$$

where $U_{m,\text{osc}}$ is the complex oscillatory energy stored in the mode field per unit length of the waveguide:

$$
U_{m,\text{osc}} = \frac{1}{4} \int_{S_0} \left(\epsilon \boldsymbol{\mathcal{E}}_m \cdot \boldsymbol{\mathcal{E}}_m + \mu \boldsymbol{\mathcal{H}}_m \cdot \boldsymbol{\mathcal{H}}_m \right) dS
\tag{5.30}
$$

If the mode is purely propagating ($\gamma_m = j\beta_m$, which requires the waveguide to be lossless), then (5.29) becomes

$$\hat{P}_{m,\text{osc}} = v_p U_{m,\text{osc}} \tag{5.31}$$

where v_p is the phase velocity of the mode. In some cases, the energy form of computing the normalization is more convenient than the direct computation of the power. Furthermore, by (J.3), the oscillatory energy can also be written in either of two alternative forms (J.7) which can help to simplify calculations even more.

5.2.5   EXAMPLE: THE PARALLEL-PLATE WAVEGUIDE

When the normalization condition is imposed on the fields of the parallel-plate waveguide treated earlier in the chapter, the constants E_0, E_1 and H_1 are determined. Here, the "cross-section" of the waveguide is just the interval from $x = 0$ to $x = a$, and our so-called powers are in fact powers per unit width in the y-direction, since this two-dimensional guide has infinite extent in y. The TEM mode whose fields are given by (5.1) is subjected to:

$$\frac{1}{2}\int_0^a \boldsymbol{\mathcal{E}} \times \boldsymbol{\mathcal{H}} \cdot \mathbf{u}_z \, dx = \frac{aE_0^2}{2\zeta} = 1$$

or

$$E_0 = \sqrt{\frac{2\zeta}{a}} \tag{5.32}$$

and we have finally

$$\boldsymbol{\mathcal{E}} = \mathbf{u}_x\sqrt{\frac{2\zeta}{a}}; \quad \boldsymbol{\mathcal{H}} = \mathbf{u}_y\sqrt{\frac{2}{a\zeta}} \tag{5.33}$$

The transverse $\boldsymbol{\mathcal{E}}$ field of a TE mode is given by (5.8); we can use (5.18) and (5.23) to obtain $\mathcal{H}_z$ and $\boldsymbol{\mathcal{H}}_T$. We arrive at the result

$$E_1 = -j\sqrt{\frac{\zeta_{\text{TE}_m}}{a}}$$

where

$$\zeta_{\text{TE}_m} = -\frac{\mathcal{E}_y}{\mathcal{H}_x} = \frac{j\omega\mu}{\gamma_m}$$

is the so-called intrinsic wave impedance for the TE$_m$ mode of the parallel-plate waveguide. Thence,

$$\begin{aligned}
\boldsymbol{\mathcal{E}} &= \mathbf{u}_y 2\sqrt{\frac{\zeta_{\text{TE}_m}}{a}} \sin\frac{m\pi x}{a} \\
\boldsymbol{\mathcal{H}} &= \mathbf{u}_z \frac{2jm\pi}{\omega\mu a}\sqrt{\frac{\zeta_{\text{TE}_m}}{a}} \cos\frac{m\pi x}{a} - \mathbf{u}_x \frac{2}{\zeta_{\text{TE}_m}}\sqrt{\frac{\zeta_{\text{TE}_m}}{a}} \sin\frac{m\pi x}{a}
\end{aligned} \tag{5.34}$$

Likewise the transverse $\boldsymbol{\mathcal{H}}$ field of a TM mode is given by (5.11); we use (5.20) and (5.22) to retrieve the remaining field components. We get in a similar fashion

$$H_1 = \sqrt{\frac{1}{\zeta_{\text{TM}_m} a}}$$

where

$$\zeta_{\text{TM}_m} = \frac{\mathcal{E}_x}{\mathcal{H}_y} = \frac{\gamma_m}{j\omega\epsilon}$$

is the so-called intrinsic wave impedance for the TM_m mode of the parallel-plate waveguide, so that

$$\mathcal{H} = \mathbf{u}_y 2\sqrt{\frac{1}{\zeta_{\mathrm{TM}_m} a}} \cos\frac{m\pi x}{a} \tag{5.35}$$

$$\mathcal{E} = \mathbf{u}_z \frac{2jm\pi}{\omega\epsilon a}\sqrt{\frac{1}{\zeta_{\mathrm{TM}_m} a}} \sin\frac{m\pi x}{a} + \mathbf{u}_x 2\zeta_{\mathrm{TM}_m}\sqrt{\frac{1}{\zeta_{\mathrm{TM}_m} a}} \cos\frac{m\pi x}{a}$$

if $m \neq 0$. If $m = 0$, the TM_0 mode is equivalent to the TEM mode from above, and (5.32)-(5.33) should be used.

5.3 "HOLLOW" METALLIC WAVEGUIDES OF ARBITRARY CROSS-SECTION

Consider the waveguide illustrated in Figure 5.4. A perfectly conducting metallic cylinder of cross-section S_0 (and boundary C_0) is oriented along the z-axis as shown. The cylinder is filled with a homogeneous medium of electrical parameters ϵ and μ (as mentioned in Chapter 1, we can easily allow this medium to be lossy by replacing ϵ by the complex dielectric constant $\hat{\epsilon}$). Most often this medium will be air, and we refer to this type of waveguide as a hollow metallic waveguide.

From (5.18), (5.20), (5.22) and (5.23), a mode solution of Maxwell's equations for this guide must satisfy:

$$\mathbf{u}_z \cdot \nabla_T \times \mathcal{E}_T = -j\omega\mu\mathcal{H}_z \tag{5.36}$$

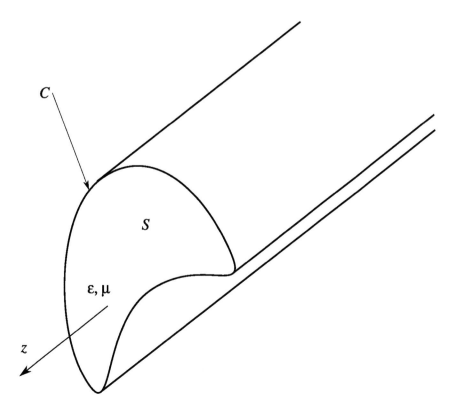

Figure 5.4 Hollow metallic waveguide.

$$\mathbf{u}_z \cdot \nabla_T \times \boldsymbol{\mathcal{H}}_T = j\omega\epsilon\mathcal{E}_z \tag{5.37}$$

$$\boldsymbol{\mathcal{E}}_T = \frac{1}{k_c^2}\left[j\omega\mu\mathbf{u}_z \times \nabla_T\mathcal{H}_z - \gamma\nabla_T\mathcal{E}_z\right] \tag{5.38}$$

$$\boldsymbol{\mathcal{H}}_T = \frac{1}{k_c^2}\left[-j\omega\epsilon\mathbf{u}_z \times \nabla_T\mathcal{E}_z - \gamma\nabla_T\mathcal{H}_z\right] \tag{5.39}$$

where we have introduced the notation

$$k_c^2 = \omega^2\mu\epsilon + \gamma^2 \tag{5.40}$$

and k_c is called the *cutoff wavenumber* for reasons which will become apparent later.

Equations for $\mathcal{E}_z$ and $\mathcal{H}_z$ can be obtained by substituting (5.38) into (5.36) and (5.39) into (5.37). Since this is a homogeneous waveguide (ϵ, μ and hence also k_c, are not functions of position), we obtain the equations

$$\left(\nabla_T^2 + k_c^2\right)\mathcal{E}_z = 0 \tag{5.41}$$

$$\left(\nabla_T^2 + k_c^2\right)\mathcal{H}_z = 0 \tag{5.42}$$

because of the vector identity $\nabla_T \times \nabla_T f \equiv 0$ for any suitable function f (see eqn. (B.14) of Appendix B). Here ∇_T^2 is the transverse Laplacian operator

$$
\begin{aligned}
\nabla_T{}^2 f &\equiv \nabla_T \cdot \nabla_T f \\
&= \frac{\partial^2 f}{\partial x^2} + \frac{\partial^2 f}{\partial y^2} \\
&= \frac{1}{\rho}\frac{\partial}{\partial\rho}\left(\rho\frac{\partial f}{\partial\rho}\right) + \frac{1}{\rho^2}\frac{\partial^2 f}{\partial\phi^2}
\end{aligned}
\tag{5.43}
$$

Naturally ∇_T^2 differs from the ordinary Laplacian ∇^2 only in the absence of the z-derivative term $\partial^2/\partial z^2$. We see that, as you may already have seen in the study of hollow rectangular waveguides, separate equations are obtained for $\mathcal{E}_z$ and $\mathcal{H}_z$ from Maxwell's equations, and the remaining field components are obtained at once from (5.38) and (5.39).

To specify the problem of finding $\mathcal{E}_z$ and $\mathcal{H}_z$ completely, we must find the appropriate boundary conditions satisfied by these quantities on the wall of the guide. Since it is tangential to the wall, which is a perfect conductor, $\mathcal{E}_z$ must vanish there:

$$\mathcal{E}_z|_{C_0} = 0 \tag{5.44}$$

However, $\mathcal{H}_z$ cannot be dealt with so immediately, since we can only relate it to a surface electric wall current, itself not yet known. We *can* obtain a second boundary condition by forcing the *other* tangential component of $\mathcal{E}$ to vanish on the wall. Denoting the *inward* normal from the wall as the unit vector $\mathbf{u}_n$, the vector $\mathbf{u}_l \equiv \mathbf{u}_z \times \mathbf{u}_n$ will be in the direction of this remaining tangential component (see Figure 5.5), so from (5.38), we require

$$
\begin{aligned}
\mathcal{E}_l|_{C_0} &= \mathbf{u}_l \cdot \boldsymbol{\mathcal{E}}_T|_{C_0} \\
&= \frac{1}{k_c^2}\left[j\omega\mu\mathbf{u}_n \cdot \nabla_T\mathcal{H}_z - \gamma\mathbf{u}_l \cdot \nabla_T\mathcal{E}_z\right]_{C_0}
\end{aligned}
\tag{5.45}
$$

But we know that

$$\mathbf{u}_l \cdot \nabla_T\mathcal{E}_z|_{C_0} = \left.\frac{\partial\mathcal{E}_z}{\partial l}\right|_{C_0} = 0$$

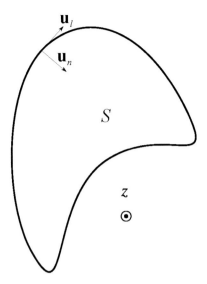

Figure 5.5   The transverse cross-section of a hollow metallic waveguide.

because the condition $\mathcal{E}_z|_{C_0} = 0$ does not change as we move *along the curve* C_0. Hence (5.45) becomes

$$\left.\frac{\partial \mathcal{H}_z}{\partial n}\right|_{C_0} = 0 \tag{5.46}$$

where $\partial/\partial n$ indicates the derivative in the direction normal to the wall at any point on the wall (i.e., $\partial f/\partial n = \mathbf{u}_n \cdot \nabla f$).

The fact that equations (5.41) and (5.44) for $\mathcal{E}_z$ and (5.42) and (5.46) for $\mathcal{H}_z$ are completely separate means that either $\mathcal{E}_z$ or $\mathcal{H}_z$ can be independently set equal to zero, *and the modes of the hollow waveguide can all be chosen either pure TE or pure TM*. Since, as we will see in Chapter 10, the TE and TM modes which are propagating along a lossless hollow waveguide carry power independently of each other, it is customary to choose the modes of these waveguides in these polarizations rather than in some other fashion.

5.4   GENERAL PROPERTIES OF TE AND TM MODES

5.4.1   TM MODES

The properties of the TM modes can be found from (5.38) and (5.39) by setting $\mathcal{H}_z \equiv 0$. Outside of being rotated by $90°$ from each other, the transverse $\mathcal{E}$ and $\mathcal{H}$ fields differ only by a proportionality constant, viz.,

$$\begin{aligned}
\mathcal{E}_T &= -\zeta_{\text{TM}} \mathbf{u}_z \times \mathcal{H}_T \\
&= -\frac{\gamma}{k_c^2} \nabla_T \mathcal{E}_z
\end{aligned} \tag{5.47}$$

The constant of proportionality ζ_{TM} is known as the *intrinsic wave impedance* for a given TM mode, and is equal to

$$\zeta_{\text{TM}} = \frac{\gamma}{j\omega\epsilon} \tag{5.48}$$

5.4.2   TE MODES

Similarly, for TE modes we have the relations

$$
\begin{aligned}
\boldsymbol{\mathcal{E}}_T &= -\zeta_{\mathrm{TE}} \mathbf{u}_z \times \boldsymbol{\mathcal{H}}_T \\
&= \frac{j\omega\mu}{k_c^2} \mathbf{u}_z \times \nabla_T \mathcal{H}_z
\end{aligned}
\tag{5.49}
$$

where the TE mode wave impedance is

$$
\zeta_{\mathrm{TE}} = \frac{j\omega\mu}{\gamma}
\tag{5.50}
$$

Of course, neither ζ_{TE} nor ζ_{TM} is known until the propagation coefficient γ is determined.

5.4.3   THE EIGENVALUES

Because of (5.40), k_c^2 can be considered as the eigenvalue for hollow guides, and γ determined from it. The problem thus reduces to solving the equation

$$
(\nabla_T^2 + k_c^2)\Psi = 0
\tag{5.51}
$$

subject to

$$
\Psi\big|_{C_0} = 0
\tag{5.52}
$$

when $\Psi = \mathcal{E}_z$, or

$$
\frac{\partial\Psi}{\partial n}\bigg|_{C_0} = 0
\tag{5.53}
$$

when $\Psi = \mathcal{H}_z$, and to simultaneously finding the permitted set of eigenvalues $k_{c,m}^2$; $m = 1, 2, \dots$. For a specific waveguide (e.g., rectangular), the mode index m may actually refer to a double index mn, as in the TE_{mn} or TM_{mn} modes.

Note that nowhere does ϵ or μ enter into the problem at this point, and it is not until we wish to find $\boldsymbol{\mathcal{E}}_T$ and $\boldsymbol{\mathcal{H}}_T$ from (5.47) or (5.49), or γ from

$$
\gamma_{\pm m} = \pm\sqrt{k_{c,m}^2 - \omega^2\mu\epsilon} \quad (m > 0)
\tag{5.54}
$$

where $\mathrm{Re}(\gamma) > 0$ (or $\mathrm{Im}(\gamma) > 0$ if $\mathrm{Re}(\gamma) = 0$) that the electrical properties of the filling come in.

We will state here (without proof) some properties of the solution of (5.51) subject to (5.52) or (5.53):

(a) There is a countably infinite, discrete set of solutions (eigenfunctions) Ψ_m, $m = 1, 2, \dots$, corresponding to the eigenvalues $k_{c,m}^2$ in either the TM or TE case ($m = 0$ is used to designate a special type of TE mode as discussed below).

(b) Writing $\kappa_m = k_{c,m}^2$ for TE modes, and $\lambda_m = k_{c,m}^2$ for TM modes, all κ_m and λ_m are real and positive when C_0 is simply connected (except for $\kappa_0 = 0$), so that we can always assume that they are ordered in increasing value:

$$
\begin{aligned}
0 &= \kappa_0 < \kappa_1 \le \kappa_2 \le \dots \\
0 &< \lambda_1 \le \lambda_2 \le \lambda_3 \le \dots
\end{aligned}
$$

The lowest TE "mode" corresponding to $\kappa_0 = 0$ for a hollow guide with a simply-connected boundary is just a constant:

$$
\Psi = \mathcal{H}_z = K_0 = \text{constant}
$$

This mode (whose only nonzero field component is $\mathcal{H}_z$ and which can only exist at zero frequency) is an exceptional case which can often be ignored in applications. It is examined in detail in Appendix H. Thus, if the medium filling the guide is lossless (μ, ϵ are real), each eigenvalue has a corresponding *cutoff frequency*

$$
\begin{aligned}
\omega_{c,m} &= 2\pi f_{c,m} \\
&= \frac{k_{c,m}}{\sqrt{\mu\epsilon}}
\end{aligned}
\tag{5.55}
$$

above which the mode is a *propagating* mode whose propagation coefficient is imaginary:

$$
\gamma_m = j\beta_m = j\sqrt{\mu\epsilon}\sqrt{\omega^2 - \omega_{c,m}^2} \quad (\omega > \omega_{c,m})
\tag{5.56}
$$

and below which the mode is *evanescent* or *cut off*:

$$
\gamma_m = \alpha_m = \sqrt{\mu\epsilon}\sqrt{\omega_{c,m}^2 - \omega^2} \quad (\omega < \omega_{c,m})
\tag{5.57}
$$

The terminology of cut-off wavenumber for $k_{c,m}$ is now clear from (5.55). The TE_1 mode is often called the *dominant* or fundamental mode of this waveguide, because

$$
\kappa_1 < \lambda_1
\tag{5.58}
$$

and thus it is the mode with lowest cutoff frequency which is capable of transporting energy.

(c) One now can also give expressions for the guide wavelength, phase velocity and group velocity for modes with index $\pm m$ which are above cutoff ($\omega > \omega_{c,m}$). These follow from (1.10)-(1.12):

$$
\lambda_{g,m} = \frac{2\pi}{\sqrt{\mu\epsilon}\sqrt{\omega^2 - \omega_{c,m}^2}}
$$

$$
v_{p,m} = \frac{1}{\sqrt{\mu\epsilon}\sqrt{1 - \omega_{c,m}^2/\omega^2}}
$$

$$
v_{g,m} = \frac{\sqrt{1 - \omega_{c,m}^2/\omega^2}}{\sqrt{\mu\epsilon}}
$$

Observe that exactly at the cutoff frequency the situation is exceptional, in that guide wavelength and phase velocity become infinite, and group velocity goes to zero. Moreover, if the fields are to remain finite, either $\mathcal{E}_T$ or $\mathcal{H}_T$ must be zero. By (5.28) this means that in a lossless waveguide, modes cannot be normalized at cutoff. Analysis of problems that require normalization (such as excitation of the mode or computation of attenuation due to wall loss, as considered in Chapters 9 and 10) must be handled by calculation away from cutoff and subsequently taking the limit.

5.4.4   POWER FLOW AND STORED ENERGY

The complex power flow through the cross section S_0 of the guide is of interest in applications as well as from a conceptual standpoint. It is given by

$$
\hat{P} = \frac{1}{2}\int_{S_0} \mathcal{E} \times \mathcal{H}^* \cdot \mathbf{u}_z \, dS
$$

From (5.47) and (5.49), we can write this as

$$\hat{P} = \frac{\zeta_m}{2} \int_{S_0} \boldsymbol{\mathcal{H}}_T \cdot \boldsymbol{\mathcal{H}}_T^* \, dS \tag{5.59}$$

where ζ_m denotes the intrinsic wave impedance

$$\zeta_m = \left\{ \begin{array}{ll} \zeta_{\text{TE}} & \text{(TE modes)} \\[2mm] \zeta_{\text{TM}} & \text{(TM modes)} \end{array} \right\} \tag{5.60}$$

of the mode. When the mode is above cutoff, the impedance is real, and since the integral in (5.59) is real, this means there is net, nonzero time-average power flow in the z-direction for a mode in these conditions. On the other hand, when a single mode decaying in the $+z$-direction exists on the guide, the wave impedance is purely imaginary, and no net time-average power flow takes place. Instead, energy is stored in the fields of the mode, as in a capacitor or inductor.

The oscillatory power is similarly given by

$$\hat{P}_{\text{osc}} = \frac{\zeta_m}{2} \int_{S_0} \boldsymbol{\mathcal{H}}_T \cdot \boldsymbol{\mathcal{H}}_T \, dS \tag{5.61}$$

and this relation can be used to normalize the fields of a mode using (5.27). This equation could also be obtained with the help of (J.7).

It can be shown that cutoff TE modes store predominantly *magnetic* energy (they have an extra component of **H**) while cutoff TM modes store predominantly *electric* energy (they have an extra component of **E**) in a hollow metallic waveguide. This has consequences for the equivalent circuits of certain types of discontinuity in such waveguides (see Chapter 11). Details of the proof are left to the reader (problem p5-11).

5.5 CIRCULAR WAVEGUIDES (METHOD OF SEPARATION OF VARIABLES)

For any hollow metallic guide, our task is now to solve (5.51)-(5.53) for a particular waveguide shape. We will first illustrate the method of separation of variables and consider the circular waveguide whose cross-section is illustrated in Figure 5.6.

A perfectly conducting metal tube of radius a is assigned Cartesian and cylindrical coordinates as shown. We first seek the TM modes for this waveguide. In the method of separation of variables, we *assume* that a solution $\mathcal{E}_z$ of (5.41) can be written as the product:

$$\mathcal{E}_z(\rho, \phi) = R(\rho)\Phi(\phi) \tag{5.62}$$

where R is a function of ρ only, and Φ is a function of ϕ only. We insert this into (5.41) to get

$$\Phi R'' + \frac{\Phi}{\rho} R' + \frac{1}{\rho^2} R\Phi'' + k_c^2 R\Phi = 0 \tag{5.63}$$

where primes denote differentiation with respect to the argument (either ρ or ϕ as the case may be). If we divide this result by $R\Phi/\rho^2$ and rearrange a bit, we get

$$\rho^2 \frac{R''}{R} + \rho \frac{R'}{R} + \rho^2 k_c^2 = -\frac{\Phi''}{\Phi} \tag{5.64}$$

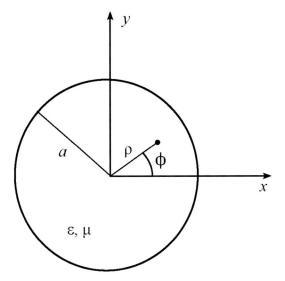

Figure 5.6 Cross-section of circular waveguide.

But this equation states that a function of ρ only (the left side) is equal to a function of ϕ only (the right side). Thus, both sides must be a constant. With a certain amount of hindsight, we call this constant ν^2. Making the right side of (5.64) equal to ν^2, we have

$$\Phi'' + \nu^2\Phi = 0$$

whose solution is well-known:

$$\Phi(\phi) = C_1 \cos \nu\phi + C_2 \sin \nu\phi \tag{5.65}$$

Here, C_1 and C_2 are some yet unknown complex constants to be determined later.

Equating the left side of (5.64) to ν^2, we have, after rearranging,

$$\rho^2 R'' + \rho R' + (k_c^2\rho^2 - \nu^2)R = 0 \tag{5.66}$$

This can be cleaned up slightly by defining a new variable $w = k_c\rho$, so that in terms of w, (5.66) becomes

$$w^2\frac{d^2R}{dw^2} + w\frac{dR}{dw} + (w^2 - \nu^2)R = 0 \tag{5.67}$$

Equation (5.67) is *Bessel's differential equation* for R as a function of w. As shown in Appendix C, the general solution of (5.67) is

$$R = C_3 J_\nu(w) + C_4 Y_\nu(w) \tag{5.68}$$

where J_ν is the Bessel function of the *first* kind, and Y_ν that of the *second* kind, of order ν and argument w. Once again we acquire two constants C_3 and C_4 that must be determined.

From (5.62), (5.65) and (5.68), then, our solution for $\mathcal{E}_z$ is

$$\mathcal{E}_z(\rho, \phi) = [C_1 \cos \nu\phi + C_2 \sin \nu\phi][C_3 J_\nu(k_c\rho) + C_4 Y_\nu(k_c\rho)] \tag{5.69}$$

It remains for us now to determine the unknown constants C_1 through C_4, the separation constant ν, and the eigenvalues k_c^2 by enforcing the boundary conditions at the waveguide wall together with certain other "common sense" restrictions on the fields. The following constraints are applied:

(a) From Appendix C, it can be seen that $Y_\nu(k_c\rho)$ blows up at $\rho = 0$:

$$|Y_\nu(k_c\rho)| \to \infty \quad \text{as } \rho \to 0$$

However, there is no physical reason for the fields to blow up at the center of the guide.[4] We conclude, therefore, that $C_4 = 0$. Equation (5.69) now becomes

$$\mathcal{E}_z(\rho, \phi) = [C_c \cos \nu\phi + C_s \sin \nu\phi] J_\nu(k_c\rho) \tag{5.70}$$

where we have combined the three remaining amplitude constants into two: $C_c = C_1 C_3$ and $C_s = C_2 C_3$.

(b) In cylindrical coordinates, the angle ϕ can vary continuously over the interval $[0, 2\pi]$ and beyond, as the wave equation holds continuously for increasing ϕ. But a given point described by $\phi = \phi_0$ is the same (for a given value of ρ) as a point with $\phi = \phi_0 + 2\pi$. We must expect however that $\mathcal{E}_z$ should be *single-valued* as a function of ϕ, because the same solution of the form (5.70) must still be valid for both values of ϕ. Thus,

$$\mathcal{E}_z(\rho, \phi) = \mathcal{E}_z(\rho, 2\pi + \phi)$$

for all ϕ. This is guaranteed only if ν in the terms $\cos \nu\phi$ and $\sin \nu\phi$ is an integer, so we set $\nu = l$ with $l = 0, 1, 2, \dots$. Equation (5.70) then becomes

$$\mathcal{E}_z(\rho, \phi) = [C_c \cos l\phi + C_s \sin l\phi] J_l(k_c\rho); \quad l = 0, 1, 2, \dots \tag{5.71}$$

Setting $\nu = l$ has fixed the indices of the Bessel functions as well. Negative values of l are unnecessary, since $\cos(-l\phi) = \cos l\phi$, and J_{-l}, Y_{-l} can be related to J_l, Y_l (Appendix C), so that no new functions result.

(c) Finally, we require that the tangential field $\mathcal{E}_z$ be zero at the guide wall $\rho = a$ (cf. (5.44)). For this requirement to be true for all angles ϕ, we must have

$$J_l(k_c a) = 0 \tag{5.72}$$

if we discard the trivial solution $C_c = C_s = 0$. From Appendix C, the Bessel functions J_l are zero only at certain values j_{lm} of their argument.[5] The subscript m denotes the ordering of the root in increasing sequence; thus

$$j_{l1} < j_{l2} < j_{l3} < \dots$$

while l denotes the index of the Bessel function involved. For example, Table C.1 of Appendix C indicates the first three zeroes of J_0 as $j_{01} \simeq 2.405$, $j_{02} \simeq 5.520$, $j_{03} \simeq 8.654$. From (5.72), we then conclude that

$$k_c a = j_{lm}$$

or

$$k_{c,\text{TM}_{lm}} = j_{lm}/a; \quad m = 1, 2, 3, \dots \tag{5.73}$$

---

[4] Infinite fields *are* in fact allowed under certain conditions provided the total field energy in a finite volume near the edge is finite. It can be shown that the stored energy of a field for which the longitudinal E-field behaves as $E_z \propto Y_\nu(k_c\rho) \cos \nu\phi \exp(-j\sqrt{k^2 - k_c^2}\,z)$ near $\rho = 0$, is infinite.

[5] No confusion should arise between $j = \sqrt{-1}$ and j_{lm}.

Substituting this result into (5.70), we get a general solution which is a linear combination of the two possible solutions

$$\mathcal{E}_z(\rho,\phi) = \left\{ \begin{array}{c} E_{lmc}\cos l\phi \\ E_{lms}\sin l\phi \end{array} \right\} J_l\left(\frac{j_{lm}}{a}\rho\right) \tag{5.74}$$

where we have renamed the remaining arbitrary constants C_c and C_s as mode amplitudes E_{lmc} and E_{lms}. These constants are determined by normalizing the oscillatory power of the mode to 1 (see below).

We can now obtain the transverse field components by substituting (5.74) into (5.47):

$$\mathcal{E}_\rho = -\frac{a\gamma_{lm}}{j_{lm}} \left\{ \begin{array}{c} E_{lmc}\cos l\phi \\ E_{lms}\sin l\phi \end{array} \right\} J_l'\left(\frac{j_{lm}}{a}\rho\right) \tag{5.75}$$

$$\mathcal{E}_\phi = \frac{l\gamma_{lm}a^2}{j_{lm}^2\rho} \left\{ \begin{array}{c} E_{lmc}\sin l\phi \\ -E_{lms}\cos l\phi \end{array} \right\} J_l\left(\frac{j_{lm}}{a}\rho\right) \tag{5.76}$$

with $\mathcal{H}_T$ obtained from

$$\mathcal{H}_T = \frac{\mathbf{u}_z \times \mathcal{E}_T}{\zeta_{\text{TM}}} \tag{5.77}$$

The modes we have just described are known as the TM$_{lm}$ modes of the circular waveguide. Indeed, there are two possible polarizations: TM$_{lmc}$ and TM$_{lms}$ (except for $l = 0$; the TM$_{0ms}$ modes, of course, do not exist, as their fields vanish identically). Thus, for example, while the TM$_{01}$ mode is invariant under rotations, the TM$_{11}$ mode can be changed into a different polarization by rotating it through $90°$. Hence, the designation TM$_{lm}$ for $l > 0$ actually refers to a pair of *degenerate*[6] modes TM$_{lmc}$ and TM$_{lms}$ having orthogonal polarizations. The same kind of thing happens in a square waveguide with the TE$_{10}$/TE$_{01}$ modes, etc.

Figure 5.7 shows the relative locations of the cutoff frequencies for the circular waveguide. For TM modes, the first few are:

$$\text{TM}_{01}: f_{c,\text{TM}_{01}} = \frac{0.383}{a\sqrt{\mu\epsilon}}$$

$$\text{TM}_{11}: f_{c,\text{TM}_{11}} = \frac{0.609}{a\sqrt{\mu\epsilon}}$$

Transverse field distributions for the TM modes can be constructed with the aid of equations (5.75)-(5.77) and Appendix I and are sketched in Figure 5.8. According to the usual conventions, solid lines denote the E-field, and dashed lines the H-field. It might be

---

[6]Two or more modes are said to be degenerate if their propagation coefficients are identical. Ordinarily this is undesirable since very little irregularity is needed to convert from one of these modes to another, and launching and receiving schemes are generally designed to deal only with one specific mode. In the presence of such a degeneracy, some type of mode filter must be devised if control over the polarization direction in the waveguide is to be maintained. However, on occasion there are devices whose design specifically utilizes the presence of degenerate modes, as in the case of Example 5.5.2.

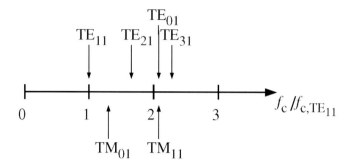

Figure 5.7 Relative cutoff frequencies for circular waveguide modes.

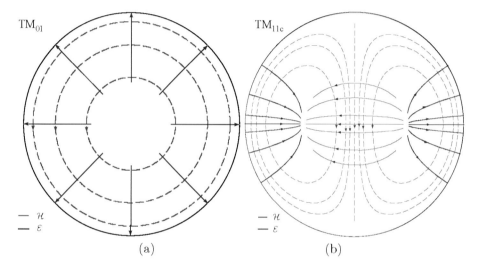

Figure 5.8 Field distributions of (a) TM_{01} and (b) TM_{11c} circular waveguide modes.

noted that the field of the TM_{11} mode bears a striking resemblance to that of the TEM mode on a two-wire transmission line (Figure 1.1(c)) and this resemblance could be used as a means of exciting this particular mode, the principles of which are presented in Chapter 10.

The TE modes are investigated in a manner almost identical to that for the TM modes. The difference is that $\mathcal{H}_z$ is being solved for instead of $\mathcal{E}_z$ (cf. (5.42)), and in step (c), we must enforce boundary condition (5.46) on $\mathcal{H}_z$. We have

$$\mathcal{H}_z(\rho, \phi) = \left\{ \begin{array}{c} H_{lmc} \cos l\phi \\ \\ H_{lms} \sin l\phi \end{array} \right\} J_l(k_c \rho) \tag{5.78}$$

and (5.46) implies

$$J_l^{'}(k_c a) = 0 \tag{5.79}$$

From Appendix C, the roots of $J_l^{'}(x)$ are simple, and can be denoted $j_{lm}^{'}$. A short table of them is given in Appendix C. We then have

$$k_{c,\mathrm{TE}_{lm}} = j_{lm}^{'}/a \tag{5.80}$$

or

$$f_{c,\text{TE}_{lm}} = \frac{j'_{lm}}{2\pi a\sqrt{\mu\epsilon}} \tag{5.81}$$

These modes (again degenerate pairs for $l > 0$) are called TE_{lm} modes. The first several cutoff frequencies are

$$\text{TE}_{11} : f_{c,\text{TE}_{11}} = \frac{0.293}{a\sqrt{\mu\epsilon}}$$

$$\text{TE}_{21} : f_{c,\text{TE}_{21}} = \frac{0.486}{a\sqrt{\mu\epsilon}}$$

$$\text{TE}_{01} : f_{c,\text{TE}_{01}} = \frac{0.609}{a\sqrt{\mu\epsilon}}$$

From Tables C.1 and C.2, we can see that the TE_{11} mode ($j'_{11} = 1.841$) actually has a lower cutoff frequency than the TE_{01} or TM_{01} modes, and is in fact the *dominant* or *fundamental* mode of this waveguide (Figure 5.7). Once again the transverse fields (of the TE_{lmc} modes) are obtained from (5.78) and (5.49):

$$\mathcal{E}_\rho = \frac{j\omega\mu la^2}{j'^2_{lm}\rho} H_{lmc} \sin l\phi J_l\left(\frac{j'_{lm}}{a}\rho\right) \tag{5.82}$$

$$\mathcal{E}_\phi = \frac{j\omega\mu a}{j'_{lm}} H_{lmc} \cos l\phi J'_l\left(\frac{j'_{lm}}{a}\rho\right) \tag{5.83}$$

$$\boldsymbol{\mathcal{H}}_T = \frac{\mathbf{u}_z \times \boldsymbol{\mathcal{E}}_T}{\zeta_{\text{TE}}} \tag{5.84}$$

Transverse field patterns for the TE_{11c}, TE_{21s} and TE_{01} modes are shown in Figure 5.9.

The transverse fields are only a part of the story of any mode, of course, and it is desirable to be able to visualize the longitudinal part of the field as well. Since for lossless guides the longitudinal fields are in phase quadrature with the transverse ones, we cannot simply take the phasor quantities and use them to plot the field lines. What is normally done is to take a "snapshot" in time of the time-varying field, and represent it either as a three-dimensional plot, or in a side view cross section, as we do for the TE_{11} mode in Figure 5.10. If we could let time increase in such a figure, we would see the entire field structure for a propagating mode moving in the direction of positive z. In addition, there are surface charge and current densities on the waveguide walls that likewise move along with the field distribution.

5.5.1 NORMALIZED MODE FIELDS

The amplitude constants E_{lms}, E_{lmc}, H_{lms} and H_{lmc} in (5.74) and (5.78) are determined from the normalization condition (5.27). The calculations are simplified somewhat by the results of problem p5-9. Thus, for TM modes, we need

$$\int_{S_0} \mathcal{E}_z^2 \, dS = \frac{2k_c^2}{j\omega\epsilon\gamma}$$

but (5.74) gives after some calculations

$$\int_{S_0} \mathcal{E}_z^2 \, dS = E_{lm}^2 \frac{2\pi a^2}{j_{lm}^2 \Delta_l} \int_0^{j_{lm}} J_l^2(u)u \, du \tag{5.85}$$

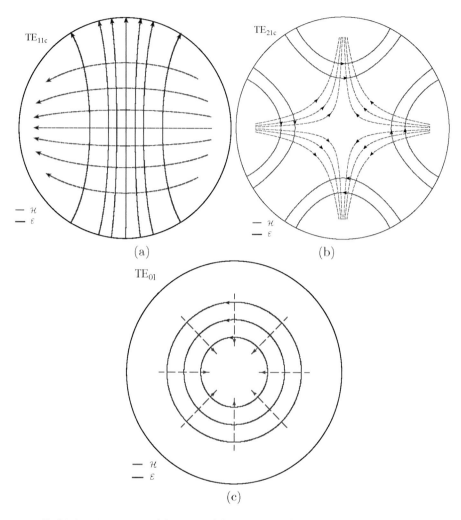

Figure 5.9 Field distributions of (a) TE_{11c}, (b) TE_{21c} and (c) TE_{01} circular waveguide modes.

where Δ_l is the so-called *Neumann factor*:

$$\Delta_l = \left\{ \begin{array}{ll} 1 & l = 0 \\ 2 & l \neq 0 \end{array} \right\} \tag{5.86}$$

From (C.40), (C.41) and (C.49), we find that the integral can be evaluated explicitly, and we have

$$\int_{S_0} \mathcal{E}_z^2 \, dS = E_{lm}^2 \frac{\pi a^2}{\Delta_l} J_l'^2(j_{lm}) \tag{5.87}$$

Thus, we find from (5.27) that

$$E_{lm} = \frac{k_c \sqrt{2\Delta_l}}{a J_l'(j_{lm}) \sqrt{j\omega\epsilon\gamma\pi}} \tag{5.88}$$

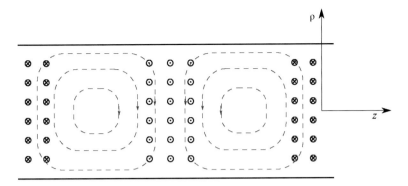

Figure 5.10 Top view of instantaneous field distribution for TE_{11c} circular waveguide mode. $\otimes$: E-field downward; $\odot$: E-field upward; dashed lines: H-field.

and an arbitrary choice of sign for E_{lm} has been made since no convention is in common use for this. Similarly we find for TE modes that

$$H_{lm} = \frac{k_c\sqrt{2\Delta_l}}{aJ_l(j'_{lm})\sqrt{j\omega\mu\gamma\pi(1 - l^2/j'^2_{lm})}} \tag{5.89}$$

5.5.2   EXAMPLE: A VARIABLE ATTENUATOR

A circular waveguide attenuator can be constructed as shown in Figure 5.11. Tapered transitions (not shown) from rectangular cross-section waveguides are used at either end together with fixed thin sheets of lossy material placed perpendicular to the electric field to assure the exclusive presence of one polarization of the TE_{11} mode (say, TE_{11c}).

To a good approximation, if the thickness of this resistive sheet is small compared to a skin depth, it can be described by the boundary condition that $\mathbf{E}_{\text{tan}}$ is continuous across the sheet (whose thickness is now approximated by zero), and

$$\mathbf{E}_{\text{tan}} = R_S \mathbf{J}_S \tag{5.90}$$

where $\mathbf{J}_S$ is the surface current in the sheet and R_S is the so-called *surface resistance* of the sheet. Eqn. (5.90) is a sort of surface version of Ohm's law. When the sheet is inserted horizontally and thus lies perpendicular to the electric field lines, little disturbance to the mode field by the sheet takes place. As would happen if the sheet were a perfect conductor, surface charges and currents appear on the sheet to properly terminate the electric and magnetic fields. The sheet is so thin that little loss is incurred in this way.

A rotatable thin sheet of lossy material occupies a diameter of the cross-section of the central portion of circular waveguide as shown. When it is horizontal, the wave coming

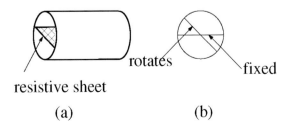

resistive sheet

(a)   (b)

Figure 5.11 A circular waveguide attenuator: (a) side view; (b) end view.

from one end of the attenuator passes through with little disturbance. On the other hand, when the sheet is oriented vertically, it represents a severe perturbation to the TE_{11c} mode, because it is parallel to the electric field lines. The modified mode which results has quite large currents $\mathbf{J}_S$ induced in the sheet, and suffers a good deal of attenuation as a result. By rotating the sheet between the horizontal and vertical positions, a variable amount of attenuation can be realized.

5.6   RECTANGULAR WAVEGUIDES (RAY DESCRIPTION OF THE MODES)

Only a few special cross-sectional shapes permit exact determination of the mode fields and eigenvalues of a hollow waveguide by separation of variables. It is desirable to have some general principles from which we can deduce some approximate characteristics of modes for any shape of waveguide, and gain some physical insight into the guiding mechanism as well. For this purpose, we will consider another method by which some types of hollow metallic waveguides can be exactly analyzed: the ray approach which was used for the parallel-plate waveguide in Section 5.1.

Since the z-dependent factor of a mode field is always $\exp(-\gamma z)$, we only need actually concern ourselves with projections of rays in the transverse (xy) plane of a waveguide. A general plane wave whose z-dependence is $\exp(-\gamma z)$ will have an electric field of the form:

$$\mathbf{E} = \mathbf{u}_e E_o e^{-\gamma z} e^{-jk_c(x\cos\theta + y\sin\theta)} \tag{5.91}$$

where $\mathbf{u}_e$ is a (constant) unit vector, E_o a constant field amplitude, and θ the angle of the projection of the ray in the transverse plane. A ray solution of

$$\left(\nabla_T^2 + k_c^2\right)\boldsymbol{\mathcal{E}} = 0 \tag{5.92}$$

will then be made up of *reduced* or *projected* plane waves of the form:

$$\boldsymbol{\mathcal{E}} = \mathbf{u}_e E_o e^{-jk_c(x\cos\theta + y\sin\theta)} \tag{5.93}$$

and the $\mathcal{H}$-field will similarly be a sum of terms of the form

$$\boldsymbol{\mathcal{H}} = \mathbf{u}_h H_o e^{-jk_c(x\cos\theta + y\sin\theta)} \tag{5.94}$$

When we speak of rays in the remainder of this section, then, we will actually mean projections of rays in the xy-plane.

Consider the hollow rectangular metallic waveguide of sides a and d whose cross-section is shown in Figure 5.12(a). A photograph of the flanged end of such a waveguide is shown in Figure 5.12(b). For (let us say) the TE modes of such a guide, it is convenient to deal with the field component $\mathcal{H}_z$ (the analysis of the parallel-plate guide could also have been carried out in terms of $\mathcal{H}_z$). A plane wave ① propagating at an angle θ between the x-axis and the projection of the ray in the xy-plane will have

$$\mathcal{H}_z^{inc} = H_1 e^{-jk_c(x\cos\theta - y\sin\theta)} \tag{5.95}$$

Upon reflecting the projection of this ray in the waveguide walls, it is evident from Figure 5.13 that three other ray directions are generated inside the waveguide. Thus, the general solution for $\mathcal{H}_z$ must be

$$\begin{aligned} \mathcal{H}_z = \quad & H_1 e^{-jk_c(x\cos\theta - y\sin\theta)} \\ & + H_2 e^{-jk_c(x\cos\theta + y\sin\theta)} \\ & + H_3 e^{-jk_c(-x\cos\theta + y\sin\theta)} \\ & + H_4 e^{-jk_c(-x\cos\theta - y\sin\theta)} \end{aligned} \tag{5.96}$$

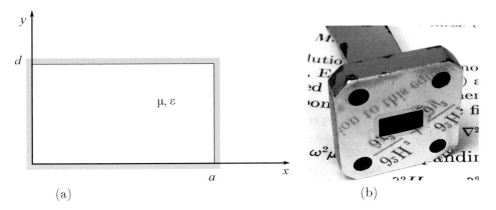

(a) (b)

Figure 5.12 (a) Cross-section of hollow rectangular waveguide; (b) flanged end of a rectangular metallic waveguide.

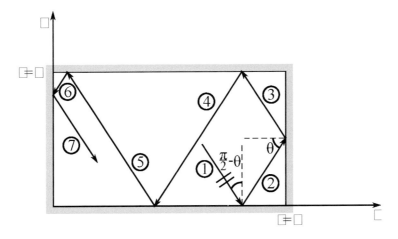

Figure 5.13 Transverse ray directions in a rectangular waveguide.

The amplitude H_i, $i = 1, 2, 3, 4$, of each plane wave corresponds to the ray $\textcircled{i}$ in Figure 5.13. It can be seen that this representation of the field $\mathcal{H}_z$ is a kind of generalized separation of variables. Each plane wave in (5.96) has the product form $X(x)Y(y)$ for some functions X and Y, and satisfies the differential equation (5.92). Unlike the ordinary method of separation of variables, however, we may require more than one such product term to be able to satisfy the boundary conditions.

To determine acceptable values for k_c and θ, we need to apply the boundary condition $\partial \mathcal{H}_z / \partial n = 0$ on each part of the wall in turn. For example, at the wall $y = 0$, $\partial/\partial n = \partial/\partial y$ and we have

$$
\begin{aligned}
0 &= \frac{\partial}{\partial y} \Big\{ H_1 e^{-jk_c(x\cos\theta - y\sin\theta)} + H_2 e^{-jk_c(x\cos\theta + y\sin\theta)} \\
&\quad + H_3 e^{-jk_c(-x\cos\theta + y\sin\theta)} + H_4 e^{-jk_c(-x\cos\theta - y\sin\theta)} \Big\}_{y=0} \\
&= jk_c \sin\theta \left\{ (H_1 - H_2)e^{-jk_c x\cos\theta} - (H_3 - H_4)e^{jk_c x\cos\theta} \right\}
\end{aligned}
$$

This condition must hold identically for all values of x, which can only happen if the terms dependent on $e^{-jk_c x \cos\theta}$ and $e^{jk_c x \cos\theta}$ each vanish separately. That is, we can examine separately the reflections ① → ② , which gives $H_1 = H_2$, and ④ → ③ , which gives $H_3 = H_4$.

Likewise, from the reflections at $x = 0$, ③ → ② and ④ → ① we get

$$H_3 = H_2 \quad \text{and} \quad H_4 = H_1$$

From the wall $x = a$:

$$H_3 e^{jk_c a \cos\theta} = H_2 e^{-jk_c a \cos\theta}$$

$$H_4 e^{jk_c a \cos\theta} = H_1 e^{-jk_c a \cos\theta}$$

and finally, from the wall $y = d$:

$$H_4 e^{jk_c d \sin\theta} = H_3 e^{-jk_c d \sin\theta}$$

$$H_1 e^{jk_c d \sin\theta} = H_2 e^{-jk_c d \sin\theta}$$

From these conditions it is not difficult to show that

$$
\begin{aligned}
H_1 = H_2 &= H_3 = H_4 \\
k_c a \cos\theta &= m\pi; \quad m = 0, 1, 2, \dots \\
k_c d \sin\theta &= n\pi; \quad n = 0, 1, 2, \dots
\end{aligned}
\tag{5.97}
$$

Hence, the solution (5.96) in terms of four bouncing plane waves can be recombined to yield

$$
\begin{aligned}
\mathcal{H}_z &= 4H_1 \cos\left(\frac{m\pi x}{a}\right) \cos\left(\frac{n\pi y}{d}\right); \\
k_{c,mn} &= \left[\left(\frac{m\pi}{a}\right)^2 + \left(\frac{n\pi}{d}\right)^2\right]^{1/2}; \quad (m, n \text{ not both zero }) \\
\gamma_{m,n} &= \left[k_{c,mn}^2 - \omega^2 \mu\epsilon\right]^{1/2} = j\left[\omega^2 \mu\epsilon - k_{c,mn}^2\right]^{1/2}
\end{aligned}
\tag{5.98}
$$

which is, of course, the well-known result for TE modes of a rectangular waveguide. The remaining transverse fields could now be obtained from (5.49), if desired. Figure 5.14 shows the relative locations of the cutoff frequencies for a rectangular waveguide with $a = 2d$.

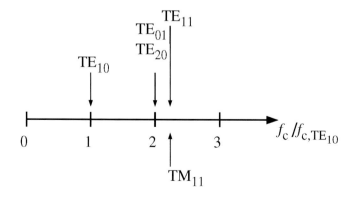

Figure 5.14   Relative cutoff frequencies for rectangular waveguide modes $(a = 2d)$.

Comparison with the distribution of cutoff frequencies for the circular waveguide shown in Figure 5.7 shows that the larger aspect ratio of the rectangular waveguide has the effect of shifting the cutoff frequencies of the lowest TM mode and the next highest TE mode higher relative to that of the fundamental TE mode.

Note that the second and third of equations (5.97) represent what are commonly called *transverse resonance* conditions: the requirement that any of the plane waves comprising the mode be in phase with all re-reflected plane waves propagating in the same direction (cf. rays ① and ⑦ in Figure 5.13). This method can also be used for the exact solution of the modes of triangular waveguides in the cases of vertex angles: (a) $45°$-$45°$-$90°$; (b) $60°$-$60°$-$60°$; (c) $30°$-$60°$-$90°$. Details are left as an exercise.

5.6.1 NORMALIZED MODE FIELDS

Whether the TE and TM modes of the hollow rectangular metallic waveguide can be found by the ray method as above, or by the method of separation of variables, as is done in many introductory texts in electromagnetics, the fields of a forward TE_{mn} mode are found to be:

$$
\left.
\begin{aligned}
\mathcal{H}_z &= H_{mn} \cos \tfrac{m\pi x}{a} \cos \tfrac{n\pi y}{d} \\[2mm]
\mathcal{E}_x &= \tfrac{j\omega\mu}{k_c^2} \tfrac{n\pi}{d} H_{mn} \cos \tfrac{m\pi x}{a} \sin \tfrac{n\pi y}{d} \\[2mm]
\mathcal{E}_y &= -\tfrac{j\omega\mu}{k_c^2} \tfrac{m\pi}{a} H_{mn} \sin \tfrac{m\pi x}{a} \cos \tfrac{n\pi y}{d} \\[2mm]
\mathcal{H}_x &= -\mathcal{E}_y/\zeta_{\mathrm{TE}} \\[2mm]
\mathcal{H}_y &= \mathcal{E}_x/\zeta_{\mathrm{TE}}
\end{aligned}
\right\}
\tag{5.99}
$$

where

$$
\left.
\begin{aligned}
\zeta_{\mathrm{TE}} &= \tfrac{j\omega\mu}{\gamma_{mn}} \\[2mm]
\gamma_{mn} &= \sqrt{k_c^2 - \omega^2 \mu\epsilon} \\[2mm]
k_c^2 &= k_{c,mn}^2 = (\tfrac{m\pi}{a})^2 + (\tfrac{n\pi}{d})^2
\end{aligned}
\right\}
\tag{5.100}
$$

and m and n range through $0, 1, 2, \cdots$, but are not both zero. The amplitude H_{mn} is now determined from (5.27):

$$
1 = \frac{1}{2} \int_0^b \int_0^a \mathcal{E} \times \mathcal{H} \cdot \mathbf{u}_z \, dx \, dy
\tag{5.101}
$$

Upon carrying out the integral in (5.101), we find that

$$
H_{mn} = \frac{jk_c\sqrt{2\Delta_m\Delta_n}}{\sqrt{-j\omega\mu\gamma_{mn}ad}}
\tag{5.102}
$$

where Δ_n is the Neumann factor defined in (5.86). As noted in previous examples, H_{mn} is only determined up to a $\pm$ sign. The choice is arbitrary, and as no standard convention exists for these modes, we have made our selection in eqn. (5.102) so as to make $\mathcal{E}_y$ for the normalized TE_{10} mode, for example, point in the $+y$ direction.

Likewise, the TM_{mn} forward mode fields are given by

$$
\left.
\begin{aligned}
\mathcal{E}_z &= E_{mn} \sin \tfrac{m\pi x}{a} \sin \tfrac{n\pi y}{d} \\[4pt]
\mathcal{E}_x &= -\tfrac{\gamma_{mn}}{k_c^2} \tfrac{m\pi}{a} E_{mn} \cos \tfrac{m\pi x}{a} \sin \tfrac{n\pi y}{d} \\[4pt]
\mathcal{E}_y &= -\tfrac{\gamma_{mn}}{k_c^2} \tfrac{n\pi}{d} E_{mn} \sin \tfrac{m\pi x}{a} \cos \tfrac{n\pi y}{d} \\[4pt]
\mathcal{H}_x &= -\mathcal{E}_y/\zeta_{\mathrm{TM}} \\[4pt]
\mathcal{H}_y &= \mathcal{E}_x/\zeta_{\mathrm{TM}}
\end{aligned}
\right\}
\tag{5.103}
$$

where γ_{mn} and k_c^2 are the same as in (5.100), while

$$
\zeta_{\mathrm{TM}} = \frac{\gamma_{mn}}{j\omega\epsilon}
\tag{5.104}
$$

and m and n range through $1, 2, \cdots$. The normalization condition (5.27) gives

$$
E_{mn} = \frac{2\sqrt{2}jk_c}{\sqrt{-j\omega\epsilon\gamma_{mn}ad}}
\tag{5.105}
$$

where an arbitrary selection of the $\pm$ sign has once again been made.

5.7 DIELECTRIC LOSSES IN HOLLOW WAVEGUIDES

Losses occur in real waveguides due to the imperfect conductivity of the walls and also because of the presence of losses in the dielectric filling the guide. The latter are easily accounted for by the foregoing theory simply by allowing $\epsilon \to \hat{\epsilon} = \epsilon' - j\epsilon''$ to be complex. Instead of (5.56) or (5.57), we find that γ_m is now neither purely real nor purely imaginary; from (5.54) we have

$$
\gamma_m \equiv \alpha_m + j\beta_m = \sqrt{k_{c,m}^2 - \omega^2\mu\epsilon' + j\omega^2\mu\epsilon''}
\tag{5.106}
$$

Instead of a sharp transition from cutoff to propagating mode, there is always *some* attenuation present, since solving for α_m and β_m gives

$$
\alpha_m = \left[\frac{k_{c,m}^2 - \omega^2\mu\epsilon' + \sqrt{(k_{c,m}^2 - \omega^2\mu\epsilon')^2 + (\omega^2\mu\epsilon'')^2}}{2} \right]^{1/2}
\tag{5.107}
$$

$$
\beta_m = \left[\frac{-(k_{c,m}^2 - \omega^2\mu\epsilon') + \sqrt{(k_{c,m}^2 - \omega^2\mu\epsilon')^2 + (\omega^2\mu\epsilon'')^2}}{2} \right]^{1/2}
\tag{5.108}
$$

It can be seen that the frequency

$$
\tilde{\omega}_{c,m} = \frac{k_{c,m}}{\sqrt{\mu\epsilon'}}
\tag{5.109}
$$

serves as a kind of *de facto* cutoff frequency, and the behavior of α_m and β_m is as shown in Figure 5.15, with a smooth transition from attenuating to propagating conditions. The

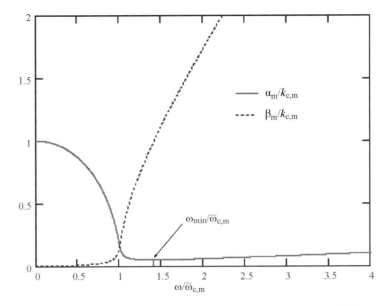

Figure 5.15 Real and imaginary parts of γ_m due to dielectric losses in hollow waveguide, for $\tan\delta = 0.05$.

attenuation coefficient α_m never becomes exactly zero, but if ϵ' and ϵ'' are substantially independent of frequency it attains a minimum value at

$$\omega_{\min} = \tilde{\omega}_{c,m}\sqrt{\frac{2}{1 + \tan^2\delta}} \tag{5.110}$$

where

$$\tan\delta = \frac{\epsilon''}{\epsilon'}$$

is the loss tangent of the dielectric. The minimum value of α_m attained at this frequency is

$$\alpha_{\min} = k_{c,m}\frac{\tan\delta}{\sqrt{1 + \tan^2\delta}} \tag{5.111}$$

and as ω increases further the attenuation subsequently increases with frequency. Eventually, the attenuation rate gets very large,

$$\alpha_m \to \frac{\omega\sqrt{\mu\epsilon'}\tan\delta}{[2 + 2\sqrt{1 + \tan^2\delta}]^{1/2}} \quad \text{as} \quad \omega \to \infty \tag{5.112}$$

although the rate of increase is fairly slow if $\epsilon'' \ll \epsilon'$ (as will be the case for a practical guide). If ϵ'' and/or ϵ' depends on frequency, the details of the behavior of α_m are rather more involved.

If ϵ'' is small enough that

$$\omega^2\mu\epsilon'' \ll |\omega^2\mu\epsilon' - k_{c,m}^2|$$

which, more than just $\epsilon'' \ll \epsilon'$, also requires the mode to be sufficiently far from the *de facto* cutoff frequency, then

$$\alpha_m \quad \simeq \quad \sqrt{k_{c,m}^2 - \omega^2\mu\epsilon'} \quad \text{if } \omega < \tilde{\omega}_{c,m}$$

$$\simeq \quad \frac{\omega^2\tan\delta}{2\tilde{\omega}_{c,m}\sqrt{\omega^2 - \tilde{\omega}_{c,m}^2}} \quad \text{if } \omega > \tilde{\omega}_{c,m} \tag{5.113}$$

$$\beta_m \simeq \frac{\omega^2 \tan \delta}{2\tilde{\omega}_{c,m}\sqrt{\tilde{\omega}_{c,m}^2 - \omega^2}} \quad \text{if } \omega < \tilde{\omega}_{c,m}$$

$$\simeq \sqrt{\omega^2 \mu \epsilon' - k_{c,m}^2} \quad \text{if } \omega > \tilde{\omega}_{c,m} \tag{5.114}$$

The treatment of wall losses in waveguides is less straightforward, and we will delay our study of them until Chapter 9 when the appropriate tools for the investigation are at our disposal.

5.8 NOTES AND REFERENCES

For general treatments of waveguide modes, see, for example:

Johnk (1975), Chapters 8 and 9;
Johnson (1965), Chapter 4;
Ramo, Whinnery and Van Duzer (1984), Chapter 8;
Mrozowski (1997);
Rozzi and Mongiardo (1997);

and also:

R. B. Adler, "Waves on inhomogeneous cylindrical structures," *Proc. IRE*, vol. 40, pp. 339-348 (1952).

Use of metallic waveguides at frequencies of several Terahertz is described in:

G. Gallot, S. P. Jamison, R. W. McGowan and D. Grischkowsky, "Terahertz waveguides," *J. Opt. Soc. Amer. B*, vol. 17, pp. 851-863 (2000).

In certain exceptional cases, waveguide modes may not be normalizable because either $\hat{P}$ or $\hat{P}_{osc}$ (or both) is zero. This can happen in a lossless waveguide for a so-called *complex mode*, where $\gamma = \alpha + j\beta$, and neither α nor β is zero. Consideration of such modes is somewhat delicate, and the interested reader is referred to the references at the end of Appendix J for more information.

The material on the general properties of uniformly filled waveguides is quite standard; see, for example:

Johnson (1965), Chapter 4;
Ramo, Whinnery and van Duzer (1984), Chapter 8;
Johnk (1988), Chapter 8;
Harrington (1961), Chapter 7;
Collin (1991), Chapter 5;
King, Mimno and Wing (1945), Chapter 3.
Milton and Schwinger (2006).

Proofs for the properties of solutions to (5.51)/(5.52) or (5.51)/(5.53) which are given in Section 5.4.3 can be found in, e.g.,

Jones (1964), Sections 4.6-4.12 and 5.1-5.2;

but see also

Courant and Hilbert (1953), Chapters 5 and 6;
Garabedian (1964), Chapter 11.

In the latter reference, the reader will find the following interesting estimates for the lowest TM and nontrivial TE eigenvalues of a hollow waveguide of arbitrary shape:

$$\kappa_1 \leq \pi(j'_{11})^2/A \simeq 10.65/A$$

$$\lambda_1 \geq \pi(j_{01})^2/A \simeq 18.17/A$$

where A is the cross-sectional area of the waveguide. It is also known that if κ_1 is degenerate, it is at most doubly so. See:

N. S. Nadirashvili, "On the multiplicity of eigenvalues of the Neumann problem," *Sov. Math. Dokl.*, vol. 33, pp. 281-282 (1986).

A useful summary of methods for computing the modes of various kinds of metallic waveguides, along with an extensive catalog of the then-known results from the literature is given in:

F. L. Ng, "Tabulation of methods for the numerical solution of the hollow waveguide problem," *IEEE Trans. Micr. Theory Tech.*, vol. 22, pp. 322-329 (1974).

Information on the use of ray diagrams and transverse resonance methods can be found in:

J. B. Keller and S. I. Rubinow, "Asymptotic solution of eigenvalue problems," *Annals of Physics*, vol. 9, pp. 24-75 (1960).

Continuous rather than discrete sets of rays can be used as the basis of approximate numerical solutions for waveguide modes. See:

H. Bateman, "On the numerical solution of linear integral equations," *Proc. Roy. Soc. London A*, vol. 100, pp. 441-449 (1922).

N. I. Platonov, "Application of plane wave expansions to the method of partial regions," *Sov. J. Commun. Technol. Electron.*, vol. 37, no. 14, pp. 20-28 (1992).

5.9 PROBLEMS

p5-1 Prove the statement following equation (5.9) that requiring the tangential electrical field to vanish at the perfect conductor is equivalent to requiring $\partial H_y/\partial x$ to vanish, for the case of parallel polarization.

p5-2 Derive expression (5.10) using the repeated reflection of plane waves produced by (5.9).

p5-3 If μ and ϵ of a medium filling a waveguide bounded by perfectly conducting walls are real functions of x and y, we know from Appendix J that γ^2 is real for modes of this waveguide for which the complex power $\hat{P} \neq 0$. Use equations (5.17)-(5.20) to show that the transverse electric field $\mathcal{E}_T$ of the mode satisfies the differential equation

$$\nabla_T \left[\frac{1}{\epsilon}\nabla_T \cdot (\epsilon\mathcal{E}_T)\right] - \mu\nabla_T \times (\frac{1}{\mu}\nabla_T \times \mathcal{E}_T) + (\omega^2\mu\epsilon + \gamma^2)\mathcal{E}_T = 0$$

along with appropriate boundary conditions. Hence, show that $\mathcal{E}_T$ may always be chosen to be real before normalization is imposed.

p5-4 Show that

$$\nabla \times [\mathcal{E}(x,y)e^{-\gamma z}] = (\nabla_T \times \mathcal{E} - \gamma\mathbf{u}_z \times \mathcal{E})e^{-\gamma z}$$

p5-5 Use the first of eqns. (5.12) and suitable vector identities from Appendix B to show that, if
$$\mathbf{u}_n \times \mathbf{E}\big|_S = 0$$
on a surface S to which the unit normal vector is $\mathbf{u}_n$, then
$$\mathbf{u}_n \cdot \mathbf{B}\big|_S = 0$$
if $\omega \neq 0$. This result means that if we enforce the boundary condition that tangential $\mathbf{E}$ is zero on a perfect electric conductor, there is no need to separately enforce the boundary condition that the normal component of $\mathbf{B}$ vanishes there. *Hint*: Use the fact that $\mathbf{u}_n = \nabla n$, where n is the normal distance of a point in space to the surface S.

p5-6 Show that for TEM, TE or TM modes of an *arbitrary* waveguide, the $\mathcal{E}$ and $\mathcal{H}$ fields are locally perpendicular, that is, $\mathcal{E} \cdot \mathcal{H} = 0$. Will this be true in general for hybrid modes?

p5-7 Give details of the derivation of (5.34) and (5.35).

p5-8 Show that the $\mathcal{E}$ lines of a TE mode obey the equations $\mathcal{H}_z(x, y(x)) = K$, where K is a constant which takes on different values for different field lines. Show that the $\mathcal{H}$ lines of a TM mode obey the equations $\mathcal{E}_z(x, y(x)) = K$. Use the method of Appendix I.

p5-9

    (a) Show that
$$\hat{P}_{\text{osc}} = \frac{j\omega\epsilon\gamma}{2k_c^4} \int_{S_0} \nabla_T \mathcal{E}_z \cdot \nabla_T \mathcal{E}_z \, dS$$

    for a TM mode, and
$$\hat{P}_{\text{osc}} = \frac{j\omega\mu\gamma}{2k_c^4} \int_{S_0} \nabla_T \mathcal{H}_z \cdot \nabla_T \mathcal{H}_z \, dS$$

    for a TE mode of a hollow metallic waveguide.

    (b) Verify the identity
$$\nabla_T \cdot (\mathcal{E}_z \nabla_T \mathcal{E}_z) = \mathcal{E}_z \nabla_T^2 \mathcal{E}_z + \nabla_T \mathcal{E}_z \cdot \nabla_T \mathcal{E}_z$$

    Use this result together with that of part (a) to prove that
$$\hat{P}_{\text{osc}} = \frac{j\omega\epsilon\gamma}{2k_c^2} \int_{S_0} \mathcal{E}_z^2 \, dS$$

    for a TM mode. Likewise, show that
$$\hat{P}_{\text{osc}} = \frac{j\omega\mu\gamma}{2k_c^2} \int_{S_0} \mathcal{H}_z^2 \, dS$$

    for a TE mode.

p5-10 Show that the complex oscillatory power carried by a TE or TM mode of a uniformly filled, lossless hollow metallic waveguide at the cutoff frequency is either zero or ∞ (if the fields have finite amplitude), and thus that it is impossible to normalize a such a mode at cutoff.

p5-11 Show that TE modes in a closed waveguide below their cutoff frequency have a greater magnetic than electric time-average stored energy per unit length ($U_{M0} > U_{E0}$), while TM modes on such a guide have $U_{E0} > U_{M0}$. Here the energy densities are as defined in (J.24) and (J.25). [*Hint:* Consider the results of problem p5-9, modified to give the complex power $\hat{P}$ instead of $\hat{P}_{\text{osc}}$.]

p5-12 Show that a TM mode of a waveguide possessing an $\mathcal{E}_z$ field that behaves as $Y_l(k_c\rho)\cos l\phi$ for ρ near 0 violates the condition that stored energy in a small cylinder about $\rho = 0$ must be finite.

p5-13 Obtain the TM mode $\mathcal{E}_z$ field and eigenvalues for the hollow semi-circular waveguide shown below, using separation of variables.

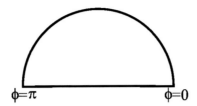

p5-14 Obtain eqns. (5.99) for the fields of TE rectangular waveguide modes by using the method of separation of variables. Likewise obtain (5.103) for the fields of the TM modes.

p5-15 A circular metallic waveguide of radius a is fitted with a perfectly conducting fin located in $\phi = 0$ as shown.

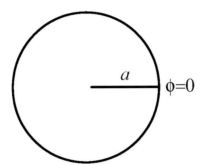

Obtain expressions for the TE mode cutoff wavenumbers k_c for this guide. Give the numerical value of $k_c a$ for the fundamental mode.

p5-16 Using separation of variables, find the $\mathcal{H}_z$ field, and the cutoff wavenumbers k_c of the TE modes of the quarter-circle air-filled metallic waveguide of radius a shown below.

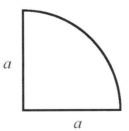

The boundary walls are at $\phi = 0$, $\phi = \pi/2$ and $\rho = a$.

p5-17 Using separation of variables, find the $\mathcal{E}_z$ field, and the cutoff wavenumbers k_c of the TM modes of the quarter-circle air-filled metallic waveguide of radius a shown in problem p5-16.

p5-18 Repeat problem p5-16, but for a waveguide whose cross-section is a $\frac{3}{4}$ sector of a circle as shown below.

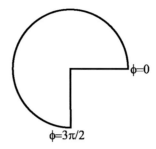

$\phi=0$

$\phi=3\pi/2$

Determine an expression for the cutoff wavenumber of the dominant TE mode, and compare it with that of the ordinary full-circle waveguide.

p5-19 Use the ray method to derive the eigenvalues and the longitudinal electric field $\mathcal{E}_z$ for the TM modes of a hollow rectangular waveguide. Show that these are identical to those obtained using separation of variables.

p5-20 Use the ray method to derive the TM mode eigenvalues for the right isosceles triangular waveguide $(45° - 45° - 90°)$ shown below:

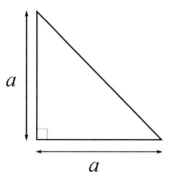

a

a

p5-21 Repeat Problem p5-20 for the TE mode eigenvalues.

p5-22 Using the ray method, obtain an expression for all the cutoff wavenumbers k_c of the TE modes of the hollow metallic waveguide whose wall is in the form of an equilateral triangle of side $2a$ shown below.

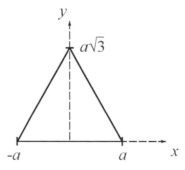

y

$a\sqrt{3}$

$-a$ a x

Use the coordinate axes shown in the figure. Make sure you include all possible ray directions in your expression for the $\mathcal{H}_z$ field.

p5-23 Repeat problem p5-22 for the TM mode case.

p5-24 The transverse electric fields of a TE_{mn} mode of a rectangular hollow waveguide are given by (5.99). Using the method of Appendix I, show that the $\mathcal{E}_T$-field lines for this mode are given by

$$\left(\cos \frac{m\pi x}{a}\right)\left(\cos \frac{n\pi y}{d}\right) = K = \text{constant}$$

with different field lines generated by different values of K. Use this expression to plot the $\mathcal{E}$-lines for the TE_{11} (rectangular) mode.

p5-25 Compare the field patterns for TE and TM modes of hollow circular waveguides given in Figures 5.8 and 5.9 with those for TE and TM modes of a rectangular guide shown below:

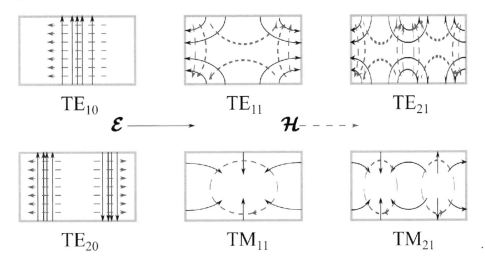

It seems clear that the TM_{01} (circular) mode is closely related to the TM_{11} (rectangular) mode, and that one would change continuously into the other if the boundary wall were gradually changed from a circular to a rectangular cross-section. It is likewise clear that a similar correspondence exists between the TE_{11} (circular) and the TE_{10} (rectangular) modes. Identify the *rectangular* waveguide modes which correspond to the following modes:

(a) TM_{11} (circular)

(b) TE_{01} (circular)

[Hint for part (b): Consider the special case of a square cross-section.]

p5-26 Show that, if $\hat{\epsilon} = \epsilon' - j\sigma/\omega$, and ϵ' and σ are independent of frequency, and $\omega > \tilde{\omega}_{c,m}$, then α_m due to dielectric losses is a monotonic function of ω, increasing if $\tilde{\omega}_{c,m} < \sigma/2\epsilon'$, and decreasing if $\tilde{\omega}_{c,m} > \sigma/2\epsilon'$. The only exception to this is if $\tilde{\omega}_{c,m} = \sigma/2\epsilon'$, when α_m is independent of frequency. Here $\tilde{\omega}_{c,m}$ is given by (5.109).

p5-27 An X-band rectangular metallic waveguide ($a = 1$ in., $d = 0.5$ in.) is uniformly filled with a lossy dielectric for which $\mu = \mu_0$, $\epsilon' = 2.25\epsilon_0$ and $\epsilon'' = 10^{-2}\epsilon_0$. These material properties are independent of frequency. Find the frequency $f_{\min}$ at which the attenuation coefficient α of the dominant TE_{10} mode of this waveguide is a minimum, and calculate the value of this attenuation coefficient. In spite of the fact that standard waveguide dimensions are given in inches, express all your results in MKS units.

p5-28 A circular metallic waveguide of radius $a = 1$ cm is uniformly filled with a lossy dielectric for which $\mu = \mu_0$, $\epsilon' = 4\epsilon_0$ and $\epsilon'' = 10^{-3}\epsilon_0$. These material properties are independent of frequency. Find the frequency $f_{\min}$ at which the attenuation coefficient α of the dominant TE_{11} mode of this waveguide is a minimum, and calculate the value of this attenuation coefficient.

6 SURFACE WAVE MODES: BASIC OPTICAL WAVEGUIDES

6.1 INTRODUCTION

A metal which is a nearly perfect reflector all the way up to microwave frequencies begins to behave differently as frequency increases beyond a value of $50 \sim 100$ GHz (the millimeter-wave band). What constitutes a "smooth" metal surface at 1 GHz might be unacceptably rough (on the scale of a wavelength) at 100 GHz. Moreover, whereas the metal may be a "good" conductor ($\epsilon''/\epsilon' = \sigma/\omega\epsilon \gg 1$) at microwave frequencies, it may exhibit completely different behavior at higher frequencies. For example, silver at $\lambda = 1$ μm has a complex relative dielectric constant of

$$\hat{\epsilon}/\epsilon_0 = \epsilon_r{}' - j\epsilon_r'' \simeq -25 - j2 \tag{6.1}$$

or, in terms of the complex refractive index $\hat{n} = (\hat{\epsilon}/\epsilon_0)^{1/2}$ (which is more commonly used at optical frequencies)[1]

$$\hat{n} \simeq 0.2 - j5 \tag{6.2}$$

The negative value of ϵ_r' indicates very different electrical behavior from that at lower frequencies. In fact, like many metals at optical frequencies, silver behaves like a *plasma* (an electrically neutral medium containing ionic free charges) which can possess permittivities with negative real parts like those of (6.1).

It is clear, then, that the changing material parameters of metals will cause traditional metallic waveguide structures to perform differently at higher frequencies than as described in Chapter 5 (this is also true of the TEM and quasi-TEM structures which we will study in Chapter 8. Moreover, the fabrication of a traditional single-mode waveguide at such high frequencies may be quite difficult: the dimensions required for a single-mode metallic rectangular waveguide at a free-space wavelength of $\lambda_0 = 1$ μm, for instance, would be 0.5 μm $< a = 2b < 1\mu$m. As sources and devices become available at such high frequencies (millimeter-wave, infrared, visible, ultraviolet and even X-ray frequencies) we need to have practical waveguides which have low losses to take over the role that traditional waveguides play at lower frequencies.

Before we start searching for new structures, however, we should first re-examine the basic physics of wave guidance. This will provide us with some principles by which useful waveguiding structures can be designed.

It will be recalled from Chapter 5 that the propagating modes of metal-boundary waveguides can be interpreted in terms of propagating plane waves which are repeatedly reflected from the conducting walls. Similarly, one may interpret TEM modes (Chapter 8) as basically axially propagating plane waves whose fields have been "warped" in order to satisfy boundary conditions at the two conductors which guide the wave, but which have the same propagation coefficient and wave impedance as the plane wave.

The mechanism of successive reflections does not require metallic boundaries. It can also take place with dielectric interfaces under appropriate conditions. A plane wave incident on a dielectric interface as shown in Figure 6.1 may be totally reflected if the angle of incidence

---

[1]As with ϵ, we will usually write $\hat{n}$ as simply n, with the understanding that it can be made complex if desired in order to describe a lossy medium.

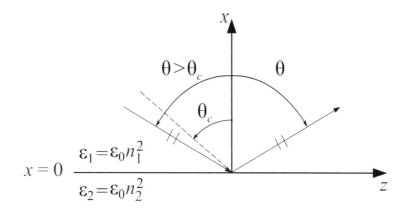

Figure 6.1 Total internal reflection at a dielectric interface.

is greater than the so-called critical angle. This critical angle exists only if $\epsilon_1 > \epsilon_2$, but suggests a variety of possible waveguiding structures using this total reflection phenomenon. Figure 6.2 illustrates a variety of guides of this type.

Figure 6.2(a) depicts the step-index optical fiber, which is perhaps the simplest realistic guide of this type. Figure 6.2(b) illustrates a *channel guide* embedded in a substrate, as is encountered in what has come to be called *integrated optics*. The *rib guide* of Figure 6.2(c) operates on slightly more complicated principles, as we will see in Chapter 7.

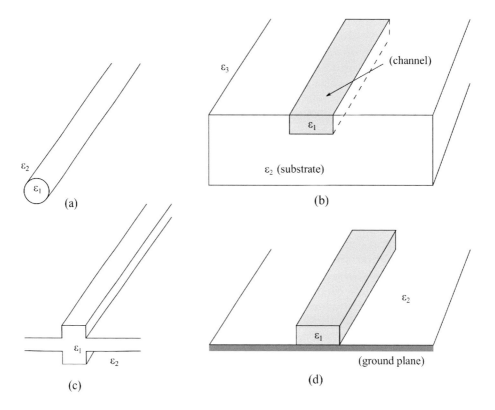

Figure 6.2 (a) Optical fiber; (b) channel (integrated optical) guide; (c) rib guide; (d) image guide.

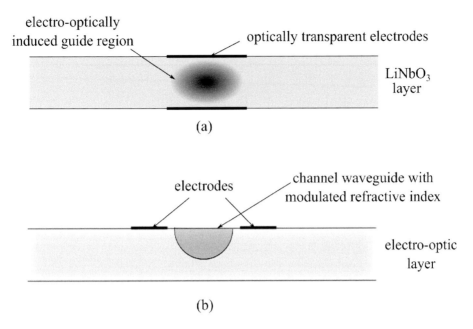

Figure 6.3 (a) Electro-optically induced waveguide; (b) electro-optic modulator.

The *image guide* shown in Figure 6.2(d) is commonly used at millimeter-wave frequencies, since its ground plane provides desirable partial shielding of the waveguide circuitry. All these waveguides are practical because smooth interfaces between dielectrics can be made, and materials are available in the mm-wave band and higher which have acceptably low losses. This type of guide is especially ideal in the optical region.

Indeed, it is even possible to induce the change in permittivity by external means, using the *electro-optic* or Pockels effect. This effect appears in certain types of materials (LiNbO$_3$ or GaAs, for instance), and manifests itself as a change in the index of refraction seen by an optical frequency field due to the presence of a much lower frequency (e. g., microwave) electric field in the material. Thus, a waveguide can be created where none was present before by impressing a DC voltage across a pair of electrodes on a piece of electro-optic material as shown in Figure 6.3(a). Another application of these materials is to change the properties of an existing waveguide by impressing a variable voltage across a pair of electrodes near the guide region. This can achieve a modulation of the phase of a signal in the waveguide, and offers the potential of utilizing the very high bandwidth that optical carrier frequencies offer (Figure 6.3(b)). Further information can be found in the references cited at the end of the chapter.

6.2 DIELECTRIC SLAB WAVEGUIDES—THE RAY DESCRIPTION

In order not to obscure the essentials of the problem, it is traditional to treat first the simple, though somewhat unrealistic, dielectric slab waveguide of Figure 6.4. Consider a dielectric slab of thickness a and permittivity ϵ_1 situated between two semi-infinite cladding regions of permittivity ϵ_2 as shown in Figure 6.4. Let us not yet make any restrictions on ϵ_1 and ϵ_2. The slab and cladding extend to infinity in both y- and z-directions, and, to obtain a purely two-dimensional problem, we seek solutions independent of y:

$$\frac{\partial}{\partial y} = 0$$

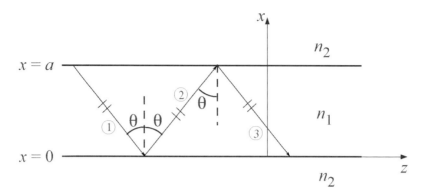

Figure 6.4 Successive ray reflections in a dielectric slab waveguide ($n_1 > n_2$).

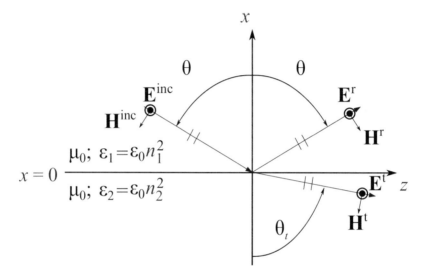

Figure 6.5 Plane-wave reflection at a dielectric interface—TE polarization.

This structure is then analogous to the parallel-plate metallic waveguide considered in Section 5.1. All media are assumed to have the same permeability $\mu = \mu_0$.

In order to treat this structure using the ray method, we consider the plane-wave reflection problem for a dielectric interface in some detail (note the change in coordinate system from the treatment in Section 1.4.1, which now aligns the z-direction with the direction of propagation of the mode). For a TE-polarized incident wave

$$\mathbf{E}^{inc} = \mathbf{u}_y E_0 e^{-jk_0 n_1(-x\cos\theta + z\sin\theta)} \qquad (x > 0) \qquad (6.3)$$

as shown in Figure 6.5, there will arise a reflected wave

$$\mathbf{E}^r = \mathbf{u}_y \rho_{\text{TE}} E_0 e^{-jk_0 n_1(x\cos\theta + z\sin\theta)} \qquad (x > 0) \qquad (6.4)$$

and, in the second dielectric, a transmitted wave

$$\mathbf{E}^t = \mathbf{u}_y \tau_{\text{TE}} E_0 e^{-jk_0 n_2(-x\cos\theta_t + z\sin\theta_t)} \qquad (x < 0) \qquad (6.5)$$

where E_0 is the amplitude of the incident wave, ρ_{TE} the *electric-field* reflection coefficient of the wave, and τ_{TE} is the *electric-field* transmission coefficient into the second region. Here, $k_0 = \omega\sqrt{\mu_0\epsilon_0} = 2\pi/\lambda_0$ is the wavenumber of free space.

If both regions are assumed to be nonmagnetic, the Fresnel reflection coefficient ρ_{TE} from (1.85) takes the form:

$$\rho_{\text{TE}} = \frac{n_1 \cos\theta - n_2 \cos\theta_t}{n_1 \cos\theta + n_2 \cos\theta_t} \tag{6.6}$$

where θ_t is related to θ by Snell's law:

$$n_2 \sin\theta_t = n_1 \sin\theta \tag{6.7}$$

If $n_1 > n_2$, there is a critical angle $\theta = \theta_c$ given by

$$\theta_c = \sin^{-1}\left(\frac{n_2}{n_1}\right) \tag{6.8}$$

such that when $\theta > \theta_c$, the transmission angle θ_t becomes complex in such a way that $\sin\theta_t > 1$ and $\cos\theta_t$ is *negative imaginary*. Under these conditions, we have from (1.89)-(1.90) that $|\rho_{\text{TE}}| = 1$; specifically,

$$\rho_{\text{TE}} = e^{j\chi_{\text{TE}}(\theta)} \qquad (\theta > \theta_c) \tag{6.9}$$

where

$$\chi_{\text{TE}}(\theta) = 2\tan^{-1}\left[\frac{\sqrt{n_1^2 \sin^2\theta - n_2^2}}{n_1 \cos\theta}\right] = \pi - 2\tan^{-1}\left[\frac{n_1 \cos\theta}{\sqrt{n_1^2 \sin^2\theta - n_2^2}}\right] = \pi - 2\sin^{-1}\left[\frac{n_1 \cos\theta}{\sqrt{n_1^2 - n_2^2}}\right] \tag{6.10}$$

Hence we have the phenomenon of *total internal reflection* when $n_1 > n_2$; all energy present in the incident wave is carried back away from the interface in the reflected wave. The transmitted wave is *evanescent*; the x-dependent factor behaves as

$$\exp\left[-k_0 n_2 |x| \frac{\sqrt{\sin^2\theta - \sin^2\theta_c}}{\sin\theta_c}\right]$$

and decays exponentially as x becomes more negative. This wave carries no energy away from the interface but merely stores it temporarily in the vicinity of $x = 0$ before it is carried away by the reflected wave.

If we compare equation (6.9) to the reflection coefficient of this wave at a perfect conductor, $e^{j\pi}$ (obtained from (6.9) and (6.10) by letting n_2 become complex and taking $|n_2| \to \infty$), we see that in our case there is an additional phase shift after reflection as compared to the perfectly-conducting boundary. It is as if the reflection had actually occurred at a perfectly conducting boundary located at a distance $d_{\text{TE}}(\theta)$ somewhat below the position of the dielectric interface. The additional phase shift is now due to further propagation through the same material as medium 1 (see Figure 6.6):

$$\begin{aligned}
d_{\text{TE}}(\theta) &= \frac{1}{k_0 n_1 \cos\theta}\left[\frac{\pi - \chi_{\text{TE}}}{2}\right] \\
&= \frac{1}{k_0 n_1 \cos\theta}\sin^{-1}\left[\frac{n_1 \cos\theta}{\sqrt{n_1^2 - n_2^2}}\right]
\end{aligned} \tag{6.11}$$

The field E_y at $x = -d_{\text{TE}}$ would be zero if such a perfect conductor were actually present. This shift in the apparent location of the reflecting boundary also can be observed for reflection of a light beam of *finite* width, in which case it is called the *Goos-Hänchen shift*,

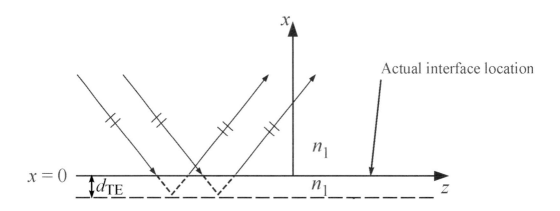

Figure 6.6 Goos-Hänchen shift modeled as reflection from displaced perfect conductor–TE polarization.

after the researchers who first detected the effect experimentally. In the limit of grazing incidence ($\theta \simeq \pi/2$), d_{TE} approaches the value

$$d_{\mathrm{TE}} \simeq \frac{1}{k_0 \sqrt{n_1^2 - n_2^2}} \tag{6.12}$$

which is independent of incidence angle. We show how d_{TE} varies with θ between grazing and critical angles in Figure 6.7. If $\cos\theta < 0.25 \cos\theta_c$, the error in using the approximate value in (6.12) is less than 1%.

We can now apply the same procedure as in the case of the parallel-plate waveguide (Chapter 5) to obtain a transverse resonance condition for the TE modes of this dielectric slab. If in $0 < x < a$, the field is assumed to be a superposition of upward and downward traveling plane waves

$$E_y = \left(E_1 e^{+jk_0 n_1 x \cos\theta} + E_2 e^{-jk_0 n_1 x \cos\theta}\right) e^{-jk_0 n_1 z \sin\theta} \tag{6.13}$$

as shown in Figure 6.4, then the reflection conditions at $x = 0$ and $x = a$ are:

$$E_2 = \rho_{\mathrm{TE}} E_1$$
$$E_1 e^{jk_0 n_1 a \cos\theta} = \rho_{\mathrm{TE}} E_2 e^{-jk_0 n_1 a \cos\theta} \tag{6.14}$$

or,

$$\exp[2jk_0 n_1 a \cos\theta] = \exp[2j\chi_{\mathrm{TE}}(\theta)] \tag{6.15}$$

Hence, the TE mode transverse resonances for the dielectric slab occur when θ satisfies the equation

$$k_0 n_1 a \cos\theta = \chi_{\mathrm{TE}}(\theta) + m\pi \quad ; \quad m = 0, 1, 2, \ldots \tag{6.16}$$

Each value of m corresponds to a mode of the slab, which we designate as the TE_m mode.

In view of (6.11), we can also write equation (6.16) as

$$k_0 n_1 (a + 2d_{\mathrm{TE}}) \cos\theta = (m+1)\pi \tag{6.17}$$

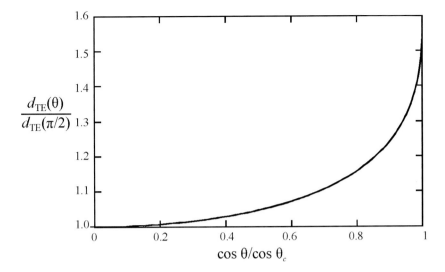

Figure 6.7 Dependence of normalized d_{TE} on angle of incidence.

Comparing this with (5.5), we can see that the TE modes of the dielectric slab can be thought of as equivalent to those of a parallel-plate waveguide of an *equivalent width* $a_{eq} = a + 2d_{TE}$ (Figure 6.8). The equivalent width depends not only on the contrast in refractive indices n_1 and n_2, and on the wavenumber k_0, but also on the angle θ at which the plane waves are incident at the interfaces. If the incidence angle is sufficiently close to grazing, we can see from (6.12) that the equivalent parallel-plate waveguide will have the same equivalent width

$$a_{eq} \simeq a + \frac{2}{k_0 \sqrt{n_1^2 - n_2^2}}$$

for any mode at a given frequency of operation. We see that $a_{eq} \to \infty$ as $k_0 \to 0$, and this suggests the possibility of a mode with zero cutoff frequency. That this does actually happen will be shown in Section 6.4.

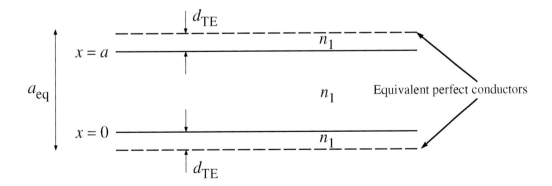

Figure 6.8 Equivalent parallel-plate waveguide for TE modes.

6.3   DIELECTRIC SLAB WAVEGUIDES—FIELD ANALYSIS

While the ray description of slab waveguide modes yields much in the way of physical insight, it is rather limited in terms of the class of waveguides which it can treat. In this section, we will rederive the TE mode eigenvalue equation and fields from a more general viewpoint.

On setting $\partial/\partial y = 0$ in equations (5.17)-(5.20), we get two separate groups of equations:

$$
\begin{aligned}
\gamma \mathcal{E}_y &= -j\omega\mu\mathcal{H}_x \\
\frac{\partial \mathcal{E}_y}{\partial x} &= -j\omega\mu\mathcal{H}_z \\
-\frac{\partial \mathcal{H}_z}{\partial x} - \gamma\mathcal{H}_x &= j\omega\epsilon\mathcal{E}_y
\end{aligned}
\tag{6.18}
$$

$$
\begin{aligned}
\gamma \mathcal{H}_y &= j\omega\epsilon\mathcal{E}_x \\
\frac{\partial \mathcal{H}_y}{\partial x} &= j\omega\epsilon\mathcal{E}_z \\
-\frac{\partial \mathcal{E}_z}{\partial x} - \gamma\mathcal{E}_x &= -j\omega\mu\mathcal{H}_y
\end{aligned}
\tag{6.19}
$$

The field components $(\mathcal{E}_y, \mathcal{H}_x, \mathcal{H}_z)$ are completely decoupled from $(\mathcal{H}_y, \mathcal{E}_x, \mathcal{E}_z)$ as far as (6.18) and (6.19) are concerned, and the boundary conditions do not couple them either. We can thus examine each set of components separately.

The fields satisfying (6.18) are TE-polarized: by eliminating $\mathcal{H}_x$ and $\mathcal{H}_z$ among the three equations, we obtain a reduced wave equation for $\mathcal{E}_y$:

$$
\left(\frac{\partial^2}{\partial x^2} + \omega^2\mu\epsilon + \gamma^2 \right) \mathcal{E}_y = 0
\tag{6.20}
$$

Of course, the x-derivatives are really ordinary derivatives, and we anticipate that $\gamma = j\beta$ for useful propagating modes. Moreover, ϵ depends on position, equaling ϵ_1 for $0 < x < a$, and ϵ_2 for $x < 0$ or $x > a$. We thus rewrite (6.20) as

$$
\left(\frac{d^2}{dx^2} + h_1^2 \right) \mathcal{E}_y = 0; \qquad (0 < x < a)
\tag{6.21}
$$

$$
\left(\frac{d^2}{dx^2} - p_2^2 \right) \mathcal{E}_y = 0; \qquad (x > a \quad \text{or} \quad x < 0)
\tag{6.22}
$$

where

$$
h_1^2 = \omega^2\mu\epsilon_1 - \beta^2 = k_1^2 - \beta^2
\tag{6.23}
$$

$$
p_2^2 = \beta^2 - \omega^2\mu\epsilon_2 = \beta^2 - k_2^2
\tag{6.24}
$$

and for definiteness, we specify that p_2 and h_1 be taken with positive real parts or purely positive imaginary. Note that h_1 and β are related to the ray angle θ of Section 6.2 by

$$
\begin{aligned}
\beta &= k_1 \sin\theta \\
h_1 &= k_1 \cos\theta
\end{aligned}
\tag{6.25}
$$

In the cladding, the solutions have the general form

$$
\begin{aligned}
\mathcal{E}_y &= Ae^{-p_2 x} + Be^{+p_2 x}; \qquad (x > a) \\
&= Ce^{-p_2 x} + De^{+p_2 x}; \qquad (x < 0)
\end{aligned}
\tag{6.26}
$$

Now, the solutions we seek are *guided* modes; if the fields are finite, we demand that they carry a finite amount of power (per unit width in the y-direction in this two-dimensional waveguide). A field solution that does not decay at infinity is not guided in this sense because the integral that gives the total power is divergent (it extends to $x = \pm\infty$). Such a field would be impossible to excite with an actual source of finite power. Thus we eliminate the growing solutions by setting $B = C = 0$;

$$
\begin{aligned}
\mathcal{E}_y &= Ae^{-p_2 x} \quad (x > a) \\
&= De^{+p_2 x} \quad (x < 0)
\end{aligned}
\tag{6.27}
$$

Within the slab, the general solution of (6.21) is equally well known:

$$
\mathcal{E}_y = E\cos(h_1 x) + F\sin(h_1 x) \quad (0 < x < a)
\tag{6.28}
$$

The remaining unknown constants A, D, E, F and β (which is, of course, embedded in h_1 and p_2) must be found by enforcing the boundary conditions at $x = 0$ and $x = a$.

At $x = 0$, $\mathcal{E}_y$ and $\mathcal{H}_z$ (the tangential field components) must be continuous. Continuity of $\mathcal{E}_y$ gives, from (6.27) and (6.28),

$$
\mathcal{E}_y|_{x=0} = D = E
\tag{6.29}
$$

On the other hand, continuity of $\mathcal{H}_z$ implies continuity of $d\mathcal{E}_y/dx$ [cf. (6.18)]; at $x = 0$ this gives

$$
\frac{d\mathcal{E}_y}{dx}\bigg|_{x=0} = p_2 D = h_1 F
\tag{6.30}
$$

At $x = a$, continuity of $\mathcal{E}_y$ gives

$$
\begin{aligned}
\mathcal{E}_y|_{x=a} &= Ae^{-p_2 a} \\
&= E\cos h_1 a + F\sin h_1 a
\end{aligned}
\tag{6.31}
$$

Using (6.29)–(6.31), we can eliminate all the unknown amplitude constants except D, and reduce (6.27) and (6.28) to

$$
\begin{aligned}
\mathcal{E}_y &= De^{p_2 x} \quad (x < 0) \\
&= D\left[\cos h_1 x + \frac{p_2}{h_1}\sin h_1 x\right] \quad (0 < x < a) \\
&= De^{-p_2(x-a)}\left[\cos h_1 a + \frac{p_2}{h_1}\sin h_1 a\right] \quad (x > a)
\end{aligned}
\tag{6.32}
$$

As always, we are left with one remaining indeterminate amplitude constant (in the absence of a specified source), but our one remaining boundary condition will give us a condition on the propagation coefficient β. Continuity of $d\mathcal{E}_y/dx$ at $x = a$ gives

$$
-p_2 D\left[\cos h_1 a + \frac{p_2}{h_1}\sin h_1 a\right] = D\left[-h_1\sin h_1 a + p_2\cos h_1 a\right]
$$

or

$$
\tan(h_1 a) = \frac{2p_2 h_1}{h_1^2 - p_2^2}
\tag{6.33}
$$

This eigenvalue equation determines the allowed values of propagation coefficient β (recall that h_1 and p_2 are both specified functions of β from (6.23) and (6.24)).

In the notation of this section, the previously obtained eigenvalue equation (6.16) has the form

$$h_1 a = m\pi + 2\tan^{-1}\left(\frac{p_2}{h_1}\right) \tag{6.34}$$

or, alternatively,

$$h_1 a = (m+1)\pi - 2\tan^{-1}\left(\frac{h_1}{p_2}\right) \tag{6.35}$$

Other forms of the eigenvalue equation are

$$\sin(h_1 a) = (-1)^m \frac{2p_2 h_1}{k_1^2 - k_2^2} \tag{6.36}$$

$$\cos(h_1 a) = (-1)^m \frac{h_1^2 - p_2^2}{k_1^2 - k_2^2} \tag{6.37}$$

It is left as an exercise for the reader to demonstrate that these are all equivalent to (6.33). We will address the solution of the eigenvalue equation in the next section; let us now look briefly at a few ways of expressing the field distribution.

Expression (6.32) for the field in the core ($0 < x < a$) can be rewritten in the form

$$\mathcal{E}_y = D\frac{\sin[h_1(x + d_{\text{TE}})]}{\sin(h_1 d_{\text{TE}})} \tag{6.38}$$

where, from (6.25), we have

$$d_{\text{TE}} = \frac{1}{h_1}\tan^{-1}(\frac{h_1}{p_2}) = \frac{1}{h_1}\sin^{-1}\left(\frac{h_1}{\sqrt{k_1^2 - k_2^2}}\right) = \frac{1}{h_1}\cos^{-1}\left(\frac{p_2}{\sqrt{k_1^2 - k_2^2}}\right) \tag{6.39}$$

With D as an arbitrary amplitude, we recognize this as a TE mode of the equivalent parallel-plate waveguide of Figure 6.8.

Because of the symmetry of the slab waveguide, its mode fields are often referred to the center of the slab rather than to one of its boundaries as we have done here. To obtain a symmetric form, we write (6.38) as

$$\mathcal{E}_y = \frac{D}{\sin(h_1 d_{\text{TE}})}\left\{\cos[h_1(\frac{a}{2} + d_{\text{TE}})]\sin[h_1(x - \frac{a}{2})]\right.$$
$$\left. + \sin[h_1(\frac{a}{2} + d_{\text{TE}})]\cos[h_1(x - \frac{a}{2})]\right\} \tag{6.40}$$

From (6.35) and (6.39), we have $h_1(\frac{a}{2} + d_{\text{TE}}) = (m+1)\frac{\pi}{2}$, so that the first term in (6.40) vanishes when m is even, and the second when m is odd. The field distributions $\mathcal{E}_y$ of the TE$_m$ modes thus share the parity of their mode index:

$$\mathcal{E}_y = \frac{(-1)^{m/2}D}{\sin(h_1 d_{\text{TE}})}\cos\left[h_1\left(x - \frac{a}{2}\right)\right] \qquad (m \text{ even})$$
$$= \frac{(-1)^{(m+1)/2}D}{\sin(h_1 d_{\text{TE}})}\sin\left[h_1\left(x - \frac{a}{2}\right)\right] \qquad (m \text{ odd}) \tag{6.41}$$

Using (6.36) and (6.37), we find that the field in $x > a$ from (6.32) can be rewritten as

$$\mathcal{E}_y = (-1)^m De^{-p_2(x-a)} \qquad (x > a) \tag{6.42}$$

Comparing this to the first line of (6.32), we see that the external fields are also even or odd according to the value of m. The form of the eigenvalue equation usually encountered when the modes are considered in this way is

$$\tan\left(\frac{h_1 a}{2}\right) = \frac{p_2}{h_1} \quad (m \text{ even})$$

$$= -\frac{h_1}{p_2} \quad (m \text{ odd}) \tag{6.43}$$

which can be obtained directly from (6.33) using the half-angle trigonometric identity.

The field amplitude D is determined by imposing the normalization condition (5.27). It is shown in problem p6-5 that

$$\frac{1}{2}\int_{-\infty}^{\infty} \boldsymbol{\mathcal{E}} \times \boldsymbol{\mathcal{H}} \cdot \mathbf{u}_z \, dx = -\frac{1}{2}\int_{-\infty}^{\infty} \mathcal{E}_y \mathcal{H}_x \, dx = \frac{\beta a_{\text{eff}}}{4} D^2 \frac{\omega(\epsilon_1 - \epsilon_2)}{h_1^2} \tag{6.44}$$

where a_{eff} is the *effective* width of the slab, defined (for a given mode at a specific frequency) as

$$a_{\text{eff}} = a + 2/p_2 \tag{6.45}$$

The effective width is the physical width of the core plus a so-called *penetration depth*, $1/p_2$, (the distance from the dielectric interface at which the field in the cladding has decayed to $1/e \simeq 37\%$ of its value at the interface) added on to each side of the core. This effective width turns out to be a more useful measure of the concentration of energy near the core than the equivalent width $a_{\text{eq}} = a + 2d_{\text{TE}}$ defined earlier. Note that in this example, as for all two-dimensional waveguides with $\partial/\partial y = 0$ (e. g., the parallel-plate waveguide in Chapter 5), we must carry out the normalization in (5.27) using an integral over x only, since the cross-section is uniform and infinite in the y-direction. Equating (6.44), which is the two-dimensional equivalent of (5.27), to 1, we have

$$D = \frac{2h_1 a}{V}\sqrt{\frac{\omega\mu}{\beta a_{\text{eff}}}} \tag{6.46}$$

where V is defined by (6.47) below.

6.4 DIELECTRIC SLAB WAVEGUIDES—THE EIGENVALUE EQUATION FOR TE MODES

The eigenvalue equation in any of its forms (6.33)–(6.35) or (6.43) cannot be solved for β explicitly; such an equation is known as a *transcendental equation*. Either graphical, numerical, or approximate methods must therefore be used to determine the propagation coefficients. Although a computer program that will find the roots of the eigenvalue equation can be constructed, the task is not altogether a straightforward one. The iteration method outlined in Appendix K is one possibility, but it requires rearranging the eigenvalue equation into a suitably convergent form. An alternate approach to the solution of the eigenvalue equation is a graphical method. from which much insight can be obtained. For this purpose it is most convenient to deal with equations (6.43). Introducing a *normalized frequency V* as

$$V = \sqrt{k_1^2 - k_2^2}\, a$$

$$= \sqrt{n_1^2 - n_2^2}\, k_0 a \tag{6.47}$$

$$= \sqrt{(p_2 a)^2 + (h_1 a)^2}$$

where n_1 and n_2 are the core and cladding refractive indices respectively, equations (6.43) become, after some rearrangement,

$$\sqrt{V^2 - (h_1 a)^2} = p_2 a \quad = \quad h_1 a \tan(h_1 a/2) \equiv F_e(h_1 a) \qquad (m \text{ even})$$
$$= \quad -h_1 a \cot(h_1 a/2) \equiv F_o(h_1 a) \qquad (m \text{ odd}) \qquad (6.48)$$

For values of $h_1 a > V$, the square root in (6.48), which is equal to $p_2 a$, becomes imaginary, and no such solutions are allowed because their fields do not decrease away from the slab [cf. (6.32)]. Similarly, imaginary values of h_1 are not allowed because in that event the left side of (6.48) would be positive real, while the right side was negative real, and no root would exist.

Figure 6.9(a) gives plots of the left and right sides of equations (6.48) for m even. The circles in the figure have radius V. We locate solutions of (6.43) for m even at the points of intersection of the curves in Figure 6.9. For $V < \pi$, the situation is as shown for the case $V = 2$. Only one intersection occurs—for $0 < h_1 a \leq \pi$—and the corresponding mode of propagation (the dominant TE mode) is called the TE_0 mode. This mode has *zero cutoff frequency*: like a TEM or quasi-TEM mode (to be studied in Chapter 8), it always propagates, and in this case, has a propagation coefficient β near that of a plane wave in the cladding, k_2. If V is increased beyond 2π (typified by the case $V = 7$ in Figure 6.9), in addition to the TE_0 mode, which continues to exist, a second solution of the eigenvalue equation for even m comes into existence between $2\pi < h_1 a \leq 3\pi$, and this is the TE_2 mode. Note that although there is an intersection of the graph of F_e with that of $p_2 a$ for $\pi < h_1 a \leq 2\pi$, this cannot be accepted as a physical solution because it has $p_2 < 0$, which would have fields that grow exponentially away from the slab. When V is increased beyond

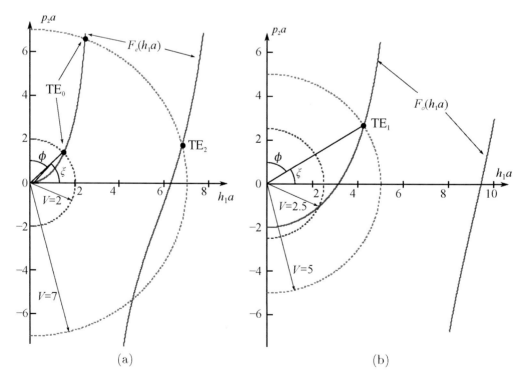

(a)   (b)

Figure 6.9   Plot of (a) $F_e(h_1 a)$ and $\sqrt{V^2 - (h_1 a)^2}$ vs. $h_1 a$ for $V = 2$ and $V = 7$; (b) $F_o(h_1 a)$ and $\sqrt{V^2 - (h_1 a)^2}$ for $V = 2.5$ and $V = 5$.

π, a mode with m odd comes into existence, being the intersection of the graph of $F_o(h_1 a)$ with that of $p_2 a$ as indicated in Figure 6.9(b). This is the TE$_1$ mode, occurring between $\pi \le h_1 a \le 2\pi$. Evidently, a new mode solution will appear each time V increases beyond $m\pi$, where m is a positive integer. Thus, the frequency

$$f_{c,m} = \frac{mc}{2a\sqrt{n_1^2 - n_2^2}} \qquad (6.49)$$

is the cutoff frequency for the TE$_m$ mode. It further follows from (6.34) and (6.35) that for this mode

$$\frac{m\pi}{a} < h_1 < \frac{(m+1)\pi}{a}.$$

Actually, the term "cutoff" could be considered slightly misleading here. In a closed metallic waveguide, a mode passes from one that propagates to one that attenuates in z at the cutoff frequency. In a slab waveguide (or indeed, any *open* waveguide—whose "boundary" is at infinity) the mode simply ceases to exist when the frequency decreases to that of mode cutoff. As can be seen from (6.46), the normalized mode amplitude D then becomes zero, because otherwise the mode field would be carrying infinite energy. Thus, at any finite frequency, only a finite number of guided modes can propagate. In this respect, dielectric waveguides differ quite substantially from their hollow metallic counterparts.

Another difference between these modes and those of hollow metallic waveguides is that the *field patterns* of these modes depend on frequency as does the propagation coefficient. Near cutoff, $p_2 a$ is very small, and $h_1 a$ is very close to its lower limit, $m\pi$. The result is a mode whose fields extend a large distance beyond the core-cladding boundaries as shown in Figure 6.10(a). Thus, for example, the fundamental TE$_0$ mode could not be used at arbitrarily low frequencies (even though it has, in principle, zero cutoff frequency) because of the large field spread in this range (Figure 6.10(a)). The effective width a_{eff} of the mode can be used as a gauge of the field spread when deciding whether a surface wave mode can be used in a given situation. Since in an actual optical waveguide the cladding does not extend to infinity, it is important to keep the value of a_{eff} well below the dimension of the cladding region.

Far from cutoff, on the other hand, $h_1 a$ approaches the limiting value of $(m+1)\pi$, and for large V we must also have large $p_2 a$. The result of this is that very little field penetrates into the cladding region ($a_{\text{eff}} \simeq a_{\text{eq}} \simeq a$). In particular, the boundary shift (6.39) is very small. The fields very nearly are zero at the core boundaries, as shown in Figure 6.10(b).

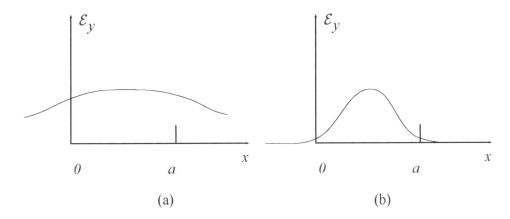

(a) (b)

Figure 6.10 TE$_0$ fields: (a) near cutoff; (b) far from cutoff.

This dependence of the field pattern on V (and thus, for fixed frequency, on a) is the basis of an end-fire antenna design. If we start with a wide ($V \gg 1$) slab waveguide on which the TE_0 mode is propagating, the slab width can then be gradually tapered by reducing a until the field spreads over a much wider region. By this point the slab is very narrow and can be allowed to disappear altogether. At this point, the field of the surface wave radiates into the uniform space in the manner of an aperture antenna, and produces a very narrow beam because of the wide extent of the aperture field.

We can use the information assembled above together with equations (6.32) and (6.41) to construct field plots for the first few modes as shown in Figure 6.11. It is seen that the mode index m refers to the number of zeroes of $\mathcal{E}_y$ within the core region. The field decays exponentially with transverse distance away from the slab, and is thus localized within the core and near the surfaces separating core from cladding. For this reason, a mode with such a field pattern is often called a *surface-wave mode*.

This kind of mode is also called a *slow-wave* mode, because its phase velocity is smaller than that of a plane wave in the surrounding (cladding) region. A uniformly filled metallic waveguide mode by contrast has a larger phase velocity than that of a plane wave in the filling medium, and could thus be referred to as a *fast-wave* mode. A fast-wave surface-wave mode is an impossibility, as the fields in the cladding medium would not decay with distance away from the core.

We can make a final practical observation at this point. Whereas the dimension of a parallel-plate metallic guide must be such that

$$k_0 a < \pi \qquad \text{or} \qquad a < \frac{\lambda_0}{2}$$

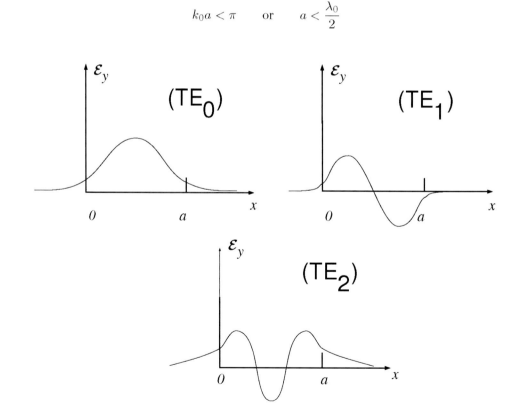

Figure 6.11 $\mathcal{E}_y$ field plots for TE_0, TE_1 and TE_2 modes.

in order to support only a single mode of propagation (TEM in this case), for the slab we need only $V < \pi$, or

$$\sqrt{n_1^2 - n_2^2}\, k_0 a < \pi \qquad \text{or} \qquad a < \frac{\lambda_0}{2\sqrt{n_1^2 - n_2^2}}$$

for only a single TE mode to propagate (as we will see a bit later, a TM mode can also propagate along the slab under these conditions). Thus if we make the dielectric contrast small (which is quite natural if core and cladding are to be made of similar materials), a considerably larger guide width a can be tolerated than in the metallic waveguide case. For example, suppose that the slab is made from a glass whose refractive index is $n_2 = 1.5$, and the core is doped slightly to increase this value by 1%. Then the quantity $\sqrt{n_1^2 - n_2^2}$ (which is sometimes referred to as the numerical aperture or NA of the guide) is about 0.21. This means that at a wavelength of $\lambda = 1~\mu$m ($f = 3 \times 10^{14}$ Hz = 300 Terahertz (THz)!), the dimension a can be as large as 2.4 μm, nearly 5 times as large as the corresponding metallic waveguide for single-mode operation.

Even though the eigenvalue equation is not solvable for β explicitly, we can use (6.35) and (6.39) to put it in a form which is solved for V in terms of hd_{TE}:

$$V \sin h_1 d_{\text{TE}} = (m+1)\pi - 2h_1 d_{\text{TE}} \tag{6.50}$$

It is thus quite easy to plot a graph of $h_1 d_{\text{TE}}$ vs. V, but it is more customary instead to plot a *confinement parameter* b:

$$b = \cos^2 h_1 d_{\text{TE}} = \left(\frac{p_2 a}{V}\right)^2 \tag{6.51}$$

The confinement parameter is zero at cutoff (when the field is not well-confined to the vicinity of the core) and approaches unity far from cutoff (when the field is highly confined near the core). In view of (6.51), it is convenient to rewrite (6.50) in terms of V and b:

$$V\sqrt{1-b} = (m+1)\pi - 2\cos^{-1}\sqrt{b} \tag{6.52}$$

or

$$V\sqrt{1-b} = m\pi + 2\sin^{-1}\sqrt{b} \tag{6.53}$$

One also sometimes sees plots of the *effective refractive index* of a mode:

$$n_{\text{eff}} \equiv \beta/k_0 \tag{6.54}$$

but since b can be expressed in terms of n_{eff}:

$$b = \frac{n_{\text{eff}}^2 - n_2^2}{n_1^2 - n_2^2} \tag{6.55}$$

and since a plot of b vs. V is a *universal* curve for TE modes (cf. (6.52) or (6.53)) independent of a, n_1, n_2 and f, it is preferable to plot b and to obtain n_{eff} from it. Figure 6.12 shows a universal plot of b vs. V for the first several TE modes. Note that the inequality $0 \le b < 1$ implies that $n_2 \le n_{\text{eff}} < n_1$ for all modes of the slab.

6.4.1 NEAR-CUTOFF AND FAR-FROM-CUTOFF APPROXIMATIONS

In parameter ranges both near and far from cutoff, it is possible to obtain explicit (though approximate) expressions for b (and thereby, for the related quantities β, n_{eff}, p or h) as a

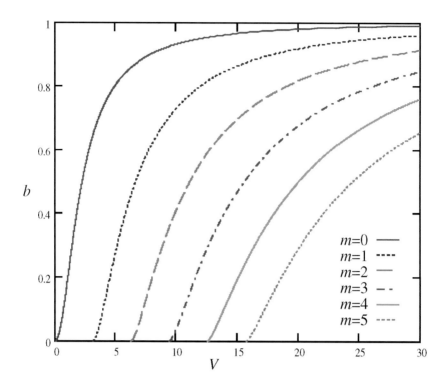

Figure 6.12 Confinement parameter b for TE_m modes of a dielectric slab.

function of V. For example, near cutoff, $b \to 0$ and $V \to m\pi$ for the TE_m mode. We put $b = \sin^2 \xi$ in (6.53) to get

$$V \cos \xi = m\pi + 2\xi \qquad (6.56)$$

The graphical meaning of ξ is shown in Figure 6.9. For small values of ξ, we can put $\cos \xi \simeq 1 - \xi^2/2$ in (6.56), and solve the resulting quadratic equation for ξ. The resulting expression for b is:

$$b \simeq \sin^2 \left\{ \frac{[4 + 2V(V - m\pi)]^{1/2} - 2}{V} \right\} \qquad \text{(near cutoff)} \qquad (6.57)$$

Far from cutoff, on the other hand, $b \to 1$ and $V\sqrt{1-b} \to (m+1)\pi$; for this limit it is more convenient to start from (6.52). We have that the angle $\phi = \pi/2 - \xi$ is small in this case; if we write (6.52) as

$$V \sin \phi = (m + 1)\pi - 2\phi \qquad (6.58)$$

we can approximate ϕ by $\sin \phi$. The resulting equation gives

$$b \simeq 1 - \left[\frac{(m+1)\pi}{V+2} \right]^2 \qquad \text{(far from cutoff)} \qquad (6.59)$$

Approximations (6.57) and (6.59) are plotted in Figure 6.13 along with the exact values of b for the TE_0 and TE_1 modes. Note that (6.59) can be written in terms of β as

$$\beta^2 \simeq k_1^2 - \left[\frac{(m+1)\pi}{a + \frac{2}{k_0 \sqrt{n_1^2 - n_2^2}}} \right]^2 \qquad (6.60)$$

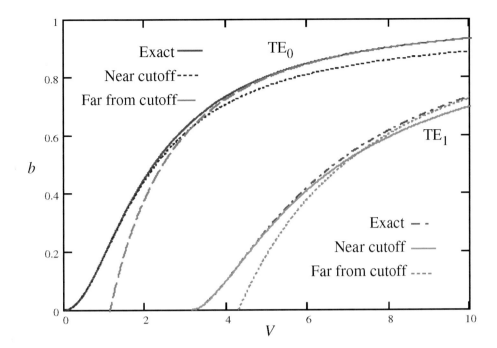

Figure 6.13 Exact, near-cutoff and far-from-cutoff approximations for the confinement parameter of the TE_0 and TE_1 dielectric slab waveguide modes.

The quantity

$$\left[\frac{(m+1)\pi}{a + \frac{2}{k_0\sqrt{n_1^2-n_2^2}}} \right]$$

is the cutoff wavenumber $\tilde{k}_c$ of the TM_{m+1} mode of a metallic parallel-plate waveguide whose plate separation is $a + 2/k_0\sqrt{n_1^2 - n_2^2}$, which is the grazing-incidence limit of the effective width a_{eq} shown in Figure 6.8 (cf. (6.12)), and also of the effective width a_{eff} defined in (6.45).

6.5 TM MODES OF THE DIELECTRIC SLAB

The TM modes can be analyzed in almost precisely the same way, either from ray description (cf. problem p6-1) or by proceeding from equations (6.19) to obtain expressions for $\mathcal{H}_y$. While $\mathcal{H}_y$ must be continuous at the dielectric interfaces (it is a tangential field component), the boundary condition that tangential $\mathcal{E}$ ($\mathcal{E}_z$ in this case) be continuous now forces $\frac{1}{\epsilon}\frac{\partial \mathcal{H}_y}{\partial x}$ to be continuous at the interfaces, i.e.,

$$\left. \frac{1}{\epsilon_2}\frac{d\mathcal{H}_y}{dx} \right|_{x=0^-} = \left. \frac{1}{\epsilon_1}\frac{d\mathcal{H}_y}{dx} \right|_{x=0^+} \tag{6.61}$$

The resulting eigenvalue or mode equation is somewhat different from (6.33); we can express it in the form:

$$\tan(h_1 a) = \frac{2p_2 h_1}{h_1^2(n_2^2/n_1^2) - p_2^2(n_1^2/n_2^2)} \tag{6.62}$$

or

$$V\sqrt{1-b} = m\pi + 2\sin^{-1}\left[\frac{(n_1/n_2)\sqrt{b}}{\sqrt{(n_1/n_2)^2 b + n_2^2(1-b)/n_1^2}}\right] \tag{6.63}$$

Although this latter equation is slightly different than (6.53), the general properties of its solutions are similar. Details can be found in any of the standard references at the end of this chapter. Note that (6.63) is again solved for V as a function of b, but does not generate universal curves of b vs. V, because the additional parameter n_1/n_2 appears in the equation. In the same way as for TE modes, we can obtain approximate formulas for b as a function of V which are good both near and far from cutoff (though they are somewhat more complicated in form; see problem p6-13).

6.6   THE GROUNDED DIELECTRIC SLAB

A structure related to the dielectric slab is the grounded dielectric slab shown in Figure 6.14. The slab lies on a conducting ground plane, and its thickness is denoted by d. This configuration is also an important building block in the analysis of more elaborate quasi-optical waveguides. Its solutions are easily obtained from those already obtained for the ungrounded slab.

Consider, for example, the TE modes of the ungrounded slab. The *odd* TE modes, according to (6.41), have $\mathcal{E}_y = 0$ at the center of the slab ($x = a/2$) by virtue of their symmetry. But this means that a perfectly conducting plane could be inserted at the center of the slab without disturbing any of the fields, because the boundary conditions would be properly satisfied. Thus, the TE modes of the grounded slab of thickness d have identical fields (above the ground plane) and identical propagation coefficients with those of the *odd* TE modes of an ungrounded slab of twice the thickness ($a = 2d$). In particular, its eigenvalue equation, by (6.43), is

$$\tan(h_1 d) = -\frac{h_1}{p_2} \tag{6.64}$$

and is solved by the same methods as before.

On the other hand, the analysis of the TM modes of the ungrounded slab shows that $\mathcal{E}_z = 0$ at the center of the slab for the *even* TM modes. Again, we can insert a conducting plane in the center of the slab without affecting these modes, and thereby obtain the TM

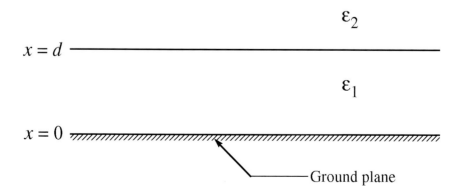

Figure 6.14   The grounded dielectric slab.

modes of the grounded slab. The resulting eigenvalue equation for these TM modes is

$$\tan(h_1 d) = \frac{\epsilon_1 p_2}{\epsilon_2 h_1} \tag{6.65}$$

Because only the *odd* TE modes and *even* TM modes of the ungrounded slab are present, the only mode of the grounded slab with no cutoff frequency is the TM_0 mode. The TE_1 mode will not exist if $\sqrt{k_1^2 - k_2^2}\, d < \pi/2$.

6.7   CIRCULAR OPTICAL FIBERS

We turn now to the rather more realistic geometry of Figure 6.15(a), the circular optical fiber. The core region, $\rho \leq a$, has a permittivity ϵ_1 which is larger than that of the cladding ($\rho \geq a$), ϵ_2. A photograph of an optical fiber is shown in Figure 6.15(b), showing not only the core and cladding regions, but also the protective external coating that is necessary

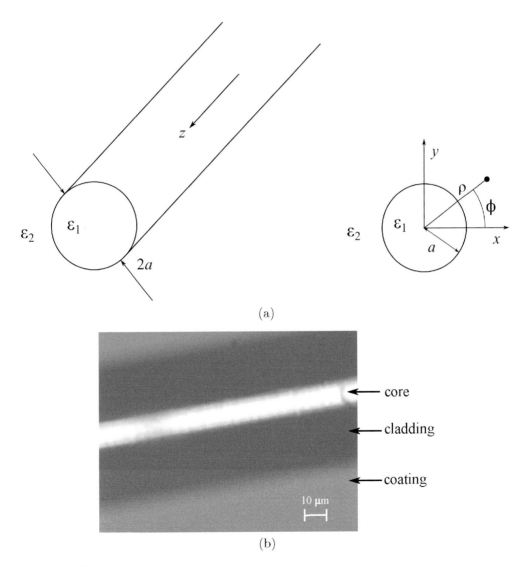

(a)

(b)

Figure 6.15   The circular optical fiber: (a) geometry; (b) photomicrograph.

for protection and practical handling of the fiber. As with the slab, we will assume that $\mu = \mu_0$ everywhere. For this waveguide, we cannot necessarily assume that only TE or TM modes will exist; we must assume that the modes will be hybrid modes, taking $\mathcal{E}_z \neq 0$ and $\mathcal{H}_z \neq 0$, and dealing with all six components of Maxwell's equations in (5.17)-(5.20). Naturally, circular cylindrical coordinates are most convenient here, and we use them to express the transverse fields in terms of $\mathcal{E}_z$ and $\mathcal{H}_z$ by means of (5.22) and (5.23).

Let us again introduce the notations

$$\gamma = j\beta$$

$$p_2 = \sqrt{\beta^2 - k_2^2}$$

$$h_1 = \sqrt{k_1^2 - \beta^2}$$

that were used in the analysis of the slab, and thus write (5.22) and (5.23) as

$$\mathcal{E}_\rho = \frac{1}{h_1^2}\left[-j\beta\frac{\partial \mathcal{E}_z}{\partial \rho} - \frac{j\omega\mu}{\rho}\frac{\partial \mathcal{H}_z}{\partial \phi}\right] \tag{6.66}$$

$$\mathcal{H}_\phi = \frac{1}{h_1^2}\left[-\frac{j\beta}{\rho}\frac{\partial \mathcal{H}_z}{\partial \phi} - j\omega\epsilon_1\frac{\partial \mathcal{E}_z}{\partial \rho}\right] \tag{6.67}$$

$$\mathcal{E}_\phi = \frac{1}{h_1^2}\left[-\frac{j\beta}{\rho}\frac{\partial \mathcal{E}_z}{\partial \phi} + j\omega\mu\frac{\partial \mathcal{H}_z}{\partial \rho}\right] \tag{6.68}$$

$$\mathcal{H}_\rho = \frac{1}{h_1^2}\left[-j\beta\frac{\partial \mathcal{H}_z}{\partial \rho} + \frac{j\omega\epsilon_1}{\rho}\frac{\partial \mathcal{E}_z}{\partial \phi}\right] \tag{6.69}$$

when $\rho < a$, and as

$$\mathcal{E}_\rho = \frac{1}{p_2^2}\left[j\beta\frac{\partial \mathcal{E}_z}{\partial \rho} + \frac{j\omega\mu}{\rho}\frac{\partial \mathcal{H}_z}{\partial \phi}\right] \tag{6.70}$$

$$\mathcal{H}_\phi = \frac{1}{p_2^2}\left[\frac{j\beta}{\rho}\frac{\partial \mathcal{H}_z}{\partial \phi} + j\omega\epsilon_2\frac{\partial \mathcal{E}_z}{\partial \rho}\right] \tag{6.71}$$

$$\mathcal{E}_\phi = \frac{1}{p_2^2}\left[\frac{j\beta}{\rho}\frac{\partial \mathcal{E}_z}{\partial \phi} - j\omega\mu\frac{\partial \mathcal{H}_z}{\partial \rho}\right] \tag{6.72}$$

$$\mathcal{H}_\rho = \frac{1}{p_2^2}\left[j\beta\frac{\partial \mathcal{H}_z}{\partial \rho} - \frac{j\omega\epsilon_2}{\rho}\frac{\partial \mathcal{E}_z}{\partial \phi}\right] \tag{6.73}$$

when $\rho > a$. These expressions should be compared with equations (5.38)-(5.39) for a homogeneously-filled metallic guide.

Substitution of (6.66)-(6.69) or (6.70)-(6.73) into (5.18) and (5.20) gives us reduced wave equations for $\mathcal{E}_z$ and $\mathcal{H}_z$:

$$\begin{aligned}(\nabla_T^2 + h_1^2)\mathcal{E}_z &= 0 \\ (\nabla_T^2 + h_1^2)\mathcal{H}_z &= 0 \end{aligned} \tag{6.74}$$

when $\rho < a$, and

$$\begin{aligned}(\nabla_T^2 - p_2^2)\mathcal{E}_z &= 0 \\ (\nabla_T^2 - p_2^2)\mathcal{H}_z &= 0 \end{aligned} \tag{6.75}$$

when $\rho > a$. The transverse Laplacian ∇_T^2 is given (in cylindrical coordinates) by

$$\nabla_T^2 f \equiv \frac{1}{\rho}\frac{\partial}{\partial\rho}\rho\frac{\partial f}{\partial\rho} + \frac{1}{\rho^2}\frac{\partial^2 f}{\partial\phi^2} \tag{6.76}$$

[cf. eqn. (5.43)].

Equations (6.74) and (6.75) are solved by separation of variables, in a manner similar to that used for the circular metallic guide in Chapter 5. In light of our experience with that problem, we expect the solution for $\mathcal{E}_z$ to take the form

$$\mathcal{E}_z(\rho,\phi) = C_1 \left\{ \begin{array}{c} \cos l\phi \\[1em] \sin l\phi \end{array} \right\} F_l(\rho) \quad ; \quad l = 0,1,2,... \tag{6.77}$$

the choice of $\cos l\phi$ corresponding to a "c" polarization and that of $\sin l\phi$ corresponding to an "s" polarization. We have included an arbitrary constant C_1 in this solution. The function $F_l(\rho)$, which is yet to be determined, must be continuous at $\rho = a$. In order to be compatible with eqns. (6.67)-(6.68) and (6.71)-(6.72) for the tangential fields $\mathcal{E}_\phi$ and $\mathcal{H}_\phi$ when we enforce continuity at $\rho = a$, it will be necessary to choose a solution for $\mathcal{H}_z$ in the form

$$\mathcal{H}_z(\rho,\phi) = C_2 \left\{ \begin{array}{c} \sin l\phi \\[1em] -\cos l\phi \end{array} \right\} G_l(\rho) \tag{6.78}$$

where C_2 is an arbitrary constant and $\sin l\phi$ and $-\cos l\phi$ correspond to the "c" and "s" polarizations, respectively. The function $G_l(\rho)$ must yet be determined, but will be continuous at $\rho = a$.

Upon substituting (6.77) and (6.78) into (6.74) and (6.75), we find that F_l and G_l must satisfy the equations

$$\rho^2 \left\{ \begin{array}{c} F_l'' \\[0.8em] G_l'' \end{array} \right\} + \rho \left\{ \begin{array}{c} F_l' \\[0.8em] G_l' \end{array} \right\} + (h_1^2\rho^2 - l^2) \left\{ \begin{array}{c} F_l \\[0.8em] G_l \end{array} \right\} = 0 \qquad (\rho < a) \tag{6.79}$$

or

$$\rho^2 \left\{ \begin{array}{c} F_l'' \\[0.8em] G_l'' \end{array} \right\} + \rho \left\{ \begin{array}{c} F_l' \\[0.8em] G_l' \end{array} \right\} - (p_2^2\rho^2 + l^2) \left\{ \begin{array}{c} F_l \\[0.8em] G_l \end{array} \right\} = 0 \qquad (\rho > a) \tag{6.80}$$

As always, we use primes to denote differentiation with respect to the argument (ρ). For $\rho < a$, we recognize that eqn. (6.79) is identical with (5.66), which we have already encountered in the study of metallic circular waveguides, except that $k_c \to h$ and $\nu \to l$. Since $\mathcal{E}_z$ and $\mathcal{H}_z$ must be finite at $\rho = 0$, we can put

$$\begin{aligned} F_l(\rho) &= J_l(h_1\rho) \\ G_l(\rho) &= J_l(h_1\rho) \end{aligned} \tag{6.81}$$

in the core region, there already being arbitrary amplitude constants present in (6.77) and (6.78).

Equation (6.80), however, is slightly different; if we make the change of variable $w = p_2\rho$, we have

$$w^2 \left\{ \begin{array}{c} \frac{d^2 F_l}{dw^2} \\ \frac{d^2 G_l}{dw^2} \end{array} \right\} + w \left\{ \begin{array}{c} \frac{dF_l}{dw} \\ \frac{dG_l}{dw} \end{array} \right\} - (w^2 + l^2) \left\{ \begin{array}{c} F_l \\ G_l \end{array} \right\} = 0 \qquad (\rho > a) \qquad (6.82)$$

From Appendix C, eqn. (C.25) ff., we see that the general solution to (6.82) is a linear combination of the *modified* Bessel functions $I_l(w)$ and $K_l(w)$. But as $\rho \to \infty$, we require that the fields decay as we move away from the core (for a surface wave). This is contrary to the behavior of $I_l(w)$ summarized in Appendix C, and so we have

$$\begin{aligned} F_l(\rho) &= C_3 K_l(p_2\rho) \\ G_l(\rho) &= C_4 K_l(p_2\rho) \qquad (\rho > a) \end{aligned} \qquad (6.83)$$

where the constants C_3 and C_4 can be obtained by requiring that the functions $F_l(\rho)$ and $G_l(\rho)$ be continuous at $\rho = a$ as noted above. Comparing (6.81) with (6.83), we see that

$$C_3 = C_4 = \frac{J_l(h_1 a)}{K_l(p_2 a)} \qquad (6.84)$$

and so $F_l(\rho)$ and $G_l(\rho)$ are actually the same function:

$$\begin{aligned} F_l(\rho) = G_l(\rho) &= J_l(h_1\rho) \qquad (\rho < a) \\ &= \frac{J_l(h_1 a)}{K_l(p_2 a)} K_l(p_2\rho) \qquad (\rho > a) \end{aligned} \qquad (6.85)$$

There now remain two amplitude constants C_1 and C_2 to determine in addition to the eigenvalue β, and two remaining boundary conditions—that $\mathcal{E}_\phi$ and $\mathcal{H}_\phi$ be continuous at $\rho = a$. Substituting (6.77) and (6.78) into (6.67)-(6.68) and (6.71)-(6.72), we obtain for these two boundary conditions that

$$\frac{1}{h_1^2} \left[\frac{j\beta}{a} C_1 l F_l(a) + j\omega\mu C_2 F_l'(a^-) \right] = \frac{1}{p_2^2} \left[-\frac{j\beta}{a} C_1 l F_l(a) - j\omega\mu C_2 F_l'(a^+) \right] \qquad (6.86)$$

$$\frac{1}{h_1^2} \left[-\frac{j\beta}{a} C_2 l F_l(a) - j\omega\epsilon_1 C_1 F_l'(a^-) \right] = \frac{1}{p_2^2} \left[\frac{j\beta}{a} C_2 l F_l(a) + j\omega\epsilon_2 C_1 F_l'(a^+) \right] \qquad (6.87)$$

Here, $F_l'(a^\pm)$ refers to the derivatives of $F_l(\rho)$ just outside and inside the core boundary, respectively:

$$F_l'(a^+) = p_2 J_l(h_1 a)\frac{K_l'(p_2 a)}{K_l(p_2 a)} \qquad (6.88)$$

$$F_l'(a^-) = h_1 J_l'(h_1 a) \qquad (6.89)$$

For the following manipulations, some simplification is gained by introducing the notations

$$\begin{aligned} \Theta_l &= \frac{F_l'(a^-)}{h_1^2 F_l(a)} + \frac{F_l'(a^+)}{p_2^2 F_l(a)} \\ &= \frac{J_l'(h_1 a)}{h_1 J_l(h_1 a)} + \frac{K_l'(p_2 a)}{p_2 K_l(p_2 a)} \end{aligned} \qquad (6.90)$$

$$\begin{aligned}
\Delta_l &= \frac{n_2^2 - n_1^2}{n_1^2} \frac{F_l'(a^+)}{p_2^2 F_l(a)} \\
&= \frac{n_2^2 - n_1^2}{n_1^2} \frac{K_l'(p_2 a)}{p_2 K_l(p_2 a)}
\end{aligned} \tag{6.91}$$

$$B = \frac{\beta}{k_1 a}\left(\frac{1}{h_1^2} + \frac{1}{p_2^2}\right) \tag{6.92}$$

(note that $\Delta_l > 0$ whenever p_2 is real and positive and that $B > 0$ if both p_2 and h_1 are real). Some rearrangement of (6.86) and (6.87) gives

$$\begin{aligned}
lBC_1 &= -\sqrt{\frac{\mu}{\epsilon_1}}\Theta_l C_2 \\
lBC_2 &= -\sqrt{\frac{\epsilon_1}{\mu}}[\Theta_l + \Delta_l]C_1
\end{aligned} \tag{6.93}$$

which are simultaneous equations to be satisfied by C_1 and C_2. If they are not both to be zero, we must have

$$l^2 B^2 = \Theta_l(\Theta_l + \Delta_l) \tag{6.94}$$

which determines the allowed values of β for the circular fiber.

Solving (6.94) for Θ_l gives two possible equations for β:

$$\Theta_l = -\frac{\Delta_l}{2} \pm \left[\left(\frac{\Delta_l}{2}\right)^2 + l^2 B^2\right]^{1/2} \tag{6.95}$$

Each of these corresponds to a particular ratio of C_1/C_2, either positive or negative, whose value is used to classify the modes of the fiber as HE or EH, according to the entries in Table 6.1.

The HE/EH nomenclature for modes of the circular fiber originated from the first applications of this structure at microwave frequencies, when a large core/cladding permittivity contrast ($\epsilon_1 \gg \epsilon_2$) was usually encountered. In this limit, the transverse fields of the HE modes are found to closely resemble those of the TE (or H) modes of the circular metallic waveguide. Similarly, fields of EH modes resemble those of TM (or E) modes in this limit. As it turns out, the second letter of the hybrid mode designation refers to the dominant longitudinal field component. But neither of these features is apparent in the case when the core and cladding refractive indices are close to each other, when only a partial resemblance

---

Table 6.1
Eigenvalue equations and C_1/C_2 ratios for HE and EH modes.

Mode Type	HE	EH
Eigenvalue equation	$\Theta_l = -\frac{\Delta_l}{2} - \left[\left(\frac{\Delta_l}{2}\right)^2 + l^2 B^2\right]^{1/2}$	$\Theta_l = -\frac{\Delta_l}{2} + \left[\left(\frac{\Delta_l}{2}\right)^2 + l^2 B^2\right]^{1/2}$
C_1 and C_2	$\dfrac{C_2}{C_1} = \sqrt{\dfrac{\epsilon_1}{\mu}} \dfrac{lB}{\frac{\Delta_l}{2}+\left[\left(\frac{\Delta_l}{2}\right)^2+l^2B^2\right]^{1/2}}$ $\left(0 \le \dfrac{C_2}{C_1} < \sqrt{\dfrac{\epsilon_1}{\mu}}\right)$	$\dfrac{C_1}{C_2} = -\sqrt{\dfrac{\mu}{\epsilon_1}} \dfrac{lB}{\frac{\Delta_l}{2}+\left[\left(\frac{\Delta_l}{2}\right)^2+l^2B^2\right]^{1/2}}$ $\left(0 \ge \dfrac{C_1}{C_2} > -\sqrt{\dfrac{\mu}{\epsilon_1}}\right)$

to hollow waveguide modes can be distinguished. Some field plots for the various modes of the fiber in such a situation will be presented in the next subsection, which should prove more useful for determining what it is that distinguishes them from a physical point of view.

6.7.1 THE WEAKLY GUIDING FIBER

For many practical applications in optical waveguides, the inequality

$$(n_1^2 - n_2^2)/n_2^2 \ll 1 \tag{6.96}$$

will be satisfied. This is known as the "small contrast" or (rather misleadingly) "weakly guiding" condition. In this situation we also have $p_2/k_2 \ll 1$ and $h_1/k_2 \ll 1$ (since $p_2^2 + h_1^2 = k_1^2 - k_2^2$). For a TE mode on a slab waveguide, from (6.18) we see that the maximum value of $|\mathcal{H}_z|$ will be much smaller than that of $|\mathcal{H}_x|$ in this case, so the weakly guiding limit produces a mode whose fields are "almost" TEM. This also holds true for the TM modes of the slab, and for modes of other types of dielectric waveguide as well. It is worth noting that for slab waveguides, (6.62) reduces to (6.33) and (6.63) to (6.53) in the weakly guided limit. The TE_m and TM_m modes of the slab are thus nearly degenerate in this limit. A similar thing occurs for other types of dielectric waveguide.

For the step-index circular fiber, the analysis of exact mode fields is quite cumbersome, and considerable simplification results in the applicable formulas if this approximation is invoked. We will thus forego the exact analysis in favor of this simpler treatment.

If (6.96) holds, then we can expect that $\Delta_l \simeq 0$ so that the eigenvalue equations in Table 6.1 for the EH and HE modes can be further simplified. By (C.40) and (C.44) they can be written as

$$\frac{J_{l-1}(h_1 a)}{h_1 a J_l(h_1 a)} = \frac{K_{l-1}(p_2 a)}{p_2 a K_l(p_2 a)} \qquad \text{(HE modes)} \tag{6.97}$$

$$\frac{J_{l+1}(h_1 a)}{h_1 a J_l(h_1 a)} + \frac{K_{l+1}(p_2 a)}{p_2 a K_l(p_2 a)} = 0 \qquad \text{(EH modes)} \tag{6.98}$$

The $l = 0$ modes are special cases. The HE_0 modes are TM modes ($C_2 = 0$ by Table 6.1), while the EH_0 modes are TE ($C_1 = 0$). Their eigenvalue equations in the weakly guiding limit are given by (6.97) and (6.98) with $l = 0$, although for TM modes it is more common to use (C.40)-(C.45) to replace J_{-1} and K_{-1} by J_1 and K_1:

$$\frac{J_1(h_1 a)}{h_1 a J_0(h_1 a)} + \frac{K_1(p_2 a)}{p_2 a K_0(p_2 a)} = 0 \qquad \text{(TM modes)} \tag{6.99}$$

The derivation of expressions for the transverse field components in the weakly guiding approximation is left as an exercise (problems p6-14 and p6-15).

When analyzing either of the characteristic equations (6.97) or (6.98), it is again convenient to deal in the dimensionless parameters b and n_{eff} given by (6.54) and (6.55). The normalized frequency of a fiber is defined as

$$V = \sqrt{k_1^2 - k_2^2}\, a = \sqrt{n_1^2 - n_2^2}\, k_0 a \tag{6.100}$$

In these terms, the eigenvalue equations are

$$\frac{J_{l-1}(V\sqrt{1-b})}{V\sqrt{1-b}\, J_l(V\sqrt{1-b})} = \frac{K_{l-1}(V\sqrt{b})}{\sqrt{b}\, K_l(V\sqrt{b})} \qquad \text{(HE or TM modes)} \tag{6.101}$$

$$\frac{J_{l+1}(V\sqrt{1-b})}{V\sqrt{1-b}\, J_l(V\sqrt{1-b})} = -\frac{K_{l+1}(V\sqrt{b})}{V\sqrt{b}\, K(V\sqrt{b})} \qquad \text{(EH or TE modes)} \tag{6.102}$$

Like the eigenvalue equations for the slab, eqns. (6.101) and (6.102) do not yield explicit solutions for β (or b) in terms of V. In fact, we cannot even find V as a function of b as we did in eqns. (6.52) and (6.53). Graphical solutions similar to those for the slab can be pursued; we note that the behavior of the left sides resembles that of $\tan(h_1 a)/h_1 a$, while the right sides vary essentially only algebraically and nonperiodically as functions of $h_1 a$. We shall not pursue the graphical solutions here (see the references at the end of the chapter), but will instead analyze the behavior of the solutions to (6.101) and (6.102) in the ranges both near and far from cutoff.

Far from cutoff, we have $V \gg 1$; in this case $b \to 1$. From the large argument behavior of the modified Bessel functions (eqn. (C.39)), we have

$$\frac{K_{l\pm1}(V\sqrt{b})}{\sqrt{b}K_l(V\sqrt{b})} \sim \frac{1}{V\sqrt{b}} \sim \frac{1}{V} \to 0 \tag{6.103}$$

as $V\sqrt{b} \to \infty$. Hence by (6.101) and (6.102), the quantity $V\sqrt{1-b}$ approaches a constant in this limit; either a root $j_{l-1,m}$ of the Bessel function J_{l-1} for the HE_{lm} mode, or a root $j_{l+1,m}$ of J_{l+1} for the EH_{lm} mode.[2] Thus, the quantity $[V\sqrt{1-b} - j_{l\pm1,m}]$ is small far from cutoff, and the Bessel functions $J_{l\pm1}(V\sqrt{1-b})$ and $J_l(V\sqrt{1-b})$ can be expanded as Taylor series:

$$\begin{aligned} J_{l\pm1}(V\sqrt{1-b}) &\simeq J_{l\pm1}(j_{l\pm1,m}) + (V\sqrt{1-b} - j_{l\pm1,m})J'_{l\pm1}(j_{l\pm1,m}) \\ &= \pm(V\sqrt{1-b} - j_{l\pm1,m})J_l(j_{l\pm1,m}) \end{aligned} \tag{6.104}$$

and

$$J_l(V\sqrt{1-b}) \simeq J_l(j_{l-1,m}) \tag{6.105}$$

The second line of (6.104) follows from eqns. (C.40) and (C.41) and the fact that $J_{l\pm1}(j_{l\pm1,m}) = 0$. Hence,

$$\frac{J_{l\pm1}(V\sqrt{1-b})}{V\sqrt{1-b}J_l(V\sqrt{1-b})} \simeq \pm\frac{V\sqrt{1-b} - j_{l\pm1,m}}{V\sqrt{1-b}} \tag{6.106}$$

Setting this equal to $1/V$ by (6.103), and solving for $V\sqrt{1-b}$, we get

$$V\sqrt{1-b} \simeq \frac{j_{l\pm1,m}}{1 + 1/V} \qquad \text{(far from cutoff)} \tag{6.107}$$

Equation (6.107) should be compared with the corresponding form (6.59) for the TE modes of the slab. These two situations have in common that h_1—their transverse wavenumber—approaches the transverse wavenumber k_c for a TM mode of the corresponding hollow metallic structure. Also, the distributions of the *transverse* fields $\mathcal{E}_x$, $\mathcal{E}_y$, $\mathcal{H}_x$ and $\mathcal{H}_y$ of the fiber modes (in the core region) approach the $\mathcal{E}_z$ field distributions of the TM modes of the metallic guide (cf. problems p6-14 and p6-15). Again we see that a surface wave mode far from cutoff tends to have nearly all of its fields concentrated within the core, almost as if an actual metal wall were present just outside the core-cladding boundary.

It is rather more difficult to obtain closed-form approximations for b in the range near cutoff. We will content ourselves here with a determination of the cutoff frequencies V_c, and obtain an approximation for b only for the fundamental mode. Since $b \to 0$ at cutoff, we have that

$$\frac{J_{l\pm1}V\sqrt{1-b})}{\sqrt{1-b}J_l(V\sqrt{1-b})} \to \frac{J_{l\pm1}(V)}{VJ_l(V)} \tag{6.108}$$

---

[2]The roots $j_{-1,m}$ are the same as j_1, m for the case of HE_{0m} (TM_m) modes.

as $b \to 0$. From (C.33) and (C.34) of Appendix C, we have that

$$\frac{K_{l+1}(V\sqrt{b})}{V\sqrt{b}K_l(V\sqrt{b})} \to -\frac{1}{V^2 b[\ln(\frac{V\sqrt{b}}{2}) + \gamma_E]} \quad (l = 0)$$

$$\to \frac{2l}{V^2 b} \quad (l \geq 1) \tag{6.109}$$

while

$$\frac{K_{l-1}(V\sqrt{b})}{V\sqrt{b}K_l(V\sqrt{b})} \to -\frac{1}{V^2 b[\ln(\frac{V\sqrt{b}}{2}) + \gamma_E]} \quad (l = 0)$$

$$\to -[\ln(\frac{V\sqrt{b}}{2}) + \gamma_E] \quad (l = 1) \tag{6.110}$$

$$\to \frac{1}{2(l-1)} \quad (l \geq 2)$$

as $b \to 0$, where γ_E is Euler's constant (C.11). By applying (6.108) and (6.109) or (6.110) to (6.101) or (6.102), we find that for TE_m or TM_m modes ($l = 0$), cutoff ($b = 0$) occurs at $V = V_c$, where

$$V_c = j_{0m} \quad (\text{TE}_m \text{ or } \text{TM}_m \text{ modes}) \tag{6.111}$$

For EH_{lm} modes ($l \geq 1$), we have

$$V_c = j_{lm} \quad (\text{EH}_{lm} \text{ modes, } l \geq 1) \tag{6.112}$$

and for HE_{lm} modes ($l \geq 2$), we find

$$V_c = j_{l-2,m} \quad (\text{HE}_{lm} \text{ modes, } l \geq 2) \tag{6.113}$$

where (C.40) and (C.41) were used to help obtain (6.113), and j_{lm} denotes the mth root of the Bessel function J_l, *not* including zero. Only the HE_{1m} class of modes allows a zero cutoff frequency; from (6.108) and (6.110) we have (as $b \to 0$):

$$b \simeq \frac{4}{V^2} e^{-2\left[\gamma_E + \frac{J_0(V)}{V J_1(V)}\right]} \tag{6.114}$$

Hence we have

$$V_c = 0 \quad (\text{HE}_{11} \text{ mode})$$

$$= j_{1,m-1} \quad (\text{HE}_{1m} \text{ modes}) \tag{6.115}$$

Thus, the HE_{11} mode can (in principle) propagate at *all* frequencies, and is the fundamental mode of the fiber. The locations of the various cutoff frequencies can be seen in Figure 6.16, where b vs. V is plotted in the weakly guiding limit for the lowest-order modes. Note the very flat part of the HE_{11} curve for about $0 \leq V < 0.7$. Even though in principle this is a guided mode (surface wave) the value of b in this range is so small (cf. eqn. (6.114)) that the fields extend for a very large distance away from the core without much decay. This is characteristic of all three-dimensional optical waveguides with homogeneous cladding, and differs markedly from the behavior of the slab shown in Figure 6.12. This means that if the fiber is to be operated in the "single-mode" regime (actually a two-mode range if both polarizations of the HE_{11} mode are accounted for), less confinement is possible ($b \simeq 0.54$) than in the case of the slab ($b \simeq 0.7$).

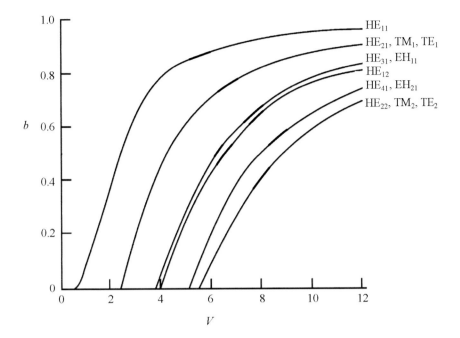

Figure 6.16 b vs. V for weakly guided fiber modes.

Figure 6.16 also illustrates the fact that in this approximation the propagation coefficients for EH_{lm} are identical to those for $HE_{l+2,m}$ modes. This can be shown directly from (6.101) and (6.102), which reduce to the same equation in this case. Thus, *in the weakly guiding limit, the $HE_{l+2,m}$ and EH_{lm} modes are approximately degenerate*. Despite the approximate nature of the degeneracy, it may be quite a long propagation distance before phase differences (due to the slight differences in β) show up, so that in practice the effect can be similar to what would be seen if exact degeneracy were present. It is common practice to combine these nearly degenerate modes in such a way that the resulting fields are *linearly polarized*, that is, of all the transverse field components only $(\mathcal{E}_x, \mathcal{H}_y)$ or $(\mathcal{E}_y, \mathcal{H}_x)$ remain. These combinations are commonly (though inaccurately) referred to as *LP modes*, with associated indices $l+1, m$. The LP_{0m} and LP_{1m} combinations arise as special cases; LP_{0m} are combinations of the two polarizations of HE_{1m} modes, while the LP_{1m} are combinations of HE_{2m}, TE_m and TM_m modes.

Figure 6.17 shows a detailed comparison of the exact dispersion curve (b vs. V) for the HE_{11} mode with the approximate values predicted by (6.107) for large V and by (6.114) for small V. Though less accurate than our corresponding approximations for the slab, these approximations are more than adequate for many design purposes.

Obtaining transverse field plots for the fiber modes is made quite easy in the weakly-guiding limit, since according to the results of problems p6-14 and p6-15, all transverse field components have the same dependence on the radial coordinate. From eqn (I.2) of Appendix I, and the field expressions given in those problems, we have the field-line equation

$$\frac{d\rho}{\rho d\phi} = -\cot l\phi \qquad (\text{``c'' polarization})$$

for HE_{lm} modes, while for EH_{lm} modes,

$$\frac{d}{\rho d\phi} = +\cot l\phi \qquad (\text{``c'' polarization})$$

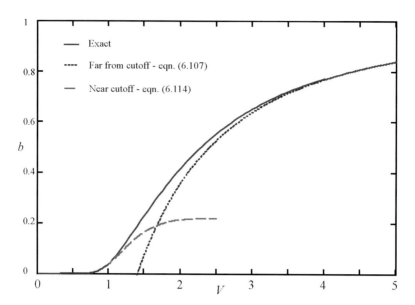

Figure 6.17 Exact, near-cutoff and far-from-cutoff calculations of b vs. V for the HE_{11} mode of the weakly guiding circular fiber.

These are easily solved and give

$$\rho = C[\sin l\phi]^{-1/l} \qquad \text{(HE modes)}$$

or

$$\rho = C[\sin l\phi]^{1/l} \qquad \text{(EH modes)}$$

By taking different values of the constant C, we generate a family of curves corresponding to the mode fields. According to the convention described in Appendix I, the density of field lines in the core can be obtained from the field amplitudes. According to problems p6-14 and p6-15, these are

$$\begin{aligned}
|\boldsymbol{\mathcal{E}}_T| &\propto A_l(\rho) \propto J_{l-1}(h_1\rho) \qquad \text{(HE modes)} \\
|\boldsymbol{\mathcal{E}}_T| &\propto B_l(\rho) \propto J_{l+1}(h_1\rho) \qquad \text{(EH modes)}
\end{aligned} \qquad (6.116)$$

For weakly guided modes, the transverse magnetic field lines are identical to the $\boldsymbol{\mathcal{E}}_T$-lines, except that they are rotated by an angle of $\pi/2l$ about the origin.

Figure 6.18 displays the $\boldsymbol{\mathcal{E}}$-field lines for the HE_{11} mode. In this case, the field lines obey

$$\rho \sin \phi = \text{const}$$

that is, $y = \text{const}$. This is therefore a linearly-polarized field, and resembles a plane wave. Sketches of HE_{21}, HE_{31}, EH_{11} and EH_{21} mode patterns are given in Figure 6.19. It can be seen that the main difference between HE and EH modes is that the field lines of HE modes tend outward and away from the center of the core, while for EH modes, the lines loop around and tend back toward the center. Figure 6.20 shows the fields for the TE_1 and TM_1 modes. The sense in which a TE mode is a special case of EH modes, and a TM mode a special case of HE modes can be seen by comparing these with those of Figure 4.21, and the special position occupied by the HE_{11} mode is quite evident from Figure 6.18.

In many (though by no means all) applications, the vector character of an optical field is irrelevant, and only the intensity distribution is of interest (i.e., $|\boldsymbol{\mathcal{E}}_T|^2$). In the weak-guidance

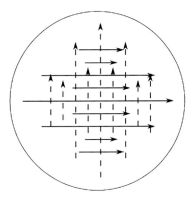

Figure 6.18 $\mathcal{E}$ and $\mathcal{H}$ fields for the HE_{11} mode of a weakly guiding fiber.

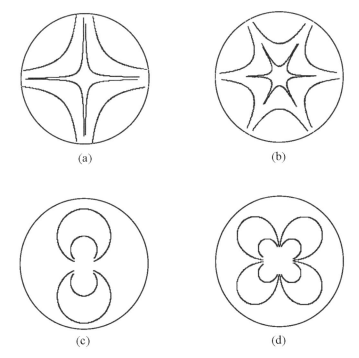

Figure 6.19 $\mathcal{E}$-lines for some higher-order modes of the circular fiber: (a) HE_{21}; (b) HE_{31}; (c) EH_{11}; (d) EH_{21}.

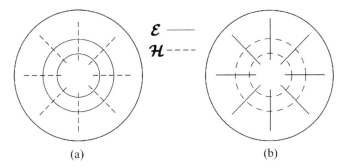

Figure 6.20 $\mathcal{E}$ and $\mathcal{H}$ lines for the (a) TE_1 mode and (b) TM_1 mode of a circular fiber.

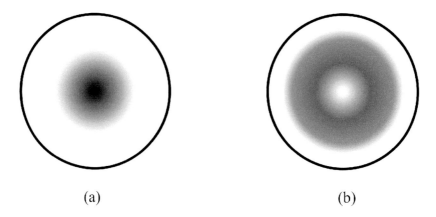

Figure 6.21   Mode intensity patterns for (a) HE_{11} and (b) TE_1, TM_1 and HE_{21} modes.

limit, these are circularly symmetric, being given by (6.116) for all modes (including TE and TM). For higher-order modes, the pattern is generally one or more rings, while for the HE_{11} mode it is simply a single circular spot centered in the core (Figure 6.21). The intensity patterns of the LP mode combinations are generally more complex, because of interference occurring between the fields of different modes.

6.8   STEP-INDEX FIBERS OF OTHER CROSS-SECTIONS

Optical fibers come in many shapes and varieties besides the circular step-index fiber. None but the circular geometry can be solved exactly in closed form, but computer numerical analyses have been done for many other types of fiber.

When the fiber is weakly guiding and a given mode on it not too close to cutoff, we can use the ideas of penetration depth and shift of the reflection plane discussed in Section 6.2 to obtain good approximate formulas for the propagation coefficient. We have already seen in eqn. (6.60) how the propagation coefficient of a slab waveguide mode can be related to that of a parallel-plate metallic guide of a larger, "effective" width. A similar result can be obtained from eqn. (6.107) using the propagation coefficients of the TM modes of a hollow circular metallic waveguide of slightly larger radius. These expressions have the form

$$\beta^2 = k_1^2 - \tilde{k}_c^2 \tag{6.117}$$

where $\tilde{k}_c$ is the cutoff wavenumber of a TM mode of a metallic circular guide of the same shape but larger radius, $a + k_0^{-1}(n_1^2 - n_2^2)^{-1/2}$:

$$\tilde{k}_c = \frac{j_{l\pm1,m}}{a + k_0^{-1}(n_1^2 - n_2^2)^{-1/2}} \tag{6.118}$$

The increase in the guide radius is simply the near-grazing limit (6.12) of the apparent shift d_{TE} of the reflecting surface. Now the limit of grazing incidence $\theta \to \pi/2$ corresponds to $\beta = k_0 n_1 \sin\theta \to k_1$, which means that the mode is far from cutoff, as indeed was the case for eqn. (6.107).

This principle can be extended to step-index fibers of any core shape, provided that we know the cutoff wavenumbers k_c for the TM modes of a correspondingly shaped hollow metallic guide. We will again obtain the result (6.117), but now $\tilde{k}_c$ will denote the cutoff wavenumber for a TM mode of a virtual hollow waveguide of a certain *effective cross-section* as shown in Figure 6.22. The effective cross-section is obtained by outwardly extending

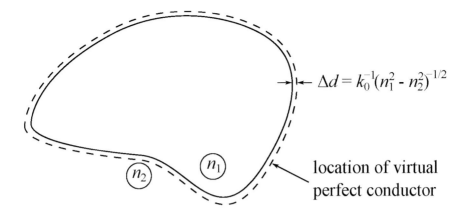

Figure 6.22   Effective cross-section for a step-index dielectric waveguide of arbitrary core shape.

the core-cladding boundary at each point in the normal direction by a distance equal to the grazing-angle shift $k_0^{-1}(n_1^2 - n_2^2)^{-1/2}$. This approximate method greatly extends the number of fibers that we can analyze, including rectangular, triangular and more. Other approximate techniques for analyzing the properties of optical waveguides do exist. Several which are based on concepts of ray theory and transverse resonance will be described in the next chapter.

6.9   THE MODE SPECTRUM OF LOSSLESS OPEN WAVEGUIDES

6.9.1   THE RADIATION MODES

All of the waveguides we have treated in this chapter have only a finite number of discrete modes at any given frequency of operation. This is true of any "open" waveguide, including those such as open microstrip, slotline, two-wire line, coplanar lines, etc. that will be treated in Chapter 8. In all these cases, there is no such thing as higher-order modes which are simply "cutoff". Instead, open waveguides possess a set of unguided "radiation modes".

It can be instructive to think of these modes as limits of modes of a closed waveguide whose metallic boundary wall recedes to infinity in the limit. Consider a dielectric slab waveguide enclosed by metallic parallel plates as shown in Figure 6.23. Although we will

Figure 6.23   Dielectric slab waveguide bounded by metallic planes.

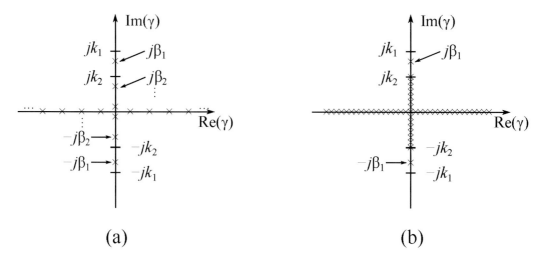

(a)   **(b)**

Figure 6.24 Location of propagation coefficients ($\times$) for modes of a shielded dielectric slab waveguide: (a) a_1 and a_2 moderate; (b) a_1 and a_2 large.

not do so in detail here, it is possible to set up expressions for the fields in the various regions of this structure, and by enforcing boundary conditions obtain a transcendental equation for the allowed values of propagation coefficient γ. It turns out that for any lossless closed waveguide, these allowed values are either pure real or pure imaginary for modes which carry nonzero complex oscillatory power, and lie at positions in the complex γ-plane as shown in Figure 6.24(a). As the distances a_1 and a_2 of the metallic plates from the slab increase, the propagation coefficients of most of the modes of this waveguide become more and more closely spaced, much the same as do the modes of a parallel-plate metallic waveguide with homogeneous filling (Figure 6.24(b). In the limit of $a_1 \to \infty$ and $a_2 \to \infty$, these modes form a *continuous spectrum* of radiation modes that have a qualitatively different nature than do the discrete, proper modes we have studied so far in this chapter.

Radiation modes describe fields which are not trapped by the guiding structure, but carry power or store energy in regions not near to the guiding structure. An example of such a field would be a plane wave in a slab waveguide whose incidence angle in the slab region was *not* greater than the critical angle, and hence produced transmitted plane waves in the cladding traveling away from the core (Figure 6.25). Unlike trapped (surface wave) modes, there is no restriction on the angle θ made by these rays with respect to the normal to the walls. Thus we have the continuous distribution of allowed values of γ as noted above,

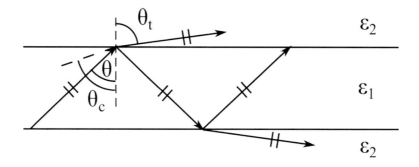

Figure 6.25 Untrapped (radiation) mode of a dielectric slab.

rather than the isolated values associated with discrete modes. For the dielectric slab, these values are:

$$\gamma = j\beta \; ; \qquad 0 \leq \beta \leq k_2$$
$$\gamma = \alpha \; ; \qquad 0 \leq \alpha < \infty$$

for "forward-traveling" continuous spectrum modes.

A single radiation mode (corresponding to a single value of γ of the continuous spectrum) cannot exist by itself if sources of finite energy content are used for excitation (a plane wave is uniform out to infinity in directions transverse to that of its propagation, and carries an infinite amount of energy). Only a superposition of such radiation modes over all possible values of γ can represent the field of such a realistic source.

Although we will not be concerned with specific applications of radiation modes in this text, the reader should be aware of their existence, and will be reminded of them from time to time in subsequent chapters. More detailed treatments can be found in the references at the end of the chapter.

6.9.2 LOCATION OF MODE PROPAGATION COEFFICIENTS

There are some principles that we can lay out that describe where the propagation coefficients of modes must be in the complex plane for a fairly general class of waveguides. For example, a qualitative analysis of the permitted range of β for bound (surface) modes of open waveguides of general type can be carried out. From equation (J.65) of Appendix J, we know that

$$\beta \leq \omega \sqrt{(\mu\epsilon)_{max}} \tag{6.119}$$

for any mode on a lossless waveguide in which $\mu > 0$ and $\epsilon > 0$ everywhere, where $(\mu\epsilon)_{max}$ represents the largest value of the product $\mu\epsilon$ over the whole guide cross-section. Furthermore, equality can be achieved in (6.119) only if the fields are identically zero except where $\mu\epsilon = (\mu\epsilon)_{max}$, which will not occur in an inhomogeneous guide (such as a dielectric slab, for example).

On the other hand, we obtain lower bounds to β by demanding that β be greater than the wavenumbers of plane waves or other wave species that can propagate out to infinity in the transverse direction. Only this guarantees exponentially decaying fields as we approach infinity in any transverse direction. For example, any waveguide of finite cross-section embedded in an infinite homogeneous region of wavenumber $k_2 = \omega\sqrt{\mu_2\epsilon_2}$ (e.g., the dielectric slab and fiber waveguides) must have $\beta > k_2$ in order that its exterior far-field (the so-called "space wave") field decay in the transverse direction. Another example is the channel guide embedded in a substrate (regarded as extending to infinity) as shown in Figure 6.2(b). For this case we require $\beta < k_1$, but $\beta > k_2$ and $\beta > k_3$ for a bound mode and hence ϵ_1 must be the largest of ϵ_1, ϵ_2, and ϵ_3 for this to happen.

When the surrounding medium of an open waveguide contains dielectric layers that extend to infinity, then not only plane waves but possibly also surface wave species may be capable of propagation away from the guiding region in the transverse direction. An example of this is given by the dielectric rib waveguide of Figure 6.2(c). For this guide, β must not only be greater than the plane wavenumber k_2 of the surrounding medium, but also greater than any of the surface wave propagation coefficients β_s along the slab itself.

Note that in all cases the upper bound on β is found from *local* properties of the permittivity distribution (i.e., where is the maximum value of $\mu\epsilon$ attained?), whereas the lower bound is due to the *global* properties of the guide (i.e., what wave species can propagate away from the local guiding region out to infinity?).

6.10 NOTES AND REFERENCES

Properties of metals at optical frequencies are discussed in

Born and Wolf (1975), Chapter 13.

The electro-optic effect and its applications are treated in

Haus (1984), Chapters 12 and 13;
R. C. Alferness, "Waveguide electro-optic modulators," *IEEE Trans. Micr. Theory Tech.*, vol. 30, pp. 1121-1137 (1982).

A nice overview of fabrication techniques for optical waveguides is given by

Lee (1986), Chapter 7.

Good introductions to the theory of optical dielectric waveguides are:

Kapany and Burke (1972);
Marcuse (1974);
Arnaud (1976);
Adams (1981);
Marcuse (1982), Chapters 8-12;
Mickelson (1993).

The derivation of the surface wave modes of a dielectric slab is to be found (in various versions) in any of these books. Figure 6.12 is adapted from

Arnaud (1976), Chapter 5.

A nice collection of reprints on the general subject of dielectric slab waveguides is:

Marcuse (1973).

Although the eigenvalue equation for the dielectric slab waveguide is not solvable in finite closed form in terms of commonly encountered functions, its many equivalent forms are encountered in numerous branches of physics. The solution of the eigenvalue equation can be expressed as a definite integral; see

C. E. Siewert, "Explicit results for the quantum-mechanical energy states basic to a finite square-well potential," *J. Math. Phys.*, vol. 19, pp. 434-435 (1978).

The solution can be expressed in terms of generalized Lambert functions as discussed in Section C.4 of Appendix C. Equation (6.58) is a special case of Kepler's equation, whose solutions describe the motion of planetary bodies. One form of its solution is as a Kapteyn series of Bessel functions:

Watson (1966), Section 1.4 and Chapter 17.

For the TE_0 mode of the dielectric slab waveguide this solution takes the form

$$b = \sin^2 \left\{ \sum_{l=0}^{\infty} \frac{(-1)^l}{l + \frac{1}{2}} J_{2l+1} \left[\left(l + \frac{1}{2} \right) V \right] \right\}$$

(although convergence of the series can be quite slow for some larger values of V). Approximate solutions similar to (6.57) and (6.59), along with further studies along these lines, were carried out by

J. F. Lotspeich, "Explicit general eigenvalue solutions for dielectric slab waveguides," *Applied Optics*, vol. 14, pp. 327-335 (1975).

M. Miyagi and S. Nishida, "An approximate formula for describing dispersion properties of optical dielectric slab and fiber waveguides," *J. Opt. Soc. Amer.*, vol. 69, pp. 291-293 (1979).

The exact analysis of the modes of a circular step-index fiber can be found in

Marcuse (1982), Chapter 8;
Arnaud (1976), Chapter 5;
Kapany and Burke (1972);
Unger (1977), Chapter 4;
Adams (1981), Chapter 7.

Elements of all of these have been incorporated into the analysis of weakly guiding fiber modes presented here. The interested reader should also consult the collection of reprints

Gloge (1976).

This contains two important articles:

D. Gloge, 'Weakly guiding fibers," *Appl. Optics*, vol. 10, pp. 2252-2258 (1971).

from which Figure 6.16 is adapted and

A. W. Snyder, "Asymptotic expressions for eigenfunctions and eigenvalues of a dielectric or optical waveguide," *IEEE Trans. Micr. Theory Tech.*, vol. 17, pp. 1130-1138 (1969).

A classic paper not included in Gloge's collection is

E. Snitzer, "Cylindrical dielectric waveguide modes," *J. Opt. Soc. Amer.*, vol. 51, pp. 491-498 (1961).

Snitzer tells how to construct the field plots illustrated in Figs. 6.18-6.20. His method was used to construct plots of the first three HE and EH mode groups in

Vzyatyshev (1970).

Other useful approximate formulas for propagation coefficients of fiber modes are given in Gloge (1971), *loc. cit.* above, but these are of limited accuracy. More accurate (though more complicated) approximations are given in Miyagi and Nishida (1979), *loc. cit.* above.

Fibers of arbitrary cross-section have been treated in the weakly guiding limit in

D. L. A. Tjaden, "First-order correction to 'weak-guidance' approximation in fibre optics theory," *Phillips J. Res.*, vol. 33, pp. 103-112 (1978).

A. W. Snyder and W. R. Young, "Modes of optical waveguides," *J. Opt. Soc. Amer.*, vol. 68, pp. 297-309 (1978).

Numerical solutions for various shapes of the fiber core are given in

C. Yeh, K. Ha, S. B. Dong and W. P. Brown, "Single-mode optical waveguides," *Applied Optics*, vol. 18, pp. 1490-1504 (1979).

The effective cross-section idea comes from

> E. F. Kuester, "Propagation coefficients for linearly polarized modes of arbitrarily-shaped optical fibers, or dielectric waveguides," *Optics Letters*, vol. 8, pp. 192-194 (1983).

and is extended to fibers which are not weakly guiding in

> E. F. Kuester, "The effective cross-section method for dielectric waveguides in or on a substrate," *Radio Science*, vol. 19, pp. 1239-1244 (1984).

Discussions of the radiation mode spectrum of optical and other open waveguides can be found in:

> Shevchenko (1971);
> Marcuse (1974);
> Marcuse (1982), Chapters 8-10;
> Vassallo (1991).

6.11 PROBLEMS

p6-1 Using the ray method of Section 6.2, and the result of problem p1-15, determine the core field $\mathcal{H}_y$ and the eigenvalue equation to be solved for β for the TM modes of the dielectric slab waveguide shown in Figure 6.4. Note: It can be somewhat more convenient to use the *current* (i.e., magnetic-field) reflection coefficient rather than the Fresnel coefficient for this problem, since the only component of the magnetic field is $\mathcal{H}_y$.

p6-2 Consider the asymmetric slab waveguide pictured below.

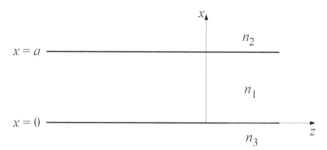

(a) Using the method of Section 6.2, give an expression for $\mathcal{E}_y$ in the core region, and derive the eigenvalue equation for the TE modes of this guide.

(b) Show that if $n_3 > n_2$, then all TE modes (even the TE_0) have a nonzero cutoff frequency, and give an expression for this frequency.

[Note: In the interest of standardized notation, put

$$p_2^2 = \beta^2 - k_0^2 n_2^2$$

$$p_3^2 = \beta^2 - k_0^2 n_3^2$$

$$V_2^2 = (n_1^2 - n_2^2)k_0^2 a^2$$

$$V_3^2 = (n_1^2 - n_3^2)k_0^2 a^2$$

Also, since no loss of generality results by doing so, assume that $n_3 > n_2$.]

p6-3 A symmetric slab dielectric waveguide is to be made from materials such that $n_2 = 3.5$ and $n_1 = 1.03n_2$. What is the largest thickness a of the core layer possible without the existence of higher order (TE_1, etc., or TM_1, etc.) modes, if the operating frequency range is to cover the range of free-space wavelengths $0.8\ \mu m \le \lambda_0 \le 1.5\ \mu m$? Suppose that one of the cladding regions is replaced by a half-space of air ($n_3 = 1$). Is it possible now to find a core layer thickness a such that only the lowest order TE and TM modes exist throughout this frequency range? If so, provide a suitable value of a.

p6-4 Prove that the time-average power flow through any plane $x = $ constant is zero for a TE surface-wave mode on a dielectric slab waveguide.

p6-5 Show that the complex power

$$\hat{P} = \frac{1}{2} \int_{y=0}^{1} \int_{x=-\infty}^{\infty} \boldsymbol{\mathcal{E}} \times \boldsymbol{\mathcal{H}}^* \cdot \mathbf{u}_z \, dxdy$$

carried through a unit width (in y) of the total cross-section of a dielectric slab waveguide for a TE mode can be expressed in terms of the effective width $a_{\text{eff}} = a + 2/p_2$ as follows:

$$\hat{P} = \frac{\beta a_{\text{eff}}}{4} |D|^2 \frac{\omega(\epsilon_1 - \epsilon_2)}{h_1^2}$$

where D is the amplitude factor from equation (6.32). Analogously, show that the complex oscillatory power is given by

$$\hat{P}_{\text{osc}} = \frac{1}{2} \int_{y=0}^{1} \int_{x=-\infty}^{\infty} \boldsymbol{\mathcal{E}} \times \boldsymbol{\mathcal{H}} \cdot \mathbf{u}_z \, dxdy = \frac{\beta a_{\text{eff}}}{4} D^2 \frac{\omega(\epsilon_1 - \epsilon_2)}{h_1^2}$$

[*Hint*: How can the trig functions $\sin(ha)$ and $\cos(ha)$ be eliminated from your expression?]

p6-6 Show that the various forms (6.33), (6.34), (6.35), (6.36), (6.37) and (6.43) for the eigenvalue equation of the TE modes of the dielectric slab are equivalent to each other.

p6-7 Consider the asymmetric slab waveguide pictured in problem p6-2.
  (a) Give an expression for $\mathcal{E}_y$ in both cladding regions as well as the core region, and derive the eigenvalue equation for the TE modes of this guide using the method of Section 6.3.
  (b) Show that if $n_2 \ne n_3$, then all TE modes (even the TE_0) have a nonzero cutoff frequency, and give an expression for this frequency.
  Use the standardized notation suggested in problem p6-2.

p6-8 Repeat problem p6-7 for the TM modes of the asymmetric slab waveguide.

p6-9 Using the result of problem p6-8, compute and plot the normalized propagation constant

$$b = \frac{n_{\text{eff}}^2 - n_3^2}{n_1^2 - n_3^2}$$

versus the normalized frequency

$$V_3 = k_0 a \sqrt{n_1^2 - n_3^2}$$

for the values $n_1 = 1.51$, $n_2 = 1.0$ and $n_3 = 1.50$. Cover the range $0 < V_3 < 20$, and plot results for all TM modes which exist in that range. Compare these results with ones obtained for the symmetric slab for the case $n_1 = 1.51$ and $n_2 = n_3 = 1.50$.

p6-10 Using the result of problem p6-7 or p6-2, compute and plot the normalized propagation constant

$$b = \frac{n_{\text{eff}}^2 - n_3^2}{n_1^2 - n_3^2}$$

versus the normalized frequency

$$V_3 = k_0 a \sqrt{n_1^2 - n_3^2}$$

for the values $n_1 = 1.51$, $n_2 = 1.0$ and $n_3 = 1.50$. Cover the range $0 < V_3 < 20$, and plot results for all TE modes which exist in that range. Compare these results with ones obtained for the symmetric slab for the case $n_1 = 1.51$ and $n_2 = n_3 = 1.50$.

p6-11 Using the method of Section 6.3, obtain expressions for the field $\mathcal{H}_y$ (both in the core and in the cladding) of the TM modes of the dielectric slab waveguide shown in Figure 6.4, along with the eigenvalue equation for determining the allowed values of β (*Note: Be careful to apply the correct boundary conditions*). Illustrate graphically the difference between the TE-mode and TM-mode eigenvalue equations.

p6-12 Suppose the cladding material of a symmetric dielectric slab waveguide has a negative real permittivity $\epsilon_2 = -|\epsilon_2|$ (or in other words, a purely negative imaginary refractive index $n_2 = -j|n_2|$). Such a material could be a metal at optical frequencies with losses neglected, an ionized plasma below its cutoff frequency, or certain kinds of metamaterial. Suppose the core region is an ordinary lossless dielectric with $\epsilon_1 \geq \epsilon_0$, and both regions are nonmagnetic.

(a) Show that there exists a TM propagating mode for $m = 0$ in this case (if $|\epsilon_2| > \epsilon_1$), that has $n_{\text{eff}} > n_1$, and that it propagates for any value of V (you may use graphical or analytical means to prove this).

(b) If $\epsilon_1/\epsilon_0 = 2$ and $\epsilon_2/\epsilon_0 = -9$, compute and plot n_{eff} versus V over the range $0 < V < 5$ for the mode described in part (a).

(c) Make a rough sketch of $\mathcal{H}_y$ versus x for this mode, and describe physically the difference in field behavior from the usual case when $\epsilon_2 > 0$.

p6-13 Show for the TM_m modes of the dielectric slab [which obey (6.63)] that

$$b \simeq \sin^2 \left[\frac{n_2^2}{n_1^2} \frac{V - m\pi}{2} \right]$$

as $b \to 0$ $(V \to m\pi)$, and

$$b \simeq 1 - \left[\frac{(m+1)\pi}{V + 2n_2^2/n_1^2} \right]^2$$

as $b \to 1$ $(V \to \infty)$.

p6-14

(a) For the HE_{lm} modes of a weakly guiding step-index circular fiber, show that (for $l \neq 0$), the transverse field components are given approximately by

$$\mathcal{E}_\rho = -jC_1 A_l(\rho) \left\{ \begin{array}{c} \cos l\phi \\ \sin l\phi \end{array} \right\}$$

$$\mathcal{E}_\phi = jC_1 A_l(\rho) \left\{ \begin{array}{c} \sin l\phi \\ -\cos l\phi \end{array} \right\}$$

$$\mathcal{H}_\rho = -\frac{j\beta}{\omega\mu} C_1 A_l(\rho) \left\{ \begin{array}{c} \sin l\phi \\ -\cos l\phi \end{array} \right\}$$

$$\mathcal{H}_\phi = -\frac{j\beta}{\omega\mu}C_1 A_l(\rho) \left\{ \begin{array}{c} \cos l\phi \\ \sin l\phi \end{array} \right\}$$

where $C_1 \simeq \sqrt{\frac{\mu}{\epsilon_1}}C_2$ and

$$A_l(\rho) = \left\{ \begin{array}{ll} \frac{\beta}{h_1}J_{l-1}(h_1\rho) & (\rho < a) \\ \frac{\beta}{p_2}\frac{J_l(h_1 a)}{K_l(p_2 a)}K_{l-1}(p_2\rho) & (\rho > a) \end{array} \right.$$

(b) For the special case $l = 0$ of the HE modes (the TM modes) show that the transverse fields are

$$\mathcal{E}_\rho = -jC_1 A_0(\rho)$$

$$\mathcal{E}_\phi = 0$$

$$\mathcal{H}_\rho = 0$$

$$\mathcal{H}_\phi = -\frac{j\beta}{\omega\mu}C_1 A_0(\rho)$$

p6-15

(a) For the EH_{lm} modes of a weakly-guiding step-index circular fiber, show that (for $l \neq 0$), the transverse field components are given approximately by

$$\mathcal{E}_\rho = jC_1 B_l(\rho) \left\{ \begin{array}{c} \cos l\phi \\ \sin l\phi \end{array} \right\}$$

$$\mathcal{E}_\phi = jC_1 B_l(\rho) \left\{ \begin{array}{c} \sin l\phi \\ -\cos l\phi \end{array} \right\}$$

$$\mathcal{H}_\rho = -\frac{j\beta}{\omega\mu}C_1 B_l(\rho) \left\{ \begin{array}{c} \sin l\phi \\ -\cos l\phi \end{array} \right\}$$

$$\mathcal{H}_\phi = \frac{j\beta}{\omega\mu}C_1 B_l(\rho) \left\{ \begin{array}{c} \cos l\phi \\ \sin l\phi \end{array} \right\}$$

where $C_1 \simeq -\sqrt{\frac{\mu}{\epsilon_1}}C_2$ and

$$B_l(\rho) = \left\{ \begin{array}{ll} \frac{\beta}{h_1}J_{l+1}(h_1\rho) & (\rho < a) \\ -\frac{\beta}{p_2}\frac{J_l(h_1 a)}{K_l(p_2 a)}K_{l+1}(p_2\rho) & (\rho > a) \end{array} \right.$$

(b) For the special case $l = 0$ of the EH modes (the TE modes), show that the transverse fields are:

$$\mathcal{E}_\rho = 0$$

$$\mathcal{E}_\phi = \frac{j\omega\mu}{\beta}C_2 B_0(\rho)$$

$$\mathcal{H}_\rho = -jC_2 B_0(\rho)$$

$$\mathcal{H}_\phi = 0$$

p6-16 A circular step-index optical fiber has a core of index $n_1 = 2.21$, and a *finite* cladding of index $n_2 = 2.20$ which extends only out to the radius c. The fiber is to be operated at a (free-space) wavelength of $\lambda_0 = 10.6\mu$m.

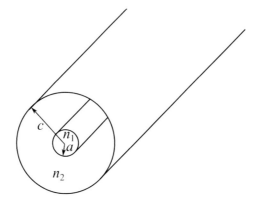

(a) What is the maximum value of a for which the fiber will support only the fundamental modes?

(b) If a is chosen to be the maximum value found in part (a), what value of c must be chosen to assure that the fields at $\rho = c$ have decayed to 0.1% of their values at $\rho = a$?

p6-17 Show that choosing $\mathcal{E}_z = C_1 F_l(\rho) \cos l\phi$ and $\mathcal{H}_z = C_2 G_l(\rho) \cos l\phi$ for the same mode leads to boundary conditions for $\mathcal{E}_\phi$ and $\mathcal{H}_\phi$ that cannot be satisfied if $l \neq 0$.

p6-18 Using the results of problems p6-14 and p6-15, show that the transverse mode fields of appropriate linear combinations of $\mathrm{HE}_{l+2,m}$ and EH_{lm} modes are linearly polarized in the core, and give approximate expressions for them.

p6-19 Use the effective cross-section method of Section 6.8 to plot b vs. $V = k_0 a \sqrt{n_1^2 - n_2^2}$ for the fundamental modes of a weakly-guiding *rectangular-core* optical fiber of dimensions $a \times d$, if $a = 2d$.

p6-20 The transverse cross-section of a dielectric strip waveguide is shown below, with $n_2 > n_3 > 1$.

air: $n = 1$

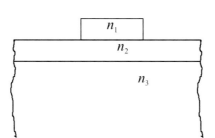

Can this structure support bound waveguide modes if

(a) $n_1 > n_2$?

(b) $n_2 > n_1 > n_3$?

(c) $n_3 > n_1$?

Explain your answers in each case.

7 TRANSVERSE RESONANCE IN GUIDED WAVE STRUCTURES

Only a very limited class of waveguides permits exact solutions for their mode fields and propagation coefficients. In fact, we have examined most of them in Chapters 5 and 6 (a few others which support TEM modes will be considered in Chapter 8). Other important waveguides (microstrip, for example) do not yield to any exact solution and must be handled either with numerical methods or by approximation. While numerical techniques are of substantial importance, they are beyond the scope of this book, in which we hope to examine more of the fundamental physics involved in guided waves.

In this chapter we will examine several types of waveguides using a very important analysis technique called the transverse resonance method. The method can be viewed as an extension of the ray methods treated in Chapter 5. In Section 7.1, it is introduced as an exact technique for the analysis of multilayer slab waveguides, whereas subsequently it will be extended as an approximate method capable of treating many types of waveguide not otherwise capable of being analyzed by other than numerical means. In specific applications other names for essentially the same approach (e.g., the effective index method in optical waveguides) are sometimes used. The technique will be introduced and developed by means of examples.

7.1 MULTILAYER SLAB WAVEGUIDES

7.1.1 TE MODES

It is of considerable interest to characterize the behavior of slab waveguides which consist of more than one central guiding layer. The layering can be employed to achieve a variety of desired physical effects. For example, thin layers of semiconducting materials with greatly differing bandgaps can be combined into so-called multiple quantum well (MQW) structures. These structures have properties which are readily exploited to make devices such as intensity modulators and optical switches. Although the methods of Chapter 6 could, in principle, be used to analyze the waveguiding properties of such multilayered slabs, the algebra involved quickly becomes much too cumbersome to handle, and is even awkward to implement in a computer program. We seek to obtain the waveguide modes by simpler means, if possible.

Consider then the multilayer slab structure of Figure 7.1(a). The TE mode fields thus satisfy (6.20) again, where now $\epsilon = \epsilon(x) = \epsilon_0 n^2(x)$. The solution within the ith layer $x_{i-1} < x < x_i$, where $n(x) = n_i$, is given as

$$\mathcal{E}_y = \tilde{A}_i e^{-p_i x} + \tilde{B}_i e^{+p_i x} \tag{7.1}$$

where

$$p_i = \sqrt{\beta^2 - k_0^2 n_i^2} = j h_i \tag{7.2}$$

and $\tilde{A}_i = A_i \sqrt{2Z_c}$ and similarly for $\tilde{B}_i$. By (6.18), the $\mathcal{H}_z$ field in this layer must be

$$\mathcal{H}_z = -\frac{1}{j\omega\mu} \frac{\partial \mathcal{E}_y}{\partial x} = \frac{p_i}{j\omega\mu} \left[\tilde{A}_i e^{-p_i x} - \tilde{B}_i e^{+p_i x} \right] \tag{7.3}$$

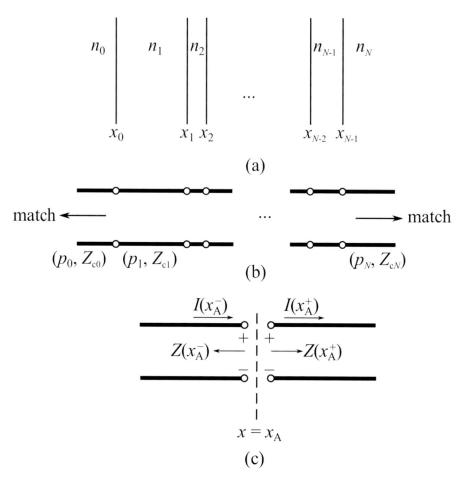

Figure 7.1 (a) Multilayered slab; (b) equivalent transverse transmission line network for TE modes; (c) decomposition into left-looking and right-looking sections.

As in Section 1.4, we compare (7.1) and (7.3) with expressions (1.7) for the voltage and current on a classical transmission line, the only difference being that here the "direction of propagation" is taken to be x instead of z. It can be seen that the behavior of the TE mode field in the transverse (x) direction is given by that of an equivalent transmission line described in Section 1.2, if the following replacements are made:

$$
\begin{aligned}
z &\rightarrow x \\
\gamma &\rightarrow p_i \\
V &\rightarrow \mathcal{E}_y \\
I &\rightarrow \mathcal{H}_z \\
Z_c &\rightarrow \frac{j\omega\mu}{p_i} = \frac{\omega\mu}{h_i}
\end{aligned}
\tag{7.4}
$$

In the cladding region $x < x_0 = 0$, we have only the wave $B_0 e^{p_0 x}$, while in $x > x_{N-1} = a$, we have only $A_N e^{-p_N x}$.

The apparatus of Section 1.2 can now be carried over for use in the solution of this multilayered waveguide. The multilayered slab is replaced by a network of equivalent transmission line sections as shown in Figure 7.1(b). We are seeking mode solutions, so no sources are present in the transmission line network. We wish to know under what conditions nonzero voltages and currents may exist in the network (i.e., nonzero fields in the slab).

Let us choose any arbitrary point $x = x_A$ in this network, and temporarily disconnect the right and left sides of the network at this point as shown in Figure 7.1(c). The impedance seen "looking right" into the right-hand section is

$$\overrightarrow{Z}(x_A) = Z(x_A^+) = \frac{V(x_A^+)}{I(x_A^+)} \tag{7.5}$$

which is the ordinary line impedance at $x = x_A^+$ as defined in (1.52). This value is easily calculated by starting at $x = x_{N-1}$ with $\overrightarrow{Z}(x_{N-1}) = Z_{cN}$ (since $\rho(x) = 0$ in region N), and repeatedly transferring this value backwards to x_{N-2}, x_{N-3} and so on, using (1.60):

$$\overrightarrow{Z}(x_i) = Z_{c,i+1} \frac{\overrightarrow{Z}(x_{i+1}) \cosh p_{i+1} d_{i+1} + Z_{c,i+1} \sinh p_{i+1} d_{i+1}}{Z_{c,i+1} \cosh p_{i+1} d_{i+1} + \overrightarrow{Z}(x_{i+1}) \sinh p_{i+1} d_{i+1}} \tag{7.6}$$

where $d_{i+1} = x_{i+1} - x_i$. In the last step to x_A, if x_A does not coincide with one of the layer boundaries x_i, we must replace d_{i+1} by $x_{i+1} - x_A$ in (7.6), as well as x_i by x_A.

Likewise, let us "look left" into the left-hand section. Here we see

$$\overleftarrow{Z}(x_A) = -Z(x_A^-) = -\frac{V(x_A^-)}{I(x_A^-)} \tag{7.7}$$

and the minus sign appears because $I(x)$ for a transmission line is always defined as positive when flowing toward increasing x. Now, $\overleftarrow{Z}$ is calculated in precisely the same way as $\overrightarrow{Z}$, starting now at $x = x_0$ with $\overleftarrow{Z}(x_0) = Z_{c0}$ and working toward x_A. Specifically,

$$\overleftarrow{Z}(x_{i+1}) = Z_{c,i+1} \frac{\overleftarrow{Z}(x_i) \cosh p_{i+1} d_{i+1} + Z_{c,i+1} \sinh p_{i+1} d_{i+1}}{Z_{c,i+1} \cosh p_{i+1} d_{i+1} + \overleftarrow{Z}(x_i) \sinh p_{i+1} d_{i+1}} \tag{7.8}$$

Reconnecting the lines at x_A once again, we require that Kirchhoff's current law $I(x_A^-) = I(x_A^+)$ and voltage law $V(x_A^-) = V(x_A^+)$ be satisfied in order for a nontrivial source-free solution to be possible. By (7.5) and (7.7) we see that this means that the *transverse resonance condition*

$$\overleftarrow{Z}(x_A) + \overrightarrow{Z}(x_A) = 0 \tag{7.9}$$

must be satisfied.

Only for certain values of β (the eigenvalues) can this occur, and so we have yet a third approach to obtaining the characteristic equation of, and physical insight into, such dielectric slab waveguides. Note that (7.9) does *not* describe an impedance match condition, since in fact both forward and reverse waves exist in all regions (line sections) except 0 and N, and the reflection coefficient $\rho(x)$ is thus not zero. Rather, (7.9) is simply a concise statement of Kirchhoff's laws for compatibility of voltage and current at the connection point x_A when the two halves of the transmission line network are connected together. This condition is similar to the condition of oscillation for a one-port oscillator. Only when this compatibility occurs is the network capable of supporting nonzero source-free voltages and currents.

It is an easy example to demonstrate that this algorithm gives the correct eigenvalue equation for the TE modes of the step-index slab. Choosing $x_A = 0$ in the slab of Figure 6.4, we have $\overleftarrow{Z}(0) = j\omega\mu_0/p_2$, while

$$\overrightarrow{Z}(0) = \frac{j\omega\mu_0}{h_1} \frac{h_1 \cos h_1 a + p_2 \sin h_1 a}{p_2 \cos h_1 a - h_1 \sin h_1 a}$$

so that

$$\frac{h_1 \cos h_1 a + p_2 \sin h_1 a}{p_2 \cos h_1 a - h_1 \sin h_1 a} + \frac{h_1}{p_2} = 0$$

which is equivalent to (6.33). We could also have chosen the matching point to be at $x_A = a/2$, in which case the impedances looking right and looking left are identical by virtue of the symmetry of the structure:

$$\overrightarrow{Z}\left(\frac{a}{2}\right) = \overleftarrow{Z}\left(\frac{a}{2}\right) = \frac{j\omega\mu_0}{h_1} \frac{h_1 \cos \frac{h_1 a}{2} + p_2 \sin \frac{h_1 a}{2}}{p_2 \cos \frac{h_1 a}{2} - h_1 \sin \frac{h_1 a}{2}}$$

The temptation is to conclude that imposition of the transverse resonance condition implies that these two impedances must equal zero. This is indeed one possibility, but not the only one. We have chosen a matching point at a plane of symmetry of the structure, and general symmetry considerations tell us that all fields of a mode of this structure must either be even or odd functions about this plane. Because of (7.3), if $\mathcal{E}_y$ is even, then $\mathcal{H}_z$ is odd, and vice versa. Setting both left- and right-looking transverse impedances equal to zero is tantamount to choosing the modes where $\mathcal{E}_y$ vanishes at the symmetry plane; those modes for which the other symmetry applies will correspond to infinite values of left- and right-looking impedances. Equivalently, we can view our procedure in terms of admittances, for which the latter case requires zero values at the symmetry plane. The result of imposing transverse resonance at $x_A = a/2$ for the dielectric slab is then

$$h_1 \cos \frac{h_1 a}{2} + p_2 \sin \frac{h_1 a}{2} = 0$$

or

$$p_2 \cos \frac{h_1 a}{2} - h_1 \sin \frac{h_1 a}{2} = 0$$

which are equivalent to (6.43).

7.1.2 TM MODES

The TM modes of the multilayered slab waveguide in Figure 7.1(a) can also be analyzed using the transverse resonance method. Instead of (7.4), we must make the correspondences

$$
\begin{aligned}
z &\rightarrow x \\
\gamma &\rightarrow p_i \\
V &\rightarrow \mathcal{E}_z \\
I &\rightarrow -\mathcal{H}_y \\
Z_c &\rightarrow \frac{p_i}{j\omega\epsilon_i} = \frac{h_i}{\omega\epsilon_i}
\end{aligned}
\tag{7.10}
$$

Other more complicated multilayered slabs can also be readily analyzed by this method. It is especially suitable for computer implementation, which is left to the reader in certain of the problems at the end of the chapter.

7.2   GRADED-INDEX PROFILES

Many methods of fabrication for optical waveguides do not result in simple abruptly changing refractive index profiles such as we have studied in Chapter 6 or in the previous section. Instead of such *step-index* distributions as in Figure 7.1(a), we more often have graded-index profiles as shown in Figure 7.2(a). The rays corresponding to the field in a step-index

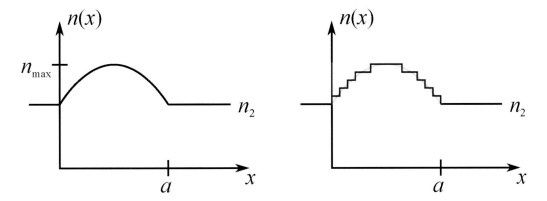

Figure 7.2 (a) Graded-index profile; (b) piecewise constant (staircase) index approximation.

guide abruptly change direction upon reflection by the boundary, while we might imagine that those of the graded-index guide do so continuously at each point within the guide. We can visualize the process approximately by imagining the continuous index profile of Figure 7.2(a) to be approximated by the piecewise constant one of Figure 7.2(b). A ray will be refracted a little bit at each jump in index until total reflection occurs, reversing the direction of transverse ray travel.

In most cases, the field equations cannot be solved exactly for continuously graded index profiles, and the piecewise constant index approximation of Figure 7.2(b), though it can be handled by the transverse resonance technique, leads to equations which are unwieldy to solve. If the change in refractive index with x is gradual enough, it is possible to use the WKB approximation of Section 1.5 to obtain an approximate transverse-resonance solution to the problem.

From the WKB approximation to the solution of the equivalent nonuniform transverse transmission line, the right-looking impedance at $x = x_i$ of an inhomogeneous section of transmission line between $x = x_i$ and x_{i+1} is given by

$$\overrightarrow{Z}(x_i) = Z_c(x_i^+) \frac{\overrightarrow{Z}(x_{i+1}) \cosh \int_{x_i}^{x_{i+1}} p(x)\, dx + Z_c(x_{i+1}^-) \sinh \int_{x_i}^{x_{i+1}} p(x)\, dx}{Z_c(x_{i+1}^-) \cosh \int_{x_i}^{x_{i+1}} p(x)\, dx + \overrightarrow{Z}(x_{i+1}) \sinh \int_{x_i}^{x_{i+1}} p(x)\, dx} \tag{7.11}$$

This can be used in place of (7.6) when the layer is inhomogeneous. When the rays in a layer approach the critical angle for one of its interfaces, the ordinary WKB approximation breaks down. This will occur whenever $p(x)$ nears zero—these values of x are turning points. The analysis must be modified near turning points as indicated in Section 1.5 and Appendix F. The result (for TE waves) is that a turning point appears to totally reflect a propagating [$p(x)$ is pure imaginary] voltage wave with a reflection coefficient of $e^{j\pi/2}$. The voltage and current decay exponentially beyond the turning point, and if a second turning point is not located too close to the first, we may ignore any penetration of the wave beyond the first turning point, and regard it as if it were isolated. In this case, a turning point located at x_t [so that $p(x_t) = 0$] in an inhomogeneous section of line with $p(x)$ imaginary for $x < x_t$ produces the right-looking impedance at $x_i < x_t$ of:

$$\overrightarrow{Z}(x_i) = Z_c(x_i^+) \frac{j \cosh \int_{x_i}^{x_t} p(x)\, dx + \sinh \int_{x_i}^{x_t} p(x)\, dx}{\cosh \int_{x_i}^{x_t} p(x)\, dx + j \sinh \int_{x_i}^{x_t} p(x)\, dx} \qquad \text{(TE polarization)} \tag{7.12}$$

This, too, can be used in place of (7.6) in the transverse resonance analysis of continuously graded slab waveguides. For TM waves, the relation corresponding to (7.12) is

$$\vec{Z}(x_i) = Z_c(x_i^+) \frac{-j \cosh \int_{x_i}^{x_t} p(x)\,dx + \sinh \int_{x_i}^{x_t} p(x)\,dx}{\cosh \int_{x_i}^{x_t} p(x)\,dx - j \sinh \int_{x_i}^{x_t} p(x)\,dx} \qquad \text{(TM polarization)} \qquad (7.13)$$

7.2.1 EXAMPLE: A DIFFUSED SLAB WAVEGUIDE

Consider a slab waveguide with an exponential permittivity profile

$$\begin{aligned}
\epsilon_r(x) &= n_0^2 \qquad (x < 0) \\
&= n_2^2 + (n_1^2 - n_2^2)e^{-x/a} \qquad (x > 0)
\end{aligned}$$

as shown in Figure 7.3, where a now represents a characteristic depth of the variation of ϵ into the substrate region $x > 0$, while n_1 is the maximum and n_2 the minimum value of refractive index in the substrate. This type of profile can result from in-diffusion of metals into previously homogeneous substrates. The WKB method will be used here to find the eigenvalue equation for the TE surface wave modes of this waveguide.

For the substrate region, we can write

$$h = h(x) = \sqrt{k_0^2 \epsilon_r(x) - \beta^2} = k_0 \sqrt{n_1^2 - n_2^2} \sqrt{e^{-x/a} - b} \qquad (7.14)$$

where again

$$b = \frac{n_{\text{eff}}^2 - n_2^2}{n_1^2 - n_2^2}$$

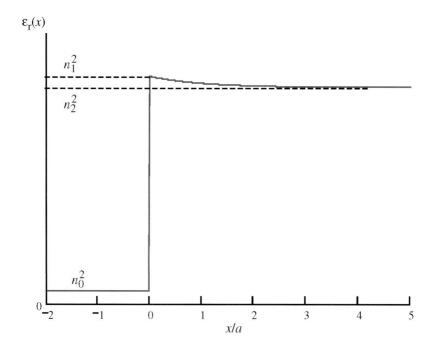

Figure 7.3 Diffused slab waveguide permittivity profile.

and

$$n_{\text{eff}} = \beta/k_0$$

The turning point x_t is located at $x_t = a \ln(1/b)$. The phase integral is evaluated by elementary functions:

$$\int_0^{x_t} p\, dx = j \int_0^{x_t} h\, dx = j2V_1[\sqrt{1-b} - \sqrt{b}\cos^{-1}\sqrt{b}] \qquad (7.15)$$

where

$$V_1 = k_0 a \sqrt{n_1^2 - n_2^2}$$

and so using (7.12) we can set up the transverse resonance condition for this waveguide:

$$\tan\left[2V_1(\sqrt{1-b} - \sqrt{b}\cos^{-1}\sqrt{b})\right] = \frac{\sqrt{n_1^2 - n_{\text{eff}}^2} + \sqrt{n_{\text{eff}}^2 - n_0^2}}{\sqrt{n_1^2 - n_{\text{eff}}^2} - \sqrt{n_{\text{eff}}^2 - n_0^2}} \qquad (7.16)$$

This equation predicts n_{eff} with high accuracy. A typical dispersion curve for this case is shown in Figure 7.4. Comparing with results for a step-index slab waveguide with the same cladding materials but having a constant index n_1 over the region $0 < x < a$ before dropping to the constant value n_2 for $x > a$, we see that both modes have a nonzero cutoff frequency, but the step-index guide has b rising much more rapidly to its high-frequency limit. This is because the diffused guide's spatial region of field concentration (where $n_{\text{eff}} > n(x)$) becomes smaller and smaller as frequency increases, while that of the step-index guide remains fixed at $0 < x < a$.

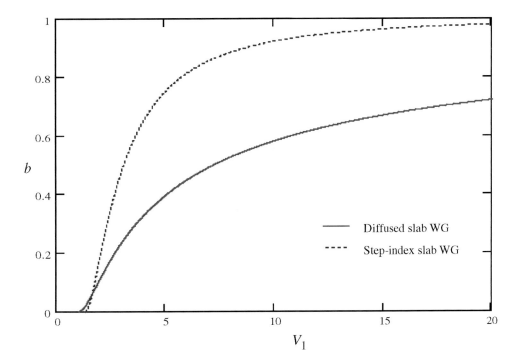

Figure 7.4 Dispersion curve for TE$_0$ mode of diffused slab waveguide with $n_0 = 1$, $n_2 = 4.0$ and $n_1 = 4.1$, with TE$_0$ mode of step-index slab waveguide of thickness a for comparison.

7.3　LOADED RECTANGULAR WAVEGUIDES

The principles of transverse resonance can be extended to other kinds of complicated waveguides if we consider reflections and transmissions of more general types of constituent waves than simple plane waves. We begin by looking at the species of wave that can exist in a uniform parallel plate region of height d as shown in Figure 7.5. Since such a uniform segment is a parallel plate waveguide, we know from Chapter 5 that a TEM wave as well as TE and TM higher order waves can be present. Let us assume for the moment that only a TEM wave exists, but that it propagates at an angle θ with respect to the x-axis as shown. Consider a rectangular metallic waveguide whose cross-section can be partitioned by one or more vertical lines into several sections of parallel-plate waveguide of height d filled with various dielectric materials. Within each section, we express the field as a pair of TEM (with respect to the *ray* directions) parallel-plate waves propagating at some angle with respect to the z-axis. The factor $\exp(-\gamma z)$ is common to all the fields, and we need deal only with the *projected* behavior of these waves in the xy-plane. The equivalent transverse transmission-line representation is readily obtained from Section 7.1—all we need do is to place perfectly conducting planes at $y = 0$ and $y = d$ in the multilayered slab geometry to obtain that of Figure 7.5. The parameters representing the equivalent transverse transmission line for each parallel-plate region i are obtained from (7.4):

$$\left.\begin{array}{rcl} z & \to & x \\ \gamma & \to & jk_{xi} = j\sqrt{k_0^2\epsilon_{ri} - \beta^2} = jk_0\sqrt{\epsilon_{ri}}\cos\theta_i \\ V & \to & \mathcal{E}_y \\ I & \to & \mathcal{H}_z \\ Z_c & \to & \frac{\omega\mu}{k_{xi}} \end{array}\right\} \quad \text{TEM waves} \qquad (7.17)$$

We have changed the notation for the propagation constant in the x-direction in order to be more adaptable for use in waveguides other than multilayer slabs. A filling medium of relative permittivity ϵ_{ri} is assumed for each parallel-plate region. If all regions have the same material parameters, then k_{xi} is the same everywhere, and is equal to the cutoff wavenumber k_c for the homogeneously filled waveguide being analyzed.

Consider as an example the rectangular metallic waveguide of Figure 5.12. While we already have a full solution for this waveguide by other means, it is instructive to see it carried out by the transverse resonance method. The equivalent transverse transmission line network is a length a of transmission line short-circuited at both ends. There being only one material filling the waveguide, we have $k_{x1} = k_c$, the cutoff wavenumber for the rectangular guide. Breaking the transverse equivalent circuit at (say) $x = 0$, and applying

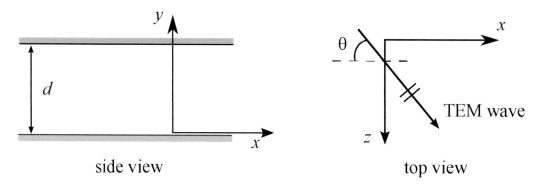

side view　　　　　　　　　　　　　　　　　　　　top view

Figure 7.5　Parallel plate region of height d.

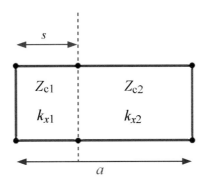

Figure 7.6 Equivalent transverse transmission line network for dielectric-loaded wave-guide.

the transverse resonance condition, we obtain $\tan k_c a = 0$, or $k_c a = m\pi$; $m = 1, 2, \ldots$. The solutions of this equation correspond to the TE_{m0} modes. Note that no other modes of this waveguide are found by the transverse resonance method, because only TEM waves have been used as building blocks for the fields.

Next, consider the effect of loading the rectangular waveguide with a nonmagnetic dielectric region of permittivity ϵ occupying $0 < x < s$, while the rest of the waveguide is air-filled. The equivalent transverse network is shown in Figure 7.6. The characteristic impedance and cutoff wavenumber are different for the two sections of transmission line:

$$Z_{c1} = \frac{\omega\mu_0}{k_{x1}}; \qquad k_{x1} = \sqrt{k^2 - \beta^2}$$

and

$$Z_{c2} = \frac{\omega\mu_0}{k_{x2}}; \qquad k_{x2} = \sqrt{k_0^2 - \beta^2}$$

where k and k_0 are the wavenumbers of the dielectric and air regions respectively. We set up the transverse resonance equation as

$$Z_{c1} \tan k_{x1}s + Z_{c2} \tan[k_{x2}(a - s)] = 0 \tag{7.18}$$

or

$$\frac{\tan k_{x1}s}{k_{x1}} + \frac{\tan[k_{x2}(a - s)]}{k_{x2}} = 0 \tag{7.19}$$

As is the case for the dielectric slab waveguide, this is a transcendental equation for the phase coefficient β, and must in general be solved numerically. This is left as an exercise for the reader.

Suppose that we wish to examine the modes of stepped cross-section, homogeneously filled metallic waveguides such as those shown in Figure 7.7. In this case, we must begin by reconsidering the species of wave that can exist in a transverse segment of the structure of uniform height d. Now not only TEM waves, but also TE and TM waves could be present, propagating (or attenuating) in any direction in the xz-plane. If $kd < \pi$, the higher order TE and TM waves are cutoff, and will quickly decay to negligible levels as we move away from the places at which they originate. If a TEM wave is incident on a junction between two parallel plate regions as shown in Figure 7.8, we expect that there will be a reflected as well as a transmitted TEM wave on the respective sides of the junction as indicated. In order to maintain consistency of the phase variation in the z-direction, the reflection and transmission angles must be the same as the incident angle, if the same medium is present on both sides of the junction. However, these waves along with the incident wave

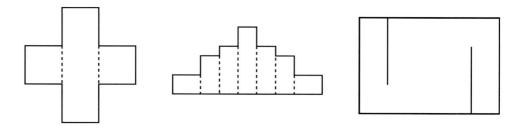

Figure 7.7 Hollow metallic guides with stepped-profile cross-sections.

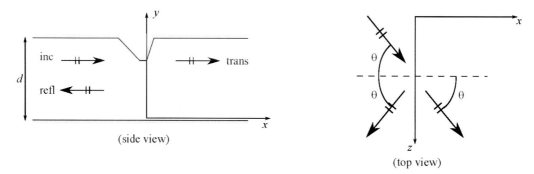

Figure 7.8 Junction between two parallel plate regions.

are not generally enough to satisfy the boundary conditions for the total field at the junction itself; it follows that higher order cutoff waves must also be generated in the vicinity of the junction. As will be seen in Chapter 11, we often wish only to consider the field due to the fundamental TEM waves, and this is essentially equal to the total field if we do not need accurate values of the field near the junction. All we require is that we be able to calculate reflection and transmission coefficients of the TEM wave from the junction, and this has been done for many types of junctions as indicated in the references at the end of the chapter.

Therefore, consider a hollow metallic waveguide whose cross-section can be partitioned by one or more vertical lines into several sections of parallel-plate waveguides of various heights, as shown in Figure 7.7. Within each section, we express the field as a pair of TEM (with respect to the *ray* directions) parallel-plate waves propagating at some angle with respect to the z-axis (we would, of course need to supplement these near the junction regions by the cutoff wave fields as discussed above). The factor $\exp(-\gamma z)$ is common to all the fields, and we need deal only with the *projected* behavior of these waves in the xy-plane.

Thus, for example, in the adjacent parallel-plate sections shown in Figure 7.9, we write

$$\mathcal{E}_z = E_1 e^{-jk_c x} + E_2 e^{+jk_c x} \qquad x < x_0 \quad \text{and} \quad 0 < y < d_1 \tag{7.20}$$

and

$$\mathcal{E}_z = E_3 e^{-jk_c x} + E_4 e^{+jk_c x} \qquad x > x_0 \quad \text{and} \quad 0 < y < d_2 \tag{7.21}$$

where $k_c^2 = k_x^2 = \omega^2 \mu \epsilon + \gamma^2 = k^2 + \gamma^2$. At such a junction between parallel-plate sections, each of the incident waves (E_1 and E_4) will partially reflect back into the same section and partially transmit into the adjacent section. We denote the reflection coefficients of the waves incident from left or right as ρ_A and ρ_B, respectively. The transmission coefficient from left to right is called τ_A, and that from right to left is called τ_B. The situation is illustrated schematically in Figure 7.10. From all this, there follow two conditions on the

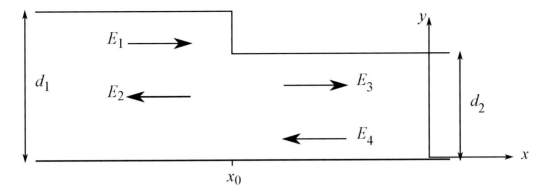

Figure 7.9 Projected TEM parallel-plate waves in adjacent parts of a cross-section.

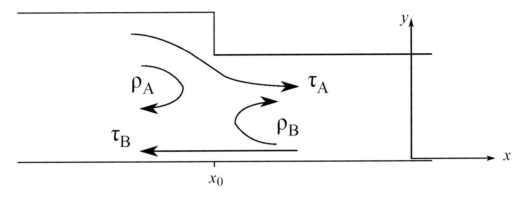

Figure 7.10 Reflection and transmission coefficients at a typical junction.

waves (7.20) and (7.21) at the junction; namely,

$$E_2 e^{jk_c x_0} = \rho_A(E_1 e^{-jk_c x_0}) + \tau_B(E_4 e^{jk_c x_0}) \tag{7.22}$$

and

$$E_3 e^{-jk_c x_0} = \rho_B(E_4 e^{jk_c x_0}) + \tau_A(E_1 e^{-jk_c x_0}). \tag{7.23}$$

Note how we must be careful to use the correct values of the incident and reflected waves in (7.22) and (7.23).

7.3.1 RIDGED, GROOVED AND FINNED RECTANGULAR METALLIC WAVEGUIDES

Exact explicit expressions for the reflection and transmission coefficients at a junction involving a diaphragm or a change in height do not exist in general. However, under certain conditions (usually that the cutoff frequency is not too large—$k_c d \leq \pi$) approximate equivalent transverse networks that include lumped circuit elements can be used to generate approximate values for these coefficients. A number of useful equivalent circuits for TEM waves at homogeneously filled parallel-plate discontinuities are shown in Table 7.1. They refer to an equivalent transverse transmission line described by (7.17). Approximations for other discontinuities can be obtained by a number of methods, notably the mode-matching technique described in Section 11.2. In this way, we can reduce the problem to one of transverse resonance on an equivalent transverse transmission line circuit appropriate to the waveguide under consideration.

Table 7.1

Transverse equivalent networks for TEM waves at parallel plate junctions.

Junction	Parameters/Equivalent Circuit
7.1.1 Symmetrical Window	$BZ_c = \frac{2k_c d}{\pi} \ln\left(\csc\frac{\pi g}{2d}\right)$ $(k_c d \le \pi)$
7.1.2 Symmetrical Diaphragm	$BZ_c = \frac{2k_c d}{\pi} \ln\left(\csc\frac{\pi g}{d}\right)$ $(k_c d \le \pi)$
7.1.3 Nonsymmetrical Window or Diaphragm	$BZ_c = \frac{4k_c d_0}{\pi} \ln\left(\csc\frac{\pi g_0}{2d_0}\right)$ $(k_c d \le \pi)$
7.1.4 Symmetrical Step in Height	$BZ_c = \frac{k_c d_2}{\pi}\left[\frac{(1-N^2)^2}{2N^2}\ln\left(\frac{1+N^2}{1-N^2}\right) + 2\ln\left(\frac{1+N^2}{2N}\right)\right];$ $N = \sqrt{d_2/d_1} \le 1;$ $(k_c d_1 \le \pi)$
7.1.5 Nonsymmetrical Step in Height	$BZ_c = \frac{2k_c d_{02}}{\pi}\left[\frac{(1-N^2)^2}{2N^2}\ln\left(\frac{1+N^2}{1-N^2}\right) + 2\ln\left(\frac{1+N^2}{2N}\right)\right];$ $N = \sqrt{d_{02}/d_{01}} \le 1;$ $(k_c d_1 \le \pi)$

We illustrate the use of Table 7.1 on the example of a symmetrically finned rectangular waveguide shown in Figure 7.11(a). The equivalent transmission line circuit for the transverse resonance method is shown in Figure 7.11(b). The side walls at $x = 0$ and $x = a$ are special cases of junction 7.1.1 as $g \to 0$, i.e. short circuits, whereas the gap in the center at $x = a/2$ is described by a shunt susceptance B as in the general form of 7.1.1. We set up the

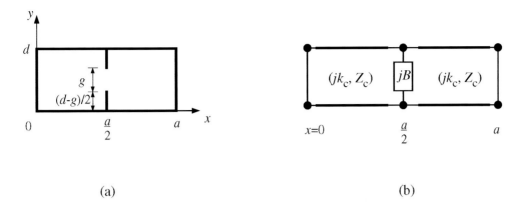

Figure 7.11 (a) Symmetrically finned rectangular waveguide; (b) equivalent transverse transmission line network.

transverse resonance condition for this guide at $x_A = \frac{a}{2}^-$, so that the shunt susceptance jB will appear in parallel with the line admittance seen looking *to the right* at $x = x_A^+$. Thus, the transverse resonance condition becomes

$$\frac{\tan \frac{k_c a}{2}\left(2 - BZ_c \tan \frac{k_c a}{2}\right)}{1 - BZ_c \tan \frac{k_c a}{2}} = 0 \tag{7.24}$$

where, by 7.1.1 of Table 7.1,

$$BZ_c = \frac{2k_c d}{\pi} \ln\left(\csc \frac{\pi g}{2d}\right) \tag{7.25}$$

One set of solutions to (7.24) has $\tan \frac{k_c a}{2} = 0$ or $k_c a = 2m\pi$, $m = 1, 2, 3, \ldots$. These are the $\mathrm{TE}_{2m,0}$ modes of the guide of width a with no fins at all. The $\mathcal{E}_y$ field for such modes is already zero at $x = a/2$ and is thus unaffected by the insertion of the fins. On the other hand, the $\mathrm{TE}_{2m+1,0}$ modes of the large guide are substantially perturbed by the presence of the fins and correspond to the other set of solutions to (7.24) which obey

$$\tan \frac{k_c a}{2} = \frac{2}{BZ_c} = \frac{\pi}{k_c d \ln\left(\csc \frac{\pi g}{2d}\right)} \tag{7.26}$$

or

$$k_c a = 2m\pi + 2 \tan^{-1}\left[\frac{\pi}{k_c d \ln\left(\csc \frac{\pi g}{2d}\right)}\right] \tag{7.27}$$

This is a transcendental equation for k_c which must be solved numerically or by approximation. If $m \neq 0$, a useful approximation is obtained by setting $k_c = 2m\pi/a$ in the right side of (7.27), giving

$$k_c a = 2m\pi + 2 \tan^{-1}\left[\frac{a}{2md \ln\left(\csc \frac{\pi g}{2d}\right)}\right] \tag{7.28}$$

while for $m = 0$ only numerical solution is sufficiently accurate. We plot results for the first three modes (along with the approximation (7.28) for the TE_{30} mode) in Figure 7.12 and corresponding E-field distributions are shown in Figure 7.13. For the $\mathrm{TE}_{2m+1,0}$ modes, we

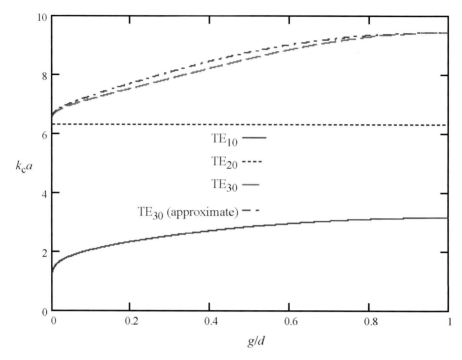

Figure 7.12 Normalized cutoff frequency $k_c a$ versus normalized gap width g/d for symmetrically finned rectangular waveguide TE_{10}, TE_{20}, and TE_{30} modes; $a = 2d$.

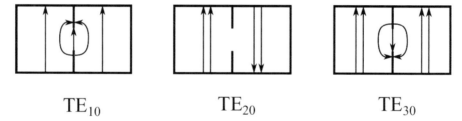

Figure 7.13 Sketches of the $\mathcal{E}$-field for TE_{10}, TE_{20}, and TE_{30} modes of symmetrically finned rectangular waveguide.

see that the cutoff frequencies are *reduced* by comparison with the unfinned guide of width $2a$ as g/d decreases. In a sense, the standing-wave distribution is allowed to "crawl down" the fins so that a larger transverse wavelength can "fit into" the guide than is possible without the fins. Subject to verification that the next higher-order modes of the finned waveguide do not undergo similar reductions in cutoff frequency, it can be seen that this guide is capable of propagating only the dominant TE_{10} mode over a wider bandwidth than is the ordinary rectangular guide.

At this point, it is well to summarize the limitations of this approach. Numerical calculations of $k_c a$ for the TE_{10} mode of the finned guide are graphically indistinguishable from those computed from (7.27) and displayed in Figure 7.12. This accuracy may not be obtained in other situations. In order to assure reasonable results, we should keep the following in mind:

(a) Only TEM parallel-plate waves have been used. If k_c is large enough, other higher-order parallel-plate waves may have to be used to adequately represent the field in each section.

(b) The method has been shown here only for TE modes. For TM modes, a similar analysis treating $\mathcal{E}_z$ must be done, which again may require the use of higher-order parallel-plate waves to represent the fields in each section.

(c) If $k_c d$ is too large (where d is the height of a section), the approximate reflection and transmission coefficients inferred from Table 7.1 are no longer accurate. The entries in this table have an error of less than 5% if $k_c d < \pi$ (where d is the largest of the heights involved in the formulas). Greater precision can be obtained at the expense of more complicated formulas.

(d) The representations (7.20) and (7.21) for the fields are not valid too close to a junction. This is because higher-order, cutoff parallel-plate waves must be present in order to provide continuity of all tangential fields at the junction (see Chapter 11). When using this method to determine k_c, we must be sure that the horizontal distance x_s separating two junctions is sufficiently large that these cutoff waves produced at one junction have decayed to a negligible value when they reach the second one. If we regard 5% as a negligible value, the criterion is roughly

$$\frac{d}{x_s} < \frac{\pi}{\sqrt{(k_c x_s)^2 + \pi^2}}$$

For the example of Figure 7.11(a), this criterion gives $d < \frac{a}{2}$ for the TE_{10} mode.

Restrictions resulting from points (c) and (d) above can only be relaxed at the cost of considerable complexity in the analysis, if at all. However, other kinds of modes can be analyzed if we use not TEM waves, but other kinds of transversely propagating waves in order to represent the fields. In order to do this, we must in each case set up an appropriate transmission line equivalence and build up a table of equivalent circuit representations for discontinuities analogous to Table 7.1. We give an example of this for dielectric waveguides later in the chapter. A further example of a waveguide which can be analyzed on the basis of TEM waves is that of quasi-optical microstrip which is treated in the next section. Other examples are given as problems at the end of the chapter.

7.4   QUASI-OPTICAL MICROSTRIP

Consider the waveguide shown in Figure 7.14. Structurally it is identical to the microstrip line we will study in Section 8.6, but it is operated in a high-frequency range such that $k_0 w \geq 1$ and so operates in a rather different way from the quasi-TEM microstrip mode. We therefore name it *quasi-optical microstrip* and will analyze it using the transverse resonance method.

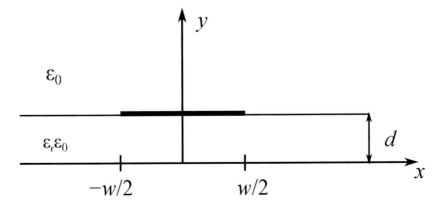

Figure 7.14   Quasi-optical microstrip (cross-section).

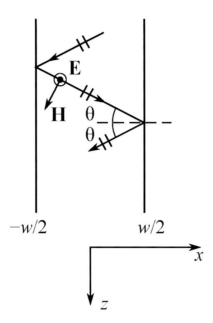

Figure 7.15 Quasi-optical microstrip (top view and ray diagram).

Once again we assume that the field in the parallel-plate region ($0 < y < d$, $|x| < w/2$) underneath the strip (but away from the edges) consists essentially only of a pair of (y-polarized) TEM waves reflecting from the edges of the strip as shown in Figure 7.15. Our total electric field $\mathcal{E}_y$ would then be

$$\mathcal{E}_y = E_1 e^{-jk_0\sqrt{\epsilon_r}x\cos\theta} + E_2 e^{+jk_0\sqrt{\epsilon_r}x\cos\theta} \tag{7.29}$$

and we can identify $k_{x1} = k_0\sqrt{\epsilon_r}\cos\theta$ in the substrate under the strip. As with the examples of Section 7.3, the edges of the strip at $x = \pm w/2$ are not simply plane reflecting walls and, in fact, will scatter an incident plane wave in a much more complicated fashion than as a single reflected plane wave. It is thus not possible to apply an exact boundary condition to (7.29) at the edges to find the relation between E_1 and E_2. Nevertheless, there will exist *some* electric field reflection coefficient for the TEM wave from a single such edge for an incidence angle of θ, which we will denote

$$\rho(\theta) = e^{j\chi(\theta)} \tag{7.30}$$

We can, however, make some general observations about the wave scattering process at the edge, and their implications about the behavior of this open waveguide. Besides a reflected TEM wave under the strip, there is also a field penetration beyond the edge, like the transmitted field into the second medium in the dielectric interface problem of Figure 6.5. The field in air will satisfy the equation

$$(\nabla_T^2 + k_T^2) \left\{ \begin{array}{c} \mathcal{E} \\ \mathcal{H} \end{array} \right\} = 0 \tag{7.31}$$

where

$$k_T^2 = k_0^2 - \beta^2 = -p^2 \tag{7.32}$$

We call this portion of the field a *space wave*, because it is not localized to the slab. As long as $\beta > k_0$, k_T^2 is negative (i.e., k_T is imaginary and p is real), and the solution of (7.31), like

(6.5) for the interface problem when $\theta > \theta_c$, decays exponentially away from the edge. But when $\beta < k_0$, k_T^2 is positive and k_T is real. This means that the solution of (7.31) will be a wave propagating away from the edge and whose energy is taken from the incident wave. For this reason only partial reflection, $|\rho(\theta)| < 1$, is to be expected.

We thus conclude that $\beta > k_0$ (or $\sin\theta > \epsilon_r^{-1/2}$) is necessary if a *guided* wave without radiation loss is to result from these repeated reflections at the edge. In fact, not only waves propagating into the air from the edge might be present, but also surface waves along the grounded dielectric slab, beyond the edge. Since the propagation factor $\exp(-\gamma z) = \exp(-j\beta z)$ is present, the surface waves whose propagation coefficients are β_{sm} (along some direction in the xz-plane), $m = 0, 1, 2, \ldots$, will have a field structure proportional to

$$e^{-p_{sm}y}e^{-j\sqrt{\beta_{sm}^2 - \beta^2}\,|x|}$$

in the air region $y > d$ beyond the edges of the strip, and to

$$\left\{ \begin{array}{c} \sin(h_{sm}y) \\ \cos(h_{sm}y) \end{array} \right\} e^{-j\sqrt{\beta_{sm}^2 - \beta^2}\,|x|}$$

in the dielectric, where p_{sm} and h_{sm} are the values of h and p corresponding to the surface wave solutions of (6.64) or (6.65). Evidently, if $\beta < \beta_{sm}$, part of the field will radiate away from the edge as a surface wave. For $\beta > \beta_{sm}$, on the other hand, this field decays exponentially in the x-direction.

We thus have a set of qualitative criteria which constrain the possible values of β for proper (bound) modes of planar quasi-optical structures. We first identify the types of wave which might carry energy away from the guiding region. In the foregoing example, the space wave (which obeys (7.31)) and the surface waves serve this role. A bound mode must then have a value of β or greater than a lower limit for each of these wave types (k_0 for a space wave and β_{sm} for a surface wave). On the other hand, Appendix J gives an upper limit of

$$\beta \le \omega\sqrt{(\mu\epsilon)_{max}} \tag{7.33}$$

where $(\mu\epsilon)_{max}$ is the maximum value of $\mu\epsilon$ over the waveguide cross-section. Thus, for our quasi-optical microstrip we have that β must be restricted to the range

$$\beta_{s,\text{TM}_0} \le \beta \le \omega\sqrt{\mu_0\epsilon_0\epsilon_r} \tag{7.34}$$

where β_{s,TM_0} is the propagation coefficient for the TM$_0$ surface wave on the grounded dielectric slab—the highest value for any of the surface waves which may exist on the slab.

If the substrate is electrically thin ($k_0\sqrt{\epsilon_r}\,d \ll 1$), it has been found that the phase $\chi(\theta)$ of the reflection coefficient can be accurately approximated by

$$\chi(\theta) \simeq$$
$$\frac{2k_0d}{\pi\epsilon_r^{1/2}\cos\theta}\left\{(1 - \epsilon_r\sin^2\theta)\left[\ln(k_0d\sqrt{\epsilon_r\sin^2\theta - 1}) + \gamma_E - 1\right]\right.$$
$$\left. - \epsilon_r\cos^2\theta\ln 2\pi - (\epsilon_r - 1)Q\left(\frac{1 - \epsilon_r}{1 + \epsilon_r}\right)\right\} \tag{7.35}$$

In (7.35), γ_E is used to denote Euler's constant (C.11), and Q is the function defined in (C.81).

Applying the reflection conditions at $x = \pm w/2$, we obtain

$$\left. \begin{array}{c} e^{j\chi(\theta)}E_1e^{-jk_0w\sqrt{\epsilon_r}\cos\theta/2} = E_2e^{jk_0w\sqrt{\epsilon_r}\cos\theta/2} \\ e^{j\chi(\theta)}E_2e^{-jk_0w\sqrt{\epsilon_r}\cos\theta/2} = E_1e^{jk_0w\sqrt{\epsilon_r}\cos\theta/2} \end{array} \right\} \tag{7.36}$$

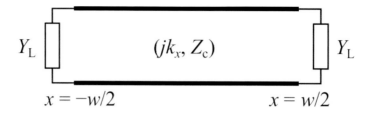

Figure 7.16 Equivalent transverse transmission-line network for quasi-optical microstrip.

whose solution yields $E_2 = \pm E_1$ and

$$k_0 w \sqrt{\epsilon_r} \cos\theta - \chi(\theta) = m\pi \quad m = 0, 1, 2, \ldots \tag{7.37}$$

As in the previous sections, we might also have obtained this result using the transverse resonance of an equivalent transverse transmission line. The near-total reflection implied by (7.30) and (7.35) is equivalent to a lumped load admittance Y_L connected at either end as shown in 7.16. In this circuit the classical transmission-line variables are given by the replacements

$$\left.\begin{array}{rcl} z & \to & x \\ \gamma & \to & jk_x \equiv jk_0\sqrt{\epsilon_r}\cos\theta \\ V & \to & \mathcal{E}_y \\ I & \to & \mathcal{H}_z \\ Z_c & \to & \frac{\omega\mu_0}{k_x} = \frac{\zeta}{\cos\theta}, \end{array}\right\} \tag{7.38}$$

and the lumped admittance Y_L for $k_0\sqrt{\epsilon_r}\,d \ll 1$ is given by

$$\begin{aligned} Y_L &= \frac{1}{Z_c}\frac{1-\rho}{1+\rho} = -\frac{j}{Z_c}\tan\frac{\chi}{2} \\ &\simeq \frac{jk_0 d}{\pi\zeta_0}\left\{\frac{p_0^2}{k_0^2}[\ln p_0 d + \gamma_E - 1] + \frac{k_x^2}{k_0^2}\ln 2\pi + (\epsilon_r - 1)Q\left(\frac{1-\epsilon_r}{1+\epsilon_r}\right)\right\} \end{aligned} \tag{7.39}$$

where $\beta = k_0\sqrt{\epsilon_r}\sin\theta$ is the yet-to-be-determined propagation coefficient of the *microstrip mode* (not of the equivalent transverse line), $k_x = \sqrt{k_0^2\epsilon_r - \beta^2}$ is the transverse wavenumber of the TEM wave under the strip, $p_0 = \sqrt{\beta^2 - k_0^2}$ and γ_E is Euler's constant. With the impedance looking left from $x = -w/2$ equal to $1/Y_L$, we have the transverse resonance condition

$$\frac{(1/Y_L)\cos k_x w + jZ_c \sin k_x w}{Z_c \cos k_x w + j(1/Y_L)\sin k_x w}Z_c + \frac{1}{Y_L} = 0, \tag{7.40}$$

which can be shown to be equivalent to (7.37).

Equation (7.37) is a transcendental equation which, like those encountered in Chapter 6, must be solved numerically or by analytic approximation. Since d/w is small under the conditions we have imposed, we can accomplish the latter as follows. We substitute (7.35) in (7.37), expressing Q in terms of

$$r_\epsilon = r\left(\frac{1-\epsilon_r}{1+\epsilon_r}\right) \tag{7.41}$$

where $r(x)$ is a special function defined in eqn. (C.80) of Appendix C. From (C.82), we have

$$r_\epsilon \simeq \sqrt{0.4052 + \frac{0.5160}{\epsilon_r} + \frac{0.0788}{\epsilon_r^2}} \tag{7.42}$$

Then, using $\epsilon_{\text{eff}} = \epsilon_r \sin^2 \theta$ instead of θ itself as the unknown, we have for the $m = 0$ mode

$$\epsilon_{\text{eff}} \left\{ 1 + \frac{2d}{\pi w} \left[\ln \left(\frac{2\pi e^{1-\gamma_E}}{k_0 \, d \sqrt{\epsilon_{\text{eff}} - 1}} \right) \right] \right\} = \epsilon_r + \frac{2d}{\pi w} \left[\ln \left(\frac{e^{1-\gamma_E}}{k_0 \, dr_\epsilon \sqrt{\epsilon_{\text{eff}} - 1}} \right) + \epsilon_r \ln(2\pi r_\epsilon) \right] \quad (7.43)$$

Now, as $d/w \to 0$, the solution to this equation approaches $\epsilon_{\text{eff}} = \epsilon_r$. The appearance of ϵ_{eff} inside the logarithm (which is why we still have a transcendental equation) provides only a weak dependence on ϵ_{eff}. For purposes of approximation, we can with acceptable accuracy replace ϵ_{eff} (there only) by ϵ_r (this is the first step of an iterative solution process, as described in Appendix K), thus obtaining an explicit approximation for ϵ_{eff}:

$$\epsilon_{\text{eff}} \cong \frac{\epsilon_r + \frac{2d}{\pi w} \left[\ln \left(\frac{e^{1-\gamma_E}}{k_0 \, dr_\epsilon \sqrt{\epsilon_r - 1}} \right) + \epsilon_r \ln(2\pi r_\epsilon) \right]}{1 + \frac{2d}{\pi w} \ln \left(\frac{2\pi e^{1-\gamma_E}}{k_0 \, d \sqrt{\epsilon_r - 1}} \right)} \quad (7.44)$$

This formula is valid so long as $k_0 d \ll 1$ but $k_0 w \geq \frac{1}{2}$. A typical example of the dispersion relation (7.44) and its comparison to numerically exact computations is shown in Figure 7.17.

A point worth noting about equation (7.35) is that, if $\sin \theta > \epsilon_r^{-1/2}$, $\chi(\theta)$ is purely real if ϵ_r is real. But, if $\sin \theta < \epsilon_r^{-1/2}$, we have

$$\ln(k_0 d \sqrt{\epsilon_r \sin^2 \theta - 1}) = \frac{\pi j}{2} + \ln(k_0 d \sqrt{1 - \epsilon_r \sin^2 \theta}),$$

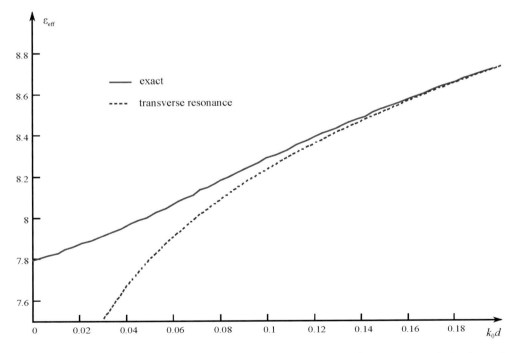

Figure 7.17 Frequency dependence of ϵ_{eff} for the fundamental mode on quasi-optical microstrip: $\epsilon_r = 9.9$, $w/d = 4.69$.

and we observe that $\left|\rho(\theta)\right| < 1$ and the reflection is less than total. In the former case, $\beta = k_0\sqrt{\epsilon_r}\sin\theta > k_0$, while in the latter case $\beta < k_0$. As discussed above, total reflection does not occur in the latter case because of the propagating (rather than exponentially decaying) space wave in the air region beyond the edge of the strip. In the approximation $k_0 d \ll 1$ which we have invoked in (7.35), the TM$_0$ mode wavenumber is approximated by (see Section 6.6 and problem p6-13):

$$
\begin{aligned}
\frac{\beta_{s,\mathrm{TM}_0}^2}{k_0^2} &\simeq 1 + k_0^2 d^2\left(\frac{\epsilon_r - 1}{\epsilon_r}\right)^2 \\
&\simeq 1
\end{aligned}
\tag{7.45}
$$

so that the left side of (7.34) becomes simply $\beta \geq k_0$ and is identical with the constraint that arises from the space wave.

7.5   DIELECTRIC RIB WAVEGUIDE: THE EFFECTIVE INDEX METHOD

Consider the dielectric rib waveguide shown in Figure 7.18. A section of dielectric slab of thickness a_i and width w is located between two semi-infinite slabs of somewhat smaller thickness a_e. (In reality, of course, these outer slabs would merely be of large but finite extent.) Let us assume that

$$
k_0 a_i \sqrt{n_1^2 - n_2^2} < \pi,
$$

so that only the TE$_0$ and TM$_0$ surface waves may exist on either of the slab sections.

Suppose we try to model certain modes of this guide such that, in the central region, $0 < x < w$, the field can be expressed as the repeated reflection of TE$_0$ surface waves between the discontinuities at $x = 0$ and $x = w$. If β_i is the surface wave propagation coefficient of the TE$_0$ wave (as found from (6.33) with $a = a_i$) and $f_{i,\mathrm{TE}}(y)$ is the characteristic field distribution of the field component $\mathcal{E}_z$ of this surface wave in the vertical direction (see (6.32) and (6.41)):

$$
\begin{aligned}
f_{i,\mathrm{TE}}(y) &= \cos h_{i1} y \quad (|y| < a_i/2) \\
&= \cos\frac{h_{i1} a_i}{2} e^{-p_{i2}(|y| - a_i/2)} \quad (|y| > a_i/2)
\end{aligned}
\tag{7.46}
$$

where $h_{i1} = \sqrt{k_1^2 - \beta_i^2}$ and $p_{i2} = \sqrt{\beta_i^2 - k_2^2}$, then we write

$$
\mathcal{E}_z = \left[E_1 e^{-j\sqrt{\beta_i^2 - \beta^2}\,x} + E_2 e^{+j\sqrt{\beta_i^2 - \beta^2}\,x}\right] f_{i,\mathrm{TE}}(y)
\tag{7.47}
$$

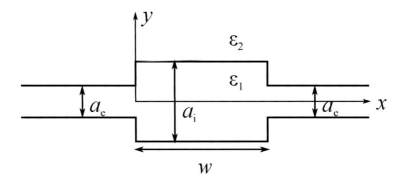

Figure 7.18   Dielectric rib waveguide.

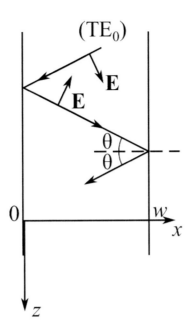

Figure 7.19 TE_0 surface waves reflecting from rib edge.

Here E_1 and E_2 are amplitude constants to be determined, while β is the (yet unknown) overall propagation coefficient of the dielectric rib waveguide mode.

If we introduce the propagation factor $\exp(-j\beta z)$ into (7.47), we note that it consists of two surface waves, propagating not in the z-direction nor in the x-direction purely, but at some angle in the xz-plane, as suggested in Figure 7.19. Indeed, the angle θ obeys

$$\begin{aligned}
\beta &= \beta_i \sin\theta \\
\sqrt{\beta_i^2 - \beta^2} &= \beta_i \cos\theta
\end{aligned} \tag{7.48}$$

and it is clear that a component of $\mathcal{E}$ in the x-direction must be present as well, due to the oblique propagation of these surface waves. We do not, however, require it for the present analysis, and we do have that $\mathcal{E}_y \equiv 0$ in the approximation we are making.

In the outer slab regions the fields, which are still associated with the same propagation factor $\exp(-j\beta z)$, will behave as $\exp\left[-j\sqrt{\beta_e^2 - \beta^2}\,|x|\right]$, where β_e is the TE_0 surface wave propagation coefficient for these outer slabs. If the mode of the rib waveguide is to be bound to the vicinity of the central region, we will have to have $\beta > \beta_e$.

In order to set up a transverse resonance analysis of this waveguide, we need to have a supply of information on equivalent circuit representations for TE_0 slab surface wave discontinuities analogous to Table 7.1. The transmission line analogy is summarized as

$$\left.\begin{aligned}
z &\to x \\
\gamma &\to j\sqrt{\beta_i^2 - \beta^2} \\
V : \quad \mathcal{E}_z &= V(x) f_{i,\text{TE}}(y)
\end{aligned}\right\} \quad TE_0 \text{ slab surface waves.} \tag{7.49}$$

where we omit specification of I and Z_c, since only one value of Z_c appears in the class of structures we will deal with here. Appropriate equivalent networks are given in Table 7.2.

We are now in a position to analyze the so-called E^x modes of the rib waveguide (there being no y-component of $\mathcal{E}$ in the constituent TE_0 surface waves and approximately none overall). The transverse equivalent transmission-line network is again as in Figure 7.16,

Table 7.2

Transverse equivalent networks for TE$_0$ surface waves at dielectric slab junctions.

Junction	Parameters/Equivalent Circuit
7.2.1 Symmetrical Step in Height	$BZ_c = \dfrac{\sqrt{\beta_i^2 - \beta^2}}{\sqrt{\beta^2 - \beta_e^2}} \dfrac{\beta_e^2}{\beta_i^2}$ $(0 \neq a_e < a_i)$
	$(Z_c, \sqrt{\beta_i^2 - \beta^2})$ $Y = jB$
7.2.2 Truncated Slab	$BZ_c = \dfrac{\sqrt{\beta_i^2 - \beta^2}}{\tilde{p}} \dfrac{k_2^2}{\beta_i^2}$; where
	$\tilde{p}^2 = \beta^2 - k_2^2 + h_{i1}^2/(4 + 8/p_{i2}a_i)$

except that h is replaced by $\sqrt{\beta_i^2 - \beta^2}$, and Y_L is given by jB from entry 7.2.1 in Table 7.2. Hence, in analogy with (7.40) we obtain the following eigenvalue equation for β:

$$\tan\sqrt{\beta_i^2 - \beta^2}\, w = \frac{2Z_c B}{Z_c^2 B^2 - 1}$$

or

$$\tan h_{si} w = \frac{2 h_{si} p_{se}}{h_{si}^2 (\beta_e^2/\beta_i^2) - p_{se}^2 (\beta_i^2/\beta_e^2)} \tag{7.50}$$

where we denote

$$\begin{aligned} h_{si} &= \sqrt{\beta_i^2 - \beta^2} \\ p_{se} &= \sqrt{\beta^2 - \beta_e^2} \end{aligned} \tag{7.51}$$

The E^y-modes of rib waveguides can similarly be analyzed as the transverse resonances of TM$_0$ slab modes. We let $f_{i,\mathrm{TM}}(y)$ denote the characteristic field distribution of the $\mathcal{E}_y$ component of the TM$_0$ wave on the central slab, and now we let β_i and β_e denote the surface wave propagation coefficients of the TM$_0$ waves on the central and outer slabs, respectively. We write

$$\mathcal{E}_y = \left[E_1 e^{-j\sqrt{\beta_i^2 - \beta^2}\, x} + E_2 e^{+j\sqrt{\beta_i^2 - \beta^2}\, x} \right] f_{i,\mathrm{TM}}(y) \tag{7.52}$$

where E_1, E_2, and β are to be determined by transverse resonance. Our transmission-line analogy is

$$
\left.
\begin{array}{ccc}
z & \to & x \\
\gamma & \to & j\sqrt{\beta_i^2 - \beta^2} \\
V: & & \mathcal{E}_y = V(x) f_{i,\text{TM}}(y)
\end{array}
\right\} \qquad \text{TM}_0 \text{ slab surface waves.} \qquad (7.53)
$$

Some useful equivalent networks for this case are given in Table 7.3. The setup and solution of the eigenvalue equation for the E^y modes of the rib guide as well as for other types of guides can now be done just as before. The details are left as an exercise for the reader.

Note the close similarity between (7.50) and the TM-mode eigenvalue equation (6.62) for a dielectric slab. Indeed, (7.50) would result if we analyzed the TM modes of a dielectric slab whose core refractive index were $\beta_i/k_0 = n_{1\text{eff}}$, the TE$_0$ mode effective index of the central slab region in Figure 7.18, while the cladding index was $\beta_e/k_0 = n_{2\text{eff}}$, the TE$_0$ mode effective index for the outer slabs in Figure 7.18. It must be kept in mind that we must first solve equation (6.33) to obtain β_i and β_e before we can proceed to solve (7.50) for β itself. The observed analogy between (7.50) and (6.62) has led to the term "effective index method" for the application of transverse resonance techniques to this kind of waveguide. But such a close analogy as this results only because of the specific form of the reflection coefficient of the constituent TE$_0$ wave at the rib boundary (i.e., of B) as inferred from 7.2.1 of Table 7.2. However, this expression is merely an approximation which has proven sufficiently accurate in practice to justify its use. In fact, the other entry in Table 7.2

Table 7.3

Transverse equivalent networks for TM$_0$ surface waves at dielectric slab junctions.

Junction	Parameters/Equivalent Circuit
7.3.1 Symmetrical Step in Height A	$BZ_c = \dfrac{\sqrt{\beta^2 - \beta_e^2}}{\sqrt{\beta_i^2 - \beta^2}} \dfrac{\beta_i^2}{\beta_e^2}; \quad (0 \neq a_e < a_i)$ $(Z_c, \sqrt{\beta_i^2 - \beta^2}) \qquad Y = jB$ A
7.3.2 Truncated Slab A	$BZ_c = \dfrac{\tilde{p}}{\sqrt{\beta_i^2 - \beta^2}} \dfrac{\beta_i/\omega}{d\beta_i/d\omega};$ where $\tilde{p}^2 = \beta^2 - k_2^2 +$ $\dfrac{h_{i1}^2}{4} \dfrac{(h_{i1}^2 - 2p_{i2}/a_i)(n_2^2/n_1^2) + 2p_{i2}/a_i + p_{i2}^2(n_1^2/n_2^2)}{h_{i1}^2(1 + 2/p_{i2}a_i)(n_2^2/n_1^2) + 2p_{i2}/a_i + p_{i2}^2(n_1^2/n_2^2)}$ $(Z_c, \sqrt{\beta_i^2 - \beta^2}) \qquad Y = jB$ A

and those in Table 7.3 do not result in such simple interpretations of effective-index slab structures, and the name "effective-index" is really a misnomer in such cases. Strictly, the "effective-index method" refers to a means of approximating B in Tables 7.2 and 7.3 and then to a straightforward application of transverse resonance techniques based upon these approximations.

7.6   NOTES AND REFERENCES

Discussions of the transverse-resonance/equivalent transmission-line model for multilayer slabs are to be found in

Felsen and Marcuvitz (1973), Section 2.4;
Solimeno, *et al.* (1986), Chapter 3.

Multiple quantum well structures are discussed in

T. H. Wood, "Multiple quantum well (MQW) waveguide modulators," *J. Lightwave Technol.*, vol. 6, pp. 743-757 (1988).
H. Yakamoto, M. Asada and Y. Suematsu, "Theory of refractive index variation in quantum well structure and related intersectional optical switch," *J. Lightwave Technol.*, vol. 6, pp. 1831-1840 (1988).

Applications of the WKB method to inhomogeneous dielectric slab waveguides (including the exponential diffused profile example) are given in:

A. Gedeon, "Comparison between rigorous theory and WKB-analysis of modes in graded-index waveguides," *Opt. Commun.*, vol. 12, pp. 329-332 (1974).
G. B. Hocker and W. K. Burns, "Modes in diffused optical waveguides of arbitrary index profile," *IEEE J. Quantum Electron.*, vol. 11, pp. 270-276 (1975).
J. Janta and J. Čtyroký, "On the accuracy of WKB analysis of TE and TM modes in planar graded-index waveguides," *Opt. Commun.*, vol. 25, pp. 49-52 (1978).
V. Ramaswamy and R. K. Lagu, "Numerical field solution for an arbitrary asymmetrical graded-index planar waveguide," *J. Lightwave Technol.*, vol. 1, pp. 408-417 (1983).

The exact solution for the exponential index profile is given in

Sodha and Gathak (1977), Section 7.2

and

E. M. Conwell, "Modes in optical waveguides formed by diffusion," *Appl. Phys. Lett.*, vol. 23, pp. 328-329 (1973).

Perhaps the earliest application of approximate transverse resonance methods was in

S. B. Cohn, "Properties of ridge waveguide," *Proc. IRE*, vol. 35, pp. 783-788 (1947).

Other applications to metallic waveguides are to be found in

L. O. Goldstone and A. A. Oliner, "Leaky-wave antennas I. Rectangular Waveguides," *IRE Trans. Ant. Prop.*, vol. 7, pp. 307–319 (1959).
J. R. Pyle, "The cutoff wavelength of the TE_{10} mode in ridged rectangular waveguide of any aspect ratio," *IEEE Trans. Micr. Theory Tech.*, vol. 14, pp. 175–183 (1966).

P. J. B. Clarricoats and P. E. Green, "Waveguide structures for double-beam leaky-wave antennas," *Proc. IEE (London)*, vol. 114, pp. 604–610 (1967).

E. V. Orleanskaya, "Broadband device for rotating the plane of polarization," *Telecommun. Radio Eng.*, vol. 22/23, no. 8, pp. 75-80 (1968).

E. V. Orleanskaya, "Application of the methods of network theory to problems of waveguides partially filled with dielectric," *Telecommun. Radio Eng.*, vol. 24/25, no. 3, pp. 86-91 (1970).

A. A. Oliner and P. Lampariello, "The dominant mode properties of open groove guide: An improved solution," *IEEE Trans. Micr. Theory Tech.*, vol. 33, pp. 755–764 (1985).

The equivalent networks in Table 7.1 are obtained from

Marcuvitz (1965), Chapter 5.

Details of the analysis of quasi-optical microstrip can be found in

D. C. Chang and E. F. Kuester, "Total and partial reflection from the end of a parallel-plate waveguide with an extended dielectric slab," *Radio Science*, vol. 16, pp. 1-13 (1981).

E. F. Kuester, R. T. Johnk and D. C. Chang, "The thin-substrate approximation for reflection from the end of a slab-loaded parallel-plate waveguide with application to microstrip patch antennas," *IEEE Trans. Ant. Prop.*, vol. 30, pp. 910-917 (1982).

The "exact" (numerical) solution given in Figure 7.17 is due to

R. H. Jansen, "High speed computation of single and coupled microstrip parameters including dispersion, high-order modes, loss and finite strip thickness," *IEEE Trans. Micr. Theory Tech.*, vol. 26, pp. 75-82 (1978).

The original paper on effective-index methods was

R. M. Knox and P. P. Toulios, "Integrated circuits for the millimeter through optical frequency range," in *Proc. Symp. Submillimeter Waves*, Brooklyn: Polytechnic Press, 1970, pp. 497-510,

and since than a number of refinements and extensions have been made. See

W. V. McLevige, T. Itoh and R. Mittra, "New waveguide structures for millimeter-wave and optical integrated circuits," *IEEE Trans. Micr. Theory Tech.*, vol. 23, pp. 788-794 (1975).

S.-T. Peng and A. A. Oliner, "Guidance and leakage properties of a class of open dielectric waveguides: Part I—Mathematical formulations," *IEEE Trans. Micr. Theory Tech.*, vol. 29, pp. 843-855 (1981).

A. A. Oliner, S.-T. Peng, T.-I. Hsu and A. Sanchez, "Guidance and leakage properties of a class of open dielectric waveguides: Part II—New physical effects," *IEEE Trans. Micr. Theory Tech.*, vol. 29, pp. 855-869 (1981).

F. P. Payne, "A new theory of rectangular optical waveguides," *Opt. Quantum Electron.*, vol. 14, pp. 525-537 (1982).

S. Zhendong and L. Jiang, "Effective dielectric parameter method—a new method for analyzing propagation characteristics of dielectric waveguides," *Micr. Opt. Technol. Lett.*, vol. 2, pp. 427-431 (1989).

A nice summary of these methods is to be found in

Adams (1981), Chapter 6.

We follow here the treatments of

Vassallo (1985), Tome 2, Chapter 7,
Koshiba (1992), Chapter 5.

from which most of the entries of Tables 7.2 and 7.3 are obtained, except for entry 7.3.1 which is due to unpublished work by D. C. Chang and L. H. Xiong.

The papers of Oliner, Peng *et al.* in particular discuss the effects of considering transformations between TE and TM modes during the reflection process at the edge of a dielectric rib. It may happen that neglecting this coupling predicts a perfectly bound mode (consisting of TE surface waves only), while the presence of a small but nonzero constituent of TM surface waves actually causes the mode to be leaky, in the manner discussed in Section 7.4. Such qualitative considerations should always be borne in mind when applying approximate transverse resonance methods.

7.7 PROBLEMS

p7-1 A multilayer slab waveguide consists of an inner core of index n_1, two coating layers of index n_3, and a cladding of index n_2 with

$$n_1 > n_2 > n_3$$

as indicated.

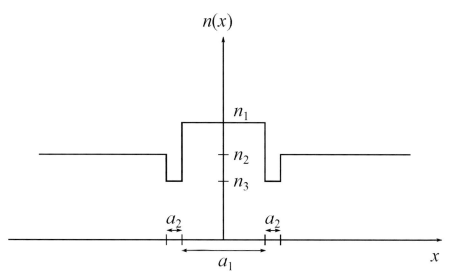

(a) Obtain the eigenvalue equation for the TE modes of this waveguide.
(b) Write a computer program to calculate n_{eff} vs. $V = k_0 a_1 \sqrt{n_1^2 - n_2^2}$ for the TE_0 mode. Plot your numerical results for the case when $n_1 = 2.0$, $n_2 = 1.9$ and $n_3 = 1.5$, with $a_2/a_1 = 0.1$, 0.3 or 1.0 successively.

p7-2 A multilayer slab waveguide consists of two core regions of index n_1, two outer cladding regions of index $n_2 < n_1$ and a central cladding region, also of index n_2, as shown.

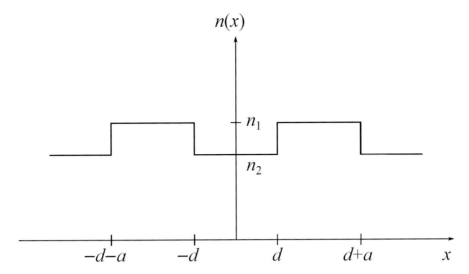

The core regions have width a each, while they are separated by a distance $2d$ as shown.

(a) Obtain the eigenvalue equation whose solutions are the propagation constants β for the TE modes of this system of coupled dielectric slab waveguides.

(b) Write a computer program to calculate b vs. V (both defined in the same way as for the single slab waveguide) for the TE_0 and TE_1 modes. Plot your numerical results for the case when $n_1 = 1.4$ and $n_2 = 1.37$, with $d/a = 0.1$, 0.3 or 1.0 successively.

p7-3 Show that the correspondences (7.10) are the correct ones to use for analyzing the TM modes of a multilayered slab waveguide by the transverse resonance method.

p7-4 Derive the transverse resonance equation for the TM modes of the slab waveguide with exponential permittivity profile described in Section 7.2. Plot b vs. V_1 for the fundamental TM mode using the values $n_1 = 5$, $n_2 = 4.9$ and $n_0 = 1$. Use the result of problem p7-3.

p7-5 A slab waveguide has a refractive index profile of

$$n^2(x) = n_2^2 \qquad (|x| > a)$$
$$= n_2^2 + (1 - \frac{x^2}{a^2})(n_1^2 - n_2^2) \qquad (|x| < a)$$

and $\mu = \mu_0$ everywhere.

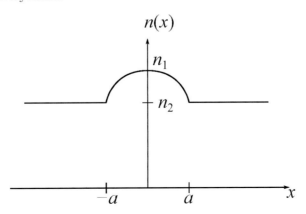

Use the WKB method to calculate and plot the values of

$$b = \frac{n_{\text{eff}}^2 - n_2^2}{n_1^2 - n_2^2}$$

vs. normalized frequency $V_1 \equiv k_0 a \sqrt{n_1^2 - n_2^2}$ for the fundamental TE mode of this guide. You may ignore the penetration of the fields beyond the turning points (when is this a good approximation?).

p7-6 A slab waveguide has a refractive index profile of

$$n^2(x) = n_2^2 + \frac{n_1^2 - n_2^2}{\cosh^2(x/a)}$$

and $\mu = \mu_0$ everywhere. Use the WKB method to calculate and plot the values of

$$b = \frac{n_{\text{eff}}^2 - n_2^2}{n_1^2 - n_2^2}$$

vs. normalized frequency $V_1 \equiv k_0 a \sqrt{n_1^2 - n_2^2}$ for the fundamental TE mode of this guide. You may ignore the penetration of the fields beyond the turning points (when is this a good approximation?).
[Hints: The integral

$$\int \sqrt{\frac{1}{\cosh^2 u} - b}\, du$$

may be usefully transformed using the successive changes of variable $\cosh u = 1/y$ and $y = \sqrt{\sin^2 \theta + b \cos^2 \theta}$. Then you may find the integral

$$\int_0^{\pi/2} \frac{d\theta}{\sin^2 \theta + b \cos^2 \theta} = \frac{\pi}{2\sqrt{b}}; \qquad b > 0$$

of some use in this problem.]

p7-7 For the partially filled waveguide of Figure 7.6, let $a = 0.9$ in, $d = 0.4$ in, $h = 0.1$ in, and let the relative permittivity of the dielectric layer be $\epsilon_r = 10$. Find the value of the cutoff frequency f_c for the fundamental mode of this waveguide (it occurs when $\gamma = \alpha + j\beta = 0$). Numerically solve the transcendental equation (7.19) for β, covering at least the range of frequencies from f_c to $3f_c$. Compare these results with those for an unloaded (air-filled) rectangular waveguide of the same overall dimensions.

p7-8 Using a TEM wave description as in Section 7.3, construct an eigenvalue equation for determining the propagation coefficients of the lowest-order TE mode (predominantly E^y-polarized) of the symmetric uniformly filled double-ridge metallic waveguide shown.

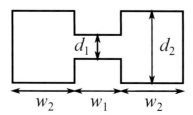

Solve this equation numerically for $k_c w_1$ for the dominant mode if $w_1 = w_2 = d_2$ and $d_1/d_2 = 0.5$. Sketch what you believe the transverse electric field distribution for this mode looks like. What advantages does this guide have over an ordinary metallic rectangular waveguide of width $3w_1$ and height w_1?

p7-9 For the symmetric uniformly filled double-ridge metallic waveguide shown in problem p7-8, let $w = 2w_2 + w_1$, $w_2 = 2w_1$, $d_2 = w/2$ and $d_1 = d_2/4$. Construct an eigenvalue equation for determining the propagation coefficient of the lowest-order TE mode (predominantly E_y-polarized). Numerically solve for the normalized cutoff wavenumber $k_c w$ of the dominant mode, and sketch what you believe the transverse electric field distribution for this mode looks like. Compare this waveguide to an ordinary metallic rectangular waveguide of width w and height d_2, stating the advantages and disadvantages of each.

p7-10 A rectangular metallic waveguide whose height $d = a/2$ is half the width a is fitted with a pair of fins whose height is $d/2$ each as shown below.

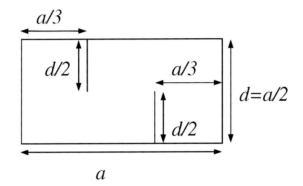

Use the transverse resonance method to obtain an equation for determining the eigenvalues k_c for the modes of this waveguide that are formed from TEM parallel plate waves. Use the software of your choice to determine the values of $k_c a$ for the first three such modes, in order of increasing cutoff wavenumber, and compare these values with those of the corresponding modes of an empty rectangular waveguide.

p7-11 Show that equation (7.40) is equivalent to equation (7.37).

p7-12 Using the methods of Sections 7.3 and 7.4, construct an eigenvalue equation for determining the propagation coefficients of E^y-polarized modes of the ridged quasi-optical microstrip shown.

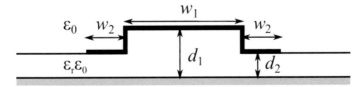

These modes are formed by the transverse resonance of the appropriate TEM parallel-plate waves.

p7-13 Plot dispersion curves of $b = (\beta^2/k_0^2 - n_2^2)/(n_1^2 - n_2^2)$ versus $V = k_0 w\sqrt{n_1^2 - n_2^2}$ for the lowest-order solution of (7.50) if $n_1^2 = 2.25$, $n_2^2 = 1$, $k_0 a_1 \sqrt{n_1^2 - n_2^2} = 1.0$, and $k_0 a_2 \sqrt{n_1^2 - n_2^2} = 0.95$. Cover the range $0 \le V \le 100$. Also display the value of $V_{\text{eff}} \equiv k_0 w\sqrt{n_{1,\text{eff}}^2 - n_{2,\text{eff}}^2}$ along the horizontal axis for comparison with V.

8 TEM AND QUASI-TEM MODES: BASIC PLANAR TRANSMISSION LINES

8.1 INTRODUCTION

In solid-state circuits operating at microwave (or higher) frequencies, especially if low power levels are involved, it is convenient to be able to fabricate intricate transmission-line networks by the same processes (photolithography, etc.) which have proved so successful in conventional circuit design, and have made printed circuits and integrated circuits commonplace. The use of transmission lines involving essentially only planar (strip) conductors reduces manufacturing problems in many situations. In this chapter, we will examine the quasistatic mode properties of such striplines, and discuss two important classes of modes in the study of such transmission lines: TEM and quasi-TEM modes.

Striplines which are filled with an electrically homogeneous medium (μ and ϵ are independent of x and y) belong to the TEM class of waveguide structures, examples of which are shown in Figure 8.1. Also belonging to this class are the coaxial and two-wire transmission lines shown in Figures 1.1(b) and (c). Common to all TEM waveguides is the presence of two or more separate conducting boundaries which allows a nontrivial TEM mode to exist.

As will be noted from these examples, some types of transmission lines have cross sections which are (at least conceptually) of infinite extent, while others are partially or completely shielded by one of the conducting boundaries. Generally, unshielded versions of planar transmission lines are suitable for use at frequencies up to about 30 or 40 GHz. Above this range, the tendency of planar circuits toward unwanted radiation dictates that shields be employed. A potential drawback to a shield is that it involves more metal surface area for the waveguide structure, which can result in increased conductor loss if a significant surface current flow is present there. In practice, shielded lines are used well into the millimeter wave band, up to 140 GHz.

Perhaps the simplest of all striplines is a thin, perfectly-conducting strip located above and parallel to a conducting ground plane of infinite extent. This idealized transmission line is sometimes referred to as *open, unshielded,* or *unbalanced* stripline (Figure 8.1(a)). A variation on this is the completely shielded line of Figure 8.1(b), called *shielded stripline* or *rectangular coaxial transmission line (RCTL)*, which is used in measurement applications to form a *TEM cell.* Another version is the partially-shielded structure of Figure 8.1(c), which is called *triplate, semi-open* or *balanced* stripline, and has two infinitely extended ground planes.

Striplines using air dielectrics must have their center conductors supported at the ends of a section of line, and this may not be adequate if close dimensional tolerances must be maintained. A homogeneous dielectric filling might be provided for the purpose of supporting the strip, but covering the strip complicates the fabrication process, makes external circuit connections more difficult, and the presence of losses in the dielectric can lead to unwanted attenuation on the line. Commonly, then, we see the use of striplines partially filled with dielectric material, so that support is provided, and yet manufacture is easy (the strips are simply deposited on the surface of the dielectric) and the total amount of potentially

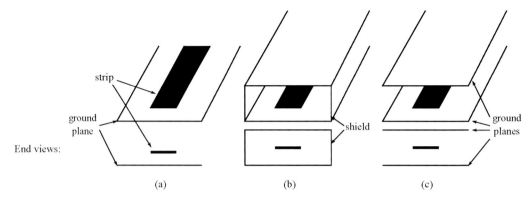

Figure 8.1 TEM-type stripline structures: (a) open stripline; (b) shielded stripline; (c) balanced stripline.

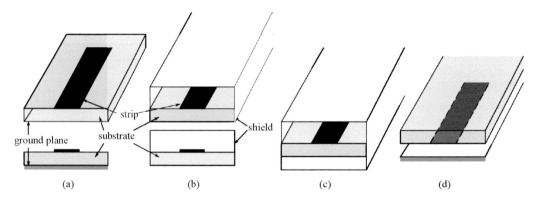

Figure 8.2 Quasi-TEM stripline structures: (a) microstrip; (b) shielded microstrip; (c) elevated microstrip; (d) inverted microstrip.

lossy dielectric is reduced. Strips deposited directly on the surface of a dielectric layer thus facilitate the manufacture of microwave integrated circuits as mentioned before.

As we will see, such structures of *inhomogeneous* cross-sectional properties (that is, where $\epsilon = \epsilon(x, y)$ and/or $\mu = \mu(x, y)$ are functions of transverse position) cannot generally support pure TEM modes, except in the trivial case of zero frequency. However, there can exist modes which are (in a sense to be made more precise later on) "close" to TEM, and are called *quasi-TEM* modes. In contrast to TEM modes, quasi-TEM modes are dispersive (phase velocity depends on frequency) and admit only an approximate description in terms of voltage and current. The quasi-TEM nature of these modes is generally present only when the characteristic transverse dimensions of the structure are small compared to a wavelength. Once frequency becomes sufficiently large, they behave more like a TE or TM mode of a hollow waveguide, or like a surface wave on a dielectric waveguide (cf. Chapter 6).

Several quasi-TEM stripline structures are depicted in Figure 8.2. Figure 8.2(a) is known as *microstrip*, or more specifically, *open microstrip*; it uses a dielectric substrate such as alumina to support the strip conductor above the ground plane. Figure 8.2(b) illustrates *shielded microstrip*, while Figures 8.2(c) and (d) show two variants of *elevated microstrip*, one shielded and one open. When the strip is located beneath the substrate but over the ground plane, we have an example of *inverted microstrip* as shown in Figure 8.2(d).

While striplines are generally (but not always) transmission lines where the dielectric, when present, is located *between* the two line conductors, other planar structures are used as

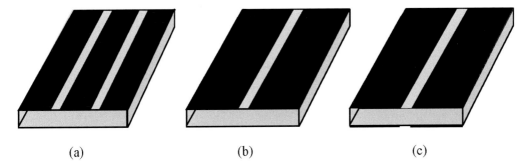

(a) (b) (c)

Figure 8.3 (a) Coplanar waveguide (CPW); (b) slotline; (c) bilateral slotline (open fin-line).

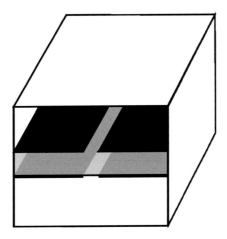

Figure 8.4 Shielded bilateral finline.

well. Figure 8.3 shows some *coplanar lines*, which have in common that all line conductors are deposited on the same side of the substrate. In this class are the *coplanar* waveguide of Figure 8.3(a) and the basic *slotline* of Figure 8.3(b). A *bilateral slotline* is shown in Figure 8.3(c); as discussed in Section 8.10, it has a tendency to radiative leakage of energy and is normally surrounded by a shield (Figure 8.4). This is an example of a *finline* which is basically a shielded slotline. Although finlines do not possess two or more separate conducting boundaries as do the other quasi-TEM structures considered in this chapter, they do share many of the same characteristics and are usually associated with them rather than with closed metallic waveguides. Each of these structures has its own advantages and weaknesses which tailor it to a particular type of application.

We will begin this chapter by discussing the general properties of TEM modes, and then proceed to discuss the characteristics of some quasi-TEM modes. The chapter concludes with some applications to specific types of planar transmission-lines.

8.2 TEM MODES

Let us begin by considering a waveguide with a doubly-connected (two-piece) perfectly conducting walls as shown in Figure 8.5. If the guide is homogeneously filled, then the conclusions reached in Chapter 5 apply to this guide as well: it supports an infinite number of TE and TM modes; each mode has some positive cutoff frequency, etc. Let us suppose for now that our waveguide is *inhomogeneous*, which means that $\epsilon(x,y)$ and/or $\mu(x,y)$

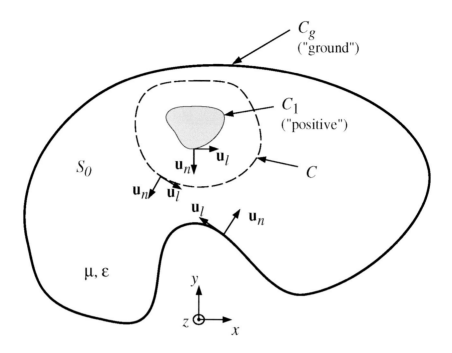

Figure 8.5 Cross-section of TEM waveguide.

are nonconstant functions of the transverse variables. There will still be modes of such waveguides analogous to the TE and TM modes of uniformly filled metallic waveguides, in that they possess positive cutoff frequencies, although now the modes are generally hybrid, as were the modes of the dielectric waveguides studied in Chapter 6.

This doubly-connected waveguide, as we will see, is also capable of propagating a mode with zero cutoff frequency. An important particular situation when this occurs is that of a TEM (transverse electromagnetic) mode. We saw in Section 5.1 the example of a parallel-plate metallic waveguide, which supports a TEM mode. This was basically the confinement between two parallel plates of a plane wave, which is also an example of a TEM wave. We devote this section to a study of the general properties of TEM modes.

For a TEM mode, $\mathcal{E}_z$ and $\mathcal{H}_z$ will vanish. Thus, if a TEM mode is possible, equations (5.22) and (5.23) are not going to be of much use to us. Thus we return to (5.17)-(5.20). Setting $\mathcal{E}_z = 0$ and $\mathcal{H}_z = 0$, we get

$$-\gamma \mathbf{u}_z \times \mathcal{E}_T = -j\omega\mu\mathcal{H}_T \tag{8.1}$$

$$\nabla_T \times \mathcal{E}_T = 0 \tag{8.2}$$

$$-\gamma \mathbf{u}_z \times \mathcal{H}_T = j\omega\epsilon\mathcal{E}_T \tag{8.3}$$

$$\nabla_T \times \mathcal{H}_T = 0 \tag{8.4}$$

Equations (8.1) and (8.3) tell us that unless $\mathcal{E}_T$ and $\mathcal{H}_T$ are both zero (the trivial case), we must have

$$\frac{\gamma}{j\omega\mu} = \frac{j\omega\epsilon}{\gamma}$$

or,

$$\gamma = \pm jk \tag{8.5}$$

where

$$k = \omega\sqrt{\mu\epsilon} \tag{8.6}$$

is the wavenumber of a plane wave in the medium filling the transmission line. We can conclude that unless $\mu\epsilon = $ constant, there can be no pure TEM mode on a transmission line. Notice, however, that this does not necessarily require μ and ϵ individually to be constant. We could imagine (with the use of magnetic materials such as ferrites) that the waveguide cross-section is filled with different media for which $\mu\epsilon$ is always the same.

Then, (8.1) and (8.3) become

$$\boldsymbol{\mathcal{H}}_T = \pm\frac{\mathbf{u}_z \times \boldsymbol{\mathcal{E}}_T}{\zeta_{\text{TEM}}(x, y)} \tag{8.7}$$

with

$$\zeta_{\text{TEM}} = \sqrt{\frac{\mu(x, y)}{\epsilon(x, y)}} = \zeta(x, y) \tag{8.8}$$

where $\zeta(x, y)$ is the wave impedance $[\mu(x, y)/\epsilon(x, y)]^{1/2}$ of a plane wave in the filling medium. The $\pm$ signs in (8.7) correspond to those in (8.5). Hereafter, we will confine ourselves to forward modes only, and take $+$ signs in (8.5) and (8.7).

If the transverse divergence is taken of eqns. (8.1) and (8.3), and use made of identity (B.13) and eqns. (8.2) and (8.4), our TEM mode field equations decouple into independent ones for $\boldsymbol{\mathcal{E}}_T$ and $\boldsymbol{\mathcal{H}}_T$ (if $\omega \neq 0$):

$$\nabla_T \cdot (\epsilon\boldsymbol{\mathcal{E}}_T) = 0 \quad ; \qquad \nabla_T \times \boldsymbol{\mathcal{E}}_T = 0 \tag{8.9}$$

$$\nabla_T \cdot (\mu\boldsymbol{\mathcal{H}}_T) = 0 \quad ; \qquad \nabla_T \times \boldsymbol{\mathcal{H}}_T = 0 \tag{8.10}$$

Equations (8.9) and (8.10) are in fact the two-dimensional ($\partial/\partial z \equiv 0$) source-free, *static* field equations for $\boldsymbol{\mathcal{E}}_T$ and $\boldsymbol{\mathcal{H}}_T$. The boundary condition for (8.9) is that the tangential field component $\mathcal{E}_l = \mathbf{u}_l \cdot \boldsymbol{\mathcal{E}}_T$ (cf. Figure 8.5) vanishes at the walls; for (8.10), that the normal component $\mathcal{H}_n = \mathbf{u}_n \cdot \boldsymbol{\mathcal{H}}_T$ vanishes there.

It must be possible to express $\boldsymbol{\mathcal{E}}_T$ as the transverse gradient of a scalar potential[1] :

$$\boldsymbol{\mathcal{E}}_T = -\nabla_T\Phi(x, y) \tag{8.11}$$

Substituting (8.11) into the first of (8.9), we get

$$\nabla_T \cdot [\epsilon\nabla_T\Phi(x, y)] = 0 \tag{8.12}$$

or, if ϵ is independent of x and y,

$$\nabla_T^2\Phi = 0 \tag{8.13}$$

which is Laplace's equation for the potential function $\Phi(x, y)$. Hence, if as in Chapter 2 we denote the boundary wall by the contour C_0 (which is made up of two parts, C_1 and C_g as shown in Figure 8.5), then (8.11) and the boundary condition on $\mathcal{E}$ implies that $\partial\Phi/\partial l = 0$ on C_g, or in other words,

$$\Phi\big|_{C_g} = \text{constant} \tag{8.14}$$

---

[1]To avoid the problematical introduction of a "script Φ", we simply use capital Φ for the present purpose. We will later use a script $\mathcal{Q}$ to avoid similarly a "script ρ". No conflict should arise from this usage.

Were the second conductor absent, the uniqueness theorem for Laplace's equation (or the more general equation (8.12)) tells us that Φ would be constant throughout the whole cross-section of the waveguide, and thus $\mathcal{E}_T$ (and with it $\mathcal{H}_T$) would vanish identically. Hence, *there can be no TEM mode in a waveguide with a simply-connected boundary.*

When two or more conducting boundaries are present, however, we can have Φ assume *different* constant values on each separate portion of the boundary, and nontrivial solutions for $\mathcal{E}_T$ and $\mathcal{H}_T$ can be found. By convention, we take C_1 to be the "positive" conductor with respect to C_g (which is sometimes referred to as "ground" and taken to have potential zero). For the two-conductor case shown in Figure 8.5, we thus put

$$
\begin{aligned}
\Phi|_{C_1} &= \Phi^{(1)} = \text{constant} \\
\Phi|_{C_g} &= 0
\end{aligned}
\tag{8.15}
$$

with $\Phi^{(1)} \neq 0$. Once we have solved for Φ throughout the entire cross-section of the waveguide, equations (8.5)-(8.8) and (8.11), in principle, tell us all we need to know about this mode.

Since TEM structures, often referred to as *transmission lines*, are usually connected between localized circuits, it is useful to associate some circuit concepts with the TEM mode. In this way, they can be related to the classical transmission line described in Section 1.2, and described by the same equations. Because of the boundary conditions on normal **E** and tangential **H** at the conducting surfaces C_1 and C_g, surface charge and current densities will be present on both conductors:

$$
\begin{aligned}
\epsilon \mathbf{u}_n \cdot \mathbf{E}|_{C_1} &= \rho_S^{(1)} \\
\mathbf{u}_n \times \mathbf{H}|_{C_1} &= \mathbf{J}_S^{(1)}
\end{aligned}
\tag{8.16}
$$

$$
\begin{aligned}
\epsilon \mathbf{u}_n \cdot \mathbf{E}|_{C_g} &= \rho_{Sg} \\
\mathbf{u}_n \times \mathbf{H}|_{C_g} &= \mathbf{J}_{Sg}
\end{aligned}
\tag{8.17}
$$

where ρ_{Sg} and $\mathbf{J}_{Sg}$ are the surface charge and current densities on conductor C_g, and $\rho_S^{(1)}$ and $\mathbf{J}_S^{(1)}$ are the surface charge and current densities on C_1. Of course, these surface densities both depend on the coordinate z as e^{-jkz}, so we can write

$$
\begin{aligned}
\rho_{Sg} &= \mathcal{Q}_{Sg}(l)e^{-jkz}; & \rho_S^{(1)} &= \mathcal{Q}_S^{(1)}(l)e^{-jkz} \\
\mathbf{J}_{Sg} &= \mathcal{J}_{Sg}(l)e^{-jkz}; & \mathbf{J}_S^{(1)} &= \mathcal{J}_S^{(1)}(l)e^{-jkz}
\end{aligned}
\tag{8.18}
$$

where the $\mathcal{Q}_S$ and $\mathcal{J}_S$ depend only on the coordinate l, which is the arc length along either C_1 or C_g. Furthermore, since $\mathcal{H}_z = 0$, there can be no transverse component of the surface current, and the $\mathcal{J}_S$ are thus totally z-directed:

$$
\mathcal{J}_{Sg} = \mathbf{u}_z \mathcal{J}_{zg}(l); \qquad \mathcal{J}_S^{(1)} = \mathbf{u}_z \mathcal{J}_z^{(1)}(l)
\tag{8.19}
$$

The *electrostatic* behavior of the configuration of Figure 8.5 can be summarized by the total charge per unit length (or linear charge density) on the conductor C_1:

$$
\mathcal{Q}_l^{(1)} = \oint_{C_1} \mathcal{Q}_S^{(1)}(l)\, dl
\tag{8.20}
$$

and the potential difference $\Phi^{(1)}$ between C_1 and C_g:

$$
\Phi^{(1)} = -\int_g^1 \mathcal{E}_T \cdot d\mathbf{l}
\tag{8.21}
$$

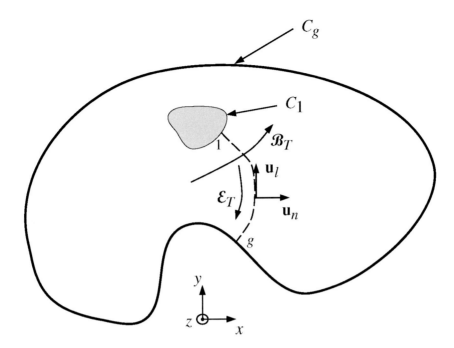

Figure 8.6 Contour for calculating potential and magnetic flux between C_1 and C_g.

Because $\mathcal{E}_T$ is the gradient of a scalar function, the path of integration in (8.21) can be any contour between a point 1 on C_1 and a point g on C_g (see Figure 8.6). Likewise, because of the first of (8.9) and the two-dimensional divergence theorem, the path of integration in (8.20) may be taken to be any closed contour C lying outside of C_1 but inside of C_g (see Figure 8.5).

The total charge per unit length on C_g is simply the negative of that on C_1, as we can show using (8.16), (8.17), (8.20), the two-dimensional divergence theorem, and (8.9):

$$
\begin{aligned}
\mathcal{Q}_{lg} &= \oint_{C_g} \mathcal{Q}_{Sg}\, dl \\
&= \oint_{C_g} \epsilon \mathbf{u}_n \cdot \mathcal{E}_T\, dl \\
&= -\oint_{C_1} \epsilon \mathbf{u}_n \cdot \mathcal{E}_T\, dl - \int_{S_0} \nabla_T \cdot (\epsilon \mathcal{E}_T)\, dS \\
&= -\mathcal{Q}_l^{(1)}
\end{aligned}
\tag{8.22}
$$

The electrostatic characteristics of the line are succinctly described by defining the distributed capacitance c per unit length as

$$
\mathrm{c} \equiv \mathcal{Q}_l^{(1)}/\Phi^{(1)}
\tag{8.23}
$$

As stated above, we can write the linear charge density as

$$
\mathcal{Q}_l^{(1)} = \oint_C \epsilon \mathcal{E}_n\, dl
\tag{8.24}
$$

for an arbitrary contour C between C_1 and C_g, keeping in mind that n represents the normal *away from* C. Thus, if we know the solution $\boldsymbol{\mathcal{E}}_T$ to (8.9) for some transmission line configuration, (8.21), (8.24) and (8.23) will allow us to determine the capacitance c. It is to be emphasized that these electrostatic characteristics are determined completely by the function $\epsilon(x, y)$ and the geometry of the line, and not by the magnetic properties $\mu(x, y)$ at all.

The magnetostatic problem (8.10) is characterized by the current flowing in conductor 1 (that we will denote by $I^{(1)}$) and the magnetic flux per unit length $\Psi_l^{(1)}$ which passes between the two conductors of the line. Referring to Figure 8.6, we find the flux by integrating $\boldsymbol{\mathcal{B}}_T \cdot \mathbf{u}_n$ along any path connecting C_1 to C_g (for instance, we may use the same one used before in (8.21) to evaluate the voltage $\Phi^{(1)}$), where $\mathbf{u}_n$ here is the unit vector normal to this path and to $\mathbf{u}_z$ as well. We obtain

$$\Psi_l^{(1)} = \int_g^1 \boldsymbol{\mathcal{B}}_T \cdot \mathbf{u}_n \, dl \tag{8.25}$$

The current $I^{(1)}$ flowing in the $+z$ direction on conductor C_1 is given by integrating the current density over C_1:

$$\begin{aligned} I^{(1)} &= \oint_{C_1} \mathcal{J}_z^{(1)}(l) \, dl \\ &= \oint_{C_1} \boldsymbol{\mathcal{H}}_T \cdot d\boldsymbol{l} \end{aligned} \tag{8.26}$$

As with the charge density, the path of integration in (8.26) can be taken to be any closed contour C between C_1 and C_g. It is readily shown by methods analogous to those used in (8.22) (problem p8 - 1) that an equal but opposite current must flow in the other conductor C_g:

$$\begin{aligned} I_g &= \oint_{C_g} \boldsymbol{\mathcal{H}}_T \cdot d\boldsymbol{l} \\ &= -I^{(1)} \end{aligned} \tag{8.27}$$

The magnetostatic characteristics of the line are then succinctly described in terms of the distributed inductance l per unit length of the line, defined as

$$l \equiv \Psi_l^{(1)} / I^{(1)} \tag{8.28}$$

We emphasize that the magnetostatic characteristics of the TEM mode are determined only by the magnetic properties $\mu(x, y)$ and geometry of the line.

We now establish the connection between the voltage and current associated with this mode, which will turn out in fact to be the telegrapher's equations as found in Chapter 1 for the classical transmission line. Indeed, we can take the voltage of forward mode of the classical transmission line corresponding to this TEM mode to be

$$V(z) = \Phi^{(1)} e^{-jkz} \tag{8.29}$$

while the current of this mode is

$$I(z) = I^{(1)} e^{-jkz} \tag{8.30}$$

Take the dot product of (8.3) with $\mathbf{u}_n$ on C_1, and integrate over C_1. From (8.24) and (8.26), the result can be written as:

$$-\gamma I^{(1)} = -j\omega \mathcal{Q}_l^{(1)} \tag{8.31}$$

which is a statement of conservation of surface charge on C_1. If this is combined with (8.23), we have

$$-\gamma I^{(1)} = -j\omega c \Phi^{(1)} \tag{8.32}$$

which is equivalent to the second telegrapher's equation in (1.2) in the absence of shunt losses g. Likewise, taking the dot product of (8.1) with $\mathbf{u}_n$ on the path between points g and 1 in Figure 8.6 and integrating along the path, the result is (using (8.21) and (8.25)):

$$-\gamma \Phi^{(1)} = -j\omega \Psi_l^{(1)} \tag{8.33}$$

Combining this result with (8.28), we get

$$-\gamma \Phi^{(1)} = -j\omega l I^{(1)} \tag{8.34}$$

which is equivalent to the first telegrapher's equation in (1.2) in the absence of series losses r. We have thus justified the use of the distributed-network model of a transmission line in the absence of losses for TEM modes, noting that $j\omega$ corresponds to $\partial/\partial t$ and $-\gamma$ corresponds to $\partial/\partial z$ for this forward mode. As in Chapter 1, having computed l and c for a given configuration of a uniform TEM mode structure, we have that

$$\gamma = j\omega \sqrt{lc} \tag{8.35}$$

and

$$Z_c = \sqrt{\frac{l}{c}} \tag{8.36}$$

For the TEM mode, however, we have already seen that γ is determined by the properties of the filling medium of the waveguide, eqn. (8.5). Comparison with (8.35) thus shows that the line parameters are intimately related in this case:

$$lc = \mu\epsilon \qquad \boxed{\text{TEM modes only}} \tag{8.37}$$

We see that the distributed inductance and distributed capacitance are *not* independent for TEM lines. There are thus several alternative expressions for Z_c in terms of l, c, and the product $\mu\epsilon$ of the medium parameters:

$$Z_c = \frac{\sqrt{\mu\epsilon}}{c} = \frac{l}{\sqrt{\mu\epsilon}} \qquad \boxed{\text{TEM modes only}} \tag{8.38}$$

8.2.1 POWER, ENERGY AND NORMALIZATION OF TEM MODES

The result of problem p8-2 shows that the complex power carried by a TEM mode can also be expressed in terms of the voltage and current associated with the mode:

$$\hat{P} = \frac{1}{2}\Phi^{(1)} I^{(1)*} \qquad \boxed{\text{TEM modes only}} \tag{8.39}$$

and the oscillatory power is equal to

$$\hat{P}_{\text{osc}} = \frac{1}{2}\Phi^{(1)} I^{(1)} \qquad \boxed{\text{TEM modes only}}$$

Since for a forward TEM mode $\Phi^{(1)} = Z_c I^{(1)}$, it can be seen that the normalization condition (5.27) is equivalent to setting the voltage and current to the particular values:

$$\Phi^{(1)} = \sqrt{2Z_c} \quad ; \quad I^{(1)} = \sqrt{2/Z_c} \tag{8.40}$$

for a TEM mode, in agreement with what was found in (1.26) for classical transmission lines.

The line parameters l and c of a TEM mode can alternatively be given in terms of energy, rather than the forms presented above. For the electrostatic problem, we can write

$$
\begin{aligned}
c[\Phi^{(1)}]^2 = \frac{[Q_l^{(1)}]^2}{c} &= \Phi^{(1)} Q_l^{(1)} = \Phi^{(1)} \oint_{C_1} Q_S^{(1)}(l)\, dl = \oint_{C_1 + C_g} \Phi \epsilon \boldsymbol{\mathcal{E}}_T \cdot \mathbf{u}_n\, dl \\
&= -\int_{S_0} \nabla_T \cdot (\Phi \epsilon \boldsymbol{\mathcal{E}}_T)\, dS = -\int_{S_0} \nabla_T \Phi \cdot \epsilon \boldsymbol{\mathcal{E}}_T\, dS \\
&= \int_{S_0} \epsilon \boldsymbol{\mathcal{E}}_T^2\, dS
\end{aligned}
\tag{8.41}
$$

where we have used (8.9) and (B.11), and $\boldsymbol{\mathcal{E}}_T^2 = \boldsymbol{\mathcal{E}}_T \cdot \boldsymbol{\mathcal{E}}_T$. Hence as an alternative to (8.23), we could also have defined c without explicit reference to the voltage $\Phi^{(1)}$, but using only the linear charge density and the energy per unit length[2] stored in the electric field:

$$
\frac{1}{c} = \frac{1}{[Q_l^{(1)}]^2} \int_{S_0} \epsilon \boldsymbol{\mathcal{E}}_T^2\, dS \qquad \boxed{\text{TEM modes only}}
\tag{8.42}
$$

Alternatively, an expression for c in terms of energy and voltage only can be obtained. Likewise, the line inductance can also be expressed without explicit reference to the flux:

$$
l = \frac{1}{[I^{(1)}]^2} \int_{S_0} \mu \boldsymbol{\mathcal{H}}_T^2\, dS \qquad \boxed{\text{TEM modes only}}
\tag{8.43}
$$

where $\boldsymbol{\mathcal{H}}_T^2 = \boldsymbol{\mathcal{H}}_T \cdot \boldsymbol{\mathcal{H}}_T$. Since

$$
\mu \boldsymbol{\mathcal{H}}_T^2 = \epsilon \boldsymbol{\mathcal{E}}_T^2 \qquad \boxed{\text{TEM modes only}}
\tag{8.44}
$$

by (8.7), we have that

$$
\frac{[Q_l^{(1)}]^2}{c} = l[I^{(1)}]^2
\tag{8.45}
$$

Then by (8.31) we obtain (8.35) without the need to utilize information about the voltage or magnetic flux. On the other hand, using (8.23) and (8.28), we could also obtain energy forms which depend on $\Phi^{(1)}$ and $\Psi_l^{(1)}$, but not explicitly upon the linear charge density or the total current. These energy forms have application to the calculation of wall losses using the incremental inductance rule (see Section 9.5.2 and Appendix M), and forms of a similar type will prove crucial to the modeling of quasi-TEM and pseudo-TEM modes later on.

8.2.2 MULTICONDUCTOR TEM MODES

TEM modes can also exist if the waveguide cross section has more than two separate conductors. Keeping C_g as ground ($\Phi|_{C_g} = 0$), we now suppose there are M other distinct conducting contours, $C_1 \ldots, C_M$. Since the potential on the other conductors may take on different values:

$$
\Phi|_{C_i} = V^{(i)}
$$

---

[2]Since there is no complex conjugate taken in the following expressions, the surface integrals below are proportional to the *oscillatory* energy, rather than the time-average energy. For lossless transmission lines, these are essentially the same thing.

there will be more than one TEM mode (M of them, all degenerate, in fact). The previous arguments once again require that $\mu\epsilon$ be constant, and that $\beta = \omega\sqrt{\mu\epsilon}$. By superposition, the general solution for the potential will be

$$\Phi = \sum_{i=1}^{M} V^{(i)}\Phi_i \tag{8.46}$$

where Φ_i is the solution of (8.12) or (8.13) subject to:

$$\Phi_i|_{C_i} = 1; \qquad \Phi_i|_{C_k, C_g} = 0 \quad (k \neq i)$$

We denote the (purely transverse) field corresponding to each of these potentials by $\boldsymbol{\mathcal{E}}_{wi} = -\nabla_T \Phi_i$. Note that these fields are not normalized to unit power; the subscript w is meant to indicate that each field is associated with nonzero potential only on one of the wires. In a straightforward manner, we can generalize equations (8.7), (8.24), (8.25), (8.26), (8.31) and (8.33) to the multiconductor case as follows:

$$\boldsymbol{\mathcal{H}}_{wi} = \pm\frac{\mathbf{u}_z \times \boldsymbol{\mathcal{E}}_{wi}}{\zeta_{\text{TEM}}(x, y)} \tag{8.47}$$

$$Q_l^{(i)} = \oint_{C_i} \epsilon\mathcal{E}_n \, dl \tag{8.48}$$

$$\Psi_l^{(i)} = \int_g^i \boldsymbol{\mathcal{B}}_T \cdot \mathbf{u}_n \, dl \tag{8.49}$$

$$I^{(i)} = \oint_{C_i} \boldsymbol{\mathcal{H}}_T \cdot d\mathbf{l} \tag{8.50}$$

$$-\gamma I^{(i)} = -j\omega Q_l^{(i)} \tag{8.51}$$

$$-\gamma V^{(i)} = -j\omega \Psi_l^{(i)} \tag{8.52}$$

where i in the path integral for $\Psi_l^{(i)}$ denotes a point on C_i.

From the foregoing, the charge per unit length on C_i can be written:

$$Q_l^{(i)} = \sum_{k=1}^{M} V^{(k)} \oint_{C_i} \epsilon\boldsymbol{\mathcal{E}}_{wk} \cdot \mathbf{u}_n \, dl \tag{8.53}$$

whence from (8.51) we get:

$$-\gamma I^{(i)} = -j\omega \sum_{k=1}^{M} V^{(k)} \oint_{C_i} \epsilon\boldsymbol{\mathcal{E}}_{wk} \, dl \tag{8.54}$$

If we form the column vectors

$$[I] = \begin{bmatrix} I^{(1)} \\ I^{(2)} \\ \vdots \\ I^{(M)} \end{bmatrix}; \qquad [V] = \begin{bmatrix} V^{(1)} \\ V^{(2)} \\ \vdots \\ V^{(M)} \end{bmatrix} \tag{8.55}$$

we can write (8.54) as

$$-\gamma [I] = -j\omega [c] [V] \tag{8.56}$$

where the elements of the capacitance matrix

$$[c] = \begin{bmatrix} c_{11} & -c_{12} & \dots & -c_{1M} \\ -c_{21} & c_{22} & \dots & -c_{2M} \\ \vdots & \vdots & \ddots & \vdots \\ -c_{M1} & -c_{M2} & \dots & c_{MM} \end{bmatrix} \tag{8.57}$$

are given by

$$c_{ii} = \oint_{C_i} \epsilon \boldsymbol{\mathcal{E}}_{wi}\, dl; \qquad c_{ik} = -\oint_{C_i} \epsilon \boldsymbol{\mathcal{E}}_{wk}\, dl \tag{8.58}$$

Although it is not immediately obvious, we can prove the reciprocity relation $c_{ik} = c_{ki}$ by a method analogous to that used in (8.41) above. For $k \neq i$ (the relation is trivial for $k = i$), we have

$$\begin{aligned} c_{ik} &= -\oint_{C_i} \epsilon \boldsymbol{\mathcal{E}}_{wk}\, dl = -\oint_{C_i} \Phi_i \epsilon \boldsymbol{\mathcal{E}}_{wk}\, dl = -\oint_{C_g+C_1+\dots+C_M} \Phi_i \epsilon \boldsymbol{\mathcal{E}}_{wk}\, dl \\ &= \int_{S_0} \nabla_T \cdot (\Phi_i \epsilon \boldsymbol{\mathcal{E}}_{wk})\, dS = \int_{S_0} \nabla_T \Phi_i \cdot \epsilon \boldsymbol{\mathcal{E}}_{wk}\, dS = -\int_{S_0} \epsilon \boldsymbol{\mathcal{E}}_{wi} \cdot \boldsymbol{\mathcal{E}}_{wk}\, dS \end{aligned} \tag{8.59}$$

which is the same if we interchange the indices i and k, so reciprocity is proved. In like manner, we obtain

$$c_{ii} = \int_{S_0} \epsilon \boldsymbol{\mathcal{E}}_{wi} \cdot \boldsymbol{\mathcal{E}}_{wi}\, dS \tag{8.60}$$

which is an extension of (8.41). Thus, it can be seen that (8.56) is equivalent to the second of (3.27) in the case $M = 2$.

The other set of multiconductor transmission line equations is a rather simple consequence of the results obtained above. Defining the column vector of fluxes between each conductor and ground by

$$[\Psi_l] = \begin{bmatrix} \Psi_l^{(1)} \\ \Psi_l^{(2)} \\ \vdots \\ \Psi_l^{(M)} \end{bmatrix} \tag{8.61}$$

we have from (8.52), (8.56) and $\gamma = j\beta = \omega\sqrt{\mu\epsilon}$ that

$$[c] [\Psi_l] = \mu\epsilon [I] \tag{8.62}$$

This leads to a definition of the inductance matrix $[l]$ by

$$[\Psi_l] = [l] [I] \tag{8.63}$$

where

$$[l] = \mu\epsilon [c]^{-1} \tag{8.64}$$

Combining (8.63) and (8.52) gives

$$-\gamma [V] = -j\omega [l] [I] \tag{8.65}$$

which is equivalent to the first of (3.27) when $M = 2$.

Equation (8.64) leads to a generalization of (8.37) for TEM modes:

$$[l] [c] = \mu\epsilon [\mathbf{1}] \qquad \boxed{\text{TEM modes only}} \tag{8.66}$$

Reciprocity of the inductances ($l_{ik} = l_{ki}$) is an easy consequence of (8.66).

8.2.3 DIELECTRIC LOSSES OF TEM MODES

It should be noted that if ϵ is complex, the quantity defined in (8.23) will also be complex, and represents $c - jg/\omega$ instead of c. Likewise, a complex value of μ implies that (8.28) is equal to $l - jr/\omega$. In the case when μ and ϵ are constants, putting $\hat{\epsilon} = \epsilon' - j\epsilon''$, we have from (8.5) and (8.6) that

$$\gamma_{\text{TEM}} = \alpha_{\text{TEM}} + j\beta_{\text{TEM}} = j\omega\sqrt{\mu(\epsilon' - j\epsilon'')} \tag{8.67}$$

where

$$\alpha_{\text{TEM}} = \omega\sqrt{\mu}\left[\frac{\sqrt{(\epsilon')^2 + (\epsilon'')^2} - \epsilon'}{2}\right]^{1/2} \tag{8.68}$$

and

$$\beta_{\text{TEM}} = \omega\sqrt{\mu}\left[\frac{\sqrt{(\epsilon')^2 + (\epsilon'')^2} + \epsilon'}{2}\right]^{1/2} \tag{8.69}$$

This, of course, is the same complex propagation coefficient we would get for a plane wave in the lossy dielectric. It is identical to (5.106)-(5.108) in the special case $k_{c,m}^2 = 0$. Note that if $\epsilon'' \ll \epsilon'$, (8.68) and (8.69) become approximately

$$\alpha_{\text{TEM}} \simeq \frac{\omega\epsilon''}{2}\sqrt{\frac{\mu}{\epsilon'}} \tag{8.70}$$

$$\beta_{\text{TEM}} \simeq k_0\sqrt{\frac{\mu\epsilon'}{\mu_0\epsilon_0}} \tag{8.71}$$

8.3 SOME TEM STRUCTURES

In this section, we will give some examples of transmission lines supporting TEM waves, and formulas for their characteristic impedances. In all these examples, μ and ϵ are taken to be independent of x and y.

8.3.1 THE COAXIAL TRANSMISSION LINE

Figure 8.7 shows a familiar TEM structure, the coaxial transmission line. The field solutions for this structure are well known (they can be found using separation of variables, for example):

$$\boldsymbol{\mathcal{E}} = \mathbf{u}_\rho\frac{\Phi^{(1)}}{\rho\ln(b/a)} \tag{8.72}$$

$$\boldsymbol{\mathcal{H}} = \mathbf{u}_\phi\frac{\Phi^{(1)}}{\zeta\rho\ln(b/a)} \tag{8.73}$$

where $\zeta = (\mu/\epsilon)^{1/2}$, and a and b are the inner and outer radii of the line, respectively. From (8.73),

$$I^{(1)} = \frac{2\pi\Phi^{(1)}}{\zeta\ln(b/a)}$$

and hence,

$$Z_c = \frac{\zeta}{2\pi}\ln\frac{b}{a} \tag{8.74}$$

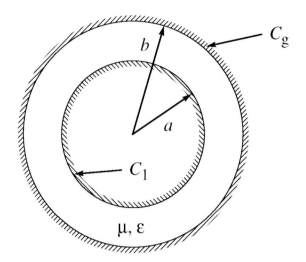

Figure 8.7   The coaxial transmission line.

To normalize the mode field, we must determine the amplitude constant $\Phi^{(1)}$ which enters into these equations. We have

$$
\frac{1}{2}\int_{S_0}\mathcal{E}_1\times\mathcal{H}_1\cdot\mathbf{a}_z\,dS = \frac{1}{2}\int_0^{2\pi}\int_a^b\frac{[\Phi^{(1)}]^2}{\zeta\rho^2}\frac{\rho\,d\rho\,d\phi}{[\ln(b/a)]^2}
$$

$$
= \frac{\pi[\Phi^{(1)}]^2}{\zeta\ln(b/a)} \tag{8.75}
$$

According to (5.27), this must be equal to 1, and we find that

$$
\Phi^{(1)} = \sqrt{\frac{\zeta}{\pi}\ln(b/a)} = \sqrt{2Z_c} \tag{8.76}
$$

in accordance with (8.40). We could equally well have chosen $\Phi^{(1)}$ to be the negative of this value, but this would not be in agreement with the conventional definition of voltage and current polarity for such a transmission line (inner conductor is positive).

8.3.2   TWO-WIRE TRANSMISSION LINE

Another familiar example is the two-wire transmission line shown in Figure 8.8. The wires are identical, with radii a and their centers are separated by a distance $b = 2h$ as shown. The fields of the TEM mode here are also well-known:

$$
\mathcal{E} = \frac{\sqrt{2Z_c}}{2\ln[(b+d)/2a]}\left\{\mathbf{u}_x\left[\frac{x-d/2}{(x-d/2)^2+y^2}-\frac{x+d/2}{(x+d/2)^2+y^2}\right]\right.
$$

$$
\left.+\mathbf{u}_y\left[\frac{y}{(x-d/2)^2+y^2}-\frac{y}{(x+d/2)^2+y^2}\right]\right\} \tag{8.77}
$$

$$
\mathcal{H} = \frac{\sqrt{2Z_c}}{2\zeta\ln[(b+d)/2a]}\left\{\mathbf{u}_y\left[\frac{x-d/2}{(x-d/2)^2+y^2}-\frac{x+d/2}{(x+d/2)^2+y^2}\right]\right.
$$

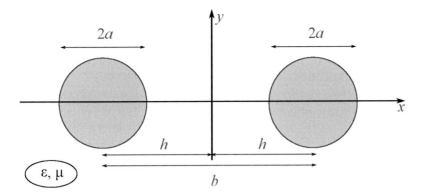

Figure 8.8 The two-wire transmission line.

$$- \mathbf{u}_x \left[\frac{y}{(x - d/2)^2 + y^2} - \frac{y}{(x + d/2)^2 + y^2} \right] \right\}$$ (8.78)

where

$$d = \sqrt{b^2 - 4a^2}$$ (8.79)

Integration of the magnetic field to obtain the current I results in the following expression for the characteristic impedance:

$$Z_c = \frac{\zeta}{\pi} \ln \frac{b + d}{2a} = \frac{\zeta}{\pi} \cosh^{-1} \frac{b}{2a}$$ (8.80)

By image theory, an infinite perfectly conducting plane could be placed at $x = 0$ without disturbing the fields of this structure. Thus, the solution for the two-wire line can also be used for a single wire whose center is a distance h away from and parallel to this ground plane. The capacitance per unit length is twice that of the two-wire line, while the inductance per unit length and characteristic impedance are reduced by a factor of two.

8.3.3 OPEN STRIPLINE

Another example where the region between the two conductors extends to infinity is the open stripline of Figure 8.9. An infinitely thin metal strip of width w is suspended at a height h above a ground plane. Unlike the previous two examples, the potentials and fields for even this most simple of the striplines cannot be written down in closed form. Various

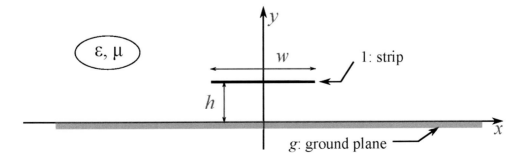

Figure 8.9 Open stripline.

approximations can be made, or numerical solutions attempted (see the references at the end of the chapter), but we will not pursue any detailed derivation here. Let us rather present some results, and concentrate on their physical importance.

If the strip were *wide* compared to its height above the ground plane ($w \gg h$), we would expect $\mathcal{E}_T$ to look as illustrated in Figure 8.10. For a given voltage $\Phi^{(1)}$ between the strip and ground plane, the surface charge on the strip is mostly on the bottom side in this case, and nearly uniform except very near the edges. Here, the field fringes, disturbing the uniformity which prevails near the center of the strip, and causing there to be *more* surface charge near the edges than would be the case if the field were assumed uniform and vertical there. The neglect of the fringing fields is commonly made in elementary discussions of parallel-plate capacitors, and corresponds to the field shown in Figure 8.11. The capacitance per unit length in this parallel-plate approximation is well-known:

$$c_{\text{pp}} = \epsilon \frac{w}{h} \tag{8.81}$$

However, even for very wide strips, this simple model can cause a significant error in the value of the capacitance: nearly 30% when $w/h = 10$. Moreover, for narrow strips, the model completely breaks down.

When the strip is narrow ($w \ll h$), the field $\mathcal{E}_T$ behaves as in Figure 8.12(a). It is quite similar to that of a wire of radius a located at the same height h above a ground plane, Figure 8.12(b). The capacitance of the latter is well-known (related by image theory to that of the two-wire line considered above):

$$c_{\text{wire}} = \frac{2\pi\epsilon}{\ln(2h/a)} \tag{8.82}$$

for $a \ll h$. If a circular wire were chosen with a radius such that the charge on it were the same as that on the strip of Figure 8.12(a) (for the same voltage $\Phi^{(1)}$ on each), we could use (8.82) to approximate the capacitance of this narrow strip. In fact, it has been found

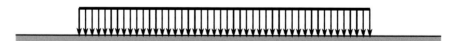

Figure 8.10 Field lines for wide stripline.

Figure 8.11 Stripline fields, neglecting fringing.

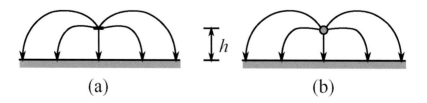

Figure 8.12 Field lines for (a) narrow stripline and (b) wire above ground plane.

that a thin circular wire of *equivalent radius*:

$$a_{eq} = w/4 \qquad (8.83)$$

does support the same charge as the strip for a given voltage under these conditions. The concept of equivalence between thin conductors of various shapes and circular wires is widely encountered in electromagnetics, and enables us to use design information for transmission lines or antennas made up of circular wires for the analysis of structures using conductors of arbitrary cross-section. The concept is applicable whenever the dimensions of the conductor cross-section are small by comparison with the distance of the conductor from any other object. Equivalent radii for conductors of other cross-sections are given in the references cited at the end of this chapter.

The following formula has been found to give c for the stripline of Figure 8.9 for any value of w/h, with an error of less than 1%:

$$c \simeq \frac{4\pi\epsilon}{\ln\left\{1 + 32\frac{h}{w}\left[\frac{h}{w} + \sqrt{(\frac{h}{w})^2 + (\frac{\pi}{8})^2}\right]\right\}} \qquad (8.84)$$

This expression has the advantage of being "reversible", i.e., it can be solved for the ratio w/h in terms of c, allowing design of a stripline with a desired value of Z_c:

$$\frac{w}{h} = \frac{4\sqrt{\pi^2 + 4[\exp(4\pi\epsilon/c) - 1]}}{\exp(4\pi\epsilon/c) - 1} \qquad (8.85)$$

8.4 DISTRIBUTED-CIRCUIT MODEL OF NON-TEM MODES

The transmission lines illustrated in Figures 8.2 and 8.3 do not fall into the class of waveguides we have analyzed in the previous sections, but have instead arbitrary inhomogeneous variations of ϵ in x and y. Some types of planar (and, for that matter, nonplanar) transmission lines involve the use of ferrite substrates which introduce variations of μ as well. We have seen that a TEM mode can exist on structures of this type only if $\mu\epsilon$ is constant throughout the cross-section.

A mode of a two-conductor line that is not TEM can nevertheless still be described in distributed-circuit terms similar to those we used for the TEM mode, with several important differences.[3] Let us examine Maxwell's equations (5.17)-(5.20) to see how they imply different behavior for the fields than the satisfaction of (8.9) and (8.10) implied for TEM mode fields. For the moment, we will continue to assume that the boundaries C_1 and C_g are perfect conductors. Eqns. (5.18) and (5.20) provide the most natural generalizations of the second parts (the curl equations) of (8.9) and (8.10) for non-TEM modes:

$$\nabla_T \times \boldsymbol{\mathcal{E}}_T = -j\omega\mu\mathbf{u}_z\mathcal{H}_z \qquad (8.86)$$

$$\nabla_T \times \boldsymbol{\mathcal{H}}_T = j\omega\epsilon\mathbf{u}_z\mathcal{E}_z \qquad (8.87)$$

The algebraic relationship between $\boldsymbol{\mathcal{E}}_T$ and $\boldsymbol{\mathcal{H}}_T$ now involves the longitudinal field components. An appropriate generalization of (8.1) and (8.3) is obtained from (5.21) and taking $\mathbf{u}_z \times$ (5.19):

$$\nabla_T \mathcal{H}_z = -\gamma\boldsymbol{\mathcal{H}}_T + j\omega\epsilon\mathbf{u}_z \times \boldsymbol{\mathcal{E}}_T \qquad (8.88)$$

---

[3]The following development is relatively general, and can be adapted (as is done for example in problem p8-8) to apply to virtually *any* mode of a waveguide with a metallic boundary—even a single-piece boundary.

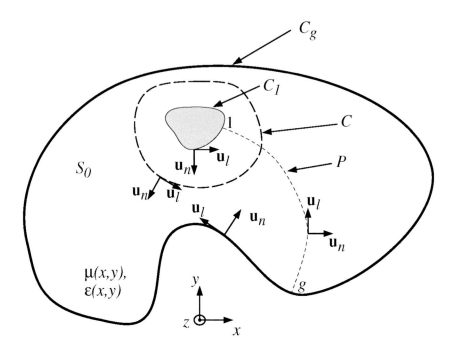

Figure 8.13   Cross-section of a non-TEM transmission line.

$$\nabla_T \mathcal{E}_z \;=\; -\gamma \mathcal{E}_T - j\omega\mu \mathbf{u}_z \times \mathcal{H}_T \tag{8.89}$$

Taking the transverse curl of (8.88) and (8.89), and then using (8.86)-(8.87) together with some vector identities yields generalizations of the first parts (the divergence equations) of (8.9) and (8.10) for non-TEM modes:

$$\nabla_T \cdot (\epsilon \mathcal{E}_T) = \gamma \epsilon \mathcal{E}_z \quad ; \qquad \nabla_T \cdot (\mu \mathcal{H}_T) = \gamma \mu \mathcal{H}_z \tag{8.90}$$

but it should be kept in mind that these are consequences of (8.88)-(8.89), and not independent equations.

Refer to the cross-section shown in Figure 8.13. We again regard C_1 as the "positive" conductor of the line (with respect to the "ground" C_g). In contrast to the TEM case, we cannot define either a potential difference $\Phi^{(1)}$ or a magnetic flux per unit length Ψ_l between C_1 and C_g in a unique fashion. For, suppose P is some path between C_1 and C_g (Figure 8.13). We can define a voltage Φ_P *associated with that path* as

$$\Phi_P = -\int_P \mathcal{E}_T \cdot d\boldsymbol{l} \tag{8.91}$$

in analogy with the definition for a TEM mode.

If we choose a second (different) path P' between C_1 and C_g, then a surface $S_{PP'}$ will be enclosed by P, P' and segments of the conductors C_1 and C_g as shown in Figure 8.14(a). If we integrate $\mathcal{E}_T \cdot d\boldsymbol{l}$ around the closed path bounding $S_{PP'}$, use Stokes' theorem, and note that $\mathcal{E}_T \cdot \mathbf{u}_l = 0$ on C_1 and C_g, we obtain

$$\Phi_P - \Phi_{P'} = \int_{S_{PP'}} \mathbf{u}_z \cdot \nabla_T \times \mathcal{E}_T \, dS \tag{8.92}$$

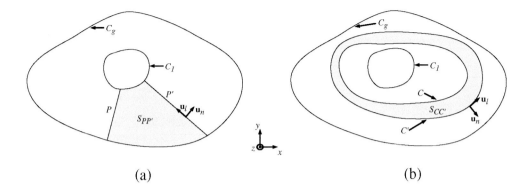

Figure 8.14 (a) Two paths P and P', and the surface $S_{PP'}$ bounded by them. (b) Two contours C and C', and the surface $S_{CC'}$ bounded by them.

or, from (8.86),

$$\Phi_{P'} = \Phi_P + j\omega \int_{S_{PP'}} \mu \mathcal{H}_z \, dS \tag{8.93}$$

i.e., $\Phi_{P'}$ differs from Φ_P by $j\omega$ times the longitudinal magnetic flux passing between P and P'. In an analogous fashion, we can integrate the second of equations (8.90) over $S_{PP'}$ and use the two-dimensional divergence theorem to obtain

$$\Psi_{lP'} = \Psi_{lP} + \gamma \int_{S_{PP'}} \mu \mathcal{H}_z \, dS \tag{8.94}$$

where

$$\Psi_{lP} = \int_P \mu \mathcal{H}_T \cdot \mathbf{u}_n \, dl \tag{8.95}$$

is the magnetic flux per unit length past P (which is likewise dependent on the path P).

Thus, we fix some choice for the path P in order to define the voltage and flux unambiguously. These can then be related in the same way as for TEM modes. We proceed by integrating (8.89) along P:

$$
\begin{aligned}
\int_P \nabla_T \mathcal{E}_z \cdot dl + \gamma \int_P \mathcal{E}_T \cdot dl &= -j\omega \int_P (\mathbf{u}_z \times \mu \mathcal{H}_T) \cdot dl \\
&= -j\omega \int_P \mu \mathcal{H}_T \cdot \mathbf{u}_n \, dl
\end{aligned}
\tag{8.96}
$$

where $\mathbf{u}_n$ is the unit vector normal to P. The first integral on the left side of (8.96) is equal to $\mathcal{E}_z|_{C_1} - \mathcal{E}_z|_{C_g} = 0$. The second term, by (8.91), is equal to $-\Phi_P$, while the integral on the right side is Ψ_P, so we have:

$$-\gamma \Phi_P = -j\omega \Psi_{lP} \tag{8.97}$$

similar to (8.33) in the TEM case. This relation holds true no matter what path P is chosen.

Similarly, we can compute the conduction current $I^{(1)}$ on C_1 by (8.26) once again, and the linear free charge density $\mathcal{Q}_l^{(1)}$ by (8.24). Now, however, that because (8.9) and (8.10) no longer hold, the current and charge are no longer the same if we use an arbitrary contour

C between C_1 and C_g to compute them, in contrast to the situation for a TEM mode. In fact, let us denote

$$Q_{lC} = \oint_C \epsilon \mathcal{E}_n \, dl \qquad (8.98)$$

and

$$I_C = \oint_C \mathcal{H}_T \cdot d\mathbf{l} \qquad (8.99)$$

Now consider any two possible closed contours C and C' such that C is outside of C_1 but inside of C', which in turn is inside of C_g (Figure 8.14(b)). Integrating the first of (8.90) over the portion of the cross-section $S_{CC'}$ between C and C', and using the two-dimensional divergence theorem gives

$$Q_{lC'} = Q_{lC} + \gamma \int_{S_{CC'}} \epsilon \mathcal{E}_z \, dS \qquad (8.100)$$

Likewise, doing the same for the z-component of (8.87) and using Stokes' theorem gives

$$I_{C'} = I_C + j\omega \int_{S_{CC'}} \epsilon \mathcal{E}_z \, dS \qquad (8.101)$$

Thus, just as the voltage and flux can depend on the path for a non-TEM mode, the current and charge lose something of their unambiguous character as well—they now depend on which contour is used to define them. In particular, the current $I_{C'}$ differs from I_C by the longitudinal displacement current through the portion of the cross section between the two contours. Finally, the charge and current on the ground conductor C_g are no longer necessarily the negatives of those on C_1. We have

$$Q_{lg} = -Q_l^{(1)} - \gamma \int_S \epsilon \mathcal{E}_z \, dS \qquad (8.102)$$

while doing the same for the z-component of (8.87) and using Stokes' theorem gives

$$I_g = -I^{(1)} - j\omega \int_S \epsilon \mathcal{E}_z \, dS \qquad (8.103)$$

We now choose some suitable contour C and integrate (8.88) around it, giving

$$\oint_C \nabla_T \mathcal{H}_z \cdot d\mathbf{l} + \gamma \oint_C \mathcal{H}_T \cdot d\mathbf{l} = j\omega \oint_C (\mathbf{u}_z \times \epsilon \mathcal{E}_T) \cdot d\mathbf{l}$$

$$= j\omega \oint_C \epsilon \mathcal{E}_T \cdot \mathbf{u}_n \, dl \qquad (8.104)$$

The first integral vanishes because $\mathcal{H}_z$ is single-valued; the second is equal to I_C by (8.99). The integral on the right is Q_{lC} by (8.98), so that

$$-\gamma I_C = -j\omega Q_{lC} \qquad (8.105)$$

which reduces to (8.31) in the TEM case. Note that (8.105) holds for an arbitrary choice of the contour C, although most frequently it is chosen to be C_1.

8.4.1   DEFINITIONS OF LINE PARAMETERS FOR NON-TEM MODES

As mentioned above, a mode that is not TEM does not have fields which are found exactly from (8.9) and (8.10), but from the full, coupled Maxwell's equations (5.17)-(5.20) (or (8.86)-(8.89)). If we wish to make (8.105) and (8.97) fully correspond to the telegrapher's equations, we have some freedom as to how to proceed. Perhaps the most obvious approach would be to define inductance and capacitance per unit length through the "mixed" equations

$$c_M(\omega, \gamma) = \frac{Q_{lC}}{\Phi_P} \tag{8.106}$$

and

$$l_M(\omega, \gamma) = \frac{\Psi_{lP}}{I_C} \tag{8.107}$$

which lead to the telegrapher's equations

$$-\gamma I_C = -j\omega c_M \Phi_P \tag{8.108}$$

and

$$-\gamma \Phi_P = -j\omega l_M I_C \tag{8.109}$$

and in particular the expression

$$\gamma = \pm j\omega \sqrt{l_M(\omega, \gamma) c_M(\omega, \gamma)} \tag{8.110}$$

for the propagation constant.[4] The line parameters defined in this way depend on the choices of both path P and contour C, and for certain kinds of non-TEM modes the choice of one or the other may not be very obvious or convenient, as in the case of pseudo-TEM modes discussed in Section 8.9. We therefore consider two alternative definitions of the line parameters that require only a choice of P or C, but not both.

We may define an inductance and capacitance per unit length in terms of oscillatory field energies and the current and linear charge density on C, in analogy with (8.42) and (8.43):

$$\frac{1}{c_C(\omega, \gamma)} = \frac{1}{Q_{lC}^2} \int_{S_0} \left(\epsilon \mathcal{E}_T^2 + \mu \mathcal{H}_z^2 \right) dS \tag{8.111}$$

and

$$l_C(\omega, \gamma) = \frac{1}{I_C^2} \int_{S_0} \left(\mu \mathcal{H}_T^2 + \epsilon \mathcal{E}_z^2 \right) dS \tag{8.112}$$

In these equations, we have used the forms of oscillatory energy from (J.7) that most closely resemble the forms for the TEM case. It can be judged that there is a certain natural connection between the pairs of fields $(\mathcal{E}_T, \mathcal{H}_z)$ and $(\mathcal{H}_T, \mathcal{E}_z)$ because of (8.86) and (8.87). If the longitudinal fields are small (as will be the case for the quasi-TEM modes discussed in the next section), (8.111) and (8.112) reduce to (8.42) and (8.43) from the TEM case. The definitions (8.111) and (8.112) can be viewed in terms of a potential Φ_C and magnetic flux per unit length Ψ_{lC} defined as

$$\Phi_C = \frac{1}{Q_{lC}} \int_{S_0} \left(\epsilon \mathcal{E}_T^2 + \mu \mathcal{H}_z^2 \right) dS \tag{8.113}$$

and

$$\Psi_{lC} = \frac{1}{I_C} \int_{S_0} \left(\mu \mathcal{H}_T^2 + \epsilon \mathcal{E}_z^2 \right) dS \tag{8.114}$$

Since (J.7) shows that the two forms of oscillatory energy used above are equal:

$$\int_{S_0} \left(\epsilon \mathcal{E}_T^2 + \mu \mathcal{H}_z^2 \right) dS = \int_{S_0} \left(\mu \mathcal{H}_T^2 + \epsilon \mathcal{E}_z^2 \right) dS, \tag{8.115}$$

---

[4]Equation (8.110), which in general is a transcendental equation for γ, might be a means to compute the propagation constant, or in some cases simply be an identity that only holds once γ has been determined.

we have from (8.105) that

$$-\gamma\Phi_C = -j\omega\Psi_{lC} \tag{8.116}$$

and from this follow the telegrapher's equations

$$-\gamma I_C = -j\omega\mathsf{c}_C\Phi_C \tag{8.117}$$

and

$$-\gamma\Phi_{lC} = -j\omega\mathsf{l}_C I_C \tag{8.118}$$

and in particular the result of classical transmission line theory:

$$\gamma = \pm j\omega\sqrt{\mathsf{l}_C(\omega,\gamma)\mathsf{c}_C(\omega,\gamma)} \tag{8.119}$$

No use is made of the path P in this formalism.

An alternative method based on the field energies for characterizing non-TEM modes is to use the path-dependent voltage and magnetic flux together with the oscillatory energy to define path-dependent line capacitance and inductance as follows:

$$\mathsf{c}_P(\omega,\gamma) = \frac{1}{\Phi_P^2}\int_{S_0}\left(\epsilon\boldsymbol{\mathcal{E}}_T^2 + \mu\boldsymbol{\mathcal{H}}_z^2\right)dS \tag{8.120}$$

and

$$\frac{1}{\mathsf{l}_P(\omega,\gamma)} = \frac{1}{\Psi_{lP}^2}\int_{S_0}\left(\mu\boldsymbol{\mathcal{H}}_T^2 + \epsilon\boldsymbol{\mathcal{E}}_z^2\right)dS \tag{8.121}$$

This approach is equivalent to defining a current and a charge per unit length as

$$I_P = \frac{1}{\Psi_{lP}}\int_{S_0}\left(\mu\boldsymbol{\mathcal{H}}_T^2 + \epsilon\boldsymbol{\mathcal{E}}_z^2\right)dS \tag{8.122}$$

and

$$\mathcal{Q}_{lP} = \frac{1}{\Phi_P}\int_{S_0}\left(\epsilon\boldsymbol{\mathcal{E}}_T^2 + \mu\boldsymbol{\mathcal{H}}_z^2\right)dS \tag{8.123}$$

We have

$$-\gamma I_P = -j\omega\mathcal{Q}_{lP} \tag{8.124}$$

and thus the telegrapher's equations

$$-\gamma I_P = -j\omega\mathsf{c}_P\Phi_P \tag{8.125}$$

and

$$-\gamma\Phi_{lP} = -j\omega\mathsf{l}_P I_P \tag{8.126}$$

and again in particular the result of classical transmission line theory, in the form

$$\gamma = \pm j\omega\sqrt{\mathsf{l}_P(\omega,\gamma)\mathsf{c}_P(\omega,\gamma)} \tag{8.127}$$

No use is made of the contour C in this formalism.

Let us pause to note several things at this point. First, although in general $\mathsf{c}_M(\omega,\gamma) \neq \mathsf{c}_P(\omega,\gamma) \neq \mathsf{c}_C(\omega,\gamma)$ and $\mathsf{l}_M(\omega,\gamma) \neq \mathsf{l}_P(\omega,\gamma) \neq \mathsf{l}_C(\omega,\gamma)$, the values of γ obtained from (8.110), (8.119) and (8.127) must be the same. We will thus use the notations $\mathsf{l}(\omega,\gamma)$ and $\mathsf{c}(\omega,\gamma)$ to refer to the set $\mathsf{l}_M(\omega,\gamma)$ and $\mathsf{c}_M(\omega,\gamma)$, $\mathsf{l}_C(\omega,\gamma)$ and $\mathsf{c}_C(\omega,\gamma)$ or $\mathsf{l}_P(\omega,\gamma)$ and $\mathsf{c}_P(\omega,\gamma)$ as might be applicable in a given context. Since our mode is *not* a TEM mode, we *cannot* immediately calculate γ without knowing $\mathsf{l}(\omega,\gamma)$ and $\mathsf{c}(\omega,\gamma)$ since (8.1) and (8.3), and therefore also (8.5) do not hold.[5] As a matter of fact, equations (8.86) through (8.127) might well apply to essentially *any* mode of this waveguide, or any other waveguide for which appropriate definitions for Φ_P and I_C can be made. If a mode below cutoff were under consideration, either $\mathsf{l}(\omega, \gamma)$ or $\mathsf{c}(\omega, \gamma)$ would have to be negative. Since l and c may depend on γ, equations (8.119) and (8.127) are generally only implicit equations which must be solved to determine γ. An example of how complicated this can make things will be seen in the example of the slotline, later on in this chapter.

The propagation constant γ for any waveguide mode is a well-defined quantity that is determined by Maxwell's equations, independent of whatever choice of path P or contour C is made (or even if no such path and contour are defined at all). The question of defining a characteristic impedance for a non-TEM mode is rather more problematical. Undoubtedly, one possible choice would be to set

$$Z_{c,VI} = \frac{\Phi_P}{I_C} = \sqrt{\frac{\mathsf{l}_M(\omega, \gamma)}{\mathsf{c}_M(\omega, \gamma)}} \tag{8.128}$$

as follows from (8.108)-(8.109), but this definition has a number of drawbacks. First, it depends on the choice of both the path P and contour C on which the voltage Φ_P and current I_C are defined. Second, for modes in general, the oscillatory power carried through the cross-section of the line is *not* related to Φ_P and I_C as it is for TEM modes:

$$P_{\mathrm{osc}} = \frac{1}{2} \int_{S_0} \boldsymbol{\mathcal{E}}_T \times \boldsymbol{\mathcal{H}}_T \cdot \mathbf{u}_z \, dS \neq \frac{1}{2} \Phi_P I_C \tag{8.129}$$

We can, however, show from (J.4)-(J.7), (8.112) and (8.119) that

$$P_{\mathrm{osc}} = \frac{1}{2} \Phi_C I_C = \frac{1}{2} Z_{c,PI} I_C^2 \tag{8.130}$$

where we have defined a "power-current" characteristic impedance as

$$Z_{c,PI} = \sqrt{\frac{\mathsf{l}_C(\omega, \gamma)}{\mathsf{c}_C(\omega, \gamma)}} \tag{8.131}$$

but in this case, the "voltage" Φ_C defined in (8.113) that would have to be used along with the current I_C in a conventional transmission-line theory is only an artificial construct with no physical meaning. In a similar way, we could define a "power-voltage" characteristic impedance in such a way that

$$P_{\mathrm{osc}} = \frac{1}{2} \Phi_P I_P = \frac{\Phi_P^2}{2 Z_{c,PV}} \tag{8.132}$$

where

$$Z_{c,PV} = \sqrt{\frac{\mathsf{l}_P(\omega, \gamma)}{\mathsf{c}_P(\omega, \gamma)}} \tag{8.133}$$

In this case we must deal with a "current" I_P from (8.122) that has no physical meaning.

In general, there is no relationship between these three possible definitions of Z_c. We are forced to admit that, except for situations where voltage and current are unambiguous

---

[5]Equation (8.127) does imply though that the product $\mathsf{l}_P \mathsf{c}_P$ must be independent of the path P just as (8.119) implies that the product $\mathsf{l}_C \mathsf{c}_C$ must be independent of the contour C and $\mathsf{l}_M \mathsf{c}_M$ is independent of both.

concepts (such as for quasi-TEM modes as discussed in Section 8.5.1), the definition of Z_c is fairly arbitrary, and is a matter of convenience for the analysis of concrete problems using the language of circuit theory. A model of equivalent transmission lines is introduced in Chapter 10 suitable for use with any waveguide, and some value of Z_c is needed to make the analogy to a classical transmission line complete. We might find that a particular definition turns out to give simpler equivalent circuits for waveguide discontinuities or excitation schemes than another. That must be the eventual basis for choosing any definition for Z_c, provided that we can be self-consistent in the matter. However, it is only the connection with circuit theoretical concepts that requires Z_c at all, and in principle, we can bypass the the concept of it completely by dispensing with the notions of voltage and current, and speaking only of the fields on the transmission line.

If the conductors C_1 and C_g are not perfect, but instead have a finite conductivity, the preceding discussion has to change somewhat. In particular, even though we can still define a path-associated voltage (8.91), the following derivation (8.92)-(8.97) is no longer true, because it assumed that the tangential components of $\mathcal{E}$ were zero on the surfaces of the conductors. However, the derivations (8.98)-(8.105) and (8.111)-(8.119) remain valid with only minor changes, as we discuss below. In addition to allowing imperfect conductors, let us also assume that there may be losses in the region between the conductors. Both conditions are accounted for by allowing the permittivity ϵ and/or the permeability μ to be complex functions of position, including "inside" the conductors C_1 and C_g. The derivation (8.98)-(8.119) remains true if we extend the surface integrals to include also the cross-sections S_1 and S_g of the conductors, and if the line parameters are modified to include shunt conductance and series resistance per unit length:

$$\frac{1}{j\omega \mathrm{c}_C(\omega,\gamma) + \mathrm{g}_C(\omega,\gamma)} = \frac{1}{j\omega Q_{lC}^2} \int_{S_0+S_1+S_g} \left(\epsilon \mathcal{E}_T^2 + \mu \mathcal{H}_z^2\right) dS \qquad (8.134)$$

and

$$j\omega \mathrm{l}_C(\omega,\gamma) + \mathrm{r}_C(\omega,\gamma) = \frac{j\omega}{I_C^2} \int_{S_0+S_1+S_g} \left(\mu \mathcal{H}_T^2 + \epsilon \mathcal{E}_z^2\right) dS \qquad (8.135)$$

whence the propagation constant obeys

$$\gamma = \pm\sqrt{[j\omega \mathrm{l}_C(\omega,\gamma) + \mathrm{r}_C(\omega,\gamma)][j\omega \mathrm{c}_C(\omega,\gamma) + \mathrm{g}_C(\omega,\gamma)]} \qquad (8.136)$$

Once again, the results agree formally with those of classical transmission line theory. The corresponding generalizations of (8.120)-(8.127) are

$$\mathrm{g}_P(\omega,\gamma) + j\omega \mathrm{c}_P(\omega,\gamma) = \frac{j\omega}{\Phi_P^2} \int_{S_0} \left(\epsilon \mathcal{E}_T^2 + \mu \mathcal{H}_z^2\right) dS \qquad (8.137)$$

$$\frac{1}{\mathrm{r}_P(\omega,\gamma) + j\omega \mathrm{l}_P(\omega,\gamma)} = \frac{1}{j\omega \Psi_P^2} \int_{S_0} \left(\mu \mathcal{H}_T^2 + \epsilon \mathcal{E}_z^2\right) dS \qquad (8.138)$$

and

$$\gamma = \pm\sqrt{[j\omega \mathrm{l}_P(\omega,\gamma) + \mathrm{r}_P(\omega,\gamma)][j\omega \mathrm{c}_P(\omega,\gamma) + \mathrm{g}_P(\omega,\gamma)]} \qquad (8.139)$$

8.5 QUASI-TEM MODES

8.5.1 "APPROXIMATELY" TEM MODES AND THE QUASI-TEM LIMIT

We must admit that even for a two-conductor transmission line with an inhomogeneously filled cross section, in the static limit ($\omega \to 0$, $\gamma \to 0$), there will be nontrivial solutions $\mathcal{E}_T$ and $\mathcal{H}_T$ to the field equations. These are the static, purely transverse fields of z-independent charge and longitudinal current distributions on the conductors of the line, like those of true TEM modes. It seems reasonable to expect that at "sufficiently low" frequencies some kind of "almost" TEM mode could exist, for which $\mathcal{E}_z$ and $\mathcal{H}_z$ were "small" compared to the corresponding transverse fields.

To say more precisely what "small" frequency means, call L a typical length over which the field varies significantly in the transverse direction (roughly speaking, this can be thought of as one of the transverse dimensions of the waveguide, e. g., its diameter). It should be noted that some of the conductors of the transmission line may be of infinite transverse extent, provided that the *field* is effectively confined to a region of dimension L. Then operating on a field with the transverse gradient, curl or divergence operators will produce a quantity which is of the order of magnitude of that field divided by L. If γ can be thought of as being the same order of magnitude as $k \equiv \omega\sqrt{\mu\epsilon}$, then if $kL \ll 1$, (8.88)-(8.89) provides the estimates

$$\mathcal{H}_z \sim |\mathcal{H}_T|O(kL) + |\mathcal{E}_T/\zeta|O(kL)$$

$$\mathcal{E}_z \sim |\mathcal{E}_T|O(kL) + |\mathcal{H}_T\zeta|O(kL)$$

where $\zeta = \sqrt{\mu/\epsilon}$ is the local wave impedance, and $O(kL)$ denotes terms which are no larger in magnitude than some constant times kL. The right sides of (8.86) and (8.87) are then $O(k^2L^2)$ smaller than the left sides, and can be set approximately equal to zero. We then revert to eqns. (8.9) and (8.10). Physically, this *quasi-TEM* approximation means that the transverse dimensions of the structure must be small compared to the wavelength of a plane wave in whatever material may constitute the waveguide structure. The transverse fields will then approximately satisfy (8.9) and (8.10) with μ and ϵ now functions of x and y, but not (8.1) or (8.3) any longer ((8.88)-(8.89) hold instead, and approximately determine $\mathcal{E}_z$ and $\mathcal{H}_z$).

The most important application of the formalism of Section 8.4 is to the quasi-TEM mode of the transmission line which approaches the static, two-dimensional field configuration for small values of kL. In many such situations, the difference between Φ_P and $\Phi_{P'}$ for any two paths P and P' becomes negligible because of (8.93) and the smallness of ω and $\mathcal{H}_z$, and Φ_P approaches a unique frequency-independent value $\Phi^{(1)}$, obtained from the *static* field $\mathcal{E}_{T0}$ satisfying the equations (8.9). Likewise, it becomes possible to replace Ψ_P by the frequency-independent value Ψ_l, obtained from the static field $\mathcal{H}_{T0}$ that satisfies (8.10). As a result, l_P and c_P become independent of both ω and γ. In the same way, the difference between I_C and I'_C for any two contours C and C' becomes negligible under these conditions because of (8.101). We can thus replace I_C by the (approximately) frequency-independent value $I^{(1)}$ obtained from the magnetostatic field $\mathcal{H}_{T0}$, and analogously $\mathcal{Q}_{lC} \to \mathcal{Q}_{l1}$. The definitions of $l_C(\omega, \gamma)$ and $c_C(\omega, \gamma)$ become independent of C, and also approach constants as frequency approaches zero. Similar statements hold true for $l_M(\omega, \gamma)$ and $c_M(\omega, \gamma)$. In this limit, we say that the mode is a quasi-TEM mode, and we can put

$$\begin{aligned}
l_C(\omega, \gamma) &\simeq l_P(\omega, \gamma) \simeq l_M(\omega, \gamma) \simeq l \\
c_C(\omega, \gamma) &\simeq c_P(\omega, \gamma) \simeq c_M(\omega, \gamma) \simeq c
\end{aligned} \tag{8.140}$$

where l and c are the static inductance and capacitance per unit length of line computed just as they were in the TEM case, from (8.9), (8.10), (8.23) and (8.28), μ and ϵ now being regarded as functions of position (x, y) in the cross section.[6] In the case of a loss-free

---

[6] Again, for certain configurations such as the slotline, l and c may fail to approach finite nonzero limits. In such cases, it is necessary to refine our treatment somewhat to calculate quantities $l(\omega, \gamma)$ and $c(\omega, \gamma)$ which do not approach as static limit. Such modes will be called pseudo-TEM modes.

quasi-TEM mode, then, we have from (8.119) or (8.127) that

$$\gamma = j\beta \simeq j\omega\sqrt{lc} \qquad (8.141)$$

for a forward mode. We emphasize that β cannot now generally be expressed directly in terms of particular values of μ and ϵ of the line in any simple way. It is, however, possible to show that the value of the product lc will lie somewhere between the minimum and maximum values of the product $\mu\epsilon$ over the cross-section of the line (cf. Appendix J). All of the definitions for characteristic impedance given in Section 8.4.1 tend to the same limit in the quasi-TEM regime:

$$Z_c = \sqrt{\frac{l}{c}} \qquad (8.142)$$

and Z_c may be used in all the same ways as for TEM modes. Thus the entire body of transmission-line analysis may be applied here just as to a TEM line. It is important to remember, though, that *only* (8.142), and none of the other alternative forms in (8.38) can be used to calculate Z_c when the mode is not pure TEM.

In the case when the conductors C_1 and C_g are not perfect, or there are losses in the material filling the waveguide cross-section, we include also the line parameters $g_C(\omega,\gamma)$ and $r_C(\omega,\gamma)$ as indicated at the end of Section 8.4. For sufficiently low frequencies, g and r become independent of C and of γ, though not generally independent of ω (especially in the case of r, due to skin effect—see Section 9.5). If the conductivity of the conductors is high, $\mathcal{E}_T$ inside them will be very small, while $\mathcal{E}_z$, while also small, will contribute to a significant longitudinal current density, which is the major contribution to r for quasi-TEM modes. In this book, we will only evaluate these effects of loss by perturbation methods (Sections 9.4-9.6), but in some cases a full solution of Maxwell's equations is necessary to accurately model them.

8.5.2   QUASI-TEM MODES ON LINES HAVING AIR AND ONE OTHER DIELECTRIC

Many important quasi-TEM lines have a dielectric which is piecewise constant. Indeed, the dielectric constant is homogeneous ($\epsilon = \epsilon_r\epsilon_0$) within a substrate region, and the remainder of the cross section is air ($\epsilon = \epsilon_0$), as shown in Figure 8.15. Suppose that the entire waveguide is made of nonmagnetic material ($\mu = \mu_0$). Let us emphasize the dependence of c on the relative permittivity ϵ_r of the "substrate" region by writing:

$$c = C(\epsilon_r) \qquad (8.143)$$

Evidently, $C(1)$ is the capacitance of the corresponding TEM mode structure obtained from the same configuration of conductors C_1 and C_g, but uniformly filled with air. The inductance l of the original quasi-TEM line must be the same as that of the air-filled line, since ϵ has no effect on the quasi-TEM inductance, which is completely determined by the solution of (8.10) which involves μ but not ϵ. Then, from (8.37) we have

$$l = \frac{\mu_0\epsilon_0}{C(1)} \qquad (8.144)$$

and (8.141) and (8.142) can be expressed in the forms

$$\beta \simeq k_0\sqrt{\frac{C(\epsilon_r)}{C(1)}} \qquad (8.145)$$

and

$$Z_c(\epsilon_r) = Z_c(1)\sqrt{\frac{C(1)}{C(\epsilon_r)}} \qquad (8.146)$$

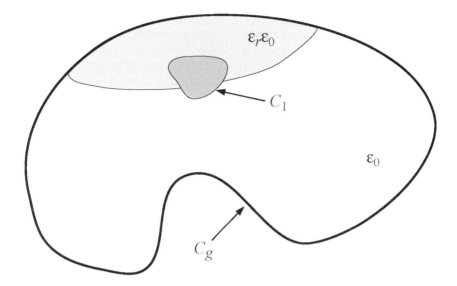

Figure 8.15   Transmission line containing two homogeneous dielectric regions.

for quasi-TEM modes, where $k_0 = \omega\sqrt{\mu_0\epsilon_0}$ and

$$Z_c(1) = \frac{\sqrt{\mu_0\epsilon_0}}{C(1)}$$

is the characteristic impedance of the corresponding air-filled line.

These expressions suggest the definition of an *effective dielectric constant* ϵ_{eff} for quasi-TEM modes according to

$$\epsilon_{\text{eff}} = \frac{C(\epsilon_r)}{C(1)} \tag{8.147}$$

so that

$$\beta \simeq k_0\sqrt{\epsilon_{\text{eff}}} \tag{8.148}$$

and

$$Z_c(\epsilon_r) = Z_c(1)/\sqrt{\epsilon_{\text{eff}}} \tag{8.149}$$

The physical significance of ϵ_{eff} is thus that a line homogeneously filled with a medium of relative permittivity ϵ_{eff} and permeability μ_0 would have the same β and Z_c as given by (8.148) and (8.149). It therefore suffices for us to obtain an expression for $C(\epsilon_r)$ as a function of ϵ_r to completely characterize the mode under quasi-TEM conditions.

8.6   EXAMPLE: OPEN MICROSTRIP

Consider the open microstrip whose cross-section is shown in Figure 8.16(a). It is identical to the open stripline of Figure 8.9, except that the region $0 < y < h$ is occupied by a substrate of relative dielectric constant ϵ_r. The region above the strip is taken to be air, and all regions are assumed to be nonmagnetic. The line is thus of the two-dielectric type studied in the previous section. A photograph showing a top view of some microstrip circuitry is shown in Figure 8.16(b), in which a branch-line hybrid (see problem p2-16) is located at the center.

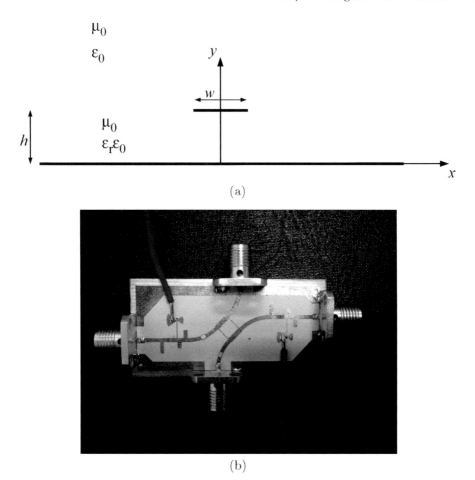

Figure 8.16 (a) Open microstrip cross section; (b) microstrip circuitry containing a branch-line hybrid.

As was the case with open stripline, an exact formula for $C(\epsilon_r)$ for microstrip cannot be given. It is possible to give one that is valid for arbitrary values of w/h with an accuracy of 2% or better:

$$C_{\mathrm{micr}}(\epsilon_r) \simeq \frac{2\pi\epsilon_0(\epsilon_r + 1)}{\ln\left\{1 + \frac{32h}{w}\left[\frac{h}{w}r_\epsilon^2 + \sqrt{\left(\frac{hr_\epsilon^2}{w}\right)^2 + \left(\frac{\pi(\epsilon_r+1)}{16\epsilon_r}\right)^2}\right]\right\}} \qquad (8.150)$$

where by (7.42)

$$r_\epsilon \simeq \sqrt{0.4052 + \frac{0.5160}{\epsilon_r} + \frac{0.0788}{\epsilon_r^2}}$$

In conjunction with (8.146) and (8.147), equation (8.150) can be used to calculate Z_c and ϵ_{eff}. Since the use of (8.150) in (8.146) results in a somewhat messy expression, the following formula is more often used to find Z_c:

$$Z_c \simeq \frac{\zeta_0}{4\pi}\sqrt{\frac{2}{\epsilon_r + 1}} \ln\left\{1 + \frac{32h}{w}\left[\frac{h}{w}r_\epsilon + \sqrt{\left(\frac{hr_\epsilon}{w}\right)^2 + \frac{\epsilon_r + 1}{2\epsilon_r}\left(\frac{\pi}{8}\right)^2}\right]\right\} \qquad (8.151)$$

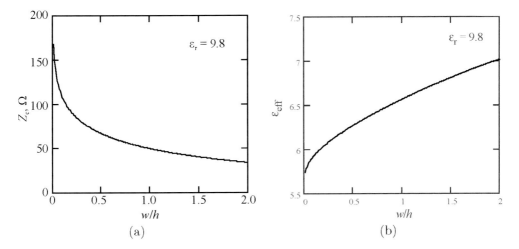

Figure 8.17 Quasi-TEM values for (a) characteristic impedance and (b) effective permittivity of microstrip.

Formula (8.151) is reversible: It can be solved for w/h in terms of Z_c, and used to design a microstrip with a desired characteristic impedance:

$$\frac{w}{h} = \frac{1}{K}\sqrt{\frac{8\pi^2(\epsilon_r+1)}{\epsilon_r}+64Kr_\epsilon} \qquad (8.152)$$

where

$$K = \exp\left[\frac{4\pi Z_c}{\zeta_0}\sqrt{\frac{\epsilon_r+1}{2}}\right]-1 \qquad (8.153)$$

Typical values of Z_c and ϵ_{eff} are shown in Figure 8.17.

Some examination of the preceding formulas shows that $C(\epsilon_r) > C(1)$ when $\epsilon_r > 1$. This is due to the fact that a portion of the field now resides in a region of higher dielectric constant (the fields also rearrange themselves slightly by comparison to the air-filled line, but this is a relatively smaller effect). It is thus possible to achieve lower values of characteristic impedance (say, 50 Ω) using narrower strips than in the case of an air-filled stripline by using an electrically dense ($\epsilon_r \sim 10$) substrate such as alumina.

The charge and current distributions are present on both the top and bottom sides of the strip, with more density on the bottom side the wider the strip. The density is often given as a total density, adding the densities from the top and bottom sides at a given point x. The total current density can be given quite accurately by:

$$\mathcal{J}_{z1}(x) \simeq \frac{\text{const}}{\sqrt{\cosh^2(\frac{\pi w}{8h})-\cosh^2(\frac{\pi x}{4h})}} \qquad (|x| < \frac{w}{2}) \qquad (8.154)$$

where $x = 0$ is the center of the strip, and the constant is chosen to make the total current on the strip equal to $I^{(1)}$. This current distribution is singular at the edges of the strip ($x = \pm w/2$) because the strip has been assumed to be infinitely thin, and to have perfectly sharp edges. Plots of (8.154) are given in Figure 8.18 for a very wide and a very narrow strip. It is evident that for wide strips, the current is very uniformly distributed (except near the edges), as would be expected from a "parallel-plate" approximation that neglects fringing of the fields. The increase in current density near the strip edges causes the attenuation of

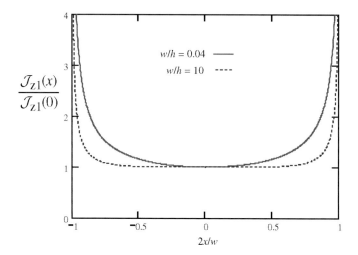

Figure 8.18 Normalized total surface current density $\mathcal{J}_{z1}(x)$ on open microstrip with $\mu = \mu_0$.

the mode due to the finite conductivity of the strip metal to be larger than if the current were uniformly distributed. This point will be examined in more detail in Section 9.5.

An accurate approximation to the charge distribution on open microstrip has been found to be:

$$Q_{S1}(x) \simeq \frac{\text{const}}{\sqrt{\cosh^2\left(\frac{\pi \epsilon_r w}{4(\epsilon_r+1)h}\right) - \cosh^2\left(\frac{\pi \epsilon_r x}{2(\epsilon_r+1)h}\right)}} \qquad (|x| < w/2) \qquad (8.155)$$

where the constant is now chosen to make the total charge per unit length equal to Q_{S1}. For wide strips and large values of ϵ_r, the charge distribution will be somewhat flatter than the current distribution (Figure 8.19). This demonstrates (from the principle of charge

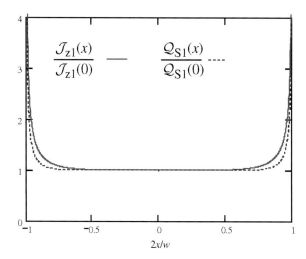

Figure 8.19 Normalized charge and current distributions for open microstrip with $\epsilon_r = 10$ and $w/h = 10$.

conservation) that at least some *transverse* current density $\mathcal{J}_{x1}(x)$ must also be present on the strip for this quasi-TEM mode.

8.7 MORE EXAMPLES: COPLANAR LINES

In this section, we present the quasi-TEM properties of two other commonly encountered planar transmission lines. In all cases, the inductance $\mathfrak{l}$ is given in terms of the expression for capacitance evaluated at $\epsilon_r = 1$, as in (8.144).

8.7.1 OPEN COPLANAR WAVEGUIDE (CPW)

A further commonly used planar transmission line is the coplanar waveguide (CPW) shown in Figure 8.20. Its positive conductor is a strip on a dielectric substrate, like the microstrip, but unlike the microstrip, its ground plane is in two parts and lies on the same side of the substrate as the strip. Thus, CPW is formed from a substrate with a ground plane on one side into which two identical slots have been cut, separated by a distance s. The mode of interest here has the two outer conductors at ground potential, while the center conductor is the positive reference, resulting in the field distribution shown in Figure 8.21. Coplanar waveguide combines the advantages of microstrip with additional ones: both series and shunt lumped elements can easily be connected to the line without disturbing the substrate.

As was the case for microstrip, exact expressions for the line parameters of CPW are not available. The following formula for $\mathfrak{c}$ has been found to track well with numerically calculated results for a wide range of substrate thicknesses h, particularly if $\epsilon_r \gg 1$ (it is exact either for $\epsilon_r = 1$ or as $h \to \infty$):

$$\mathfrak{c} = C_{\text{CPW}}(\epsilon_r) = 4\epsilon_0 \left[\frac{K(k)}{K(k')} + \frac{\epsilon_r - 1}{2} \frac{K(k_1)}{K(k_1')} \right] \qquad (8.156)$$

where $K(k)$ is the complete elliptic integral of the first kind, and is discussed in Appendix C where simple approximate formulas for it are given. The moduli k and k_1 appearing in

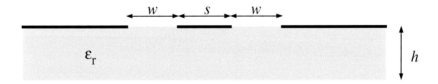

Figure 8.20 The open coplanar waveguide (CPW).

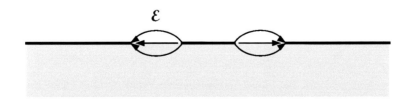

Figure 8.21 Transverse electric field distribution of the fundamental (quasi-TEM) mode in coplanar waveguide.

(8.156) are

$$k = \frac{s}{s + 2w}$$

$$k_1 = \frac{\sinh(\pi s/4h)}{\sinh[\pi(s + 2w)/4h]} \tag{8.157}$$

while the complementary moduli are $k' = \sqrt{1 - k^2}$ and $k_1' = \sqrt{1 - k_1^2}$. As $h \to \infty$, we have $k_1 \to k$, and in this limit as well as $\epsilon_r \to 1$, the expression (8.156) becomes exact for the quasi-TEM case.

In this approximation, the dependence of c on ϵ_r is much simpler than was $C_{\mathrm{micr}}(\epsilon_r)$. We have ϵ_{eff} and Z_c given by

$$\epsilon_{\mathrm{eff}} = 1 + \frac{\epsilon_r - 1}{2} \frac{K(k')}{K(k)} \frac{K(k_1)}{K(k_1')} \tag{8.158}$$

and

$$Z_c = \frac{\zeta_0}{4\sqrt{\epsilon_{\mathrm{eff}}}} \frac{K(k')}{K(k)} \tag{8.159}$$

based on (8.156). Some typical behavior of Z_c and ϵ_{eff} vs. $s/(s+2w)$ is shown in Figure 8.22.

Notice that since there are two ratios of length dimensions for us to adjust, we can specify *both* Z_c and ϵ_{eff} for a design with a given value of ϵ_r. This is in contrast to the cases of slotline or microstrip, in which only one of these can be independently specified. Equations (C.70) and (C.72) of Appendix C allow us to calculate the value of the modulus k given the value of the ratio of $K(k')/K(k)$. Thus, we could proceed as follows. With Z_c and ϵ_{eff} given, we find the value of $K(k')/K(k)$ and therefore k from (8.159). The value of $K(k_1')/K(k_1)$ and thus k_1 is then found from (8.158). The value of s/w is then found from the first of eqns. (8.157), and s/h or w/h from the second of these equations. In the latter case, the equation is not explicitly solvable for the desired quantity, and some form of numerical or approximate method (e.g., the iteration method outlined in Appendix K) must be used to obtain its value.

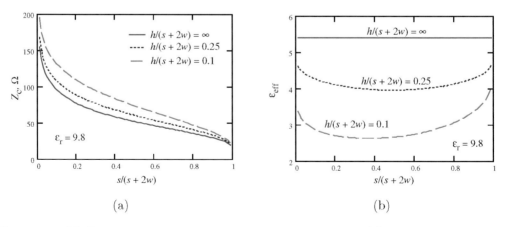

(a) (b)

Figure 8.22 (a) Characteristic impedance of coplanar waveguide, and (b) effective dielectric constant of coplanar waveguide or coplanar strip line.

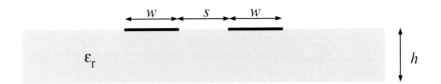

Figure 8.23 Coplanar strip line (CPS).

8.7.2 COPLANAR STRIP LINE (CPS)

A structure complementary to CPW which is sometimes used as a traveling-wave electrode structure for electro-optic devices is the coplanar strip line (CPS) shown in Figure 8.23. Here again, both series and shunt connections of lumped circuit elements may easily be made to this line.

Corresponding formulas to (8.156)-(8.159) for the CPW are also available for CPS. With the moduli k and k_1 again given by (8.157), we have

$$\mathsf{c} = C_{\text{CPS}}(\epsilon_r) = \mathsf{c}_0 \left[1 + \frac{\epsilon_r - 1}{2} \frac{\mathsf{c}_0}{\epsilon_0} \frac{K(k_1)}{K(k'_1)} \right] \tag{8.160}$$

where c_0 is the value of c when $\epsilon_r = 1$:

$$\mathsf{c}_0 = C_{\text{CPS}}(1) = \epsilon_0 \frac{K(k')}{K(k)} \tag{8.161}$$

We can now give the formulas for ϵ_{eff} and Z_c as

$$\epsilon_{\text{eff}} = 1 + \frac{\epsilon_r - 1}{2} \frac{K(k')}{K(k)} \frac{K(k_1)}{K(k'_1)} \tag{8.162}$$

(the same as for CPW), while

$$Z_c = \frac{\zeta_0}{\sqrt{\epsilon_{\text{eff}}}} \frac{K(k)}{K(k')} \tag{8.163}$$

Typical plots of Z_c for CPS are shown in Figure 8.24; reference can be made to Figure 8.22(b) for values of ϵ_{eff}.

Other design data and formulas for quasi-TEM planar transmission lines can be found in the references at the end of the chapter.

8.8 DISPERSION

When the frequency of operation increases beyond the quasi-TEM range, β begins to display a nonlinear dependence on ω (that is, ϵ_{eff} depends upon frequency). The effective dielectric constant $\epsilon_{\text{eff}} \equiv \beta^2/k_0^2$ can be shown to *increase* with frequency (see eqn. (J.70)). This phenomenon is called *dispersion*, and is in fact an inherent property of non-TEM modes.[7]

---

[7]This is evident from (J.68), wherein the inequality is strict whenever there are nonzero z-components of the field present. When these field components are zero or negligible (as at low frequencies), dispersion will be absent, as in the case of TEM or quasi-TEM modes. Dispersion also becomes negligible in the limit of very high frequency, since, as shown by (J.65), ϵ_{eff} cannot increase beyond the upper limit of ϵ_r, in which case the z-components of the field once again become small and the field tends to concentrate almost exclusively within the region of highest $\mu\epsilon$. We can say that a mode which is essentially nondispersive at a given frequency is behaving as a plane wave, which shares these properties.

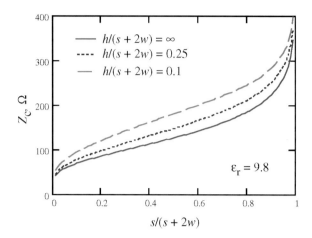

Figure 8.24 Characteristic impedance of coplanar strip line.

The reason for this (in the case of microstrip, for example) is that a larger fraction of the electric fields become concentrated in the substrate, hence making the wave "see" more of the denser medium, and behave more like a plane wave would in that medium. This is true quite generally for all waveguides, but the proof involves some subtlety and is left as an exercise (problem p8-10). The nonlinear dependence of β on frequency causes distortion of pulses transmitted along the line (see Chapter 4), and so will impose a limit on the rate at which information can be sent. It is thus important to be able to predict the variation of ϵ_{eff} with frequency.

8.8.1 DISPERSION IN MICROSTRIP

Actual computation or approximation of ϵ_{eff} as a function of ω can be carried out numerically. The process can be rather sophisticated, but considerable headway has been made since about 1970. Typical *dispersion characteristics* (ϵ_{eff} vs. frequency) for open microstrip are shown in Figure 8.25. It is clear that for a given substrate permittivity ϵ_r and line aspect ratio w/h, a thinner substrate allows for lower dispersion up to much higher frequencies.

There have also been efforts to provide empirical design formulas giving accurate approximations for the dispersive behavior of open microstrip. Most accurate of existing formulas seems to be:

$$\epsilon_{\text{eff}}(f) = \epsilon_r - \frac{\epsilon_r - \epsilon_{\text{eff}}(0)}{1 + (f/f_h)^m} \tag{8.164}$$

where $\epsilon_{\text{eff}}(0)$ is the zero-frequency limit of ϵ_{eff}, and

$$f_h = \frac{f_T}{0.75 + (0.75 - 0.332/\epsilon_r^{1.73})(w/h)} \tag{8.165}$$

$$f_T = \frac{\tan^{-1}\left[\epsilon_r \sqrt{\frac{\epsilon_{\text{eff}}(0)-1}{\epsilon_r - \epsilon_{\text{eff}}(0)}}\right]}{2\pi h \sqrt{\mu_0 \epsilon_0 (\epsilon_r - \epsilon_{\text{eff}}(0))}} \tag{8.166}$$

$$m = m_0 m_c \tag{8.167}$$

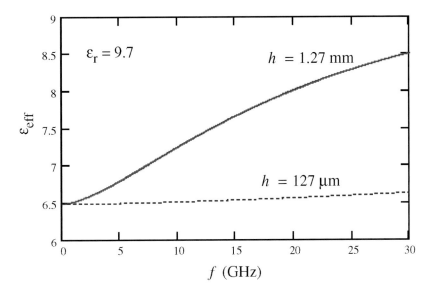

Figure 8.25 Effective dielectric constant as a function of frequency for open microstrip: $\mu = \mu_0$; $\epsilon_r = 9.7$; $w/h = 0.96$.

$$m_0 = 1 + \frac{1}{1 + \sqrt{w/h}} + 0.32(1 + \sqrt{w/h})^{-3} \qquad (8.168)$$

$$
\begin{aligned}
m_c &= 1 + \frac{1.4}{1 + w/h}[0.15 - 0.235\exp(-0.45f/f_h)] \qquad (w/h \leq 0.7) \\
&= 1 \qquad (w/h \geq 0.7)
\end{aligned}
\qquad (8.169)
$$

8.8.2 DISPERSION IN COPLANAR TRANSMISSION LINES

Dispersion likewise occurs in other quasi-TEM transmission lines. Details about these for several important cases can be found in the references cited at the end of the chapter. Here we present empirical formulas for the dispersion of coplanar transmission lines. The formula for effective permittivity of either CPW or CPS as a function of frequency is slightly different than (8.164):

$$\sqrt{\epsilon_{\text{eff}}(f)} = \sqrt{\epsilon_{\text{eff}}(0)} + \frac{\sqrt{\epsilon_r} - \sqrt{\epsilon_{\text{eff}}(0)}}{1 + (f/f_h)^m} \qquad (8.170)$$

in which the parameters are given by

$$m = -1.8 \qquad (8.171)$$

$$f_h = \frac{A^{0.56}}{4h\sqrt{\mu_0\epsilon_0(\epsilon_r - 1)}} \qquad (8.172)$$

$$A = \exp\left[u\ln\left(\frac{s}{w}\right) + v\right] \qquad (8.173)$$

$$u = 0.54 - 0.64 \ln \left(\frac{s}{h} \right) + 0.015 \left[\ln \left(\frac{s}{h} \right) \right]^2 \tag{8.174}$$

$$v = 0.43 - 0.86 \ln \left(\frac{s}{h} \right) + 0.54 \left[\ln \left(\frac{s}{h} \right) \right]^2 \tag{8.175}$$

8.9 PSEUDO-TEM MODES

Some waveguides which might appear at first glance to support a quasi-TEM mode turn out not to do so, because of the failure of the line parameters $\mathsf{l}(\omega, \gamma)$ and/or $\mathsf{c}(\omega, \gamma)$ to approach a finite, nonzero limit as $\omega \to 0$. The dominant modes of such waveguides often turn out to be sorts of hybrids between quasi-TEM modes and surface waves or other waveguide mode types. Like surface-wave and closed waveguide modes, these hybrids will be essentially dispersive at all frequencies. Nevertheless if certain important dimensions of the guide cross section are small compared to a wavelength, it will still make sense to use a formalism similar to that used for quasi-TEM modes for their description. We can call such modes *pseudo-TEM* modes, several examples of which are shown in Figs. 8.26 and 8.27. In Figure 8.28, we show some structures whose fundamental mode is quasi-TEM at low frequencies, but becomes pseudo-TEM when the frequency becomes large enough.

In Figure 8.26, the waveguides have a small central conductor on which a current and linear charge density can be defined fairly unambiguously, but where no unique voltage exists because the second conductor is electrically very far from the first or is nonexistent. The current on the central conductor produces a predominantly TM field, and we call this type of pseudo-TEM mode *quasi-TM*. On the other hand, in Figure 8.27, the waveguides have a small gap between different parts of the conducting wall across which a voltage and magnetic flux can be reasonably defined, but where no unique current or charge density exists because there is only one piece of the conducting wall overall, or the conductors have infinite extent. The field of this kind of mode is predominantly TE, and we call this type of pseudo-TEM mode *quasi-TE*. In the following we consider one example of each type in some detail.

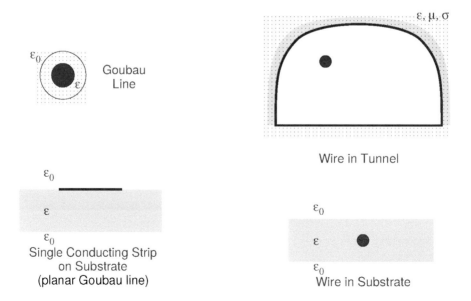

Figure 8.26 Waveguides supporting pseudo-TEM modes of quasi-TM type.

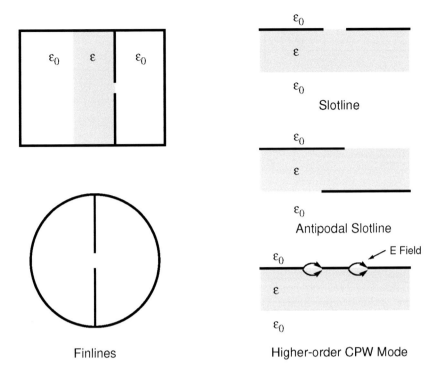

Figure 8.27 Waveguides supporting pseudo-TEM modes of quasi-TE type.

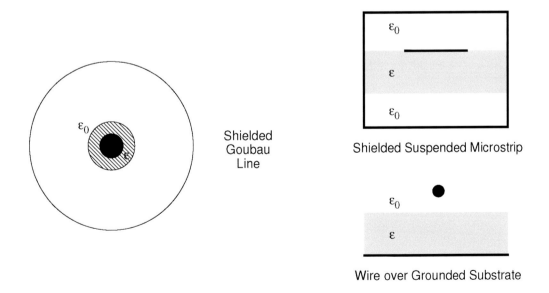

Figure 8.28 Waveguides supporting modes that are quasi-TEM at low enough frequencies and pseudo-TEM of quasi-TM type at sufficiently high frequencies.

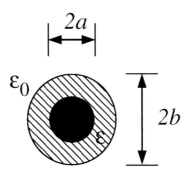

Figure 8.29   Goubau line.

8.9.1   THE GOUBAU LINE

The Goubau line is a perfectly conducting wire of radius a surrounded by a dielectric layer occupying $a < \rho < b$ (Figure 8.29). This waveguide supports surface waves similar to those of the dielectric fiber (cylinder) considered in Section 6.7. Unlike the pure dielectric cylinder, however, there is (for some modes at least) substantial current flow in the wire which can be connected to transmission lines as a form of excitation and reception. This line can be used in antenna structures and as a microwave transmission line that does not require a second (return) conductor.

Proceeding as in Section 6.7, we consider the TM_1 mode of this line. The azimuthally independent $(\partial/\partial\phi \equiv 0)$ electric field has the form

$$\mathcal{E}_z = E_1 F_0(\rho) \tag{8.176}$$

where

$$
\begin{aligned}
F_0(\rho) &= J_0(h_1\rho) - J_0(h_1 a)\frac{Y_0(h_1\rho)}{Y_0(h_1 a)} \qquad (a < \rho < b) \\
&= \left[J_0(h_1 b) - J_0(h_1 a)\frac{Y_0(h_1 b)}{Y_0(h_1 a)} \right]\frac{K_0(p_2\rho)}{K_0(p_2 b)} \qquad (\rho > b)
\end{aligned}
\tag{8.177}
$$

while $\mathcal{H}_\phi$ and $\mathcal{E}_\rho$ are obtained from (6.66) and (6.67) or (6.70) and (6.71):

$$
\begin{aligned}
\mathcal{E}_\rho &= -\frac{j\beta}{h_1^2}E_1 F_0'(\rho) \quad (a < \rho < b); \\
&= \frac{j\beta}{p_2^2}E_1 F_0'(\rho) \quad (\rho > b)
\end{aligned}
\tag{8.178}
$$

and

$$\mathcal{H}_\phi = \frac{\omega\epsilon}{\beta}\mathcal{E}_\rho \quad (a < \rho < b); \qquad \mathcal{H}_\phi = \frac{\omega\epsilon_0}{\beta}\mathcal{E}_\rho \quad (\rho > b) \tag{8.179}$$

Here

$$h_1 = \sqrt{k_0^2\epsilon_r - \beta^2}; \qquad p_2 = \sqrt{\beta^2 - k_0^2}$$

and $\epsilon_r = \epsilon/\epsilon_0$ is the relative permittivity of the wire coating. The exact equation for determining the propagation constant β can be found in the same way as for the circular dielectric fiber in Section 6.7, by enforcing continuity of tangential $\mathbf{H}$ at $\rho = a$. The result is

$$\frac{J_1(h_1 b)Y_0(h_1 a) - J_0(h_1 a)Y_1(h_1 b)}{J_0(h_1 b)Y_0(h_1 a) - J_0(h_1 a)Y_0(h_1 b)} = -\frac{h_1}{\epsilon_r p_2}\frac{K_1(p_2 b)}{K_0(p_2 b)} \tag{8.180}$$

Alternatively, we could use these fields to compute c_C and l_C from (8.111) and (8.112). If the result is approximated for an electrically thin wire and coating ($h_1 b \ll 1$ and $p_2 b \ll 1$), we arrive at

$$l(\omega, \beta) = \frac{\mu_0}{2\pi}\left(\ln\frac{2}{e^{\gamma_E}p_2 a}\right) \tag{8.181}$$

$$\frac{1}{c(\omega, \beta)} = \frac{1}{2\pi\epsilon_0}\left(\frac{1}{\epsilon_r}\ln\frac{b}{a} + \ln\frac{2}{e^{\gamma_E}p_2 b}\right) \tag{8.182}$$

where γ_E denotes Euler's constant (C.11). These expressions are then inserted into (8.119) to get a transcendental equation for determining β:

$$\frac{\beta^2}{k_0^2} = \frac{\ln\frac{2}{e^{\gamma_E}p_2 a}}{\frac{1}{\epsilon_r}\ln\frac{b}{a} + \ln\frac{2}{e^{\gamma_E}p_2 b}} \quad \Rightarrow \quad \left(\frac{p_2 b}{V}\right)^2 = \frac{B}{(\epsilon_r - 1)\ln\frac{2}{e^{\gamma_E}p_2 b} + B} \tag{8.183}$$

where

$$B = \left(1 - \frac{1}{\epsilon_r}\right)\ln\frac{b}{a} \quad \text{and} \quad V = k_0 b\sqrt{\epsilon_r - 1} \tag{8.184}$$

Equation (8.180) can also be shown to reduce to (8.183) in the limit of an electrically thin wire.

Numerical solutions of this equation are left to the reader as an exercise, but the solution of (8.183) can be expressed explicitly in terms of a special function known as Lambert's W-function (see Section C.4 of Appendix C). To do this, we put

$$A = -\frac{k_0^2 a^2 B}{2}e^{2(\gamma_E + B)}; \qquad \theta = 2\left(B - \ln\frac{2}{e^{\gamma_E}p_2 a}\right) \tag{8.185}$$

and thereby reduce (8.183) to the form

$$\theta e^\theta = A \tag{8.186}$$

From Section C.4 we then obtain

$$\theta = W_{-1}(A) \quad \Rightarrow \quad p_2 a = 2e^{-(B + \gamma_E) + \theta/2} \quad \Rightarrow \quad \epsilon_{\text{eff}} = 1 + \left(\frac{p_2 a}{k_0 a}\right)^2 \tag{8.187}$$

The function W_{-1} is accurately approximated using (C.75), and a sample plot of $\epsilon_{\text{eff}} = \beta^2/k_0^2$ vs. V from both (8.180) and (8.183) is shown in Figure 8.30. The approximation (8.183) is seen to be quite accurate for $V < 0.5$, but becomes progressively less accurate for larger V, and for V greater than about 1.1 it no longer gives a real solution for ϵ_{eff}.

8.9.2 THE SLOT LINE

The slotline consists of a grounded substrate of thickness h the ground plane of which has been interrupted by a slot of width w (Fig. 8.31). Note that there is no conductor on the bottom of the substrate. The slotline has the advantage that shunt elements can be easily connected across the line, a connection which is quite cumbersome and destructive on microstrip. On the other hand, series connections are troublesome because the current and charge densities on the conductors are spread out over a relatively large distance in the transverse direction. Slotline has been employed mainly in short lengths for intermediate

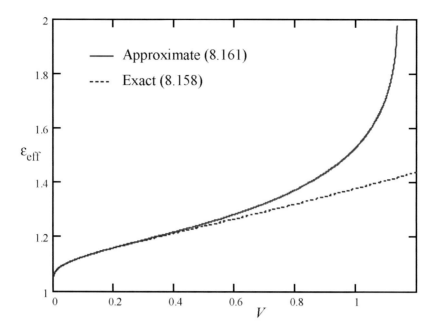

Figure 8.30   Effective permittivity of a Goubau line: $\epsilon_r = 2.2$, $b/a = 2.5$.

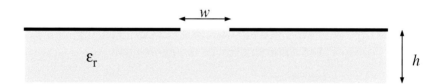

Figure 8.31   Slotline.

transitions between other lines where impedance transformations or baluns are needed, or as slot antennas.

If we attempt to carry out a DC calculation of l and c for this line, we find that $l = 0$ and $c = \infty$. This is basically because any voltage difference between the conductors produces an infinite charge on each of them, because they are both infinitely wide.[8] The mode is thus a pseudo-TEM mode of quasi-TE type. Choosing the path P for defining the voltage and magnetic flux of the slotline to be a straight line across the slot, and a contour C that "hugs" the top and bottom sides of one of the conductors, we compute the mixed line parameters $l_M(\omega, \beta)$ and $c_M(\omega, \beta)$ as described in Section 8.4 for any mode of an arbitrary waveguide.

Unless w/h is impractically large (≥ 10), the value of $c_M(\omega, \beta)$ can be expressed (accurate to a few percent) in terms of the capacitance $C_{micr}(\epsilon_r)$ from (8.150) which describes the microstrip structure having the same values of w and h:

$$c_M(\omega, \beta) = C_{\text{slot}}(\epsilon_r, \omega, \beta) = \frac{2\epsilon_0}{\pi} \ln\left(\frac{8}{e^{\gamma_E} w \sqrt{\beta^2 - k_0^2}}\right) + \frac{\epsilon_0^2(\epsilon_r^2 - 1)}{C_{micr}(\epsilon_r)} \tag{8.188}$$

---

[8] If the conductors are very wide but finite, l is very small and c is very large, and because of the large dimensions dispersion will begin to occur at much lower frequencies.

where γ_E refers to Euler's constant ($e^{\gamma_E} = 1.78107\ldots$), and $k_0^2 = \omega^2 \mu_0 \epsilon_0$. The line inductance $l_M(\omega, \beta)$ is related to the line capacitance for $\epsilon_r = 1$, just as is the case for quasi-TEM modes (eqn. (8.144)):

$$l_M(\omega, \beta) = \frac{\mu_0 \epsilon_0}{C_{\text{slot}}(1, \omega, \beta)} \tag{8.189}$$

Notice now that equation (8.110) for this line,

$$\beta^2 = \omega^2 l_M(\omega, \beta) c_M(\omega, \beta) = k_0^2 \frac{C_{\text{slot}}(\epsilon_r, \omega, \beta)}{C_{\text{slot}}(1, \omega, \beta)}, \tag{8.190}$$

is an implicit definition of β; we must first solve for β and afterwards define ϵ_{eff} for the slotline by

$$\epsilon_{\text{eff}} = \beta^2 / k_0^2 \tag{8.191}$$

The solution of (8.190) can be carried out by graphical or numerical means. In particular, we may do so by the iteration described in Appendix K, since the right side of (8.190) is a relatively slowly varying function of the unknown β (it appears inside a logarithm). We choose an initial guess at β, call it $\beta^{(0)}$, say, and substitute into the right side of (8.190). the resulting value is the square of a new estimate, $\beta^{(1)}$ for the unknown, which can be substituted repeatedly in this way into the right side of (8.190). This sequence of approximations converges rapidly to the value of β.

Once we know β, the voltage-current version (8.128) of the characteristic impedance can be computed from:

$$Z_{c,VI}(\omega) = \sqrt{\frac{l_M(\omega, \beta)}{c_M(\omega, \beta)}} = \sqrt{\mu_0 \epsilon_0} \frac{\sqrt{\epsilon_{\text{eff}}}}{C_{\text{slot}}(\epsilon_r, \omega, \beta)} \tag{8.192}$$

Since the slotline possesses a fairly well-defined voltage, but an ill-defined current (especially at low frequencies where the surface wave character of the mode becomes pronounced and the current density spreads out extensively over the conducting planes), the power-voltage definition (8.133) of characteristic impedance is also frequently used, and can differ considerably from $Z_{c,VI}$.

Typical behaviors of ϵ_{eff} and $Z_{c,VI}$ vs. ω for a slotline are shown in Figure 8.32. We see that the effective permittivity is rather sensitive to the ratio of slot width to substrate

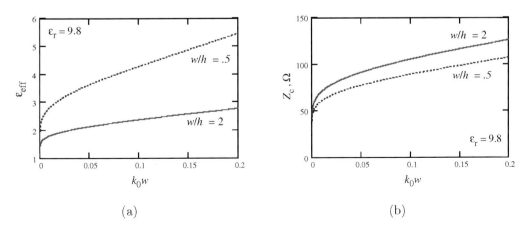

Figure 8.32 (a) Effective dielectric constant for a slotline with $\epsilon_r = 9.8$, $w/h = 0.5$ and 2 vs. normalized frequency $k_0 w$; (b) $Z_{c,VI}$ for this slotline.

thickness, while the characteristic impedance is much less dependent on this ratio, and is difficult to reduce below about 50 Ω in practical situations.

8.10 QUALITATIVE ANALYSIS OF OPEN PLANAR WAVEGUIDES

As we have done in Section 6.9.2, as well as in Section 7.4 for quasi-optical microstrip and in Section 7.5 for dielectric rib waveguide, a qualitative analysis of the permitted range of β for bound (surface) modes of open planar waveguides of general type can be carried out. Thus, as before, eqn. (6.119) is valid for the proper modes of any open lossless waveguide. On the other hand, in an open planar guide, we obtain lower bounds to β by identifying the species of wave which can propagate out to infinity and demand that β be greater than the wavenumbers characterizing each of these species. Only this guarantees exponentially decaying fields as we approach infinity in any transverse direction, and therefore that the mode carries a finite amount of energy.

This statement would seem to prohibit the existence of a TEM mode on an open waveguide (one whose cross section extends to infinity). Indeed, if we consider the coaxial transmission line of Section 8.3.1 and let $b \to \infty$ so that we are left with a single wire in infinite space, we would have the fields

$$\mathcal{E}_\rho = \frac{Q_{l1}}{2\pi\epsilon\rho}; \qquad \mathcal{H}_\phi = \frac{I^{(1)}}{2\pi\rho} \qquad \text{(for } \rho > a\text{)}$$

and the total (oscillatory) power carried by the mode would be

$$\hat{P}_{\mathrm{osc}} = \frac{Q_{l1} I^{(1)}}{4\pi} \int_a^\infty \frac{d\rho}{\rho} = \infty$$

Thus, in the absence of the outer shield, the TEM mode of this degenerate coaxial line does not exist. On the other hand, if both conductors are of finite extent and carry charge per unit length and current that are equal but opposite on the two conductors, the field can carry finite power without exponential decay as the transverse coordinates approach infinity. In this case, the fields resemble those of the two-wire transmission line of Section 8.3.2. Putting $x = \rho \cos \phi$ and $y = \rho \sin \phi$ in (8.77) and letting $\rho \gg (a, b)$, we find

$$\mathcal{E} \simeq \frac{\sqrt{2Z_c}}{2 \ln \left(\frac{b+d}{2a}\right)} \frac{1}{\rho^2} \left(\mathbf{u}_x \cos 2\phi + \mathbf{u}_y \sin 2\phi\right) \tag{8.193}$$

while $\mathcal{H} = (\mathbf{u}_z \times \mathcal{E})/\zeta$. The Poynting vector $\mathcal{E} \times \mathcal{H}$ thus behaves like const/ρ^4 as $\rho \to \infty$, leaving an integral of the form

$$\int^\infty \frac{d\rho}{\rho^3}$$

which is finite. Thus, even without exponential field decay in the transverse direction, it is still possible for a TEM (or quasi-TEM, for that matter) mode to exist on an open waveguide. The strict inequality $\beta > k$ is not necessary in such cases, but the finite energy condition must be checked whenever exponential field decay is not guaranteed.

Consider the examples of slotline, both with and without a ground plane as shown in Figure 8.33. With no ground plane, Figure 8.33(a), both a space wave (wavenumber k_0) and one or more surface waves of the grounded slab (maximum wavenumber β_{s,TM_0}) can propagate to infinity. From this and from (6.119), we find

$$\beta_{s,\mathrm{TM}_0} < \beta \leq k_0 \sqrt{\epsilon_r},$$

Figure 8.33   (a) Open slotline; (b) slotline with ground plane.

which was also true of microstrip as shown in Section 7.4. However, with a ground plane present (Figure 8.33(b)), there can now exist a TEM parallel-plate wave (wavenumber $k_0\sqrt{\epsilon_r}$) going out to infinity on this cross-section, and so we would have to have

$$k_0\sqrt{\epsilon_r} < \beta \le k_0\sqrt{\epsilon_r},$$

which is impossible with the first inequality being a strict one. Hence, no bound modes can exist on slotline which has a ground plane unless we can somehow arrange for the TEM parallel-plate wave constituent to be absent from the total field.

Now, we ought to note that just because a particular wave species traveling to infinity *might* exist in a waveguide cross-section does not necessarily mean that it will form a constituent of *every* mode which that waveguide might support. Consider, for instance, the open symmetric finline shown in Figure 8.34. Due to the symmetry of this structure, it seems possible that a mode might exist having the electric field lines shown in either Figure 8.34(a) or(b). However, the fields of Figure 8.34(b) are evidently settling down to a TEM wave as we move into the parallel plate regions away from the gaps and so, according to our earlier discussions, could not be a proper bound mode. On the other hand, the fields shown in Figure 8.34(a) have an opposite symmetry to those of Figure 8.34(b) and can contain no TEM wave at all. Therefore, for those modes of the finline which by symmetry contain no constituent TEM waves, there may be bound modes which can exist such that

$$\sqrt{k_0^2\epsilon_r - \pi^2/d^2} = \beta_1 < \beta < k_0\sqrt{\epsilon_r}$$

where β_1 is the propagation coefficient of TE_1 and TM_1 parallel-plate waves.

Unfortunately, these kinds of modes can be rather unstable since small errors in dimensional tolerances can cause deviations from symmetry in an actual guide. This means in general that *some* TEM wave will be present in practice, causing what would have been a purely bound mode to be "leaky." This will make it necessary even in such cases of apparent symmetry to use a shield or "grounding" connections to assure the existence of the desired modes without leakage. In Figure 8.35(a), a shield is added to a slotline with ground plane to suppress the TEM wave. The plates of the slotline are maintained at the same potential as the shield, making a TEM wave impossible: no wave species can propagate to infinity in the transverse direction, and the only lower bound on β based on the foregoing considerations is zero. In a case like this, however, any modes which do exist are not particularly well

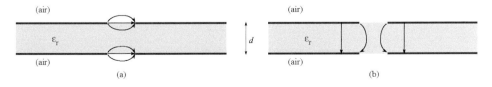

Figure 8.34   Open symmetric finline and its E-field lines: (a) proper mode; (b) improper mode.

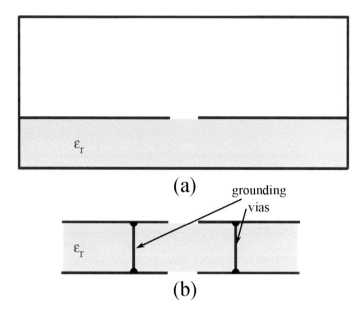

Figure 8.35 Schemes to eliminate leaky modes in slotlines: (a) shielded slotline; (b) use of vias to suppress TEM waves in bilateral slotline.

localized to the neighborhood of the slot, but have field distributions which tend to extend throughout the entire cross-section within the shield. Moreover, the fundamental mode no longer has zero cutoff frequency. This loss of field localization could be a disadvantage in some applications. Figure 8.35(b) shows an alternative method using vias (wires connecting the upper and lower plates through the substrate at periodic intervals along the length of the line).

Perhaps the most commonly encountered example of this phenomenon is that of the conductor-backed coplanar waveguide (CBCPW) shown in Figure 8.36(a). Without some means of suppressing the TEM wave, significant lateral leakage of energy is likely to occur. "Channelizing" the structure as shown in Figure 8.36(b) not only suppresses the TEM wave, but also ties together the outer ground planes, suppressing the higher-order pseudo-TEM CPW mode as well and eliminating the need for an air bridge to accomplish this task.

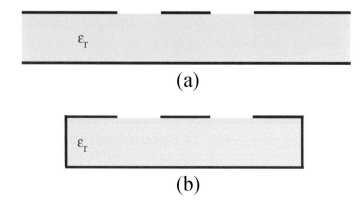

Figure 8.36 (a) Conductor-backed coplanar waveguide; (b) channelized coplanar waveguide.

8.11   NOTES AND REFERENCES

The standard development of the theory of TEM modes can be found in

Johnk (1975), Chapter 9;
Ramo, Whinnery and van Duzer (1984), Chapter 8;
Johnson (1965), Chapter 4;
Collin (1991), Chapter 4.

Limiting expressions for stripline and microstrip capacitance as $w/h \to 0$ or $w/h \to \infty$ are to be found in:

E. A. Shchapoval, "Capacitance, inductance and effective relative permittivity of microstrip line," *Electron. Lett.*, vol. 11, pp. 225-226 (1975).

D. C. Chang and E. F. Kuester, "An analytic theory for narrow open microstrip," *Arch. Elek. Übertragungstech.*, vol. 33, pp. 199-206 (1979).

E. F. Kuester and D. C. Chang, "Closed-form expressions for the current or charge distribution on parallel strips or microstrip," *IEEE Trans. Micr. Theory Tech.*, vol. 28, pp. 254-259 (1980).

W. C. Chew and J. A. Kong, "Asymptotic formula for the capacitance of two oppositely charged discs," *Math. Proc. Camb. Phil. Soc.*, vol. 89, pp. 373-384 (1981).

S. Y. Poh, W. C. Chew and J. A. Kong, "Approximate formulas for the line capacitance and characteristic impedance of microstrip line," *IEEE Trans. Micr. Theory Tech.*, vol. 29, pp. 135-142 (1981).

The unified approximations (8.84), (8.150) and (8.151) were first obtained by H. A. Wheeler:

H. A. Wheeler, "Transmission-line properties of a strip on a dielectric sheet on a plane," *IEEE Trans. Micr. Theory Tech.*, vol. 25, pp. 631-647 (1977).

This paper includes formulas accounting for the nonzero thickness of the strip as well. Expression (7.41) for r_ϵ is an improvement on that in Wheeler's formula. See:

E. F. Kuester, "Accurate approximations for a function appearing in the analysis of microstrip," *IEEE Trans. Micr. Theory Tech.*, vol. 32, pp. 131-133 (1984).

Equivalent radii (sometimes also called *external conformal radii* or *outer radii*) for conductors of various shapes are given in:

Pólya and Szegö (1951), pp. 251-254, 273;
Yu. Ya. Iossel' and L. S. Perel'man, "Calculation of the capacitance of wires," *Elektrichestvo*, no. 9, pp. 38-43 (1986).

Approximate equivalent radii for some other conductor shapes are given in:

C. W. H. Su and J. P. German, "The equivalent radius of noncircular antennas," *Microwave J.*, vol. 9, no. 4, pp. 64-67 (1966).

Some methods for obtaining stripline capacitance are given in

Collin (1991), Sections 4.5 and 4.6;

while formulas and tables for line parameters of many types of stripline are given in:

Saad (1971), vol. 1, pp. 115-151;
Gunston (1972);
Hilberg (1979).

The approach of Section 8.4 to the definition of transmission-line parameters for non-TEM modes is a modification of that given in:

J. R. Brews, "Transmission line models for lossy waveguide interconnections in VLSI," *IEEE Trans. Electron Devices*, vol. 33, pp., 1356-1365 (1986).

R. B. Marks and D. F. Williams, "A general waveguide circuit theory," *J. Res. Nat. Inst. Standards Technol.*, vol. 97, pp. 533-562 (1992).

This method is distinguished by the use of energy and power-loss integrals over the entire cross-section of the waveguide to define the line parameters. Other authors have used line integrals analogous to those used for TEM modes in Section 8.2; see

S. A. Schelkunoff, "Impedance concept in wave guides," *Quart. Appl. Math.*, vol. 2, pp. 1-15 (1944).

R. A. Waldron, "Characteristic impedances of waveguides," *Marconi Rev.*, vol. 30, pp. 125-136 (1967).

Both methods yield the same results in the case of a quasi-TEM or pure TEM mode, but in other cases give different values for the line voltage and current, and thus for the characteristic impedance. Quasi-TEM modes are discussed from a number of viewpoints in:

C. Snow, "Alternating current distribution in cylindrical conductors," *Sci. Papers Bureau of Standards*, no. 509, pp. 277-338 (1925) [also in *Proceedings of the Seventh International Congress of Mathematicians*, vol. II. Toronto: University of Toronto Press, 1928, pp. 157-218].

G. K. Grunberger, V. Keine and H. H. Meinke, "Longitudinal field components and frequency-dependent phase velocity in the microstrip transmission line," *Electron. Lett.*, vol. 6, pp. 683-685 (1970).

A. F. dos Santos and J. P. Figanier, "The method of series expansion in the frequency domain applied to multidielectric transmission lines," *IEEE Trans. Micr. Theory Tech.*, vol. 23, pp. 753-756 (1975).

M. Aubourg, "Onde quasi-TEM des guides cylindriques blindés," *Ann. Telecomm.*, vol. 34, pp. 45-51 (1979).

A. Magos, "Calculation of quasi-TEM waves by series expansion in powers of frequency," *Periodica Polytechnica (Elec. Eng.)*, vol. 23, pp. 7-25 (1979).

Ye. N. Korshunova, A. N. Sivov and A. D. Shatrov, "Quasistatic theory of metal-magnetodielectric waveguide structures," *Sov. J. Commun. Technol. Electron.*, vol. 37, no. 7, pp. 75-83 (1992).

In the case of non-TEM modes, particularly for those of closed metallic waveguides, a variety of approaches have been taken for the definition of characteristic impedance. See:

E. D. Farmer, "Junction admittance between waveguides of arbitrary cross-sections," *Proc. IEE (London) pt. C*, vol. 103, pp. 145-152 (1956).

W. K. McRitchie, and M. M. Z. Kharadly, "Properties of interface between homogeneous and inhomogeneous waveguides," *Proc. IEE (London)*, vol. 121, pp. 1367-1374 (1974).

E. F. Kuester, D. C. Chang and L. Lewin, "Frequency-dependent definitions of microstrip characteristic impedance," *URSI Symp. EM Theory*, Munich, 1980, pp. B335/1-B335/3.

R. H. Jansen and N. H. L. Koster, "New aspects concerning the definition of microstrip characteristic impedance as a function of frequency," *IEEE MTT Symposium*, 1982, pp. 305-307.

R. H. Jansen and M. Kirschning, "Arguments and an accurate model for the power-current formulation of microstrip characteristic impedance," *Arch. Elek. Übertrag.*, vol. 37, pp. 108-112 (1983).

L. Zhu and K. Wu, "Revisiting characteristic impedance and its definition of microstrip line with a self-calibrated 3-D MoM scheme," *IEEE Micr. Guided-Wave Lett.*, vol. 8, pp. 87-89 (1998).

F. Mesa and D. R. Jackson, "A novel approach for calculating the characteristic impedance of printed-circuit lines," *IEEE Micr. Wireless Comp. Lett.*, vol. 15, pp. 283-285 (2005).

R. Rodríguez-Berral, F. Mesa and D. R. Jackson, "A high-frequency circuit model for the gap excitation of a microstrip line," *IEEE Trans. Micr. Theory Tech.*, vol. 54, pp. 4100-4110 (2006).

Often the definitions are made so that the proper reflection coefficient at the junction between two waveguides can be calculated solely in terms of the characteristic impedances. Other times, the choice is based on the type of source excitation applied to the waveguide. But in neither case does this take into account the effect of parasitics due to higher-order cutoff modes near the junction (see Chapter 11). In fact, different values of characteristic impedance can be accommodated by varying the equivalent circuit of these parasitics, so in the end the characteristic impedance cannot be a uniquely defined quantity except in certain limiting cases, and choosing one definition over another is essentially a matter of convenience (e.g., to simplify the equivalent circuit of the junction).

More information on microstrip and other types of planar quasi-TEM lines is given in:

Garg, Bahl and Bozzi (2103);

Gupta and Singh (1974), Chapter 3;

R. Mittra and T. Itoh, "Analysis of microstrip transmission lines," in *Advances in Microwaves*, vol. 8 (L. Young, ed.). New York: Academic Press, 1974, pp. 67-141;

Nefedov and Fialkovskii (1980);

Edwards (1981);

Hoffmann (1987);

Rozzi and Mongiardo (1997).

These references contain discussion and further references on dispersion, losses, effects of nonzero strip thickness, and so on. In particular, we mention that although the strip conductors in planar transmission lines are usually very thin, they do have nonzero thickness. This can result in capacitance and inductance per unit length that are noticeably different from the values given by formulas applicable to the zero thickness case. Further, the implications for the computation of attenuation due to conductor loss (as done in Section 9.5) can be quite substantial. If the strips are located in a homogeneous medium, correction formulas for capacitance per unit length when the conductor thickness is small compared to all other dimensions in the structure have been given in:

S. B. Cohn, "Thickness corrections for capacitive obstacles and strip conductors," *IRE Trans. Micr. Theory Tech.*, vol. 8, pp. 638-644 (1960).

Kh. L. Garb, I. B. Levinson and P. Sh. Fridberg, "Effect of wall thickness in slot problems of electrodynamics," *Radio Eng. Electron. Phys.*, vol. 13, pp. 1888-1896 (1968).

When a material substrate is present, no universal correction formulas exist. For microstrip, such formulas have been given by Wheeler, and enhanced by later researchers:

H. A. Wheeler, "Transmission-line properties of a strip on a dielectric sheet on a plane," *IEEE Trans. Micr. Theory Tech.*, vol. 25, pp. 631-647 (1977).

I. J. Bahl and R. Garg, "Simple and accurate formulas for a microstrip with finite strip thickness," *Proc. IEEE*, vol. 65, pp. 1611-1612 (1977).

The coplanar waveguide was introduced by

C. P. Wen, "Coplanar waveguide: A surface strip transmission line suitable for nonreciprocal gyromagnetic device application," *IEEE Trans. Micr. Theory Tech.*, vol. 17, pp. 1087-1090 (1969).

Formulas (8.156)-(8.159) for CPW are a special case of a result due to

E. S. Kochanov, "Capacitance of a planar strip line allowing for the dielectric substrate thickness," *Telecommun. Radio Eng.*, vol. 29/30, no. 1, pp. 127-128 (1975).

Formulas (8.160)-(8.163) for CPS were given by

G. Ghione and C. Naldi, "Analytical formulas for coplanar lines in hybrid and monolithic MICs," *Electron. Lett.*, vol. 20, pp. 179-181 (1984).

The method used to obtain the CPW and CPS parameters is a quite general one, capable of treating a variety of two-dielectric lines; see:

E. S. Kochanov, "Computation of electric capacitance of wires in two-layer structures," *Elektrichestvo*, no. 7, pp. 81-84 (1977).

Kochanov also gave a CPS formula by this method. Although it is less accurate than that of Ghione and Naldi, it contains the case of unequal strip widths:

E. S. Kochanov, "Parasitic capacitances in printed wiring of radio equipment," *Telecommun. Radio Eng.*, vol. 21/22, no. 7, pp. 129-132 (1967).

Synthesis formulas analogous to (8.152) for the design of CPW's with desired values of characteristic impedance (but not desired values of ϵ_{eff}) are given in:

T. Q. Deng, M. S. Leong and P. S. Kooi, "Accurate and simple closed-form formulas for coplanar waveguide synthesis," *Electron. Lett.*, vol. 31, pp. 2017-2019 (1995).

and for CPS in:

T. Q. Deng, M. S. Leong, P. S. Kooi and T. S. Yeo, "Synthesis formulas for coplanar lines in hybrid and monolithic MICs," *Electron. Lett.*, vol. 32, pp. 2253-2254 (1996).

Conductor-backed coplanar waveguide (CBCPW) is a coplanar waveguide with a ground plane on the opposite side of the substrate from the strip and gaps. Approximate formulas for its line parameters are given in

G. Ghione and C. Naldi, "Parameters of coplanar waveguides with lower ground plane," *Electron. Lett.*, vol. 19, pp. 734-735 (1983).

G. Ghione and C. Naldi, "Coplanar waveguides for MMIC applications: Effect of upper shielding, conductor backing, finite-extent ground planes, and line-to-line coupling," *IEEE Trans. Micr. Theory Tech.*, vol. 35, pp. 260-267 (1987).

Note that this structure requires mitigation of potential leakage mechanisms by methods like those shown in Figure 8.35. Many other design formulas are given in

Simons (2001), Chapter 2;
Wolff (2006), Chapter 2.

Calculation of the dispersion of quasi-TEM lines generally requires quite an elaborate mathematical analysis. See, for example,

G. Kowalski and R. Pregla, "Dispersion characteristics of single and coupled microstrips," *Arch. Elek. Übertragungstech*, vol. 26, pp. 276-280 (1972).

from which Figure 8.25 is adapted. See also Chang and Kuester (1979) above and

E. F. Kuester and D. C. Chang, "Theory of dispersion in microstrip of arbitrary width," *IEEE Trans. Micr. Theory Tech.*, vol. 28, pp. 259-265 (1980).

For engineering applications, it is often best to make use of empirical closed-form results such as (8.164), which is due to

M. Kobayashi, "A dispersion formula satisfying recent requirements in microstrip CAD," *IEEE Trans. Micr. Theory Tech.*, vol. 36, pp. 1246-1250 (1988).

Formulas for the frequency dependence of the characteristic impedance of microstrip can be found in

Hoffmann (1987), pp. 175-179.

while (8.170) for CPW and CPS was given in:

G. Hasnain, A. Dienes and J. R. Whinnery, "Dispersion of picosecond pulses in coplanar transmission lines," *IEEE Trans. Micr. Theory Tech.*, vol. 34, pp. 738-741 (1986).

Other dispersion formulas for CPW and CPS can be found in:

M. Y. Frankel, S. Gupta, J. A. Valdmanis and G. A. Mourou, "Terahertz attenuation and dispersion characteristics of coplanar transmission lines," *IEEE Trans. Micr. Theory Tech.*, vol. 39, pp. 910-916 (1991).

S. Gevorgian, T. Martinsson, A. Deleniv, E. Kollberg and I. Vendik, "Simple and accurate dispersion expression for the effective dielectric constant of coplanar waveguides," *IEE Proc. pt. H*, vol. 144, pp. 145-148 (1997).

A. K. Rastogi and S. Mishra, "Coplanar waveguide characterization with thick metal coating," *Int. J. Inf. Millimeter Waves*, vol. 20, pp. 505-520 (1999).

Pseudo-TEM modes on wires located above and within a finitely-conducting earth have been studied in:

Sunde (1968), Chapters 5 and 8.

The pseudo-TEM mode guided by a wire in a tunnel is discussed in:

E. F. Kuester and D. B. Seidel, "Low-frequency behavior of the propagation constant along a thin wire in an arbitrarily shaped mine tunnel," *IEEE Trans. Micr. Theory Tech.*, vol. 27, pp. 736-741 (1979).

The Goubau line is treated in many texts; see for example

Collin (1991), Section 11.6.

A solution for the low-frequency behavior of the propagation constant of the Goubau line in terms of the Lambert W-function was carried out in

D. C. Jenn, "Applications of the Lambert W function in electromagnetics," *IEEE Ant. Prop. Mag.*, vol. 44, no. 3, pp. 139-142 (2002).
D. Jaisson, "Simple formula for the wave number of the Goubau line," *Electromagnetics*, vol. 34, pp. 85-91 (2014).

A variety of open transmission lines based on metalized dielectric cylinders is discussed in:

Shestopalov (1997).

The slotline was introduced by S. Cohn in 1969, and studied intensively in the decade that followed:

S. B. Cohn, "Slot line on a dielectric substrate," *IEEE Trans. Micr. Theory Tech.*, vol. 17, pp. 768-778 (1969).
E. A. Mariani, C. P. Heinzman, J. P. Agrios and S. B. Cohn, "Slot line characteristics," *IEEE Trans. Micr. Theory Tech.*, vol. 17, pp. 1091-1096 (1969).
J. B. Knorr and K.-D. Kuchler, "Analysis of coupled slots and coplanar strips on dielectric substrate," *IEEE Trans. Micr. Theory Tech.*, vol. 23, pp. 541-548 (1975).
R. Garg and K. C. Gupta, "Expressions for wavelength and impedance of a slotline," *IEEE Trans. Micr. Theory Tech.*, vol. 24, p. 532 (1976).
C. M. Krowne, "Approximations to hybrid mode slot line behaviour," *Electron. Lett.*, vol. 14, pp. 258-259 (1978).
R. Janaswamy and D. H. Schaubert, "Dispersion characteristics for wide slotlines on low-permittivity substrates," *IEEE Trans. Micr. Theory Tech.*, vol. 33, pp. 723-726 (1985).
R. Janaswamy and D. H. Schaubert, "Characteristic impedance of a wide slotline on low-permittivity substrates," *IEEE Trans. Micr. Theory Tech.*, vol. 34, pp. 900-902 (1986).

Summaries of this and other work on slotlines are found in

Hoffmann (1987), Chapter 14;
Garg *et al.*, Chapter 5.

The results (8.188) and (8.190) for slotline are from unpublished work of E. F. Kuester and D. C. Chang, who improved the results of:

L. A. Vainshtein, N. I. Lesik and V. Kondrat'ev, "Quasi-static theory of the dominant mode in a slot line," *Radio Eng. Electron. Phys.*, vol. 22, no. 9, pp. 36-43 (1977).

Finally, finlines are comprehensively treated in

Bhat and Koul (1987).

Conductor-backed coplanar waveguide is treated in

Simons (2001), Chapter 3.

Discussions of the general question of whether a structure supports bound modes in certain ranges of β, can be found in

A.-M. A. El-Sherbiny, "Exact analysis of shielded microstrip lines and bilateral fin lines," *IEEE Trans. Micr. Theory Tech.*, vol. 29, pp. 669-675 (1981).

R. W. Jackson, "Considerations in the use of coplanar waveguide for millimeter-wave integrated circuits," *IEEE Trans. Micr. Theory Tech.,*, vol. 34, pp. 1450-1456 (1986).

H. Shigesawa, M. Tsuji and A. A. Oliner, "Conductor-backed slotline and coplanar waveguide: Dangers and full-wave analysis," *1988 IEEE MTT Symposium Digest*, vol. 1, New York, NY, pp. 199–202 (May 1988).

H. Shigesawa, M. Tsuji and A. A. Oliner, "Dominant mode power leakage from printed-circuit waveguides," *Radio Science*, vol. 26, pp. 559-564 (1991).

These papers also treat the related question of so-called "leaky" modes. Leaky modes display attenuation along the direction of propagation due to radiation loss into one or more of the wave species that can propagate in the transverse direction if β is less than their wavenumber. Although not proper bound modes, (they are in fact superpositions of continuous spectrum waves concentrated near a particular value of β), leaky wave modes can behave for many purposes as if they were.

8.12   PROBLEMS

p8-1 Show that the current in conductor C_g (the "ground") of a TEM transmission line is equal and opposite to the current flowing on C_1; i.e., prove (8.27).

p8-2 The complex power carried by a mode on a waveguide is

$$\hat{P} = \frac{1}{2} \int_{S_0} \boldsymbol{\mathcal{E}} \times \boldsymbol{\mathcal{H}}^* \cdot \mathbf{u}_z \, dS$$

where S_0 is the cross-section of the guide. When the mode is a TEM mode, show that

$$\hat{P} = \frac{1}{2} \Phi^{(1)} I^{(1)*}$$

which is identical with the corresponding result (1.19) for the distributed constant model of a classical transmission line. Here $I^{(1)}$ and $\Phi^{(1)}$ are defined by (8.26) and (8.21) respectively. Likewise show that the complex oscillatory power

$$\hat{P}_{\text{osc}} = \frac{1}{2} \int_{S_0} \boldsymbol{\mathcal{E}} \times \boldsymbol{\mathcal{H}} \cdot \mathbf{u}_z \, dS$$

is equal to $\Phi^{(1)} I^{(1)}/2$.

p8-3 Any magnetic field can be expressed as $\mathbf{B} = \nabla \times \mathbf{A}$, where $\mathbf{A}$ is known as the vector potential of the field; the purely transverse magnetic mode field $\boldsymbol{\mathcal{B}}_T(x,y)$ of a TEM mode that does not depend on z can be expressed as

$$\boldsymbol{\mathcal{B}}_T = \nabla_T \times [\mathbf{u}_z \mathcal{A}_z(x,y)] = -\mathbf{u}_z \times \nabla_T \mathcal{A}_z$$

where $\mathcal{A}_z(x,y)$ is the z-component of a vector potential.

(a) Since $\boldsymbol{\mathcal{B}}_T \cdot \mathbf{u}_n = 0$ at the surface of a perfect conductor (PEC), show that we must have $\mathcal{A}_z = $ constant on a PEC wall. In fact, if we choose that constant to be 0 on the ground conductor C_g, show that $\mathcal{A}_z = \Psi_l^{(1)}$ on conductor C_1, where $\Psi_l^{(1)}$ is the magnetic flux per unit length defined by (8.25).

(b) Using $\nabla_T \times \boldsymbol{\mathcal{H}}_T = 0$ together with any appropriate vector identities, show that $\mathcal{A}_z$ obeys the following differential equation in the cross section of the waveguide:

$$\nabla_T \cdot \left(\frac{1}{\mu} \nabla_T \mathcal{A}_z \right) = 0$$

(c) Using the results of parts (a) and (b), prove equation (8.43) using a method similar to that used to prove (8.42).

p8-4 For a multiconductor TEM line with $M = 2$, show that

$$l_{11} = \frac{\mu \epsilon \mathsf{c}_{22}}{\mathsf{c}_{11}\mathsf{c}_{22} - \mathsf{c}_{12}^2}; \qquad l_{12} = \frac{\mu \epsilon \mathsf{c}_{12}}{\mathsf{c}_{11}\mathsf{c}_{22} - \mathsf{c}_{12}^2}; \qquad l_{22} = \frac{\mu \epsilon \mathsf{c}_{11}}{\mathsf{c}_{11}\mathsf{c}_{22} - \mathsf{c}_{12}^2}$$

p8-5 Design a coaxial cable that will have a characteristic impedance of $Z_c = 72\,\Omega$, using a dielectric of polyethylene ($\epsilon_r = 2.26$) and an inner conductor of AWG 20 gauge solid copper wire (diameter $= 0.8118$ mm).

p8-6 Consider an open stripline as in Figure 8.9 with a strip width $w = 1$ mm. The dielectric medium is nonmagnetic, with $\epsilon = 6\epsilon_0$. What values of the strip height h above the ground plane would be necessary to achieve the following characteristic impedances: $Z_c = 10\ \Omega$, $Z_c = 30\ \Omega$, $Z_c = 50\ \Omega$, $Z_c = 100\ \Omega$, $Z_c = 300\ \Omega$ and $Z_c = 500\ \Omega$? Comment on the practical feasibility of each of these transmission lines.

p8-7 A *triplate* stripline is a homogeneously filled stripline that has two ground planes symmetrically located above and below the strip conductor as shown below.

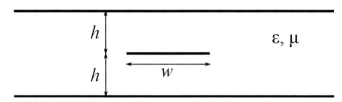

If the strip conductor has zero thickness, the capacitance per unit length can be shown to be given by the expression

$$\mathsf{c} = 4\epsilon \frac{K(k')}{K(k)}$$

where ϵ is the permittivity of the medium between the conductors, K is the complete elliptic integral of the first kind, the modulus is

$$k = \operatorname{sech}\frac{\pi w}{4h}$$

and the complementary modulus is $k' = \sqrt{1 - k^2}$. Assuming $\epsilon = 2.2\epsilon_0$, compute and plot the characteristic impedance Z_c for the TEM mode of this structure as a function of w/h over the range $0.1 < w/h < 10$. If $w \gg h$, what is the "parallel-plate" approximation (neglecting the fringing fields) for Z_c? How much error does this approximation have when $w/h = 10$? Derive a second approximation valid when $w \ll h$ and assess its accuracy.

p8-8 Suppose that for a TE mode of a lossless hollow metallic waveguide with a simply connected wall, we define a voltage Φ_P as

$$\Phi_P = - \int_P \boldsymbol{\mathcal{E}}_T \cdot d\boldsymbol{l}$$

where P is a path from a point 1 to a second point 2 on the waveguide wall as shown. Assume that the path P is such that $\Phi_P \neq 0$, and that the mode is above cutoff ($k > k_c$).

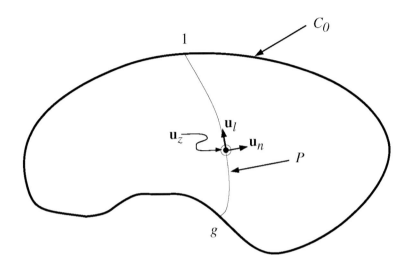

(a) Show that for this mode

$$\mathsf{l}_P(\omega, \beta) = \mu K_H$$

and

$$\mathsf{c}_P(\omega, \beta) = \frac{\epsilon}{K_H} \frac{\beta^2}{k^2}$$

where K_H is a dimensionless, frequency-independent parameter defined by:

$$K_H \equiv \frac{\left[\int_P (\partial \mathcal{H}_z / \partial n)\, dl \right]^2}{k_c^2 \int_{S_0} \mathcal{H}_z^2\, dS}$$

The results of problem p5-9 may be used in your solution without proving them.

(b) Show that the path-dependent "characteristic impedance" of this mode is

$$Z_{c,P} \equiv \sqrt{\frac{\mathsf{l}_P(\omega, \beta)}{\mathsf{c}_P(\omega, \beta)}} = K_H \zeta_{\mathrm{TE}}$$

Sketch a graph of $Z_{c,P}$ vs. frequency, showing its general form.

(c) Show that the oscillatory power is given by

$$\hat{P}_{\mathrm{osc}} = \frac{\Phi_P^2}{2 Z_{c,P}}$$

no matter what path P is chosen.

(d) Apply the results of parts (a) and (b) to the TE_{10} mode of a rectangular metallic waveguide, choosing P to be a vertical line from $y = 0$ to $y = d$ located at an arbitrary position $x = x_0$ along the x-axis. Find the quantities l_P, c_P and K_H, and determine $Z_{c,P}$. If $x_0 = a/2$, show that $K_H = 2d/a$.

p8-9 Suppose that for a TM mode of a lossless hollow metallic waveguide with a simply connected wall, we define a current I_C as

$$I_C = \int_C \boldsymbol{\mathcal{H}}_T \cdot d\boldsymbol{l}$$

where C is a portion of the conducting wall between a point 1 and a second point 2 on the waveguide wall as shown. Assume that C is such that $I_C \neq 0$, and that the mode is above cutoff ($k > k_c$).

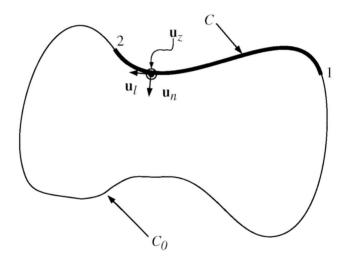

(a) Show that for this mode

$$\mathsf{c}_C(\omega, \beta) = \epsilon K_E$$

and

$$\mathsf{l}_C(\omega, \beta) = \frac{\mu}{K_E} \frac{\beta^2}{k^2}$$

where K_E is a dimensionless, frequency-independent parameter defined by:

$$K_E \equiv \frac{\left[\int_C (\partial \mathcal{E}_z / \partial n)\, dl \right]^2}{k_c^2 \int_{S_0} \mathcal{E}_z^2\, dS}$$

The results of problem p5-9 may be used in your solution without proving them.

(b) Show that the "characteristic impedance" of this mode (dependent on the choice of C) is

$$Z_{c,C} \equiv \sqrt{\frac{\mathsf{l}_C(\omega, \beta)}{\mathsf{c}_C(\omega, \beta)}} = \frac{\zeta_{\mathrm{TM}}}{K_E}$$

Sketch a graph of $Z_{c,C}$ vs. frequency, showing its general form.

(c) Show that the oscillatory power is given by

$$\hat{P}_{\mathrm{osc}} = \frac{1}{2} Z_{c,C} I_C^2$$

no matter what segment C is chosen.

(d) Apply the results of parts (a) and (b) to the TM_{11} mode of a rectangular metallic waveguide, choosing C to be the bottom wall of the waveguide: $y = 0$ from $x = 0$ (point 1) to $x = a$ (point 2). Find the quantities l_C, c_C and K_E, and determine $Z_{c,C}$.

p8-10 Consider a propagating forward mode ($\gamma = j\beta$, $\beta > 0$) on an arbitrary lossless waveguide whose permeability is that of free space μ_0 while the permittivity ϵ may be inhomogeneous (a function of (x,y)) but not dispersive (ϵ is not a function of ω). The permittivity is assumed to have a maximum value of $\epsilon_{\max}$. Demonstrate the following properties as $\omega \to \infty$:

(a) $\epsilon_{\mathrm{eff}} \to \epsilon_{\mathrm{eff}}^{(\infty)}$, where $\epsilon_{\mathrm{eff}}^{(\infty)}$ is a constant less than or equal to $\epsilon_{\max}/\epsilon_0$.

(b) $v_p \to v_g$.

(c) In a certain sense, $|\mathcal{E}_z/\mathcal{E}_T|$ and $|\mathcal{H}_z/\mathcal{H}_T|$ approach zero.

(d) In fact, $\epsilon_{\mathrm{eff}}^{(\infty)} = \epsilon_{\max}/\epsilon_0$, and both $\boldsymbol{\mathcal{E}}$ and $\boldsymbol{\mathcal{H}}$ approach zero at all points except where $\epsilon = \epsilon_{\max}$.

[Hint: Make use of the results of Appendix J, especially (J.33), (J.35), (J.39) and (J.70), as well as (5.17) and (5.19).]

p8-11 Provide details of the derivation of (8.100) and (8.101).

p8-12 A coaxial line is made up of an inner conductor of radius a, a dielectric layer of permittivity ϵ_1 in $a < \rho < c$, a dielectric layer of different permittivity ϵ_2 in $c < \rho < b$, and an outer conductor of radius b as shown.

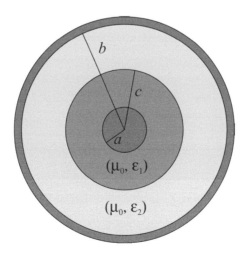

Obtain the quasi-TEM line parameters l, c, Z_c and β for this line. (Hint: cf. for example Johnk (1975), example 4-4.)

p8-13 Repeat problem p8-12 for the case where $\epsilon(\rho)$ and $\mu(\rho)$ are arbitrary functions of the radial coordinate ρ, but are independent of ϕ. Obtain expressions for the quasi-TEM line parameters l, c, Z_c and β for this line.

p8-14 A coaxial line is made up of an inner conductor of radius a, a material layer of permittivity ϵ_1 and permeability μ_0 in $a < \rho < c$, a material layer of permittivity ϵ_0 in $c < \rho < b$, and an outer conductor of radius b as shown.

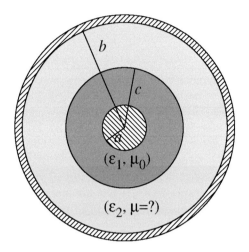

(a) What must be the permeability of the outer material layer in $c < \rho < b$ for a TEM mode to exist on this structure?

(b) Obtain expressions for the line parameters l (in terms of permeabilities μ only), c (in terms of permittivities ϵ only), Z_c and β for this line.

p8-15 Suppose a quasi-TEM line is formed with air and a homogeneous substrate with relative *permeability* $\mu_r \neq 1$, as well as relative permittivity $\epsilon_r \neq 1$.

(a) Let the inductance per unit length of this line be given by the function $L(\mu_r)$. Show that
$$L(\mu_r) = \frac{\mu_0 \epsilon_0}{C(\frac{1}{\mu_r})}$$

Hence show that the propagation coefficient is

$$\beta = \omega \sqrt{\mu_0 \epsilon_0} \sqrt{\frac{C(\epsilon_r)}{C(\frac{1}{\mu_r})}}$$

and that the characteristic impedance is

$$Z_c = \zeta_0 \frac{\epsilon_0}{\sqrt{C(\epsilon_r) C(\frac{1}{\mu_r})}}$$

(b) Are the values of β and Z_c increased or decreased by an increase in the value of μ_r? Why?

(c) Define an *effective permeability* μ_{eff} in analogy with ϵ_{eff}, and express β and Z_c in terms of μ_{eff} and ϵ_{eff}. Can you give your definition a clear physical interpretation?

p8-16 Design microstrip lines according to each of the following specifications:
   (a) $Z_c = 10\,\Omega$; use a 50 mil (1.27 mm) alumina substrate ($\epsilon_r = 9.7$).
   (b) $Z_c = 300\,\Omega$; use a 50 mil (1.27 mm) alumina substrate ($\epsilon_r = 9.7$).
   (c) $Z_c = 100\,\Omega$; use a 2 mm thick Teflon substrate ($\epsilon_r = 2.1$).

p8-17 Plot the characteristic impedance Z_c of a microstrip line versus w/h for several values of ϵ_r (say, 1.0, 2.5 and 10), using (8.151).

p8-18 Some microstrip circuitry is laid out on a Teflon substrate ($\epsilon_r = 2.1$, $h = 2$ mm) and operates at a frequency of $f = 1$ GHz. We need to design a matching section between two portions of line, one of which has a characteristic impedance of 50 Ω, and the other 75 Ω as shown.

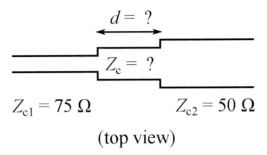

(top view)

Assuming that junction fringing effects can be neglected, find the width w for each of the three sections of microstrip, and the length d of the matching section.

p8-19 A 100 Ω load impedance is connected to the end of a 50 Ω microstrip line as shown.

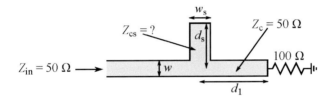

The microstrip is to be fabricated on a 50 mil (1.27 mm) alumina substrate ($\epsilon_r = 9.7$). An open-ended microstrip stub of length d_s with characteristic impedance Z_{cs} is to be used as a single-stub tuner to match the line to the load at $f = 2$ GHz. Find the stub lengths d_s and strip widths w_s needed to achieve this match for the three cases $Z_{cs} = 20\,\Omega$, $50\,\Omega$ and $100\,\Omega$. In each case, use the solution with the smallest value of d_1. Field fringing effects at the junction, the load, and the open end of the stub are to be ignored, but you may find it instructive (optionally) to compare your analytical results with those obtained from a microwave circuit analysis program. Use d_1 and d_s measured from the centers of the strips as shown in the figure if you do make this comparison.

p8-20 A CPW is to be fabricated on a substrate with $\epsilon_r = 6.2$ and $h = 5$mm. If Z_c is to be 75 Ω, plot a graph showing w vs. s such that this value of characteristic impedance is achieved.

p8-21 If k and k_1 for a CPW are given by (8.157), show that $k_1 \leq k$, with equality happening only if $h \to \infty$. As a result, show that

$$\epsilon_{\text{eff}} \leq \frac{\epsilon_r + 1}{2}$$

for a CPW.

p8-22 A coplanar stripline as in Figure 8.23 is to have $Z_c = 50\,\Omega$, while $\epsilon_{\text{eff}} = 4$. If $\epsilon_r = 17$, obtain values of w/h and s/h such that these line parameters are achieved. A transcendental equation has to be solved numerically to obtain the solution.

p8-23 It can be observed from Figure 8.22 that for given values of ϵ_r and $h/(s + 2w)$, there is a certain value of $s/(s+2w)$ for which ϵ_{eff} for a CPW becomes a minimum.

    (a) If $\epsilon_r = 12$ and $h/(s + 2w) = 1$, find the value of $s/(s + 2w)$ at which the minimum of ϵ_{eff} occurs, give its minimum value, and calculate the characteristic impedance Z_c at this minimum point.

    (b) Explain physically why this minimum occurs in general.

p8-24 A one-sided CPW is obtained from the ordinary CPW by removing one of the ground planes as shown below.

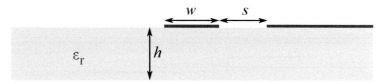

This design eliminates the problem of maintaining both ground planes at the same potential (which requires air bridges), but is an unbalanced line (like the coaxial line) instead of a balanced line (like the two-wire line). An approximate formula for its capacitance per unit length has been found to be[9]

$$C(\epsilon_r) = 2\epsilon_0 \left[\frac{K(k)}{K(k')} + \frac{\epsilon_r - 1}{2} \frac{K(k_1)}{K(k_1')} \right]$$

where

$$k = \sqrt{\frac{w}{s + w}}; \qquad k_1 = e^{-\pi s/4h} \sqrt{\frac{\sinh \frac{\pi w}{2h}}{\sinh \frac{\pi(s+w)}{2h}}}$$

For $\epsilon_r = 9.8$ and $(s + w)/h = 0.5$, compute and plot the characteristic impedance and effective dielectric constant of this line as a function of $\frac{w}{s}$ over the range $0.2 < \frac{w}{s} < 5$. Do you think a practical line of this type with $Z_c = 50\,\Omega$ can be realized? Give your reasons.

p8-25 A substrate of yttrium iron garnet, or YIG ($\epsilon_r = 16$, $h = .055$ in $= 1.4$ mm) is to be used in the design of a 72 Ω slotline. If the operating frequency is $f = 3$ GHz, what is the required slot width w? What will the corresponding value of ϵ_{eff} be? Note that a transcendental equation has to be solved numerically for this problem.

p8-26 Give upper and lower bounds for the propagation coefficients β of the CPW and CPS coplanar structures considered in Section 8.7. Can either of these support bound modes if an infinite ground plane is placed on the bottom side of the substrate? If so, are any special shielding or grounding arrangements required?

p8-27 A groove guide is a pair of metallic parallel-plates with a bulge or groove at the center as shown in cross-section below.

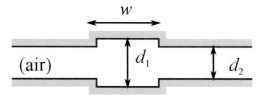

[9] S. S. Gevorgian and 1. G. Mironenko, "Asymmetric coplanar-strip transmission lines for MMIC and integrated optic applications," *Electron. Lett.*, vol. 26, pp. 1916-1918 (1990).

Could a bound mode exist on this waveguide? Explain your reasoning. Does the answer change if vias are allowed? If so, how should they be connected?

p8-28 An open planar Goubau line as shown below is essentially a microstrip without a ground plane.

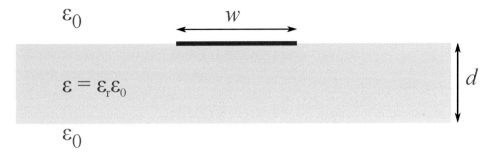

Discuss whether a bound mode can exist on this structure, and explain your reasoning. If a mode does exist, will it be of quasi-TEM or pseudo-TEM type?

9 ORTHOGONALITY, POWER FLOW AND WAVEGUIDE LOSSES

9.1 RECIPROCITY AND REACTION

One of the important tools in electromagnetic theory is the reciprocity theorem of Lorentz. We will derive it in this section and follow up with a number of applications, including the orthogonality of waveguide modes and the approximate calculation of waveguide losses. Further applications will be made in the succeeding chapters.

When dealing with sources which are small or physically remote from the points where the field is of interest, it is occasionally useful to consider the effect of fictitious *magnetic* currents[1] $\mathbf{M}$ in addition to that of the *electric* currents $\mathbf{J}$. If such fictitious magnetic currents could exist, they would enter into Maxwell's equations as

$$\nabla \times \mathbf{E} = -j\omega\mu\mathbf{H} - \mathbf{M} \tag{9.1}$$

$$\nabla \times \mathbf{H} = j\omega\epsilon\mathbf{E} + \mathbf{J} \tag{9.2}$$

In Chapter 10, we will examine in more detail the ways in which magnetic currents can arise, specifically in the context of field equivalence principles.

Consider a set of sources $\mathbf{J}^a$ and $\mathbf{M}^a$ which produces the fields $\mathbf{E}^a$, $\mathbf{H}^a$ in some linear isotropic medium whose parameters are ϵ and μ (which may be functions of position). We have

$$\left.\begin{array}{rcl} \nabla \times \mathbf{E}^a &=& -j\omega\mu\mathbf{H}^a - \mathbf{M}^a \\ \nabla \times \mathbf{H}^a &=& j\omega\epsilon\mathbf{E}^a + \mathbf{J}^a \end{array}\right\} \tag{9.3}$$

The fields $\mathbf{E}^a$, $\mathbf{H}^a$ produced by the currents $\mathbf{J}^a$, $\mathbf{M}^a$ constitute a *state "a"* of our system. If we remove $\mathbf{J}^a$ and $\mathbf{M}^a$ and replace them with a *different* set of sources $\mathbf{J}^b$ and $\mathbf{M}^b$ (but place them in the *same* medium and boundary geometry), we observe a different field $\mathbf{E}^b$, $\mathbf{H}^b$ which obeys

$$\left.\begin{array}{rcl} \nabla \times \mathbf{E}^b &=& -j\omega\mu\mathbf{H}^b - \mathbf{M}^b \\ \nabla \times \mathbf{H}^b &=& j\omega\epsilon\mathbf{E}^b + \mathbf{J}^b \end{array}\right\} \tag{9.4}$$

and constitutes a *state "b"*. Our next steps resemble the derivation of Poynting's theorem for oscillatory energy as done in Appendix D.

We take the dot product of the first of equations (9.3) with $\mathbf{H}^b$ and the second of equations (9.4) with $\mathbf{E}^a$ and subtract the results. Using (B.13), we get

$$\begin{array}{rl} & \mathbf{H}^b \cdot \nabla \times \mathbf{E}^a - \mathbf{E}^a \cdot \nabla \times \mathbf{H}^b \\ = & \nabla \cdot (\mathbf{E}^a \times \mathbf{H}^b) \\ = & -j\omega\mu\mathbf{H}^a \cdot \mathbf{H}^b - j\omega\epsilon\mathbf{E}^a \cdot \mathbf{E}^b - \mathbf{M}^a \cdot \mathbf{H}^b - \mathbf{J}^b \cdot \mathbf{E}^a \end{array} \tag{9.5}$$

---

[1]The use of $\mathbf{M}$ to denote fictitious magnetic currents is not universal and is in fact fairly recent. Many texts on electromagnetics use $\mathbf{J}_m$ to denote this current, while $\mathbf{M}$ is used to signify the *magnetization density* of dipole moment per unit volume. Since we will not use this latter quantity [which would be given by $\mathbf{M}/(j\omega\mu)$ in our notation], the relatively simple notation $\mathbf{M}$ will be followed here.

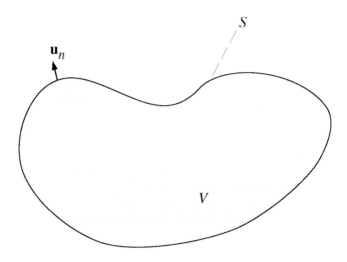

Figure 9.1 Volume and surface used for Lorentz reciprocity theorem.

A second relation can be obtained by interchanging the superscripts a and b:

$$\nabla \cdot (\mathbf{E}^b \times \mathbf{H}^a) = -j\omega\mu\mathbf{H}^a \cdot \mathbf{H}^b - j\omega\epsilon\mathbf{E}^a \cdot \mathbf{E}^b - \mathbf{M}^b \cdot \mathbf{H}^a - \mathbf{J}^a \cdot \mathbf{E}^b \tag{9.6}$$

Finally, subtracting (9.6) from (9.5) gives

$$\nabla \cdot (\mathbf{E}^a \times \mathbf{H}^b - \mathbf{E}^b \times \mathbf{H}^a) = -(\mathbf{E}^a \cdot \mathbf{J}^b - \mathbf{H}^a \cdot \mathbf{M}^b) + (\mathbf{E}^b \cdot \mathbf{J}^a - \mathbf{H}^b \cdot \mathbf{M}^a) \tag{9.7}$$

This is the differential form of the *Lorentz Reciprocity Theorem*. We can obtain the integral form of the theorem by integrating (9.7) over a volume V bounded by the closed surface S (Figure 9.1) and applying the divergence theorem:

$$\oint_S (\mathbf{E}^a \times \mathbf{H}^b - \mathbf{E}^b \times \mathbf{H}^a) \cdot \mathbf{u}_n \, dS$$
$$= -\int_V (\mathbf{E}^a \cdot \mathbf{J}^b - \mathbf{H}^a \cdot \mathbf{M}^b) \, dV + \int_V (\mathbf{E}^b \cdot \mathbf{J}^a - \mathbf{H}^b \cdot \mathbf{M}^a) \, dV \tag{9.8}$$

Here $\mathbf{u}_n$ is the normal unit vector pointing outward from the surface S. In (9.8) we require only that the medium parameters ϵ and μ be the same *inside* V for states "a" and "b." Outside of V, the material environments can be completely different.

If the volume V is extended to cover *all* the sources "a" and "b," then the *right* side of (9.8) no longer depends on the choice of V. We denote

$$\langle a, b \rangle \equiv \int_{V^b} (\mathbf{E}^a \cdot \mathbf{J}^b - \mathbf{H}^a \cdot \mathbf{M}^b) \, dV$$
$$\langle b, a \rangle \equiv \int_{V^a} (\mathbf{E}^b \cdot \mathbf{J}^a - \mathbf{H}^b \cdot \mathbf{M}^a) \, dV \tag{9.9}$$

in this case, where V^a contains all the "a" sources and V^b contains all the "b" sources. The quantity $\langle a, b \rangle$ is called the *reaction* of the field "a" on the sources "b". From (9.8), then,

$$\oint_S (\mathbf{E}^a \times \mathbf{H}^b - \mathbf{E}^b \times \mathbf{H}^a) \cdot \mathbf{u}_n \, dS = -\langle a, b \rangle + \langle b, a \rangle \tag{9.10}$$

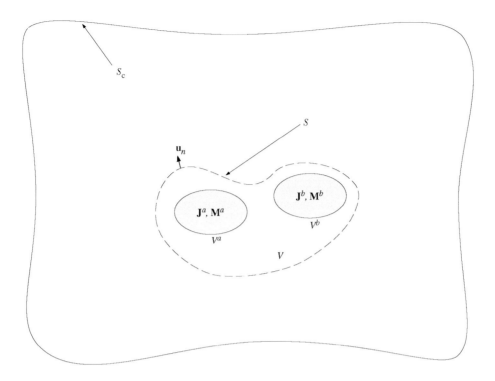

Figure 9.2 Two sets of sources in a region enclosed by a perfect conductor S_c.

where S is any surface completely enclosing both V^a and V^b, and the left side of (9.10) is independent of the choice of S within these constraints. If for the moment we suppose that a perfectly conducting surface S_c completely surrounds all the sources in the problem, then by *choosing* S to be the same as S_c (see Figure 9.2), we find that this constant must be zero:

$$\oint_{S_c} (\mathbf{E}^a \times \mathbf{H}^b - \mathbf{E}^b \times \mathbf{H}^a) \cdot \mathbf{u}_n \, dS = \oint_{S_c} [\mathbf{u}_n \times \mathbf{E}^a \cdot \mathbf{H}^b - \mathbf{u}_n \times \mathbf{E}^b \cdot \mathbf{H}^a] \, dS = 0$$

because $\mathbf{u}_n \times \mathbf{E} = 0$ on the surface of a perfectly conducting body. If we imagine unbounded space as the limiting case where S_c recedes to infinity, we have shown that the surface integral is zero in this general case as well.[2] Hence, we conclude that

$$\langle a, b \rangle = \langle b, a \rangle \tag{9.11}$$

when the sources are of finite extent.

A direct physical interpretation of the reaction is rather subtle in general. In the case when $\mathbf{M} = 0$, we find that the so-called *self-reaction* $\langle a, a \rangle$ is equal to

$$\langle a, a \rangle = \int_{V^a} \mathbf{E}^a \cdot \mathbf{J}^a \, dV \tag{9.12}$$

---

[2]The actual rigorous proof is somewhat more complicated.

and by the results of Appendix D, we can relate this to the oscillatory part of the power being dissipated in or generated by the sources of the "a" field. In more general circumstances, however, when both electric and magnetic currents are present, we can probably best think of $\langle a, b \rangle$ roughly as a measure of how well correlated are the "a" fields with the "b" sources, or as a sort of "measured" value of the "a" field with the "b" sources acting as a probe. We will expand on this idea below, but it was Spike Milligan who, presumably not knowing anything about the subject of reciprocity, put the concept most succinctly:

—*Intelligence has established that the people attacking us are the enemy.*
—*Do the enemy realize that you have this information?*
—*Oh no, we got 'em fooled, they think that* we're *the enemy.*

When some or all of the sources are confined to surfaces or lines, the reaction can be computed by writing the volume current distributions in terms of appropriate Dirac delta functions. Let n be the normal distance from a surface S^b that contains the surface currents $\mathbf{J}_S^b$ and $\mathbf{M}_S^b$, measured as positive on the side of S^b toward which the unit vector $\mathbf{u}_n$ points. Then the corresponding *volume* densities are

$$\mathbf{J}^b = \mathbf{J}_S^b \delta_{S^b}(n)$$

$$\mathbf{M}^b = \mathbf{M}_S^b \delta_{S^b}(n)$$

where $\delta_{S^b}(n)$ is the Dirac delta function concentrated on the surface S^b. Then if a field $\mathbf{E}^a$, $\mathbf{H}^a$ is continuous at S^b, the properties of the delta function give

$$\langle a, b \rangle \equiv \int\limits_{V^b} (\mathbf{E}^a \cdot \mathbf{J}^b - \mathbf{H}^a \cdot \mathbf{M}^b) \, dV = \int\limits_{S^b} (\mathbf{E}^a \cdot \mathbf{J}_S^b - \mathbf{H}^a \cdot \mathbf{M}_S^b) \, dS \qquad (9.13)$$

Similar definitions apply to the case of line sources and point sources (dipoles).

9.1.1 EXAMPLE

As a simple example of the application of the reciprocity theorem, consider the situation shown in Figure 9.3(a). On the surface S of a perfect conductor is placed a tangential electric surface current $\mathbf{J}_S^a$, resulting in the field $\mathbf{E}^a, \mathbf{H}^a$ over all space. We can use the reciprocity theorem to show that this field is identically zero. Let a "b" state be produced by either an electric Hertz dipole

$$\mathbf{J}^b = j\omega \mathbf{p} \delta(x - x_0) \delta(y - y_0) \delta(z - z_0) \qquad (9.14)$$

or a magnetic Hertz dipole

$$\mathbf{M}^b = j\omega \mu \mathbf{m} \delta(x - x_0) \delta(y - y_0) \delta(z - z_0) \qquad (9.15)$$

which we will use as a probe for the "a" state field (see Appendix D for the properties of Hertz dipoles). We have either

$$\langle a, b \rangle = j\omega \mathbf{p} \cdot \mathbf{E}^a(x_0, y_0, z_0) \qquad (9.16)$$

in the case of (9.14), or

$$\langle a, b \rangle = -j\omega \mu \mathbf{m} \cdot \mathbf{H}^a(x_0, y_0, z_0) \qquad (9.17)$$

in the case of (9.15), so that $\langle a, b \rangle$ is a measure of some component of the "a" state field. But $\langle a, b \rangle = \langle b, a \rangle$ and, by (9.13),

$$\langle b, a \rangle = \int\limits_S \mathbf{E}^b \cdot \mathbf{J}_S^a \, dS = 0 \qquad (9.18)$$

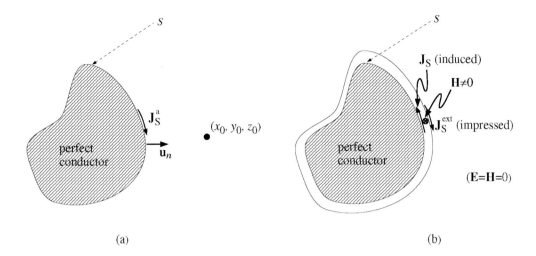

Figure 9.3 (a) A tangential impressed surface current on a perfect conductor; (b) Zero external field produced by electric currents at the surface of a perfect conductor.

because $\mathbf{E}^b_{tan}$ on the perfect conductor S is identically zero. By (9.16) and (9.17), this implies that $\mathbf{E}^a, \mathbf{H}^a$ are both zero at any point in space. We might say that the conductor "shorts" out the source $\mathbf{J}^a_S$ and results in no net exterior field.

If we imagine a small space (which is reckoned eventually to approach zero in the limit) between an impressed source $\mathbf{J}^{ext}_S$ and the perfect conductor as shown in Figure 9.3(b), we can gain a little further insight into what is happening here. An equal and opposite *induced* current $\mathbf{J}_S$ appears on the conductor in a way similar to that predicted by image theory (even though the conductor is not a plane surface). The resultant $\mathbf{E}$ field becomes zero everywhere as the surface S containing the impressed sources approaches the conductor. The $\mathbf{H}$-field inside the conductor and outside S also becomes zero, but the $\mathbf{H}$-field in the infinitesimal layer between them continues to support a nonzero $\mathbf{H} = \mathbf{J}_S \times \mathbf{u}_n = \mathbf{u}_n \times \mathbf{J}^{ext}_S$. In circuit theory, this is analogous to an ideal current generator which has a short circuit connected across it. Current flows around the loop, but currents and voltages outside this loop remain zero.

This result is used in an approximate way to reduce the electromagnetic interference (EMI) produced by high-speed digital currents flowing on printed circuit boards (PCBs). If a large ground plane is placed next to the plane(s) in which the current-carrying conducting traces lie (separated from the traces by a thin dielectric substrate), then the fields produced outside the PCB are nearly zero, because of being approximately short-circuited by the ground plane.

9.2 ORTHOGONALITY

The reciprocity theorem can be used to deduce a number of properties of waveguide modes, many of which are presented in Appendix J. One of the most important such properties is the *orthogonality* between fields of different modes of the same waveguide. Orthogonality is a kind of statement that, in a uniform section of waveguide, each mode behaves independently of all the others, in a sense to be specified below. The property is used extensively in later chapters when discussing excitation, scattering and coupling of waveguide modes.

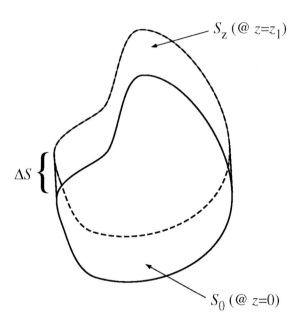

S_z (@ $z=z_1$)

ΔS

S_0 (@ $z=0$)

Figure 9.4   Slice of waveguide between $z = 0$ and $z = z_1$.

Consider the application of the Lorentz reciprocity theorem to the fields of two modes of the same waveguide. Take

$$
\begin{aligned}
\mathbf{E}^a &= \boldsymbol{\mathcal{E}}_m e^{-\gamma_m z} \; ; \quad \mathbf{H}^a = \boldsymbol{\mathcal{H}}_m e^{-\gamma_m z} \\
\mathbf{E}^b &= \boldsymbol{\mathcal{E}}_n e^{-\gamma_n z} \; ; \quad \mathbf{H}^b = \boldsymbol{\mathcal{H}}_n e^{-\gamma_n z}
\end{aligned}
\tag{9.19}
$$

and apply (9.8) to a source-free region of the waveguide. Specifically, let the volume V be a slice cut out of the waveguide cross-section (which we can, as in Figure 9.2, temporarily assume to be bounded by a perfectly conducting wall and then, if desired, allow the wall to recede to infinity to allow for cases—such as open transmission lines or optical waveguides— where the waveguide cross-section is unbounded) as shown in Figure 9.4. The surfaces at $z = z_1$ (S_z) and $z = 0$ (S_0) make up the front and back parts of the bounding surface S, while the lateral surface at the wall of the waveguide is denoted ΔS. Because the region is assumed source-free, the right side of (9.8) will be zero. Since the waveguide wall is perfectly conducting, we have on ΔS

$$
\mathbf{E}^a \times \mathbf{H}^b \cdot \mathbf{u}_n = \mathbf{u}_n \times \mathbf{E}^a \cdot \mathbf{H}^b = 0
$$

along with a similar expression with $a \leftrightarrow b$, so the contribution to the surface integral from ΔS vanishes. Equation (9.8) now becomes

$$
e^{-(\gamma_m + \gamma_n)z_1} \int_{S_z} (\boldsymbol{\mathcal{E}}_m \times \boldsymbol{\mathcal{H}}_n - \boldsymbol{\mathcal{E}}_n \times \boldsymbol{\mathcal{H}}_m) \cdot \mathbf{u}_z \, dS - \int_{S_0} (\boldsymbol{\mathcal{E}}_m \times \boldsymbol{\mathcal{H}}_n - \boldsymbol{\mathcal{E}}_n \times \boldsymbol{\mathcal{H}}_m) \cdot \mathbf{u}_z \, dS = 0 \tag{9.20}
$$

Now $\boldsymbol{\mathcal{E}}$ and $\boldsymbol{\mathcal{H}}$ depend only on x and y and not on z, so the integral over S_z is the same as the integral over S_0, and thus

$$
[e^{-(\gamma_m + \gamma_n)z_1} - 1] \int_{S_0} (\boldsymbol{\mathcal{E}}_m \times \boldsymbol{\mathcal{H}}_n - \boldsymbol{\mathcal{E}}_n \times \boldsymbol{\mathcal{H}}_m) \cdot \mathbf{u}_z \, dS = 0 \tag{9.21}
$$

or, since (9.21) must hold for *any* value of z_1,

$$\int_{S_0} (\mathcal{E}_m \times \mathcal{H}_n - \mathcal{E}_n \times \mathcal{H}_m) \cdot \mathbf{u}_z \, dS = 0 \qquad (\text{if } \gamma_m \neq -\gamma_n) \tag{9.22}$$

This is known as the *orthogonality* property between fields of modes having other than exactly opposite propagation coefficients. It should be noted that only the *transverse* components of $\mathcal{E}$ and $\mathcal{H}$ enter into the orthogonality relation (9.22). Sometimes orthogonality is expressed in a somewhat different form than (9.22). If we let $n \to -n$ in (9.22) and apply (5.24), we obtain a second result which can be combined with (9.22) to yield

$$\int_{S_0} \mathcal{E}_m \times \mathcal{H}_n \cdot \mathbf{u}_z \, dS = 0 \qquad (\text{if } \gamma_m^2 \neq \gamma_n^2) \tag{9.23}$$

For certain waveguides, we have seen that more than one mode can sometimes have the same propagation coefficient γ. Modes with this property are said to be *degenerate*. In a lossless waveguide, if they are not already orthogonal, it is always possible to form combinations of degenerate modes into a new subset of modes so that (9.22) holds under the sole requirement that $m \neq -n$. Alternatively, we could imagine a slight perturbation being introduced to the waveguide (a change in its wall shape, for instance) which destroys the degeneracy in practice. In any event, we will always assume that (9.22) holds for $m \neq -n$ in all our applications.

Thus, with the normalization condition (5.27) taken into account, we finally have in place of (9.23)

$$
\begin{aligned}
\frac{1}{2} \int_{S_0} \mathcal{E}_m \times \mathcal{H}_n \cdot \mathbf{u}_z \, dS \;\; &= \;\; 0 \qquad (\text{if } m^2 \neq n^2) \\
&= \;\; 1 \qquad (\text{if } +m \text{ or } -m = n > 0; \text{ forward mode}) \\
&= \;\; -1 \qquad (\text{if } +m \text{ or } -m = n < 0; \text{ reverse mode})
\end{aligned}
\tag{9.24}
$$

These properties are the analogs of those obtained for multiconductor transmission lines in terms of mode voltages and currents in Section 3.3.2—equations (3.48)-(3.53).

When continuous or radiation modes can exist on a waveguide structure, the orthogonality condition takes a somewhat different form. If $\mathcal{E}(\kappa)$ and $\mathcal{H}(\kappa)$ are the fields of a continuous mode whose propagation coefficient is $\gamma(\kappa)$, and if $\gamma(\kappa_1) \neq \gamma(\kappa_2)$ for $\kappa_1 \neq \kappa_2$, and $\gamma(-\kappa) = -\gamma(\kappa)$, then we can normalize the continuous mode fields such that

$$\int_{S_0} \left(\mathcal{E}(\kappa_1) \times \mathcal{H}(\kappa_2) - \mathcal{E}(\kappa_2) \times \mathcal{H}(\kappa_1) \right) \cdot \mathbf{u}_z \, dS = 4\delta(\kappa_1 + \kappa_2) \tag{9.25}$$

where δ is the Dirac delta-function. Moreover, a discrete guided mode is always orthogonal to a continuous mode:

$$\int_{S_0} \left(\mathcal{E}(\kappa) \times \mathcal{H}_m - \mathcal{E}_m \times \mathcal{H}(\kappa) \right) \cdot \mathbf{u}_z \, dS = 0 \tag{9.26}$$

9.2.1 ORTHOGONALITY OF MODES ON LOSSLESS WAVEGUIDES

Equation (9.24) can be put in a somewhat different form for most modes of a lossless waveguide. In such a situation, if the propagation coefficient γ_m is either pure real or pure imaginary (i.e., the mode is not a complex mode), then $\mathcal{E}_{mT}$ can be chosen purely real before normalization is imposed (as shown in problem p5-3). From the orthogonality property (9.24) we then have

$$\int_{S_0} \boldsymbol{\mathcal{E}}_{mT} \times \boldsymbol{\mathcal{H}}_{m'T}^* \cdot \mathbf{u}_z \, dS = \left\{ \int_{S_0} \boldsymbol{\mathcal{E}}_{mT} \times \boldsymbol{\mathcal{H}}_{m'T} \cdot \mathbf{u}_z \, dS \right\}^* = 0 \tag{9.27}$$

if $m \neq \pm m'$. Multiplication of the fields by a complex constant will obviously not change this result, so it is true also for the normalized mode fields.

We further have from Section J.2 of Appendix J that γ_m is either pure real or pure imaginary (corresponding to cutoff or propagating modes respectively) provided that the complex power $\hat{P}$ carried by the mode is not zero, and further that $\hat{P}_m$ is either real or imaginary as found in (5.26). Then, by (5.28), eqn. (9.24) for $m > 0$ and $m' > 0$ can be written in the form

$$\begin{aligned}
\frac{1}{2} \int_{S_0} \boldsymbol{\mathcal{E}}_{mT} \times \boldsymbol{\mathcal{H}}_{m'T}^* \cdot \mathbf{u}_z \, dS &= 0 & m^2 \neq m'^2 \\
&= 1 & m^2 = m'^2, \, \gamma_m = j\beta_m \\
&= ju_m & m^2 = m'^2, \, \gamma_m = \alpha_m
\end{aligned} \tag{9.28}$$

where $u_m = +1$ if the mode stores more magnetic than electric energy per unit length, and $u_m = -1$ if the mode stores more electric than magnetic energy per unit length (cf. (J.23)). This has implications for power transport in waveguides as discussed in the next section.

A final note about the fields of a propagating mode on a lossless waveguide: if $\boldsymbol{\mathcal{E}}_{mT}$ and $\boldsymbol{\mathcal{H}}_{mT}$ are chosen to be real, it is evident that $\boldsymbol{\mathcal{E}}_{mz}$ and $\boldsymbol{\mathcal{H}}_{mz}$ will be pure imaginary. Thus, reverse mode fields can be expressed as

$$\begin{aligned}
\boldsymbol{\mathcal{E}}_{-m} &= \boldsymbol{\mathcal{E}}_m^* \\
\boldsymbol{\mathcal{H}}_{-m} &= -\boldsymbol{\mathcal{H}}_m^*
\end{aligned} \tag{9.29}$$

9.3   POWER FLOW IN LOSSLESS WAVEGUIDES: EQUIVALENT TRANSMISSION LINE MODEL

There is a close analogy which exists between any waveguide mode and the voltage and current on a classical transmission line. Let us develop the notion of this equivalence here in a systematic way.

Suppose that we somehow set up sources on either side of a source-free section $z_1 \leq z \leq z_2$ of a uniform waveguide such that only $\pm z$-traveling modes with given indices $\pm m$ are excited, as shown in Figure 9.5. Then the field in this region will be given by:

$$\begin{aligned}
\mathbf{E} &= a_m \boldsymbol{\mathcal{E}}_m(x,y) e^{-\gamma_m z} + b_m \boldsymbol{\mathcal{E}}_{-m}(x,y) e^{\gamma_m z} \\
\mathbf{H} &= a_m \boldsymbol{\mathcal{H}}_m(x,y) e^{-\gamma_m z} + b_m \boldsymbol{\mathcal{H}}_{-m}(x,y) e^{\gamma_m z}
\end{aligned} \tag{9.30}$$

where a_m, b_m are wave amplitudes associated with the $\pm m$ modes. We will see how to compute them in Chapter 10. Because of relations (5.24) between components of the $\pm m$ mode fields, we can obtain separate expressions for the longitudinal and transverse components of the fields:

$$\left. \begin{aligned}
\mathbf{E}_T &= \boldsymbol{\mathcal{E}}_{mT}(x,y)[a_m e^{-\gamma_m z} + b_m e^{\gamma_m z}] \\
\mathbf{H}_T &= \boldsymbol{\mathcal{H}}_{mT}(x,y)[a_m e^{-\gamma_m z} - b_m e^{\gamma_m z}]
\end{aligned} \right\} \tag{9.31}$$

$$\left. \begin{aligned}
E_z &= \mathcal{E}_{mz}(x,y)[a_m e^{-\gamma_m z} - b_m e^{\gamma_m z}] \\
H_z &= \mathcal{H}_{mz}(x,y)[a_m e^{-\gamma_m z} + b_m e^{\gamma_m z}]
\end{aligned} \right\} \tag{9.32}$$

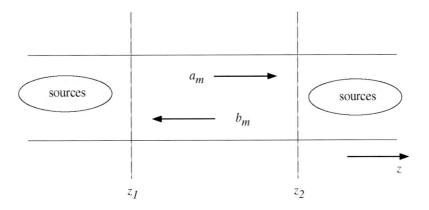

Figure 9.5 Sources producing modes $\pm m$.

The z-dependent factor in the expressions for $\mathbf{E}_T$ and H_z is formally the same as the expression for the z-dependent voltage on a classical transmission line, eqn. (1.7), if we identify a_m (b_m) as the forward- (reverse-) traveling wave amplitude, and γ_m as the propagation coefficient. Within a constant factor (the characteristic impedance), the z-dependent factor in the expressions for $\mathbf{H}_T$ and E_z is formally identical to that for the current on this same line.

Now let us assume that the transverse fields in a lossless waveguide are given as a sum of all possible such $\pm z$-directed mode pairs from (9.31):

$$\mathbf{E}_T = \sum_{m>0} [a_m e^{-\gamma_m z} + b_m e^{\gamma_m z}] \boldsymbol{\mathcal{E}}_{mT}(x,y)$$

$$\mathbf{H}_T = \sum_{m>0} [a_m e^{-\gamma_m z} - b_m e^{\gamma_m z}] \boldsymbol{\mathcal{H}}_{mT}(x,y) \qquad (9.33)$$

The total complex power being carried through the cross-section is then

$$\hat{P}(z) = \frac{1}{2} \int_{S_z} (\mathbf{E}_T \times \mathbf{H}_T^*) \cdot \mathbf{u}_z \, dS$$

$$= \frac{1}{2} \sum_{m>0} \sum_{m'>0} [a_m e^{-\gamma_m z} + b_m e^{\gamma_m z}] \qquad (9.34)$$

$$\times [a_{m'}^* e^{-\gamma_{m'}^* z} - b_{m'}^* e^{\gamma_{m'}^* z}] \int_{S_0} (\boldsymbol{\mathcal{E}}_{mT} \times \boldsymbol{\mathcal{H}}_{m'T}^*) \cdot \mathbf{u}_z \, dS$$

From the orthogonality condition (9.27), the double sum in (9.34) then reduces to a single sum of the form:

$$\hat{P}(z) = \frac{1}{2} \sum_{m>0} [a_m e^{-\gamma_m z} + b_m e^{\gamma_m z}][a_m^* e^{-\gamma_m^* z} - b_m^* e^{\gamma_m^* z}] \int_{S_0} (\boldsymbol{\mathcal{E}}_{mT} \times \boldsymbol{\mathcal{H}}_{mT}^*) \cdot \mathbf{u}_z \, dS \quad (9.35)$$

It can be inferred from (9.35) that the complex power carried by a collection of modes existing in a lossless waveguide is equal to the sum of the complex powers associated with each forward/reverse ($\pm z$-directed) pair of modes individually. This is a generalization of the result (3.82) obtained for power carried by propagating modes on lossless multiconductor transmission lines.

When we take the real part of (9.35) to obtain the time-average power carried through the guide, we then obtain using (9.28) that

$$
\begin{aligned}
P_{\text{av}}(z) &= \operatorname{Re} P(z) = \sum_{m>0;\gamma_m=j\beta_m} \left\{ |a_m|^2 - |b_m|^2 \right\} \\
&+ \sum_{m>0;\gamma_m=\alpha_m} 2u_m \operatorname{Im}(a_m b_m^*)
\end{aligned}
\tag{9.36}
$$

This is a generalization of relation (1.22) for a lossless classical transmission line. The forward and reverse propagating modes thus all carry power independently of each other. The cutoff modes can carry power only in the presence of *both* forward and reverse attenuating modes; if only one type is present, then a_m or b_m is zero, and their contribution to the total power flow vanishes.

9.4   DIELECTRIC LOSSES IN WAVEGUIDES

For homogeneously filled waveguides, we have previously seen that accounting for the effect of dielectric losses on the modes—especially on their propagation coefficients—is fairly straightforward. We have only to make ϵ (or μ) appropriately complex, and use the formulas already derived for the modes. When the waveguide is inhomogeneously filled, the problem becomes more difficult.

The effect of dielectric losses in the medium filling an arbitrary waveguide can be determined by an approximation procedure if the losses in the medium are small. Suppose for simplicity that the filling material is piecewise homogeneous, with relative permittivity ϵ_r taking on the values ϵ_{r1}, ϵ_{r2}, and so on, in the subareas S_1, S_2, etc., of the waveguide cross section S_0 (which may extend to infinity in the case of an open waveguide). Depending on the context, we may define an effective permittivity or effective index for a particular mode as

$$
\epsilon_{\text{eff}} = n_{\text{eff}}^2 = \beta^2/k_0^2
\tag{9.37}
$$

and emphasize its functional dependence on the relative dielectric constants

$$
\epsilon_{r1} = n_1^2 \quad ; \quad \epsilon_{r2} = n_2^2 \quad ; \dots
\tag{9.38}
$$

of the various sections by writing

$$
\epsilon_{\text{eff}} = \epsilon_{\text{eff}}(\epsilon_{r1}, \epsilon_{r2}, \dots)
\tag{9.39}
$$

The presence of small losses ϵ_1'', ϵ_2'', ..., in the materials causes

$$
\epsilon_{r1,2} \rightarrow \epsilon_{r1,2} - j\frac{\epsilon_{1,2}''}{\epsilon_0}
$$

The complex propagation coefficient γ can be evaluated by taking the first-order terms in a Taylor series expansion about real values of the ϵ_{ri}:

$$
\gamma = \alpha + j\beta \simeq \gamma|_{\epsilon_{r1},\epsilon_{r2},\dots=\text{real}} - j\sum_i \frac{\epsilon_i''}{\epsilon_0} \frac{\partial\gamma}{\partial\epsilon_{ri}}\bigg|_{\epsilon_{r1},\epsilon_{r2},\dots=\text{real}}
$$

The value of $\gamma = j\beta$ and its derivatives at real values of the filling material permittivities can be expressed in terms of, say, ϵ_{eff}, by means of (9.37):

$$
\beta|_{\epsilon_{r1},\epsilon_{r2},\dots=\text{real}} = k_0 \sqrt{\epsilon_{\text{eff}}|_{\epsilon_{r1},\epsilon_{r2},\dots=\text{real}}}
$$

$$\left.\frac{\partial \beta}{\partial \epsilon_{ri}}\right|_{\epsilon_{r1},\epsilon_{r2},...=\text{real}} = \frac{k_0}{2\sqrt{\epsilon_{\text{eff}}\big|_{\epsilon_{r1},\epsilon_{r2},...=\text{real}}}} \left.\frac{\partial \epsilon_{\text{eff}}}{\partial \epsilon_{ri}}\right|_{\epsilon_{r1},\epsilon_{r2},...=\text{real}}$$

We now define *filling factors* q_1, q_2, etc., as

$$q_i = \left.\frac{\partial \epsilon_{\text{eff}}}{\partial \epsilon_{ri}}\right|_{\epsilon_{r1},\epsilon_{r2}=...=\text{real}} \tag{9.40}$$

Then to a first approximation, α and β are given by

$$\beta \simeq k_0 n_{\text{eff}}$$
$$\alpha \simeq \frac{k_0}{2n_{\text{eff}}} \left(\sum_i q_i \frac{\epsilon_i''}{\epsilon_0}\right) \tag{9.41}$$

where n_{eff} refers to the value of n_{eff} when the ϵ_{ri} are real. From (9.41), we can see that β is virtually unaffected by small dielectric losses. The effect upon α of losses in different regions depends on the magnitude of the filling factor for those regions.

In Appendix J, we show how the filling factors can be expressed in terms of fractional parts of stored energy or power carried in the various dielectrics of the waveguide, as well as the group velocity. This physical interpretation is useful in gaining insight into the loss mechanisms of a waveguide. We have that

$$q_i = \frac{c^2/\epsilon_{ri}}{v_p v_g} \frac{\int_{S_i} \epsilon \boldsymbol{\mathcal{E}}_m \cdot \boldsymbol{\mathcal{E}}_m^* \, dS}{\int_{S_0} \epsilon \boldsymbol{\mathcal{E}}_m \cdot \boldsymbol{\mathcal{E}}_m^* \, dS} \tag{9.42}$$

where c is the speed of light in vacuum, and v_p and v_g are the phase and group velocities of the mode of the corresponding lossless guide. For a metallic waveguide uniformly filled with dielectric, we have from (J.40) and (9.42) that $q = 1$, and (9.41) reduces to (5.113) and (5.114) of Chapter 5.

9.4.1 EXAMPLE: DIELECTRIC LOSS FOR QUASI-TEM MODES

For a quasi-TEM mode on a structure containing only air and one other uniform dielectric region (the substrate, with subscript $i = 1$ which will be dropped here), we have

$$\alpha \simeq \frac{\omega}{2}\sqrt{\frac{\mu_0}{\epsilon_0 \epsilon_{\text{eff}}}} q \epsilon'' \tag{9.43}$$

where

$$q = \frac{\partial \epsilon_{\text{eff}}}{\partial \epsilon_r} \tag{9.44}$$

and ϵ_{eff} is to be evaluated at (the real value of) ϵ_r. Equation (9.43) should be compared to the small loss limit of the attenuation coefficient (8.70) of a TEM mode:

$$\alpha_{\text{TEM}} \simeq \frac{\omega}{2}\sqrt{\frac{\mu}{\epsilon'}} \epsilon'' \tag{9.45}$$

For microstrip for instance, the wider the line is, the closer to 1 the filling factor will be, and the larger the attenuation coefficient for a given value of ϵ''. Qualitatively, however, dielectric losses for quasi-TEM modes are very similar to those of pure TEM modes.

9.4.2 EXAMPLE: DIELECTRIC LOSS IN STEP-INDEX, WEAKLY-GUIDING WAVEGUIDES

Consider a step-index dielectric waveguide which is weakly guiding. It is known that the waveguide can be described completely by two normalized quantities: the normalized frequency

$$V = k_0 d \sqrt{\epsilon_{r1} - \epsilon_{r2}} \tag{9.46}$$

where d is some convenient length parameter of the cross-section, ϵ_{r1} is the relative permittivity of the core and ϵ_{r2} is that of the cladding; and the confinement parameter

$$b = \frac{\epsilon_{\text{eff}} - \epsilon_{r2}}{\epsilon_{r1} - \epsilon_{r2}} \tag{9.47}$$

which implies the relation

$$\epsilon_{\text{eff}} = \epsilon_{r2} + b(\epsilon_{r1} - \epsilon_{r2}) \tag{9.48}$$

In this case, b is a function of k_0, ϵ_{r1}, ϵ_{r2} and d only through the parameter V:

$$b = f(V) \tag{9.49}$$

Now, from (9.46) we have that partial derivatives of V with respect to ϵ_{r1}, ϵ_{r2} and even the frequency ω are related to each other:

$$\frac{\partial V}{\partial \epsilon_{r1}} = -\frac{\partial V}{\partial \epsilon_{r2}} = \frac{k_0 d}{2\sqrt{\epsilon_{r1} - \epsilon_{r2}}} \tag{9.50}$$

and

$$\frac{\partial V}{\partial \omega} = \frac{2(\epsilon_{r1} - \epsilon_{r2})}{\omega} \frac{\partial V}{\partial \epsilon_{r1}} = \frac{d}{c}\sqrt{\epsilon_{r1} - \epsilon_{r2}} \tag{9.51}$$

where $c = (\mu_0 \epsilon_0)^{-1/2}$ is the velocity of light in vacuum.

Although we may not know an explicit form of f, we do know from (9.48) that

$$q_1 = \frac{\partial \epsilon_{\text{eff}}}{\partial \epsilon_{r1}} = b + (\epsilon_{r1} - \epsilon_{r2})\frac{\partial b}{\partial \epsilon_{r1}} = b + (\epsilon_{r1} - \epsilon_{r2})f'(V)\frac{\partial V}{\partial \epsilon_{r1}} \tag{9.52}$$

and

$$q_2 = \frac{\partial \epsilon_{\text{eff}}}{\partial \epsilon_{r2}} = 1 - b + (\epsilon_{r1} - \epsilon_{r2})\frac{\partial b}{\partial \epsilon_{r2}} = 1 - b + (\epsilon_{r1} - \epsilon_{r2})f'(V)\frac{\partial V}{\partial \epsilon_{r2}} \tag{9.53}$$

so it follows immediately from (9.50) that

$$q_1 + q_2 = 1 \tag{9.54}$$

This relation holds *only* for weakly-guiding step-index dielectric waveguides.

There is also a relation between the filling factors (and thus the attenuation coefficient) and the group velocity of a mode on a weakly guiding step-index guide. It is more usual to express this relation in terms of the so-called *group-index* of the mode:

$$n_g = \frac{c}{v_g} = c\frac{d\beta}{d\omega} = \frac{d}{d\omega}(\omega n_{\text{eff}}) = n_{\text{eff}} + \omega\frac{dn_{\text{eff}}}{d\omega} \tag{9.55}$$

Taking the derivative of (9.48) with respect to ω and using (9.49), we get

$$2n_{\text{eff}}\frac{dn_{\text{eff}}}{d\omega} = (\epsilon_{r1} - \epsilon_{r2})f'(V)\frac{\partial V}{\partial \omega} \tag{9.56}$$

Substituting (9.51), (9.52) and (9.55) into this equation, we obtain after some algebra:

$$q_1 = \frac{n_{\mathrm{eff}} n_g - n_2^2}{n_1^2 - n_2^2} \tag{9.57}$$

The comparison of q_1 as given by (9.57) with b as given by (6.55) is interesting. Equations (9.57) and (9.54) allow us to compute the dielectric losses for a weakly guiding fiber completely from a dispersion curve (β vs. ω) for each mode, if we can accurately read the slope at the desired frequency.

9.5 WALL LOSSES IN WAVEGUIDES

The effect of wall losses due to finite conductivity of the boundary is not so easy to account for as that of dielectric losses. If we were to let the walls be treated as a finitely conducting medium, the waveguide would no longer be electrically homogeneous, and any simplifications which were possible in the case of ideal walls (such as the existence of pure TE and TM modes) will be lost. Besides, if (as would be reasonable to assume in practice) the walls are *highly* conducting, there should be very little field penetration into the walls at all. In fact, recalling the behavior of a plane wave being refracted into a highly conducting medium (Figure 9.6), we might guess that the fields vary almost purely normally away from the interface and rapidly decay in that same direction within a few skin depths δ_c, where

$$\delta_c = \sqrt{\frac{2}{\omega \mu_c \sigma_c}} \tag{9.58}$$

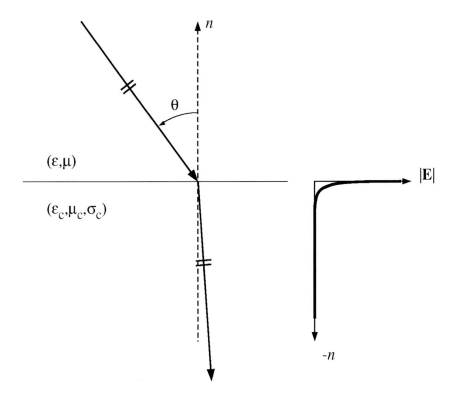

Figure 9.6 Plane-wave penetration into a good conductor for $(\epsilon_c - j\sigma_c/\omega)/\epsilon \ll 1$.

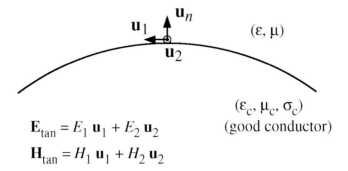

$$\mathbf{E}_{\tan} = E_1 \mathbf{u}_1 + E_2 \mathbf{u}_2$$

$$\mathbf{H}_{\tan} = H_1 \mathbf{u}_1 + H_2 \mathbf{u}_2$$

Figure 9.7   Geometry for impedance boundary conditions ($\mathbf{u}_1$ and $\mathbf{u}_2$ are unit vectors tangent to the interface).

This characteristic behavior of fields within a good conductor makes it possible to formulate an approximate boundary condition at its surface which *eliminates the need to calculate fields within the conductor at all.* Boundary conditions of this type are called *impedance boundary conditions*[3] and take the form

$$\mathbf{E}_{\tan} = Z_S \mathbf{u}_n \times \mathbf{H}_{\tan} \tag{9.59}$$

where $\mathbf{u}_n$ is the unit normal vector pointing away from the conductor at the interface, and $\mathbf{E}_{\tan}$ and $\mathbf{H}_{\tan}$ are the components of the fields *outside* the conductor, *tangential* to the interface (Figure 9.7). The quantity Z_S is known as the *surface impedance* associated with the given interface. It may be a function of position along the surface.

In Appendix L, we show that, for a smooth, highly conducting surface,

$$Z_S = \zeta_c = \sqrt{\frac{\mu_c}{\epsilon_c - j\sigma_c/\omega}} \tag{9.60}$$

where ζ_c is the complex wave impedance for a plane wave in the conducting medium.[4] If, as is usually the case, $\sigma_c/\omega\epsilon_c \gg 1$, we have approximately

$$Z_S = (1+j)\sqrt{\frac{\omega\mu_c}{2\sigma_c}} = \frac{1+j}{\delta_c\sigma_c} \equiv R_S + jX_S \tag{9.61}$$

Unfortunately, even the use of the simpler boundary condition (9.59) is not enough to allow exact solution of the mode fields for most cross-sectional shapes (even rectangular). Thus, we will develop here an approximate method to determine the effect of wall losses on the *propagation coefficient* by assuming that the *fields* are very nearly what they would be if the walls were perfectly conducting. To do this, we apply the reciprocity theorem to the fields

$$
\begin{aligned}
\mathbf{E}^a &= \boldsymbol{\mathcal{E}}_{m0} e^{-\gamma_{m0}z}; \quad \mathbf{H}^a = \boldsymbol{\mathcal{H}}_{m0} e^{-\gamma_{m0}z} \\
\mathbf{E}^b &= \boldsymbol{\mathcal{E}}_{-m} e^{\gamma_m z}; \quad \mathbf{H}^b = \boldsymbol{\mathcal{H}}_{-m} e^{\gamma_m z}
\end{aligned}
\tag{9.62}
$$

---

[3]Though (9.59) is superficially similar to the resistive sheet boundary condition (5.90), it differs in that $\mathbf{H}_{\tan}$ is nonzero only on one side ($n > 0$) of the boundary.

[4]Many types of surface (rough, covered with dielectric layers, etc.) can also be characterized by surface impedances Z_S given by other expressions.

where $\mathcal{E}_{m0}$, $\mathcal{H}_{m0}$, and γ_{m0} are associated with a mode of the *perfect* guide, and $\mathcal{E}_{-m}$, $\mathcal{H}_{-m}$, and $-\gamma_m$ with a corresponding mode of the guide with imperfect walls propagating (or attenuating) in the opposite direction. We apply reciprocity to the volume of Figure 9.4. The manipulations are very similar to those in the proof of orthogonality, except that the integral over ΔS does not vanish because $\mathbf{u}_n \times \mathcal{E}_m$ is not zero on the guide wall but obeys (9.59). Since

$$\int_{\Delta S} = \int_0^{z_1} dz \oint_{C_0} dl,$$

we obtain

$$\gamma_m - \gamma_{m0} = \frac{\oint_{C_0} Z_S \mathcal{H}_{m0(tan)} \cdot \mathcal{H}_{-m(tan)}\, dl}{\int_{S_0} (\mathcal{E}_{m0} \times \mathcal{H}_{-m} - \mathcal{E}_{-m} \times \mathcal{H}_{m0}) \cdot \mathbf{u}_z\, dS} = -\frac{\oint_{C_0} Z_S \mathcal{H}_{m0(tan)} \cdot \mathcal{H}_{-m(tan)}\, dl}{\int_{S_0} (\mathcal{E}_{m0} \times \mathcal{H}_m + \mathcal{E}_m \times \mathcal{H}_{m0}) \cdot \mathbf{u}_z\, dS}$$

(9.63)

Though (9.63) is exact, we can do little with it as it stands since $\mathcal{E}_m$ and $\mathcal{H}_m$ are not known. However, we reason that, if Z_S is small enough, we should have

$$\mathcal{E}_{-m} \simeq \mathcal{E}_{-m0} \quad \text{and} \quad \mathcal{H}_{-m} \simeq \mathcal{H}_{-m0}$$

(9.64)

so that

$$\gamma_m - \gamma_{m0} \simeq \frac{\oint_{C_0} Z_S \mathcal{H}_{m0} \cdot \mathcal{H}_{-m0}\, dl}{-2 \int_{S_0} \mathcal{E}_{m0} \times \mathcal{H}_{m0} \cdot \mathbf{u}_z\, dS}$$

(9.65)

Only the components of $\mathcal{H}$ tangential to the wall enter into the line integral in the numerator of (9.65); these would be the scalar components $\mathcal{H}_{m0l}$ and $\mathcal{H}_{m0z}$ in the coordinate systems of Figures 5.5 or 8.5.

If the medium filling the guide is lossless, the real part of (9.65) is interpretable as one half the power lost into the walls per unit length divided by the power carried through the cross-section S_0 of the guide. Indeed, the real part of (9.65) gives the attenuation coefficient of the mode, and if the unperturbed guide is lossless, then for a propagating mode (see (9.29)),

$$\alpha_m \simeq \frac{\oint_{C_0} R_S \mathcal{H}_{m0} \cdot \mathcal{H}_{-m0}\, dl}{-2 \int_S \mathcal{E}_{m0} \times \mathcal{H}_{m0} \cdot \mathbf{u}_z\, dS} = \frac{\oint_{C_0} R_S \mathcal{H}_{m0} \cdot \mathcal{H}_{m0}^*\, dl}{2 \int_{S_0} \mathcal{E}_{m0} \times \mathcal{H}_{m0}^* \cdot \mathbf{u}_z\, dS}$$

(9.66)

For a TM mode (and also, as a special case, for a TEM mode) of a homogeneously filled and otherwise loss-free waveguide whose walls have uniform surface resistance, we have from (9.61), (9.66), (5.47), and (5.48) that, for $\omega > \omega_{c,m}$,

$$\alpha_m \simeq \frac{R_S}{2\zeta} \frac{1}{\sqrt{1 - \omega_{c,m}^2/\omega^2}} \frac{\oint_{C_0} |\mathcal{H}_{m0l}|^2\, dl}{\int_{S_0} \mathcal{H}_{m0T} \cdot \mathcal{H}_{m0T}^*\, dS}$$

(9.67)

which is the attenuation (or additional attenuation) due to finite conductivity of the walls. Here ζ is the wave impedance of the medium *filling* the guide, while R_S is the surface resistance of the walls (eqn. (9.61)).

For TE modes of homogeneously filled guides, a similar but more complicated expression follows from (5.49) and (5.50) and problem p5-9:

$$
\alpha_m \simeq \frac{R_S}{2\zeta} \left\{ \sqrt{1 - \omega_{c,m}^2/\omega^2} \frac{\oint_{C_0} |\mathcal{H}_{m0l}|^2 \, dl}{\int_{S_0} \mathcal{H}_{m0T} \cdot \mathcal{H}_{m0T}^* \, dS} + \frac{\omega_{c,m}^2/\omega^2}{\sqrt{1 - \omega_{c,m}^2/\omega^2}} \frac{\oint_{C_0} |\mathcal{H}_{m0z}|^2 \, dl}{\int_{S_0} |\mathcal{H}_{m0z}|^2 \, dS} \right\} \quad (9.68)
$$

The ratios of the line and surface integrals in (9.67) and (9.68) are independent of frequency, so it is evident that the attenuation α_{TM} of a TM mode due to wall losses goes as

$$
\alpha_{\text{TM}} \sim A_m \sqrt{\omega} \qquad \text{as } \omega \to \infty
$$
$$
\alpha_{\text{TE}} \sim B_m \sqrt{\omega} + C_m \omega^{-3/2} \quad \text{as } \omega \to \infty,
$$

where A_m, B_m, and C_m are mode-dependent constants. Clearly the attenuation of all modes increases for high enough frequency except for TE modes (if any exist) for which $B_m = 0$— i.e., for which $|\mathcal{H}_{m0l}|^2$ integrates to zero around the boundary C_0. We will see below that the TE_{0m} mode of the hollow circular waveguide is one such, which (along with the fact that, unlike the TE_{11} mode, it is nondegenerate) makes it attractive to use in practical applications, especially at millimeter-wave, submillimeter-wave and even infrared frequencies.

9.5.1   EXAMPLE: THE TE$_{01}$ MODE OF A HOLLOW CIRCULAR WAVEGUIDE

Consider the TE_{0m} modes of a hollow circular waveguide as studied in Chapter 5. Its $\mathcal{E}$ and $\mathcal{H}$ fields are given by (5.78) and (5.82)-(5.84) with $n = 0$. It will be noticed that the transverse component of $\mathcal{H}$ which is tangential to the wall, $\mathcal{H}_l = \mathcal{H}_\phi$, is identically zero. The first term in (9.68) thus vanishes, and the second term is evaluated by substitution of the field expressions from Chapter 5 into the remaining term and carrying out the integrals. With the help of (C.40), (C.41) and (C.49) we find that

$$
\alpha_{\text{TE}_{0m}} \simeq \frac{R_S}{\zeta a} \frac{f_c^2/f^2}{\sqrt{1 - f_c^2/f^2}} \quad (9.69)
$$

where $f_c = f_{c,\text{TE}_{0m}}$ is the cutoff frequency of the mode. The decay of this expression as frequency increases can be seen.

In Figure 9.8, a plot of the theoretical (eqn. (9.69)) and measured values of the attenuation coefficient of the TE_{01} mode are presented. The actual losses are seen to decrease with frequency less rapidly than theoretically predicted; the deviation between theory and experiment is attributable to an increase of Z_S with frequency when the roughness of the conductor surface is taken into account. This effect will eventually negate the inherent low-loss properties of this mode, but there will be a broad range of frequencies for which very low attenuation can be achieved in practice.

The TE_{01} mode was extensively researched during the 1950s and 1960s as a practical communications channel. The Bell System put into operation the WT-4 millimeter waveguide system using this mode. The system has the following parameters:

Repeater spacing: 20 miles
Bandwidth: 40-110 GHz
Capacity: 60 2-way channels @ 282 M Bits/sec/channel = 240,000 2-way voice circuits.

In order to use the TE_{01} mode effectively, ways had to be found to suppress the other modes which can propagate along with it: at the very least, the TE_{11}, TE_{21} and TM_{11} (each in

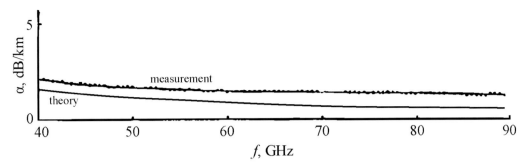

Figure 9.8 Theoretical and measured values of attenuation coefficient α_c due to wall losses for the TE$_{01}$ mode of a circular metallic waveguide: $a = 25.5$ mm, $\sigma_c = 5.8 \times 10^7$ S/m, $\mu_c = \mu_0$.

both polarizations) and the TM$_{01}$ modes. Mode filters and converters can be devised for this purpose; for example, a wire helix can be used to line the waveguide wall, which disturbs the TE$_{01}$ mode very little, but tends to suppress the other undesired modes.

9.5.2 THE INCREMENTAL INDUCTANCE RULE

In some circumstances, it is possible to express the wall loss in a different form which is easier to compute than the ratio of field integrals (9.65). This is accomplished by the *incremental inductance rule* of H. A. Wheeler, which we will derive below. Assume for present purposes that μ and ϵ do not change (or change negligibly) with frequency. Then the results (J.3) and (J.19) from Appendix J show that

$$\frac{\partial \gamma_{m0}}{\partial \omega} = -j \frac{\int_{S_0} \mu \boldsymbol{\mathcal{H}}_{m0} \cdot \boldsymbol{\mathcal{H}}_{-m0} \, dS}{\int_{S_0} \boldsymbol{\mathcal{E}}_{m0} \times \boldsymbol{\mathcal{H}}_{m0} \cdot \mathbf{u}_z \, dS} \tag{9.70}$$

Using this expression in (9.65), we find that (again, for an otherwise lossless mode)

$$\gamma_m - \gamma_{m0} \simeq \frac{(1+j)\omega}{2} \frac{\partial \beta_{m0}}{\partial \omega} q_c \tag{9.71}$$

where a sort of "magnetic filling factor" has been defined by

$$q_c \equiv \frac{\oint_{C_0} \delta n(l) \boldsymbol{\mathcal{B}}_{m0} \cdot \boldsymbol{\mathcal{H}}_{m0}^* \, dl}{\int_{S_0} \boldsymbol{\mathcal{B}}_{m0} \cdot \boldsymbol{\mathcal{H}}_{m0}^* \, dS} \tag{9.72}$$

with the distance $\delta n(l)$ defined as

$$\delta n(l) = \frac{R_S}{\omega \mu} \simeq \frac{\mu_c}{\mu} \frac{\delta_c}{2} \tag{9.73}$$

the latter approximation holding if R_S is given by (9.61). Here R_S, ϵ and μ may be functions of position. Note that q_c is analogous to the dielectric filling factors (9.42) of the previous section, and that

$$\gamma_m - \gamma_{m0} \simeq \frac{(1+j)k_0}{2n_{\text{eff}}} q_c \tag{9.74}$$

which is completely analogous to (9.41).

The integral around C_0 in (9.72) is shown in Appendix M to be equal to $\delta l/l_0$, the incremental fractional change in inductance per unit length due to indentation of the conducting boundaries $C_0 = C_1 \cup C_2$ by the small distance $\delta n(l)$ given in (9.73). Thus, from (M.18), we have

$$\gamma_m - \gamma_{m0} \simeq \frac{(1+j)\omega}{2} \frac{\partial \beta_{m0}}{\partial \omega} \frac{\delta l}{l_0} \tag{9.75}$$

This, in its most general form, is the *incremental inductance rule*. If μ, μ_c and δ_c (and thus also δn) are independent of position, we can write (9.75) in the form

$$\gamma_m - \gamma_{m0} \simeq \frac{(1+j)\omega}{2} \frac{\partial \beta_{m0}}{\partial \omega} \frac{\delta n}{l_0} \frac{\delta l}{\delta n} \tag{9.76}$$

where $\delta l/\delta n$ is calculable by direct differentiation of l_0 with respect to the geometrical dimensions of the line, without the need to perform any integrations or to know the values of the fields at all.

The incremental inductance rule simplifies under the conditions for a quasi-TEM mode. If l_0 and c_0 are independent of ω as $\omega \to 0$, then $\beta_{m0} \simeq \omega\sqrt{l_0 c_0}$ and thus $\partial \beta_{m0}/\partial \omega \simeq \sqrt{l_0 c_0}$, and from (9.76) we have:

$$\gamma_m - \gamma_{m0} \simeq \frac{(1+j)R_S}{2Z_{c0}} \frac{1}{\mu} \frac{\delta l}{\delta n} \tag{9.77}$$

where $Z_{c0} = \sqrt{l_0/c_0}$ is the characteristic impedance of the original lossless line. Of interest is the real part of (9.77), which gives the attenuation coefficient of the mode due to wall conduction losses:

$$\alpha_m \simeq \frac{R_S}{2\mu Z_{c0}} \frac{\delta l}{\delta n} \tag{9.78}$$

This is the most commonly encountered form of the incremental inductance rule.

We can also cast the incremental inductance rule in terms of equivalent transmission line parameters: the series impedance per unit length $r + j\omega l$ and the shunt admittance per unit length $g + j\omega c$. According to classical transmission-line theory, (equation (1.9)), we have

$$\gamma_m = \sqrt{(r + j\omega l)(g + j\omega c)} \tag{9.79}$$

Correspondingly, for the unperturbed (lossless) line we have

$$\gamma_{m0} = j\beta_{m0} = j\omega\sqrt{l_0 c_0} \tag{9.80}$$

If $r/(\omega l_0)$ and $(l - l_0)/l_0$ are small compared to one, while $g = 0$ and $c = c_0$, (i. e., the finite wall conductivity does not change the series line parameters r and l much and the shunt parameters not at all), then

$$\gamma_m - \gamma_{m0} \simeq j\beta_{m0} \left[\frac{l - l_0}{2l_0} + \frac{r}{2j\omega l_0} \right] \tag{9.81}$$

Comparing this to equation (9.75), we find that we can obtain the same expression if we take

$$l \simeq l_0 + \delta l \tag{9.82}$$

and

$$r \simeq \omega \delta l \tag{9.83}$$

The inductance of the line with imperfect conductors is thus the same as that of a lossless line whose walls are indented by a distance (9.73). The additional inductance δl is often referred to as the *internal inductance* of the conductors of the line. It is worth noting that for the lossy line $lc \neq \mu\epsilon$; even if the unperturbed mode was pure TEM, the perturbed mode will only be quasi-TEM.

A word of caution is in order regarding the application of (9.77) to strip-type transmission lines. It is common practice to model the strips of such lines as infinitely thin, which does not cause any essential error when computing l and c so long as the actual thickness of the strip is small compared to the width of the strip and the substrate thickness. The infinitely thin strip, however, is an inadequate model when considering conductor losses, because the strip thickness is not large compared to a skin depth, and this violates the validity of our surface impedance concept. More than this, the inverse square-root singularity in the current at the edges of an infinitely thin strip (cf. Figure 8.18) causes the line integral in the numerator of (9.63) to diverge. In order to compute conductor loss effects for thin strip conductors, a way must be found to correctly account for the influence of the nonzero thickness of the strip as well as the detailed shape of the strip edge. Methods to do this are found in the references at the end of the chapter.

9.5.3 EXAMPLE: CONDUCTOR LOSS IN COAXIAL LINE

Consider once again the case of a coaxial line as shown in Figure 8.7. We have

$$l_0 = \sqrt{\mu\epsilon}\, Z_{c0} = \frac{\mu}{2\pi} \ln\left(\frac{b}{a}\right) \tag{9.84}$$

If both inner and outer conductors have the same conductivity σ_c and permeability μ_c, we can simply compute

$$\frac{\delta l}{\delta n} = \frac{\partial l_0}{\partial b} - \frac{\partial l_0}{\partial a} = \frac{\mu}{2\pi}\left(\frac{1}{b} + \frac{1}{a}\right) \tag{9.85}$$

Note that we place a minus sign in front of the $\partial/\partial a$ term in (9.85) because a recession of the inner conductor means a *decrease* in a, whereas the corresponding motion of the outer conductor means an *increase* in b. From (9.82) and (9.83), we have

$$l \simeq \frac{\mu}{2\pi}\left\{\ln\left(\frac{b}{a}\right) + \frac{\delta_c\mu_c}{2\mu}\left(\frac{1}{b} + \frac{1}{a}\right)\right\} \tag{9.86}$$

$$r \simeq \frac{\omega\delta_c\mu_c}{4\pi}\left(\frac{1}{b} + \frac{1}{a}\right), \tag{9.87}$$

and by (9.78) the attenuation coefficient is

$$\begin{aligned}
\alpha &\simeq \frac{\omega\mu_c\delta_c}{4\zeta \ln(b/a)}\left(\frac{1}{b} + \frac{1}{a}\right) \\
&\simeq \frac{1}{2\zeta\delta_c\sigma_c \ln(b/a)}\left(\frac{1}{b} + \frac{1}{a}\right).
\end{aligned}$$

If α is plotted as a function of a for a fixed value of b, the result (in arbitrary units) is as shown in Figure 9.9. The attenuation is minimized when $\partial\alpha/\partial a = 0$, or

$$\frac{b}{a} = \frac{1}{\ln\left(\frac{b}{a}\right) - 1}$$

whose numerical solution gives an optimum outer to inner conductor radius ratio of $(b/a)_{\text{opt}} = 3.591$. For this value, the characteristic impedance has the value

$$Z_{c,\text{opt}} = \frac{1}{\sqrt{\epsilon_r}} \frac{\zeta_0}{2\pi} \ln\left(\frac{b}{a}\right)_{\text{opt}} = \frac{76.71\,\Omega}{\sqrt{\epsilon_r}}$$

If $\epsilon_r = 1$, the optimum characteristic impedance is approximately $75\,\Omega$, while for polyethylene ($\epsilon_r = 2.26$), the optimum value is close to $50\,\Omega$.

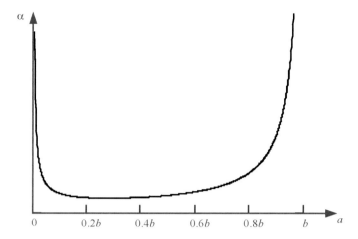

Figure 9.9 Variation of coaxial transmission line attenuation constant versus a for fixed b.

9.6 CONCLUSION: LOSS EFFECTS ON ORTHOGONALITY AND POWER FLOW

The orthogonality properties (9.22) and (9.24) hold even for the modes of a lossy waveguide, whose dielectric constant is complex or whose walls impose an impedance boundary condition. Only (9.28) is restricted to the case of lossless waveguides. This does mean that the conclusions of Section 9.3 on power flow must be modified for a lossy waveguide, however, since the "power orthogonality" condition (9.28) is no longer true. A lossy waveguide on which several modes exist transfers a power which generally consists not merely of the sum of the powers in individual modes, but is a complicated expression involving cross terms between all the various modes. In a waveguide with low losses, the conclusions arrived at for lossless guides can be expected to hold true to a good approximation, but the limitations of our earlier conclusions should be kept in mind.

Likewise, our derivation of the effect of wall losses was done initially without regard for whether the corresponding guide with ideally conducting walls was lossless or not. Thus, (9.65) is valid for any unperturbed guide, whether or not it has dielectric losses. This unperturbed guide's dielectric losses can be computed as in Section 9.4, however, and from the form of (9.65) we see that in this approximation the dielectric losses and wall losses have an effect on the attenuation coefficient of a mode that is simply additive:

$$\alpha_m = \alpha_{m,c} + \alpha_{m,d} \tag{9.88}$$

Any attenuation $\alpha_{m,d}$ due to dielectric losses from the material filling the guide must be added to the wall loss attenuation $\alpha_{m,c}$ to give the total attenuation coefficient of the mode.

9.7 NOTES AND REFERENCES

The discussion of reciprocity, reaction and orthogonality is inspired mostly by

Harrington (1961), Chapters 3,7.

In Harrington's book, the variational or stationary character of integral formulas derived from the reciprocity theorem is discussed; this means that if approximate expressions for the fields are used in such a formula, the error in the integral is of smaller order of magnitude than the error in the fields. This observation was originally due to Rumsey:

V. H. Rumsey, "Reaction concept in electromagnetic theory," *Phys. Rev.*, vol. 94, pp. 1483-1491 (1954).

The reader may also consult

Collin (1991), Sections 1.10, 5.2 and 11.7;
Vainshtein (1988), Chapter 14;
Olyslager (1999);

for other points of view about reciprocity. The quotation near the end of Section 9.1 comes from:

S. Milligan, *More Goon Show Scripts*. London: Sphere Books Ltd., 1974, p. 22.

Equation (9.43) for the dielectric loss in a microstrip is due to:

M. V. Schneider, "Dielectric loss in hybrid integrated circuits," *Proc. IEEE*, vol. 57, pp. 1206-1207 (1969).

The reader should exercise caution in using formulas involving the filling factors q of planar lines which are given in the literature, since not all authors use the same definition for q. Equation (9.43) is completely accurate to first order in ϵ'', regardless of the geometry of the line involved. Some authors quote the formula

$$\epsilon_{\text{eff}} \simeq (1 - q) + q\epsilon_r \tag{9.89}$$

implying

$$q \simeq \frac{\epsilon_{\text{eff}} - 1}{\epsilon_r - 1} \tag{9.90}$$

for the filling factor. This would be exactly true if q were independent of ϵ_r—that is, if ϵ_{eff} were a linear function of ϵ_r. Another widely used formula for the filling factor of microstrip is

$$q \simeq \frac{1}{2}[1 + (1 + 10h/w)^{-1/2}] \tag{9.91}$$

but the accuracy of this formula is not quite as good as that of (9.90), as is checked in the solution of problem p9-15(b). Expression (9.90) can also be used with fairly good accuracy for CPW and CPS, but not necessarily for other planar lines, especially the slotline.

The connection of dielectric losses with the group velocity for weakly-guiding fibers seems to have been first noted by

D. Gloge, "Propagation effects in optical fibers," *IEEE Trans. Micr. Theory Tech.*, vol. 23, pp. 106-120 (1975).

for the case of the circular fiber; this paper is also reprinted in Gloge (1976). The approximate expression $q_1 \simeq V^2/(V^2 + 2)$ for q_1 of the TE_0 mode of a dielectric slab was given by:

D. Botez, "Analytical approximation of the radiation confinement factor for the TE_0 mode of a double heterojunction laser," *IEEE J. Quantum Electron.*, vol. 14, pp. 230-232 (1978).

who called it the radiation confinement factor. The expression was shown to be accurate for small V as well as for large V, and was used by

K.-L. Chen and S. Wang, "An approximate expression for the effective refractive index in symmetric DH lasers," *IEEE J. Quantum Electron.*, vol. 19, pp. 1354-1356 (1983).

to obtain the approximate relation

$$b \simeq 1 - \frac{2}{V^2} \ln(1 + V^2/2)$$

for the TE_0 mode for any value of V.

For the discussion of wall losses, see

Johnson (1965), pp. 153–162;
Collin (1991), Section 5.3.

Information on the Bell WT-4 waveguide system can be found in:

Martin (1969), Chapters 8 and 9;
Martin (1971), Chapter 14.

Figure 9.8 is based on experimental data given in:

S. Hatano and F. Nihei, "Measurement of surface resistance in oversized circular waveguide at millimeter wavelengths," *IEEE Trans. Micr. Theory Tech.*, vol. 24, pp. 886-887 (1976).

The incremental inductance rule was first given in

H. A. Wheeler, "Formulas for the skin effect," *Proc. IRE*, vol. 30, pp. 412-424 (1942).

For a rigorous proof, see

V. Alessandrini, *et al.*, "The skin effect in multiconductor systems," *Int. J. Electron.*, vol. 40, pp. 57–63 (1976).

One way of computing conductor losses for planar transmission lines is to use formulas for the inductance per unit length for strips of *finite* thickness, and then apply the incremental inductance rule. Such formulas are more complicated than those given in Chapter 8, and for microstrip are given in

M. V. Schneider, "Microstrip lines for microwave integrated circuits," *Bell Syst. Tech. J.*, vol. 48, pp. 1422-1444 (1969).
H. A. Wheeler, "Transmission-line properties of a strip on a dielectric sheet on a plane," *IEEE Trans. Micr. Theory Tech.*, vol. 25, pp. 631-647 (1977).

See also

Gupta, Garg and Chadha (1981), Chapter 3.

An alternative to this approach which requires only information about the properties of the line with infinitely thin strip conductors was developed independently in:

L. Lewin, "A method of avoiding the edge current divergence in perturbation loss calculations," *IEEE Trans. Micr. Theory Tech.*, vol. 32, pp. 717-719 (1984).
L. A. Vainshtein and S. M. Zhurav, "Strong skin effect at the edges of metal plates," *Sov. Tech. Phys. Lett.*, vol. 12, pp. 298-299 (1986).

and has been extended and refined by

S. M. Zhurav, "Edge loss in metal plates of rectangular cross section," *Sov. Tech. Phys. Lett.*, vol. 13, pp. 147-148 (1987).

G. Ya. Slepyan, "More precise impedance conditions for calculating the heat losses in a thin open metal screen," *Sov. J. Commun. Technol. Electron.*, vol. 33, no. 4, pp. 172-175 (1988).

E. L. Barsotti, E. F. Kuester and J. M. Dunn, "Strip edge shape effects on conductor loss calculations using the Lewin/Vainshtein method," *Electron. Lett.*, vol. 26, pp. 983-985 (1990).

E. L. Barsotti, E. F. Kuester and J. M. Dunn, "Effect of metallization edge shape on conductor loss of open coplanar waveguide," *Micr. Opt. Technol. Lett.*, vol. 3, pp. 389-391 (1990).

C. L. Holloway and E. F. Kuester, "Edge shape effects and quasi-closed form expressions for the conductor loss of microstrip lines," *Radio Science*, vol. 29, pp. 539-559 (1994).

C. L. Holloway and E. F. Kuester, "A quasi-closed form expression for the conductor loss of CPW lines, with an investigation of edge shape effects," *IEEE Trans. Microwave Theory and Techniques*, vol. 43, pp. 2695-2701 (1995).

9.8 PROBLEMS

p9-1 Let a "b" state be produced by an electric current loop situated along a closed contour C:

$$\mathbf{J}^b = I\mathbf{u}_l \delta_C(x, y, z)$$

where $\mathbf{u}_l$ is the unit vector tangent to the contour C at any point on the contour, and the delta function δ_C is to be understood as concentrated at the loop in the following sense: for any function $f(x, y, z)$, we have

$$\int_V f\delta_C \, dV = \int_C f \, dl$$

if the volume V completely contains the loop C. The "b" state possesses no magnetic current sources. Let the "a" state field be produced by any sources at all, provided that $\mathbf{M}^a = 0$ for points at the loop.

(a) Express the reaction $\langle a, b \rangle$ as a line integral around C.

(b) Use Stokes' theorem to re-express $\langle a, b \rangle$ as an integral over a surface S spanning C.

(c) Suppose this surface S contracts to zero and the current I on the loop goes to infinity in such a way that $I \int_S \mathbf{u}_n \, dS \to \mathbf{m}$, where $\mathbf{m}$ is a constant vector, and $\mathbf{u}_n$ is a unit vector normal to S. Show that $\langle a, b \rangle$ approaches an expression of the form (9.17) in this limit. [*Note:* $\mathbf{m}$ *is called the magnetic dipole moment of the loop.*]

p9-2 It is known that an arbitrary time-harmonic electric field $\mathbf{E}$ can be expressed in terms of a scalar potential Φ and a vector potential $\mathbf{A}$ as follows:

$$\mathbf{E} = -\nabla\Phi - j\omega\mathbf{A}$$

Show that, in the absence of magnetic currents ($\mathbf{M} \equiv 0$), the reaction $\langle a, b \rangle$ can also be expressed as

$$\langle a, b \rangle = -j\omega \int_{V^b} (\Phi^a \rho^b + \mathbf{A}^a \cdot \mathbf{J}^b) \, dV$$

where V^b is a volume which contains the "b" sources and on whose surface $\mathbf{u}_n \cdot \mathbf{J}^b = 0$, while ρ^b is the volume (electric) charge density corresponding to $\mathbf{J}^b$.

p9-3   Let $S(z)$ be a surface in the plane $z = \text{constant}$ bounded by the closed curve $C(z)$. The shapes of $S(z)$ and $C(z)$ do not change as z is varied; only a shift of their location is performed.

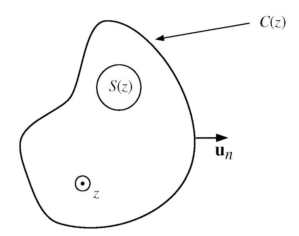

Prove the following consequence of the reciprocity theorem:

$$-W(a, b; z) + W(b, a; z)$$
$$= \frac{d}{dz} \int_{S(z)} (\mathbf{E}^a \times \mathbf{H}^b - \mathbf{E}^b \times \mathbf{H}^a) \cdot \mathbf{u}_z \, dS$$
$$+ \oint_{C(z)} (\mathbf{E}^a \times \mathbf{H}^b - \mathbf{E}^b \times \mathbf{H}^a) \cdot \mathbf{u}_n \, dl$$

where $\mathbf{u}_n$ is the outward unit normal vector to $C(z)$ and

$$W(a, b; z) = \int_{S(z)} (\mathbf{E}^a \cdot \mathbf{J}^b - \mathbf{H}^a \cdot \mathbf{M}^b) \, dS$$

is a sort of reaction density per unit length in the z-direction. [Hint: Integrate this expression from 0 to z and compare with equation (9.8).]

p9-4   Consider two states with fields $\mathbf{E}^{a,b}$ and $\mathbf{H}^{a,b}$ in the surface $S(z)$ and on the curve $C(z)$ as defined in problem p9-3. Use (9.5) to show that

$$\frac{d}{dz} \int_{S(z)} \mathbf{E}^a \times \mathbf{H}^b \cdot \mathbf{u}_z \, dS + \oint_{C(z)} \mathbf{E}^a \times \mathbf{H}^b \cdot \mathbf{u}_n \, dl =$$
$$-j\omega \int_{S(z)} (\epsilon \mathbf{E}^a \cdot \mathbf{E}^b + \mu \mathbf{H}^a \cdot \mathbf{H}^b) \, dS - \int_{S(z)} (\mathbf{J}^b \cdot \mathbf{E}^a + \mathbf{M}^a \cdot \mathbf{H}^b) \, dS$$

p9-5   A perfect magnetic conductor (PMC) is defined as a surface on which the total tangential magnetic field is equal to zero ($\mathbf{u}_n \times \mathbf{H} = 0$). Use the reciprocity theorem to prove that a surface-concentrated magnetic current density $\mathbf{M}_S$ located at and tangential to a PMC surface S produces zero field exterior to that surface.

p9-6 Without directly invoking the reciprocity theorem, show that for a pair of TE modes on a hollow metallic waveguide the properties

$$\int_{S_0} \nabla_T \mathcal{H}_{z1} \cdot \nabla_T \mathcal{H}_{z2} \, dS = 0 \qquad (\gamma_1^2 \neq \gamma_2^2)$$

$$\int_{S_0} \mathcal{H}_{z1} \mathcal{H}_{z2} \, dS = 0 \qquad (\gamma_1^2 \neq \gamma_2^2)$$

follow from orthogonality property (9.24).

p9-7 Without directly invoking the reciprocity theorem, show that for a pair of TM modes on a hollow metallic waveguide the properties

$$\int_{S_0} \nabla_T \mathcal{E}_{z1} \cdot \nabla_T \mathcal{E}_{z2} \, dS = 0 \qquad (\gamma_1^2 \neq \gamma_2^2)$$

$$\int_{S_0} \mathcal{E}_{z1} \mathcal{E}_{z2} \, dS = 0 \qquad (\gamma_1^2 \neq \gamma_2^2)$$

follow from orthogonality property (9.24).

p9-8 Without directly invoking the reciprocity theorem, show that a TE mode is always orthogonal to a TM mode in the sense of (9.23) for a hollow waveguide, regardless of whether or not $\gamma_m^2 = \gamma_n^2$.

p9-9 In equations (5.75)–(5.77), expressions for $\mathcal{E}_T$ and $\mathcal{H}_T$ of the TM_{11c} and TM_{11s} circular waveguide modes are given. Show by direct calculation of the integral that (9.22) holds for $m = +\mathrm{TM}_{11c}$ and $n = -\mathrm{TM}_{11s}$, even though $\gamma_m = -\gamma_n$ for these degenerate modes.

p9-10 Obtain an "energy orthogonality" property for modes on an arbitrary waveguide by proceeding as follows. In (9.5), let the a state be the field of mode m of the waveguide, and the b state be mode n. Integrate over the slice of waveguide shown in Figure 9.4. Show that this results in the orthogonality relation (for normalized waveguide modes):

$$\frac{1}{4} \int_{S_0} (\epsilon \mathcal{E}_m \cdot \mathcal{E}_n + \mu \mathcal{H}_m \cdot \mathcal{H}_n) \, dS \; = \; 0 \qquad (\text{if } m \neq n)$$

$$= \; \frac{\gamma_m}{j\omega} \qquad (\text{if } m = n > 0; \text{ forward mode})$$

$$= \; -\frac{\gamma_m}{j\omega} \qquad (\text{if } m = n < 0; \text{ reverse mode})$$

Note the connection with (5.29)-(5.30).

In the same way, obtain the energy orthogonality relation corresponding to (9.28) for a lossless waveguide.

p9-11 Suppose there exist two modes $\mathcal{E}_1, \mathcal{H}_1$ and $\mathcal{E}_2, \mathcal{H}_2$ in a waveguide which have the same propagation coefficient $\gamma_1 = \gamma_2 = \gamma$. Suppose, too, that the mode fields $\mathcal{E}_1$, $\mathcal{H}_1$ and $\mathcal{E}_{-2}, \mathcal{H}_{-2}$ are not orthogonal to each other. That is,

$$\int_{S_0} (\mathcal{E}_1 \times \mathcal{H}_{-2} - \mathcal{E}_{-2} \times \mathcal{H}_1) \cdot \mathbf{u}_z \, dS \neq 0.$$

Find the constant A in order that the combination fields

$$\mathcal{E}_a = \mathcal{E}_1 + A\mathcal{E}_2$$

$$\mathcal{H}_a = \mathcal{H}_1 + A\mathcal{H}_2$$

are orthogonal to $\mathcal{E}_1$, $\mathcal{H}_1$. In other words, give A in terms of $\mathcal{E}_1$, $\mathcal{H}_1$ and $\mathcal{E}_2$, $\mathcal{H}_2$ such that

$$\int_{S_0} (\mathcal{E}_1 \times \mathcal{H}_{-a} - \mathcal{E}_{-a} \times \mathcal{H}_1) \cdot \mathbf{u}_z \, dS = 0.$$

p9-12 Suppose that instead of a perfectly conducting boundary wall, a waveguide has an impedance wall, at which the fields obey (9.59). Show that even in this more general case, orthogonality relations (9.22) and (9.23) still hold.

p9-13

(a) Obtain the following "complex power"-type of reciprocity relation:

$$\nabla \cdot (\mathbf{E}^a \times \mathbf{H}^{b*} + \mathbf{E}^{b*} \times \mathbf{H}^a) = -\mathbf{E}^{b*} \cdot \mathbf{J}^a - \mathbf{H}^{b*} \cdot \mathbf{M}^a - \mathbf{E}^a \cdot \mathbf{J}^{b*} - \mathbf{H}^a \cdot \mathbf{M}^{b*}$$

valid where ϵ and μ are real quantities. If ϵ and μ are complex, we may still use this result if we absorb the effects of their imaginary parts ϵ'' and μ'' into the sources as $\mathbf{J}^{a,b} = \omega\epsilon''\mathbf{E}^{a,b}$, and $\mathbf{M}^{a,b} = \omega\mu''\mathbf{H}^{a,b}$. We can even allow ϵ'' and μ'' to be different for states a and b if their effect is included into the source terms in this fashion.

(b) Use the result of part (a) to obtain the power-type orthogonality relation

$$\int_{S_0} (\mathcal{E}_m \times \mathcal{H}_n^* + \mathcal{E}_n^* \times \mathcal{H}_m) \cdot \mathbf{u}_z \, dS = 0$$

valid for a lossless waveguide if $\gamma_m \neq -\gamma_n^*$.

(c) Use the result of part (b) to provide an independent proof of the first line of (9.28).

p9-14 A source excites a rectangular metallic waveguide, all of whose modes have a cutoff frequency above the operating frequency of the source. One end of the waveguide opens out into empty space as shown, and a measurement shows the presence of a nonzero radiated power coming from the end of the guide.

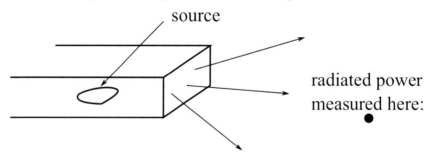

Explain how this cutoff waveguide can deliver power from the source into space as described. Specifically, what conditions on the wave amplitudes of the cutoff modes in the waveguide must hold?

p9-15

(a) Using (9.40) and the limiting forms of (8.150), show that the filling factor q for a microstrip line is approximately $\frac{1}{2}$ for narrow strips ($w/h \ll 1$) and approximately 1 for a very wide strip ($w/h \gg 1$).

(b) Plot the filling factor q vs. w/h for several values of ϵ_r (say, 1.0, 2.5 and 10), using (8.150). Compare these values with those obtained from (9.90) and (9.91), and determine which has better accuracy.

p9-16 Use the results in Section 8.7.1 to obtain a formula for the filling factor q of the quasi-TEM mode of CPW. Plot q vs. $s/(s + 2w)$ for the cases $(s + 2w)/h = 0.2$, 1, and 4.

p9-17 Use Figure 6.17 and the results of Section 9.4 to find the attenuation coefficient α of the TE_0 mode in a slab waveguide with $a = 2\mu m$, $n_1 = 2.01$, $n_2 = 2.0$, $\lambda_0 = 0.5\mu m$, $\epsilon_1''/\epsilon_0 = 0.001$, and $\epsilon_2''/\epsilon_0 = 0.002$.

p9-18 For a weakly guiding step-index optical waveguide of arbitrary cross-sectional shape, show that

$$q_1 = b + \frac{V}{2}\frac{db}{dV}$$

Plot q_1 vs. V for the TE_0 mode of a dielectric slab waveguide, and compare it to a plot of b vs. V for this same mode.

p9-19 Use the result of problem p9-13(a) and the fields

$$\mathbf{E}^a = \boldsymbol{\mathcal{E}}_{m0}e^{-\gamma_{m0}z}; \qquad \mathbf{H}^a = \boldsymbol{\mathcal{H}}_{m0}e^{-\gamma_{m0}z}$$

$$\mathbf{E}^b = \boldsymbol{\mathcal{E}}_{m}e^{-\gamma_{m}z}; \qquad \mathbf{H}^b = \boldsymbol{\mathcal{H}}_{m}e^{-\gamma_{m}z}$$

where $\gamma_{m0} = j\beta_{m0}$ is the propagation coefficient of a mode of a lossless waveguide whose dielectric constant is ϵ' (state a) and $\gamma_m = \alpha_m + j\beta_m$ is that of a waveguide whose dielectric constant has the slightly lossy value of $\epsilon' - j\epsilon''$ (state b) to obtain the approximation

$$\alpha_m = \gamma_{m0} + \gamma_m^* \simeq \frac{\int_{S_0} \omega\epsilon'' \boldsymbol{\mathcal{E}}_{m0} \cdot \boldsymbol{\mathcal{E}}_{m0}^* \, dS}{2\int_{S_0} \boldsymbol{\mathcal{E}}_{m0} \times \boldsymbol{\mathcal{H}}_{m0}^* \cdot \mathbf{u}_z \, dS}$$

for the attenuation coefficient of the lossy waveguide mode. Show that this result is compatible with (9.41)-(9.42).

p9-20 Obtain from (9.67) an expression for the attenuation coefficient of a TM_{11} (rectangular) waveguide mode due to finite wall conductivity. The $\boldsymbol{\mathcal{H}}$-field of this mode is

$$\boldsymbol{\mathcal{H}} = H_{11}\left\{\mathbf{u}_x\frac{\pi}{d}\sin\frac{\pi x}{a}\cos\frac{\pi y}{d} - \mathbf{u}_y\frac{\pi}{a}\cos\frac{\pi x}{a}\sin\frac{\pi y}{d}\right\}.$$

p9-21 There are many cases in which we desire to calculate wall losses of a mode which is degenerate (or nearly so) with another mode of the same waveguide. One special example of this is at or near cutoff, where both the forward and reverse modes' propagation coefficients approach 0. In such cases, it works out that the approximation of $\boldsymbol{\mathcal{E}}_{-m}$ and $\boldsymbol{\mathcal{H}}_{-m}$ which led from (9.24) to (9.25) is inadequate and a more refined approach is required.

(a) Suppose that the mode $(\gamma_{m0}, \boldsymbol{\mathcal{E}}_{m0}, \boldsymbol{\mathcal{H}}_{m0})$ is nearly degenerate with the mode $(\gamma_{n0}, \boldsymbol{\mathcal{E}}_{n0}, \boldsymbol{\mathcal{H}}_{n0})$. Use the approximation

$$\boldsymbol{\mathcal{E}}_{-m} \simeq A\boldsymbol{\mathcal{E}}_{-m0} + B\boldsymbol{\mathcal{E}}_{-n0}$$

$$\boldsymbol{\mathcal{H}}_{-m} \simeq A\boldsymbol{\mathcal{H}}_{-m0} + B\boldsymbol{\mathcal{H}}_{-n0}$$

in (9.63) to obtain a relationship between γ_m and the yet unknown constants A and B.

(b) Repeat the derivation of part (a), using

$$\mathbf{E}^a = \boldsymbol{\mathcal{E}}_{n0}e^{-\gamma_{n0}z} ; \quad \mathbf{H}^a = \boldsymbol{\mathcal{H}}_{n0}e^{-\gamma_{n0}z}$$

for the "a" state fields instead of the fields indicated in (9.62). Use the approximation for $\boldsymbol{\mathcal{E}}_{-m}$ and $\boldsymbol{\mathcal{H}}_{-m}$ suggested in part (a) above to obtain a second relationship between γ_m, A, and B. Eliminate A and B between this result and that of part (a).

(c) Now assume that $(\gamma_{n0}, \mathcal{E}_{n0}, \mathcal{H}_{n0})$ is the reverse-traveling mode corresponding to $(\gamma_{m0}, \mathcal{E}_{m0}, \mathcal{H}_{m0})$. That is, suppose that

$$\gamma_{n0} = -\gamma_{m0}; \quad \mathcal{E}_{n0} = \mathcal{E}_{-m0}; \quad \mathcal{H}_{n0} = \mathcal{H}_{-m0}$$

From the results of parts (a) and (b) above, show that

$$\gamma_m \simeq \pm\sqrt{\gamma_{m0}^2 + 2(\alpha_z + \alpha_T)\gamma_{m0} + 4\alpha_z\alpha_T}$$

where

$$\alpha_z = \frac{Z_S \oint_{C_0} \mathcal{H}_{m0l}^2 \, dl}{2 \int_{S_0} \mathcal{E}_{m0} \times \mathcal{H}_{m0} \cdot \mathbf{u}_z \, dS}$$

$$\alpha_T = -\frac{Z_S \oint_{C_0} \mathcal{H}_{m0z}^2 \, dl}{2 \int_{S_0} \mathcal{E}_{m0} \times \mathcal{H}_{m0} \cdot \mathbf{u}_z \, dS}.$$

p9-22 Use the techniques of problem p9-21 to derive a similar type of perturbation expression for dielectric losses for a waveguide mode which may be near a cutoff frequency.

p9-23 Obtain from (9.68) a formula for the attenuation coefficient α due to finite wall conductivity σ of the TE$_{11}$ and TE$_{01}$ modes of an air-filled circular metallic waveguide of radius a. The unperturbed mode fields are given in (5.82)-(5.84).

p9-24 A rectangular metallic waveguide of width a and height d has top and bottom walls ($y = 0$ and $y = d$) made of one metal, whose surface impedance is Z_{S1}, and side walls ($x = 0$ and $x = a$) made of another (Z_{S2}). Obtain a formula for the attenuation constant due to wall loss of the TE$_{10}$ mode of such a waveguide. If the top and bottom walls are made of silver ($\sigma = 6.17 \times 10^7$ S/m) and the sidewalls are made of an alloy of copper and zinc for which $\sigma = 2.56 \times 10^7$ S/m, find the value of α for a WR-90 size waveguide ($a = 0.9$ in, $d = 0.4$ in) operating at $f = 10$ GHz. Compare this to the case when the same waveguide has all walls made of the copper-zinc alloy.

p9-25 Consider the two-wire transmission line shown in Figure 8.8. Using the incremental inductance rule, obtain a formula for the attenuation constant α of the TEM mode of this line due to finite conductivity of the wires.

(a) If the medium surrounding the wires is assumed to be air, determine the value of a/b for which this attenuation is minimum (regard the separation b between wire centers as fixed while a is varied). What is the characteristic impedance of the line under this minimum attenuation condition?

(b) Let the conductors be made from copper ($\epsilon_c = \epsilon_0$, $\mu_c = \mu_0$, $\sigma_c = 5.8 \times 10^7$ S/m), and take $a = 1$ mm and $b = 2$ cm. Find the attenuation coefficient of the TEM mode of this line at the frequencies $f = 3$, 30 and 300 MHz.

p9-26 Sometimes a transmission line is made from a single conducting wire located above the earth, with the earth serving as the return conductor for the current flowing along the wire (a so-called "earth-return" transmission line). Let the geometry of such a line be as shown in the figure below.

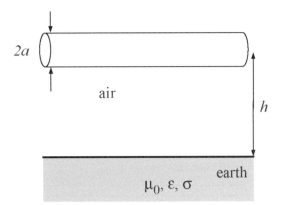

Use the results from Section 8.3.2 to obtain the line parameters of the lossless line (i. e., when the earth is a perfect conductor). Obtain a formula for the attenuation constant of the TEM mode of this structure, assuming that the wire and the earth are good conductors, have different conductivities, σ_{Cu} and σ_{soil} respectively, and that both the wire and the earth are nonmagnetic.

(a) Obtain an equation whose solution determines the value of a/h for which the attenuation constant of this mode is minimum (h remaining fixed). If $\sigma_{\mathrm{wire}} = 10^5 \sigma_{\mathrm{soil}}$, determine this value of a/h. What is the characteristic impedance of the mode under this condition?

(b) Let the wire have a radius $a = 1$ mm and its center be located at a height $h = 5$ m above the earth. The wire is made of copper, whose permittivity and permeability are those of free space, while its conductivity is $\sigma_{\mathrm{Cu}} = 5.8 \times 10^7$ S/m. The electromagnetic properties of the soil vary strongly with its composition and moisture content; here we will assume that $\epsilon_{\mathrm{soil}} = 10\epsilon_0$ and $\sigma_{\mathrm{soil}} = 10^{-2}$ S/m. If the operating frequency is 5 MHz, find the attenuation constant α of this line. For what range of frequencies will the use of the impedance boundary condition in this problem be justified?

10 EXCITATION OF WAVEGUIDES

Now that we have seen several examples of waveguides and the types of modes they can support, it is relevant to ask how we can go about setting up mode fields in a guide. This question is answered by specifying the sources to be used to set up the fields. The current sources may be electric or magnetic; they may be actual currents flowing in a thin wire or equivalent currents distributed over some surface. In this chapter, we will use the orthogonality property of waveguide modes to determine the amplitudes with which modes are excited by a given set of sources.

10.1 IMPRESSED SOURCES VERSUS INDUCED CURRENTS FOR ELECTROMAGNETIC FIELDS

Up to this point, we have been concerned only with solutions of Maxwell's equations in regions where no sources are present. Of course, some kind of externally generated currents and/or charges must be supplied somewhere in order to produce our mode solutions with nonzero amplitude, just as such impressed sources are required to energize a circuit. In a time-harmonic problem, it is enough to specify the current density, since the charge density will then follow by the law of conservation of charge.

It is well to be clear about what we mean by the term "externally generated" currents. One way of writing Maxwell's equations in a material medium is (for time-harmonic fields)

$$\left. \begin{array}{rcl} \nabla \times \mathbf{E} & = & -j\omega \mathbf{B} \\ \nabla \times \left(\dfrac{\mathbf{B}}{\mu_0} \right) & = & j\omega\epsilon_0 \mathbf{E} + \mathbf{J} \end{array} \right\} \tag{10.1}$$

However, we must understand that the current density $\mathbf{J}$ in (10.1) includes, in addition to currents we can "control" (generators, antennas, etc.), other currents caused by the presence of the electromagnetic field in the medium. These are the *conduction* currents

$$\mathbf{J}_{\text{cond}} = \sigma \mathbf{E} \tag{10.2}$$

the *polarization* currents

$$\mathbf{J}_p = j\omega(\epsilon - \epsilon_0)\mathbf{E} \tag{10.3}$$

and the *magnetization* currents

$$\mathbf{J}_m = \nabla \times \left[\left(\frac{1}{\mu_0} - \frac{1}{\mu} \right) \mathbf{B} \right] \tag{10.4}$$

On the surface of a perfect conductor as shown in Figure 10.1, surface currents

$$\mathbf{J}_S = \mathbf{u}_n \times \frac{\mathbf{B}}{\mu} \tag{10.5}$$

will also result as a limiting form of (10.2). The current densities (10.2)–(10.5) are ones we cannot "control" directly because they depend on the field $\mathbf{E}$, $\mathbf{H}$ present in the medium or on the conductor surface. This in turn must be solved for from (10.1), which requires that we know $\mathbf{J}$, which depends on $\mathbf{E}$ and $\mathbf{H}$, which were unknown in the first place.

To avoid such a vicious circle, it is common practice to account for (10.2)–(10.4) by way of material parameters μ, σ and ϵ and to rewrite (10.1) in the form

$$\left. \begin{array}{rcl} \nabla \times \mathbf{E} & = & -j\omega\mu \mathbf{H} \\ \nabla \times \mathbf{H} & = & j\omega\hat{\epsilon} \mathbf{E} + \mathbf{J}_{\text{ext}} \end{array} \right\} \tag{10.6}$$

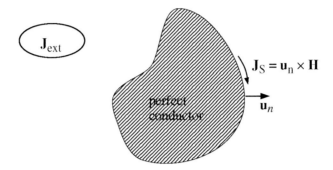

Figure 10.1   Induced surface currents on a perfect conductor.

where the **H** field is defined using the permeability μ:

$$\mathbf{B} = \mu\mathbf{H}$$

and

$$\mathbf{J}_{\text{ext}} = \mathbf{J} - \mathbf{J}_{\text{cond}} - \mathbf{J}_p - \mathbf{J}_m \tag{10.7}$$

are called *externally impressed* (or simply *impressed*) currents whose value we know or control, even before we obtain a knowledge of **E** and **H** by solving Maxwell's equations. Hereafter we will drop the subscript "ext" on impressed electric or magnetic currents **J** or **M** except when necessary for clarity or emphasis.

Of course, no such current sources can be achieved in practice because the environment in which we place a real source will always affect it to some degree. In this sense, the concept of an impressed source is very much like that of an ideal voltage or current generator in circuit theory. We will refer to currents such as (10.2)–(10.5), which disappear when the impressed sources (and hence the resultant fields) are turned off, as *induced* currents. An example illustrating the distinction between impressed currents and induced currents will be studied in some detail in the next section in connection with Love's equivalence theorem.

10.2   EQUIVALENCE PRINCIPLES

Equivalence principles analogous to those of circuit theory play an important role in electromagnetics, and come in many varieties. For example, to see how magnetic currents might be useful, consider a field **E**, **H** that is a solution of the "ordinary" Maxwell's equations (D.3). Now consider a hypothetical field $\mathbf{E}_1$, $\mathbf{H}_1$ given by

$$
\begin{aligned}
\mathbf{E}_1 &= \mathbf{E} + \frac{\mathbf{J}}{j\omega\epsilon} \\
\mathbf{H}_1 &= \mathbf{H}
\end{aligned}
\tag{10.8}
$$

(Note that this field is identical to **E**, **H** wherever $\mathbf{J} = 0$, i. e., outside the region containing the sources.) Evidently, $\mathbf{E}_1$, $\mathbf{H}_1$ satisfy

$$
\begin{aligned}
\nabla \times \mathbf{E}_1 &= -j\omega\mu\mathbf{H}_1 + \nabla \times \left(\frac{\mathbf{J}}{j\omega\epsilon}\right) \\
\nabla \times \mathbf{H}_1 &= j\omega\epsilon\mathbf{E}_1
\end{aligned}
\tag{10.9}
$$

That is (see (D.4)), these fields act as though they were produced by the equivalent, purely *magnetic* current density

$$\mathbf{M}_{eq} = -\nabla \times \left(\frac{\mathbf{J}}{j\omega\epsilon} \right) \tag{10.10}$$

So long as $\mathbf{J}$ and our fictitious $\mathbf{M}_{eq}$ are zero at a given observation point, the original field $\mathbf{E}$, $\mathbf{H}$ is precisely that which would be produced by $\mathbf{M}_{eq}$ instead of by $\mathbf{J}$. Often, the equivalent magnetic current given by (10.10) proves to be simpler to describe than the original $\mathbf{J}$. In such cases, it proves more convenient to use $\mathbf{M}_{eq}$ directly. This result is an *equivalence principle*. Such principles are often used in the formulation of electromagnetic problems to simplify concepts or computations and to increase our insight and understanding of the physics involved.

Next, consider a set of sources $\mathbf{J}$ and $\mathbf{M}$ located inside a closed surface S as shown in Figure 10.2(a). Let $\mathbf{E}$, $\mathbf{H}$ be the fields produced by these sources. Let $\mathbf{E}_1$, $\mathbf{H}_1$ be defined as

$$\mathbf{E}_1 = \begin{cases} \mathbf{E} & \text{at points exterior to } S; \\ 0 & \text{at points interior to } S \end{cases}$$

$$\mathbf{H}_1 = \begin{cases} \mathbf{H} & \text{at points exterior to } S; \\ 0 & \text{at points interior to } S. \end{cases}$$

Now $\mathbf{E}_1$ and $\mathbf{H}_1$ satisfy the source-free Maxwell equations both inside and outside of S but are discontinuous across S. If they are to represent a legitimate electromagnetic field, there must be surface currents (along with appropriate surface charges) present on S to support this discontinuity. From Section D.1 of Appendix D, these surface currents must be

$$\begin{aligned} \mathbf{J}_{S,eq} &= \mathbf{u}_n \times \mathbf{H}\big|_{S+} \\ \mathbf{M}_{S,eq} &= \mathbf{E} \times \mathbf{u}_n\big|_{S+} \end{aligned} \tag{10.11}$$

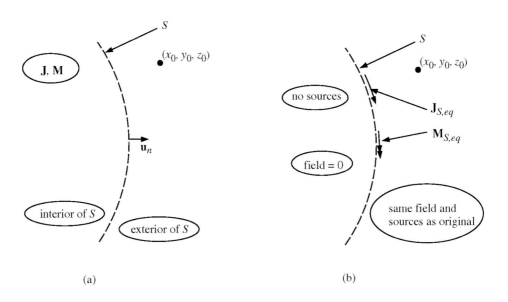

(a)   (b)

Figure 10.2   (a) Original sources; (b) equivalent sources for observation point (x_0, y_0, z_0) outside of S.

Hence, *the electric or magnetic field produced by the sources* $\mathbf{J}$ *and* $\mathbf{M}$ *inside* S *is exactly the same as that produced by the equivalent sources (10.11) on* S, *observed at any point outside* S. This statement is known as *Love's equivalence principle*.

The fields $\mathbf{E}_1$ and $\mathbf{H}_1$ of the equivalent problem (Figure 10.2(b)) are identically zero throughout the whole interior of S. Hence, Love's equivalence principle will be most useful to us when there is a surface S enclosing the original sources, inside which we do not need to know the fields and on which we know, at least approximately, the fields produced by the original sources, so that the equivalent sources (10.11) can be treated as if they were impressed ones. In general, even though the original sources may be pure *electric* currents, we must expect that tangential components of both $\mathbf{E}$ and $\mathbf{H}$ will exist on S, and so both electric *and* magnetic equivalent sources will be required.

We can see that instead of choosing $\mathbf{E}_1$ and $\mathbf{H}_1$ to be zero inside of S, we could have made them equal to any convenient functions. In this case, we would require generally different distributions of surface current on S to support the field discontinuities, as well as volume source distributions $\mathbf{J}_1$ and $\mathbf{M}_1$ to support the nonzero field:

$$\left. \begin{array}{rcl} \mathbf{J}_1 &=& \nabla \times \mathbf{H}_1 - j\omega\hat{\epsilon}\mathbf{E}_1 \\ \mathbf{M}_1 &=& -\nabla \times \mathbf{E}_1 - j\omega\mu\mathbf{H}_1 \end{array} \right\} \quad \text{interior to } S \qquad (10.12)$$

A great degree of flexibility is thus possible in the use of equivalence principles. We will explore some variations on this theme by way of an example.

10.2.1 EXAMPLE: IMPRESSED AND INDUCED SOURCES AT A PERFECTLY CONDUCTING SURFACE

In Figure 10.2(b), it can make no difference what medium is located within S (where the original sources were), since in the equivalent problem the field is zero there. In particular, let us put a perfectly conducting surface just inside of S. If we do this, we still have the *impressed* current $\mathbf{M}_S = \mathbf{E} \times \mathbf{u}_n$ as before, but we shall see that the surface electric current on S can be taken either as the *impressed* current $\mathbf{J}_S^{\text{ext}} = \mathbf{J}_S = \mathbf{u}_n \times \mathbf{H}$ that is present in Figure 10.2(b) or the *induced* current $\mathbf{J}_S = \mathbf{u}_n \times \mathbf{H}$ that results when *only* a magnetic current $\mathbf{M}_S^{\text{ext}} = \mathbf{M}_S$ is impressed at S.

Thus, the situation in Figure 10.3(a) is obtained by placing a perfect conductor just inside S, which already had zero fields in the situation of Figure 10.2(b). Both $\mathbf{J}_S^{\text{ext}}$ and $\mathbf{M}_S^{\text{ext}}$ are still impressed sources, and the total field resulting from these sources is the superposition of the fields resulting from $\mathbf{J}_S^{\text{ext}}$ and $\mathbf{M}_S^{\text{ext}}$ separately. We have already shown in Section 9.1.1 that an impressed current $\mathbf{J}_S^{\text{ext}}$ on a perfectly conducting surface produces no external fields. That result is similar to, though distinct from, image theory (Appendix D), which is itself a form of equivalence principle. Thus it must be that whatever fields are present in Figure 10.3(a) must be produced by $\mathbf{M}_S^{\text{ext}}$ alone. However, the boundary conditions and uniqueness theorems for the electromagnetic field (Appendix D) still require that the discontinuity in tangential $\mathbf{H}$ field from the right side of S to the interior of the conductor (where it is zero) must be compensated by a surface electric current. Since this current is no longer impressed on S, it must be induced in the surface of the perfect conductor, as shown in Figure 10.3(b). That is, the surface current is now a response rather than an excitation. By the uniqueness theorem for the fields, this induced current $\mathbf{J}_S$ must be the same as the equivalent impressed current $\mathbf{J}_S^{\text{ext}}$ in Figure 10.3(a). This result, known as the Schelkunoff Induction Theorem, is the electromagnetic analogy of the situation in the circuit problem of Figure 3.2. The electric impressed sources, like the current generator I, have no effect in the region in which we are interested.

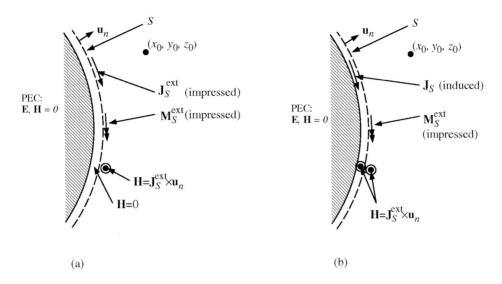

Figure 10.3 (a) Impressed electric currents on S producing zero field between S and perfect conductor; (b) induced surface currents on perfect conductor (produced by $\mathbf{M}_S^{\text{ext}}$ on S alone) with nonzero H-field between S and perfect conductor.

10.2.2 EXAMPLE: THEVENIN AND NORTON TYPES OF EQUIVALENCE THEOREMS FOR ELECTROMAGNETIC FIELDS

Suppose that a thin perfectly electrically conducting (PEC) sheet is placed at the surface S in Figure 10.2(a), and that the sources $\mathbf{J}$, $\mathbf{M}$ acting inside it produce the "short-circuit" surface current density $\mathbf{J}_{S,sc}$ in the sheet, as shown in Figure 10.4(a). Then the field out-

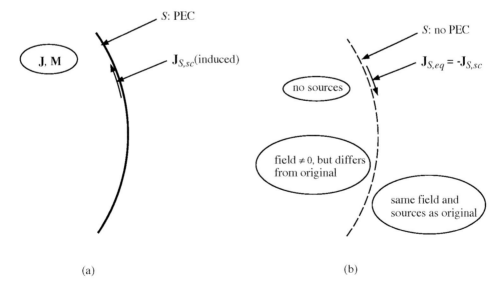

Figure 10.4 (a) Original currents inside S radiating with S "short-circuited" by a PEC; (b) Norton equivalent surface electric currents radiating at S without PEC present.

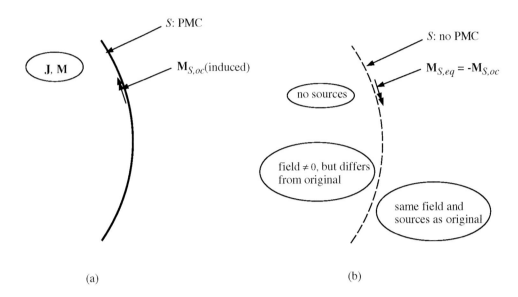

Figure 10.5 (a) Original currents inside S radiating with S "open-circuited" by a PMC; (b) Thévenin equivalent surface electric currents radiating at S without PMC present.

side of S in the original problem is the same as that produced by the equivalent electric surface currents $\mathbf{J}_{S,\text{eq}} = -\mathbf{J}_{S,sc}$ as shown in Figure 10.4(b) in the exterior of S, without the PEC present and with the original sources deactivated. This is the analog of the Norton equivalence theorem in circuit theory.

On the other hand, suppose that a sheet which is a perfect magnetic conductor (PMC) (that is, on which $\mathbf{u}_n \times \mathbf{H} = 0$) is placed at S, resulting in the "open-circuit" magnetic surface current density $\mathbf{M}_{S,oc}$ on the sheet, as shown in Figure 10.5(a). The field exterior to S is now the same as that produced by the equivalent magnetic surface current source $\mathbf{M}_{S,\text{eq}} = -\mathbf{M}_{S,oc}$ located at S, without the PMC present and with the original sources deactivated. This is the electromagnetic analog of Thévenin's equivalence theorem.

Note that, in contrast with the Love equivalence principle, the fields on the interior side of S are *not* generally zero for the Norton and Thévenin equivalents, nor are they the same as in the original problem. It is thus not generally justified to make changes in media (such as inserting conducting bodies) inside S after replacing the original sources by these equivalents, as it was for the case of Love equivalent sources. The unchanged medium inside S is the analog of the Thévenin or Norton equivalent impedance from circuit theory.

10.2.3   EXAMPLE: LARGE APERTURES AND THE KIRCHHOFF APPROXIMATION

When an electromagnetic wave is incident at an aperture which is large in extent (compared to a wavelength and to all other characteristic dimensions of the problem), it is sometimes possible to utilize the *Kirchhoff approximation* to simplify the analysis. To be more specific, consider the situation shown in Figure 10.6. In addition to the large aperture, there may be obstacles such as the dielectric shown in the figure which also affect the field. When making the Kirchhoff approximation, we assume that we can neglect the backscattered field behind the aperture as compared to $\mathbf{E}^{\text{inc}}$, $\mathbf{H}^{\text{inc}}$ as far as computing the aperture field $\mathbf{E}_0$, $\mathbf{H}_0$ is concerned. Because $\mathbf{E}_0$ and $\mathbf{H}_0$ are being approximated, any fields we compute in front of the aperture using Love's equivalent sources will also be approximate. In fact, if we now fill in the aperture with (say) a perfect electric conductor, use of approximate Kirchhoff

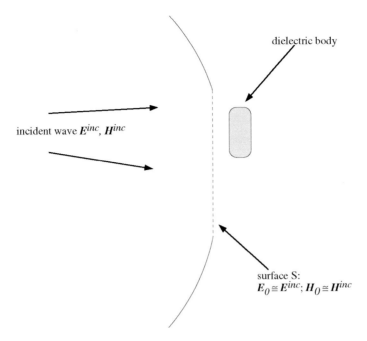

Figure 10.6 Kirchhoff approximation for a wave incident at a large aperture.

equivalent sources will in general lead to a somewhat different field computed outside the aperture. The discrepancy between the two computed fields is one measure of the error we commit by making the Kirchhoff approximation. We will illustrate the use of the Kirchhoff approximation in Section 10.5.1.

10.2.4   EXAMPLE: TANGENTIAL MAGNETIC AND NORMAL ELECTRIC SURFACE SOURCES

Consider a surface concentration of *normal* electric current (i. e., a continuous sheet of electric dipoles oriented perpendicular to the sheet):

$$\mathbf{J} = J_{Sn}\mathbf{u}_n\delta(n) \tag{10.13}$$

According to (10.10), one equivalent to this source would be the (tangential) magnetic surface sources

$$\mathbf{M}_{\text{eq}} = -\nabla \times \left(\frac{J_{Sn}\mathbf{u}_n\delta(n)}{j\omega\epsilon}\right) = \mathbf{u}_n \times \nabla_t\left(\frac{J_{Sn}}{j\omega\epsilon}\right)\delta(n) \tag{10.14}$$

or in other words,

$$\mathbf{M}_{S,\text{eq}} = \mathbf{u}_n \times \nabla_t\left(\frac{J_{Sn}}{j\omega\epsilon}\right) \tag{10.15}$$

This equivalent source has the disadvantage that we must take a derivative of J_{Sn} to compute it, and if the normal surface current is confined to a relatively small region, its derivative is likely to be quite large in that region. Even if a reasonable approximation to J_{Sn} can be found, the derivative of this approximation will not necessarily be close to the derivative of the exact distribution, so we will seek an alternate set of equivalent sources that does not suffer from this shortcoming.

To do this, suppose that the electric field has been expressed in terms of scalar and vector potentials as:

$$\mathbf{E} = -\nabla\Phi - j\omega\mathbf{A}$$

Suppose also that we have used the Love equivalence principle and the induction theorem as in Section 10.2.1 to arrive at an equivalent surface magnetic source

$$\mathbf{M}_{S,\text{eq}} = \mathbf{E}_t \times \mathbf{u}_n\big|_{n=0^+} = \mathbf{u}_n \times \nabla_t\Phi\big|_{n=0^+} - j\omega\mathbf{A}_t \times \mathbf{u}_n\big|_{n=0^+} \qquad (10.16)$$

in front of a PEC. The term due to the scalar potential has the same form as (10.14), and so it can be further made equivalent to a surface distribution of normal electric currents. The final set of equivalent sources is then

$$J_{Sn,\text{eq}} = j\omega\epsilon\Phi\big|_{n=0^+} \qquad \mathbf{M}_{S,\text{eq}} = -j\omega\mathbf{A}_t \times \mathbf{u}_n\big|_{n=0^+} \qquad (10.17)$$

in which expressions no differentiations occur, provided that Φ and $\mathbf{A}_t$ are known.

10.2.5   EXAMPLE: EQUIVALENT DIPOLES FOR ELECTRICALLY SMALL OBJECTS

If an electrically small object (that is, an object whose dimensions are small compared to a wavelength) is placed in an incident electromagnetic field, charges and currents will be induced on the object, due either to conduction, polarization or magnetization depending on the nature of the object. Just as in the case of microscopic objects like atoms or molecules, the primary effect of these induced charges and currents will be to create electric and magnetic dipole moments $\mathbf{p}$ and $\mathbf{m}$ that will be proportional to the incident field. These dipole moments now serve as secondary (i.e., dependent) sources

$$\mathbf{J}_{\text{eq}} \simeq j\omega\mathbf{p}\,\delta(x - x_0)\delta(y - y_0)\delta(z - z_0) \qquad (10.18)$$

$$\mathbf{M}_{\text{eq}} \simeq j\omega\mu\mathbf{m}\,\delta(x - x_0)\delta(y - y_0)\delta(z - z_0) \qquad (10.19)$$

located at the coordinates (x_0, y_0, z_0) of the scattering object, and produce an additional field. The proportionality between the incident field and the dipole moments is expressed through dyadic quantities called *polarizabilities* that depend on the geometry and electromagnetic material properties of the object. If the object is surrounded by free space, then

$$\mathbf{p} = \epsilon_0 \overset{\leftrightarrow}{\alpha}_e \cdot \mathbf{E}^{\text{inc}} \qquad (10.20)$$

$$\mathbf{m} = \frac{1}{\mu_0} \overset{\leftrightarrow}{\alpha}_m \cdot \mathbf{B}^{\text{inc}} \qquad (10.21)$$

The polarizabilities (whose dimensions are those of volume) are known in closed form for only a few simple shapes, and must generally be computed numerically. A sphere of radius a, relative permittivity ϵ_r and relative permeability μ_r has the electric polarizability

$$\overset{\leftrightarrow}{\alpha}_e = 4\pi a^3 \frac{\epsilon_r - 1}{\epsilon_r + 2}\overset{\leftrightarrow}{I} \qquad (10.22)$$

and the magnetic polarizability

$$\overset{\leftrightarrow}{\alpha}_m = 4\pi a^3 \frac{\mu_r - 1}{\mu_r + 2}\overset{\leftrightarrow}{I} \qquad (10.23)$$

where $\overset{\leftrightarrow}{I}$ is the identity dyadic (see Appendix B). For a perfectly conducting sphere, we get the polarizabilities by letting $\epsilon_r \to \infty$ in (10.22) and $\mu_r \to 0$ in (10.23).

10.2.6   EXAMPLE: EQUIVALENT DIPOLES FOR ELECTRICALLY SMALL APERTURES

The usefulness of the equivalence principle in any of its variants depends upon the extent to which we may know accurately the values of electric or magnetic fields on a given surface. For electrically large apertures, a Kirchhoff-type approximation may be appropriate, wherein some simple incident field is used instead of the exact field (incident plus scattered) at the surface containing the aperture. When the aperture is small electrically, the fields near the aperture tend to be strongly dependent upon the shape of the aperture as well as upon the incident field.

Small aperture coupling has many important applications in the design of resonators, directional couplers and measurement devices. The analysis of small apertures is carried out using Bethe's small-aperture theory, in which the equivalent sources $\mathbf{M}_S$ and $\mathbf{J}_S$ at the aperture are replaced by an appropriately chosen set of Hertzian dipoles.

Consider the situation shown in Figure 10.7. A perfectly conducting plane is located at $n = 0$, where n is the normal distance from the plane as shown. The region $n < 0$ under the plane has material parameters ϵ_- and μ_-, while above it $(n > 0)$ the parameters are ϵ_+ and μ_+. The plane itself is described by the Cartesian coordinates (or other arc-length coordinates) (x_1, x_2) so that (x_1, x_2, n) form a right-handed coordinate system. Some given impressed sources are present which produce what we will call the "short-circuit" field $\mathbf{E}^{sc}$, $\mathbf{H}^{sc}$ in the presence of the plane but with no aperture cut into it (that is, the aperture is metallized or "shorted out").

When an electrically small aperture A is cut into the plane, the field produced by these impressed sources is perturbed and is now equal to

$$\mathbf{E}^{\text{tot}} = \mathbf{E}^{sc} + \mathbf{E}^A; \qquad \mathbf{H}^{\text{tot}} = \mathbf{H}^{sc} + \mathbf{H}^A \tag{10.24}$$

where $\mathbf{E}^A$ and $\mathbf{H}^A$ are due to the presence of the aperture. The qualitative behavior of these fields is illustrated in Figure 10.8. Figure 10.8(a) shows a short-circuit electric field which is nonzero only on the left side of the screen $(n < 0)$. $\mathbf{E}^{sc}$ is, of course, normal to the screen. After the aperture is cut into the wall, the electric field fringes and penetrates through the screen, roughly as shown in Figure 10.8(b). A small tangential electric field is present in the aperture, and by the equivalence theorem we could express $\mathbf{E}^A$ and $\mathbf{H}^A$ in $n < 0$ (or $n > 0$) by an appropriate surface magnetic current impressed at $n = 0^-$ (or $n = 0^+$, respectively) and with the aperture closed off by a perfect conductor.

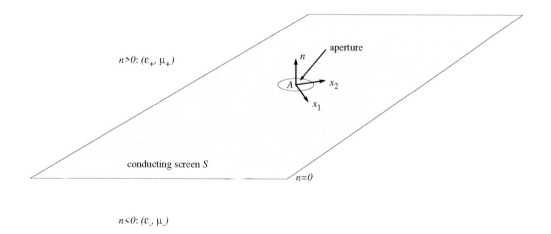

Figure 10.7   Aperture in a conducting plane.

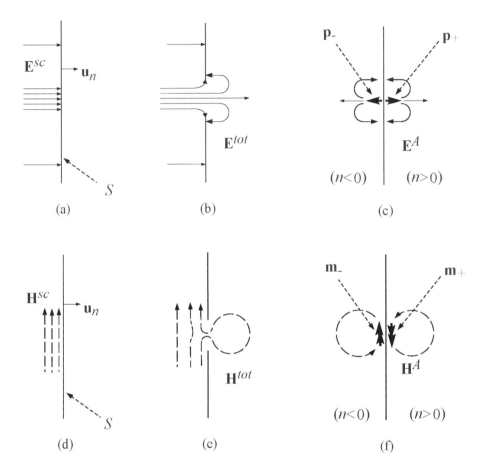

Figure 10.8   Field fringing at a small aperture in a conducting wall.

This tangential aperture electric field, however, is rather hard to determine in simple form, and it is more convenient to use a set of equivalent sources like those of Section 10.2.4 in order to compute $\mathbf{E}^A$ and $\mathbf{H}^A$. Upon examining Figure 10.8(b), we see that $\mathbf{E}^A$ on either side of the screen closely resembles the electric field that would be produced by a Hertzian electric dipole (cf. Appendix D) normal to the aperture, with the aperture closed off once again ("metallized") behind it, as indicated in Figure 10.8(c). Likewise, a short-circuit magnetic field in $n < 0$ (Figure 10.8(d)) will fringe as shown in Figure 10.8(e) when the aperture is cut. The field $\mathbf{H}^A$ which results on either side is nearly the same as the magnetic field of a Hertzian magnetic dipole with the aperture metallized behind it, as in Figure 10.8(f).

To justify the use of dipoles as equivalent sources, we suppose that $J_{Sn,\text{eq}}$ and $\mathbf{M}_{S,\text{eq}}$ have been obtained as in (10.17), with the potentials chosen so that $\Phi = 0$ and $\mathbf{A}_t = 0$ on the screen S, so that the equivalent sources differ from zero only in the aperture A. If the linear dimensions of the aperture are small compared to a wavelength, the potentials near the aperture can be found approximately from the static field equations. Moreover, if our observation point is not too close to the aperture, the distributed equivalent sources can be approximately lumped into dipoles concentrated at some point (x_{10}, x_{20}) in the middle of the aperture. At $n = 0^+$, we put

$$\mathbf{M}_{S,\text{eq}} \simeq j\omega\mu_+\mathbf{m}_+\delta(x_1 - x_{10})\delta(x_2 - x_{20}) \tag{10.25}$$

$$\mathbf{J}_{S,\text{eq}} \simeq j\omega\mathbf{p}_+\delta(x_1 - x_{10})\delta(x_2 - x_{20}) \tag{10.26}$$

as given in equations (D.16) and (D.17). The dipole moments $\mathbf{p}_+$ and $\mathbf{m}_+$ are chosen so as to give the same integrated values of $\mathbf{J}_{S,\text{eq}}$ and $\mathbf{M}_{S,\text{eq}}$ over the plane $n = 0$. Hence

$$j\omega\mu_+\mathbf{m}_+ = \int_A \mathbf{M}_{S,\text{eq}}\,dS = j\omega \int_A \mathbf{u}_n \times \mathbf{A}_t\,dS \tag{10.27}$$

or

$$\mathbf{m}_+ = \frac{1}{\mu_+} \int_A \mathbf{u}_n \times \mathbf{A}_t\,dS \tag{10.28}$$

Similarly,

$$\mathbf{p}_+ = \epsilon_+\mathbf{u}_n \int_A \Phi\,dS \tag{10.29}$$

We need dipoles on *both* sides of the closed-off aperture (at $n = 0^\pm$) in order to produce the field $\mathbf{E}^A$, $\mathbf{H}^A$ in both $n > 0$ and $n < 0$. At $n = 0^+$, we place the dipoles $\mathbf{p}_+$ and $\mathbf{m}_+$, while at $n = 0^-$, we place the dipoles $\mathbf{p}_-$ and $\mathbf{m}_-$ which are found from the values of Φ and $\mathbf{A}_t$ in the aperture similarly to $\mathbf{p}_+$ and $\mathbf{m}_+$. Because the potentials must be continuous through the aperture, the dipole strengths on the two sides turn out to be related by

$$\frac{\mathbf{p}_-}{\epsilon_-} = -\frac{\mathbf{p}_+}{\epsilon_+} \tag{10.30}$$

and

$$\mu_-\mathbf{m}_- = -\mu_+\mathbf{m}_+ \tag{10.31}$$

Most generally, the short-circuit field will exist on both sides of the screen. The strengths of the equivalent dipoles depend on these short-circuit field values as follows:

$$\begin{aligned}
\frac{\mathbf{p}_+}{\epsilon_+} &= -\frac{2}{\epsilon_+ + \epsilon_-}\alpha_E\,[\mathbf{D}^{\text{sc}}]_{n=0^-}^{0^+} \\
\mu_+\mathbf{m}_+ &= \frac{2\mu_+\mu_-}{\mu_+ + \mu_-}\overset{\leftrightarrow}{\alpha}_M \cdot [\mathbf{H}^{\text{sc}}]_{n=0^-}^{0^+}
\end{aligned} \tag{10.32}$$

while $\mathbf{p}_-$ and $\mathbf{m}_-$ are given by (10.30) and (10.31). The jumps $[\mathbf{D}^{\text{sc}}]$ and $[\mathbf{H}^{\text{sc}}]$ of the short-circuit field across the screen are related to the surface charge and current densities which exist when the aperture is metalized:

$$\rho_S^{\text{sc}} = \mathbf{u}_n \cdot [\mathbf{D}^{\text{sc}}]_{n=0^-}^{0^+} \quad ; \quad \mathbf{J}_S^{\text{sc}} = \mathbf{u}_n \times [\mathbf{H}^{\text{sc}}]_{n=0^-}^{0^+}$$

Removal of the metal (and its associated charge and current densities) at the aperture, along with the redistribution of these densities on the metal surrounding the aperture, is what causes the fields $\mathbf{E}^A$ and $\mathbf{H}^A$.

The coefficient α_E in (10.32) is known as the electric polarizability of the aperture, while the dyadic (or tensor) (essentially a 2×2 matrix here) $\overset{\leftrightarrow}{\alpha}_M$ is called the magnetic polarizability of the aperture. We use capital letter subscripts $_E$ and $_M$ to distinguish these aperture polarizabilities from those for scattering objects as used in (10.20) and (10.21). Like those polarizabilities, the dimensions of these are those of volume; in a sense they are effective volumes—volumes of equivalent scatterers which can produce the same polarization and magnetization (approximately) as the dipoles (10.30)-(10.32). The polarizabilities of various aperture shapes can be calculated from the corresponding static field problems. We give a brief list of some of these in Table 10.1. Values of α_E and $\overset{\leftrightarrow}{\alpha}_M$ for other apertures can be found in the references at the end of the chapter.

The dipoles (10.32), (10.30) and (10.31) succeed only approximately in determining $\mathbf{E}^A$ and $\mathbf{H}^A$. They are strictly valid only for apertures in infinite perfectly conducting planes and

Table 10.1

Polarizabilities for electrically small apertures in an infinite plane screen of zero thickness located at $n = 0$**.**

Aperture shape	α_E	$\overset{\leftrightarrow}{\alpha}_M = \mathbf{u}_1\mathbf{u}_1\alpha_M^{11} + \mathbf{u}_2\mathbf{u}_2\alpha_M^{22}$
10.1.1 Circle of radius r_0	$\frac{2}{3}r_0^3$	$\alpha_M^{11} = \alpha_M^{22} = \frac{4}{3}r_0^3$
10.1.2 Ellipse with semiaxes r_1 and r_2	$\dfrac{\pi(r_1r_2)^{3/2}}{3}\dfrac{2k'}{2E(k) - k'^2K(k)}$ $\left(k = \dfrac{r_1 - r_2}{r_1 + r_2}\right)$	$\alpha_M^{11} =$ $\dfrac{\pi(r_1r_2)^{3/2}}{6}\dfrac{k'}{2(1+k)K(k) - E(k)}$ $\alpha_M^{22} =$ $\dfrac{\pi(r_1r_2)^{3/2}}{6}\dfrac{k'}{2(1-k)K(k) - E(k)}$
10.1.3 Square	$\alpha_{E\square} = 0.1138w^3$	$\alpha_M^{11} = \alpha_M^{22} = \alpha_{M\square} = 0.2600w^3$
10.1.4 Rectangle	$\dfrac{w^2l^2}{(w^2 + l^2)^{3/2}}\left[Cwl + \frac{\pi}{16}(l - w)^2\right]$ $C = 0.3219$	$\alpha_M^{11} = \dfrac{\pi C_1 l^3}{\ln\left(1 + C_1\frac{l}{w} + C_2\frac{l^2}{w^2}\right)}$ $\alpha_M^{22} = \dfrac{\pi C_1 w^3}{\ln\left(1 + C_1\frac{w}{l} + C_2\frac{w^2}{l^2}\right)}$ $C_1 = 1.328; \quad C_2 = 0.4077$
10.1.5 Greek cross	$\dfrac{w^2l(2l - w)\left[2\sqrt{l^2 + w^2} + l\sqrt{2}\right]}{12(l^2 + 2w^2)}$	$\alpha_M^{11} = \alpha_M^{22} =$ $\dfrac{2w}{9l}\dfrac{l^3 + w^2l - w^3}{\sinh^{-1}\frac{w}{l} + \frac{w}{l}\left[\sinh^{-1}\frac{l}{w} - A\right]}$ where $A = \ln(1 + \sqrt{2}) \simeq 0.8814$

Note: Expressions given for the circle and ellipse are exact [K and E are the elliptic integrals given by (C.57) and (C.58)], values for the square are from numerical simulation and the remainder are the best available approximate expressions.

for static fields. They do not accurately predict the perturbing field very close (within a few aperture diameters) to the aperture. Nevertheless, we may use the Bethe theory for a more general surface provided that the aperture is located at a place where the radius of curvature is large compared to the aperture size and away from corners and bends in the surface. We may use the theory for nonzero frequency provided that the largest linear dimension (i.e., diameter) of the aperture is small compared to a wavelength. The approximate nature of (10.32) becomes apparent when conservation of energy is examined. As the reader may verify, e.g., for the simple example of plane wave excitation studied below, the field $\mathbf{E}^A$, $\mathbf{H}^A$ will generally not satisfy conservation of energy exactly, the discrepancy tending to zero as the ratio of aperture diameter to wavelength goes to zero. In some applications it may be desirable to remedy this defect (when calculating quality factors of resonators, for example), and a means to do so is given in Appendix N.

10.2.7 ELECTRICALLY SMALL APERTURES IN A THICK CONDUCTOR

It often happens that the nonzero thickness of the conducting screen in which an electrically small aperture is cut is not negligible in comparison with the aperture dimension. In this case, the determination of equivalent dipole moments on both sides of the screen is more complicated. Let the surfaces of the screen be located at $n = \pm h/2$, where h is the thickness of the screen. For simplicity, we assume that the material constants (μ, ϵ) are the same within and on both sides of the aperture. Instead of (10.30)-(10.32), the electric dipoles located at $n = \pm h/2$ with the aperture filled in have the form

$$
\begin{aligned}
\mathbf{p}_+ &= -\alpha_E^s \left[\mathbf{D}^{sc}\right]_{n=-h/2}^{h/2} - \alpha_E^a \left[\mathbf{D}^{sc}_{n=-h/2} + \mathbf{D}^{sc}_{n=h/2}\right] \\
\mathbf{p}_- &= +\left[\mathbf{D}^{sc}\right]_{n=-h/2}^{h/2} - \alpha_E^a \left[\mathbf{D}^{sc}_{n=-h/2} + \mathbf{D}^{sc}_{n=h/2}\right]
\end{aligned}
\tag{10.33}
$$

where α_E^s is the *symmetric* electric polarizability of the aperture, and α_E^a is its *antisymmetric* electric polarizability. Likewise, the magnetic dipoles at $n = \pm h/2$ with the aperture filled in are

$$
\begin{aligned}
\mathbf{m}_+ &= +\overset{\leftrightarrow s}{\alpha}_M \cdot \left[\mathbf{H}^{sc}\right]_{n=-h/2}^{h/2} + \overset{\leftrightarrow a}{\alpha}_M \cdot \left[\mathbf{H}^{sc}_{n=-h/2} + \mathbf{H}^{sc}_{n=h/2}\right] \\
\mathbf{m}_- &= -\overset{\leftrightarrow s}{\alpha}_M \cdot \left[\mathbf{H}^{sc}\right]_{n=-h/2}^{h/2} + \overset{\leftrightarrow a}{\alpha}_M \cdot \left[\mathbf{H}^{sc}_{n=-h/2} + \mathbf{H}^{sc}_{n=h/2}\right]
\end{aligned}
\tag{10.34}
$$

where $\overset{\leftrightarrow s}{\alpha}_M$ is the *symmetric* magnetic polarizability dyadic of the aperture, and $\overset{\leftrightarrow a}{\alpha}_M$ is its *antisymmetric* magnetic polarizability dyadic.

Closed-form expressions for the polarizabilities of apertures in a thick screen are not available for most aperture shapes, but for a circular hole of radius r_0 in a conductor of thickness h, we have

$$
\alpha_E^s \simeq \frac{2r_0^3}{3} \frac{1}{1 + 1.33 \tanh\left(\frac{2.405h}{2r_0}\right)}; \qquad \alpha_E^a \simeq \frac{2r_0^3}{3} \frac{1}{1 + 1.33 \coth\left(\frac{2.405h}{2r_0}\right)}
\tag{10.35}
$$

and

$$
\overset{\leftrightarrow s}{\alpha}_M \simeq \frac{4r_0^3}{3} \frac{\mathbf{u}_1\mathbf{u}_1 + \mathbf{u}_2\mathbf{u}_2}{1 + 1.817 \tanh\left(\frac{1.841h}{2r_0}\right)}; \qquad \overset{\leftrightarrow a}{\alpha}_M \simeq \frac{4r_0^3}{3} \frac{\mathbf{u}_1\mathbf{u}_1 + \mathbf{u}_2\mathbf{u}_2}{1 + 1.817 \coth\left(\frac{1.841h}{2r_0}\right)}
\tag{10.36}
$$

For other shapes, the polarizabilities must be computed numerically.

10.2.8   EXAMPLE: PLANE WAVE INCIDENT AT A SMALL APERTURE

To illustrate the use of (10.30)-(10.32), consider the situation of Figure 10.9. A normally incident plane wave in $z < 0$ is reflected by a perfectly conducting plane at $z = 0$, in which is cut a small aperture centered at $x = 0$, $y = 0$. The media on both sides of the plane are supposed to be free space. With the aperture closed off, the short-circuit field is the sum of the incident wave $\mathbf{E}^i, \mathbf{H}^i$ and the reflected wave $\mathbf{E}^r, \mathbf{H}^r$:

$$\left.\begin{array}{rcl} \mathbf{E}^{sc} & = & \mathbf{E}^i + \mathbf{E}^r = \mathbf{u}_x E_0 (e^{-jk_0 z} - e^{+jk_0 z}) \quad (z < 0) \\ \mathbf{H}^{sc} & = & \mathbf{H}^i + \mathbf{H}^r = \mathbf{u}_y (E_0/\zeta_0)(e^{-jk_0 z} + e^{+jk_0 z}) \quad (z < 0) \end{array}\right\} \tag{10.37}$$

The short-circuit fields are equal to zero in $z > 0$. From these equations, $[\mathbf{D}^{sc}]_{n=0-}^{0+} = 0$ while

$$[\mathbf{H}^{sc}]_{n=0-}^{0+} = -2\mathbf{u}_y E_0/\zeta_0 \tag{10.38}$$

where we have taken $n = z$. By (10.32), therefore, there will be no electric dipoles, while the magnetic dipoles

$$\mathbf{m}_+ = \overset{\leftrightarrow}{\alpha}_M \cdot [\mathbf{H}^{sc}]_{n=0-}^{0+} = -2\frac{E_0}{\zeta_0}\left(\mathbf{u}_x \alpha_{Mxy} + \mathbf{u}_y \alpha_{Myy}\right) \quad (\text{at } z = 0^+) \tag{10.39}$$

and

$$\mathbf{m}_- = 2\frac{E_0}{\zeta_0}\left(\mathbf{u}_x \alpha_{Mxy} + \mathbf{u}_y \alpha_{Myy}\right) \quad (\text{at } z = 0^-) \tag{10.40}$$

radiate in the presence of the conducting plane with the aperture closed off. The fields $\mathbf{E}^A$ and $\mathbf{H}^A$ due to the aperture are thus found by using image theory (a dipole of twice the strength radiates in infinite free space) as described in Appendix D.

Other examples of the application of small aperture theory will be given later in this chapter.

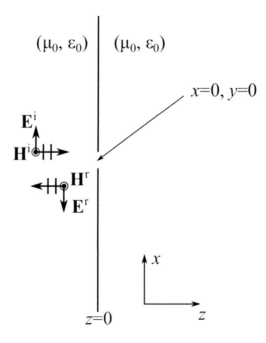

Figure 10.9   Plane wave normally incident at a small aperture in a planar conductor.

10.3 EQUIVALENT TRANSMISSION LINES—NORMALIZATION AND CHARACTERISTIC IMPEDANCE

The analogy between waveguide modes and equivalent transmission lines introduced in Section 9.3 is incomplete in one respect: there is no definition for the characteristic impedance of an arbitrary waveguide mode. If a suitable definition for a characteristic impedance Z_{cm} for *any* mode of a waveguide could be made, the discussion above could be extended to define a voltage and current associated with the mode, just as is done for the TEM case. The analogy between each waveguide mode and a classical transmission line suggested in eqns. (9.31) and (9.32) would then be complete. The trouble is, that there is no obviously unique way to define Z_{cm} for a non-TEM mode.

In spite of this, some constraints do exist. From Section J.2 of Appendix J, we have that $\gamma_m \hat{P}$ for any normalizable mode of a lossless waveguide is pure imaginary, and therefore if an expression analogous to (8.39) for complex power is to hold for our equivalent transmission line model of a mode, along with condition (9.28), we must have that Z_{cm} is real if $\gamma_m = j\beta_m$ (a propagating mode), or imaginary if $\gamma_m = \alpha_m$ (a cutoff mode). Note that this implies that a single cutoff mode carries no time-average power in the z-direction, since the complex power will be imaginary, as was shown in another way in Section 9.3.

Other than this, the precise value of Z_{cm} is essentially unrestricted for non-TEM modes. There are, however, certain choices for Z_{cm} which are traditionally made in specific cases, and these are shown in Table 10.2.

Considering the field representations (9.31) and (9.32), let us make some choice for Z_{cm}, and take it to be the characteristic impedance of an equivalent classical transmission line. We can thus rewrite (9.31) and (9.32) as:

$$\left.\begin{array}{rcl}
\mathbf{E}_T &=& \boldsymbol{\mathcal{E}}_{mT}(x,y)\tilde{V}_m(z) \\
\mathbf{H}_T &=& \boldsymbol{\mathcal{H}}_{mT}(x,y)\tilde{I}_m(z) \\
E_z &=& \mathcal{E}_{mz}(x,y)\tilde{I}_m(z) \\
H_z &=& \mathcal{H}_{mz}(x,y)\tilde{V}_m(z)
\end{array}\right\} \tag{10.41}$$

where following Section 1.2 for the classical transmission line,

$$\tilde{V}_m(z) = \frac{V_m(z)}{\sqrt{2Z_{cm}}} = a_m e^{-\gamma_m z} + b_m e^{\gamma_m z} \tag{10.42}$$

$$\tilde{I}_m(z) = \sqrt{\frac{Z_{cm}}{2}}\, I_m(z) = a_m e^{-\gamma_m z} - b_m e^{\gamma_m z} \tag{10.43}$$

Table 10.2

Common choices of Z_{cm} for waveguide modes.

Type of mode	Z_{cm}
10.2.1 TE mode of metallic waveguide	ζ_{TE}
10.2.2 TM mode of metallic waveguide	ζ_{TM}
10.2.3 TEM or quasi-TEM modes	Z_c

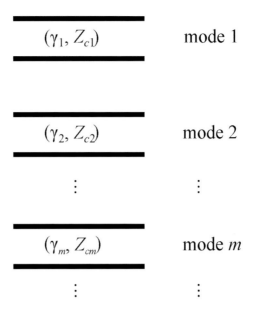

Figure 10.10 Equivalent transmission-line representation of a waveguide.

The constant factors involving Z_{cm} have been introduced into (10.43) so that, as suggested by (5.27), we will have

$$\frac{1}{2}\int_{S_0} \mathbf{E} \times \mathbf{H} \cdot \mathbf{u}_z \, dS = \frac{1}{2} V_m I_m$$

and when (9.28) holds as well, we will have

$$\frac{1}{2}\int_{S_0} \mathbf{E} \times \mathbf{H}^* \cdot \mathbf{u}_z \, dS = \frac{1}{2} V_m I_m^*$$

as would be the case for TEM modes.

Thus, we have a complete, formal analogy between each set of forward/reverse ($\pm z$-traveling) modes on a uniform section of waveguide, and an equivalent transmission line with parameters Z_{cm} and γ_m. The entire waveguide is thus modeled by an infinite set of completely decoupled equivalent transmission lines, as shown in Figure 10.10. By differentiating (10.42) and (10.43), we can easily obtain a set of telegrapher's equations for these equivalent lines, in the form (cf. (1.4)):

$$\frac{dV_m(z)}{dz} = -\gamma_m Z_{cm} I_m(z) \tag{10.44}$$

$$\frac{dI_m(z)}{dz} = -\frac{\gamma_m}{Z_{cm}} V_m(z) \tag{10.45}$$

Clearly, we may identify $\gamma_m Z_{cm}$ as a *series impedance* per unit length of the equivalent line, while γ_m/Z_{cm} serves as a *shunt admittance* per unit length. The voltage $V_m(z)$ and current $I_m(z)$ may almost be regarded as quantities of physical significance, since the fields may be obtained from them by using (10.41).

10.4   EXPANSION OF THE FIELD AS A SUM OF MODES

We are now in a position to calculate the excitation amplitudes of the modes in a general waveguide in a manner analogous to that for the classical transmission line in Section 3.1.

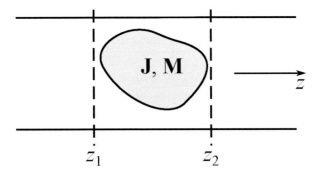

Figure 10.11 Sources in a waveguide.

Consider the situation of Figure 10.11. A collection of impressed current sources $\mathbf{J}$ and $\mathbf{M}$ is distributed in a waveguide over a volume contained between the planes $z = z_1$ and $z = z_2$. Outside the source region, i.e., in $z \le z_1$ or $z \ge z_2$, it seems reasonable that any possible field can be expressed as a sum of mode fields (which are themselves source-free solutions of Maxwell's equations). If there are no sources at infinity, then to the right of z_2 we expect to have only forward modes which propagate or attenuate in the $+z$ direction:

$$\left. \begin{aligned} \mathbf{E} &= \sum_{m>0} a_{m0} \boldsymbol{\mathcal{E}}_m(x,y) e^{-\gamma_m z} \\ \mathbf{H} &= \sum_{m>0} a_{m0} \boldsymbol{\mathcal{H}}_m(x,y) e^{-\gamma_m z} \end{aligned} \right\} \quad (z > z_2) \tag{10.46}$$

The amplitudes of $\boldsymbol{\mathcal{H}}_m$ are the same as those for $\boldsymbol{\mathcal{E}}_m$ since Maxwell's equations would otherwise not be satisfied. We sum over only *positive* m to indicate that only forward modes are present. Similarly, to the left of z_1, the most general field should be a sum of reverse modes which propagate or attenuate in the $-z$ direction:

$$\left. \begin{aligned} \mathbf{E} &= \sum_{m>0} b_{m0} \boldsymbol{\mathcal{E}}_{-m}(x,y) e^{\gamma_m z} \\ \mathbf{H} &= \sum_{m>0} b_{m0} \boldsymbol{\mathcal{H}}_{-m}(x,y) e^{\gamma_m z} \end{aligned} \right\} \quad (z < z_1) \tag{10.47}$$

What we will assume more generally here is the so-called *completeness* property of the modes to describe the transverse components of the total field in a uniform waveguide with sources of bounded extent (Figure 10.11). This assumption is a generalization of our assumption (3.10) for the classical transmission line, and of representation (10.41) for the transverse field components in terms of equivalent transmission lines:

$$\mathbf{E}_T = \sum_{m>0} \tilde{V}_m(z) \boldsymbol{\mathcal{E}}_{mT}(x,y) \tag{10.48}$$

$$\mathbf{H}_T = \sum_{m>0} \tilde{I}_m(z) \boldsymbol{\mathcal{H}}_{mT}(x,y) \tag{10.49}$$

where

$$\left. \begin{aligned} \tilde{V}_m(z) &= \frac{V_m(z)}{\sqrt{2Z_{cm}}} = a_m(z) e^{-\gamma_m z} + b_m(z) e^{\gamma_m z} \\ \tilde{I}_m(z) &= \sqrt{\frac{Z_{cm}}{2}} I_m(z) = a_m(z) e^{-\gamma_m z} - b_m(z) e^{\gamma_m z} \end{aligned} \right\} \tag{10.50}$$

From (10.46) and (10.47) it is clear that

$$a_m(z) \equiv 0 \quad ; \quad b_m(z) \equiv b_{m0} \qquad \text{for } z \leq z_1 \tag{10.51}$$

$$b_m(z) \equiv 0 \quad ; \quad a_m(z) \equiv a_{m0} \qquad \text{for } z \geq z_2 \tag{10.52}$$

The completeness property (10.48)-(10.49) for the transverse fields is far from obvious, especially within the source region. In cases where this property has been proved, the proof is quite difficult; we will not attempt to describe such a proof here. There are, in fact, many waveguides for which the completeness property has not been proved at all. Nevertheless, the property is routinely assumed to be true anyway on the grounds that it is physically reasonable.

In cases like microstrip or dielectric waveguides, some or all of whose "sidewall" boundaries extend to infinity in the transverse plane, the discrete modes of propagation are generally *not* complete by themselves (as we have mentioned already in Chapter 6), and must be supplemented by a spectrum of continuous or radiation modes. Although such modes can be treated (at least formally) in a manner similar to the discrete modes, the details are rather delicate. To get a feel for what is involved in the extension to continuous modes, observe that in place of the sums (10.46) and (10.47), we have to represent the fields as integrals over the continuous spectrum of radiation modes in addition to a sum over the discrete modes, e.g.:

$$
\begin{aligned}
\mathbf{E} &= \sum_{m>0} a_{m0} \boldsymbol{\mathcal{E}}_m(x,y) e^{-\gamma_m z} \\
&+ \int_0^{k_2} a_0(j\beta) \boldsymbol{\mathcal{E}}(j\beta; x, y) e^{-j\beta z} \, d\beta + \int_0^\infty a_0(\alpha) \boldsymbol{\mathcal{E}}(\alpha; x, y) e^{-\alpha z} \, d\alpha
\end{aligned}
$$

where $\boldsymbol{\mathcal{E}}(\gamma; x, y)$ is the field of a radiation mode of propagation coefficient γ as discussed in Sections 6.9 and 9.2, and $a_0(\gamma)$ is the wave amplitude of that spectral component of the field. Further details of this procedure can be found in the references at the end of the chapter.

We will now apply Lorentz' reciprocity theorem to the volume V illustrated in Figure 10.12. The surface S bounding V consists of the back end plane S_b at z_b and the front end plane S_f at z_f, together with the sidewall of the guide for $z_b < z < z_f$. Here z_b and z_f are two values of z chosen for convenience; we will specify them below. For the "a" state, we take the actual fields produced in the guide by the given sources; their transverse components are given by (10.48)-(10.49). For the second state, which we will call the "n" state, we take the field of a single mode of the guide with index n:

$$
\begin{aligned}
\mathbf{E}^n &= \boldsymbol{\mathcal{E}}_n(x,y) e^{-\gamma_n z} \\
\mathbf{H}^n &= \boldsymbol{\mathcal{H}}_n(x,y) e^{-\gamma_n z}
\end{aligned} \tag{10.53}
$$

We insert these fields into (9.8). The surface integral over the sidewall vanishes since $\mathbf{u}_n \times \mathbf{E}^a = \mathbf{u}_n \times \mathbf{E}^n = 0$ on this wall. Moreover, the "n" state sources are identically zero, and so we have

$$
\begin{aligned}
\sum_{m>0} &\left\{ \left[\tilde{V}_m(z_f) \int_{S_0} \boldsymbol{\mathcal{E}}_m \times \boldsymbol{\mathcal{H}}_n \cdot \mathbf{u}_z \, dS - \tilde{I}_m(z_f) \int_{S_0} \boldsymbol{\mathcal{E}}_n \times \boldsymbol{\mathcal{H}}_m \cdot \mathbf{u}_z \, dS \right] e^{-\gamma_n z_f} \right. \\
&\left. - \left[\tilde{V}_m(z_b) \int_{S_0} \boldsymbol{\mathcal{E}}_m \times \boldsymbol{\mathcal{H}}_n \cdot \mathbf{u}_z \, dS - \tilde{I}_m(z_b) \int_{S_0} \boldsymbol{\mathcal{E}}_n \times \boldsymbol{\mathcal{H}}_m \cdot \mathbf{u}_z \, dS \right] e^{-\gamma_n z_b} \right\} \\
&= \int_{z_b}^{z_f} e^{-\gamma_n z} \, dz \int_{S(z)} [\boldsymbol{\mathcal{E}}_n \cdot \mathbf{J}(x,y,z) - \boldsymbol{\mathcal{H}}_n \cdot \mathbf{M}(x,y,z)] \, dS \tag{10.54}
\end{aligned}
$$

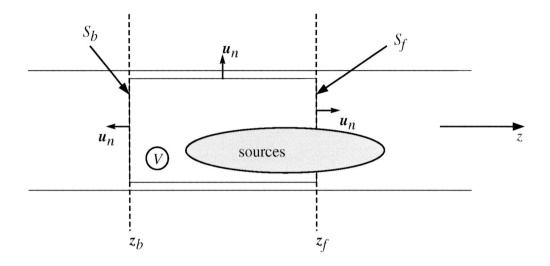

Figure 10.12 Application of Lorentz reciprocity to a waveguide containing impressed sources.

where S_0 is a generic cross-section of the waveguide, while $S(z)$ is the cross-section at the plane z.

Now by (9.24), the cross-section integrals on the left side of (10.54) will vanish by orthogonality unless $m = |n|$. The sums in (10.54) reduce to single terms, and these nonzero cross-section integrals are evaluated from the normalization condition (5.27). As a result, (10.54) becomes:

$$\int_{z_b}^{z_f} e^{-\gamma_n z}\, dz \int_{S(z)} [\boldsymbol{\mathcal{E}}_n \cdot \mathbf{J}(x,y,z) - \boldsymbol{\mathcal{H}}_n \cdot \mathbf{M}(x,y,z)]\, dS$$
$$= \quad 4[b_{|n|}(z_f) - b_{|n|}(z_b)] \qquad \text{if } n > 0$$
$$= \quad -4[a_{|n|}(z_f) - a_{|n|}(z_b)] \qquad \text{if } n < 0 \qquad (10.55)$$

where (10.50) has been used. Consider the case where $z_f = z_2$ for $n > 0$ and where $z_b = z_1$ for $n < 0$; we have

$$b_{|n|}(z_f) = a_{|n|}(z_b) = 0$$

by (10.51) and (10.52). Then taking $z_b = z$ for $n > 0$ and $z_f = z$ for $n < 0$, we have for the functions $a_m(z)$ and $b_m(z)$:

$$a_m(z) =$$
$$-\frac{1}{4} \int_{z_1}^{z} e^{\gamma_m z'} \left\{ \int_{S(z')} [\boldsymbol{\mathcal{E}}_{-m} \cdot \mathbf{J}(x,y,z') - \boldsymbol{\mathcal{H}}_{-m} \cdot \mathbf{M}(x,y,z')]\, dS \right\}\, dz'$$
$$(z > z_1) \qquad (10.56)$$

$$b_m(z) =$$
$$-\frac{1}{4} \int_{z}^{z_2} e^{-\gamma_m z'} \left\{ \int_{S(z')} [\boldsymbol{\mathcal{E}}_{m} \cdot \mathbf{J}(x,y,z') - \boldsymbol{\mathcal{H}}_{m} \cdot \mathbf{M}(x,y,z')]\, dS \right\}\, dz'$$
$$(z < z_2) \qquad (10.57)$$

Eqns. (10.56) and (10.57) are seen to be fully equivalent to (3.13), (3.15) and (3.16) for the classical transmission line if we identify the equivalent distributed sources $E^m_{ext}(z)$ and $H^m_{ext}(z)$ for the $\pm m$ mode pair as:

$$E^m_{ext}(z) = \sqrt{\frac{Z_{cm}}{2}} \int_{S(z)} (\mathcal{E}_{mz} J_z - \mathcal{H}_{mT} \cdot \mathbf{M}_T) \, dS$$

$$H^m_{ext}(z) = \frac{1}{\sqrt{2Z_{cm}}} \int_{S(z)} (\mathcal{H}_{mz} M_z - \mathcal{E}_{mT} \cdot \mathbf{J}_T) \, dS \qquad (10.58)$$

along with the identifications $\gamma \to \gamma_m$ and $Z_c \to Z_{cm}$. Inserting (10.56) and (10.57) into (10.50), and thence into (10.48) and (10.49) gives us a complete solution for the transverse fields due to any impressed sources $\mathbf{J}$ and $\mathbf{M}$, in terms of all the modes of the waveguide. The description of each mode amplitude is fully analogous to an equivalent classical transmission line.

10.4.1   REPRESENTATION OF THE LONGITUDINAL FIELD

The completeness property (10.48)-(10.49) which generally holds for transverse fields is not always true for the z-components of $\mathbf{E}$ and $\mathbf{H}$. This is related to the fact that $\mathcal{E}_z$ and $\mathcal{H}_z$ do not enter into the orthogonality property (9.24) which is used to obtain the amplitudes in (10.48) and (10.49). We must instead infer expressions for E_z and H_z from (10.48)-(10.49) and Maxwell's equations. From the z-components of the Maxwell curl equations (9.1) and (9.2), we obtain

$$E_z = -\frac{J_z}{j\omega\epsilon} + \frac{1}{j\omega\epsilon} \mathbf{u}_z \cdot \nabla_T \times \mathbf{H}_T \qquad (10.59)$$

$$H_z = -\frac{M_z}{j\omega\mu} - \frac{1}{j\omega\mu} \mathbf{u}_z \cdot \nabla_T \times \mathbf{E}_T \qquad (10.60)$$

We insert the expansions (10.48) and (10.49) for $\mathbf{E}_T$ and $\mathbf{H}_T$ into (10.59) and (10.60), and make use of (5.18) and (5.20). Our result is:

$$E_z = -\frac{J_z}{j\omega\epsilon} + \sum_{m>0} \tilde{I}_m(z) \mathcal{E}_{mz}(x, y) \qquad (10.61)$$

$$H_z = -\frac{M_z}{j\omega\mu} + \sum_{m>0} \tilde{V}_m(z) \mathcal{H}_{mz}(x, y) \qquad (10.62)$$

In other words, the z-components, in addition to containing a sum of mode fields as in (10.41), must also contain terms corresponding to the z-components of the impressed currents. Outside of the source region, these extra terms are naturally zero, and a sum of modes is sufficient to represent the entire field, as we should expect.

In some situations, however, especially when equivalent sources are used in the formulation of a problem, the extra terms in (10.61) and (10.62) may be important. As an example of the application of this result, suppose that a cylindrical region of altered material properties $\epsilon + \Delta\epsilon$ and $\mu + \Delta\mu$, occupying the cross-sectional area ΔS is inserted into the waveguide parallel to the z-axis. The original waveguide's material parameters are ϵ and μ, and these as well as $\Delta\epsilon$ and $\Delta\mu$ can be functions of x and y. It may be desirable for some reason to represent the fields of this new (perturbed) waveguide in terms of the modes of the original (unperturbed) waveguide—for example, because the unperturbed waveguide modes ($\mathcal{E}_m$, $\mathcal{H}_m$, γ_m) are simpler than those of the perturbed guide. Equations (10.48)-(10.49) can still be used to represent the transverse fields (although the z-dependences will no longer be simple sums of exponential functions). However, the longitudinal fields will be different, as we now show. The cylindrical region ΔS must be regarded as carrying the equivalent volume current sources

$$\mathbf{J}_{\mathrm{eq}} = j\omega\Delta\epsilon\mathbf{E}; \qquad \mathbf{M}_{\mathrm{eq}} = j\omega\Delta\mu\mathbf{H} \tag{10.63}$$

acting in the environment μ, ϵ of the unperturbed waveguide. Introducing the z-components of these expressions into (10.61)-(10.62) gives

$$E_z = -\frac{\Delta\epsilon}{\epsilon}E_z + \sum_{m>0}\tilde{I}_m(z)\mathcal{E}_{mz}(x,y) \tag{10.64}$$

$$H_z = -\frac{\Delta\mu}{\mu}H_z + \sum_{m>0}\tilde{V}_m(z)\mathcal{H}_{mz}(x,y) \tag{10.65}$$

or

$$E_z = \frac{\epsilon}{\epsilon+\Delta\epsilon}\sum_{m>0}\tilde{I}_m(z)\mathcal{E}_{mz}(x,y) \tag{10.66}$$

$$H_z = \frac{\mu}{\mu+\Delta\mu}\sum_{m>0}\tilde{V}_m(z)\mathcal{H}_{mz}(x,y) \tag{10.67}$$

This result will find application in the calculation of coupling between parallel dielectric waveguides in Chapter 12.

10.4.2   OPTIMIZING EXCITATION OF A GIVEN MODE

Examination of the formulas for excitation of a waveguide reveals some guidelines for enhancing the excitation of a particular mode relative to that of other possible modes. A given mode is excited primarily in the $+z$ rather than $-z$ direction using the methods discussed in Section 3.2, together with eqns. (10.58). The source distributions $\mathbf{J}$ and $\mathbf{M}$ should be chosen to maximize E_{ext}^m and/or H_{ext}^m for the desired mode while minimizing those for undesired modes.

As an example, consider the excitation of a rectangular metallic waveguide by a current probe—a wire carrying a filamentary electric current I (Figure 10.13). Examining (10.56)-(10.57) or (10.58), we see that a current of given magnitude will have maximum effect in exciting a given mode if it is directed parallel to the electric field and located at a position where the magnitude of that field is maximum. For a TE$_{10}$ mode of the rectangular guide, this would be a vertically oriented wire located at the center of the guide cross section as shown in Figure 10.13(a). If a second wire carrying an equal but oppositely directed current were similarly positioned in the guide at a distance l separated from the first in the z-direction as shown in Figure 10.13(b), then from (10.56), the total amplitude with which the mode is excited would differ from that of the single wire excitation by a factor of $(1 - e^{j\beta l})$. This becomes zero if l is an integer multiple of a guide wavelength λ_g, but is maximized if l is an odd number of half-guide wavelengths. Since the configuration of Figure 10.13(b) is equivalent to that of Figure 10.13(c) (where the original probe radiates a distance $l/2$ in front of a short-circuiting plane) by image theory, it can be seen that a single wire can be made to launch the dominant mode of this guide quite efficiently by placing it $\lambda_g/4$ in front of such a short circuit.

If the wire is the inner conductor of a coaxial line, we appear to have obtained a suitable means of transition between the coax and the rectangular waveguide as shown in Figure 10.14. In fact, positioning the coaxial probe exactly $\lambda_g/4$ from the waveguide short circuit is not a good choice from the point of view of an impedance match, but fortunately a

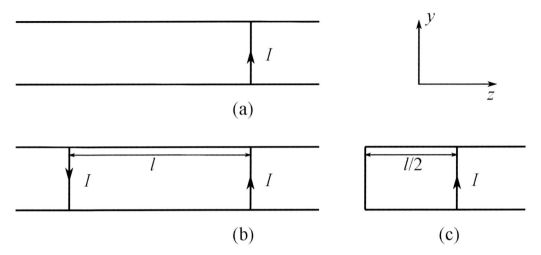

Figure 10.13 Current probe excitation of rectangular metallic waveguide: (a) single probe; (b) two probes separated by a distance l; (c) Image equivalent of (b).

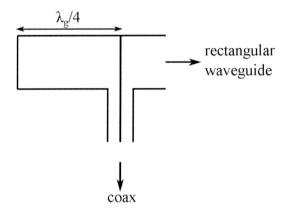

Figure 10.14 Coaxial-to-rectangular waveguide transition.

small deviation of this position from $\lambda_g/4$ does not change the mode excitation very much, and allows the reflection from this junction to be reduced by adjusting it properly.

Similar considerations apply to excitation of other kinds of waveguide. Examples will be studied later in the chapter.

10.5   EXCITATION BY GIVEN ARBITRARY APERTURE FIELDS

If the excitation currents are confined to a transverse plane (the cross-section $z = 0$, for instance), then the volume integrals (10.56) and (10.57) are replaced by surface integrals. For instance,

$$a_m(z) \equiv a_{m0} = -\frac{1}{4} \int_{S_0} (\boldsymbol{\mathcal{E}}_{-m} \cdot \mathbf{J}_S - \boldsymbol{\mathcal{H}}_{-m} \cdot \mathbf{M}_S)\, dS \qquad (z > 0) \qquad (10.68)$$

Suppose now that we take these to be the *equivalent* surface currents corresponding to fields produced by sources radiating in $z < 0$. By the Love equivalence theorem,

$$\mathbf{J}_S = \mathbf{u}_z \times \mathbf{H}_0; \qquad \mathbf{M}_S = \mathbf{E}_0 \times \mathbf{u}_z$$

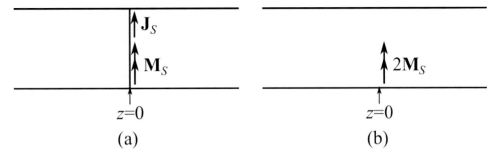

Figure 10.15 (a) Perfect conductor placed behind equivalent sources at $z = 0$; (b) equivalent source plus image radiating in infinite guide.

where $\mathbf{E}_0$ and $\mathbf{H}_0$ are the actual fields present at $z = 0$ (the so-called *aperture field*). Then from (10.68),

$$a_{m0} = -\frac{1}{4} \int_{S_0} (\mathbf{E}_0 \times \mathcal{H}_{-m} - \mathcal{E}_{-m} \times \mathbf{H}_0) \cdot \mathbf{u}_z \, dS \tag{10.69}$$

Eqns. (10.69), (10.48)-(10.50) and (10.61)-(10.62) completely describe the field produced in $z > 0$.

The field produced by these equivalent currents is zero for $z < 0$ as discussed in Section 10.2. We are thus free to insert a perfectly conducting plane at $z = 0^-$, just behind the equivalent sources, without disturbing the field in $z > 0$ (Figure 10.15(a)). From Section 10.2, we know that the effect of $\mathbf{J}_S$ is "shorted out" by the perfect conductor, and we can consider the effect of $\mathbf{M}_S$ only. But by image theory, the effect of $\mathbf{M}_S$ together with the conductor at $z = 0$ is the same as that of a source $2\mathbf{M}_S$ with the conductor removed. Hence, we can also write (10.69) as

$$\begin{aligned} a_{m0} &= -\frac{1}{2} \int_{S_0} \mathbf{E}_0 \times \mathcal{H}_{-m} \cdot \mathbf{u}_z \, dS \\ &= \frac{1}{2} \int_{S_0} \mathbf{E}_0 \times \mathcal{H}_m \cdot \mathbf{u}_z \, dS \end{aligned} \tag{10.70}$$

This is none other than the Thévenin equivalent source of Section 10.2.2.

In a similar way, we could short out the effect of $\mathbf{M}_S$ by placing a magnetic wall at $z = 0$, and after invoking image theory we would obtain

$$a_{m0} = \frac{1}{2} \int_{S_0} \mathcal{E}_m \times \mathbf{H}_0 \cdot \mathbf{u}_z \, dS \tag{10.71}$$

which is the expression for excitation by the Norton equivalent sources of Section 10.2.2. We thus need only to know the transverse $\mathbf{E}$ *or* $\mathbf{H}$ field at $z = 0$ in order to completely determine the strength of each mode in $z > 0$.

If $\mathbf{E}_0$ or $\mathbf{H}_0$ is chosen appropriately (say $\mathbf{E}_0 = \mathcal{E}_m$ or $\mathbf{H}_0 = \mathcal{H}_m$), then $a_m = 1$, and all other mode amplitudes vanish by orthogonality. Naturally, it may be difficult if not impossible to realize a given desired aperture field in practice. Even so, this result gives a goal to strive for if maximum and exclusive excitation of a particular mode is desired.

Suppose now that an aperture S_w is cut in the *side-wall* of a waveguide: a wall which in the absence of the hole would be a perfect conductor. Introduce the right-handed coordinate system (n, l, z) in the aperture, where the unit vector $\mathbf{u}_n$ normal to the sidewall points *into* the interior of the waveguide, and $\mathbf{u}_z \times \mathbf{u}_n = \mathbf{u}_l$. If a tangential electric field

$$\mathbf{E}_0 = \mathbf{u}_l E_{0l} + \mathbf{u}_z E_{0z} \tag{10.72}$$

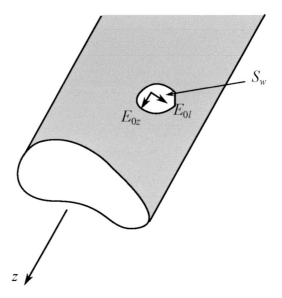

Figure 10.16 Side-wall aperture in a waveguide.

is applied at S_w as shown in Figure 10.16, then according to the Love equivalence principle of Section 10.2, this may be modeled by the equivalent magnetic surface current

$$\mathbf{M}_S = \mathbf{E}_0 \times \mathbf{u}_n\big|_{S_w} = E_{0z}\mathbf{u}_l - E_{0l}\mathbf{u}_z \tag{10.73}$$

impressed at S_w. The aperture may be closed up with a perfectly conducting wall just behind these currents, and the equivalent electric surface currents at S_w ignored because they produce no field in the waveguide. From (10.58) we have

$$E_{\text{ext}}^m(z) = -\sqrt{\frac{Z_{cm}}{2}} \oint_{C(z) \cap S_w} \mathcal{H}_{ml} E_{0z}\, dl \tag{10.74}$$

$$H_{\text{ext}}^m(z) = -\frac{1}{\sqrt{2Z_{cm}}} \oint_{C(z) \cap S_w} \mathcal{H}_{mz} E_{0l}\, dl \tag{10.75}$$

where $C(z)$ is the boundary of the waveguide at z, and the boundary integrals are taken over the intersection of this contour with S_w.

In an open waveguide with no metallic sidewall, the "aperture" must be taken to extend for the whole length of the guide, or at least over those places where the "aperture" fields are significantly different from zero. Moreover, since we do not metallize the aperture when applying the equivalence principle in this case, we will have to use both electric and magnetic equivalent surface sources over whatever "aperture" we use.

10.5.1   EXAMPLE: PLANE WAVE EXCITATION OF A DIELECTRIC SLAB

Consider a semi-infinite dielectric slab waveguide illuminated by the normally incident plane wave

$$\begin{aligned}
\mathbf{E}^{\text{inc}} &= \mathbf{u}_y E_i e^{-jk_0 z} \\
\mathbf{H}^{\text{inc}} &= -\mathbf{u}_x \frac{E_i}{\zeta_0} e^{-jk_0 z}
\end{aligned} \tag{10.76}$$

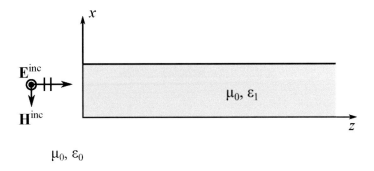

Figure 10.17 Excitation of a dielectric slab by an incident plane wave.

from free space as shown in Figure 10.17. If the resulting aperture field $\mathbf{E}_0$, $\mathbf{H}_0$ in the plane $z = 0$ were known, we could use any of eqns. (10.69), (10.70) or (10.71) to calculate the excitation amplitude of any of the TE surface waves on the slab. If we can assume that $\mathbf{E}_0$, $\mathbf{H}_0$ are approximately given by (10.76) (this is the Kirchhoff approximation described in Section 10.2.3), then our calculations can proceed.

From (10.71), for example, along with eqns. (6.46), (6.32) and (6.43), we have (putting $\int dS \rightarrow \int dx$)

$$
\begin{aligned}
a_{m0} &\simeq -\frac{1}{2} \int_{-\infty}^{\infty} \mathcal{E}_y H_x^{\text{inc}} \, dx \bigg|_{z=0} \\
&= \frac{1}{2} \frac{E_i}{\zeta_0} \int_{-\infty}^{\infty} \mathcal{E}_y \, dx \\
&= \frac{2E_i}{p_2 \sqrt{\zeta_0 n_{\text{eff}} a_{\text{eff}}}} \frac{V}{h_1 a} \quad (m \text{ even})
\end{aligned}
\tag{10.77}
$$

while $a_{m0} = 0$ for m odd. From (10.70) on the other hand, we get

$$
\begin{aligned}
a_{m0} &\simeq -\frac{1}{2} \int_{-\infty}^{\infty} E_y^{\text{inc}} \mathcal{H}_x \, dx \bigg|_{z=0} \\
&= -n_{\text{eff}} \int_{-\infty}^{\infty} \mathcal{E}_y H_x^{\text{inc}} \, dx \bigg|_{z=0} \\
&= \frac{2E_i \sqrt{n_{\text{eff}}}}{p_2 \sqrt{\zeta_0 a_{\text{eff}}}} \frac{V}{h_1 a} \quad (m \text{ even})
\end{aligned}
\tag{10.78}
$$

and again $a_{m0} = 0$ for m odd. Thus, the two approximations differ only by the factor n_{eff}, and our result can be expected to be accurate if n_{eff} is close to 1. The result of (10.69) will be the arithmetic mean of these two values, and in the absence of other evidence of accuracy might be chosen as the best of the three approximations. This of course has to be checked by numerical calculations.

Naturally, a better excitation of the TE_0 mode could be obtained by illuminating the end of the slab with a field whose profile more closely matches that of the mode. A Gaussian beam is such a field: a well-collimated beam which is capable of propagating relatively long distances without significant loss of amplitude due to diffraction (spreading of the energy in the direction transverse to that of propagation). The mathematical details of the calculation are in this case rather more cumbersome.

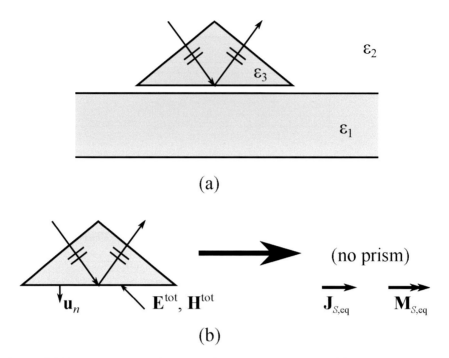

Figure 10.18 (a) Excitation of a slab waveguide by a prism coupler; (b) plane wave inside prism and equivalent currents.

10.5.2 PRISM COUPLER

Another way of exciting the surface wave modes of a dielectric waveguide is from the side, as shown in Figure 10.18(a). A plane wave enters a dielectric prism at an angle such that total internal reflection would occur at the bottom face of the prism if no waveguide were present underneath (Figure 10.18(b)). The sources and material of the prism can, by Love's equivalence principle, be replaced by a uniform material of the same properties as the cladding, and equivalent surface electric and magnetic currents placed where the bottom face of the prism was, and these can be used to calculate the excitation of the waveguide. Since the field below the bottom of the prism is evanescent when no guide is present, we assume that the field at the bottom of the prism is nearly the same even when the guide is present for purposes of calculating these equivalent currents.

Though the detailed evaluation of the mode amplitudes will not be done here, we can note that the phase variation of the plane wave field in the prism should be the same as that of the desired surface wave mode $(e^{-j\beta z})$ in order to maximize that particular mode amplitude. The arguments are similar to those given at the end of Section 3.2.

10.5.3 EXAMPLE: EXCITATION OF MICROSTRIP BY A SLOT IN THE GROUND PLANE

Consider next a microstrip line which is excited by a slot in the ground plane transverse to the strip, and across which a voltage V_{slot} is applied. A side view of the situation is shown in Figure 10.19. We will assume that the slot width g is so small electrically that the electric field in the slot can be reckoned as quasistatic, giving this applied voltage a well-defined meaning. The quasistatic field is written as:

$$\mathbf{E}_0\big|_{y=0} = \begin{cases} -\mathbf{u}_z E_{0z}(z) & (|z| < \frac{g}{2}) \\ 0 & (|z| > \frac{g}{2}) \end{cases} \tag{10.79}$$

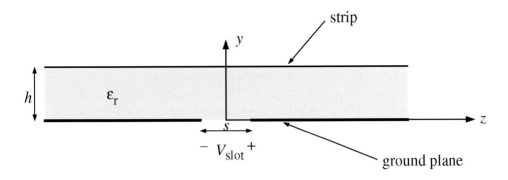

Figure 10.19 Slot excitation of microstrip.

where $E_{0z}(z)$ satisfies

$$\int_{-g/2}^{g/2} E_{0z}(z)\,dz = V_{\text{slot}} \tag{10.80}$$

For this "sidewall" aperture, we have $\mathbf{u}_n = \mathbf{u}_y$, $\mathbf{u}_l = -\mathbf{u}_x$ and hence $\mathcal{H}_{ml} = -\mathcal{H}_{mx}$. By (10.75), $H_{\text{ext}}^m \equiv 0$, while by (10.74)

$$
\begin{aligned}
E_{\text{ext}}^m(z) &= \sqrt{\frac{Z_{cm}}{2}} E_{0z}(z) \int_{-\infty}^{\infty} \mathcal{H}_{mx}\,dx \Bigg|_{y=0} \\
&\simeq \begin{cases} -E_{0z}(z) & (|z| < \frac{g}{2}) \\ 0 & (|z| > \frac{g}{2}) \end{cases}
\end{aligned}
\tag{10.81}
$$

We have made use of the fact that for the quasistatic limit of the quasi-TEM mode,

$$\int_{-\infty}^{\infty} \mathcal{H}_{mx}\,dx \Bigg|_{y=0} = I \simeq \sqrt{\frac{2}{Z_{cm}}} \tag{10.82}$$

[compare (8.40)]. The minus sign in (10.81) reflects the fact that the voltage source has been inserted into the ground conductor of this transmission line instead of the positive side. By (3.23) and (10.80), we see as expected that the slot is equivalent to a single series-connected lumped voltage source as in Figure 3.10, with $V_0 = -V_{\text{slot}}$.

10.6 EXCITATION BY ELECTRIC AND MAGNETIC DIPOLES AND RELATED SCATTERING PROBLEMS

An elementary electric dipole

$$\mathbf{J} = j\omega\mathbf{p}\delta(x - x_0)\delta(y - y_0)\delta(z - z_0) \tag{10.83}$$

concentrated at the point (x_0, y_0, z_0) produces, according to (10.46) and (10.56), the forward mode amplitudes

$$a_{m0} = -\frac{j\omega}{4}\mathbf{p} \cdot \boldsymbol{\mathcal{E}}_{-m}(x_0, y_0)e^{\gamma_m z_0} \tag{10.84}$$

for $z > z_0$, and by (10.47) and (10.57) the reverse mode amplitudes

$$b_{m0} = -\frac{j\omega}{4}\mathbf{p} \cdot \boldsymbol{\mathcal{E}}_m(x_0, y_0)e^{-\gamma_m z_0} \tag{10.85}$$

in $z < z_0$. By (10.58) and (3.24), this is equivalent to what the lumped sources

$$
\begin{aligned}
V_0^{(m)} &= j\omega\sqrt{\frac{Z_{cm}}{2}}\mathcal{E}_{mz}(x_0, y_0)p_z \\
I_0^{(m)} &= -j\omega\frac{1}{\sqrt{2Z_{cm}}}\mathcal{E}_{mT}(x_0, y_0)\cdot\mathbf{p}_T
\end{aligned} \tag{10.86}
$$

located at $z = z_0$ would excite. Similarly, the elementary magnetic dipole

$$
\mathbf{M} = j\omega\mu\mathbf{m}\delta(x - x_0)\delta(y - y_0)\delta(z - z_0) \tag{10.87}
$$

produces the mode amplitudes

$$
a_{m0} = \frac{j\omega\mu}{4}\mathbf{m}\cdot\mathcal{H}_{-m}(x_0, y_0)e^{\gamma_m z_0} \tag{10.88}
$$

in $z > z_0$ and

$$
b_{m0} = \frac{j\omega\mu}{4}\mathbf{m}\cdot\mathcal{H}_m(x_0, y_0)e^{-\gamma_m z_0} \tag{10.89}
$$

in $z < z_0$, equivalent to the lumped sources

$$
\begin{aligned}
V_0^{(m)} &= -j\omega\mu\sqrt{\frac{Z_{cm}}{2}}\mathcal{H}_{mT}(x_0, y_0)\cdot\mathbf{m}_T \\
I_0^{(m)} &= j\omega\mu\frac{1}{\sqrt{2Z_{cm}}}\mathcal{H}_{mz}(x_0, y_0)m_z
\end{aligned} \tag{10.90}
$$

These results can also be used in conjunction with the small object scattering theory of Section 10.2.5 or the small aperture theory of Section 10.2.6 to obtain mode amplitudes for waveguides containing electrically small obstacles or apertures, as shown in examples 10.6.2 or 10.6.3 below.

10.6.1 EXAMPLE: DIPOLE EXCITATION OF A COAXIAL CABLE

Let us consider the effect of an elementary radially-directed electric dipole $\mathbf{p} = \mathbf{u}_\rho p$ located at a radius $\rho = \rho_0$ and at $z = 0$ in the coaxial line described in Section 8.3.1. According to (D.18) of Appendix D, this dipole is approximately equivalent to an electrically short wire antenna of length l carrying a current I, provided that

$$
p = \frac{Il}{j\omega}
$$

According to (10.84), we have in this case

$$
\begin{aligned}
\left.\begin{array}{c} a_{10} \\ b_{10} \end{array}\right\} &= -\frac{1}{4}Il\mathbf{u}_\rho\cdot\mathcal{E}_{\mp 1}(\rho_0, \phi_0) \\
&= -\frac{\sqrt{\zeta}Il}{\sqrt{\pi\ln(b/a)}4\rho_0}
\end{aligned} \tag{10.91}
$$

where the index $m = 1$ denotes the TEM mode.

The power carried away from the dipole by the TEM wave (in both directions) is

$$
\begin{aligned}
P_{\text{rad}} &= |a_{10}|^2 + |b_{10}|^2 \\
&= \frac{\zeta I^2 l^2}{8\pi\rho_0^2\ln(b/a)}
\end{aligned} \tag{10.92}
$$

If the dimensions of the coax are such that all higher-order modes are cutoff, any of these which may be excited will carry no power since Z_{cm} for a cutoff mode is purely imaginary. In that case, we can define a *radiation resistance* R_{rad} of the dipole in the coaxial line as

$$R_{\text{rad}} \equiv \frac{2P_{\text{rad}}}{|I|^2} = \frac{\zeta l^2}{4\pi\rho_0^2 \ln(b/a)} = Z_c \left(\frac{l}{2\rho_0}\right)^2 \tag{10.93}$$

This should be compared with the radiation resistance of a dipole in free space, which is

$$R_{\text{rad}} = \frac{\zeta_0 k_0^2 l^2}{6\pi} \tag{10.94}$$

Unlike (10.94), (10.93) does not vanish with frequency, and so this means of excitation is equally efficient (or inefficient) at all frequencies.

10.6.2 EXAMPLE: SCATTERING BY A SMALL OBSTACLE IN A RECTANGULAR WAVEGUIDE

Suppose there is an electrically small object centered at the coordinates (x_0, y_0) in the plane $z = 0$ of an air-filled rectangular metallic waveguide (Figure 10.20). We wish to compute the amplitudes of the TE_{10} modes transmitted past and reflected back from this obstacle when a TE_{10} mode of unit amplitude is incident from the left. Without the obstacle present, the field is merely that of the incident wave

$$\begin{aligned} \mathbf{E}^{\text{inc}} &= \boldsymbol{\mathcal{E}}_1 e^{-j\beta_1 z} \\ \mathbf{H}^{\text{inc}} &= \boldsymbol{\mathcal{H}}_1 e^{-j\beta_1 z} \end{aligned} \tag{10.95}$$

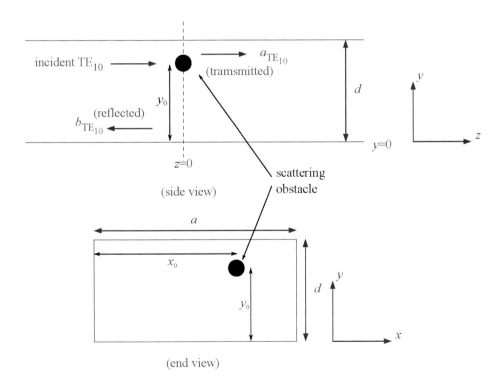

Figure 10.20 An electrically small obstacle in a rectangular waveguide.

which will act on the obstacle to induce equivalent electric and magnetic dipoles. Here, $\mathcal{E}_{\pm1}, \mathcal{H}_{\pm1}$ and β_1 are the normalized TE_{10} mode fields and propagation coefficient as given in Section 5.6.1:

$$\mathcal{E}_{\pm1} = \mathbf{u}_y E_{10} \sin \frac{\pi x}{a} \tag{10.96}$$

$$\mathcal{H}_{\pm1} = \frac{E_{10}}{Z_{c1}} \left[\mp \mathbf{u}_x \sin \frac{\pi x}{a} + \mathbf{u}_z \frac{j\pi}{a\beta_1} \cos \frac{\pi x}{a} \right]$$

where by (5.102),

$$E_{10} = -\frac{j\omega\mu_0 a}{\pi} H_{10} = 2\sqrt{\frac{Z_{c1}}{ad}} \tag{10.97}$$

and

$$Z_{c1} = \zeta_{TE_{10}} = \frac{\omega\mu_0}{\beta_1}$$

is the characteristic impedance of the mode.

By (10.20)-(10.21) the induced dipole moments of the obstacle are

$$\mathbf{p} = \epsilon_0 \overset{\leftrightarrow}{\alpha}_e \cdot \mathcal{E}_1 = E_{10} \sin \frac{\pi x_0}{a} \left(\overset{\leftrightarrow}{\alpha}_e \cdot \mathbf{u}_y \right) \tag{10.98}$$

$$\mathbf{m} = \overset{\leftrightarrow}{\alpha}_m \cdot \mathcal{H}_1 = E_{10} \left[-\frac{\beta_1}{\omega\mu_0} \sin \frac{\pi x_0}{a} \left(\overset{\leftrightarrow}{\alpha}_m \cdot \mathbf{u}_x \right) + \frac{j\pi}{\omega\mu_0 a} \cos \frac{\pi x_0}{a} \left(\overset{\leftrightarrow}{\alpha}_m \cdot \mathbf{u}_z \right) \right] \tag{10.99}$$

For simplicity, we will assume that the polarizability dyadics are diagonal; that is,

$$\overset{\leftrightarrow}{\alpha}_e = \mathbf{u}_x\mathbf{u}_x\alpha_{e,xx} + \mathbf{u}_y\mathbf{u}_y\alpha_{e,yy} + \mathbf{u}_z\mathbf{u}_z\alpha_{e,zz}; \qquad \overset{\leftrightarrow}{\alpha}_m = \mathbf{u}_x\mathbf{u}_x\alpha_{m,xx} + \mathbf{u}_y\mathbf{u}_y\alpha_{m,yy} + \mathbf{u}_z\mathbf{u}_z\alpha_{m,zz}$$

Thus, by superposition of (10.84) and (10.88), we obtain the transmitted wave amplitude in $z > 0$ to be:

$$a_1 \simeq 1 - \frac{j\omega}{4}\mathbf{p} \cdot \mathcal{E}_{-1}(x_0, y_0) + \frac{j\omega\mu_0}{4}\mathbf{m} \cdot \mathcal{H}_{-1}(x_0, y_0) = 1 - j\Delta_+ \tag{10.100}$$

where

$$\Delta_+ = \frac{1}{\beta_1 ad} \left[\left(k_0^2\alpha_{e,yy} + \beta_1^2\alpha_{m,xx} \right) \sin^2 \frac{\pi x_0}{a} + \frac{\pi^2}{a^2}\alpha_{m,zz} \cos^2 \frac{\pi x_0}{a} \right] \tag{10.101}$$

Note that the expression for a_1 includes the original incident wave in addition to what is excited by the dipoles. In a similar way, the reflected wave $b_1 = a_{-1}$ in $z < 0$ is found to be

$$b_1 \simeq -\frac{j\omega}{4}\mathbf{p} \cdot \mathcal{E}_1(x_0, y_0) + \frac{j\omega\mu_0}{4}\mathbf{m} \cdot \mathcal{H}_1(x_0, y_0) = -j\Delta_- \tag{10.102}$$

where

$$\Delta_- = \frac{1}{\beta_1 ad} \left[\left(k_0^2\alpha_{e,yy} - \beta_1^2\alpha_{m,xx} \right) \sin^2 \frac{\pi x_0}{a} + \frac{\pi^2}{a^2}\alpha_{m,zz} \cos^2 \frac{\pi x_0}{a} \right] \tag{10.103}$$

The careful reader will have detected a slight anomaly in this solution. If all modes except the dominant TE_{10} mode are cutoff, then only that is capable of transferring power in the infinitely long sections of waveguide. Now, the power carried toward the obstacle by the incident wave is 1, since the wave has unit amplitude. Likewise, the reflected and transmitted waves carry the powers

$$|b_1|^2$$

and
$$|a_1|^2$$
away from the obstacle. The total of these should equal the total power incident on the obstacle, i. e.,
$$|a_1|^2 + |b_1|^2 = 1 \tag{10.104}$$
A glance at (10.100) and (10.102) shows that our solution gives
$$|a_1|^2 + |b_1|^2 = 1 + \Delta_+^2 + \Delta_-^2 \tag{10.105}$$

The problem is that small obstacle polarizability theory as presented in Section 10.2.5 and used here is correct only to first order in the small quantities $\overleftrightarrow{\alpha}_e$ and $\overleftrightarrow{\alpha}_m$, and is, in fact, in error by terms of order $\left(\overleftrightarrow{\alpha}_e\right)^2$ and $\left(\overleftrightarrow{\alpha}_m\right)^2$. This is precisely what we observe in eqn. (10.105).

We may fix this difficulty by using, instead of $\mathbf{p}$ and $\mathbf{m}$ as given by eqns. (10.98) and (10.99),
$$\mathbf{p}' = \frac{\mathbf{p}}{1 + \delta_{ap}}; \qquad \mathbf{m}' = \frac{\mathbf{m}}{1 + \delta_{ap}} \tag{10.106}$$

for the equivalent dipole of the obstacle, where δ_{ap} is a complex constant which has the same order of magnitude as Δ, and is to be determined by the condition of conservation of complex power. When this change is made, we find instead of (10.100) that
$$a_1 = 1 - \frac{j\Delta_+}{1 + \delta_{ap}} \tag{10.107}$$

and instead of (10.102) that
$$b_1 = -\frac{j\Delta_-}{1 + \delta_{ap}} \tag{10.108}$$

Upon enforcing condition (10.104), we find that
$$\delta_{ap} = \delta_r + j\frac{\Delta_+^2 + \Delta_-^2}{2\Delta_+}$$

where the real part δ_r is not determined by balance of the real part of the complex power (10.104), but must be obtained from other considerations, as shown in Appendix N. For now, we set $\delta_r = 0$ so that we finally have the expressions
$$
\begin{aligned}
a_1 &= 1 - \frac{j\Delta_+}{1 + j\frac{\Delta_+^2 + \Delta_-^2}{2\Delta_+}} \\
b_1 &= -\frac{j\Delta_-}{1 + j\frac{\Delta_+^2 + \Delta_-^2}{2\Delta_+}}
\end{aligned}
\tag{10.109}
$$

which have been corrected to satisfy conservation of energy on the time average.

10.6.3 EXAMPLE: DIAPHRAGM WITH A SMALL APERTURE IN A RECTANGULAR WAVEGUIDE

We next consider the case of a rectangular waveguide with a perfectly conducting diaphragm at $z = 0$, into which a small aperture centered at x_0 and y_0 has been cut (Figure 10.21). We wish to compute the amplitudes of the TE_{10} modes transmitted through and reflected back from this diaphragm when a TE_{10} mode of unit amplitude is incident from the left.

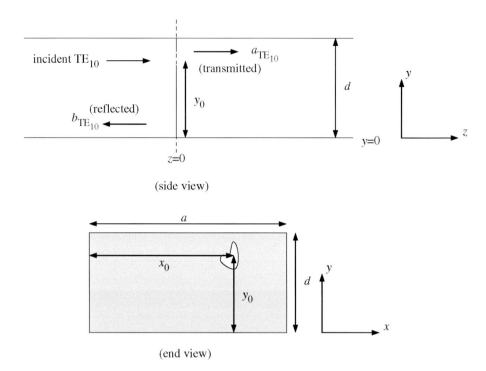

Figure 10.21 Diaphragm with small aperture in a rectangular waveguide.

If there were no aperture present, the incident wave would be totally reflected at the diaphragm, resulting in the short-circuit field

$$\begin{aligned} \mathbf{E}^{sc} &= [\boldsymbol{\mathcal{E}}_1 e^{-j\beta_1 z} - \boldsymbol{\mathcal{E}}_{-1} e^{j\beta_1 z}] \quad \text{(no aperture)} \\ \mathbf{H}^{sc} &= [\boldsymbol{\mathcal{H}}_1 e^{-j\beta_1 z} - \boldsymbol{\mathcal{H}}_{-1} e^{j\beta_1 z}] \end{aligned} \tag{10.110}$$

in $z < 0$, with the mode fields given by (10.96)-(10.97) above. The short-circuit field is zero for $z > 0$. The jump in the short-circuit field across $(x_0, y_0, 0)$ at the diaphragm is thus

$$\begin{aligned} [\mathbf{D}^{sc}]_{(x_0,y_0,0^-)}^{(x_0,y_0,0^+)} &= 0 \\ [\mathbf{H}^{sc}]_{(x_0,y_0,0^-)}^{(x_0,y_0,0^+)} &= \mathbf{u}_x \frac{2E_{10}}{Z_{c1}} \sin \frac{\pi x_0}{a} \end{aligned} \tag{10.111}$$

By Section 10.2.6, the fields in $z > 0$ resulting from the aperture being opened up are approximately equal to those produced by the magnetic dipole

$$\begin{aligned} \mathbf{m}_+ &= \overset{\leftrightarrow}{\alpha}_M \cdot [\mathbf{H}^{sc}]_{(x_0,y_0,0^-)}^{(x_0,y_0,0^+)} \\ &= 2\frac{\beta_1}{\omega\mu} E_{10} \sin \frac{\pi x_0}{a} (\overset{\leftrightarrow}{\alpha}_M \cdot \mathbf{u}_x) \end{aligned} \tag{10.112}$$

located at $(x_0, y_0, 0^+)$ with the aperture closed off. The fields in $z < 0$ are the sum of (i) the incident wave and reflected wave as given by (10.110), and (ii) a field produced by the magnetic dipole

$$\mathbf{m}_- = -2\frac{\beta_1}{\omega\mu} E_{10} \sin \frac{\pi x_0}{a} (\overset{\leftrightarrow}{\alpha}_M \cdot \mathbf{u}_x) \tag{10.113}$$

located at $(x_0, y_0, 0^-)$ with the aperture closed off.

By image theory, the dipole (10.112) radiating in the presence of the diaphragm is (for $z > 0$) equivalent to a dipole of twice the strength acting without the diaphragm (e. g., in the infinite waveguide). The amplitude of the forward-traveling TE_{10} mode in $z > 0$ is therefore, by (10.88) and (10.112)

$$a_1 \simeq \frac{1}{2} j\omega\mu \mathbf{m}_+ \cdot \boldsymbol{\mathcal{H}}_{-1}(x_0, y_0) = j\Delta \tag{10.114}$$

where

$$\Delta = \frac{4\beta_1}{ad} \alpha_{M,xx} \sin^2 \frac{\pi x_0}{a} \tag{10.115}$$

and

$$\alpha_{M,xx} = \mathbf{u}_x \cdot \overset{\leftrightarrow}{\alpha}_M \cdot \mathbf{u}_x$$

is the xx-component of the tensor $\overset{\leftrightarrow}{\alpha}_M$. The total amplitude of the $-z$ traveling TE_{10} mode in $z < 0$ is, similarly,

$$b_1 \simeq -1 + j\Delta \tag{10.116}$$

which includes the reflected part of the field $\mathbf{E}^{\mathrm{sc}}$, $\mathbf{H}^{\mathrm{sc}}$ from (10.110).

As with the small obstacle problem considered in Section 10.6.2, we observe that this solution does not satisfy power conservation exactly. As was done in that case, we fix this difficulty by using, instead of $\mathbf{m}_\pm$ as given by eqns. (10.112) and (10.113),

$$\mathbf{m}'_\pm = \frac{\mathbf{m}_\pm}{1 + \delta_{\mathrm{ap}}} \tag{10.117}$$

for the equivalent dipole of the aperture, where $\delta_{\mathrm{ap}} = j\Delta$ (a more accurate calculation that determines the real part of δ_{ap} can be carried out using the method of Appendix N). We end up with the expressions

$$\begin{aligned} a_1 &= \frac{j\Delta}{1 + j\Delta} \\ b_1 &= -1 + \frac{j\Delta}{1 + j\Delta}, \end{aligned} \tag{10.118}$$

corrected to satisfy conservation of energy on the time average. We will see in Chapter 11 that when δ_r is specified in this way, the simplest possible equivalent circuit for the diaphragm results.

10.7 NOTES AND REFERENCES

Field equivalence principles are discussed in

Harrington (1961), Chapter 3;
Collin (1991), Chapter 1;

as well as

S. A. Schelkunoff, "Kirchhoff's formula, its vector analogue, and other field equivalence theorems," *Commun. Pure Appl. Math.*, vol. 4, pp. 43-59 (1951).

V. H. Rumsey, "Reaction concept in electromagnetic theory," *Phys. Rev.*, vol. 94, pp. 1483-1491 (1954); erratum; *ibid.*, vol. 95, p. 1705 (1954).

V. H. Rumsey, "Some new forms of Huygens' principle," *IRE Trans. Ant. Prop.*, vol. 7, pp. S103-S116 (1959).

B. S. Svetov and V. P. Gubatenko, "Equivalence of systems of internal electric and magnetic currents," *Sov. J. Commun. Technol. Electron.*, vol. 30, no. 7, pp. 67-72 (1985).

J. Appel-Hansen, "Comments on field equivalence principles," *IEEE Trans. Ant. Prop.*, vol. 35, pp. 242-244 (1987).

Schelkunoff's paper above includes electromagnetic analogs of the Thévenin and Norton equivalence theorems from circuit theory; they are close in spirit to the static field equivalence theorem given by Helmholtz:

H. L. F. von Helmholtz, "Ueber einige Gesetze der Vertheilung elektrischer Ströme in körperlichen Leitern mit Anwendung auf die thierisch-elektrischen Versuche," *Ann. Phys. Chem.*, vol. 89, pp. 211-233, 353-377 (1853) [also in H. L. F. von Helmholtz, *Wissenschaftliche Abhandlungen*, vol. 1. Leipzig: Johann Ambrosius Barth, 1882, pp. 475-519].

a summary of which (in English) can be found in:

Koenigsberger (1965), pp. 99-103.

The general theory of small apertures and their polarizabilities is given in

Collin (1991), Sections 7.3-7.5.

Montgomery, Dicke and Purcell (1965), pp. 176-179.

Mashkovtsev *et al.* (1966), Chapter 4.

L. Jülke, "Der Durchgriff elektromagnetischer Felder durch kleiner Löcher in einem metallischen Schirm," *Hochfreq. Elektroakust.*, vol. 76, pp. 68-76 (1967).

J. Van Bladel, "Small-hole coupling of resonant cavities and waveguides," *Proc. IEE (London)*, vol. 117, pp. 1098-1104 (1970).

J. Van Bladel, "Small holes in a waveguide wall," *Proc. IEE (London)*, vol. 118, pp. 43-50 (1971).

Collin (2001), Section 4.13.

Sporleder and Unger (1979), pp. 47-55.

Figure 10.8 is adapted from one in Collin (2001). Note that not all authors use the same definition for the polarizabilities. Some include the effect of image theory by adding the image dipole moment to that of the original; our treatment does not include the image term. Methods of computing or measuring the polarizabilities, numerical values and approximations for various aperture shapes are given in

S. B. Cohn, "Determination of aperture parameters by electrolytic-tank measurements," *Proc. IRE*, vol. 39, pp. 1416-1421 (1951). Correction, *ibid.*, vol. 40, p. 33 (1952).

S. B. Cohn, "The electric polarizability of apertures of arbitrary shape," *Proc. IRE*, vo. 40, pp. 1069-1071 (1952).

R. F. Fikhmanas and P. Sh. Fridberg, "Theory of diffraction at small apertures: Computation of upper and lower bounds of polarizabilities," *Radio Eng. Electron. Phys.*, vol. 18, pp. 824-829 (1973).

R. De Meulenaere and J. Van Bladel, "Polarizability of some small apertures," *IEEE Trans. Ant. Prop.*, vol. 25, pp. 198-205 (1977).

R. De Smedt and J. Van Bladel, "Magnetic polarizability of some small apertures," *IEEE Trans. Ant. Prop.*, vol. 28, pp. 703-707 (1980).

E. E. Okon and R. F. Harrington, "The polarizabilities of electrically small apertures of arbitrary shape," *IEEE Trans. Electromag. Compat.*, vol. 23, pp. 359-366 (1981).

E. Arvas and R. F. Harrington, "Computation of the magnetic polarizability of conducting disks and the electric polarizability of apertures," *IEEE Trans. Ant. Prop.*, vol. 31, pp. 719-725 (1983).

Kh. L. Garb, R. S. Meierova, G. V. Pocikaev and P. Sh. Fridberg, "Polarizability coefficients of a rectangular aperture with parallel screens placed in its vicinity," *Sov. Phys. Collection*, vol. 24, no. 2, pp. 51-54 (1984).

N. A. McDonald, "Polynomial approximations for the electric polarizabilities of some small apertures," *IEEE Trans. Micr. Theory Tech.*, vol. 33, pp. 1146-1149 (1985).

N. A. McDonald, "Polynomial approximations for the transverse magnetic polarizabilities of some small apertures," *IEEE Trans. Micr. Theory Tech.*, vol. 35, pp. 20-23 (1987).

V. I. Fabrikant, "Magnetic polarizability of small apertures: analytical approach," *J. Phys. A: Math. Gen.*, vol. 20, pp. 323-338 (1987).

V. I. Fabrikant, "Electrical polarizability of small apertures: analytical approach," *Int. J. Electron.*, vol. 62, pp. 533-545 (1987).

N. A. McDonald, "Simple approximations for the longitudinal magnetic polarizabilities of some small apertures," *IEEE Trans. Micr. Theory Tech.*, vol. 36, pp. 1141-1144 (1988).

R. E. English, Jr. and N. George, "Diffraction from a small square aperture: approximate aperture fields," *J. Opt. Soc. Amer. A*, vol. 5, pp. 192-199 (1988).

The approximate expressions for α_E and $\overset{\leftrightarrow}{\alpha}_M$ of the rectangular aperture in Table 10.1 are modified versions of those given in McDonald's papers. Those for the Greek cross are adapted from Fabrikant's papers, where their accuracy is also assessed.

Various estimates and bounds for the polarizabilities of apertures in zero-thickness screens are available. For "star-shaped" apertures with respect to an origin inside A, the inequality

$$\alpha_E \leq \frac{1}{\pi} \int_A \rho \, dS \tag{10.119}$$

where ρ is the distance from that origin, was obtained by

L. E. Payne, "Isoperimetric inequalities and their applications," *SIAM Review*, vol. 9, pp. 453-488 (1967).

For convex apertures whose aspect ratio is not too great, the further inequalities,

$$\frac{4A^2}{3\pi P} \leq \alpha_E \leq \frac{2}{3} \left(\frac{A}{\pi} \right)^{3/2} \tag{10.120}$$

and

$$\frac{4}{3} \left(\frac{A}{\pi} \right)^{3/2} \leq \frac{\alpha_{M,uu} + \alpha_{M,vv}}{2} \leq \frac{4}{3} \left(\frac{P}{2\pi} \right)^3 \tag{10.121}$$

where A is the area of the aperture and P its perimeter, were conjectured in

D. L. Jaggard and C. H. Papas, "On the application of symmetrization to the transmission of electromagnetic waves through small convex apertures of arbitrary shape," *Appl. Phys.*, vol. 15, pp. 21-25 (1978).

Lastly, for an aperture of any shape we have

$$\frac{1}{\alpha_E} \geq \frac{1}{\alpha_{M,uu}} + \frac{1}{\alpha_{M,vv}} \tag{10.122}$$

if the axes u and v are chosen so that $\alpha_{M,uv} = \alpha_{M,vu} = 0$ (these are the so-called *principal axes*). This inequality was proved in

> K. S. H. Lee and L. K. Warne, "A general relation between electric polarizabilities of a plane aperture," *Radio Sci.*, vol. 39, art. RS6007, 2004.

The theory of electrically small apertures in screens of nonzero thickness has been carried out in

> A. N. Akhiezer, "On the inclusion of the effect of the thickness of the screen in certain diffraction problems," *Sov. Phys. Tech. Phys.*, vol. 2, pp. 1190-1196 (1957).
>
> A. N. Akhiezer, "On the coupling of rectangular waveguides by means of an aperture in the wide wall," *Sov. Phys. Tech. Phys.*, vol. 5, pp. 802-805 (1961).
>
> N. A. McDonald, "Electric and magnetic coupling through small apertures in shield walls of any thickness," *IEEE Trans. Micr. Theory Tech.*, vol. 20, pp. 689-695 (1972).
>
> F. Sporleder, *Erweiterte Theorie der Lochkopplung.* Dr.-Ing. dissertation, Technische Universität Braunschweig, 1976.
>
> Sporleder and Unger (1979), pp. 51-52.
>
> R. L. Gluckstern and J. A. Diamond, "Penetration of fields through a circular hole in a wall of finite thickness," *IEEE Trans. Micr. Theory Tech.*, vol. 39, pp. 274-279 (1991).

from which we have adapted the presentation given here.

A detailed consideration of the validity of the completeness assumption is given for hollow metallic waveguides in

> Kurokawa (1969).
>
> Milton and Schwinger (2006), Chapter 10.

A treatment of completeness which covers waveguides with piecewise constant filling (e.g., microstrip) is given in

> Yu. G. Smirnov, "Use of operator-beam method in the problem of the natural wave modes of a partially filled waveguide," *Sov. Phys. Dokl.*, vol. 35, pp. 430-431 (1990).

Mode expansions outside of the source region are treated in

> Collin (1991), Chapter 7 and Section 11.8;
>
> Van Bladel (2007), Chapter 15;
>
> Johnson (1965), pp. 195-202.

A very lucid discussion on the choice of Z_{cm} for equivalent transmission lines (and its arbitrary nature) is given in

> S. A. Schelkunoff, "Impedance concept in wave guides," *Quart. Appl. Math.* vol.2, pp. 1-15 (1944).

Other approaches to the choice of Z_{cm} for non-TEM modes are to be found in:

> A. Magos, "Characteristic impedance of waveguides used in quasi-TEM mode," *Periodica Polytechnica Elec. Eng.* vol. 30, pp. 17-26 (1986).
>
> J. R. Brews, "Characteristic impedance of microstrip lines," *IEEE Trans. Micr. Theory Tech.* vol. 35, pp. 30-34 (1987).

For the treatment of mode expansions within the source region, we have loosely followed Vainshtein's classic Russian text on electromagnetic waves,

Vainshtein (1988), Chapter 14.

Some discussion of the expressions for the longitudinal field components (10.61)-(10.62) in the English language is to be found in

G. S. Kino, "Mode properties of passive transmission systems," in *Electronic Waveguides*. Brooklyn: Polytechnic Press, 1958, pp. 269-281.

Felsen and Marcuvitz (1973), Chapter 8.

R. E. Collin, "On the incompleteness of E and H modes in wave guides," *Canad. J. Phys.* vol. 51, pp.1135-1140 (1973).

C. Vassallo, "On the expansion of axial field components in terms of normal modes in perturbed waveguides," *IEEE Trans. Micr. Theory Tech.* vol. 23, pp. 264-265 (1975).

The example given in Section 10.4.1 is based on

Barybin and Dmitriev (2002), Chapter 6.

The extension of mode expansions to open waveguides (including the continuous spectrum) is found in

A. B. Manenkov, "The excitation of open uniform waveguides," *Radiophys. Quantum Electron.* vol.13, pp. 578-586 (1970).

The prism coupler for dielectric waveguides is treated in

P. K. Tien, R. Ulrich and R. J. Martin, "Modes of propagating light waves in thin deposited semiconductor films," *Appl. Phys. Lett.*, vol. 14, pp. 291-294 (1969).

The radiation resistance of dipoles in waveguides is discussed in

Johnson (1965), pp. 198-201;
Collin (1991), Sect. 7.1.

The scattering from a conducting sphere in a rectangular waveguide is considered in detail in

L. B. Felsen and C. L. Ren, "Scattering by obstacles in a multimode waveguide," *Proc. IEE (London)*, vol. 113, pp. 16-26 (1966).

10.8   PROBLEMS

p10-1 A small battery forces a voltage V to exist across a small gap of width g between two sections of perfectly conducting wire as shown.

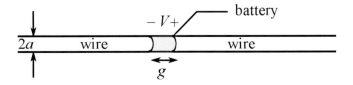

(a) Sketch the electric field lines that will exist near the gap (explain!).

(b) What equivalent current source(s) $\mathbf{J}_S$ and/or $\mathbf{M}_S$ on the *surface* of the wire and/or battery ($\rho = a$) can be used to model this battery? Sketch the directions of the current lines.

(c) If what is connected to the far ends of the wire is unknown, is there a system of equivalent sources which uses only (essentially) known sources?

p10-2 Suppose that instead of a perfectly conducting boundary wall, a waveguide has an impedance wall, at which the fields obey (9.59). Show that even in this more general case, if impressed sources as shown in Figure 10.12 act in this waveguide, the mode excitation amplitudes are still given by (10.56)-(10.57), and (10.58) remains true.

p10-3 A quasi-TEM mode is propagating on a wide ($w/h \gg 1$) microstrip line as shown, with an amplitude a_m in the $+z$-direction.

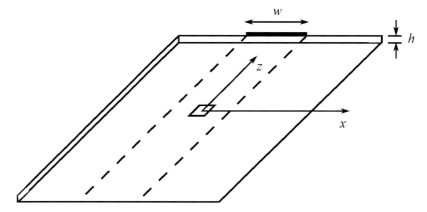

A small square hole of side a is cut into the ground plane $y = 0$ centered at $x = z = 0$. Using an approximation for the fields of this mode that neglects fringing at the edges of the strip, find expression(s) for the moments of the equivalent dipole(s) radiating in the presence of the closed-off aperture that, according to Bethe small-aperture theory, produce the fields $\mathbf{E}^A$ and $\mathbf{H}^A$ caused by the aperture.

p10-4 At $z = 0$, a vertical strip carrying the impressed surface electric current density

$$\mathbf{J}_S = \frac{I}{w}\mathbf{u}_y$$

is connected between the line and ground plane of a stripline as shown.

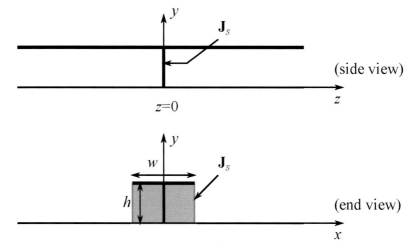

The strip has width w and height h. Find the externally impressed sources appearing in eqns. (10.58) for this configuration, and draw an equivalent circuit for this source, if m denotes the index of the fundamental TEM mode. Calculate the mode amplitudes a_{m0} and b_{m0} for this mode.

p10-5 Suppose a voltage source like that of problem p10-1 is placed in the inner conductor of a coaxial line at $z = 0$ as shown.

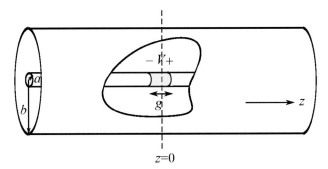

$z{=}0$

Use an equivalent *magnetic* current representation for this source as found in problem p10-1 to find the equivalent sources $E_m^{\text{ext}}(z)$ and $H_m^{\text{ext}}(z)$ of this generator in the equivalent transmission line for the TEM mode of this waveguide. From this result, obtain an equivalent circuit of the form of Figure 3.10.

p10-6 Repeat problem p10-5 for the case of the TEM mode of a two-wire transmission line with generator inserted in one of the conductors as shown.

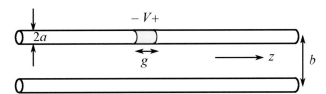

p10-7 Consider a general TEM waveguide mode on the structure of Figure 8.5. A current-carrying wire is located on an arbitrary path between conductors C_1 and C_g as shown.

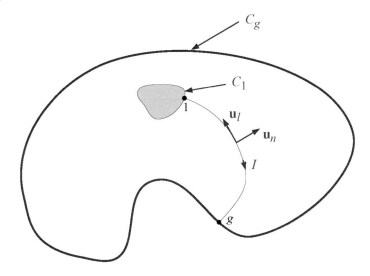

The corresponding impressed current density is

$$\mathbf{J} = -\mathbf{u}_l I \delta(z) \delta(n)$$

where l and n are arc length and normal coordinates on the wire path, analogous to those on the path "$g-1$" shown in Figure 8.6.

    (a) Show that the equivalent distributed sources (10.58) for the $m = $ TEM mode of this waveguide are

$$E_{\text{ext}}^m(z) = 0$$

$$H_{\text{ext}}^m(z) = -I\delta(z)$$

    (b) Obtain expressions for the forward and reverse mode amplitudes $a_m(z)$ and $b_m(z)$ of (10.56) and (10.57) for this mode.

    (c) Explain the negative signs in front of your answers for parts (a) and (b).

p10-8 Rework example 10.5.1 for the case of an *obliquely* incident plane wave

$$\mathbf{E}^{\text{inc}} = \mathbf{u}_y E_i e^{-jk_0(z\cos\theta + x\sin\theta)}$$

$$\mathbf{H}^{\text{inc}} = (-\mathbf{u}_x\cos\theta + \mathbf{u}_z\sin\theta)\frac{E_i}{\zeta_0}e^{-jk_0(z\cos\theta + x\sin\theta)}$$

using the Kirchhoff approximation. If m denotes the TE_0 mode, sketch the dependence of a_{m0} vs. θ for some reasonable value of V in the single-mode range.

p10-9 A slotline of width w is excited by a transversely oriented strip of width s carrying a total current I_0 (distributed as $J_{Sx}(z)$) and situated on the opposite side of the substrate from the slot as shown.

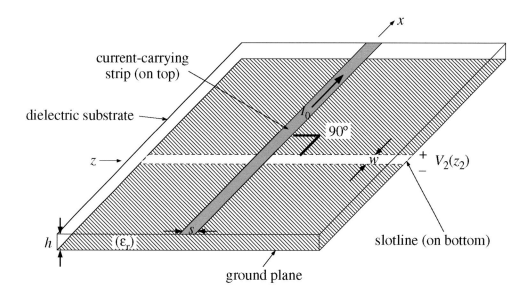

Assume that s is small compared to a wavelength. Find the equivalent distributed sources $E_{\text{ext}}^m(z)$ and $H_{\text{ext}}^m(z)$ that drive the fundamental mode of the slotline under pseudo-TEM conditions. Show that these distributed sources are equivalent to a lumped shunt current source I_0 located at the point of crossover between the slotline and the strip. Be sure to indicate the polarity of the sources; that of the voltage of the normalized slotline mode is indicated in the figure.

p10-10 A perfectly conducting sphere of radius r_0 is placed with its center at the location $(x_0, y_0, 0)$ in a rectangular metallic waveguide. If all modes of the waveguide except the TE_{10} are cutoff, find an expression for the position of the obstacle that will result in zero reflected wave from an incident TE_{10} mode.

p10-11 An electrically small circular coupling hole of radius r_0 is located in the common wall of two rectangular waveguides at $z = 0$ as shown.

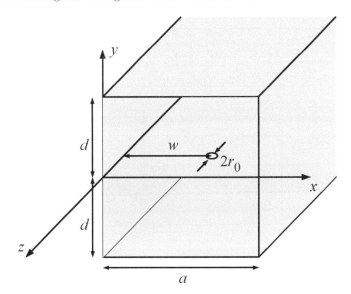

If a TE_{10} mode (index $= m$) traveling in the $+z$-direction with amplitude a_m^u exists in the upper guide, what are the amplitudes a_m^l and b_m^l of the $\pm z$-traveling TE_{10} modes produced in the lower guide due to the presence of this hole?

p10-12 What are the total amplitudes of the $\pm z$-traveling TE_{10} modes produced in the upper guide in problem p10-11 due to the presence of the coupling hole? Is conservation of energy satisfied by this solution? If not, can you suggest a modification by means of which power conservation *is* satisfied?

p10-13 Rework example 10.6.3 for the case when the right-hand waveguide (in $z > 0$) has dimensions a', d' instead of a, d.

p10-14 Rework example 10.6.3 for the case when the right-hand waveguide (in $z > 0$) is uniformly filled with a medium of relative permittivity ϵ_r and permeability μ_r.

p10-15 Rework example 10.6.3 for the case when the aperture is a circle of radius r_0 in a thick conducting diaphragm that extends from $z = -h/2$ to $z = h/2$. For what value of h is the transmission coefficient S_{21} reduced to half its value for a diaphragm of zero thickness?

p10-16 The inner conductor of a coaxial line is to serve as a probe for the excitation of a hollow rectangular waveguide as shown.

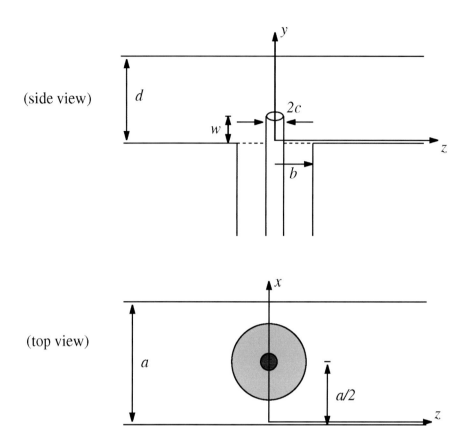

The coaxial line has inner radius c, outer radius b, and the inner conductor extends a distance w down from the center of the top wall of the rectangular guide as shown. Assume that the current $I_2(y)$ in the y-direction on the probe end is the remnant of a standing wave on the coaxial line:

$$I_2(y) = -2j\frac{V_2^+}{Z_2} \sin \beta_2(y - w) \qquad (0 < y < w)$$

where V_2^+ is the voltage of the incident wave on the coaxial line (relative to the end of the probe at $y = w$), and Z_2 and β_2 are the characteristic impedance and propagation constant of the coaxial line. Assume further that the equivalent magnetic current due to the electric field at the annular aperture in the top wall makes a negligible contribution to the excitation. Find the amplitudes $a_{\mathrm{TE}_{10}}$ and $b_{\mathrm{TE}_{10}}$ of the TE_{10} modes excited by this probe in the rectangular waveguide. Assume that c is sufficiently small that it can approximately be regarded as a current filament.

p10-17 For the situation of problem p10-3, suppose that a plane wave with electric field amplitude E_0 at the center of the square aperture is incident from the air region below the ground plane ($y < 0$) instead of having the quasi-TEM mode of the microstrip incident at the aperture. If the plane wave is normally incident to the ground plane, what are the amplitudes of the $+z$- and $-z$-traveling quasi-TEM microstrip modes excited by the equivalent dipoles of the aperture? Is it possible, by having the plane wave be *obliquely* incident to the ground plane and/or by changing the shape or orientation of the aperture, to suppress the excitation of the $-z$-traveling quasi-TEM mode? If so, how?

11 NETWORK THEORY FOR GUIDED WAVES

11.1 INTRODUCTION

Up to this point, we have (in accordance with the definition in Chapter 1) dealt only with waveguides infinite in length and uniform along the guiding (z) direction. In real life, lengths of waveguide start somewhere, must negotiate bends and other types of irregularity, and finally terminate. The exact field solution for a typical waveguide run is usually hopelessly complex, and not worth the trouble. Much useful insight for design and analysis can be obtained by using the formal equivalence between waveguide modes and classical transmission lines which was outlined in the previous chapter.

For one thing, we now know that a modal representation of waveguide fields is possible not only in an infinitely long waveguide (for which mode solutions were originally obtained), but for any finite section of uniform guide as well. Consider the complicated waveguide system of Figure 11.1(a). The fields produced in the uniform section of waveguide between z_1 and z_2 by the sources outside this section obey Maxwell's equations everywhere, together with the appropriate boundary conditions. In principle, these fields could be determined to a desired degree of accuracy using suitable mathematical software, such as a finite-difference, finite-element or moment-method tool. Only in rare circumstances could they be determined analytically. In any event, they could also be thought of as produced by a set of equivalent surface currents located at z_1 and z_2, as indicated in Figure 11.1(b), which radiate into an infinitely long waveguide. By the equivalence theorem, once the fields at z_1 and z_2 are specified, the fields throughout the uniform section can be found from the mode expansion (10.48)-(10.49) and (10.61)-(10.62) of the previous chapter.

Now that the mode representation has been shown to be valid in the uniform waveguide section, it suffices to find only the forward and reverse wave amplitudes $a_m(z)$ and $b_m(z)$ (or equivalently the voltages $V_m(z)$ and currents $I_m(z)$) associated with each mode at any plane z, in order to determine the total field there (recall that it is the orthogonality property of the waveguide modes that makes determination of a_m and b_m possible). But at a cross-sectional plane of a waveguide which adjoins a nonuniform region, a relationship among the wave amplitudes (or currents and voltages) at this plane must exist which is determined by the properties of the nonuniform region itself. Specifically, consider a waveguide feeding a source-free termination as shown in Figure 11.2(a). We define the *terminal plane A* at some constant value $z = z_A$ within the uniform waveguide as shown. The uniform waveguide section can be modeled (à la Chapter 10) as an infinite sequence of equivalent transmission lines with associated wave amplitudes $a_m(z)$ and $b_m(z)$. If the termination (that is, the region beyond the terminal plane) contains only *linear* material in addition to containing no sources, then by the linearity of Maxwell's equations there must be a *linear* relationship between the a_m's and b_m's at $z = z_A$ of the following type:

$$\begin{bmatrix} b_1(z_A) \\ b_2(z_A) \\ \vdots \end{bmatrix} = \begin{bmatrix} S^I_{G,11} & S^I_{G,12} & \cdots \\ S^I_{G,21} & S^I_{G,22} & \cdots \\ \vdots & \vdots & \ddots \end{bmatrix} \begin{bmatrix} a_1(z_A) \\ a_2(z_A) \\ \vdots \end{bmatrix} \tag{11.1}$$

where $S^I_{G,ij}$ are constants independent of $a_m(z_A)$. In matrix notation this is simply $[b(z_A)] = [S^I_G][a(z_A)]$, where the matrix $[S^I_G]$ is made up of the elements $S^I_{G,ij}$. Of course, this is like

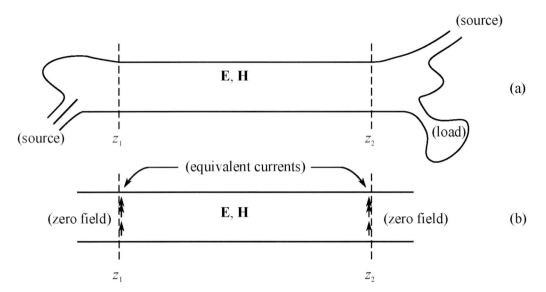

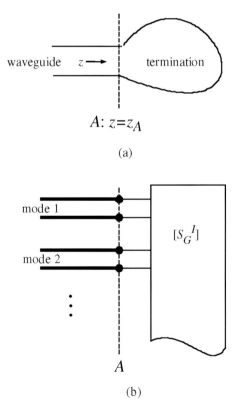

Figure 11.1 (a) Complex waveguide system containing uniform section $z_1 \leq z \leq z_2$; (b) equivalent currents producing the same fields.

Figure 11.2 (a) Waveguide junction; (b) equivalent network.

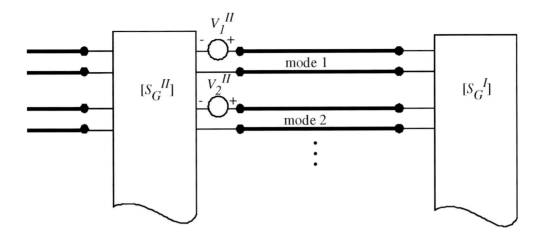

Figure 11.3 Equivalent circuit of waveguide with two terminations.

the scattering matrix representation of Chapter 2, but with a port for each mode, whether or not the mode is above cutoff: $[S_G^I]$ is called the Generalized Scattering Matrix (GSM) of the termination.

In terms of the equivalent transmission line model, eqn. (11.1) is nothing more than a mathematical representation of a lumped network terminating the transmission lines. Thus we have an equivalent circuit as shown in Figure 11.2(b). This representation is mathematically rigorous, since all modes (whether propagating or not) are accounted for. However, once this waveguide is connected to other networks (some with with sources, represented for example by a Thévenin equivalent network as shown in Figure 11.3), an infinite set of linear equations will need to be solved to determine rigorously the amplitudes of all the modes. In practice, however, we normally have only one (or a few) of the modes propagating, and all the rest are cutoff. Here, we will assume for simplicity that the only propagating mode is mode 1. To give the equivalent transmission line concept practical value, we stipulate that *no two nonuniformities in a waveguide system are to occur so close together that a cutoff mode excited at one of them has appreciable amplitude at the position of the other*. Quantitatively, this means that $|\alpha_2 l|$, $|\alpha_3 l|$, and so on, must be large compared to one, where l is the minimum distance between nonuniformities, and α_2, α_3 are the real parts of γ_2, γ_3, ...

This condition means that a terminal plane of a termination can always be chosen such that only the amplitude of mode 1 is significant at that plane, simply by moving the terminal plane away from the termination by a sufficient distance. The equivalent lines for modes 2, 3, etc., now appear effectively semi-infinite looking from the equivalent network of the termination, since any amount of the cutoff mode excited on one of these lines decays essentially to zero before encountering any discontinuity, and thus the reflection from such a discontinuity can be ignored. These semi-infinite lines can thus be replaced by lumped impedances equal to their characteristic impedances Z_{c2}, Z_{c3}, ... , and the transmission line for mode 1 will simply see a reflection coefficient of $S_{11} = S_{G,11}^I$, and thus an equivalent scalar impedance

$$Z_{11} = Z_{c1}\frac{1 + S_{11}}{1 - S_{11}}$$

as the entire effect of the termination (Figure 11.4). We will assume hereafter that the terminal planes of a junction can always be chosen so that only propagating modes need be considered, and the resulting equivalent network will have one port for each propagating

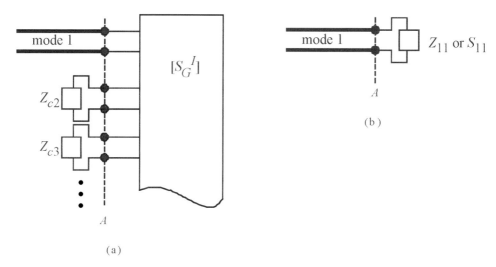

Figure 11.4 (a) Cutoff mode equivalent transmission lines replaced by their characteristic impedances; (b) equivalent load impedance for mode 1.

mode of each waveguide connected to the junction. With this understanding, the equivalent network of Figure 11.2(a) is thus the *one-port network* consisting of the scalar impedance Z_{11} of Figure 11.4(b).

Two things should be emphasized at this point. First, despite the fact that we seem to be considering only the fundamental mode in this model, the higher order modes do play a role in determining the values of Z_{11} and S_{11}. These modes store energy in a localized region near the junction, and in a lossless waveguide contribute reactive effects not accounted for by the fundamental mode. We will have more to say about this in Sections 11.3-11.7. Second, Z_{11} has meaning only with respect to a specifically chosen characteristic impedance Z_{c1} associated with mode 1, and therefore with this port of terminal plane A. To each port of any equivalent network, therefore, we must associate a particular value of characteristic impedance—that of the waveguide mode which constitutes the given port.

In waveguide network theory, one assumes that a given nonuniformity in the waveguide can be characterized by a lumped equivalent network in the manner we have just described, even though actual calculation of the network parameters does require solution of a field problem, and calculations of this type (even approximate ones) can be quite formidable. Although a comprehensive treatment of the advanced numerical and analytical methods used to address these problems is beyond the scope of this book (see the references at the end of this chapter), we will give a brief introduction to one important technique in the next section, to give a flavor of what is involved. Once the network is found, however, all of the relatively elementary methods of circuit theory and transmission line theory can be used to analyze complicated waveguide junction problems.

For general multiport networks involving more than one waveguide, we find it useful to adopt the following conventions, in addition to those from Section 2.1.4 used for multiport networks containing lumped elements and transmission lines:

(i) With respect to the junction, separate z-axes for each waveguide are defined such that the $+z$-direction points *into* the junction, as shown in Figure 11.5.

(ii) For propagating modes on lossless waveguides, both the normalization of the mode fields to unit power transfer (9.28) as well as the usual normalization condition (5.27) hold, while Z_{cm} is taken to be real.

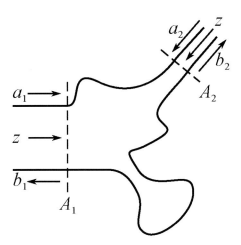

Figure 11.5 Sign conventions for multiport networks.

(iii) The index m will hereafter be used to denote not only a particular mode of a waveguide, but also will designate which waveguide (i.e., which port of the network) the mode exists on.

(iv) At the terminal plane A_m ($z = z_{A_m}$) of the mth port, we define a terminal voltage and current as those of the associated waveguide mode:

$$V_m(z_{A_m}) \equiv V_m \qquad (11.2)$$
$$I_m(z_{A_m}) \equiv I_m \qquad (11.3)$$

We also define a *forward* (incident) *wave amplitude* a_m and a *reverse* (reflected) *wave amplitude* b_m at this port as:

$$a_m = \frac{V_m + Z_{cm}I_m}{\sqrt{8Z_{cm}}} = \frac{\tilde{V}_m + \tilde{I}_m}{2}$$

$$b_m = \frac{V_m - Z_{cm}I_m}{\sqrt{8Z_{cm}}} = \frac{\tilde{V}_m - \tilde{I}_m}{2} \qquad (11.4)$$

These amplitudes are related to the wave amplitudes $a_m(z)$ and $b_m(z)$ defined in (10.50) by the relations:

$$a_m = a_m(z_{A_m})e^{-\gamma_m z_{A_m}}$$
$$b_m = b_m(z_{A_m})e^{\gamma_m z_{A_m}} \qquad (11.5)$$

The factor involving $Z_{cm}^{-1/2}$ is introduced into (11.4) so that whenever Z_{cm} is real, the expression $\frac{1}{2}\text{Re}(V_m I_m^*) = |a_m|^2 - |b_m|^2$ represents the time-average power entering port m, similar to what has been done in Chapters 1, 2 and 10. This is always true for propagating modes on lossless waveguides as discussed in (ii) above. This normalization is perhaps the most widely used one in the literature, but is not the only such possible. It does simplify certain network properties to be derived later on.

11.2 THE MODE MATCHING METHOD

Determination of the scattering parameters of a waveguide junction requires in general the solution of a boundary-value problem for Maxwell's equations, often by numerical methods

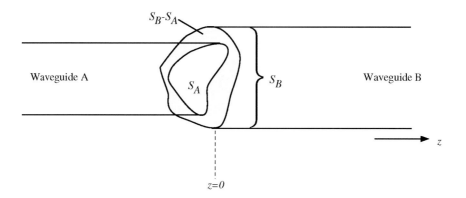

Figure 11.6 Step discontinuity between two waveguides.

that are outside the scope of this book. However, one technique exists that is capable of either high accuracy as a numerical method, or moderate accuracy from relatively simple formulas. This method is known as the *mode-matching method*, and is based on the expansion of the total field as a sum of waveguide modes that are subjected to appropriate boundary conditions at the junction.

Rather than describe the mode-matching technique in full generality, we will illustrate its use for a particular class of waveguide junctions. Changes required for other structures are usually obvious, but in any case can be found in the references at the end of the chapter. Consider a step discontinuity between two semi-infinite waveguides A and B as shown in Figure 11.6. The cross sections S_A and S_B of the waveguides are arbitrary except that we assume (to avoid certain complications) that the cross section of waveguide B completely contains that of waveguide A: $S_B \supset S_A$. We will also assume that the boundaries of both waveguides are perfect conductors, as is the portion of S_B at the junction plane $z = 0$ that is not common with S_A, which we denote $S_B - S_A$.

In waveguide A ($z < 0$), we represent the transverse fields by (10.48) and (10.49):

$$\mathbf{E}_T = \sum_{m>0} \tilde{V}_{mA}(z)\boldsymbol{\mathcal{E}}_{mAT}(x,y) \tag{11.6}$$

$$\mathbf{H}_T = \sum_{m>0} \tilde{I}_{mA}(z)\boldsymbol{\mathcal{H}}_{mAT}(x,y) \tag{11.7}$$

where

$$\tilde{V}_{mA}(z) = a_{mA}e^{-\gamma_{mA}z} + b_{mA}e^{\gamma_{mA}z} \tag{11.8}$$

$$\tilde{I}_{mA}(z) = a_{mA}e^{-\gamma_{mA}z} - b_{mA}e^{\gamma_{mA}z} \tag{11.9}$$

In waveguide B ($z > 0$) on the other hand,

$$\mathbf{E}_T = \sum_{m>0} \tilde{V}_{mB}(z)\boldsymbol{\mathcal{E}}_{mBT}(x,y) \tag{11.10}$$

$$\mathbf{H}_T = \sum_{m>0} \tilde{I}_{mB}(z)\boldsymbol{\mathcal{H}}_{mBT}(x,y) \tag{11.11}$$

where

$$\tilde{V}_{mB}(z) = b_{mB}e^{-\gamma_{mB}z} + a_{mB}e^{\gamma_{mB}z} \tag{11.12}$$

$$\tilde{I}_{mB}(z) = b_{mB}e^{-\gamma_{mB}z} - a_{mB}e^{\gamma_{mB}z} \tag{11.13}$$

Note the change of position of a_{mB} and b_{mB} in formulas (11.12)-(11.13). This is due to the convention that b_{mB} represents a scattered wave, while a_{mB} is an incident wave at the junction. If the waveguides are open (unbounded in the transverse direction), continuous mode spectra must be included in the expressions above by replacing portions of the summations by integrals (see Sections 6.9 and 10.4).

Relationships between the amplitudes a_{mA}, b_{mA}, a_{mB} and b_{mB} (that is, the GSM) are to be determined by the conditions:

(i) The tangential fields $\mathbf{E}_T$ and $\mathbf{H}_T$ are continuous across the surface S_A at $z = 0$.
(ii) The tangential field $\mathbf{E}_T$ is zero on the perfectly conducting surface $S_B - S_A$ at $z = 0$.[1]

Mathematically, we express these as

$$\sum_{m'>0} \tilde{V}_{m'B}(0)\boldsymbol{\mathcal{E}}_{m'BT}(x,y) = \sum_{m'>0} \tilde{V}_{m'A}(0)\boldsymbol{\mathcal{E}}_{m'AT}(x,y) \qquad \text{(on } S_A\text{)}$$
$$= 0 \qquad \text{(on } S_B - S_A\text{)} \tag{11.14}$$

and

$$\sum_{m'>0} \tilde{I}_{m'B}(z)\boldsymbol{\mathcal{H}}_{m'BT}(x,y) = \sum_{m'>0} \tilde{I}_{m'A}(z)\boldsymbol{\mathcal{H}}_{m'AT}(x,y) \qquad \text{(on } S_A\text{)} \tag{11.15}$$

Imposing these conditions at all points of the junction plane $z = 0$ would prove unwieldy, so instead we will weight the conditions by a suitable mode field and integrate over an appropriate cross-sectional plane. For conditions (11.14), the best choice turns out to be to take $\boldsymbol{\mathcal{H}}_{mBT}$ for each of the modes of the *larger* waveguide, form the product $(-)\times\boldsymbol{\mathcal{H}}_{mBT}\cdot\mathbf{u}_z$ with both sides, and to integrate over S_B.[2] The orthogonality property (9.23) can be used to eliminate all but one term from the left side of this result:

$$\tilde{V}_{mB}(0) = \sum_{m'>0} C_{mm'}\tilde{V}_{m'A}(0) \qquad \text{(for all } m > 0\text{)} \tag{11.16}$$

where $C_{mm'}$ is an *overlap integral* defined by

$$C_{mm'} = \frac{1}{2}\int_{S_A} \boldsymbol{\mathcal{E}}_{m'AT} \times \boldsymbol{\mathcal{H}}_{mBT} \cdot \mathbf{u}_z \, dS \tag{11.17}$$

The values of $C_{mm'}$ are known (can be computed) if the mode fields of both waveguides are known. The extent to which the values of $C_{mm'}$ differ from one is a measure of how much mismatch there is between the transverse fields of the two modes involved. Two identical mode field patterns will result in $C_{mm'} = 1$.

To impose these constraints exactly would require an infinite number of equations in the infinite number of wave amplitudes (or voltages and currents) for all the modes of both waveguides. Thus, we content ourselves with an approximation in which only a finite number of modes M_A is retained in the series (11.6)-(11.7), and M_B terms in the series (11.10)-(11.11). Under this approximation, we replace the infinite sum on the right side of (11.16)

---

[1]We do not enforce a boundary condition for $\mathbf{H}_T$ on $S_B - S_A$, because it is related to the surface current there, which is unknown at this stage.

[2]If we had used $\boldsymbol{\mathcal{H}}_{mAT}$ and only integrated over S_A, we would have failed to enforce the condition on $S_B - S_A$ even approximately.

with one that is truncated to M_A terms only, and ask that the equation hold only for the first M_B values of m. Forming the column vectors

$$[\tilde{V}_A] = \begin{bmatrix} \tilde{V}_{1A}(0) \\ \vdots \\ \tilde{V}_{M_A,A}(0) \end{bmatrix} \tag{11.18}$$

$$[\tilde{V}_B] = \begin{bmatrix} \tilde{V}_{1B}(0) \\ \vdots \\ \tilde{V}_{M_B,B}(0) \end{bmatrix} \tag{11.19}$$

and the matrix

$$[C] = \begin{bmatrix} C_{11} & C_{12} & \cdots \\ C_{21} & \ddots & \cdots \\ \vdots & \vdots & C_{M_B,M_A} \end{bmatrix} \tag{11.20}$$

we can express (11.16) in the form

$$[\tilde{V}_B] = [C][\tilde{V}_A] \tag{11.21}$$

which is a kind of generalized Kirchhoff Voltage Law (KVL) for the mode voltages at this junction.

As a weighting procedure for (11.15), we form the product $\mathcal{E}_{mAT} \times (-) \cdot \mathbf{u}_z$ with both sides of (11.15), and integrate over S_A. Using orthogonality of the modes in waveguide A, we arrive at

$$[\tilde{I}_A] = [C]^T [\tilde{I}_B] \tag{11.22}$$

where

$$[\tilde{I}_A] = \begin{bmatrix} \tilde{I}_{1A}(0) \\ \vdots \\ \tilde{I}_{M_A,A}(0) \end{bmatrix} \tag{11.23}$$

$$[\tilde{I}_B] = \begin{bmatrix} \tilde{I}_{1B}(0) \\ \vdots \\ \tilde{I}_{M_B,B}(0) \end{bmatrix} \tag{11.24}$$

which is a Kirchhoff Current Law (KCL) for the mode currents at the junction.

The matrix whose transpose appears in (11.22) is the same as the one appearing in (11.21), so no new computations of matrix elements are required in setting up this second equation. In fact, if we assign a characteristic impedance $Z_{cA,m}$ to each mode of waveguide A, and $Z_{cB,m}$ to each mode of waveguide B, and form the characteristic impedance matrices

$$[Z_{cA}] = \begin{bmatrix} Z_{cA,1} & 0 & \cdots \\ 0 & Z_{cA,2} & \cdots \\ \vdots & \vdots & \ddots \end{bmatrix} ; \qquad [Z_{cB}] = \begin{bmatrix} Z_{cB,1} & 0 & \cdots \\ 0 & Z_{cB,2} & \cdots \\ \vdots & \vdots & \ddots \end{bmatrix} \qquad (11.25)$$

analogous to (2.21), we can convert $[\tilde{I}_A]$, $[\tilde{I}_B]$, $[\tilde{V}_A]$ and $[\tilde{V}_B]$ to unnormalized quantities via:

$$[\tilde{V}_A] = \frac{1}{\sqrt{2}}[Z_{cA}]^{-1/2}[V_A]; \qquad [\tilde{I}_A] = \frac{1}{\sqrt{2}}[Z_{cA}]^{1/2}[I_A] \qquad (11.26)$$

with similar expressions for $[V_B]$ and $[I_B]$. Inserting these into eqns. (11.21) and (11.22), we obtain

$$[V_B] = [N][V_A]; \qquad [I_A] = [N]^T[I_B] \qquad (11.27)$$

where the elements of $[N] = [Z_{cB}]^{1/2}[C][Z_{cA}]^{-1/2}$ are given by

$$N_{mm'} = \sqrt{\frac{Z_{cB,m}}{Z_{cA,m'}}} C_{mm'} \qquad (11.28)$$

Equation (11.27) expresses a multiport ideal transformer relationship between the voltages $[V_A]$ and $[V_B]$, and currents $[I_A]$ and $[I_B]$, as shown in Figure 11.7(a). If only the fundamental mode ($m' = 1$) of waveguide A, and the fundamental mode ($m = 1$) of waveguide B are above cutoff, we can use the viewpoint of Figure 11.4 together with that of Figure 11.7(a) to get an equivalent circuit applicable for just these two modes as shown in Figure 11.7(b). Here port A denotes the fundamental mode of waveguide A, while port B denotes the fundamental mode of waveguide B.

These results can be understood physically in the following way. Equation (11.21) says that any mode in waveguide B is excited by (i.e., couples to) all the modes of waveguide A, unless by some accident of symmetry the overlap integral between a given pair of modes is zero. Conversely, equation (11.22) says that any mode in waveguide A is generally excited by all the modes of waveguide B. This means that an incident mode from waveguide A at the junction excites all other modes of waveguide A in general, indirectly through the excitation of modes in waveguide B that couple back to those of waveguide A. Usually, all but a few modes of each waveguide will be cutoff (and thus have purely imaginary characteristic impedances), and form a near field at the junction that acts indirectly to affect the reflection and transmission coefficients of the propagating modes.

11.2.1   GENERALIZED SCATTERING MATRIX (GSM)

The incident wave amplitudes in waveguide A are $[a_A] = ([\tilde{V}_A] + [\tilde{I}_A])/2$, while the scattered wave amplitudes are $[b_A] = ([\tilde{V}_A] - [\tilde{I}_A])/2$. Likewise in waveguide B, remembering the convention about directions, the incident wave amplitudes are $[a_B] = ([\tilde{V}_B] - [\tilde{I}_B])/2$, while the scattered wave amplitudes are $[b_B] = ([\tilde{V}_B] + [\tilde{I}_B])/2$ (note the reference direction for $[\tilde{I}_B]$ in Figure 11.7(a)). We now define the GSM $[S_G]$ through

$$\begin{bmatrix} [b_A] \\ [b_B] \end{bmatrix} = [S_G] \begin{bmatrix} [a_A] \\ [a_B] \end{bmatrix} \qquad (11.29)$$

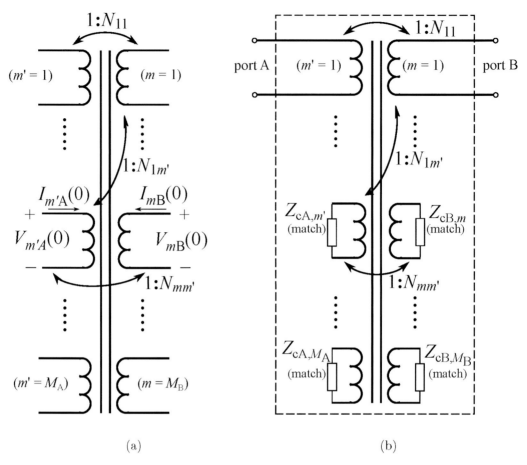

(a)   (b)

Figure 11.7 Waveguide step discontinuity: (a) multi-turn ideal transformer equivalent circuit for mode-matching equations; (b) equivalent circuit for fundamental modes only.

where $[S_G]$ in partitioned form is

$$[S_G] = \begin{bmatrix} [S_{AA}] & [S_{AB}] \\ [S_{BA}] & [S_{BB}] \end{bmatrix} \tag{11.30}$$

so that

$$\begin{aligned} [b_A] &= [S_{AA}][a_A] + [S_{AB}][a_B] \\ [b_B] &= [S_{BA}][a_A] + [S_{BB}][a_B] \end{aligned} \tag{11.31}$$

Some algebra based on the relationships between wave amplitudes and voltages and currents, together with the generalized KVL (11.21) and KCL (11.22) leads to

$$[b_A] = [a_A] + 2[C]^T[a_B] - [C]^T[C]([a_A] + [b_A]) \tag{11.32}$$

and

$$[b_B] = -[a_B] + 2[C][a_A] - [C][C]^T([b_B] - [a_B]) \tag{11.33}$$

from which we infer that

$$[S_{AA}] = \left([\mathbf{1}] + [C]^T[C]\right)^{-1} \left([\mathbf{1}] - [C]^T[C]\right) \tag{11.34}$$

$$[S_{AB}] = 2 \left([\mathbf{1}] + [C]^T[C]\right)^{-1} [C]^T \tag{11.35}$$

$$[S_{BA}] = 2 \left([\mathbf{1}] + [C][C]^T\right)^{-1} [C] \tag{11.36}$$

$$[S_{BB}] = - \left([\mathbf{1}] + [C][C]^T\right)^{-1} \left([\mathbf{1}] - [C][C]^T\right) \tag{11.37}$$

By inspection we can see that the block matrices $[S_{AA}]$ and $[S_{BB}]$ are symmetric. It can also be shown that $[S_{BA}] = [S_{AB}]^T$, so it follows that the GSM is symmetric, just as is the case with an ordinary reciprocal scattering matrix. If only one mode ($m = 1$, say) is above cutoff on each side of the junction, we can recover the ordinary scattering matrix for the junction by taking only the "11" entry from each of the block matrices:

$$[S] = \begin{bmatrix} S_{AA,11} & S_{AB,11} \\ \\ S_{BA,11} & S_{BB,11} \end{bmatrix} \tag{11.38}$$

If only one mode on either side of the junction is retained ($M_A = M_B = 1$), in this case $[C]$ and all the block matrices $[S_{AA}]$ etc. are 1×1 matrices, i. e., scalars. We have

$$S_{AA} = -S_{BB} = \frac{1 - C_{11}^2}{1 + C_{11}^2}; \qquad S_{AB} = S_{BA} = \frac{2C_{11}}{1 + C_{11}^2} \tag{11.39}$$

In general, this approximation will be too crude to give accurate results, but if each of the modes can be associated with a characteristic impedance ($Z_{cA,1}$ or $Z_{cB,1}$), equation (11.28) and Figure 11.7 allow us to represent the discontinuity by a simple $1 : N_{11}$ ideal transformer connected between the equivalent transmission lines in a qualitatively correct way. Note that reflection and transmission from the junction, even in this lowest-order approximation, will usually involve more than just the characteristic impedances (as would be the case with classical transmission lines), but the turns ratio N_{11} as well.

For (11.38) to give sufficient accuracy, large enough numbers of modes M_A and M_B on each side of the junction must be used. In practice, if only one mode in each waveguide is above cutoff, usually no more than about 20 modes are needed on each side in order for the mode-matching technique to give highly accurate values for $[S]$. A criterion has been developed for choosing an optimal truncation (M_A, M_B) of the numbers of modes retained. Once one value (M_A, say) has been chosen, we choose the other such that the maximum propagation constants of the highest modes retained on each side of the junction are approximately equal (i. e., $\gamma_{M_A A} \simeq \gamma_{M_B B}$), in order that the corresponding mode fields possess approximately the same spatial resolution in the transverse coordinates (analogous to the truncation of a Fourier series). More information on convergence of the method and criteria for truncation of the mode set can be found in the references cited at the end of the chapter.

11.2.2   EXAMPLE: PARALLEL-PLATE WAVEGUIDE STEP IN HEIGHT

Consider the step discontinuity in an air-filled parallel-plate waveguide as shown in Figure 11.8. We consider only the case of TM modes (all overlap integrals between TE and

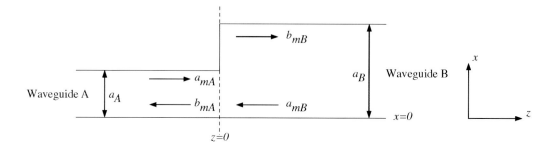

Figure 11.8 Step in height for parallel-plate waveguides.

TM modes of this waveguide are identically zero). The transverse fields of the modes have only one component each, and can be written as

$$\mathcal{E}_{mBx} = \zeta_{mB} H_{mB} \cos\left(\frac{m\pi x}{a_B}\right) \tag{11.40}$$

and

$$\mathcal{H}_{mBy} = H_{mB} \cos\left(\frac{m\pi x}{a_B}\right) \tag{11.41}$$

and similarly for waveguide A by letting $B \to A$ everywhere in the above. Note that by convention the index $m = 0$ is used to denote the TEM mode of the parallel-plate guide. In this example, therefore, all summations will start at $m = 0$ instead of $m = 1$. We have defined

$$\zeta_{mB} = \frac{\gamma_{mB}}{j\omega\epsilon_0} \tag{11.42}$$

$$\gamma_{mB} = j\sqrt{k_0^2 - \left(\frac{m\pi}{a_B}\right)^2} \tag{11.43}$$

$$H_{mB} = \begin{cases} \sqrt{\dfrac{2}{a_B \zeta_0}} & (m = 0) \\[2mm] \dfrac{2}{\sqrt{a_B \zeta_{mB}}} & (m \neq 0) \end{cases} \tag{11.44}$$

with analogous definitions having $B \to A$. The TEM mode of the parallel-plate waveguide can be characterized by its transmission line parameters as follows. Suppose for a moment that the width of the waveguide in the y direction is not infinite, but merely a very large value w. Then the voltage between the plates is $\Phi_1 = \mathcal{E}_x a_{A,B}$ and the current on the positive plate is $I_1 = \mathcal{H}_y w$. The characteristic impedance is thus

$$Z_{cA,B} = \frac{\Phi_1}{I_1} = \zeta_0 \frac{a_{A,B}}{w} \tag{11.45}$$

The overlap integrals are given by

$$C_{mm'} = \frac{1}{2}\int_0^{a_A} \mathcal{H}_{mBy}\mathcal{E}_{m'Ax}\,dx \tag{11.46}$$

$$= \frac{\zeta_{m'A}}{4} H_{mB} H_{m'A} \left\{ \frac{\sin\left[a_A\left(\frac{m\pi}{a_B} + \frac{m'\pi}{a_A}\right)\right]}{\frac{m\pi}{a_B} + \frac{m'\pi}{a_A}} + \frac{\sin\left[a_A\left(\frac{m\pi}{a_B} - \frac{m'\pi}{a_A}\right)\right]}{\frac{m\pi}{a_B} - \frac{m'\pi}{a_A}} \right\} \tag{11.47}$$

If $m/a_B = m'/a_A$, the second term on the right side of (11.47) will need to be evaluated as an appropriate limit. By direct calculation, we find that the first several elements of the $[C]$ matrix are:

$$C_{00} = \sqrt{\frac{a_A}{a_B}} \tag{11.48}$$

$$C_{01} = 0 \tag{11.49}$$

$$C_{10} = \frac{\sqrt{2}}{\pi} \sqrt{\frac{a_B}{a_A}} \sqrt{\frac{\zeta_0}{\zeta_{1B}}} \sin \frac{\pi a_A}{a_B} \tag{11.50}$$

$$C_{11} = \frac{2}{\pi} \sqrt{\frac{a_A}{a_B}} \sqrt{\frac{\zeta_{1A}}{\zeta_{1B}}} \frac{a_A a_B}{a_B^2 - a_A^2} \sin \frac{\pi a_A}{a_B} \tag{11.51}$$

$$C_{20} = \frac{\sqrt{2}}{\pi} \sqrt{\frac{a_B}{a_A}} \sqrt{\frac{\zeta_0}{\zeta_{2B}}} \sin \frac{2\pi a_A}{a_B} \tag{11.52}$$

and

$$C_{21} = \frac{4}{\pi} \sqrt{\frac{a_A}{a_B}} \sqrt{\frac{\zeta_{1A}}{\zeta_{2B}}} \frac{a_A a_B}{a_B^2 - 4a_A^2} \sin \frac{2\pi a_A}{a_B} \tag{11.53}$$

with the limiting case

$$C_{21} = \sqrt{\frac{\zeta_{1A}}{2\zeta_{2B}}} \qquad (\text{if } a_B = 2a_A)$$

If only one mode on either side of the junction is retained, equation (11.39) gives

$$S_{AA} = -S_{BB} = \frac{a_B - a_A}{a_B + a_A}; \qquad S_{AB} = S_{BA} = \frac{2\sqrt{a_B a_A}}{a_B + a_A} \tag{11.54}$$

and so by (11.45) our S matrix in this lowest order approximation can be written

$$[S] = \begin{bmatrix} \frac{Z_{cB} - Z_{cA}}{Z_{cB} + Z_{cA}} & \frac{2\sqrt{Z_{cB} Z_{cA}}}{Z_{cB} + Z_{cA}} \\[2mm] \frac{2\sqrt{Z_{cB} Z_{cA}}}{Z_{cB} + Z_{cA}} & \frac{Z_{cA} - Z_{cB}}{Z_{cB} + Z_{cA}} \end{bmatrix} \tag{11.55}$$

which is exactly what classical transmission line theory predicts (observe that by (11.28) the turns ratio of the ideal transformer between the TEM modes is $N_{00} = 1$, so the transformer can be omitted from the equivalent circuit of the junction). A more general version of this result is obtained in problem p11-3.

To examine what happens if we retain more modes on each side of the junction, consider the particular case where $a_B/a_A = 2$. According to the criterion given at the end of Section 11.2.1, if we include two modes ($m = 0$ and $m = 1$) in guide A, we should then include three modes ($m = 0, 1$ and 2) in guide B. Using (11.34)-(11.38), we can evaluate the S-parameters shown in Figures 11.9. We can see that even including just a few cut-off modes on either side of the junction gives very accurate results for the S-parameters when compared to a numerical (finite-element) solution, demonstrating the efficiency of the mode-matching method for this problem. Retention of only a single mode on each side only gives adequate accuracy for very small values of $k_0 a_B$ (i.e., very low frequency).

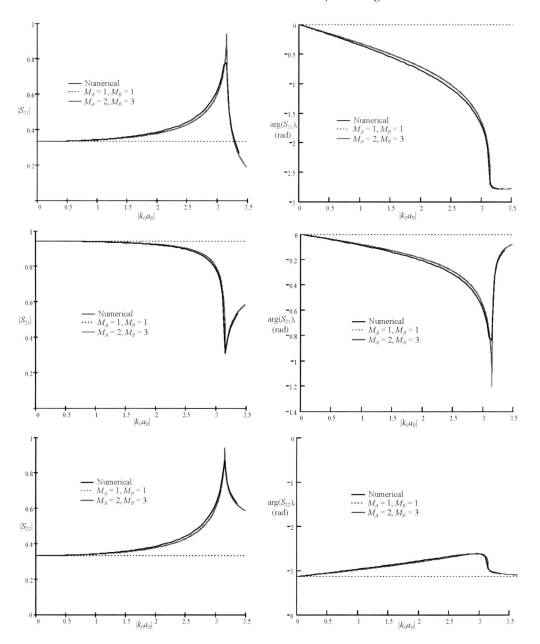

Figure 11.9 S-parameters for a step in height of a parallel-plate waveguide via mode-matching and numerical methods: $a_B/a_A = 2$.

In fact, for this junction we find that the impedance parameters are all equal, and can be cast in the form

$$Z_{11} = Z_{12} = Z_{21} = Z_{22} = -j\frac{1}{\omega C_J},$$

where C_J is a shunt junction capacitance (with w arbitrarily chosen to be 1 m), defined using the characteristic impedances (11.45). Such a capacitance is known as a *parasitic* capacitance, which is in excess to the per-unit-length capacitance associated with the parallel-plate TEM mode. The stored energy in this capacitance corresponds to the energy stored in the

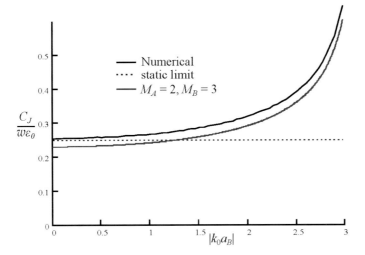

Figure 11.10 Equivalent junction capacitance of a step in height of a parallel-plate waveguide via mode-matching and numerical methods: $a_B/a_A = 2$.

cutoff mode fields near the junction. The normalized value of this capacitance is plotted in Figure 11.10. The junction capacitance is frequency dependent, but approaches a static limit as frequency nears zero. This limit is slightly underestimated by the mode-matching method when only a few cutoff modes are included. We see, however, that these first few cutoff modes account for the majority of the capacitive effect of the junction.

11.2.3 EXAMPLE: RECTANGULAR WAVEGUIDE STEP IN WIDTH

A second example illustrates some more complicated aspects of waveguide discontinuity problems. We consider a symmetric step in width for hollow rectangular metallic waveguides, as shown in Figure 11.11. Here the height d of both guides is the same, and the x-coordinate of each waveguide has been recast to one centered on each waveguide: $x_0 = x - a_A/2$ or $x_0 = x - a_B/2$ as appropriate, so that a common coordinate system can apply to both. If a TE_{10} mode is incident from either side of the junction, invariance of the structure in the y-direction dictates that only TE_{m0} modes will be excited by the discontinuity. Denoting

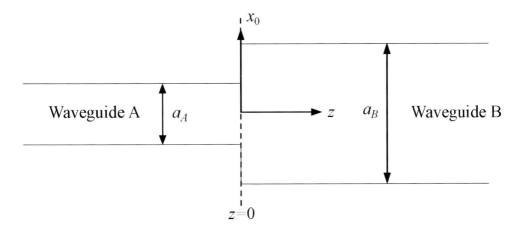

Figure 11.11 Symmetric step in width for rectangular metallic waveguides (top view).

such modes by the index m, elementary if tedious calculations using the mode fields given in Section 5.6.1 show that the relevant overlap integrals are $C_{mm'} = 0$ if one index is odd and the other is even, while

$$C_{mm'} = (-1)^{(m-1)/2} \frac{4}{\pi} \sqrt{\frac{\beta_m^B a_B}{\beta_{m'}^A a_A}} \frac{m' a_A a_B \cos \frac{m\pi a_A}{2a_B}}{(m' a_B)^2 - (m a_A)^2} \tag{11.56}$$

if m and m' are both odd. Here the propagation constants are given by

$$\beta_m^{A,B} = \sqrt{k_0^2 - \left(\frac{m\pi}{a_{A,B}}\right)^2} \tag{11.57}$$

Although, as modes that are not TEM, the TE modes of the rectangular waveguide do not have uniquely defined characteristic impedances, most commonly used for them is the power-voltage definition described in Problem p8-8 of Chapter 8. From the result of that problem, we have that the characteristic impedances for the TE_{10} modes are given by

$$Z_{cA,B} = \zeta_1^{\mathrm{TE}} K_H^{A,B} \tag{11.58}$$

where $K_H^{A,B} = \frac{2d}{a_{A,B}}$, and the wave impedance of the TE_{m0} mode in guide A or B is

$$\zeta_m^{A,B} = \frac{\omega \mu_0}{\beta_m^{A,B}} \tag{11.59}$$

These characteristic impedances are frequency dependent, unlike that of a TEM mode, and in fact become infinite at the cutoff frequency. From (11.28),

$$C_{11} = \sqrt{\frac{Z_{cA,m'}}{Z_{cB,m}}} N_{11} \tag{11.60}$$

where the turns ratio of the ideal transformer joining the fundamental modes is

$$N_{11} = \frac{4}{\pi} \frac{a_A a_B \cos \frac{\pi a_A}{2a_B}}{a_B^2 - a_A^2} \tag{11.61}$$

and by l'Hôpital's rule, $N_{11} \to 1$ as $a_A \to a_B$. Unlike C_{00} in the parallel-plate example of the previous section, C_{11} for the present case cannot be expressed only in terms of the ratio of the characteristic impedances of the two fundamental modes. The turns ratio N_{11} that must also be taken into account expresses a kind of mismatch of the mode fields; here it is independent of frequency and depends specifically on the geometry of the junction.

We illustrate the effect of these various aspects of this discontinuity in Figure 11.12. As with the parallel-plate example above, good agreement is obtained if only one cutoff mode is retained on both sides of the junction. With no cutoff modes retained, the phases of S_{11} and S_{21} are not predicted very accurately at all, and their magnitudes are not quite as accurate either. We can see that including the effect of N_{11} is crucial, because the magnitudes of the S-parameters suffer significant loss of precision without it. We can see that attempts to use only the "characteristic impedances" of the fundamental waveguide modes (when they are ill-defined) to predict reflection and transmission at a junction will in general not be very successful.

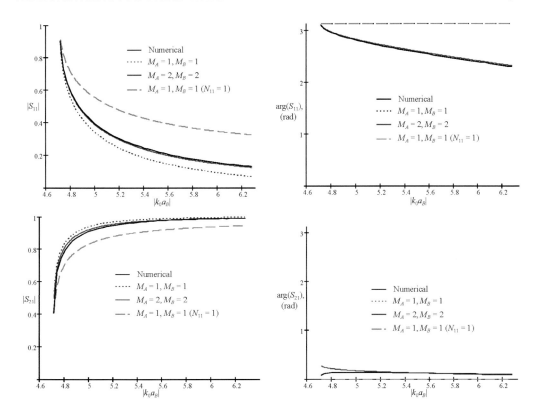

Figure 11.12 S-parameters for a step in width of a rectangular waveguide via mode-matching and numerical methods: $a_B/a_A = 1.5$. The curves denoted with $N_{11} = 1$ are computed by using that value in (11.60) in place of the exact C_{11}.

11.3   ONE-PORT EQUIVALENT NETWORKS

We can say a few general things about the impedance or admittance of the one-port termination of Figure 11.2(a) when the waveguide is lossless. Let us first consider the case when the termination is closed (occupies a bounded volume V surrounded by perfectly conducting walls (or, more generally, walls that have a nonzero surface impedance Z_S). The termination is thus bounded by the terminal plane A, and some sections of walls S_w as shown in Figure 11.13.

Applying Poynting's theorem to the complex Poynting vector $\frac{1}{2}\mathbf{E} \times \mathbf{H}^*$ over the volume V of the termination, we have

$$\frac{1}{2}\int_A \mathbf{E} \times \mathbf{H}^* \cdot \mathbf{u}_n \, dS = \tag{11.62}$$

$$-\frac{j\omega}{2}\int_V [\mu \mathbf{H} \cdot \mathbf{H}^* - \hat{\epsilon}^* \mathbf{E} \cdot \mathbf{E}^*] \, dV - \frac{1}{2}\int_{S_w} \mathbf{E} \times \mathbf{H}^* \cdot \mathbf{u}_n \, dS$$

However, the terminal plane A has been chosen so that only the propagating mode is present at that plane, so the complex power through the plane A will be:

$$\frac{1}{2}\int_A \mathbf{E} \times \mathbf{H}^* \cdot \mathbf{u}_n \, dS = -\frac{1}{4}V_1 I_1^* \int_A \boldsymbol{\mathcal{E}}_{1T} \times \boldsymbol{\mathcal{H}}_{1T}^* \cdot \mathbf{u}_z \, dS$$

$$= -\frac{V_1 I_1^*}{2} \tag{11.63}$$

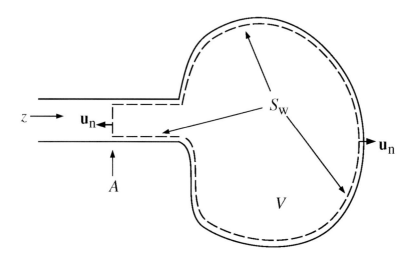

Figure 11.13 Nonradiating one-port termination.

by (9.28), (10.48) and (10.49), if the mode is, as mentioned before, propagating. The surface integral over S_w, on the other hand, is simply the complex power directed into the walls (if any). The real part is the energy loss due to finite wall conductivity, for example:

$$P_{c,r} = \frac{1}{2}\mathrm{Re}\int_{S_w} \mathbf{E}\times\mathbf{H}^*\cdot\mathbf{u}_n\,dS \quad ; \qquad P_{c,i} = \frac{1}{2}\mathrm{Im}\int_{S_w} \mathbf{E}\times\mathbf{H}^*\cdot\mathbf{u}_n\,dS$$

while the imaginary part can be associated with stored energy in the walls (Appendix D). On the right hand side of (11.62), we find the stored time-average electric and magnetic energies W_E and W_M interior to V, as well as the (dielectric) power loss P_d dissipated in the material inside V:

$$w_{E,\mathrm{av}} = \frac{1}{4}\mathrm{Re}\int_V \hat{\epsilon}\mathbf{E}\cdot\mathbf{E}^*\,dV \tag{11.64}$$

$$w_{M,\mathrm{av}} = \frac{1}{4}\int_V \mu\mathbf{H}\cdot\mathbf{H}^*\,dV \tag{11.65}$$

$$P_d = -\frac{\omega}{2}\mathrm{Im}\int_V \hat{\epsilon}\mathbf{E}\cdot\mathbf{E}^*\,dV \tag{11.66}$$

Thus, (11.62) becomes

$$\frac{V_1 I_1^*}{2} = P_L + j\left[2\omega(w_{M,\mathrm{av}} - w_{E,\mathrm{av}}) + P_{c,i}\right] \tag{11.67}$$

where $P_L = P_{c,r} + P_d$ is the total power dissipated in both filling medium and the walls. Since, however, $V_1 = Z_{11}I_1$, where Z_{11} is the equivalent impedance of the junction, we have

$$Z_{11} = \frac{2}{|I_1|^2}\{P_L + 2j\omega W_{\mathrm{excess}}\} \tag{11.68}$$

where

$$W_{\mathrm{excess}} = w_{M,\mathrm{av}} - w_{E,\mathrm{av}} + \frac{P_{c,i}}{2\omega} = \frac{1}{4}\mathrm{Re}\int_V (\mu\mathbf{H}\cdot\mathbf{H}^* - \hat{\epsilon}\mathbf{E}\cdot\mathbf{E}^*)\,dV$$

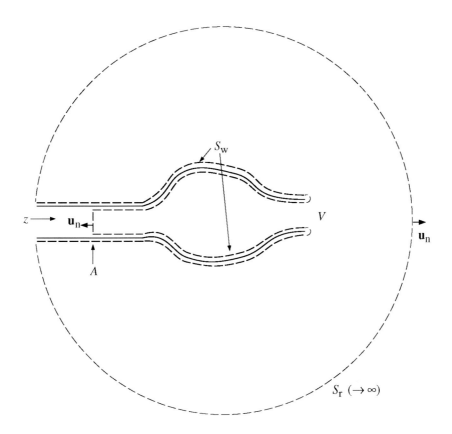

Figure 11.14   Radiating one-port termination.

is the excess of stored magnetic energy over stored electric energy (both in the volume V and in the walls). The real part of Z_{11} is due to ohmic power losses in the dielectric and in the walls, as expected physically. If $P_{c,i}$ is negligible, the sign of the imaginary part X_{11} of Z_{11} (i. e., the reactance) is determined by whether or not $w_{M,\mathrm{av}}$ is greater than $w_{E,\mathrm{av}}$. If $w_{M,\mathrm{av}} > w_{E,\mathrm{av}}$, the reactance is positive (inductive termination); otherwise, the reactance is negative (capacitive termination). This result is the generalization to the field description of that obtained by circuit theory in Section 2.3.

Now suppose that the one-port is capable of radiating energy as shown in Figure 11.14. The volume V of the termination is now infinite, and its bounding surface must include a portion S_r that extends to infinity. The derivation leading to (11.67) and (11.68) is changed only by the addition of an extra surface integral over S_r:

$$P_r = \frac{1}{2} \int_{S_r} \mathbf{E} \times \mathbf{H}^* \cdot \mathbf{u}_n \, dS$$

which is the (real) time average power loss due to radiation from the one-port, and the necessity of carrying out the volume integrals in (11.66) for the stored energies over a volume of infinite extent.

The calculation of P_r involves only the far fields of the termination, and is a result of the termination acting as an antenna, for better or for worse. For radiating fields such as these, $w_{M,\mathrm{av}}$ and $w_{E,\mathrm{av}}$ as given by (11.66) are always infinite; however, their difference (the excess stored energy W_{excess}) remains finite as we extend S_r to infinity. It is also true that the far field of a radiating structure becomes locally a plane wave, for which it is known that the electric and magnetic energy densities are equal, and so the quantity W_{excess} from which X_{11} is computed depends only on the near field of the termination, which is often quasi-static in nature. We may thus replace $w_{M,\text{av}}$ and $w_{E,\text{av}}$ by the so-called "near-field" or "observable" parts of the stored energies for purposes of this calculation (see the references at the end of the chapter for how this can be done in a unique fashion). Thus, for a radiating termination, (11.68) should be modified by augmenting the heat losses P_L by the radiated power loss P_r. Otherwise, the interpretation of this equation remains the same.

11.4   EXAMPLES OF ONE-PORT TERMINATIONS

We present here three examples of one-port terminations. Resonant cavities represent a case of special interest which will be investigated separately in Chapter 13. Further examples of one-port terminations are to be found in the references at the end of the chapter.

11.4.1   THE OPEN-ENDED TWO-WIRE LINE

A simple example of a one-port equivalent circuit is provided by the open-ended two-wire transmission line shown in Figure 11.15(a). We can readily argue that the admittance of this open end must be capacitive rather than inductive, based on the behavior of the fields near the end of the line. For, exactly at the end, the wire current must be near zero, while the voltage has built up to near its maximum value (Figure 11.15(b)). We should expect, consequently, that the **H**-field of the TEM mode of this line is nearly zero at $z = 0$, while the **E**-field is relatively large. Higher-order modes, whether cutoff modes or radiation modes, must be present in order to ensure continuity of **E** and **H** in the vicinity of the open end,

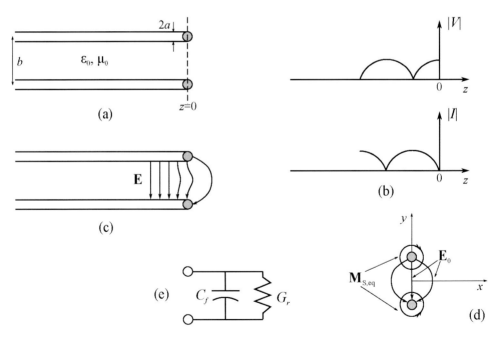

Figure 11.15   (a) Open-ended two-wire transmission line; (b) approximate voltage and current distributions near the end; (c) static fringing field near the end; (d) aperture field and equivalent currents at $z = 0$; (e) terminal admittance of open end at $z = 0$.

but the relative magnitudes of $\mathbf{E}$ and $\mathbf{H}$ should not change much in this region. We thus have $w_{E,\text{av}} > w_{M,\text{av}}$, and the reactance or susceptance will be capacitive.

A more detailed study of the electric field near the ends of the wires shows a fringing as in Figure 11.15(c). Similarly to the fringing at the edges of a stripline (Figure 8.10), this extra field produces an additional capacitance C_f at the end of the line, over and above that accounted for by

$$c = \frac{\pi \epsilon_0}{\ln(b/a)} \qquad (b \gg a) \qquad (11.69)$$

(the capacitance per unit length of the line). A solution of the static field problem of Figure 11.15(c) gives an approximate value of

$$C_f = \frac{b}{2\pi} \frac{c^2}{\epsilon_0} \qquad (b \gg a) \qquad (11.70)$$

for this fringing capacitance.

Beyond this, an examination of the aperture field at $z = 0$ (Figure 11.15(d)) shows that a number of magnetic current loops appear there to radiate into the half-space $z > 0$. As a result, a certain amount of power is lost from the transmission line, and can be accounted for by adding an equivalent radiation conductance G_r to the equivalent circuit at $z = 0$ (Figure 11.15(e)). The value of G_r is found by calculating the power radiated from the end of the line using standard methods of antenna theory, and is given by

$$G_r = \frac{k_0^2 b^2}{8\pi} \frac{\zeta_0}{Z_c^2} \qquad (b \gg a, \ k_0 b \ll 1) \qquad (11.71)$$

where Z_c is the characteristic impedance of the two-wire line:

$$Z_c = \frac{\zeta_0}{\pi \ln(b/a)} \qquad (b \gg a) \qquad (11.72)$$

For a power-line frequency of $f = 60$ Hz, and $a = 0.5$ mm, $b = 1.0$ cm (values similar to those of household power outlets in North America), we have from (11.72) that

$$Z_c = 359.5\,\Omega$$

and thus

$$G_r = 1.83 \times 10^{-20}\text{S}$$

A voltage of 115 volts RMS (163 volts peak) at the open end thus radiates a power of

$$\frac{1}{2}(G_r)163^2 = 2.43 \times 10^{-16}\text{watts}$$

from the open end. This provides an answer for James Thurber's mother-in-law, who was in some confusion about the matter:

> *She came naturally by her confused and groundless fears, for her own mother lived the latter years of her life in the horrible suspicion that electricity was dripping invisibly all over the house. It leaked, she contended, out of empty sockets if the wall switch had been left on. She would go around screwing in bulbs, and if they lighted up she would hastily and fearfully turn off the wall switch and go back to her* Pearson's or Everybody's, *happy in the satisfaction that she had stopped not only a costly but a dangerous leakage. Nothing could ever clear this up for her.*

So she was right, however slightly.

11.4.2 A MICROSTRIP PATCH ANTENNA

A more practical example of a one-port termination is the microstrip patch antenna shown in Figure 11.16. Here, we show the patch fed by a microstrip of characteristic impedance Z_c at one of its edges—one of a number of possible feed methods. An analysis of this structure shows that patch presents the feed line with a complex impedance Z_{in} that varies rapidly in a resonant fashion with frequency. The magnitude of the resulting input reflection coefficient is shown for a particular example in Figure 11.17. As the frequency increases from 2.7 GHz to 3.0 GHz, we observe an almost perfect match at a resonance frequency of about 2.86 GHz. Nearly all of the power incident from the microstrip line is "dissipated" in the real part of the input impedance Z_{in} of the patch (in reality, of course, this power would be that radiated by the patch). Observe that at the resonant frequency, the width w_1 of the patch is a little less than half a wavelength. If ordinary transmission-line theory were applied to

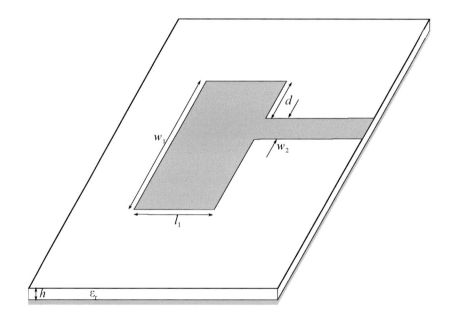

Figure 11.16 Rectangular microstrip patch antenna fed by a microstrip line.

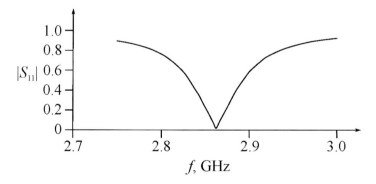

Figure 11.17 Magnitude of the input reflection coefficient for the microstrip-line fed patch antenna of Figure 11.16; $\epsilon_r = 1.0$, $h = 1.0$ mm, $w_1 = 5.0$ cm, $d = 1.375$ cm, $l_1 = 6.0$ cm, $w_2 = 5.0$ mm, $Z_c = 48.8$ Ω. (© 1990 IEEE. Used with permission.)

the patch, the width of the patch would be expected to be exactly $\lambda/2$; excess parasitic capacitance due to fringing of the electric field from the lateral (radiating) edges of the patch make it appear longer than its physical dimensions.

The microstrip patch structure finds wide application because it is relatively compact and easy to fabricate to close tolerances. Since the feed itself is a microstrip transmission line, we have a completely integrated waveguide/antenna structure eminently suited for use in satellite communications systems. Further information on microstrip patch antennas can be found in the references at the end of the chapter.

11.4.3 TRUNCATED DIELECTRIC SLAB WAVEGUIDE

As illustrated in the first example above, truncated TEM or quasi-TEM lines tend to reflect nearly all of the energy incident at the truncation. By contrast, surface waves on dielectric waveguides "want" to continue propagation in the forward direction. Figure 11.18 shows some reflection coefficients for the TE_0 and TM_0 surface waves at a truncation of the slab into free space. This structure is a model of the end face of a semiconductor laser. A normally incident plane wave in a medium with refractive index $n_1 = 3.6$ would have $|S_{11}|^2 = 0.32$ as calculated from the Fresnel reflection coefficient, and we see that the TE reflection is virtually always larger than this, while TM reflection is smaller. This phenomenon is associated with the presence of radiation modes near the end face. As the quantity $1 - |S_{11}|^2$ represents the power radiated from the end of the guide, it can be seen that this structure is very suitable for use as an endfire antenna, as most of the incident power is indeed radiated, more so for TM polarization than for TE. In either case, reflection can be minimized by placing a layer of dielectric over the end face of the waveguide, in a similar way that a quarter-wave layer can minimize plane-wave reflection coefficients.

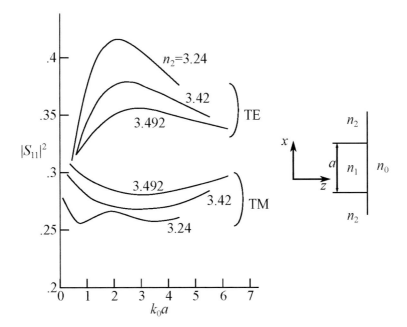

Figure 11.18 Magnitude squared of the reflection coefficient of TE_0 and TM_0 modes at the truncated end of a dielectric slab waveguide: $n_1 = 3.6$, $n_0 = 1$. (© 1993 IEEE. Used with permission.)

11.5 TWO-PORT AND MULTIPORT NETWORKS

A multiport junction of waveguides can be viewed as a multiport equivalent network, described by the formalism of Sections 2.1 and 2.2. Each port is connected to the transmission line which is equivalent to one of the waveguide modes; thus, a two-port junction may arise from the connection of two single-mode waveguides, or from the termination of a single guide supporting two propagating modes. The network description will be the same in either case, although the parameters of the equivalent network can be expected to be different.

By similar arguments to those used for one-port networks, we know that a linear relationship must exist among the terminal variables of a multiport network. In matrix form as in Chapter 2:

$$[V] = [Z][I] \tag{11.73}$$

Not all the elements of the impedance matrix $[Z]$ are independent if the material filling the junction obeys reciprocity. In fact, consider any two possible fields $(\mathbf{E}^a, \mathbf{H}^a)$ and $(\mathbf{E}^b, \mathbf{H}^b)$ which can exist in the case of a two-port junction. To these there will correspond the terminal variables $(V_1^a, V_2^a, I_1^a, I_2^a)$ and $(V_1^b, V_2^b, I_1^b, I_2^b)$ respectively. Applying Lorentz reciprocity to the volume V bounded by the junction walls and terminal planes A_1 and A_2 (Figure 11.19), we find from (9.8) that

$$\int_{A_1} (\mathbf{E}^a \times \mathbf{H}^b - \mathbf{E}^b \times \mathbf{H}^a) \cdot \mathbf{u}_n \, dS = -\int_{A_2} (\mathbf{E}^a \times \mathbf{H}^b - \mathbf{E}^b \times \mathbf{H}^a) \cdot \mathbf{u}_n \, dS \tag{11.74}$$

because the surface integrals over the walls of the junction vanish, and because for a passive junction there are no sources in V. Using (5.27), (10.48), (10.49) and the fact that $\mathbf{u}_n = -\mathbf{u}_z$ on each of the terminal planes, we find that

$$(V_1^a I_1^b - V_1^b I_1^a) = (-V_2^a I_2^b + V_2^b I_2^a) \tag{11.75}$$

Eqn. (2.4) can now be expressed in matrix-vector notation as:

$$[I^b]^T [V^a] = [I^a]^T [V^b] \tag{11.76}$$

But since (2.3) holds for any state of this junction, we have as in Chapter 2 that $[Z]$ is a symmetric matrix, provided the material inside the junction is such that the reciprocity theorem is valid. Indeed, all the apparatus developed in Chapter 2 to describe the behavior of multiport lumped-element or classical transmission-line networks can be carried over without modification to the description of any waveguide multiport junction.

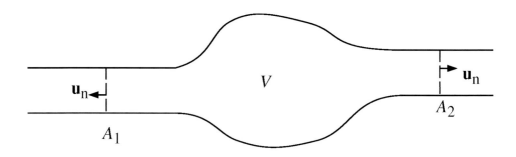

Figure 11.19 Geometry for application of Lorentz reciprocity to a two-port junction.

As with one-port equivalent circuits, the capacitive or inductive nature of multiport networks is dependent upon whether predominantly electric or magnetic energy is stored in the junction. It is more convenient to express the general relations for this in terms of the scattering matrix, rather than $[Z]$ or $[Y]$. We start by applying Poynting's theorem to the volume occupied by the junction, in the manner of eqn. (11.62). The complex power leaving the junction through the ports is now

$$-\frac{1}{2}[I]^{\dagger}[V]$$

In place of (11.67) we now have

$$\frac{1}{2}[I]^{\dagger}[V] = P + 2j\omega W_{\text{excess}} \tag{11.77}$$

where $P = P_r + P_L$ is the sum of the radiated and ohmic power losses from the junction (cf. eqns. (11.62) and (11.66)), and once again W_{excess} is the excess of stored magnetic energy over stored electric energy in the volume V. Using the wave amplitudes $[a]$ and $[b]$ defined in (2.25), we have

$$\{[a] - [b]\}^{\dagger}\{[a] + [b]\} = P + 2j\omega W_{\text{excess}} \tag{11.78}$$

Again introducing the scattering matrix $[S]$ as in (2.23), we have from (11.78):

$$[a]^{\dagger}\left\{[1] + [S] - [S]^{\dagger} - [S]^{\dagger}[S]\right\}[a] = P + 2j\omega W_{\text{excess}} \tag{11.79}$$

Separating real and imaginary parts, we get (2.90) and (2.91) as in Chapter 2, where the powers and stored energies are now expressed in terms of field integrals instead of summations over circuit elements.

In the example of scattering from a "thin" obstacle in a hollow metallic waveguide, such an obstacle can be represented by a two-port network, whose equivalent circuit is a single shunt impedance element Z_{12}. In this case the scattering matrix takes the form of (2.97), if both ports are referred to the same characteristic impedance Z_0. Therefore (2.96) holds, and we conclude, just as we did for a one-port equivalent network, that for this simple shunt element, $\text{Im}(Z_{12})$ is positive (inductive) if $w_{M,\text{av}} > w_{E,\text{av}}$ and negative (capacitive) if $w_{M,\text{av}} < w_{E,\text{av}}$.

Since cutoff TE modes store predominantly *magnetic* energy and cutoff TM modes store predominantly *electric* energy (problem p5-11) in a hollow metallic waveguide, if we can (by using considerations of symmetry or some other feature of the junction) determine that only *one* kind (TE or TM) of cutoff mode exists in the vicinity of a "thin" obstacle, then we can determine the sign of the shunt reactance of the equivalent circuit. If a two-port is lossless, then $P = 0$ in (2.90), and we must have again that $[S]$ is unitary as in (2.94). This then implies (2.95).

11.6 EXAMPLES OF TWO-PORT NETWORKS

11.6.1 TRANSVERSE DIAPHRAGM WITH SMALL APERTURE IN A RECTANGULAR WAVEGUIDE

A first example of a two-port junction is provided by a vertical conducting diaphragm in a rectangular waveguide. The diaphragm is located at $z = 0$, with only an electrically small aperture centered at (x_0, y_0) allowing coupling through it. This is the structure we looked at in example 10.6.3 using small aperture theory. We display it again in Figure 11.20. We will take the terminal planes for both ports 1 and 2 to be at $z = 0$ (plane A) as shown.

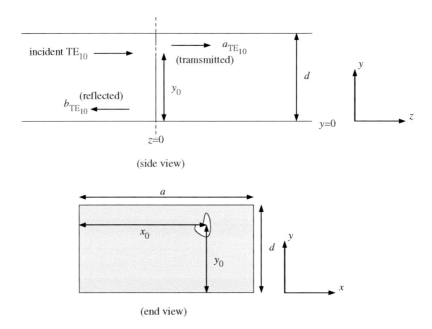

Figure 11.20 Diaphragm with small aperture in a rectangular metallic waveguide.

From our results (10.118) together with definition (2.23) for $[S]$ and (11.4) for $[a]$ and $[b]$, we conclude that

$$
\begin{aligned}
S_{12} &= \frac{j\Delta}{1 + j\Delta} \\
S_{11} &= -1 + S_{12} = -\frac{1}{1 + j\Delta}
\end{aligned}
\tag{11.80}
$$

where, from (10.115) we have

$$
\Delta = \frac{4\beta_1}{ad}\alpha_{M,xx}\sin^2\frac{\pi x_0}{a}
\tag{11.81}
$$

By reflection symmetry about $z = 0$, we have $S_{22} = S_{11}$ and $S_{12} = S_{21}$ (which also follows from reciprocity).

Now, this $[S]$-matrix has the form of (2.97) appropriate to a "thin" obstacle, with

$$
\bar{Z}_{12} = Z_{12}/Z_0 = j\Delta/2
\tag{11.82}
$$

Thus, $\bar{Z}_{12}$ is positive imaginary, and represents a pure shunt inductance across the equivalent transmission line at $z = 0$. According to (2.96), there is more magnetic energy stored near the diaphragm than there is electric energy. We infer that primarily TE modes are present near the terminal plane.

11.6.2 CAPACITIVE DIAPHRAGM; PARTITIONING OF EQUIVALENT NETWORKS

It sometimes happens that an equivalent network can be partitioned into two or more sections, each of which has its own significance as a building block in making up equivalent networks for totally new junctions. We will illustrate this idea with the example of a capacitive diaphragm in a rectangular metallic waveguide, as shown in Figure 11.21(a). This "thin"

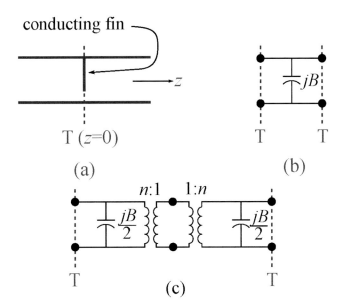

Figure 11.21 (a) Capacitive diaphragm in a rectangular metallic waveguide; (b) equivalent circuit; (c) alternate equivalent circuit for Figure 11.21, with two "building-block" sections.

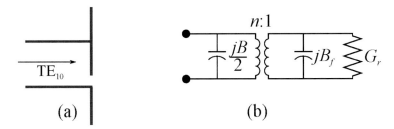

Figure 11.22 (a) Capacitively diaphragmed waveguide radiating into free space; (b) equivalent circuit.

obstacle presents a capacitive susceptance B to the equivalent transmission line as indicated in Figure 11.21(b). However, the equivalent circuit could be redrawn in an infinite number of ways. One such way is shown in Figure 11.21(c), where the shunt susceptance has been divided into two parts, and these parts further isolated using a pair of ideal transformers whose turns ratio n is chosen to transfer the characteristic impedance of the waveguide to some other (but as yet unspecified) value. If we turn now to a different problem wherein the right side of the diaphragm opens onto free space and radiates as an aperture antenna (Figure 11.22(a)), we have fringing and radiation fields to the right of the diaphragm, but the fields to the left of the diaphragm remain largely unaltered. The resulting one-port could be modeled as a single lumped impedance Z_{11}, but it is more instructive to show the equivalent circuit as in Figure 11.1(b). Here, the fields of cutoff modes to the left of the diaphragm show up as one of the half-sections of Figure 11.21(c), while the fringing susceptance B_f and radiation conductance G_r appear to the right of this half-section, modeling the effect of the fields in this half-space. The ideal transformer's turns ratio n can be selected in such a way that $B/2$ depends only on properties of the diaphragm in the waveguide and *not* on those of the half-space into which this aperture radiates, while B_f and G_r do not depend on the properties of the fields inside the waveguide.

This same "radiation network" $G_r + jB_f$ could be connected to a different "half-network" if the aperture were fed by some other waveguide. The interchangeability of component parts of an equivalent circuit is contingent upon the coupling mechanism between the partitioned sections being sufficiently weak (i.e., the aperture must be small enough). Nonetheless, this concept is often useful in the modeling and design of various kinds of equivalent networks.

11.6.3 RIGHT-ANGLE BEND IN MICROSTRIP

It is necessary in microstrip and other planar circuits to change the direction of the line to accommodate specific design goals (Figure 11.23(a)). This causes a discontinuity whose effects must be modeled by an appropriate two-port. Often the bend is *chamfered* as shown in Figure 11.23(b) by trimming off a portion of the sharp exterior corner in order to help minimize reflection from the bend. Further improvement can be expected if we split the bend into two gentler ones as shown in Figure 11.23(c). Some typical data for the *S*-parameters of these bends are shown in Figure 11.24. It will be observed that an almost reflectionless bend can be obtained over a wide frequency range by means of chamfering or double-bending, whereas the uncompensated bend's performance is seriously degraded as frequency increases. In fact, the double bend's performance is achieved in part by destructive interference between the two partial bends, as evidenced by the ripple in the scattering parameters that minimizes the reflection at about 28 GHz.

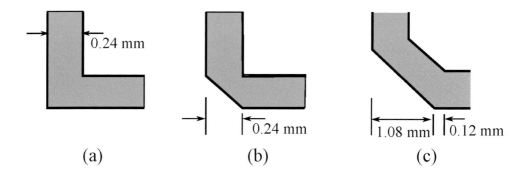

Figure 11.23 (a) Right-angle microstrip bend; (b) 45° chamfered bend; (c) double bend.

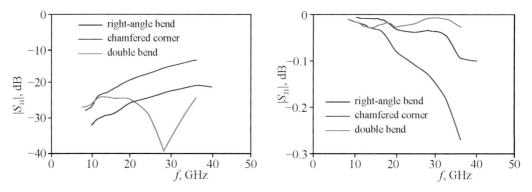

Figure 11.24 Magnitudes of S_{11} and S_{21} for bends of Figure 11.23: $h =$ substrate thickness $= 0.24$ mm, $\epsilon_r = 9.8$. (© 1990 IEEE. Used with permission.)

11.7 EXAMPLES OF MULTIPORT NETWORKS

11.7.1 USE OF PHYSICAL SYMMETRY–THE E-PLANE AND H-PLANE T–JUNCTIONS IN A RECTANGULAR WAVEGUIDE

Consider the E-plane and H-plane T-junctions in rectangular waveguides as shown in Figure 11.25. All the ports are assumed to have identical cross-sections and to support only the TE_{10} mode above cutoff. Both are therefore 3-port networks.

In Chapter 2, we saw how the symmetry of a network composed of lumped elements and transmission lines could be used to simplify the derivation of its scattering parameters. The same kinds of symmetry considerations can be applied to the electromagnetic field in order to find the S-parameters of waveguide junctions. In the present case, from Figure 11.26(a), we can see that the E-field in the E–plane junction will fringe in such a way that $S_{13} = -S_{23}$. From Figure 11.26(b) on the other hand, it is clear that $S_{13} = +S_{23}$ for the H-plane tee. From the physical symmetry of the junctions, it should also be evident that $S_{11} = S_{22}$ in both cases.

As a result, we conclude that the symmetric $[S]$ matrix for the E-plane tee must have the form

$$[S] = \begin{bmatrix} S_{11} & S_{12} & S_{13} \\ S_{12} & S_{11} & -S_{13} \\ S_{13} & -S_{13} & S_{33} \end{bmatrix} \tag{11.83}$$

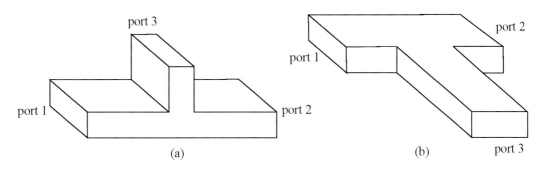

Figure 11.25 (a) E-plane T-junction; (b) H-plane T-junction.

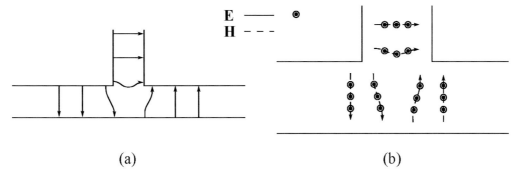

Figure 11.26 Field fringing in (a) E-plane tee and (b) H-plane tee junctions of rectangular waveguides.

and for the H-plane tee, we must have

$$[S] = \begin{bmatrix} S_{11} & S_{12} & S_{13} \\ S_{12} & S_{11} & S_{13} \\ S_{13} & S_{13} & S_{33} \end{bmatrix} \tag{11.84}$$

In both cases, only 4 complex parameters remain to be determined. If the junctions are lossless in addition, the following *real* relationships follow from the unitary property (2.94) of the $[S]$ matrix for the E-plane tee:

$$|S_{11}|^2 + |S_{12}|^2 + |S_{13}|^2 = 1 \tag{11.85}$$
$$2|S_{13}|^2 + |S_{33}|^2 = 1 \tag{11.86}$$
$$S_{11}^* S_{12} + S_{12}^* S_{11} - |S_{13}|^2 = 0 \tag{11.87}$$
$$S_{11}^* S_{13} - S_{12}^* S_{13} + S_{13}^* S_{33} = 0 \tag{11.88}$$

These constraints mean that only 4 independent *real* parameters would now need to be found to completely specify the behavior of this junction. Similar constraints can also be obtained for the H-plane junction.

11.7.2 90° MICROSTRIP-SLOTLINE CROSSOVER

In Section 10.5.3 and problem p10-9, we considered the excitation of a microstrip line by a voltage-excited slot in the ground plane, and that of a slotline by a transverse current-carrying strip on the opposite side of the substrate. If the respective sources of excitation in these two problems are regarded as a slotline and a microstrip carrying suitable voltage and current in the vicinity of the crossover point $z_1 = z_2 = 0$ of a 90° microstrip-slotline transition (see Figure 11.27), then we can use the results to obtain an equivalent circuit for the crossover. From the results of the two excitation problems mentioned above, we

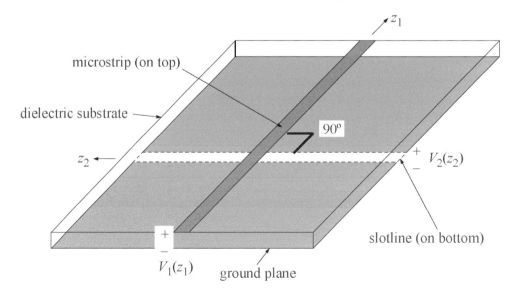

Figure 11.27 90° microstrip-slotline transition.

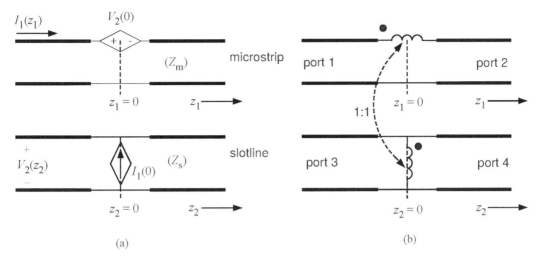

Figure 11.28 (a) Dependent generator equivalent circuit for Figure 11.27; (b) ideal transformer equivalent circuit.

can extract the equivalent circuit of Figure 11.28(a). From eqn. (2.11), we recognize that the behavior of this equivalent circuit is the same as that of an ideal 1:1 transformer as connected in Figure 11.28(b), and thus this too is an equivalent circuit for the crossover, valid at least for low enough frequencies that all phenomena involved can be regarded as quasistatic.

The elements of the scattering matrix can be determined by placing matched loads at various ports and exciting the remaining ones in certain ways. For example, if Z_{cm} is the characteristic impedance of the microstrip and Z_s is that of the slotline, placing matched loads of Z_s at ports 3 and 4 means that the 1:1 transformer in the microstrip line presents a lumped series impedance of $Z_s/2$ in the line. Choosing the terminal planes for ports 1 and 2 at $z_1 = 0$ (and likewise for ports 3 and 4 at $z_2 = 0$), we reduce to the two-port problem as shown in Figure 11.29. If, further, port 2 is connected to a matched load of Z_{cm} and port 1 is driven with (say) a voltage of $V_1 = 1$, then we can easily evaluate the resulting voltage V_2 and currents I_1 and I_2 at ports 1 and 2. Wave amplitudes a_1, a_2, b_1 and b_2 are then found from (2.25), and we find:

$$S_{11} = \left.\frac{b_1}{a_1}\right|_{a_2=a_3=a_4=0} = \frac{Z_s}{Z_s + 4Z_{cm}} \tag{11.89}$$

$$S_{21} = \left.\frac{b_2}{a_1}\right|_{a_2=a_3=a_4=0} = \frac{4Z_{cm}}{Z_s + 4Z_{cm}} \tag{11.90}$$

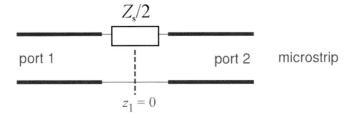

Figure 11.29 Equivalent network for Figure 11.27 if ports 3 and 4 are matched.

From the physical symmetry of the junction, we have $S_{11} = S_{22}$ and $S_{12} = S_{21}$. The other elements of $[S]$ can be found in a similar way.

11.7.3   A MICROSTRIP COUPLER

As has been seen in Section 3.3, the presence of distributed coupling (capacitive and/or inductive) between two or more transmission lines can result in significant transfer of energy between the two lines. A more detailed treatment of this coupling mechanism by analysis of the electromagnetic fields can be carried out using the method presented in Chapter 12. We present here only some typical results for a microstrip coupler—the four-port shown in Figure 11.30. The magnitudes and phases of the scattering parameters S_{i1} are shown in Figure 11.31. Notice how significant coupling in both the "forward" (S_{31}) as well as "reverse" (S_{41}) directions occurs, depending on the frequency of operation. At approximately 13 and 26 GHz, nearly all incident power at port 1 is delivered either to port 2 or port 3, meaning that essentially only codirectional coupling occurs. The amount of power as well as the ports to which it is delivered can be varied by adjusting the parameters of the coupler.

11.7.4   INTERSECTING DIELECTRIC WAVEGUIDES

In order to achieve significant densities of circuit elements in integrated optics, ways have to be found to allow dielectric waveguides to cross each other without undesired crosstalk or radiation loss. One way of doing this is simply to let two such guides intersect each other as shown in Figure 11.32. This configuration results in coupling, not only between the surface

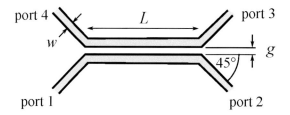

Figure 11.30   Microstrip coupler.

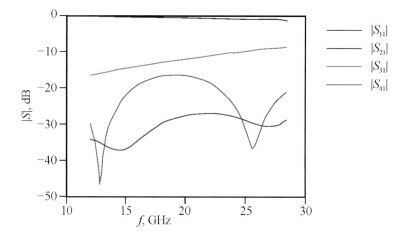

Figure 11.31 S-parameters for a microstrip coupler: $\epsilon_r = 12.9$; $w = 0.074$ mm; $g = 0.074$ mm; $L = 4.00$ mm; $h = 0.10$ mm.

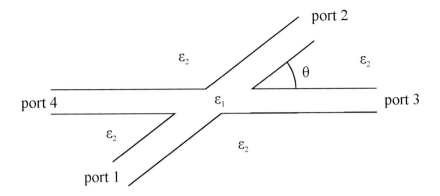

Figure 11.32 Intersecting dielectric slab waveguides.

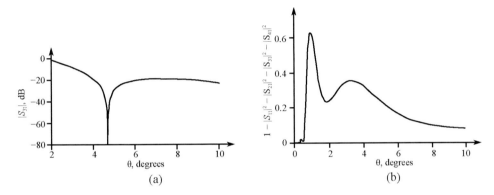

Figure 11.33 (a) Crosstalk ($|S_{31}|^2$), and (b) radiation loss ($1 - \sum_i |S_{i1}|^2$) as functions of angle of intersection for dielectric slab waveguides. (© 1987 AIP. Used with permission.)

waves of the two guides, but also between the surface waves and radiation modes of the two guides. There is a complex interference pattern due to the superposition of these effects, and the resulting structure exhibits radiated power loss, as well as an S_{13} that goes to zero at a certain value of intersection angle between the guides (Figure 11.33). The existence of this so-called "magic" angle suggests a promising way of crossover for integrated optical or millimeter wave dielectric waveguides, with relatively minor radiation losses.

11.8   NOTES AND REFERENCES

In a very short period of time, waveguide network theory has become well enough studied to be regarded as classical. More details can be found in the treatments of

Altman (1964);
Collin (2001);
Ghose (1963);
Kerns and Beatty (1967);
Montgomery, Dicke and Purcell (1965).

Shorter treatments are available in

Harrington (1961), Chapter 8;
Ramo, Whinnery and van Duzer (1984), Chapter 11;

Johnson (1965), Chapter 14;
Van Bladel (2007), Chapter 15.

Mode-matching techniques are covered in

Mittra and Lee (1971), Chapters 2 and 4;
Vassallo (1985), Section 8.3;
Uher *et al.* (1993), Section 2.1;
Conciauro *et al.* (2000), Chapter 3

as well as in

R. Mittra, "Relative convergence of the solution of a doubly infinite set of equations," *J. Res. Nat. Bureau Stand. D*, vol. 67, pp. 245-254 (1963).

J. R. Pace and R. Mittra, "Generalized scattering matrix analysis of waveguide discontinuity problems," in *Proceedings of the Symposium on Quasi-Optics*. Brooklyn, NY: Polytechnic Press, 1964, pp. 177-197.

G. I. Veselov and V. M. Temnov, "The solution of some systems of equations in electrodynamics and the 'relative convergence' phenomenon," *Radio Eng. Electron. Phys.*, vol. 26, no. 10, pp. 13-20 (1981).

Y. C. Shih, "The mode-matching method," in *Numerical Techniques for Microwave and Millimeter-Wave Passive Structures* (T. Itoh, ed.). New York: Wiley, 1989, Chapter 9.

A. S. Il'inskii and E. Yu. Fomenko, "Investigation of infinite-dimensional systems of linear algebraic equations of the second kind in waveguide diffraction problems," *Comp. Math. Math. Phys.*, vol. 31, no. 3, pp 1-11 (1991).

G. V. Eleftheriades, A. S. Omar, L. P. B. Katehi and G. M. Rebeiz, "Some important properties of waveguide generalized scattering matrices in the context of the mode matching technique," *IEEE Trans. Micr. Theory Tech.*, vol. 42, pp. 1896-1903 (1994).

A. Morini and T. Rozzi, "On the definition of the generalized scattering matrix of a lossless multiport," *IEEE Trans. Micr. Theory Tech.*, vol. 49, pp. 160-165 (2001).

P. Russer, M. Mongiardo and L. B. Felsen, "Electromagnetic field representations and computations in complex structures III: Network representations of the connection and subdomain circuits," *Int. J. Num. Model.: Elec. Netw., Dev. Fields*, vol. 15, pp. 127-145 (2002).

The papers by Mittra, Veselov and Temnov, and Il'inskii and Fomenko above discuss the so-called "relative convergence" phenomenon, and optimal ways of truncating the mode series in this method. The concept of multi-winding ideal transformers as circuit elements is discussed in the papers by Morini and Rossi, and by Russer *et al.* above, as well as in

Newcomb (1966), Section 1-2.

Other theoretical techniques for evaluating circuit parameters for waveguide and transmission-line discontinuities are given in

Montgomery, Dicke and Purcell (1965);
Collin (1991);
Ghose (1963);
Harrington (1961);
Johnson (1965);

Van Bladel (2007);
Vainshtein (1969b);
Marcuvitz (1965);
Schwinger and Saxon (1968);
Jones (1964), Chapter 5;
Mittra and Lee (1971);
Lewin (1975);
Katsenelenbaum *et al.* (1998);
Milton and Schwinger (2006), Chapter 13.

and

F. E. Borgnis and C. H. Papas, "Electromagnetic waveguides and resonators," in *Handbuch der Physik*, vol. 16 (S. Flügge, ed.). Berlin: Springer-Verlag, 1958, pp. 285-422.

Although a network formalism may, in principle, be used to treat discontinuities in optical or other dielectric waveguides just as for any other type of waveguide, impedance concepts are not generally invoked in this context. The scattering matrix is, however, sometimes used, and analysis techniques for discontinuities in such open waveguides are to be found in

Marcuse (1982), Chapter 9;
Marcuse (1974);
Shevchenko (1971).

Discontinuities involving curvature and other changes in direction of the waveguide axis are given in

Lewin (1975), Chapter 4;
Lewin, Chang and Kuester (1977);
Londergan, Carini and Murdock (1999)

A highly useful collection of formulas and graphs for junctions and discontinuities in hollow rectangular and circular waveguides as well as coaxial transmission lines is found in:

Marcuvitz (1965).

The expression of the stored energies of radiating junctions in terms of "observable" or near-field stored energies has been treated from differing points of view. See:

D. R. Rhodes, "Observable stored energies of electromagnetic systems," *J. Franklin Inst.*, vol. 302, pp. 225-237 (1976).
D. R. Rhodes, "A reactance theorem," *Proc. Roy. Soc. London*, vol. A353, pp. 1-10 (1977).
I. M. Polishchuk, "The Q-factor and energy center of antennas," *Radio Eng. Electron. Phys.*, vol. 28, no. 6, pp. 1-5 (1983).

See also the Notes and References at the end of Chapter 13.

Chapter 5 of

King (1965)

discusses equivalent circuits of many types of discontinuities in coaxial and two-wire lines. Equation (11.70) can be found there, as well as in

Vainshtein (1969b)

where expression (11.71) is also given. An extension of these results to transmission lines of more general geometrical configuration is given in

R. Ianconescu and V. Vulfin, "Radiation from free space TEM transmission lines," *IET Microw. Ant. Prop.*, vol. 13, pp. 2242-2255 (2019).

R. Ianconescu and V. Vulfin, "Radiation from quasi-TEM insulated transmission lines," *IET Microw. Ant. Prop.*, vol. 13, pp. 761-773 (2019).

The quotation at the end of Section 11.4.1 is from:

J. Thurber, *My Life and Hard Times*. New York: Bantam Books, 1961, p. 33.

In recent years, full-wave numerical simulation techniques for the determination of the scattering parameters of waveguide junctions have become widely available, in both freely available and commercial versions. Software based on mode-matching methods can be extremely fast, but is usually limited to waveguide geometries whose modes can be expressed in analytical closed form. More general structures can be handled by finite-element, finite-difference, Galerkin (moment) or other methods. For example, a moment method solution of the integral equation describing microstrip circuits was developed at the University of Colorado Boulder by B. L. Brim, D. I. Wu, J.-X. Zheng and D. C. Chang. See:

D. I. Wu, D. C. Chang and B. L. Brim, "Accurate numerical modeling of microstrip junctions and discontinuities," *Int. J. Micr. Millim.-Wave Comp.-Aided Eng.*, vol. 1, pp. 48-58 (1991).

D. I. Wu and D. C. Chang, "A review of the electromagnetic properties and the full-wave analysis of the guiding structures in MMIC," *Proc. IEEE*, vol. 79, pp. 1529-1537 (1991).

D. C. Chang and J.-X. Zheng, "Electromagnetic modeling of passive circuit elements in MMIC," *IEEE Trans. Micr. Theory Tech.*, vol. 40, pp. 1741-1747 (1992).

The results of Figure 11.17 are taken from:

J.-X. Zheng and D. C. Chang, "Computer-aided design of electromagnetically-coupled and tuned, wide band microstrip patch antennas," *IEEE Ant. Prop. Symp.*, vol. 3, pp. 1120-1123 (1990).

those of Figure 11.24 are from:

J.-X. Zheng and D. C. Chang, "Numerical modelling of chamfered bends and other microstrip junctions of general shape in MMICS," *IEEE Micr. Theory Tech. Symp.*, vol. 2, pp. 709-712 (1990).

and those of Figure 11.31 are from:

J.-X. Zheng, "Electromagnetic modeling of microstrip circuit discontinuities and antennas of arbitrary shape," *MIMICAD Technical Report No. 7*, University of Colorado Boulder (1991).

Equivalent circuits and their parameters for a variety of microstrip discontinuities are given in

Garg, Bahl and Bozzi (2013), Chapters 3 and 4.

Similar data for coplanar waveguides are given in

Wolff (2006), Chapters 3-5.

For more on the microstrip-slotline crossover of Section 11.7.2, see

J. B. Knorr, "Slot-line transitions," *IEEE Trans. Micr. Theory Tech.* vol. 22, pp. 548-554 (1974).

H.-Y. Yang and N. G. Alexopoulos, "A dynamic model for microstrip-slotline transition and related structures," *IEEE Trans. Micr. Theory Tech.* vol. 36, pp. 286-293 (1988).

T. Uwano, T. Itoh and R. Sorrentino, "Characterization of microstrip-to-slotline transition discontinuities by transverse resonance analysis," *Alta Frequenza* vol. 57, pp. 183-191 (1988).

The open-ended dielectric waveguide is treated by many authors. The plots in Figure 11.18 were adapted from

P. C. Kendall, D. A. Roberts, P. N. Robson, M. J. Adams, and M. J. Robertson, "New formula for semiconductor laser facet reflectivity," *IEEE Photon. Technol. Lett.*, vol. 5, pp. 148-150 (1993).

See also

P. Kaczmarski and P. E. Lagasse, "Bidirectional beam propagation method," *Electron. Lett.*, vol. 24, pp. 675-676 (1988).

I. G. Tigelis and A. B. Manenkov, "Scattering from an abruptly terminated asymmetrical slab waveguide," *J. Opt. Soc. Amer. A*, vol. 16, pp. 523-532 (1999).

The crossover of two dielectric slab waveguides has been examined in

N. Agrawal, L. McCaughan and S. R. Seshadri, "A multiple scattering interaction analysis of intersecting waveguides," *J. Appl. Phys.*, vol. 62, pp. 2187-2193 (1987).

L. McCaughan and N. Agrawal, "Novel physical effects in intersecting waveguides," *Appl. Phys. Lett.*, vol. 51, pp. 1389-1391 (1987).

from the second of which the data in Figure 11.33 are taken.

11.9   PROBLEMS

p11-1 Consider the step-junction of two dielectric slab waveguides as shown.

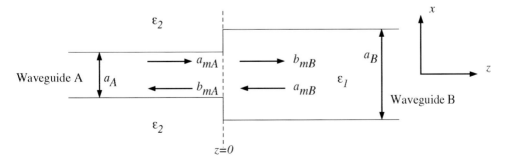

Using only the TE_0 mode on each waveguide, perform a simple mode-matching procedure to obtain expressions for the S-parameters of this junction. Multiply both sides of the equation for continuity of the tangential E-field by the modal $\mathcal{H}$ field of waveguide B, and the equation for tangential H by the $\mathcal{E}$ field of waveguide A to obtain your results.

p11-2 Show that $[S_{BA}] = [S_{AB}]^T$, where the matrices are given in (11.35) and (11.36).

p11-3 If m' denotes a TEM mode on waveguide A, and m denotes a TEM mode on waveguide B, show that the overlap integral (11.17) for these two modes is equal to

$$C_{mm'} = \sqrt{\frac{Z_{cA}}{Z_{cB}}}$$

no matter what the detailed structure of either waveguide, where $Z_{cA,B}$ are the characteristic impedances of the TEM modes on guides A and B respectively. Hence, the turns ratio of the ideal transformer directly coupling these modes is $N_{mm'} = 1$, and only parasitics due to higher order modes will alter the reflection/transmission properties of this junction.

p11-4 Consider a mode "m" of a uniformly filled metallic waveguide (it may be either TE or TM). When the medium filling the waveguide has material parameters ϵ_1 and μ_1, we call the waveguide "guide 1" and denote the mode fields $\mathcal{E}_m^{(1)}$, $\mathcal{H}_m^{(1)}$, the propagation constant $\gamma_m^{(1)}$, and the intrinsic wave impedance of the mode (5.60) as $\zeta_m^{(1)}$. If the material filling the guide has ϵ_2, μ_2, we call the guide "2" and replace the superscripts $^{(1)}$ in the previous notations by $^{(2)}$.

(a) Let mode m be incident from guide 1 to guide 2, the junction between the two guides being a plane interface located at $z = 0$ as shown. Mode m in guide 1 is assumed to be above cutoff $(\gamma_m^{(1)} = j\beta_m^{(1)})$.

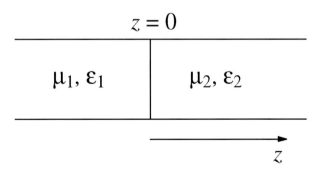

Show that *only* a reflected mode m in guide 1 and a transmitted mode in guide 2 are produced, and that if the transverse electric and magnetic fields are given by

$$\mathbf{E}_T = (e^{-j\beta_m^{(1)}z} + \rho e^{j\beta_m^{(1)}z}) \frac{\mathcal{E}_{mT}^{(1)}}{\sqrt{\zeta_m^{(1)}}} \qquad (z < 0)$$

$$\mathbf{H}_T = (e^{-j\beta_m^{(1)}z} - \rho e^{j\beta_m^{(1)}z}) \frac{\mathcal{H}_{mT}^{(1)}}{\sqrt{\zeta_m^{(1)}}} \qquad (z < 0)$$

$$\mathbf{E}_T = \tau \frac{\mathcal{E}_{mT}^{(2)} e^{-j\beta_m^{(2)}z}}{\sqrt{\zeta_m^{(2)}}} \qquad (z > 0)$$

$$\mathbf{H}_T = \tau \frac{\mathcal{H}_{mT}^{(2)} e^{-j\beta_m^{(2)} z}}{\sqrt{\zeta_m^{(2)}}} \qquad (z > 0)$$

then the reflection and transmission coefficients are given by

$$\rho = \frac{\zeta_m^{(2)} - \zeta_m^{(1)}}{\zeta_m^{(2)} + \zeta_m^{(1)}}; \qquad \tau = \frac{2\zeta_m^{(2)}}{\zeta_m^{(2)} + \zeta_m^{(1)}}$$

(b) Suppose a second interface back to the material of guide 1 is placed at $z = l$ as shown.

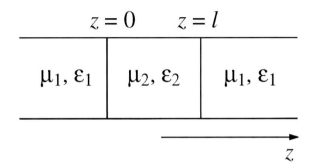

Find expressions for the total $\mathbf{E}_T$ and $\mathbf{H}_T$ in $0 < z < l$, and find the time-average power passed in the positive z-direction for any z in this region. Give two separate expressions for the power, depending on whether mode m in guide 2 is propagating or cutoff.

p11-5 Consider the magic-tee shown below, where all 4 ports are rectangular metallic waveguides of identical cross section supporting only the TE_{10} mode above cutoff.

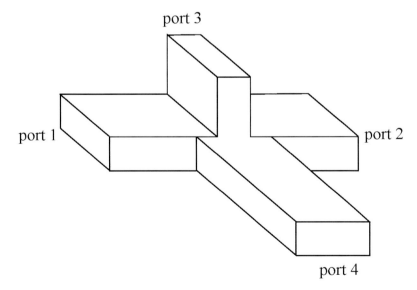

Using the properties of reciprocity, losslessness and physical symmetry of this junction, find as many relationships as you can between the elements of the scattering matrix for this junction. Look especially for any property which might explain the name "magic-tee" which is used for this structure—are any ports isolated from each other?

p11-6 Find the remaining elements of the scattering matrix for the 4-port network representing the microstrip-slotline transition described in example 11.7.2. The terminal planes for ports 1 and 2 are to be taken at $z_1 = 0$, while those for ports 3 and 4 are to be taken at $z_2 = 0$.

p11-7 Suppose that port 2 of a microstrip-slotline $90°$ crossover is terminated in a short circuit, and port 4 is terminated in an open circuit.

    (a) Find the scattering matrix of the 2-port which results from the remaining ports 1 and 3.

    (b) What further modification(s) to this structure are needed to realize a reflectionless microstrip-to-slotline transition if $Z_{c1} \neq Z_{c2}$?

    (c) Suggest ways to achieve the idealized short and open circuit terminations at ports 2 and 4, in view of the fact that at the physical locations of these terminal planes there are likely to be significant amounts of higher-order modes present.

12 COUPLED-MODE THEORY

The basic concepts describing the coupling between two waveguides were laid out in Section 3.3, for the case of transmission lines modeled as distributed networks. What is missing from this model is the connection with the electromagnetic field of the coupled waveguides. In this chapter, we will provide this missing piece of the picture, constructing a theory that can be applied to any kind of waveguide, not just to conventional transmission lines. In this model, each waveguide acts as a distributed source exciting the other, and in this way energy is continuously exchanged between the two guides. The first four sections of this chapter will focus on instances of *co-directional coupling*, wherein a mode on one waveguide is coupled to a mode on a second waveguide that propagates in the same direction. Generally speaking, substantial coupling between modes propagating in opposite directions (*contra-directional coupling*) requires a nonuniform coupling mechanism, and this is treated in the final section of the chapter.

Coupled waveguides can, of course, be analyzed as an entire waveguide, rather than separate but coupled ones. If nothing else, this can be carried out by numerical simulation. However, our analyses in this chapter will be based on various kinds of perturbation approximation. That is, the coupling is assumed to be small enough that the individual, uncoupled waveguides approximately retain their distinct character. In this way, we can gain insight into the behavior of coupled modes in waveguides.

12.1 COUPLING OF CLOSED METALLIC WAVEGUIDES

We have seen in Chapter 10 that one way of exciting the modes of a waveguide is via discrete coupling mechanisms such as electrically small apertures in common walls, or loop or post-type probes. If more than a small amount of power transfer is desired, there can be large mismatches that require compensation, and this compensation will in general introduce unwanted frequency dependence into the circuit with attendant reduction in bandwidth. Somewhat more control over the coupling process between waveguides, and better performance, can be attained if the coupling is done in a gradual, continuous or nearly continuous manner. For example, open waveguides could be placed in parallel proximity to each other. For closed waveguides, a series of very small holes could be placed in the common wall between two waveguides, as shown in Figure 12.1. Since each hole presents only a very small disturbance to the waveguides, spurious reflections are minimized, while the large number of holes enables a substantial amount of coupling to occur.

To examine how this kind of gradual coupling process works, suppose that identical apertures in the common wall between two closed metallic waveguides are spaced periodically in the z-direction with separation distance p, and at the position (x_c, y_c) in the transverse plane. We assume that the spacing p is small compared to a wavelength, but large compared to the aperture dimensions so that near-field coupling between adjacent apertures can be neglected. For simplicity, we will assume that the common wall between the waveguides has zero thickness, and that the magnetic polarizability dyadic is diagonal, like those given in Table 10.1. These limitations could be readily removed if desired. Also, it will be assumed that, if the coupling apertures were absent, only one mode would be above cutoff in each isolated guide; we designate this as mode 1 in the first waveguide, and mode 2 in the other. When the coupling apertures are present, we will assume [following (10.48)-(10.49)] that in guide 1 or 2 respectively we have the transverse fields

$$\mathbf{E}_T(x, y, z) \quad = \quad \tilde{V}_{1,2}(z)\boldsymbol{\mathcal{E}}_{(1,2)T}(x, y) \tag{12.1}$$

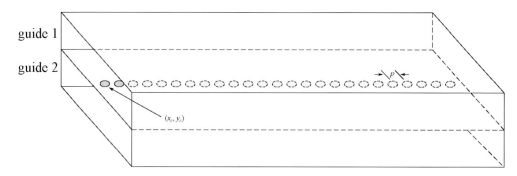

Figure 12.1 Distributed coupling between closed waveguides via a series of small apertures in the common wall (rectangular waveguides and circular apertures shown as an example).

$$\mathbf{H}_T(x, y, z) \quad = \quad \tilde{I}_{1,2}(z)\boldsymbol{\mathcal{H}}_{(1,2)T}(x, y)$$

and [following (10.61)-(10.62)] the longitudinal fields

$$E_z(x, y, z) \quad = \quad \tilde{I}_{1,2}(z)\mathcal{E}_{(1,2)z}(x, y) \tag{12.2}$$
$$H_z(x, y, z) \quad = \quad \tilde{V}_{1,2}(z)\mathcal{H}_{(1,2)z}(x, y)$$

where $\tilde{V}_{1,2}(z)$ and $\tilde{I}_{1,2}(z)$ are normalized mode voltages and currents of the modes in each guide.

To start with, consider the case when only a single aperture is present, located at $z = z_0$ (cf. problem p10-11). Denote the material parameters in guide 1 at (x_c, y_c) by μ_1, ϵ_1, and those in guide 2 at this location by μ_2, ϵ_2. Using (10.32), (10.86) and (10.90) with (12.1)-(12.2) as the short-circuit fields, we find that the mode in guide 1 is excited by the following impressed equivalent lumped sources at $z = z_0$:

$$V_0^{(1)} = \sqrt{2Z_{c1}}\tilde{V}_0^{(1)}; \qquad I_0^{(1)} = \sqrt{\frac{2}{Z_{c1}}}\tilde{I}_0^{(1)} \tag{12.3}$$

where Z_{c1} is the characteristic impedance assigned to the fundamental mode of guide 1,

$$\tilde{V}_0^{(1)}(z_0) \quad = \quad -j\omega\frac{\mu_1\mu_2\alpha_M^{ll}}{\mu_1 + \mu_2}\mathcal{H}_{1l}\left[\tilde{I}_1(z_0)\mathcal{H}_{1l} - \tilde{I}_2(z_0)\mathcal{H}_{2l}\right] \tag{12.4}$$

$$\tilde{I}_0^{(1)}(z_0) \quad = \quad j\omega\left\{\frac{\mu_1\mu_2\alpha_M^{zz}}{\mu_1 + \mu_2}\mathcal{H}_{1z}\left[\tilde{V}_1(z_0)\mathcal{H}_{1z} - \tilde{V}_2(z_0)\mathcal{H}_{2z}\right]\right.$$
$$\left. + \frac{\epsilon_1\alpha_E}{\epsilon_1 + \epsilon_2}\mathcal{E}_{1n}\left[\tilde{V}_1(z_0)\epsilon_1\mathcal{E}_{1n} - \tilde{V}_2(z_0)\epsilon_2\mathcal{E}_{2n}\right]\right\},$$

all mode fields are evaluated at the location (x_c, y_c) of the coupling hole, and (n, l) denote the transverse coordinates normal and tangential to the waveguide wall respectively, as shown in Figure 5.5. Here n points *into* waveguide 1, but away from waveguide 2. From (10.50), (3.21) and the differential equation (3.6), we can now obtain

$$\frac{d\tilde{V}_1}{dz} \quad = \quad -j\beta_1\tilde{I}_1 + \tilde{V}_0^{(1)}(z_0)\delta(z - z_0)$$

$$\frac{d\tilde{I}_1}{dz} \quad = \quad -j\beta_1\tilde{V}_1 + \tilde{I}_0^{(1)}(z_0)\delta(z - z_0) \tag{12.5}$$

Now, if there are multiple coupling holes instead of just one, closely spaced periodically in the z-direction with a distance p between their centers, we can approximate this situation by "smearing out" the discrete excitations represented by the delta functions in (12.5):

$$\frac{d\tilde{V}_1}{dz} = -j\beta_1\tilde{I}_1 + \frac{\tilde{V}_0^{(1)}(z)}{p}$$

$$\frac{d\tilde{I}_1}{dz} = -j\beta_1\tilde{V}_1 + \frac{\tilde{I}_0^{(1)}(z)}{p} \qquad (12.6)$$

The justification for this approximation is discussed in more detail in Section 12.5.2 below. From (12.4), we can put (12.6) into the form

$$\frac{d\tilde{V}_1}{dz} = -j\beta_1\tilde{I}_1 - j\Delta\tilde{l}_{11}\tilde{I}_1 - j\tilde{l}_{12}\tilde{I}_2$$

$$\frac{d\tilde{I}_1}{dz} = -j\beta_1\tilde{V}_1 - j\Delta\tilde{c}_{11}\tilde{V}_1 + j\tilde{c}_{12}\tilde{V}_2 \qquad (12.7)$$

wherein

$$\Delta\tilde{l}_{11} = \frac{\omega\mu_1\mu_2}{\mu_1 + \mu_2}\frac{\alpha_M^{ll}}{p}\mathcal{H}_{1l}^2$$

$$\tilde{l}_{12} = -\frac{\omega\mu_1\mu_2}{\mu_1 + \mu_2}\frac{\alpha_M^{ll}}{p}\mathcal{H}_{1l}\mathcal{H}_{2l}$$

$$\Delta\tilde{c}_{11} = -\frac{\omega\epsilon_1^2}{\epsilon_1 + \epsilon_2}\frac{\alpha_E}{p}\mathcal{E}_{1n}^2 - \frac{\mu_1\mu_2}{\mu_1 + \mu_2}\frac{\alpha_M^{zz}}{p}\mathcal{H}_{1z}^2$$

$$\tilde{c}_{12} = -\frac{\omega\epsilon_1\epsilon_2}{\epsilon_1 + \epsilon_2}\frac{\alpha_E}{p}\mathcal{E}_{1n}\mathcal{E}_{2n} - \frac{\mu_1\mu_2}{\mu_1 + \mu_2}\frac{\alpha_M^{zz}}{p}\mathcal{H}_{1z}\mathcal{H}_{2z} \qquad (12.8)$$

are equivalent normalized (excess) self and mutual inductances and capacitances for the coupled waveguides. In a completely analogous way, we also have

$$\frac{d\tilde{V}_2}{dz} = -j\beta_2\tilde{I}_2 - j\Delta\tilde{l}_{22}\tilde{I}_2 - j\tilde{l}_{12}\tilde{I}_1$$

$$\frac{d\tilde{I}_2}{dz} = -j\beta_2\tilde{V}_2 - j\Delta\tilde{c}_{22}\tilde{V}_2 + j\tilde{c}_{12}\tilde{V}_1 \qquad (12.9)$$

with

$$\Delta\tilde{l}_{22} = \frac{\omega\mu_1\mu_2}{\mu_1 + \mu_2}\frac{\alpha_M^{ll}}{p}\mathcal{H}_{2l}^2$$

$$\Delta\tilde{c}_{22} = -\frac{\omega\epsilon_2^2}{\epsilon_1 + \epsilon_2}\frac{\alpha_E}{p}\mathcal{E}_{2n}^2 - \frac{\mu_1\mu_2}{\mu_1 + \mu_2}\frac{\alpha_M^{zz}}{p}\mathcal{H}_{2z}^2 \qquad (12.10)$$

Except for the normalizing factors contained in $\tilde{V}_m$ and $\tilde{I}_m$, we see that these are identical with the matrix telegrapher's equations (3.26) that describe the voltages and currents on a multiconductor transmission line. The normalized line parameters are related to those of Chapter 3 by

$$\beta_1 + \Delta\tilde{l}_{11} \equiv \tilde{l}_{11} = \frac{\omega l_{11}}{Z_{c1}}$$

$$\beta_2 + \Delta\tilde{l}_{22} \equiv \tilde{l}_{22} = \frac{\omega l_{22}}{Z_{c2}}$$

$$\beta_1 + \Delta\tilde{c}_{11} \equiv \tilde{c}_{11} = \omega c_{11} Z_{c1}$$

$$\beta_2 + \Delta\tilde{c}_{22} \equiv \tilde{c}_{22} = \omega c_{22} Z_{c2}$$

$$\tilde{l}_{12} = \frac{\omega l_{12}}{\sqrt{Z_{c1} Z_{c2}}} \qquad \tilde{c}_{12} = \omega c_{12} \sqrt{Z_{c1} Z_{c2}} \qquad (12.11)$$

All of the theory of Section 3.3 can thus be applied to this pair of coupled waveguides, even though there are no actual wires carrying currents or supporting voltages with respect to some ground conductor. This is merely an equivalent circuit representation. Clearly these results could be extended to the more general lossy case in which we have line parameters z_{il}, y_{il} and γ_i as in Chapter 3.

If instead of a periodic string of apertures, the two waveguides are coupled by a long, continuous slot of width g, then the terms $\alpha_{E,M}/p$ in the above expressions can be replaced (in two cases) by the polarizabilities per unit length

$$\frac{\alpha_E}{p} \to \alpha_{El} = \frac{\pi g^2}{16}; \qquad \frac{\alpha_M^{ll}}{p} \to \alpha_{Ml}^{ll} = \frac{\pi g^2}{16} \qquad (12.12)$$

which are obtained by putting $w \to g$ in entry 10.1.3 of Table 10.1, dividing by l and letting $l \to \infty$. The term $\frac{\alpha_M^{zz}}{p}$ cannot be treated in this way, and problems where this term appears must be handled by other techniques (for example, using the transverse equivalent networks of Table 7.1). If the center of the slot is located at a position on the wall where $\mathcal{H}_{1z} = \mathcal{H}_{2z} = 0$, then we have in place of (12.8) and (12.10):

$$\Delta\tilde{l}_{11} = \frac{\omega\mu_1\mu_2}{\mu_1 + \mu_2} \frac{\pi g^2}{16} \mathcal{H}_{1l}^2$$

$$\Delta\tilde{c}_{11} = -\frac{\omega\epsilon_1^2}{\epsilon_1 + \epsilon_2} \frac{\pi g^2}{16} \mathcal{E}_{1n}^2$$

$$\tilde{l}_{12} = -\frac{\omega\mu_1\mu_2}{\mu_1 + \mu_2} \frac{\pi g^2}{16} \mathcal{H}_{1l}\mathcal{H}_{2l}$$

$$\tilde{c}_{12} = -\frac{\omega\epsilon_1\epsilon_2}{\epsilon_1 + \epsilon_2} \frac{\pi g^2}{16} \mathcal{E}_{1n}\mathcal{E}_{2n}$$

$$\Delta\tilde{l}_{22} = \frac{\omega\mu_1\mu_2}{\mu_1 + \mu_2} \frac{\pi g^2}{16} \mathcal{H}_{2l}^2$$

$$\Delta\tilde{c}_{22} = -\frac{\omega\epsilon_2^2}{\epsilon_1 + \epsilon_2} \frac{\pi g^2}{16} \mathcal{E}_{2n}^2 \qquad (12.13)$$

12.1.1 EXAMPLE: RECTANGULAR WAVEGUIDES COUPLED THROUGH A CONTINUOUS SLOT IN THE BROAD WALL

We can apply the foregoing theory to the problem of two identical air-filled rectangular waveguides, each supporting a TE_{10} mode, coupled by a centered slot of width g in the common broad wall between them, as shown in Figure 12.2. The n and l transverse coordinates at the slot are indicated there for the upper waveguide (guide 1). Because $\mathcal{H}_z$ of the TE_{10} mode is zero at the center of this slot, we may use (12.13). From Section 5.6.1, we have

$$\mathcal{H}_{l1} = \mathcal{H}_{l2} = -\mathcal{H}_x\left(\frac{a}{2}, 0\right) = -2\sqrt{\frac{\beta_{10}}{\omega\mu_0 ad}}; \qquad \mathcal{E}_{n1} = \mathcal{E}_{n2} = \mathcal{E}_y\left(\frac{a}{2}, 0\right) = 2\sqrt{\frac{\omega\mu_0}{\beta_{10} ad}} \quad (12.14)$$

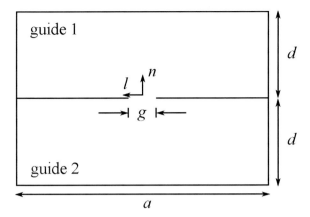

Figure 12.2 Cross section of coupled rectangular waveguides coupled by a centered slot in the common broad wall.

where $\beta_{10} = \beta_1 = \beta_2 = \sqrt{k_0^2 - \left(\frac{\pi}{a}\right)^2}$ is the propagation constant of the uncoupled TE_{10} modes. From (12.13) therefore,

$$\Delta\tilde{l}_{11} = \Delta\tilde{l}_{22} = -\tilde{l}_{12} = \beta_{10}\Delta$$
$$\Delta\tilde{c}_{11} = \Delta\tilde{c}_{22} = \tilde{c}_{12} = -\frac{k_0^2}{\beta_{10}}\Delta \tag{12.15}$$

where

$$\Delta = \frac{\pi g^2}{8ad} \tag{12.16}$$

is a measure of the coupling strength, usually $\ll 1$ because the gap width g is small.

Using (12.15) and (12.11) together with the results of example 3.3.7, we can find the propagation constants of the even and odd normal modes of these coupled waveguides. We have

$$K_C = \frac{\tilde{c}_{12}}{\tilde{c}_{11}} = -\frac{\frac{k_0^2}{\beta_{10}^2}\Delta}{1 - \frac{k_0^2}{\beta_{10}^2}\Delta}; \qquad K_L = \frac{\tilde{l}_{12}}{\tilde{l}_{11}} = -\frac{\Delta}{1+\Delta} \tag{12.17}$$

and

$$\hat{\beta}_1 = \hat{\beta}_2 = \omega\sqrt{(l_{11}+\Delta l_{11})(c_{11}+\Delta c_{11})} = \beta_{10}\sqrt{(1+\Delta)\left(1 - \frac{k_0^2}{\beta_{10}}\Delta\right)} \tag{12.18}$$

so by (3.96) we have

$$\beta_e = \beta_{10}$$
$$\beta_o = \beta_{10}\sqrt{(1+2\Delta)\left(1 - 2\frac{k_0^2}{\beta_{10}}\Delta\right)} \tag{12.19}$$

These results are of course approximate, being based on small aperture theory. As we have already seen from examples 10.6.2 and 10.6.3 in Chapter 10, we may need to appeal to other information about the problem in order to obtain a better approximation. In the present case, since the structure of Figure 12.2 is an air-filled closed waveguide, it must be true that the propagation constants of the even and odd modes have the forms

$$\beta_e = \sqrt{k_0^2 - k_{ce}^2} \qquad k_o = \sqrt{k_0^2 - k_{co}^2} \tag{12.20}$$

where k_{ce} and k_{co} are the cutoff wavenumbers for these modes, and are independent of the operating frequency. Evidently the cutoff wavenumber of the even mode is unchanged from that of the uncoupled waveguides: $k_{ce} = \pi/a$. However, that of the odd mode is changed; equating the expressions for β_o in (12.19) and (12.20) to each other, expanding and neglecting the term in Δ^2, we find that the odd-mode cutoff wavenumber has increased:

$$k_{co} \simeq \frac{\pi}{a}\sqrt{1+2\Delta} \simeq \frac{\pi}{a}(1+\Delta) \tag{12.21}$$

for $\Delta \ll 1$.

From (3.106), we can find that the smallest coupling length for complete power transfer from guide 1 to guide 2 is

$$l_c \simeq \frac{\beta_{10}a^2}{\pi\Delta} = \frac{8a^3\beta_{10}d}{\pi^2 g^2} \tag{12.22}$$

The narrower the slot, the longer will be the coupling length. As an example, if $a = 2$ cm, $d = 1$ cm, $g = 2$ mm and $f = 10$ GHz, then $l_c = 2.25$ m; if the slot width is doubled to $g = 4$ mm, then the coupling length is reduced to $l_c = 56$ cm, a more manageable size.

Rough sketches of the electric field lines of the even and odd normal modes are shown in Figure 12.3. We can see that the fields of the even mode are unaffected by the presence of the slot, explaining why its propagation is unchanged by the opening of the slot. However, fringing of the fields near the slot occurs in the odd mode (there would otherwise be a discontinuity in the normal component of $\epsilon\mathcal{E}$ at the slot, and there is no conductor present to support the surface charge density that would be required). This fringing is associated with the slight increase in the cutoff wavenumber for the odd mode exhibited in (12.21)—the field is "pinched" toward the side walls, making the width of the guide appear slightly smaller for this mode $[a_{\text{eq}} = a/(1+\Delta)]$.

Consider a section of these coupled waveguides with length l connected to four uncoupled rectangular waveguides as shown in Figure 12.4. Interfacing the uncoupled guides to the coupled guides at each end are transducers that convert uncoupled modes to the normal modes of the coupled section. If these transducers are simply direct connections between the coupled and uncoupled waveguides, then ideally the transducer on the left side would relate the wave amplitudes of modes 1 and 2 of the uncoupled guides with those of the even

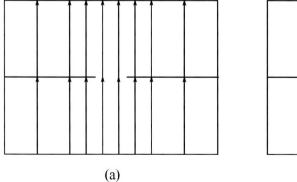
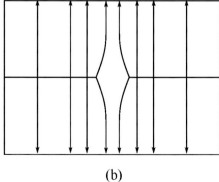

(a)   (b)

Figure 12.3 Electric field lines of the (a) even mode and (b) odd mode of the slot-coupled rectangular waveguides.

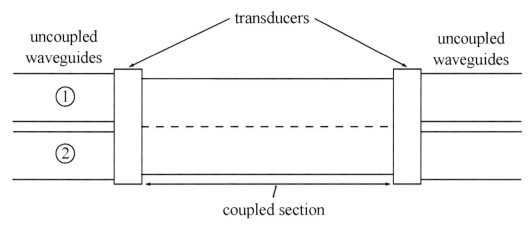

Figure 12.4 Directional coupler formed by the junction of uncoupled and coupled rectangular waveguides.

and odd modes of the coupled section by $[b] = [S][a]$, where

$$
[b] = \begin{bmatrix} b_1 \\ b_2 \\ b_e \\ b_o \end{bmatrix} \quad ; \quad [a] = \begin{bmatrix} a_1 \\ a_2 \\ a_e \\ a_o \end{bmatrix} \tag{12.23}
$$

and the scattering matrix of the transducer is:

$$
[S] = \frac{1}{\sqrt{2}} \begin{bmatrix} 0 & 0 & 1 & 1 \\ 0 & 0 & 1 & -1 \\ 1 & 1 & 0 & 0 \\ 1 & -1 & 0 & 0 \end{bmatrix} \tag{12.24}
$$

However, the presence of fringing fields in the odd mode means that the overlap integrals (11.17) will not exactly lead to the values given in (12.24), and there may also in fact be mode conversion at this transducer, leading to excitation of higher order modes and resulting parasitics. It is necessary to account for these effects in the design of directional couplers based on this mechanism. One way to minimize the nonideal behavior is to taper the slot gradually to zero width near the ends, rather than to truncate it abruptly. More details can be found in the references at the end of the chapter.

12.2 COUPLED-WAVE EQUATIONS

An alternative representation for the coupled transmission-line equations of Chapter 3 is sometimes used, which can often be a more natural way of thinking about waveguide coupling.

We construct the following *pseudo*-wave amplitudes for each hypothetical uncoupled waveguide mode:

$$a_i(z) = \frac{1}{\sqrt{8\hat{Z}_{ci}}}\left[V^{(i)}(z) + \hat{Z}_{ci}I^{(i)}(z)\right]; \qquad b_i(z) = \frac{1}{\sqrt{8\hat{Z}_{ci}}}\left[V^{(i)}(z) - \hat{Z}_{ci}I^{(i)}(z)\right] \quad (12.25)$$

by analogy with (10.42)-(10.43). These are not to be confused with modal wave amplitudes, as defined for example in the case of a multiconductor transmission line by (3.67). The telegrapher's equations (3.30) can now be rewritten in terms of the column vectors

$$[a(z)] = \begin{bmatrix} a_1(z) \\ a_2(z) \end{bmatrix}; \qquad [b(z)] = \begin{bmatrix} b_1(z) \\ b_2(z) \end{bmatrix} \quad (12.26)$$

of wave amplitudes as

$$\begin{aligned} \frac{d[a(z)]}{dz} &= -([\hat{\gamma}] + [v])\,[a(z)] + [v_r]\,[b(z)] \\ \frac{d[b(z)]}{dz} &= ([\hat{\gamma}] + [v])\,[b(z)] - [v_r]\,[a(z)] \end{aligned} \quad (12.27)$$

where

$$[\hat{\gamma}] = \begin{bmatrix} \hat{\gamma}_1 & 0 \\ 0 & \hat{\gamma}_2 \end{bmatrix} \quad (12.28)$$

is the propagation constant matrix,

$$[v] = \begin{bmatrix} 0 & v_{12} \\ v_{12} & 0 \end{bmatrix} \quad (12.29)$$

is the codirectional coupling matrix, and

$$[v_r] = \begin{bmatrix} 0 & v_{r12} \\ v_{r12} & 0 \end{bmatrix} \quad (12.30)$$

is the contradirectional coupling matrix, with

$$v_{12} = \frac{\sqrt{\hat{\gamma}_1\hat{\gamma}_2}}{2}\left(K_Z + K_Y\right); \qquad v_{r12} = \frac{\sqrt{\hat{\gamma}_1\hat{\gamma}_2}}{2}\left(K_Z - K_Y\right) \quad (12.31)$$

If we assume a propagating mode on lossless coupled lines, the uncoupled lines are described as in Section 3.3.3. Then the foregoing equations take the form

$$\begin{aligned} \frac{d[a(z)]}{dz} &= -j\left([\hat{\beta}] + [\kappa]\right)[a(z)] + j\,[\kappa_r]\,[b(z)] \\ \frac{d[b(z)]}{dz} &= j\left([\hat{\beta}] + [\kappa]\right)[b(z)] - j\,[\kappa_r]\,[a(z)] \end{aligned} \quad (12.32)$$

where

$$[\hat{\beta}] = \begin{bmatrix} \hat{\beta}_1 & 0 \\ 0 & \hat{\beta}_2 \end{bmatrix} \tag{12.33}$$

is the propagation constant matrix,

$$[\kappa] = \begin{bmatrix} 0 & \kappa_{12} \\ \kappa_{12} & 0 \end{bmatrix} \tag{12.34}$$

is the codirectional coupling matrix, and

$$[\kappa_r] = \begin{bmatrix} 0 & \kappa_{r12} \\ \kappa_{r12} & 0 \end{bmatrix} \tag{12.35}$$

is the contradirectional coupling matrix, with

$$\kappa_{12} = \frac{\sqrt{\hat{\beta}_1 \hat{\beta}_2}}{2}(K_L - K_C); \qquad \kappa_{r12} = \frac{\sqrt{\hat{\beta}_1 \hat{\beta}_2}}{2}(K_L + K_C) \tag{12.36}$$

Equations (12.27) or (12.32) are known as *coupled-wave equations*. Their physical meaning is that the wave amplitudes on hypothetical uncoupled lines propagate with uncoupled propagation constants $\hat{\gamma}_i$ or $\hat{\beta}_i$, but are also modified by the presence of forward or reverse waves on the other line, with coupling strengths between waves propagating in the same direction given by υ_{12} or κ_{12}, and between waves propagating in opposite directions by υ_{r12} or κ_{r12}, respectively.

The coupled-mode equations are often reduced to simpler forms, under the assumption that coupling between certain modes is so weak as to be negligible. For example, waves traveling on a lossless system in the forward direction can be thought of as distributed externally impressed voltage and current sources, varying as $e^{-j\hat{\beta}_{1,2}z}$. As is demonstrated in the solution to problem p3-6, the phase variation of this source is so different from that of either of the backward traveling waves that little excitation of the reverse waves occurs. In fact, it can be shown that distributed contradirectional coupling (that is, between the a_i and b_i) will be negligibly small provided that

$$|\hat{\beta}_i + \hat{\beta}_k| \gg |\kappa_{rik}| \quad \Rightarrow \quad |K_L + K_C| \ll 2 \left| \sqrt{\frac{\hat{\beta}_1}{\hat{\beta}_2}} + \sqrt{\frac{\hat{\beta}_2}{\hat{\beta}_1}} \right|; \qquad (i \neq k)$$

which will be true if $K_L \ll 1$ and $K_C \ll 1$. On the other hand, codirectional coupling is negligible only if

$$|\hat{\beta}_i - \hat{\beta}_k| \gg |\kappa_{ik}| \quad \Rightarrow \quad |K_L - K_C| \ll 2 \left| \sqrt{\frac{\hat{\beta}_1}{\hat{\beta}_2}} - \sqrt{\frac{\hat{\beta}_2}{\hat{\beta}_1}} \right|; \qquad (i \neq k)$$

and this is less likely to be true because $\hat{\beta}_i$ and $\hat{\beta}_k$ are often very close to each other numerically. The reduced coupled-mode equations are:

$$\frac{d[a(z)]}{dz} = -j\left([\hat{\beta}] + [\kappa]\right)[a(z)]$$

$$\frac{d[b(z)]}{dz} = j\left([\hat{\beta}] + [\kappa]\right)[b(z)] \tag{12.37}$$

and are decoupled 2×2 sets of differential equations, rather than the fully coupled 4×4 system (12.32). These simplified equations are most often used to model coupling between uniform, parallel dielectric waveguides for which the conditions of validity of this approximation are most strongly satisfied.

It is interesting to compare the exact solution for the symmetric coupled-line example of Section 3.3.7 with that obtained from (12.37) when contradirectional coupling is neglected. We seek even or odd modes of the pseudo-wave amplitudes as

$$[a_e(z)] = \begin{bmatrix} a_e^{(1)} \\ a_e^{(1)} \end{bmatrix} e^{-j\beta_e z}; \qquad [a_o(z)] = \begin{bmatrix} a_o^{(1)} \\ -a_o^{(1)} \end{bmatrix} e^{-j\beta_o z} \tag{12.38}$$

Substituting these into (12.37), we obtain

$$\beta_e = \hat{\beta}_1 + \kappa_{12} = \hat{\beta}_1\left[1 + \frac{K_L - K_C}{2}\right]; \qquad \beta_o = \hat{\beta}_1 - \kappa_{12} = \hat{\beta}_1\left[1 - \frac{K_L - K_C}{2}\right] \tag{12.39}$$

But if $K_L \ll 1$ and $K_C \ll 1$, we have

$$\sqrt{1 \pm K_{L,C}} \simeq 1 \pm \frac{K_{L,C}}{2}$$

and so (12.39) can also be obtained from (3.96) in this approximation, observing that the error terms in the former are of the order of K_L^2, K_C^2 and $K_L K_C$.

12.3   FIELD ANALYSIS OF PROXIMITY-COUPLED WAVEGUIDES

Open waveguides such as microstrip transmission lines or optical fibers can be coupled simply by placing their "guiding regions" (i.e., the ungrounded conductors or the dielectric cores) in proximity with each other, rather than creating holes in a common metal wall. The fields of each waveguide then overlap each other—one can think of equivalent sources (such as those in the Love equivalence principle) created by the fields of a mode of one waveguide continuously exciting modes on the other waveguide, and vice versa. For such problems, removal of the coupling is not as simple as closing off coupling holes or slots—it requires complete removal of each guide from the presence of the other. In this section, we will present a framework for studying the coupling of adjacent open waveguides from a field-theoretic viewpoint by means of a perturbation method. Application to the specific case of coupled adjacent dielectric waveguides will be made in Section 12.4, while the treatment of coupled open transmission lines is left as an exercise for the reader (problem p12-11).

Consider a general configuration of two waveguides whose guiding regions are contained within the boundaries C_1 and C_2, and which may also possess a ground conductor C_g as shown in Figure 12.5. The ground conductor may be closed, as shown, or open (extending to infinity, like the ground plane in microstrip, for example), or even absent (as for a dielectric waveguide). For simplicity, we will assume that all regions are nonmagnetic, with permeability μ_0. The interior cross-sectional surface of C_1 (the guiding region of waveguide 1) is denoted by S_1, and that of C_2 (the guiding region of waveguide 2) by S_2. The guiding regions are embedded in a "background" medium whose permittivity is ϵ_c, occupying the common cross-section S_c of the coupled waveguides. The interior of each guiding region may be a conductor (in the case of a transmission line) or a different dielectric medium (in the case of a surface-wave structure). The guiding region S_1 will be characterized by the permittivity ϵ_1, and S_2 by ϵ_2. These parameters may be complex if the guiding region is conducting, and a perfect conductor will be obtained by taking the limit $\epsilon_{1,2} \to -j\infty$. The entire cross section of the waveguide will be denoted as $S = S_c \cup S_1 \cup S_2$. The permittivity as a function of position will be denoted by ϵ:

$$\epsilon = \begin{cases} \epsilon_1 & \text{in } S_1 \\ \epsilon_2 & \text{in } S_2 \\ \epsilon_c & \text{in } S_c \end{cases} \tag{12.40}$$

Note that ϵ_1, ϵ_2 and ϵ_c can themselves be functions of (x, y) within S_1, S_2 and S_c respectively.

If either guiding region S_1 or S_2 is removed and replaced by the background medium, we are left with a single waveguide as shown in Figure 12.6. The corresponding permittivities as functions of position are denoted by

$$\tilde{\epsilon}_1 = \begin{cases} \epsilon_1 & \text{in } S_1 \\ \epsilon_c & \text{in } S_c \cup S_2 \end{cases} \qquad \tilde{\epsilon}_2 = \begin{cases} \epsilon_2 & \text{in } S_2 \\ \epsilon_c & \text{in } S_c \cup S_1 \end{cases} \tag{12.41}$$

Let us assume that each of the individual guides supports a propagating mode of some kind with mode fields $(\boldsymbol{\mathcal{E}}_1, \boldsymbol{\mathcal{H}}_1)$ or $(\boldsymbol{\mathcal{E}}_2, \boldsymbol{\mathcal{H}}_2)$, propagation constant β_1 or β_2 and, if appropriate, characteristic impedance Z_{c1} or Z_{c2} respectively. All these quantities are assumed to be known. We would like to determine, at least approximately, the interaction of the individual waveguide modes when both guiding regions are present simultaneously as in Figure 12.5.

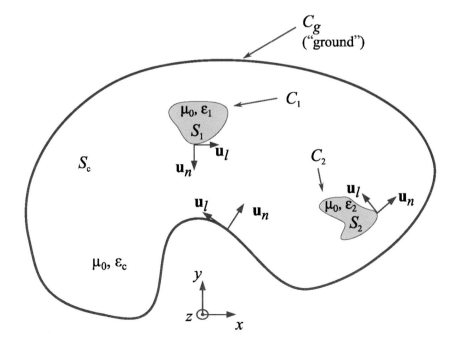

Figure 12.5 Coupled waveguides with a common ground conductor.

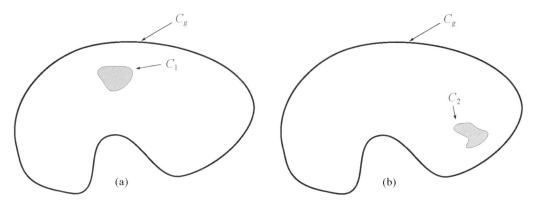

Figure 12.6 Individual waveguides 1 and 2 having the same "background" environment and ground conductor: (a) guide 1; (b) guide 2.

We will adopt a perturbation method for the study of these coupled waveguides, similar to that used in Section 9.5 to calculate waveguide attenuation due to wall losses. Let $\mathbf{E}(x, y, z)$ and $\mathbf{H}(x, y, z)$ be an exact solution of Maxwell's equations (not necessarily a single mode) in a region of the structure of Figure 12.5 that is free of impressed sources; this field obeys the boundary condition $\mathbf{u}_n \times \mathbf{E} = 0$ on the ground plane C_g. These fields will be taken as the "a" state in the reciprocity theorem, for which we will use the form obtained as the result of problem p9-3 instead of the artifice of integrating over a "slice" of the waveguide. For the "b" state let us choose the field of either the forward or reverse mode of the structure of Figure 12.6(a):

$$\mathbf{E}^b = \boldsymbol{\mathcal{E}}_{\pm 1} e^{\mp j\beta_1 z}; \qquad \mathbf{H}^b = \boldsymbol{\mathcal{H}}_{\pm 1} e^{\mp j\beta_1 z} \tag{12.42}$$

This field also obeys $\mathbf{u}_n \times \mathbf{E} = 0$ on the ground plane C_g. The reciprocity theorem requires that both states exist in the same medium, so we will apply the reciprocity theorem only to the region S_c that is common to the coupled waveguides shown in Figure 12.5 and to the individual guides shown in Figure 12.6.

After some simplification, use of these fields in the reciprocity identity of problem p9-3 results in

$$\left(\frac{d}{dz} \mp j\beta_1\right) \int_{S_c} [\mathbf{E}(x, y, z) \times \boldsymbol{\mathcal{H}}_{\pm 1} - \boldsymbol{\mathcal{E}}_{\pm 1} \times \mathbf{H}(x, y, z)] \cdot \mathbf{u}_z \, dS$$
$$= \oint_{C_1 \cup C_2} [\mathbf{E}(x, y, z) \times \boldsymbol{\mathcal{H}}_{\pm 1} - \boldsymbol{\mathcal{E}}_{\pm 1} \times \mathbf{H}(x, y, z)] \cdot \mathbf{u}_n \, dl \tag{12.43}$$

For some purposes it is convenient to use the results of adding and subtracting the equations in (12.43), which yield respectively:

$$\frac{d}{dz} \int_{S_c} \boldsymbol{\mathcal{E}}_{1T} \times \mathbf{H}_T(x, y, z) \cdot \mathbf{u}_z \, dS = -j\beta_1 \int_{S_c} \mathbf{E}_T(x, y, z) \times \boldsymbol{\mathcal{H}}_{1T} \cdot \mathbf{u}_z \, dS$$
$$+ \oint_{C_1 \cup C_2} [\boldsymbol{\mathcal{E}}_{1T} \times \mathbf{u}_z H_z(x, y, z) - \mathbf{E}_T(x, y, z) \times \mathbf{u}_z \mathcal{H}_{1z}] \cdot \mathbf{u}_n \, dl \tag{12.44}$$

and

$$\frac{d}{dz} \int_{S_c} \mathbf{E}_T(x, y, z) \times \boldsymbol{\mathcal{H}}_{1T} \cdot \mathbf{u}_z \, dS = -j\beta_1 \int_{S_c} \boldsymbol{\mathcal{E}}_{1T} \times \mathbf{H}_T(x, y, z) \cdot \mathbf{u}_z \, dS$$
$$+ \oint_{C_1 \cup C_2} [\mathbf{u}_z E_z(x, y, z) \times \boldsymbol{\mathcal{H}}_{1T} - \mathbf{u}_z \mathcal{E}_{1z} \times \mathbf{H}_T(x, y, z)] \cdot \mathbf{u}_n \, dl \tag{12.45}$$

By interchanging the indices 1 and 2, we arrive at the additional relations

$$\left(\frac{d}{dz} \mp j\beta_2\right) \int_{S_c} [\mathbf{E}(x,y,z) \times \boldsymbol{\mathcal{H}}_{\pm 2} - \boldsymbol{\mathcal{E}}_{\pm 2} \times \mathbf{H}(x,y,z)] \cdot \mathbf{u}_z \, dS$$

$$= \oint_{C_1 \cup C_2} [\mathbf{E}(x,y,z) \times \boldsymbol{\mathcal{H}}_{\pm 2} - \boldsymbol{\mathcal{E}}_{\pm 2} \times \mathbf{H}(x,y,z)] \cdot \mathbf{u}_n \, dl \qquad (12.46)$$

$$\frac{d}{dz} \int_{S_c} \boldsymbol{\mathcal{E}}_{2T} \times \mathbf{H}_T(x,y,z) \cdot \mathbf{u}_z \, dS = -j\beta_2 \int_{S_c} \mathbf{E}_T(x,y,z) \times \boldsymbol{\mathcal{H}}_{2T} \cdot \mathbf{u}_z \, dS$$

$$+ \oint_{C_1 \cup C_2} [\boldsymbol{\mathcal{E}}_{2T} \times \mathbf{u}_z H_z(x,y,z) - \mathbf{E}_T(x,y,z) \times \mathbf{u}_z \mathcal{H}_{2z}] \cdot \mathbf{u}_n \, dl \qquad (12.47)$$

and

$$\frac{d}{dz} \int_{S_c} \mathbf{E}_T(x,y,z) \times \boldsymbol{\mathcal{H}}_{2T} \cdot \mathbf{u}_z \, dS = -j\beta_2 \int_{S_c} \boldsymbol{\mathcal{E}}_{2T} \times \mathbf{H}_T(x,y,z) \cdot \mathbf{u}_z \, dS$$

$$+ \oint_{C_1 \cup C_2} [\mathbf{u}_z E_z(x,y,z) \times \boldsymbol{\mathcal{H}}_{2T} - \mathbf{u}_z \mathcal{E}_{2z} \times \mathbf{H}_T(x,y,z)] \cdot \mathbf{u}_n \, dl \qquad (12.48)$$

Although identities (12.43)-(12.48) are exact, we must make a reasonable approximation of $\mathbf{E}$ and $\mathbf{H}$ in order to make practical use of them. We will make the assumption that the diameters of the boundaries C_1 and C_2 are sufficiently small compared to their separation distance that no appreciable redistribution of the fields over S_1 or S_2 takes place when they are both present, as compared to when only one of them is. Such redistribution is called the *proximity effect*, and will be neglected in our treatment. Exactly how we approximate the total fields will depend on what kinds of waveguides are being coupled; we will limit ourselves here to examining the case of coupled dielectric waveguides that support surface waves in the next section.

12.4   DIELECTRIC WAVEGUIDE COUPLING

Typically, there is no conducting shield in a dielectric waveguide; if that is the case, we regard the outer boundary C_g in Figure 12.5 as having receded to infinity. The cladding region is S_c, while the materials inside S_1 and S_2 serve as core or guiding regions for each guide in isolation. These isolated guides are once again as depicted in Figure 12.6. We will make use of the continuity of tangential $\mathbf{E}$ and $\mathbf{H}$ at the contours C_1 and C_2 to transform the boundary integrals in (12.44)-(12.48) to integrals over S_1 and S_2 using the divergence theorem before approximating the total fields.

Consider for example the integral

$$\oint_{C_2} \mathbf{E}(x,y,z) \times \boldsymbol{\mathcal{H}}_{\pm 1} \cdot \mathbf{u}_n \, dl$$

that appears in (12.43). Applying the two-dimensional divergence theorem (B.26), we have

$$\oint_{C_2} \mathbf{E}(x,y,z) \times \boldsymbol{\mathcal{H}}_{\pm 1} \cdot \mathbf{u}_n \, dl = \int_{S_2} \nabla_T \cdot [\mathbf{E}(x,y,z) \times \boldsymbol{\mathcal{H}}_{\pm 1}] \, dS \qquad (12.49)$$

$$= \int_{S_2} [\boldsymbol{\mathcal{H}}_{\pm 1} \cdot \nabla_T \times \mathbf{E}(x,y,z) - \mathbf{E}(x,y,z) \cdot \nabla_T \times \boldsymbol{\mathcal{H}}_{\pm 1}] \, dS$$

$$= -j\omega \int_{S_2} [\mu_0 \boldsymbol{\mathcal{H}}_{\pm 1} \cdot \mathbf{H}(x,y,z) + \epsilon_c \boldsymbol{\mathcal{E}}_{\pm 1} \cdot \mathbf{E}(x,y,z)] \, dS$$

$$- \left(\frac{d}{dz} \mp j\beta_1\right) \int_{S_2} \mathbf{E}_T(x,y,z) \times \boldsymbol{\mathcal{H}}_{\pm 1 T} \cdot \mathbf{u}_z \, dS$$

where we have used various vector identities and Maxwell's equations along the way, including (5.16). Note that in S_2, the total field obeys Maxwell's equations in the medium ϵ_2, while $\boldsymbol{\mathcal{E}}_{\pm 1}$, $\boldsymbol{\mathcal{H}}_{\pm 1}$ is a solution in medium ϵ_c. The other boundary integrals in (12.43) and (12.46) are transformed in a similar way, so that finally these equations can be written in the form

$$\left(\frac{d}{dz} \mp j\beta_1\right) \int_S [\mathbf{E}(x,y,z) \times \boldsymbol{\mathcal{H}}_{\pm 1} - \boldsymbol{\mathcal{E}}_{\pm 1} \times \mathbf{H}(x,y,z)] \cdot \mathbf{u}_z \, dS = j\omega \int_{S_2} (\epsilon_2 - \epsilon_c) \mathbf{E}(x,y,z) \cdot \boldsymbol{\mathcal{E}}_{\pm 1} \, dS$$

$$(12.50)$$

and

$$\left(\frac{d}{dz} \mp j\beta_2\right) \int_S [\mathbf{E}(x,y,z) \times \boldsymbol{\mathcal{H}}_{\pm 2} - \boldsymbol{\mathcal{E}}_{\pm 2} \times \mathbf{H}(x,y,z)] \cdot \mathbf{u}_z \, dS = j\omega \int_{S_1} (\epsilon_1 - \epsilon_c) \mathbf{E}(x,y,z) \cdot \boldsymbol{\mathcal{E}}_{\pm 2} \, dS$$

$$(12.51)$$

Note that the first two integrals in each equation are now carried out over the entire cross section $S = S_c \cup S_1 \cup S_2$ instead of just the common background region S_c. These equations are still exact, and require suitable approximations of $\mathbf{E}$ and $\mathbf{H}$ to be of practical use.

Since contradirectional coupling between uniform dielectric waveguides can normally be neglected (see Section 12.2), we will assume that only forward modes exist on the waveguides. Regarding the fields $\mathbf{E}$ and $\mathbf{H}$ as aperture fields in a plane $z = $ constant, we can use (10.69) to obtain wave amplitudes for each mode as

$$a_1(z) = -\frac{1}{4} \int_S [\mathbf{E}(x,y,z) \times \boldsymbol{\mathcal{H}}_{-1} - \boldsymbol{\mathcal{E}}_{-1} \times \mathbf{H}(x,y,z)] \cdot \mathbf{u}_z \, dS$$

$$a_2(z) = -\frac{1}{4} \int_S [\mathbf{E}(x,y,z) \times \boldsymbol{\mathcal{H}}_{-2} - \boldsymbol{\mathcal{E}}_{-2} \times \mathbf{H}(x,y,z)] \cdot \mathbf{u}_z \, dS \qquad (12.52)$$

We will approximate the transverse fields as a superposition of two uncoupled forward modes only:

$$\mathbf{E}_T(x,y,z) \simeq A_1(z)\boldsymbol{\mathcal{E}}_{1T}(x,y) + A_2(z)\boldsymbol{\mathcal{E}}_{2T}(x,y) \qquad (12.53)$$

$$\mathbf{H}_T(x,y,z) \simeq A_1(z)\boldsymbol{\mathcal{H}}_{1T}(x,y) + A_2(z)\boldsymbol{\mathcal{H}}_{2T}(x,y)$$

For the longitudinal fields, we must be a little more careful with our approximation. Analogous to the considerations that led to (10.61)-(10.62), we obtain expressions for the longitudinal fields from the z-components of the Maxwell curl equations:

$$E_z(x,y,z) = \frac{1}{j\omega\epsilon}\mathbf{u}_z \cdot \nabla_T \times \mathbf{H}_T(x,y,z); \qquad H_z(x,y,z) = -\frac{1}{j\omega\mu}\mathbf{u}_z \cdot \nabla_T \times \mathbf{E}_T(x,y,z)$$

so using (12.53), (5.18) and (5.20) we obtain the approximations:

$$E_z(x,y,z) \simeq A_1(z)\frac{\tilde{\epsilon}_1}{\epsilon}\mathcal{E}_{1z}(x,y) + A_2(z)\frac{\tilde{\epsilon}_2}{\epsilon}\mathcal{E}_{2z}(x,y) \qquad (12.54)$$

$$H_z(x,y,z) \simeq A_1(z)\mathcal{H}_{1z}(x,y) + A_2(z)\mathcal{H}_{2z}(x,y)$$

Note that the amplitudes $A_{1,2}(z)$ are not the same as the wave amplitudes $a_{1,2}(z)$. This is because in the core region S_1 the mode fields $\boldsymbol{\mathcal{E}}_2$ and $\boldsymbol{\mathcal{H}}_2$ are not zero (though they are small if the guides are not too close together because they are evanescent outside of S_2). In fact, substituting (12.53) into (12.52), we find that

$$a_1(z) = A_1(z) + N_{12}A_2(z); \qquad a_2(z) = A_2(z) + N_{12}A_1(z) \qquad (12.55)$$

where

$$N_{12} = \frac{C_{12} + C_{21}}{2} \qquad (12.56)$$

and

$$C_{21} = \frac{1}{2} \int_S \boldsymbol{\mathcal{E}}_1 \times \boldsymbol{\mathcal{H}}_2 \cdot \mathbf{u}_z \, dS; \qquad C_{12} = \frac{1}{2} \int_S \boldsymbol{\mathcal{E}}_2 \times \boldsymbol{\mathcal{H}}_1 \cdot \mathbf{u}_z \, dS \qquad (12.57)$$

are power overlap integrals between the two modes. They are small compared to 1 for weak coupling.

With only two unknowns a_1 and a_2 to determine, only two of the four equations in (12.50)-(12.51) are required, and we choose to employ only the ones with the lower signs. Substituting (12.53) and (12.54) into these and using (12.55) gives equations of the form (12.37), with

$$[\hat{\beta}] = \begin{bmatrix} \hat{\beta}_1 & 0 \\ 0 & \hat{\beta}_2 \end{bmatrix} ; \qquad [\kappa] = \begin{bmatrix} 0 & \kappa_{12} \\ \kappa_{21} & 0 \end{bmatrix} \tag{12.58}$$

where

$$\hat{\beta}_1 = \beta_1 + \frac{\omega}{2(1 - N_{12}^2)}(T_{e11} - T_{m11} - N_{12}T_{e12} + N_{12}T_{m12})$$

$$\hat{\beta}_2 = \beta_2 + \frac{\omega}{2(1 - N_{12}^2)}(T_{e22} - T_{m22} - N_{12}T_{e21} + N_{12}T_{m21})$$

$$\kappa_{12} = \frac{\omega}{2(1 - N_{12}^2)}(T_{e12} - T_{m12} - N_{12}T_{e11} + N_{12}T_{m11})$$

$$\kappa_{21} = \frac{\omega}{2(1 - N_{12}^2)}(T_{e21} - T_{m21} - N_{12}T_{e22} + N_{12}T_{m22}) \tag{12.59}$$

where the integrals defined by

$$T_{e11} = \frac{1}{2}\int_{S_2}(\epsilon_2 - \epsilon_c)\boldsymbol{\mathcal{E}}_{1T}\cdot\boldsymbol{\mathcal{E}}_{1T}\,dS \qquad T_{e12} = \frac{1}{2}\int_{S_2}(\epsilon_2 - \epsilon_c)\boldsymbol{\mathcal{E}}_{1T}\cdot\boldsymbol{\mathcal{E}}_{2T}\,dS$$

$$T_{e21} = \frac{1}{2}\int_{S_1}(\epsilon_1 - \epsilon_c)\boldsymbol{\mathcal{E}}_{1T}\cdot\boldsymbol{\mathcal{E}}_{2T}\,dS \qquad T_{e22} = \frac{1}{2}\int_{S_1}(\epsilon_1 - \epsilon_c)\boldsymbol{\mathcal{E}}_{2T}\cdot\boldsymbol{\mathcal{E}}_{2T}\,dS$$

$$T_{m11} = \frac{1}{2}\int_{S_2}(\epsilon_2 - \epsilon_c)\frac{\epsilon_c}{\epsilon_2}\mathcal{E}_{1z}\mathcal{E}_{1z}\,dS \qquad T_{m12} = \frac{1}{2}\int_{S_2}(\epsilon_2 - \epsilon_c)\mathcal{E}_{1z}\mathcal{E}_{2z}\,dS$$

$$T_{m21} = \frac{1}{2}\int_{S_1}(\epsilon_1 - \epsilon_c)\mathcal{E}_{1z}\mathcal{E}_{2z}\,dS \qquad T_{m22} = \frac{1}{2}\int_{S_1}(\epsilon_1 - \epsilon_c)\frac{\epsilon_c}{\epsilon_1}\mathcal{E}_{2z}\mathcal{E}_{2z}\,dS \tag{12.60}$$

will be called polarization overlap integrals. Note that, unlike (12.34), we have in general $\kappa_{12} \neq \kappa_{21}$. This is because we have neglected coupling between any but one mode of each waveguide, and this neglect results in coupled-mode equations that violate reciprocity, at least to some small extent.

The overlap integrals with mixed subscripts ($_{12}$ or $_{21}$) will be small of first order due to the exponential smallness of the field of one guide in the presence of the core region of the other. Those with identical subscripts ($_{11}$ or $_{22}$) will be even smaller, of second order, because we integrate the square of the field of one guide over the core of the other. Since we are using a perturbation method that is expected only to give accurate results to first order, we may neglect terms of higher order in (12.59), resulting in the simpler expressions

$$\hat{\beta}_1 = \beta_1$$

$$\hat{\beta}_2 = \beta_2$$

$$\kappa_{12} = \frac{\omega}{2}(T_{e12} - T_{m12})$$

$$\kappa_{21} = \frac{\omega}{2}(T_{e21} - T_{m21}) \tag{12.61}$$

Properties of and interrelationships between the overlap integrals are derived in Appendix O. In particular, in the case of degenerate modes ($\beta_1 = \beta_2$), we have from (O.13) that $\kappa_{12} = \kappa_{21}$, which suggests that this perturbation-based coupling model should be most accurate when β_1 is sufficiently close to β_2.

12.5   NONUNIFORM MODE COUPLING

12.5.1   ADIABATIC COUPLING

In many applications the bandwidth of a coupler such as that described in Section 12.1.1 is unacceptably narrow (see problem p12-6). In such cases we seek ways to broaden this bandwidth, as well as to ease problems caused by fabrication variations. One solution is the so-called *adiabatic coupler* (also sometimes called a *tapered-velocity coupler*). In this type of coupler, one allows the propagation constants β_1 and β_2 of the coupled waveguides (and possibly the coupling coefficients) to be functions of z within the coupling region. The principles involved are similar to those of the nonuniform transmission line studied in Section 1.5. The tapering is usually carried out in such a way that $\beta_1 \neq \beta_2$ over most of the coupler's length, with $\beta_1 \simeq \beta_2$ only in a small neighborhood of one point. This point of synchronism may change location with frequency, but should always remain within the coupling region.

When the velocity tapers are linear:

$$\beta_1 = \beta_0 + C_1 z; \qquad \beta_2 = \beta_0 + C_2 z$$

the coupled-mode equations (12.37) can either be solved exactly in terms of special functions known as parabolic cylinder functions, or solved approximately by a method analogous to the Rayleigh-Bremmer technique of Section 1.5.2 (see the references at the end of this chapter). A rough plot of a solution for $|V_2(z)|$ when $V_2(-\infty) = 0$ is shown in Figure 12.7. The most important coupling occurs near the point of synchronism ($z = 0$), and after that, the tendency for the power to transfer back to guide 1 is frustrated by the large difference between β_1 and β_2. The situation is similar to that in problem p3-3, in which a transmission line is excited by a distributed source with nonuniformly varying phase.

It can be shown that bandwidths as large as 3:1 are attainable by this technique. Tolerance problems are also eased, but the price paid is that longer coupling regions are required, in much the same way as broadband impedance matching is possible with nonuniform transmission lines if the line length is long enough (see the exponential line example in Section 1.5.2).

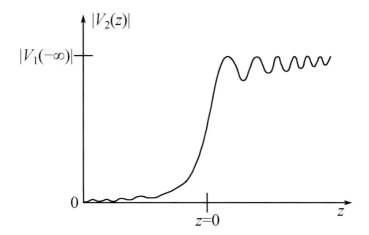

Figure 12.7   Coupled signal in an adiabatic coupler versus coupling distance z.

12.5.2   PERIODIC AND CONTRADIRECTIONAL COUPLING

Continuous uniform couplers such as the slot coupler of example 12.1.1 are often not used in practice because the fringing field present in the odd system mode will cause significant reflection at the junction with the uncoupled waveguides. A resonant slot-concentrated wave can be present as a result. This causes a degradation of the directivity and bandwidth of the coupler. A periodic arrangement of discrete coupling holes as shown in Figure 12.1 can be used instead, but rather than model this coupling by an approximation of uniform coupling as in (12.7) and (12.9), we use the coupled-wave equations (12.37) wherein the coupling matrix is a periodic function of z: $[\kappa(z+p)] = [\kappa(z)]$. Such a function can be represented as a Fourier series:

$$[\kappa(z)] = \sum_{i=-\infty}^{\infty} [\kappa^{(i)}] e^{2\pi j \frac{iz}{p}} \tag{12.62}$$

Substituting (12.62) into the first of (12.37), we get

$$j\frac{da_1}{dz} = \hat{\beta}_1 a_1 + \sum_{i=-\infty}^{\infty} \kappa_{12}^{(i)} e^{2\pi j \frac{iz}{p}} a_2$$

$$j\frac{da_2}{dz} = \hat{\beta}_2 a_2 + \sum_{i=-\infty}^{\infty} \kappa_{12}^{(i)} e^{2\pi j \frac{iz}{p}} a_1 \tag{12.63}$$

If $\hat{\beta}_1$ and $\hat{\beta}_2$ are constants, introduce new amplitudes $U_1(z)$ and $U_2(z)$ by

$$a_1(z) = e^{-j\hat{\beta}_1 z} U_1(z); \qquad a_2(z) = e^{-j\hat{\beta}_2 z} U_2(z) \tag{12.64}$$

and get

$$j\frac{dU_1}{dz} = \sum_{i=-\infty}^{\infty} \kappa_{12}^{(i)} e^{jz\left(\hat{\beta}_1 - \hat{\beta}_2 + \frac{2\pi i}{p}\right)} U_2$$

$$j\frac{dU_2}{dz} = \sum_{i=-\infty}^{\infty} \kappa_{12}^{(i)} e^{jz\left(\hat{\beta}_2 - \hat{\beta}_1 + \frac{2\pi i}{p}\right)} U_1 \tag{12.65}$$

When κ_{12} is constant (that is, when only the terms with $i = 0$ are present in (12.65)), we have argued in Section 12.2 that significant power transfer between modes 1 and 2 occurs only if $|\hat{\beta}_2 - \hat{\beta}_1| \lesssim |\kappa_{12}| = |\kappa_{12}^{(0)}|$. Similar arguments for the general case can be advanced for (12.65), indicating that only those terms of the summation for which the coefficient of jz in the exponent is small compared to the corresponding matrix element $\kappa_{12}^{(i)}$ will significantly affect the solution. That is, only when

$$\left| \hat{\beta}_1 - \hat{\beta}_2 + \frac{2\pi i}{p} \right| \lesssim |\kappa_{12}^{(i)}| \tag{12.66}$$

will these terms need to be retained.

   If the period p of the coupling is short enough, only the $i = 0$ terms have the possibility of contributing to the coupling in a substantial way. But if all the $i \neq 0$ terms are dropped from (12.65), we reduce to the case of continuous, uniform coupling as discussed in Section 12.2, with the matrix $[\kappa]$ replaced with its zeroth-order Fourier coefficient:

$$[\kappa^{(0)}] = \frac{1}{p} \int_0^p [\kappa(z)] \, dz \tag{12.67}$$

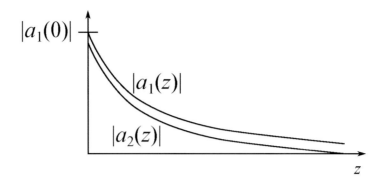

Figure 12.8 Behavior of mode amplitudes in a contradirectional coupler.

For larger values of p, we must consider the possibility that one of the conditions (12.66) might be satisfied for some $i \neq 0$, even though $\hat{\beta}_1 \neq \hat{\beta}_2$ (which would suppress coupling in the case when $|\kappa|$ is a small enough constant). In particular, if $\hat{\beta}_2 = -\hat{\beta}_1$, and if the period $p = \pi/\hat{\beta}_1$ is half of an uncoupled guide wavelength, then the condition (12.66) can be satisfied with $i = -1$. With such a value of p, significant power transfer can occur between modes traveling in *opposite* directions, and we have what is called *contradirectional* coupling. It is usually necessary in such problems to consider the simultaneous coupling among only four modes (± 1 and ± 2) to achieve sufficient accuracy. Although there is no conceptual difficulty about doing so, the algebra that results is rather cumbersome, and a two-mode approximate description of contradirectional coupling is often made. If a_1 is a forward mode and a_2 a reverse mode, the results obtained are qualitatively as shown in Figure 12.8. The amplitude $|a_1(z)|$ attenuates as z increases, and as it does so, the reverse wave amplitude $|a_2(z)|$ is building up as z decreases. Under the two-mode approximation, complete power transfer is never attainable using only contradirectional coupling. However, in practice a certain amount of co-directional coupling will inevitably take place, and it is possible to balance the various coupling processes so as to achieve almost total power transfer from the forward wave to the reverse one.

12.6   NOTES AND REFERENCES

References for the basic theory of coupled transmission-line and coupled-wave equations are given in Section 3.4.

Coupling through a slot in the common wall between two metallic waveguides is covered in:

Huang Hung-Chia, "Theory of coupled waveguides," *Scientia Sinica*, vol. 11, pp. 16-57 (1962).

Gu Fu-Nian, "On eigenvalues of coupled waveguides," *Chinese J. Phys. (Peking)*, vol. 22, pp. 613-625 (1967).

E. H. Kopp and R. S. Elliott, "Coupling between dissimilar waveguides," *IEEE Trans. Micr. Theory Tech.*, vol. 16, pp. 6-11 (1968).

P. F. Wilson and D. C. Chang, "Mode coupling by a longitudinal slot for a class of planar waveguiding structures: Part I—Theory," *IEEE Trans. Micr. Theory Tech.*, vol. 33, pp. 981-987 (1985).

P. F. Wilson and D. C. Chang, "Mode coupling by a longitudinal slot for a class of planar waveguiding structures: Part II—Applications," *IEEE Trans. Micr. Theory Tech.*, vol. 33, pp. 988-993 (1985).

and the particular example of Section 12.1.1 is considered in detail in

E. Schuon, "Eigenschaften uund Bemessung des Langschlitz-Richtkopplers," *Arch. Elek. Übertrag.*, vol. 12, pp. 237-243 (1958).

Single-hole, multi-hole and slot couplers are discussed in:

R. L. Kyhl, "Directional couplers," in Montgomery (1966), Chapter 14.
Altman (1964), Section 4.3;
A. J. Sangster, "Variational method for the analysis of waveguide coupling," *Proc. IEE (London)*, vol. 112, pp. 2171-2179 (1965).
R. Levy, "Directional couplers," in *Advances in Microwaves, vol. 1*. New York: Academic Press, 1966, pp. 115-209.
Mashkovtsev, Tsibizov and Emelin (1966), Chapters 4-6;
Sporleder and Unger (1979), Section 2.7;
Collin (2001), Chapter 6;
Pozar (2005), Section 7.4.

The classic paper on the coupled-wave equations (in contrast to the multiconductor transmission-line equations of Section 3.3) is

S. E. Miller, "Coupled wave theory and waveguide applications," *Bell Syst. Tech. J.*, vol. 33, pp. 661-719 (1954).

Other treatments can be found in:

Watkins (1958), Chapter 3;
Louisell (1960);
Johnson (1965), Chapter 9;
R. Levy, "Directional couplers," in *Advances in Microwaves, vol. 1*. New York: Academic Press, 1966, pp. 115-209.
Waldron (1969), Chapter 9;
Sporleder and Unger (1979);
Snyder and Love (1983), Chapters 27-29.
Huang Hung-Chia (1984).

Two different derivations of the condition under which coupling between a given pair of modes in coupled-mode theory can be neglected are given in

L. T. ter-Martirosian, "Evaluation of errors in the theory of coupled modes," *Radio Eng. Electron. Phys.*, vol. 14, pp. 1096-1102 (1969).

and

Sporleder and Unger (1979), Section 6.1.

The connection between the coupled-mode and telegrapher's equation formulations was discussed in

J. E. Adair and G. I. Haddad, "Coupled-mode analysis of nonuniform coupled transmission lines," *IEEE Trans. Micr. Theory Tech.*, vol. 17, pp. 746-752 (1969).

Coupling between parallel dielectric waveguides has been modeled by a variety of techniques. Different approaches have yielded expressions for the coupling coefficients that are either line integrals around certain contours in the waveguide cross-sections, or surface integrals over all or part of the cross-section. All of these methods have yielded similar results, but different numerical values for the coupling effects are observed in the case of strong coupling. See the books:

Sporleder and Unger (1979), Section 2.8.
Marcuse (1982), Chapter 10.
Vassallo (1985), tome 2, Section 1.4-2.
Barybin and Dmitriev (2002), Chapter 7.

as well as the papers

J. A. Arnaud, "Transverse coupling in fiber optics Part I: Coupling between trapped modes," *Bell Syst. Tech. J.*, vol. 53, pp. 217-224 (1974).
E. F. Kuester and D. C. Chang, "Nondegenerate surface-wave mode coupling between dielectric waveguides," *IEEE Trans. Micr. Theory Tech.*, vol. 23, pp. 877-882 (1975).
A. Hardy and W. Streifer, "Coupled mode theory of parallel waveguides," *J. Lightwave Technol.*, vol. 3, pp. 1135-1146 (1985).
W. Streifer, M. Osiński, and A. Hardy, "Reformulation of the coupled-mode theory of multiwaveguide systems," *J. Lightwave Technol.*, vol. 5, pp. 1-4 (1987).
S. L. Chuang, "A coupled mode formulation by reciprocity and a variational principle," *J. Lightwave Technol.*, vol. 5, pp. 5-15 (1987).
H. A. Haus, W. P. Huang, S. Kawakami, and N. A. Whitaker, "Coupled-mode theory of optical waveguides," *J. Lightwave Technol.*, vol. 5, pp. 16-23 (1987).
S. L. Chuang, "Application of the strongly coupled-mode theory to integrated optical devices," *IEEE J. Quant. Electron.*, vol. 23, pp. 499-509 (1987).
H.-C. Chang, "Coupled-mode equations for dielectric waveguides based on projection and partition modal amplitudes," *IEEE J. Quant. Electron.*, vol. 23, pp. 1929-1937 (1987).
C. Vassallo, "About coupled-mode theories for dielectric waveguides," *J. Lightwave Technol.*, vol. 6, pp. 294-303 (1988).
H.-C. Chang, H. S. Huang and Y.-C. Wang, "On the various forms of the coupled-mode theory for optical waveguides," *Micr. Opt. Technol. Lett.*, vol. 3, pp. 296-298 (1990).
W. P. Huang, "Coupled-mode theory for optical waveguides: an overview," *J. Opt. Soc. Amer. A*, vol. 11, pp. 963-983 (1994).
E. O. Kamenetskii, "Induced polarization effects in coupling processes of waveguide modes," *IEEE Trans. Micr. Theory Tech.*, vol. 44, pp. 572-584 (1996).
B. E. Little and W. P. Huang, "Coupled-mode theory for optical waveguides," *PIER*, vol. 10, pp. 217-270 (1995).
A. N. Kireev and T. Graf, "Vector coupled-mode theory of dielectric waveguides," *IEEE J. Quant. Electron.*, vol. 39, pp. 866-873 (2003).
J. Xu and Y. Chen, "General coupled mode theory in non-Hermitian waveguides," *Opt. Express*, vol. 23, pp. 22619-22627 (2015).

in which the variants of this technique are discussed and compared. Vassallo in particular has asserted that there is no one formulation that is "best" for all possible applications, but that in many practical situations any formulation has adequate accuracy. We have opted in Section 12.4 for a derivation that is in some sense the simplest for a clear understanding of the coupling process. Note that there is not complete agreement between the various authors on the notation for overlap integrals and coupling coefficients, so the reader should be alert when examining this literature.

The method used in Sections 12.3-12.4 to model coupled dielectric waveguides can be adapted to treat coupled transmission lines:

K. Yasumoto, "Coupled-mode formulation of multilayered and multiconductor transmission lines," *IEEE Trans. Micr. Theory Tech.*, vol. 44, pp. 585-590 (1996).

M. Matsunaga, M. Katayama and K. Yasumoto, "Coupled-mode analysis of line parameters of coupled microstrip lines," *PIER*, vol. 24, pp. 1-17 (1999).

D. Chen, Z. Shen and Y. Lu, "Coupled mode analysis of forward and backward coupling in multiconductor transmission lines," *IEEE Trans. Electromag. Compat.*, vol. 47, pp. 463-470 (2005).

and has shown good agreement with full-wave simulations of microstrip-like lines. If the fields and propagation constants of the system modes of a set of coupled transmission lines can be determined exactly, a method for calculating the per-unit-length line parameters was given in

F. Olyslager, A. Franchois, and D. De Zutter, "Reciprocity based transmission line equations for higher order eigenmodes in lossy waveguides," *Wave Motion*, vol. 41, pp. 229-238 (2005).

Adiabatic couplers are treated in:

W. H. Louisell, "Analysis of the single tapered mode coupler," *Bell Syst. Tech. J.*, vol. 34, pp. 853-870 (1955).

H. E. Rowe, "Approximate solutions for the coupled line equations," *Bell Syst. Tech. J.*, vol. 41, pp. 1011-1029 (1962).

R. A. Waldron, "The theory of coupled modes," *Quart. J. Mech. Appl. Math.*, vol. 18, pp. 385-404 (1965).

Waldron (1969), Section IX.J.

R. B. Smith, "Analytic solutions for linearly tapered directional couplers," *J. Opt. Soc. Amer.*, vol. 66, pp. 882-892 (1976).

Sporleder and Unger (1979), Section 8.2.

using either approximate or (in special cases) exact solutions of the coupled-mode equations. Discussions of periodic coupling are found in:

S. E. Miller, "On solutions for two waves with periodic coupling," *Bell Syst. Tech. J.*, vol. 47, pp. 1801-1822 (1968).

A. W. Snyder and O. J. Davies, "Asymptotic solution of coupled-mode equations for sinusoidal coupling," *Proc. IEEE*, vol. 58, pp. 168-169 (1970).

A. L. Cullen and O. J. Davies, "Periodic coupling of waveguide modes," *Proc. IEE (London)*, vol. 117, pp. 2061-2068 (1970).

A. Yariv, "Coupled-mode theory for guided-wave optics," *IEEE J. Quant. Electron.*, vol. 9, pp. 919-933 (1973).

C. O. Egalon, A. M. Buoncristiani and R. S. Rogowski, "Asymptotic approximation and first-order correction of coupled-mode equations," *Opt. Eng.*, vol. 36, pp. 1930-1934 (1997).

12.7 PROBLEMS

p12-1 In the example of Section 12.1.1, let the two waveguides be filled with different materials: the upper waveguide has parameters μ_1 and ϵ_1, while the lower one has μ_2 and ϵ_2. Find the normal modes of this system of coupled waveguides.

p12-2 Repeat problem p12-1, but for air-filled waveguides of different widths: the upper guide has width a_1, while the lower one has width a_2.

p12-3 Two air-filled rectangular waveguides, one on top of the other, are coupled by a periodically spaced string of circular apertures of radius r_0 along the center of the common broad wall between the waveguides. Obtain an expression for the coupling length l_c of this system.

p12-4 Consider a pair of adjacent identical coaxial transmission lines, coupled by a narrow slot of width g in the common shield between them as shown in cross section by the figure below.

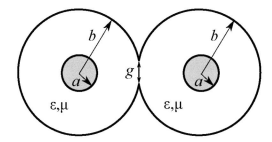

Using the normalized fields of the TEM mode of the coaxial line given in Section 8.3.1, obtain expressions for the mutual capacitance c_{12} and mutual inductance l_{12} per unit length of this coupled system, as well as the increments $\Delta c_{11} = \Delta c_{22}$ and $\Delta l_{11} = \Delta l_{22}$ in the self capacitance and inductance per unit length. From these results obtain expressions for the even and odd mode characteristic impedances of this system from equations (3.97).

p12-5 Repeat problem p12-4, but assume that the materials filling the two cross-sections are different: (ϵ_1, μ_1) or (ϵ_2, μ_2). Instead of determining the even and odd mode characteristic impedances (which do not apply in this case because of the lack of symmetry), find an expression for the coupling length l_c of this system.

p12-6 Suppose that a section of coupled rectangular waveguides as described in Section 12.1.1 has been chosen to have the coupling length $l = l_c$ from (12.22) at a certain design frequency ω_0. If the actual operating frequency ω is different from ω_0, only a part of the power initially launched in guide 1 will have been transferred to guide 2. Determine the range of frequencies (as fractions of ω_0) over which at least 90% of the power from guide 1 will be transferred to guide 2, and from this the fractional operating bandwidth of the coupler. Show that this bandwidth is independent of the slot width g.

p12-7 Consider the structure of Figure 7.11 as a pair of rectangular waveguides coupled by a slot in their common narrow wall. In the limit when the gap g is very small, how does the behavior of this system differ from that of Section 12.1.1? What practical constraints limit the performance of Figure 7.11 as a directional coupler?

p12-8 If $a_1(z)$ and $a_2(z)$ are true wave amplitudes, and contradirectional coupling is negligible so that $b_1(z) = b_2(z) = 0$ and the first line of (12.37) applies, conservation of energy for a system of lossless coupled waveguides would require that the relation

$$\frac{d}{dz}\left[|a_1(z)|^2 + |a_2(z)|^2\right] = 0$$

holds everywhere. If $\hat{\beta}_1$ and $\hat{\beta}_2$ are real, show that conservation of energy requires that $\kappa_{21} = \kappa_{12}^*$, and in particular that $\kappa_{21} = \kappa_{12}$ if they are real.

p12-9 Show that the general solution to the first line of (12.37) with parameters (12.58) is

$$a_1(z) = e^{-j\beta_{av}z}\left\{a_1(0)\cos\delta z - j\left[\kappa_{12}a_2(0) + \Delta\beta a_1(0)\right]\frac{\sin\delta z}{\delta}\right\}$$

$$a_2(z) = e^{-j\beta_{av}z}\left\{a_2(0)\cos\delta z - j\left[\kappa_{21}a_1(0) - \Delta\beta a_2(0)\right]\frac{\sin\delta z}{\delta}\right\}$$

where

$$\beta_{av} = \frac{\hat{\beta}_1 + \hat{\beta}_2}{2}; \qquad \Delta\beta = \frac{\hat{\beta}_1 - \hat{\beta}_2}{2}; \qquad \delta = \sqrt{(\Delta\beta)^2 + \kappa_{12}\kappa_{21}}$$

If $a_2(0) = 0$, what is the maximum value that $|a_2(z)|^2$ may take, and at what values of z does this occur? Under what condition does $|a_2(z)|_{max}^2 = |a_1(0)|^2$?

p12-10 Using the method of Section 12.4, compute the coupling coefficients for a pair of identical dielectric slab waveguides with core thickness a and separation distance d between the nearest faces of the slabs. Obtain expressions the propagation constants of the even and odd modes of this coupled system. For the case of $n_1 = 1.4$ and $n_2 = 1.37$, with $d/a = 0.1$, 0.3 or 1.0 successively, compute and plot n_{eff} of even and odd coupled TE$_0$ modes for $0 < V < 5$, and compare this result with that of an exact analysis using the transverse resonance method of Section 7.1 (cf. problem p7-2). Note that in this two-dimensional waveguide structure, surface integrals over the cross-section will reduce to integrals over the x-coordinate only, and integrals over the boundary contour will reduce to evaluations at the boundary points of the cores.

p12-11 Consider the multiconductor transmission line that results when the surfaces S_1 and S_2 in Figure 12.5 are occupied by perfect conductors. Imitate the technique of Section 12.4 to model the coupling between a pair of single transmission lines, but enforce the condition that the tangential components of $\mathbf{E}$ and $\mathcal{E}_{1,2}$ at the conductor surfaces C_1, C_2 and C_g are zero in (12.43)-(12.48) instead of making the transformations used at the beginning of Section 12.4. In the resulting equations, use the approximations

$$\mathbf{E}_T(x,y,z) \simeq \tilde{V}_1(z)\mathcal{E}_{1T}(x,y) + \tilde{V}_2(z)\mathcal{E}_{2T}(x,y)$$

$$\mathbf{H}_T(x,y,z) \simeq \tilde{I}_1(z)\mathcal{H}_{1T}(x,y) + \tilde{I}_2(z)\mathcal{H}_{2T}(x,y)$$

$$E_z(x,y,z) \simeq \tilde{I}_1(z)\mathcal{E}_{1z}(x,y) + \tilde{I}_2(z)\mathcal{E}_{2z}(x,y)$$

$$H_z(x,y,z) \simeq \tilde{V}_1(z)\mathcal{H}_{1z}(x,y) + \tilde{V}_2(z)\mathcal{H}_{2z}(x,y)$$

for the total fields in S_c to derive the coupled transmission line equations (3.27) relating the voltages and currents $V_{1,2}(z)$ and $I_{1,2}(z)$. Obtain expressions for the coefficients in these equations in terms of overlap integrals between the fields of the isolated transmission lines.

p12-12 Using the results of problem p12-11, obtain expressions for the inductance and capacitance parameters per unit length for the geometry shown in the figure below.

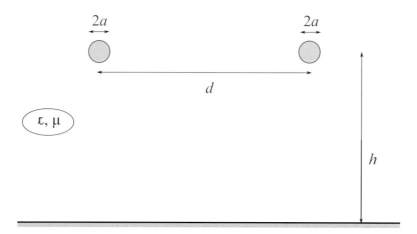

From these obtain expressions for the even and odd mode characteristic impedances.

p12-13 Consider two modes that are coupled within a certain range of z by a z-dependent coupling coefficient $\kappa(z)$, but with z-independent, identical propagation constants $\beta_1 = \beta_2 = \beta$. The mode amplitudes are governed by the system of differential equations

$$j\frac{da_1}{dz} = \beta a_1 + \kappa(z)a_2$$
$$j\frac{da_2}{dz} = \beta a_2 + \kappa(z)a_1$$

Obtain the solutions $a_1(z)$ and $a_2(z)$ of this system that satisfy $a_1(z_0) = 1$ and $a_2(z_0) = 0$. As a hint, you may wish to try the changes of variable

$$a_1(z) = e^{-j\beta z}U_1(z); \qquad a_2(z) = e^{-j\beta z}U_2(z); \qquad x = \int_{z_0}^{z} \kappa(z')\,dz'$$

Suggest some specific function $\kappa(z)$ that might alleviate the problem of needing tight dimensional tolerances to achieve complete power transfer (a 0 dB coupler). Is there any such function that can increase the bandwidth of a 0 dB coupler? If not, why not?

13 RESONANT ELEMENTS FOR CIRCUITS AND WAVEGUIDES

13.1 INTRODUCTION

A resonator (roughly defined) is a device capable of storing a large amount of energy during oscillations in which relatively little of this energy is dissipated. In waveguide and transmission-line circuits, just as in lumped-element circuits of a more conventional nature, resonant elements find a wide range of application. Filters, frequency measuring devices and oscillators all involve resonators of one sort or another, but at higher frequencies, even radiating elements (antennas) are resonant circuits. It is thus important for us to understand the properties of the resonators themselves, as well as their effect on a waveguide or transmission-line system into which they are connected. The properties of lumped-element resonators will be examined first in Section 13.2, while later in the chapter we will study cavity resonators, dielectric resonators and other such devices based on their electromagnetic field descriptions.

Before embarking on the general analysis of resonators, let us first examine the simplest and most familiar example of a resonator—the LC tank from basic circuit theory, shown in Figure 13.1. With either a nonzero initial voltage $v(0)$ present across the capacitor, or a nonzero initial current $i(0)$ flowing through the inductor, this idealized circuit has the solutions

$$v(t) = A_1 \cos \omega_0 t + A_2 \sin \omega_0 t \tag{13.1}$$

$$i(t) = \sqrt{\frac{C}{L}} \left(-A_2 \cos \omega_0 t + A_1 \sin \omega_0 t \right)$$

where A_1 and A_2 are constants determined by the initial conditions of the circuit, and

$$\omega_0 = \frac{1}{\sqrt{LC}} \tag{13.2}$$

is the *resonant frequency* of the circuit. Another way to obtain this solution is to start with the assumption of the time-harmonic voltage and current

$$v(t) = \mathrm{Re}\left(V e^{j\omega t} \right) \tag{13.3}$$

$$i(t) = \mathrm{Re}\left(I e^{j\omega t} \right)$$

and require that the circuit equations admit nontrivial solutions ($V \neq 0$ and $I \neq 0$). The result is that the phasor voltage and current are given by

$$V = A \tag{13.4}$$

$$I = \mp j A \sqrt{\frac{C}{L}}$$

where $A = A_1 - j A_2$, and $\omega = \pm \omega_0$.

These solutions are called the *free* or *natural oscillations* of the resonator of Figure 13.1, or more simply, its *modes*. They are analogous to the modes of a waveguide or transmission line, except that in those cases we assume a z-dependence of $e^{-\gamma z}$ and search for nontrivial

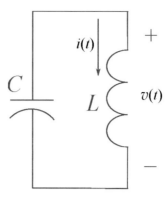

Figure 13.1 Resonant LC tank circuit.

field solutions, while here we have assumed an analogous exponential time dependence and searched for a nonzero response. The modes of a resonator are *source-free*—they exist at times t when any externally applied sources are turned off or disconnected from the circuit. In the absence of loss, as in our simple example, the modes will oscillate without decay for all time, as stored energy is merely exchanged between the inductor and the capacitor as time evolves.

In general, resonant elements have a more complex nature; they have losses, are usually driven by external sources and must be connected to external circuitry to perform any practical function. We consider resonators more generally in the remainder of this chapter.

13.2   LUMPED-ELEMENT RESONATORS

We will consider here a fairly general class of RLC networks as resonators, and afterwards illustrate by means of some examples. Consider a circuit made up only of resistors, capacitors, inductors and ideal voltage and current generators, such that we may choose a tree of the network with: (i) all tree branches consisting only of resistors, capacitors and/or current sources in parallel, and (ii) all links consisting only of resistors, inductors and voltage sources in series (see Appendix A for definitions of graph-theoretic terms relevant to circuit theory). Such a tree is called a *proper tree*. A section of such a circuit showing a typical tree branch and typical link is presented in Figure 13.2. From (A.3)-(A.4), the circuit equations

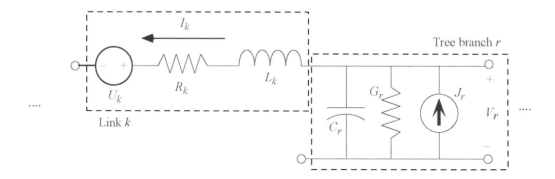

Figure 13.2 Portion of a resonator circuit showing tree branch r and link k.

for this network can be written in the form

$$[D][I_{\text{int}}] = ([G] + j\omega[C])[V_{\text{int}}] - [J] \tag{13.5}$$

$$-[D]^T[V_{\text{int}}] = ([R] + j\omega[L])[I_{\text{int}}] + [U] \tag{13.6}$$

where $[D]$ is the matrix described in Appendix A that specifies the connections in the network, $[I_{\text{int}}]$ is a column vector of internal link currents I_k, $[V_{\text{int}}]$ is a column vector of internal tree branch voltages V_r, $[J]$ is a column vector of the independent current sources J_r in the tree branches, and $[U]$ is a column vector of the independent voltage sources U_k in the links. We use the subscripts in $[V_{\text{int}}]$ and $[I_{\text{int}}]$ to distinguish these quantities internal to the resonator circuit from the voltages $[V]$ and currents $[I]$ at the ports of the network (as was done in Section 2.3). The matrices $[G]$, $[C]$, $[R]$ and $[L]$ are diagonal matrices whose entries are the values of tree branch conductance, tree branch capacitance, link resistance and link inductance, respectively.

13.2.1   NATURAL MODES OF AN ISOLATED RESONATOR

We first examine the free (or natural) modes of this network: that is, the possible nontrivial solutions that can exist in the absence of excitation (when the voltage sources are replaced by short circuits $[U] = 0$ and the current sources by open circuits $[J] = 0$). We can expect this to happen only for certain (generally complex) special values of the angular frequency ω called *natural frequencies*. For a mode designated by the index m, we denote these by $\Omega_m = \omega_m + js_m$, and the corresponding voltages $[V^{(m)}]$ and currents $[I^{(m)}]$ satisfy

$$[D][I^{(m)}] = ([G] + j\Omega_m[C])[V^{(m)}] \tag{13.7}$$

$$-[D]^T[V^{(m)}] = ([R] + j\Omega_m[L])[I^{(m)}] \tag{13.8}$$

These equations define a generalized eigenvalue problem wherein Ω_m is the eigenvalue and the set of vectors $([V^{(m)}], [I^{(m)}])$ is the generalized eigenvector.

Orthogonality properties are obtained by pre-multiplying $[V^{(n)}]^T$ by (13.7) and $[I^{(n)}]^T$ by (13.8) to give, respectively:

$$[V^{(n)}]^T[D][I^{(m)}] = [V^{(n)}]^T([G] + j\Omega_m[C])[V^{(m)}] \tag{13.9}$$

$$-[I^{(n)}]^T[D]^T[V^{(m)}] = [I^{(n)}]^T([R] + j\Omega_m[L])[I^{(m)}] \tag{13.10}$$

If the indices m and n are interchanged in (13.10), and the result added to (13.9), we obtain

$$0 = [V^{(n)}]^T([G] + j\Omega_m[C])[V^{(m)}] + [I^{(m)}]^T([R] + j\Omega_n[L])[I^{(n)}] \tag{13.11}$$

If we now interchange m and n in (13.11), and subtract the result from (13.11) itself, we arrive at the following identity:

$$(\Omega_n - \Omega_m)\left\{[I^{(m)}]^T[L][I^{(n)}] - [V^{(m)}]^T[C][V^{(n)}]\right\} = 0 \tag{13.12}$$

If $\Omega_n \neq \Omega_m$, we have the orthogonality property:

$$\left\{[I^{(m)}]^T[L][I^{(n)}] - [V^{(m)}]^T[C][V^{(n)}]\right\} = 0 \tag{13.13}$$

analogous to (3.51)-(3.53) for multiconductor transmission lines. If there are linearly independent voltage and current vectors satisfying (13.7) and (13.8) that have $\Omega_n = \Omega_m$, then these modes are said to be *degenerate*. Such cases are rare (but do include the case of critical damping in the language of circuit analysis). In any event, the degeneracy can always be removed by a small change in one or more of the component values, so we will only consider the case of nondegenerate modes from here on.

If Ω_m is a natural frequency of the resonator, then from (13.7) and (13.8) it can be seen that $-\Omega_m^* \equiv \Omega_{-m} = -\omega_m + js_m$ will also be a natural frequency, corresponding to the voltages and currents

$$[V^{(-m)}] = [V^{(m)}]^*; \qquad [I^{(-m)}] = [I^{(m)}]^* \tag{13.14}$$

If $\omega_m \neq 0$, the distinct pair of modes corresponding to $\Omega_{\pm m}$ is said to be underdamped (usually the case of most interest). In this case, the orthogonality property (13.13) together with (13.14) implies that

$$[I^{(m)}]^\dagger [L][I^{(m)}] = [V^{(m)}]^\dagger [C][V^{(m)}] \tag{13.15}$$

i.e., the time-average stored electric and magnetic energies in the circuit for this mode are equal. Putting $n = -m$ in (13.11) above, and using (13.14), we have, after some rearrangement:

$$s_m = \frac{[I^{(m)}]^\dagger [R][I^{(m)}] + [V^{(m)}]^\dagger [G][V^{(m)}]}{[I^{(m)}]^\dagger [L][I^{(m)}] + [V^{(m)}]^\dagger [C][V^{(m)}]} \tag{13.16}$$

In other words, the inverse time constant or decay constant for a mode is equal to one-half the ratio of the time-average dissipated power in the network to the time-average stored energy in the network.

If $\omega_m = 0$, so that $\Omega_m = js_m$ is purely imaginary, the mode is said to be overdamped, and it is possible to choose the amplitudes of all mode voltages and currents to be real. In this case, we adopt the convention that the indices $\pm m$ refer to the same mode.

13.2.2   INTERNAL EXCITATION OF A RESONATOR

Let next consider how independent voltage or current sources excite the modes of a resonator. We do not yet connect the resonator to an external circuit; all sources are assumed to be "inside" the resonant circuit in the sense that they appear as in Figure 13.2. Assuming that the modes form a complete set, any total voltage and current vectors can be expressed as a linear combination of them:

$$[V] = \sum_m d_m [V^{(m)}]; \qquad [I] = \sum_m d_m [I^{(m)}] \tag{13.17}$$

where d_m are complex constants to be determined. But this can be done by manipulations similar to those that led to the orthogonality property between modes. First we left multiply (13.5) and (13.6) by $[V^{(n)}]^T$ and $[I^{(n)}]^T$ respectively:

$$[V^{(n)}]^T [D][I] = [V^{(n)}]^T ([G] + j\omega[C])[V] - [V^{(n)}]^T [J]$$
$$-[I^{(n)}]^T [D]^T [V] = [I^{(n)}]^T ([R] + j\omega[L])[I] + [I^{(n)}]^T [U]$$

Similarly, we left multiply (13.7) and (13.8) (with $m \to n$) by $[V]^T$ and $[I]^T$ respectively:

$$[V]^T [D][I^{(n)}] = [V]^T ([G] + j\Omega_n[C])[V^{(n)}]$$
$$-[I]^T [D]^T [V^{(n)}] = [I]^T ([R] + j\Omega_n[L])[I^{(n)}]$$

The first and fourth of these equations are then subtracted from the sum of the second and third, giving

$$0 = j(\omega - \Omega_n)\left\{[I^{(n)}]^T [L][I] - [V^{(n)}]^T [C][V]\right\} + [I^{(n)}]^T [U] + [V^{(n)}]^T [J] \tag{13.18}$$

We now substitute the expansion (13.17) into (13.18) and use the orthogonality property of the modes, which allows us to solve for d_n:

$$d_n = \frac{j}{\omega - \Omega_n} \frac{[V^{(n)}]^T[J] + [I^{(n)}]^T[U]}{[I^{(n)}]^T[L][I^{(n)}] - [V^{(n)}]^T[C][V^{(n)}]}$$

Thus the total voltage and current vectors are given by

$$[V] = \sum_m \frac{j[V^{(m)}]}{\omega - \Omega_m} \frac{[V^{(m)}]^T[J] + [I^{(m)}]^T[U]}{[I^{(m)}]^T[L][I^{(m)}] - [V^{(m)}]^T[C][V^{(m)}]} \tag{13.19}$$

$$[I] = \sum_m \frac{j[I^{(m)}]}{\omega - \Omega_m} \frac{[V^{(m)}]^T[J] + [I^{(m)}]^T[U]}{[I^{(m)}]^T[L][I^{(m)}] - [V^{(m)}]^T[C][V^{(m)}]} \tag{13.20}$$

Note that this result is independent of any arbitrary amplitude we may have chosen for the mode voltages and currents themselves.

We see that any single mode's contribution to these sums is resonant. That is, its magnitude relative to its maximum value (which occurs at $\omega = \omega_m$—the resonant frequency of the mode) varies with frequency as

$$\frac{1}{|\omega - \Omega_m|} = \frac{1}{\sqrt{(\omega - \omega_m)^2 + s_m^2}}$$

The half-power (or $1/\sqrt{2}$-amplitude) points of this response occur when $|\omega - \omega_m| = s_m$, corresponding to a half-power bandwidth of $2s_m$. The *quality factor* or Q of the mode is then defined as the inverse of the relative bandwidth:

$$Q_m = \frac{|\omega_m|}{2s_m} \tag{13.21}$$

and provides a measure of how sharp this resonance is in the frequency response of the circuit. The Q of an overdamped mode is evidently 0.

Equations (13.19) and (13.20) also show us another way in which the excitation of a particular mode may be enhanced relative to the others. We can choose the excitation $[U]$ and $[J]$ to be orthogonal to all but one of the modes as follows:

$$[U] = [L][I^{(n)}]; \qquad [J] = [C][V^{(n)}]$$

The orthogonality property causes all the d_m to be zero except d_n.

If only one independent source is active in the resonator, we can use the foregoing to express the input impedance or admittance seen by this source. In the case when only the current source J_r in Figure 13.2 is active, it sees an impedance by (13.19) of

$$Z_{in,\,r} = \frac{V_r}{J_r} = \sum_m \frac{j(V_r^{(m)})^2}{(\omega - \Omega_m)([I^{(m)}]^T[L][I^{(m)}] - [V^{(m)}]^T[C][V^{(m)}])} \tag{13.22}$$

Likewise, when only the voltage source U_k is active, it sees an admittance by (13.20) of

$$Y_{in,\,k} = \frac{I_k}{U_k} = \sum_m \frac{j(I_k^{(m)})^2}{(\omega - \Omega_m)([I^{(m)}]^T[L][I^{(m)}] - [V^{(m)}]^T[C][V^{(m)}])} \tag{13.23}$$

Formally, eqn. (13.22) is identical to the impedance of a series string of $RLCG$ modules (each one made up of the parallel combination of a lossy inductor—an R and L in series—and

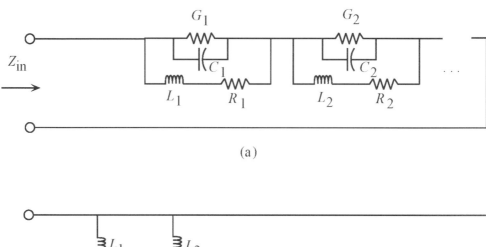

(a)

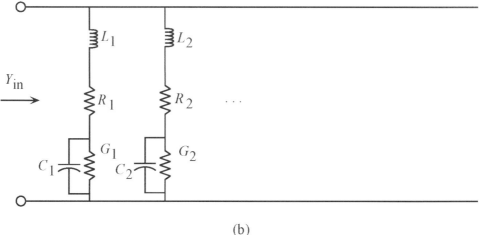

(b)

Figure 13.3 Equivalent networks of a lumped element resonant circuit: (a) first Foster canonical equivalent network; (b) second Foster canonical equivalent network.

a lossy capacitor—a G and C in parallel, and representing a single mode of the resonator) as shown in Figure 13.3(a). This is the so-called first Foster canonical equivalent network. Likewise, eqn. (13.23) is formally identical to the admittance of a parallel combination of modules each made up of the series combination of a lossy inductor and lossy capacitor as shown in Figure 13.3(b). This is the so-called second Foster canonical equivalent network. In either case, each module corresponds to the contribution from a single pair of underdamped modes of the resonator, or to that of a single overdamped mode (in which case the module degenerates into a GC or RL unit only).

We said that this equivalence was formal because it is possible that the resulting equivalent circuit might not be physically realizable, because negative values of resistance, inductance and so on would be required. For low-loss networks or modes with high enough Q, however, it can be shown that all values of R, L, G and C in these equivalent circuits are positive. We can, in any event, use the equivalent circuit as we would an actual one for most practical purposes. Note also that the component values for the two cases (Z_{in} and Y_{in}) are in general unrelated, depending as they do not only on the mode but also on the location and type of excitation of the network.

At frequencies near the isolated resonant frequency of a high-Q mode, one term of the equivalent circuit (or one term of the series (13.22) or (13.23)) will dominate all the others, and the network behaves as a simple four-element resonant circuit. For the link voltage source, we have

$$Z_{in} \simeq \frac{Z_{\max}}{1 + 2jQ_m(\omega - \omega_m)/\omega_{|m|}} \tag{13.24}$$

where

$$Z_{\max} = -\frac{(V_r^{(m)})^2}{s_m([I^{(m)}]^T[L][I^{(m)}] - [V^{(m)}]^T[C][V^{(m)}])}$$

while for the tree branch current source,

$$Y_{in} \simeq \frac{Y_{\max}}{1 + 2jQ_m(\omega - \omega_m)/\omega_{|m|}} \tag{13.25}$$

where

$$Y_{\max} = -\frac{(I_k^{(m)})^2}{s_m([I^{(m)}]^T[L][I^{(m)}] - [V^{(m)}]^T[C][V^{(m)}])}$$

This allows for considerable simplification of many resonator circuit analyses.

13.2.3 NATURAL MODES OF A LOADED RESONATOR

A resonator has no practical use unless it is connected to an external circuit. For example, filters and diplexers often employ resonators connected between a source and a load. The act of connecting a resonator to an external circuit will inherently modify its behavior, but it is usually desired to make such modification as small as possible. In this subsection, we will investigate the effect of connecting an external load to a lumped-element resonator. There are two basic ways to do this.

The first way is to create a port for connection to the resonator by making a *pliers entry* to the resonator: a branch of the resonator circuit is broken, and the two resulting terminals form a port, an example of which is shown in Figure 13.4. If a load impedance Z_L were connected to the port thus created, there would in general be a substantial perturbation to the resonant behavior of the network, unless $|Z_L|$ was sufficiently small. There are several possible ways to achieve efficient coupling to a load impedance that is not small. One way is to connect in parallel with the port a sufficiently large coupling admittance Y_{coup}, as shown in Figure 13.5(a). To avoid unnecessary losses, this coupling admittance is normally chosen to be imaginary. No matter what else is connected to the resonator's port, the resonant behavior should be close to that of the uncoupled resonator because a large admittance (close to a short circuit) will always appear across the port. Another way to provide coupling to a resonator via a pliers-entry port is to use an ideal transformer with a small turns ratio N as shown in Figure 13.5(b).

The second method of creating a port for external connection to the resonator is by *soldering-iron entry*: a pair of existing nodes of the resonator circuit is used to make the port, as shown in Figure 13.6. If an externally connected load at this port is to avoid making a large impact on the properties of the resonator, only small load admittances Y_L should

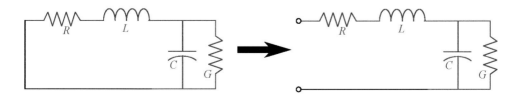

Figure 13.4 Creation of a resonator port by pliers entry.

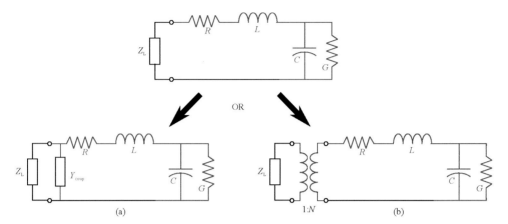

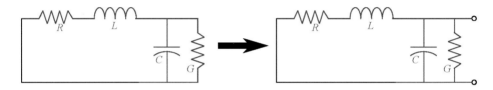

Figure 13.5 Small-perturbation coupling to a pliers-entry port: (a) using a large parallel admittance, (b) using an ideal transformer.

Figure 13.6 Creation of a resonator port by soldering-iron entry.

be connected there. Once again, we can guarantee this in a number of ways. For example, we could connect a large coupling impedance Z_{coup} in series with the port as shown in Figure 13.7(a), or connect the load through an ideal transformer with a large turns ratio N as shown in Figure 13.7(b).

It can be seen from these simple examples that the effect of creating an access port to the resonator and connecting a load will be to perturb either the series impedance of a link (in the case of pliers-entry) or the shunt admittance of a tree branch (in the case of soldering-iron entry). The change in resonant frequency of a given mode due to external loading can then be determined by perturbation theory, somewhat analogously to the determination of the change in propagation constant of a waveguide mode caused by the presence of wall losses (Section 9.5). Let the unloaded (unperturbed) resonator be described by (13.7) and (13.8), with an additional superscript zero placed on the eigenvalue and eigenvectors:

$$[D][I^{(m0)}] = ([G] + j\Omega_{m0}[C])[V^{(m0)}] \tag{13.26}$$

$$-[D]^{T}[V^{(m0)}] = ([R] + j\Omega_{m0}[L])[I^{(m0)}] \tag{13.27}$$

In general, the resonator could be loaded at any link k by adding a perturbation impedance $Z_{k,\mathrm{pert}}$ in series with that link so that

$$R_k + j\omega L_k \rightarrow R_k + j\omega L_k + Z_{k,\mathrm{pert}}$$

where, for example,

$$Z_{k,\mathrm{pert}} = \frac{1}{Y_{k,\mathrm{coup}} + \frac{1}{Z_{k,L}}}$$

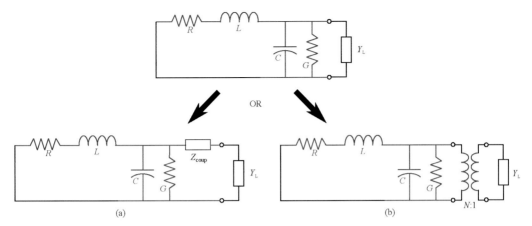

Figure 13.7 Small-perturbation to a soldering-iron-entry port: (a) using a large series admittance, (b) using an ideal transformer.

is the parallel combination of the coupling admittance in Figure 13.5(a) with the external load impedance $Z_{k,L}$. In a similar fashion, we could attach a load admittance $Y_{r,L}$ in series with the coupling impedance in the port created by soldering-iron entry to any tree branch r as in Figure 13.7(a) to get the perturbation admittance

$$Y_{r,\text{pert}} = \frac{1}{Z_{r,\text{coup}} + \frac{1}{Y_{r,L}}}$$

so that the tree branch admittance is modified to

$$G_k + j\omega C_r \rightarrow G_k + j\omega C_r + Y_{r,\text{pert}}$$

In general, we could create access ports at any place in the resonator, which would have the effect of replacing

$$[R] + j\omega[L] \rightarrow [R] + j\omega[L] + [Z_{\text{pert}}(\omega)]$$

and

$$[G] + j\omega[C] \rightarrow [G] + j\omega[C] + [Y_{\text{pert}}(\omega)]$$

in (13.26) and (13.27). Thus for the loaded resonator we have

$$[D][I^{(m)}] = \{[G] + j\Omega_m[C] + [Y_{\text{pert}}(\Omega_m)]\} [V^{(m)}] \qquad (13.28)$$

$$-[D]^T[V^{(m)}] = \{[R] + j\Omega_m[L] + [Z_{\text{pert}}(\Omega_m)]\} [I^{(m)}] \qquad (13.29)$$

A perturbation formula is constructed by multiplying (13.26) on the left by $[V^{(m)}]^T$, (13.27) by $[I^{(m)}]^T$, (13.28) by $[V^{(m0)}]^T$ and (13.29) by $[I^{(m0)}]^T$:

$$[V^{(m)}]^T[D][I^{(m0)}] = [V^{(m)}]^T([G] + j\Omega_{m0}[C])[V^{(m0)}] \qquad (13.30)$$

$$-[I^{(m)}]^T[D]^T[V^{(m0)}] = [I^{(m)}]^T([R] + j\Omega_{m0}[L])[I^{(m0)}] \qquad (13.31)$$

$$[V^{(m0)}]^T[D][I^{(m)}] = [V^{(m0)}]^T([G] + j\Omega_m[C] + [Y_{\text{pert}}(\Omega_m)])[V^{(m)}] \qquad (13.32)$$

$$-[I^{(m0)}]^T[D]^T[V^{(m)}] = [I^{(m0)}]^T([R] + j\Omega_m[L] + [Z_{\text{pert}}(\Omega_m)])[I^{(m)}] \qquad (13.33)$$

Since $[V^{(m)}]^T[D][I^{(m0)}] = [I^{(m0)}]^T[D]^T[V^{(m)}]$ and $[I^{(m)}]^T[D]^T[V^{(m0)}] = [V^{(m0)}]^T[D][I^{(m)}]$, adding (13.30) to (13.33) gives

$$0 = [V^{(m)}]^T([G] + j\Omega_{m0}[C])[V^{(m0)}] + [I^{(m0)}]^T([R] + j\Omega_m[L] + [Z_{\text{pert}}(\Omega_m)])[I^{(m)}] \quad (13.34)$$

and adding (13.31) to (13.32) gives

$$0 = [I^{(m)}]^T([R] + j\Omega_{m0}[L])[I^{(m0)}] + [V^{(m0)}]^T([G] + j\Omega_m[C] + [Y_{\text{pert}}(\Omega_m)])[V^{(m)}] \quad (13.35)$$

If (13.35) is subtracted from (13.34), since the matrices $[Z_{\text{pert}}(\omega)]$, $[Y_{\text{pert}}(\omega)]$, $[R]$, $[L]$, $[C]$ and $[G]$ are all diagonal we get after some rearrangement

$$\Omega_m - \Omega_{m0} = j\frac{[I^{(m)}]^T[Z_{\text{pert}}(\Omega_m)][I^{(m0)}] - [V^{(m)}]^T[Y_{\text{pert}}(\Omega_m)][V^{(m0)}]}{[I^{(m)}]^T[L][I^{(m0)}] - [V^{(m)}]^T[C][V^{(m0)}]} \quad (13.36)$$

which is exact. Finally, we make the perturbation approximation $[V^{(m)}] \simeq [V^{(m0)}]$, $[I^{(m)}] \simeq [I^{(m0)}]$ and $\Omega_m \simeq \Omega_{m0}$ on the right side of (13.36) to get

$$\Omega_m - \Omega_{m0} \simeq j\frac{[I^{(m0)}]^T[Z_{\text{pert}}(\Omega_{m0})][I^{(m0)}] - [V^{(m0)}]^T[Y_{\text{pert}}(\Omega_{m0})][V^{(m0)}]}{[I^{(m0)}]^T[L][I^{(m0)}] - [V^{(m0)}]^T[C][V^{(m0)}]} \quad (13.37)$$

which depends only on the properties of the unloaded resonator, and the known perturbation impedances and admittances.

13.2.4   EXTERNAL STEADY-STATE EXCITATION OF A RESONATOR

Not only is it necessary to connect external loads to resonators in order to realize some practical function, but external sources are required for the operation of devices, not only such as filters, but also oscillators and other active devices. To model such configurations, we thus need to consider a real generator (with a nonzero equivalent impedance) connected externally to the resonator, instead of just ideal current or voltage generators placed within the resonator. Because of the Thévenin and Norton equivalence theorems, it is sufficient to consider either an ideal voltage source with a series impedance, or an ideal current source with a shunt admittance as the external excitation. In the former case, the excitation is inserted by pliers entry to the resonator: a branch of the resonator circuit is broken, and the Thévenin equivalent circuit is inserted in series with this branch. In the latter case, the excitation is connected by a soldering-iron entry: the Norton equivalent circuit of the external generator is connected across a pair of existing nodes of the resonator circuit. If the external generator is to avoid making a large impact on the properties of the resonator, the generator impedance should be low for a pliers entry, and high (i. e., low admittance) for a soldering-iron entry.

13.2.5   EXAMPLE: *RLCG* RESONATOR

Consider the $RLCG$ circuit shown in Figure 13.8. If the voltage source V_G and its associated impedance Z_G (which have been inserted using a pliers entry) are removed from the circuit by replacing them with a short circuit, we have the elementary component of the equivalent circuit in Figure 13.3(a), representing a single mode of a general resonant circuit driven by a current source. If on the other hand the current source I_G and its associated admittance Y_G (which are connected by a soldering-iron entry) are removed from the circuit, we have the elementary component of the equivalent circuit in Figure 13.3(b) driven by a voltage source. There is one tree branch and one link in this circuit: the tree branch voltage is that across the capacitor $V_1 = V_C$, and the link current is that through the inductor $I_1 = I_L$.

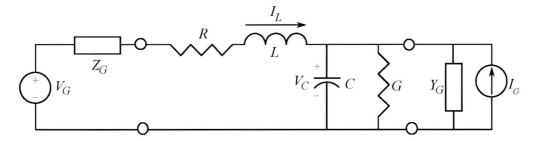

Figure 13.8 An $RLGC$ circuit.

The resonant modes satisfy (13.7) and (13.8):

$$
\begin{aligned}
I_1^{(m)} &= (G + j\Omega_m C)V_1^{(m)} \\
-V_1^{(m)} &= (R + j\Omega_m L)I_1^{(m)}
\end{aligned}
\tag{13.38}
$$

which are easily solved to give a pair of modes. If

$$
\frac{1}{LC} > \frac{1}{4}\left(\frac{G}{C} - \frac{R}{L}\right)^2
\tag{13.39}
$$

then the modes form an underdamped pair with

$$
\Omega_{\pm 1} = \pm\omega_1 + js_1 = \pm\sqrt{\frac{1}{LC} - \frac{1}{4}\left(\frac{G}{C} - \frac{R}{L}\right)^2} + \frac{j}{2}\left(\frac{G}{C} + \frac{R}{L}\right)
\tag{13.40}
$$

while if

$$
\frac{1}{LC} < \frac{1}{4}\left(\frac{G}{C} - \frac{R}{L}\right)^2
\tag{13.41}
$$

the modes are overdamped, with the imaginary resonant frequencies

$$
\Omega_{1,2} = \pm j\sqrt{\frac{1}{4}\left(\frac{G}{C} - \frac{R}{L}\right)^2 - \frac{1}{LC}} + \frac{j}{2}\left(\frac{G}{C} + \frac{R}{L}\right)
\tag{13.42}
$$

For the underdamped modes, the quality factor is

$$
Q = \frac{\omega_1}{\dfrac{G}{C} + \dfrac{R}{L}}
\tag{13.43}
$$

For high-Q modes, $\omega_1 \simeq 1/\sqrt{LC}$. In this case, the maximum admittance seen by the voltage generator V_G in the absence of the generator impedance Z_G is, from (13.25),

$$
Y_{\max} \simeq \frac{1}{R + \dfrac{LG}{C}}
\tag{13.44}
$$

and the maximum impedance seen by the current generator I_G in the absence of Y_G is by (13.24)

$$
Z_{\max} \simeq \frac{1}{G + \dfrac{CR}{L}}
\tag{13.45}
$$

The insertion of the actual voltage generator V_G and Z_G, aside from inducing voltages and currents in the resonator, also perturbs its natural frequencies. For, even if the ideal source V_G were zero, the added impedance Z_G to the RL link would result in more resistance: $R \to R + R_G$, and a change in the link reactance from ωL to $\omega L + X_G$. If the mode has a high Q, so that $|\Omega_m L| \gg R$ and $|\Omega_m C| \gg G$, and the generator impedance does not vary rapidly near the unperturbed resonant frequencies, then the effect of the generator impedance on the natural modes of the circuit is to change $R \to R + R_G$ as above and $L \to L + X_G/|\omega_m|$. This changes the real parts $\omega_{\pm 1}$ of (13.40) by small amounts, due mostly to X_G. The imaginary part s_1 may be more significantly changed, since now

$$s_1 = \frac{1}{2}\left(\frac{G}{C} + \frac{R + R_G}{L} \right)$$

In fact, the quality factor has been changed from the *unloaded* Q (or Q_U) of (13.43) to the *loaded* Q or Q_L, which is found from

$$\frac{1}{Q_L} = \frac{1}{Q_U} + \frac{1}{Q_e} \tag{13.46}$$

where Q_e is called the *external* Q, and is related to the additional resistance introduced into the network by the external generator:

$$\frac{1}{Q_e} = \frac{R_G/L}{\omega_1} \tag{13.47}$$

We define the *coupling factor*

$$\kappa = Q_U/Q_e \tag{13.48}$$

between the inserted generator and the resonator, so that

$$\frac{1}{Q_L} = \frac{1 + \kappa}{Q_U} \tag{13.49}$$

Maximum power transfer from the inserted generator to the resonator occurs when $R_G = 1/Y_{\max}$, or from (13.44), (13.47) and (13.49),

$$Q_e = Q_U \quad \text{or} \quad \kappa = 1 \tag{13.50}$$

This is known as the case of critical coupling, and is akin to the situation in circuit theory when maximum power transfer is achieved by a conjugate impedance match.

13.3   CAVITY RESONATORS FROM SECTIONS OF METALLIC WAVEGUIDE

A waveguide can be made to work as a resonator by connecting short circuits (i.e., perfectly conducting walls) at both ends of a uniform section of the waveguide with length H. For a hollow metallic waveguide, such a resonator is called a *cavity resonator*, examples of which are shown in Figure 13.9. Since each mode of an arbitrary waveguide has been shown in Section 9.3 to be equivalent to a transmission line, the modes of such cavities can be analyzed using the transmission-line equivalent circuit of Figure 13.10. The resonator of Figure 13.10 can thus be used to represent a wide variety of resonant structures. For each different waveguide mode with index m, the propagation constant of the transmission line should be replaced by that of the waveguide mode: $\gamma_m(\omega)$.

Ignoring losses for the moment, and assuming the waveguide mode to be above cutoff, we have $\gamma_m(\omega) = j\beta_m(\omega)$, where

$$\beta_m(\omega) = \sqrt{\omega^2 \mu\epsilon - k_{c,m}^2} \tag{13.51}$$

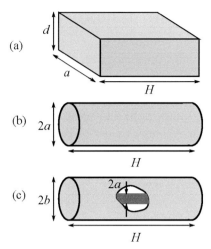

Figure 13.9 Cavity resonators formed by short-circuited sections of waveguide: (a) rectangular, (b) circular cylindrical, (c) coaxial.

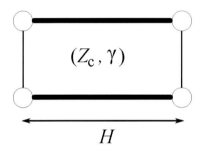

Figure 13.10 Short-circuit transmission-line resonator.

and $k_{c,m}$ is the cutoff frequency of mode m. Nontrivial voltages and currents on the source-free transmission line will occur only if the length is an integer number of half-wavelengths, so the resonant frequencies will be determined by

$$\beta_m(\omega_{mp})H = \pi p \qquad p = 0, \pm 1, \pm 2, \dots \tag{13.52}$$

Solving for ω_{mp} we obtain

$$\omega_{mp} = \frac{1}{\sqrt{\mu\epsilon}} \sqrt{k_{c,m}^2 + \left(\frac{\pi p}{H}\right)^2} \tag{13.53}$$

These cavity modes are typically denoted with the mode number of the corresponding waveguide mode, along with the longitudinal index p. Thus, for example, the TE_{10} (rectangular) waveguide mode generates the TE_{10p} modes of the cavity, whose resonant frequencies are

$$\omega_{\mathrm{TE}_{10p}} = \frac{1}{\sqrt{\mu\epsilon}} \sqrt{\left(\frac{\pi}{a}\right)^2 + \left(\frac{\pi p}{H}\right)^2} \tag{13.54}$$

In the following example, we will examine resonant modes of this type in more detail, including the effect of losses.

13.3.1 EXAMPLE: SHORT-CIRCUITED SECTION OF LOSSY TRANSMISSION LINE

Consider the example of a length H of lossy transmission line short-circuited at one end as shown in Figure 13.11. The input impedance seen at the other end is

$$Z_{\text{in}} = Z_c \tanh \gamma H \tag{13.55}$$

where Z_c and γ are given by (1.8), (1.9) and (1.5), where r, l, c and g are assumed to be frequency-independent. The resonant modes of Figure 13.10 will occur at the resonant frequencies that obey $Z_{\text{in}} = 0$, or

$$\gamma H = jp\pi \tag{13.56}$$

for any integer p.

For the special case when $p = 0$, the factors $\sqrt{g + j\omega c}$ in Z_{in} will cancel, leaving the special case $r + j\omega l = 0$, or the purely damped complex resonant frequency

$$\Omega_0 = js_0 = j\frac{r}{l} \tag{13.57}$$

that is independent of H. The damping coefficient s_0 will be recognized as the inverse of the RL time constant for the loop circuit containing a resistance rH and inductance lH, formed when the transmission line is shorted at both ends.

For $p \neq 0$, we can square both sides of (13.56) and solve the resulting quadratic equation in ω to obtain the complex resonant frequencies

$$\Omega_{\pm p} = \pm\omega_p + js_p \tag{13.58}$$

If we define the resonant frequencies in the lossless limit as

$$\omega_{0p} = \frac{p\pi}{H\sqrt{lc}} \tag{13.59}$$

then for

$$\omega_{0p} > \frac{1}{2}\left|\frac{g}{c} - \frac{r}{l}\right| \tag{13.60}$$

the mode is underdamped (compare with condition (13.39)), and

$$\omega_p = \sqrt{\omega_{0p}^2 - \frac{1}{4}\left(\frac{g}{c} - \frac{r}{l}\right)^2} \tag{13.61}$$

and

$$s_p = \frac{1}{2}\left(\frac{g}{c} + \frac{r}{l}\right) \tag{13.62}$$

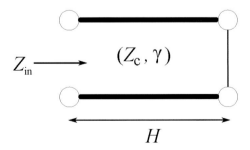

Figure 13.11 Short-circuited transmission-line section.

For high-order modes with $\omega_p \simeq \omega_{0p} \gg s_p$, the unloaded Q of the underdamped modes is

$$Q_{U,p} \simeq \frac{p\pi}{2s_p H \sqrt{lc}} \tag{13.63}$$

which increases as the mode number increases since all the s_p's are equal. If

$$\omega_{0p} < \frac{1}{2} \left| \frac{g}{c} - \frac{r}{l} \right| \tag{13.64}$$

then the mode is overdamped (compare with (13.41)), so that $\omega_p = 0$ and

$$s_p = \frac{1}{2} \left(\frac{g}{c} + \frac{r}{l} \right) \pm \sqrt{\frac{1}{4} \left(\frac{g}{c} - \frac{r}{l} \right)^2 - \omega_{0p}^2} \tag{13.65}$$

13.3.2 MIRROR RESONATORS

Lasers require resonators for optical frequencies to lock the oscillations to a particular frequency. The simplest resonator of this type is the Fabry-Perot resonator shown in Figure 13.12(a). In this structure, beams of light that are approximately plane waves are reflected back and forth between the conducting mirrors. The structure is approximately represented by the transmission line of Figure 13.10. As above, resonance occurs when the distance between the mirrors is an integer multiple p of a half-wavelength (at optical frequencies, p is typically very large). Because the optical beams are not exactly plane waves, they have a tendency to diverge (broaden their width), and losses can occur when part of the energy is lost due to a fraction of the field passing beyond the edges of the mirrors. This effect can be reduced by curving the mirrors so as to focus the beams, reducing the field amplitude at the mirror edges, as shown in Figure 13.12(b). Dielectric lenses can also be used for this purpose. In practice, the mirrors of the resonator are made to have a small transparency (a little less than perfect reflection) in order to be able to couple energy out of the resonant cavity. Alternatively, small coupling apertures are sometimes used for this purpose.

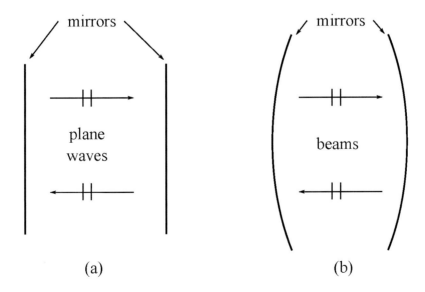

Figure 13.12 Fabry-Perot resonators: (a) plane mirrors; (b) curved mirrors.

13.3.3   NATURAL MODES OF RESONATORS COUPLED TO EXTERNAL CIRCUITS

Like that of the ideal, uniform waveguide of infinite length, the notion of a completely isolated resonator is an idealization: useful for visualizing the physical processes involved in the resonance of the structure, but quite a bit removed from the actual configuration as found in practice. This is because to be of use, the resonator must be coupled to the outside world somehow, if its frequency selective behavior is to find practical application. For lumped-element resonators, we have studied the coupling process in Section 13.2.4.

For a hollow rectangular waveguide cavity, we might cut a small aperture in one of its walls, thereby connecting the cavity to a section of waveguide as shown in Figure 13.13. In Section 11.6.1, we saw that the transmission-line equivalent of this structure for the TE_{10} mode is a small impedance connected in shunt across the transmission line at the location of the aperture plane A, so an equivalent circuit for Figure 13.13 is that shown in Figure 13.14, in which the equivalent aperture inductance is given by

$$L_a = \frac{Z_{c1}\Delta}{2\omega} = \frac{2\mu\alpha_{M,xx}}{ad}\sin^2\frac{\pi x_0}{a} \tag{13.66}$$

where Δ is given by (11.81) and $Z_{c1} = j\omega\mu/\gamma_1$ is the characteristic impedance of the transmission line equivalent to the TE_{10} (rectangular) waveguide mode. If the coupling is reduced to zero by closing off the aperture ($L_a \to 0$), the cavity to the right of A is simply the isolated resonator studied in Sections 13.3.1 and 13.3 above.

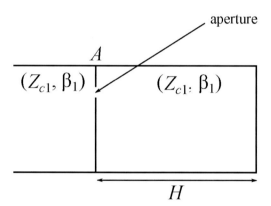

Figure 13.13 Rectangular waveguide section (cavity) coupled by a small aperture to a hollow rectangular waveguide.

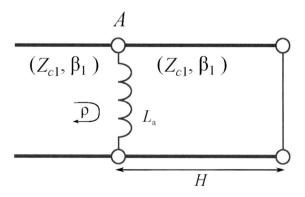

Figure 13.14 Equivalent circuit for Figure 13.13.

If the transmission line to the left of A is of infinite length (or terminated in a matched load Z_{c1}), and there are as yet no sources connected to the system, we may regard the loaded resonator of Figure 13.14 as a resonant system in its own right. We can again ask what complex resonant frequencies it possesses. To answer this question, consider the scattering matrix of the coupling inductance, in which $S_{11} = S_{22}$ and $S_{21} = S_{12}$ are given by (11.80). With no sources present, $a_1 = 0$ (cf. the figure), while a_2 and b_2 are related by traversing a round trip of length $2H$ through the waveguide section and accounting for the reflection coefficient of -1 at the short circuit on the far right:

$$a_2 = -e^{-2\gamma_1 H} b_2 \tag{13.67}$$

Enforcing the scattering matrix relation at A then gives

$$
\begin{bmatrix} b_1 \\ b_2 \end{bmatrix} =
\begin{bmatrix} S_{11} & S_{12} \\ S_{12} & S_{11} \end{bmatrix}
\begin{bmatrix} 0 \\ -e^{-2\gamma_1 H} b_2 \end{bmatrix}
\tag{13.68}
$$

In order for a nontrivial solution ($b_2 \neq 0$) of (13.68) to exist, the condition

$$1 = -S_{11} e^{-2\gamma_1 H} \tag{13.69}$$

must be satisfied. Using (11.80), we can write this as

$$\gamma_1(\omega_p) H + \frac{1}{2} \ln(1 + j\Delta) = jp\pi \tag{13.70}$$

Evidently, in the absence of coupling ($\Delta = 0$), this reduces to equation (13.56) or (13.52) for the resonant frequencies of the isolated cavity.

Since $\Delta \ll 1$ for the case of a small aperture, the logarithm in (13.70) can be expanded using the first few terms of its Taylor series expansion:

$$\ln(1 + j\Delta) \simeq j\Delta + \frac{\Delta^2}{2} \tag{13.71}$$

Expressing Δ in terms of L_a and γ_1 by means of (13.66), and using (13.71), we obtain a quadratic equation for $\gamma_1(\omega)$:

$$\gamma_1(\omega_p)(H + H_{\mathrm{ex}}) + \gamma_1^2(\omega_p) H_{\mathrm{ex}}^2 = jp\pi \tag{13.72}$$

where

$$H_{\mathrm{ex}} = \frac{L_a}{\mu} \tag{13.73}$$

is a kind of "length extension" of the resonator caused by fields penetrating through the aperture. We solve (13.72) for γ_1:

$$\gamma_1(\omega_p) = \frac{2\pi p j}{H + H_{\mathrm{ex}} + \sqrt{(H + H_{\mathrm{ex}})^2 - 4\pi p j H_{\mathrm{ex}}^2}} \tag{13.74}$$

where we have chosen the positive sign in front of the square root because the other choice will lead to a resonant frequency with a negative imaginary part; a natural mode that grows exponentially with time is impossible in a passive system. For H_{ex} sufficiently small, we can approximate (13.74) by

$$\gamma_1(\omega_p) = \frac{\pi p j}{H + H_{\mathrm{ex}}} \left[1 + \pi p j \left(\frac{H_{\mathrm{ex}}}{H + H_{\mathrm{ex}}} \right)^2 \right] \tag{13.75}$$

Using

$$\omega^2 = \frac{1}{\mu\epsilon}\left[\left(\frac{\pi}{a}\right)^2 - \gamma_1^2\right]$$

and still assuming H_{ex} to be small, we arrive at the complex resonant frequencies

$$\Omega_{\pm p} = \omega_{\pm p} + js_p \simeq \pm\frac{1}{\sqrt{\mu\epsilon}}\sqrt{\left(\frac{\pi}{a}\right)^2 + \left(\frac{p\pi}{H'}\right)^2} + j\frac{H_{\text{ex}}^2}{H'\sqrt{\mu\epsilon}}\frac{\left(\frac{p\pi}{H'}\right)^3}{\sqrt{\left(\frac{\pi}{a}\right)^2 + \left(\frac{p\pi}{H'}\right)^2}} \tag{13.76}$$

for these TE_{10p} modes ($p \neq 0$), where

$$H' = H + H_{\text{ex}} \tag{13.77}$$

is an "effective length", slightly larger than the physical length of the cavity.

Two important features of (13.76) should be noted. First, the real parts $\omega_{\pm p}$ of $\Omega_{\pm p}$ are shifted from their values for a lossless, isolated (uncoupled) cavity; the length H' appears in place of H. We may interpret this to mean that the small amount of field penetration out of the cavity through the aperture has the effect of making the cavity appear longer than its isolated counterpart. Second, the resonant frequencies are now complex, even though the cavity itself has no losses. In the transmission-line resonator example of Section 13.3.1, the appearance of an imaginary part of the resonant frequency was due to intrinsic ohmic losses (r or g) of the resonator itself. In the present case, it is due to the presence of a nonzero wave amplitude carrying power away from the resonant portion of the structure. This kind of loss is described in terms of the external quality factor Q_e of a cavity that is coupled to an external circuit. For the present configuration, since $|\omega_{\pm p}| \gg s_p$ and using (13.21) to define Q_e, we have

$$Q_e \simeq \frac{H'^2}{2p\pi H_{\text{ex}}^2}\left[1 + \left(\frac{H'^2}{pa}\right)^2\right] \tag{13.78}$$

The Q of an isolated resonator (due only to intrinsic losses) is the unloaded Q, denoted by Q_U, similar to what appears in the consideration of external excitation of a resonator in Section 13.2.4. The Q of a resonator coupled to an exterior load (with both intrinsic losses as well as power lost to the external region) is once again called the loaded Q, and is designated Q_L. As shown already in (13.46) for a special case, we can show more generally that

$$\frac{1}{Q_L} = \frac{1}{Q_U} + \frac{1}{Q_e} \tag{13.79}$$

approximately, if the quality factors are sufficiently large. Obviously, Q_e and Q_L will depend on the manner in which the cavity is coupled to the outside; Q_U on the other hand depends not at all on the coupling mechanism. Let us note in passing that high-Q modes (for which $|\omega| \simeq |\omega_p| \gg s_p$) have complex resonant frequencies of the form

$$\Omega_p \simeq \omega_p\left(1 + \frac{j}{2Q}\right) \tag{13.80}$$

where Q is either Q_U or Q_L for the case when the resonator is unloaded or loaded, as appropriate.

13.3.4   STEADY-STATE EXCITATION

The natural oscillations of loaded or unloaded resonators provide a means for describing the response of a resonant system *after* the sources have been turned off. But as in Section 13.2.4, we also need to find the steady-state response of a resonant system driven by time-harmonic sources at a real operating frequency ω. The situation is analogous to that of excitation of a waveguide or transmission line by sources distributed in space (see Sections 3.2 and 10.4). For values of z where sources are present, the z-dependence of the source-free mode fields must be different than simply $e^{\pm \gamma_m z}$. So too in a cavity, at instants of time when sources are operating, the time dependences of the natural modes must be modified from $e^{j\Omega_m t}$. In fact, for time-harmonic sources in the steady state, their time dependence $e^{j\omega t}$ must carry over to the entire field. While the general problem of excitation of cavities and other types of resonator will be treated later in the chapter, we can introduce most of the essential ideas in this problem using the simple resonator system of Figure 13.14.

Imagine that a generator is connected to the transmission line on the left side of Figure 13.14 in such a way that a wave amplitude a_1 is incident on the aperture plane A from the left. We will again allow intrinsic losses within the cavity so that we replace β_1 by $-j\gamma_1 = \beta_1 - j\alpha_1$ wherever appropriate. Thus the wave amplitudes within the cavity are related by (13.67). Now let ρ be the overall reflection coefficient in the left hand waveguide in Figure 13.14, defined by $b_1 = \rho a_1$. Instead of (13.68), the S-matrix relation at the aperture plane now takes the form

$$
\begin{bmatrix} \rho a_1 \\ b_2 \end{bmatrix} = \begin{bmatrix} S_{11} & S_{12} \\ S_{12} & S_{11} \end{bmatrix} \begin{bmatrix} a_1 \\ -e^{-2\gamma_1 H} b_2 \end{bmatrix}
\tag{13.81}
$$

For a nontrivial solution $(a_1, b_2 \neq 0)$ of (13.81), the determinant of the system (13.81) must be zero:

$$
\det \left\{ \begin{bmatrix} S_{11} & S_{12} \\ S_{12} & S_{11} \end{bmatrix} \begin{bmatrix} 1 & 0 \\ 0 & -e^{-2\gamma_1 H} \end{bmatrix} - \begin{bmatrix} \rho & 0 \\ 0 & 1 \end{bmatrix} \right\}
\tag{13.82}
$$

or explicitly,

$$
\rho = \frac{S_{11} + (S_{11}^2 - S_{12}^2)e^{-2\gamma_1 H}}{1 + S_{11}e^{-2\gamma_1 H}}
\tag{13.83}
$$

and thence from the second row of (13.81) we get

$$
b_2 = \frac{S_{12}}{1 + S_{11}e^{-2\gamma_1 H}} a_1
\tag{13.84}
$$

Our analysis will be simplified somewhat by using (13.66) to rewrite eqns. (11.80)-(11.81) in the form

$$
\begin{aligned}
S_{12} &= j\kappa e^{-2j\beta_1 L_a/\mu} \\
S_{11} &= -\sqrt{1 - \kappa^2} e^{-2j\beta_1 L_a/\mu}
\end{aligned}
\tag{13.85}
$$

where the coupling parameter κ is defined by

$$
\kappa = \frac{\Delta}{\sqrt{1 + \Delta^2}} \simeq \Delta
\tag{13.86}
$$

the last approximation being valid because of the condition $\Delta \ll 1$. Let us assume that the cavity losses are small: $2\alpha_1 H \ll 1$. Then we can write (13.83) and (13.84) as

$$
\rho = e^{-2j\beta_1 L_a/\mu} \left(\frac{e^{-2j\beta_1 H' - 2\alpha_1 H'} - \sqrt{1 - \kappa^2}}{1 - \sqrt{1 - \kappa^2} e^{-2j\beta_1 H' - 2\alpha_1 H'}} \right)
\tag{13.87}
$$

and

$$b_2 = e^{-2j\beta_1 L_a/\mu} \left(\frac{j\kappa}{1 - \sqrt{1 - \kappa^2} e^{-2j\beta_1 H' - 2\alpha_1 H'}} \right) a_1 \qquad (13.88)$$

where H' is given by (13.77) and the quantity $2\alpha_1(H' - H) = 2\alpha_1 H_{\text{ex}}$ has been considered negligible compared to 1. From (13.87) and (13.88), we can express the magnitude squared of the input reflection coefficient of the cavity as

$$|\rho|^2 = 1 - \left(1 - e^{-4\alpha_1 H'}\right) \left| \frac{b_2}{a_1} \right|^2 \qquad (13.89)$$

The magnitude of the ratio b_2/a_1 will be a maximum (i.e., the magnitude of the signal introduced into the cavity is maximized) when the frequency is such that

$$\beta_1 H' = p\pi, \qquad (13.90)$$

that is, when the (real) operating frequency ω is equal to the real part of a complex resonant frequency Ω_p as given in (13.76). This is the *resonance condition* for forced oscillations, and represents the closest we can come to the complex natural frequency of the resonator at a real operating frequency. At such frequencies,

$$\left| \frac{b_2}{a_1} \right|^2_{\text{max}} = \frac{\kappa^2}{(1 - \sqrt{1 - \kappa^2} e^{-2\alpha_1 H'})^2} \qquad (13.91)$$

and we see that at resonance the input reflection coefficient is minimized:

$$|\rho|^2_{\text{min}} = \frac{(e^{-2\alpha_1 H'} - \sqrt{1 - \kappa^2})^2}{(1 - \sqrt{1 - \kappa^2} e^{-2\alpha_1 H'})^2} \qquad (13.92)$$

or in other words provides the best match of the cavity/coupling system to the waveguide for given values of L_a and α_1. If the coupling coefficient κ can be adjusted (by choosing the size or position of the aperture, for example), then the reflection coefficient can be made equal to zero by requiring that

$$\kappa = \kappa_{\text{crit}} \equiv \sqrt{1 - e^{-4\alpha_1 H'}} \simeq 2\sqrt{\alpha_1 H'} \quad (\text{if } \alpha_1 H' \ll 1) \qquad (13.93)$$

In this case all the power of the incident wave a_1 is dissipated in the losses inside the cavity. If $\kappa > \kappa_{\text{crit}}$, the cavity is said to be *overcoupled*, while if $\kappa < \kappa_{\text{crit}}$, the cavity is *undercoupled*.

Under the critical coupling condition at resonance, we see from (13.91) that

$$\left| \frac{b_2}{a_1} \right|_{\text{max}} = \frac{1}{\kappa}, \qquad (13.94)$$

meaning that a huge standing wave exists in the cavity relative to the wave on the waveguide that feeds it. Reflection in the feed waveguide has been suppressed because the small fraction of this huge standing wave that penetrates through the aperture out into the feed waveguide is exactly canceled by the substantial portion of a_1 that is reflected by the aperture alone (in the absence of the cavity). When the cavity is undercoupled, the wave reflected from the diaphragm is dominant, while in the overcoupled case, more energy spills out of the aperture than is reflected by the diaphragm. In either case, the result is a mismatch in the feed waveguide.

Another viewpoint for this problem can be achieved by devising an equivalent circuit for the load presented by the coupled cavity to the waveguide containing the incident wave.

The factor $e^{-2j\beta_1 L_a/\mu}$ in (13.87) can be accounted for by including a short length H_{ex} of transmission line to the left of the terminal plane A. The remaining factor (call it ρ') in parentheses in (13.87) can be interpreted as the impedance

$$Z_{11} = Z_{c1}\frac{1+\rho'}{1-\rho'} = Z_{c1}\frac{1-\sqrt{1-\kappa^2}}{1+\sqrt{1-\kappa^2}}\coth(\alpha_1 H' + j\beta_1 H') \qquad (13.95)$$

The second factor in (13.95),

$$Z_{\mathrm{oc}} = Z_{c1}\coth(\alpha_1 H' + j\beta_1 H'), \qquad (13.96)$$

is the input impedance of an *open-circuited* length H' of lossy waveguide in the absence of the coupling diaphragm (which depends on the details of coupling only through the small length extension H_{ex}). Near a resonant frequency, Z_{oc} can be approximately characterized as an LCG network. To see this, assume $\alpha_1 H'$ is small and let ω be close to ω_p—the frequency at which $\beta_1 H' = p\pi$. Then

$$\beta_1(\omega) \simeq \beta_1(\omega_p) + (\omega-\omega_p)\beta_1'(\omega_p) = \frac{p\pi}{H'} + \frac{\omega-\omega_p}{v_{g,p}}$$

where $v_{g,p} = 1/\beta_1'(\omega_p)$ is the group velocity of the waveguide mode at the frequency ω_p. From (13.96), we then get

$$\begin{aligned}
Z_{\mathrm{oc}} &\simeq Z_{c1}\coth\left[\alpha_1 H' + jp\pi + j(\omega-\omega_p)\frac{H'}{v_{g,p}}\right] = Z_{c1}\coth\left[\alpha_1 H' + j(\omega-\omega_p)\frac{H'}{v_{g,p}}\right] \\
&\simeq \frac{Z_{c1}}{\alpha_1 H' + j(\omega-\omega_p)\frac{H'}{v_{g,p}}} = \frac{Z_{c1}}{\alpha_1 H'}\frac{1}{1 + j\frac{\omega-\omega_p}{v_{g,p}\alpha_1}}
\end{aligned} \qquad (13.97)$$

having used the small argument limit of the hyperbolic cotangent function. On the other hand, if we consider the parallel LCG network of Figure 13.8, with the voltage generator replaced by a short circuit and R taken to be zero, the current generator in the resulting network sees the equivalent circuit shown in Figure 13.15. The impedance of that equivalent circuit is

$$Z_{\mathrm{eq}} = \frac{j\omega L_{\mathrm{eq}}}{1 - \omega^2 L_{\mathrm{eq}}C_{\mathrm{eq}} + j\omega L_{\mathrm{eq}}G_{\mathrm{eq}}} \simeq \frac{1}{G_{\mathrm{eq}}}\frac{1}{1 + j\frac{2C_{\mathrm{eq}}}{G_{\mathrm{eq}}}\left(\omega - \frac{1}{\sqrt{L_{\mathrm{eq}}C_{\mathrm{eq}}}}\right)} \qquad (13.98)$$

if $G_{\mathrm{eq}}\sqrt{\frac{L_{\mathrm{eq}}}{C_{\mathrm{eq}}}} \ll 1$ and ω is close to $1/\sqrt{L_{\mathrm{eq}}C_{\mathrm{eq}}}$. Comparing (13.97) with (13.98), we obtain the following values for the elements in the equivalent circuit of Figure 13.15:

$$G_{\mathrm{eq}} = \frac{1}{R_{\mathrm{eq}}} = \frac{\alpha_1 H'}{Z_{c1}} \qquad (13.99)$$

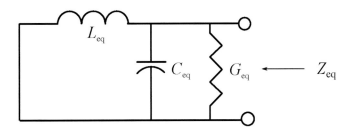

Figure 13.15   An LGC equivalent circuit.

$$C_{eq} = \frac{H'}{2Z_{c1}v_{g,p}} \tag{13.100}$$

$$L_{eq} = \frac{2Z_{c1}v_{g,p}}{H'\omega_p^2} \tag{13.101}$$

The equivalent circuit representation of a cavity mode near resonance provides us with a convenient way of determining Q_U, Q_e and Q_L of the system. For by construction,

$$Q_U = \frac{1}{G_{eq}}\sqrt{\frac{C_{eq}}{L_{eq}}} \tag{13.102}$$

On the other hand, the loaded Q of the cavity is found by connecting a matched load Z_{c1} across terminal plane A of Figure 13.14. The result is a parallel LCG circuit as in Figure 13.15, but with G_{eq} replaced by

$$G'_{eq} = G_{eq} + \frac{N^2}{Z_{c1}} = G_{eq}(1+\kappa) \tag{13.103}$$

where

$$\kappa = \frac{N^2}{G_{eq}Z_{c1}} \tag{13.104}$$

is the coupling coefficient of the cavity. The loaded Q is therefore

$$Q_L = \frac{1}{G'_{eq}}\sqrt{\frac{C_{eq}}{L_{eq}}} = \frac{Q_U}{1+\kappa} \tag{13.105}$$

The external Q is that which would result from having only the external shunt conductance N^2/Z_{c1} in the circuit of Figure 13.15, so

$$Q_e = \frac{Z_{c1}}{N^2}\sqrt{\frac{C_{eq}}{L_{eq}}} = \frac{Q_U}{\kappa} \tag{13.106}$$

It is readily seen from the foregoing that eqn. (13.79) is satisfied.

From (13.105)-(13.106), we see that the ratio

$$\frac{Q_L}{Q_e} = \frac{\kappa}{1+\kappa} \tag{13.107}$$

represents the ratio of the power dissipated in the external circuit (presumably usefully by delivery to some load) to the total power dissipated in the cavity and in the external circuit. It might seem at first that a very large value of κ (a highly overcoupled state) should be sought. This is not true, for several reasons. First, the feed waveguide would be severely mismatched, meaning that the overall power delivered to the cavity would be low. But possibly even more important is the effect that larger κ (and hence smaller Q_L) would have on the bandwidth of the system.

To see how the bandwidth is affected by the Q of a resonator, let us rewrite (13.98) using expressions for Q and the resonant frequency of this resonator:

$$Z_{eq} = -j\frac{\omega\omega_1}{G_{eq}Q}\frac{1}{\omega^2 - \omega_1^2 - j\omega\omega_0/Q} \tag{13.108}$$

Let $\omega = \omega_1 + \Delta\omega$, where $\Delta\omega \ll \omega_1$ is a small frequency shift away from the resonant frequency. Then, neglecting terms of order $\Delta\omega/\omega_1$ compared to one, but *not* terms like $Q\Delta\omega/\omega_1$, because Q can be large, (13.108) becomes approximately

$$Z_{\text{eq}} \simeq \frac{R_{\text{eq}}}{1 + 2jQ\Delta\omega/\omega_1} \tag{13.109}$$

The real part of (13.109) has its maximum value (R_{eq}) when $\Delta\omega = 0$, and falls off rapidly above and below this point (Figure 13.16). It is reduced by a factor of 2 when

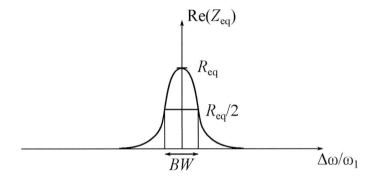

Figure 13.16 A typical resonance curve: Re (Z_{eq}) vs. $\Delta\omega/\omega_1$.

$2Q\Delta\omega/\omega_1 = \pm 1$; we use these two frequency points to define the fractional bandwidth BW of the resonator:

$$BW \equiv 2\frac{\Delta\omega}{\omega_1} = \frac{1}{Q} \tag{13.110}$$

Thus, the q-factor (loaded or unloaded) of a resonator is a direct measure of the narrowness of the frequency interval over which the resonant behavior is manifest. If a highly selective circuit is desired, we must have a large Q_L. For fixed losses in the cavity itself we must make design trade-offs between the goals of small bandwidth and maximum power delivered to a load.

A resonator may have more than one coupling port, for example, one port that accepts power from a source, and a second that supplies power to a load. This type of cavity is called a transmission cavity, and is used when highly selective transmission (narrow bandpass behavior) is desired. The analysis is similar to that carried out for a one-port cavity, and will not be presented here.

13.4   CAVITIES OF OTHER SHAPES

Other shapes (e.g., spherical) of cavity resonators that do not correspond to sections of uniform waveguide are also sometimes used. The general theory of such cavities will be outlined briefly in this section. It is analogous to the theory of metallic waveguides of arbitrary cross section, and the mathematics parallels that used in Chapter 9.

Assume that the cavity is a volume V containing a lossless material of real permittivity ϵ and real permeability μ (both positive, and possibly functions of position) and bounded by a perfectly conducting surface S. These restrictions can be relaxed at the expense of a more complicated derivation. The natural modes of this resonator are the sets of fields $\mathbf{E} = \mathbf{E}_m e^{j\Omega_m t}$ and $\mathbf{H} = \mathbf{H}_m e^{j\Omega_m t}$ that obey $\mathbf{u}_n \times \mathbf{E}_m = 0$ on S (here, as we did for waveguides, we take $\mathbf{u}_n$ to be the unit vector normal to S pointing into V). The resonant angular frequency is $\Omega_m = \omega_m + js_m$ as before. The mode fields obey

$$\nabla \times \mathbf{E}_m = -j\Omega_m \mu \mathbf{H}_m; \qquad \nabla \times \mathbf{H}_m = j\Omega_m \epsilon \mathbf{E}_m \tag{13.111}$$

For every mode m with $\Omega_m \neq 0$, there will evidently be a corresponding mode $-m$ with $\Omega_{-m} = -\Omega_m$ such that

$$\mathbf{E}_{-m} = \mathbf{E}_m; \qquad \mathbf{H}_{-m} = -\mathbf{H}_m$$

We can show that all resonant frequencies are real in the following way. From (13.111) and a vector identity we have

$$\nabla \cdot (\mathbf{E} \times \mathbf{H}^*) = -j\Omega_m \mu \mathbf{H}_m \cdot \mathbf{H}_m^* + j\Omega_m^* \epsilon \mathbf{E}_m \cdot \mathbf{E}_m^* \tag{13.112}$$

Integrating this over V, using the divergence theorem and the boundary condition on $\mathbf{E}_m$ at S, we get

$$\Omega_m \int_V \mu \mathbf{H}_m \cdot \mathbf{H}_m^* \, dV - \Omega_m^* \int_V \epsilon \mathbf{E}_m \cdot \mathbf{E}_m^* \, dV = 0 \tag{13.113}$$

Since the integrands in (13.113) are real, the imaginary part of (13.113) gives

$$s_m \int_V (\mu \mathbf{H}_m \cdot \mathbf{H}_m^* + \epsilon \mathbf{E}_m \cdot \mathbf{E}_m^*) \, dV = 0 \tag{13.114}$$

For fields not identically zero, the integral in (13.114) is positive and we conclude that $s_m = 0$, so that $\Omega_m = \omega_m$ is real for a lossless closed cavity. One consequence of this is that we can always choose the mode field $\mathbf{E}_m$ to be a real function, and if we do, then $\mathbf{H}_m$ will be purely imaginary.

In some cases we may also have electrostatic or magnetostatic modes that have $\Omega_m = \omega_m = 0$. The electrostatic modes have $\mathbf{H}_m \equiv 0$, while $\mathbf{E}_m = -\nabla \Phi_m$; the magnetostatic modes have $\mathbf{E}_m \equiv 0$, while $\mathbf{H}_m = -\nabla \Psi_m$. These modes satisfy the curl equations (13.111) trivially, and we must use the divergence equations (the Gauss laws) to fully describe them:

$$\nabla \cdot (\epsilon \nabla \Phi_m) = 0; \qquad \nabla \cdot (\mu \nabla \Psi_m) = 0 \tag{13.115}$$

along with the appropriate boundary conditions. An example of a resonator with an electrostatic mode is a cavity with a two-piece conducting boundary (such as a pair of concentric spheres). In this case, the electrostatic mode is the well-known static capacitor configuration.

13.4.1   ORTHOGONALITY AND NORMALIZATION

In a manner similar to that of Section 9.2, we can obtain orthogonality properties of the mode fields. We can by techniques that are now standard obtain the identity

$$\nabla \cdot (\mathbf{E}_m \times \mathbf{H}_{m'}) = -j\Omega_m \mu \mathbf{H}_m \cdot \mathbf{H}_{m'} - j\Omega_{m'} \epsilon \mathbf{E}_m \cdot \mathbf{E}_{m'} \tag{13.116}$$

for any pair of modes m and m'. Integrating (13.116) over the volume V and using the divergence theorem and boundary condition, we get

$$\Omega_m \int_V \mu \mathbf{H}_m \cdot \mathbf{H}_{m'} \, dV + \Omega_{m'} \int_V \epsilon \mathbf{E}_m \cdot \mathbf{E}_{m'} \, dV = 0 \tag{13.117}$$

Interchanging m and m' in this result gives

$$\Omega_{m'} \int_V \mu \mathbf{H}_m \cdot \mathbf{H}_{m'} \, dV + \Omega_m \int_V \epsilon \mathbf{E}_m \cdot \mathbf{E}_{m'} \, dV = 0 \tag{13.118}$$

If $\Omega_m \neq \pm\Omega_{m'}$, then we have the orthogonality relations

$$\int_V \epsilon\mathbf{E}_m \cdot \mathbf{E}_{m'} \, dV = \int_V \mu\mathbf{H}_m \cdot \mathbf{H}_{m'} \, dV = 0 \tag{13.119}$$

If there are distinct modes with $\Omega_m = \Omega_{m'}$ but linearly independent mode fields, the proof above does not apply. However, new linear combinations of the two mode fields can always be found such that (13.119) is valid. The proof is similar to that used for waveguide modes in problem p9-11.

When $m = m'$, we will normalize the cavity mode fields to unit stored electric oscillatory energy:

$$\frac{1}{4}\int_V \epsilon\mathbf{E}_m \cdot \mathbf{E}_m \, dV = -\frac{1}{4}\int_V \mu\mathbf{H}_m \cdot \mathbf{H}_m \, dV = 1 \tag{13.120}$$

the second equality following from (13.117) or (13.118).

13.4.2 EXCITATION OF CAVITIES BY IMPRESSED SOURCES

Now let externally impressed sources $\mathbf{J}_{\text{ext}}$ and $\mathbf{M}_{\text{ext}}$ act within the cavity. We assume the completeness property of the cavity modes so that we can expand the total field as a superposition of them:

$$\mathbf{E} = \sum_{m'} A_{m'}\mathbf{E}_{m'}; \qquad \mathbf{H} = \sum_{m'} B_{m'}\mathbf{H}_{m'} \tag{13.121}$$

for some complex amplitudes A_m and B_m, whose values are to be determined. The impressed sources operate at a frequency ω—not, in general, one of the resonant mode frequencies, so the total field must obey

$$\nabla \times \mathbf{E} = -j\omega\mu\mathbf{H} - \mathbf{M}_{\text{ext}}; \qquad \nabla \times \mathbf{H} = j\omega\epsilon\mathbf{E} + \mathbf{J}_{\text{ext}} \tag{13.122}$$

We now substitute (13.121) into the first of (13.122), take the dot product of both sides with the mode field $\mathbf{H}_m$, integrate over the volume V and make use of the orthogonality relation and (13.111). The result is

$$4j\Omega_m A_m = 4j\omega B_m - \int_V H_m \cdot \mathbf{M}_{\text{ext}} \, dV \tag{13.123}$$

A similar procedure on the second of (13.122) gives

$$4j\Omega_m B_m = 4j\omega A_m + \int_V E_m \cdot \mathbf{J}_{\text{ext}} \, dV \tag{13.124}$$

Equations (13.123) and (13.124) can be readily solved to obtain

$$\begin{aligned}
A_m &= \frac{j}{4}\frac{\omega \int_V E_m \cdot \mathbf{J}_{\text{ext}} \, dV - \Omega_m \int_V H_m \cdot \mathbf{M}_{\text{ext}} \, dV}{\omega^2 - \Omega_m^2} \\
B_m &= \frac{j}{4}\frac{\Omega_m \int_V E_m \cdot \mathbf{J}_{\text{ext}} \, dV - \omega \int_V H_m \cdot \mathbf{M}_{\text{ext}} \, dV}{\omega^2 - \Omega_m^2}
\end{aligned} \tag{13.125}$$

It will be seen that the frequency dependence of these coefficients is the same as that of (13.19) and (13.20), obtained earlier for a lumped-element resonator.

It should be noted that the integrals in (13.125) can be written in an alternate and sometimes more convenient form for the electrostatic and magnetostatic modes. For electrostatic modes, we have obviously $\int_V H_m \cdot \mathbf{M}_{\text{ext}}\, dV = 0$, while

$$\int_V E_m \cdot \mathbf{J}_{\text{ext}}\, dV = -j\omega \int_V \Phi_m \rho_{\text{ext}}\, dV - \oint_S \Phi_m \mathbf{J}_{\text{ext}} \cdot \mathbf{u}_n\, dS \tag{13.126}$$

where ρ_{ext} is the electric charge density corresponding to $\mathbf{J}_{\text{ext}}$, and is obtained from (D.7). In like manner, for a magnetostatic mode, $\int_V E_m \cdot \mathbf{J}_{\text{ext}}\, dV = 0$, while

$$\int_V H_m \cdot \mathbf{M}_{\text{ext}}\, dV = -j\omega \int_V \Psi_m \rho_{\text{m,ext}}\, dV - \oint_S \Psi_m \mathbf{M}_{\text{ext}} \cdot \mathbf{u}_n\, dS \tag{13.127}$$

and $\rho_{\text{m,ext}}$ is the magnetic charge density corresponding to $\mathbf{M}_{\text{ext}}$, and is also obtained from (D.7).

13.4.3   WALL LOSSES IN CAVITY RESONATORS

Let $\mathbf{E}_{m0}$, $\mathbf{H}_{m0}$ and $\Omega_{m0} = \omega_{m0}$ be a mode of a cavity resonator of a given geometry with a perfectly conducting wall and lossless filling material. Let $\mathbf{E}_m$, $\mathbf{H}_m$ and $\Omega_m = \omega_m + js_m$ be the corresponding mode of a resonator of the same geometry and filling material, but whose wall has finite conductivity and forces the impedance boundary condition (9.59) to hold.

We will apply a result similar to the reciprocity theorem (but not identical, since the frequencies of the two states are not the same). Using a vector identity and Maxwell's curl equations (13.111), we have

$$\nabla \cdot (\mathbf{E}_{m0} \times \mathbf{H}_m - \mathbf{E}_m \times \mathbf{H}_{m0}) = j(\Omega_m - \Omega_{m0})(\mu \mathbf{H}_m \cdot \mathbf{H}_{m0} - \epsilon \mathbf{E}_m \cdot \mathbf{E}_{m0}) \tag{13.128}$$

Next, we integrate (13.128) over the volume V, use the divergence theorem and apply the boundary conditions $\mathbf{u}_n \times \mathbf{E}_{m0} = 0$ and $\mathbf{u}_n \times \mathbf{E}_m = -Z_S \mathbf{H}_{m0(\text{tan})}$ at S, with the result that

$$\Omega_m - \Omega_{m0} = j\frac{\oint_S Z_S \mathbf{H}_{m0(\text{tan})} \cdot \mathbf{H}_{m(\text{tan})}\, dS}{\int_V (\mu \mathbf{H}_{m0} \cdot \mathbf{H}_m - \epsilon \mathbf{E}_{m0} \cdot \mathbf{E}_m)\, dV} \tag{13.129}$$

Using a perturbation approximation we put $\mathbf{E}_m \simeq \mathbf{E}_{m0}$ and $\mathbf{H}_m \simeq \mathbf{H}_{m0}$ to get an explicitly computable approximation:

$$\Omega_m - \Omega_{m0} \simeq j\frac{\oint_S Z_S \mathbf{H}_{m0(\text{tan})} \cdot \mathbf{H}_{m0(\text{tan})}\, dS}{\int_V (\mu \mathbf{H}_{m0} \cdot \mathbf{H}_{m0} - \epsilon \mathbf{E}_{m0} \cdot \mathbf{E}_{m0})\, dV} = j\frac{\oint_S Z_S \mathbf{H}_{m0(\text{tan})} \cdot \mathbf{H}_{m0(\text{tan})}\, dS}{2 \int_V \mu \mathbf{H}_{m0} \cdot \mathbf{H}_{m0}\, dV} \tag{13.130}$$

the last equality following from (13.120). Since $s_{m0} = 0$ for the lossless cavity, the real part of (13.130) gives the decay constant due to wall losses for the cavity:

$$s_m \simeq \frac{\oint_S R_S \mathbf{H}_{m0(\text{tan})} \cdot \mathbf{H}_{m0(\text{tan})}\, dS}{2 \int_V \mu \mathbf{H}_{m0} \cdot \mathbf{H}_{m0}\, dV} \tag{13.131}$$

where $R_S = \text{Re}(Z_S)$ is the surface resistance of the wall. As is the case for wall losses in a waveguide (Section 9.5), equation (13.130) has a simple physical interpretation—it is half the ratio of time-average power dissipated in the walls to the stored energy in the cavity.

13.5 DIELECTRIC RESONATORS

The ready availability of low-loss, high-permittivity ($\epsilon_r > 200$) ceramic dielectrics such as barium, magnesium, strontium or calcium titanates (or compounds/composites thereof) has allowed the development of small-size resonators at microwave frequencies. The basic mechanism for resonance of a dielectric resonator can be seen from the simple one-dimensional example of a layer of high-permittivity nonmagnetic material with relative permittivity ϵ_r occupying the region $0 < z < H$, with free-space in $z < 0$ and $z > H$. If the fields do not vary with x or y, we can apply the equivalent transmission-line model of Section 1.4 to analyze the structure, as shown in Figure 13.17. The characteristic impedances are $Z_{cd} = \zeta_0/\sqrt{\epsilon_r}$

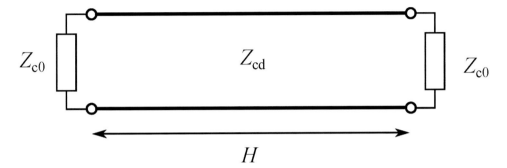

Figure 13.17 Transmission-line equivalent circuit for a dielectric slab resonator.

for the dielectric layer, and $Z_{c0} = \zeta_0$ for free space. The propagation constant in the layer is $\beta_d = \Omega\sqrt{\epsilon_r}/c$, where c is the speed of light in vacuum.

Using a transverse resonance technique like that of Chapter 7, we find that the resonant frequencies of the dielectric slab obey

$$\tan \frac{\Omega H \sqrt{\epsilon_r}}{c} = \frac{2j Z_{c0} Z_{cd}}{Z_{c0}^2 + Z_{cd}^2} = \frac{2j\sqrt{\epsilon_r}}{\epsilon_r + 1} \tag{13.132}$$

The right side of (13.132) is small in magnitude since $\epsilon_r \gg 1$, and we have, to the lowest order of approximation that $\Omega H \sqrt{\epsilon_r} = m\pi c$, which is what we would have if the ends of the transmission line were open-circuited instead of loaded with the large impedances Z_{c0}. The solutions of (13.132) are

$$\Omega_m = \frac{m\pi c}{H\sqrt{\epsilon_r}} + j\frac{c}{H\sqrt{\epsilon_r}} \ln \frac{\sqrt{\epsilon_r}+1}{\sqrt{\epsilon_r}-1}; \qquad m = 1, 2, \ldots \tag{13.133}$$

If $\epsilon_r \gg 1$, the imaginary part of Ω_m is approximately

$$s_m \simeq \frac{2c}{H\epsilon_r} \tag{13.134}$$

which is much smaller than the real part

$$\omega_m = \frac{m\pi c}{H\sqrt{\epsilon_r}} \tag{13.135}$$

Thus, not only does high permittivity provide a reduction in wavelength that shrinks the size of the resonator, but also a high reflection coefficient at the interface with air that reduces the radiation losses that account for s_m.

Because of the high value of the permittivity, at the lowest resonant frequencies a dielectric resonator will still be dimensionally small compared to a free-space wavelength. This means that coupling to these modes of the resonator can be conveniently modeled using their polarizabilities, similar to what is done for an electrically small object in Section 10.2.5. Now, however, the polarizabilities will be resonant functions of frequency. For most resonator shapes, no analytical expressions for the polarizabilities are available, but an exception is the spherical resonator of radius a, relative permittivity ϵ_r and relative permeability μ_r. If the resonator is located in free space, we have instead of (10.22) and (10.23),

$$\overset{\leftrightarrow}{\alpha}_e = 4\pi a^3 \frac{F(ka)\epsilon_r - 1}{F(ka)\epsilon_r + 2}\overset{\leftrightarrow}{I} \tag{13.136}$$

and

$$\overset{\leftrightarrow}{\alpha}_m = 4\pi a^3 \frac{F(ka)\mu_r - 1}{F(ka)\mu_r + 2}\overset{\leftrightarrow}{I} \tag{13.137}$$

which are valid under the condition that $k_0 a \ll 1$, where

$$F(\theta) = \frac{2(\sin\theta - \theta\cos\theta)}{(\theta^2 - 1)\sin\theta + \theta\cos\theta} \tag{13.138}$$

and $k = k_0\sqrt{\mu_r\epsilon_r}$. A sample plot of normalized polarizabilities is shown in Figure 13.18. Note that even though the sphere in this case is made of a nonmagnetic material, the

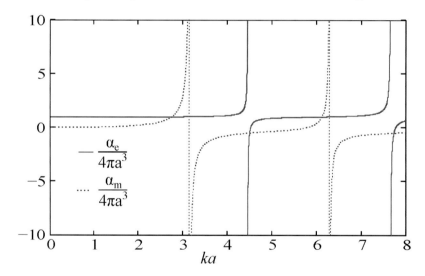

Figure 13.18   Normalized electric and magnetic polarizabilities for a spherical dielectric resonator with $\epsilon_r = 100$ and $\mu_r = 1$.

magnetic polarizability becomes different from zero as frequency increases, and in fact the first resonant frequency (where the polarizability becomes infinite in the absence of losses) is a magnetic one, at $ka = \pi$, while the first electric resonance ($\alpha_e \to \infty$) is at $ka \simeq 4.449$.

13.6   SPLIT-RING RESONATORS AND RELATED STRUCTURES

Dielectric resonators have the advantage that they are physically small compared to a wavelength in free space, allowing more compact resonator design than available from sections of ordinary transmission line. Another way to achieve size reduction without the use of special high-dielectric materials is by the use of the so-called split-ring resonator and its variants. The basic idea here is that, by wrapping the transmission-line resonator of Figure 13.10 into a circle as shown in Figure 13.19(a), the overall dimension can be reduced from a length of $\lambda/2$ to a diameter of $\lambda/2\pi$. However, the wrapping of this geometry causes interactions of the fields in various parts of the transmission line that do not exist in the unwrapped resonator. This has meant that several different structures related to the wrapped transmission line have been employed instead. Thus we have the loop-gap resonator, a single conductor bent into an almost complete circle as shown in Figure 13.19(b), which can be approximately modeled as a series combination of a capacitance associated with the gap, and an inductance associated with the loop. A pair of loop-gap resonators can be nested as shown in Figure 13.19(c) to form a split-ring resonator, which provides a stronger resonant amplitude, and therefore stronger coupling, than the single-loop version.

The equivalent circuit representations of such structural variants are generally more complicated; examples of these can be found in the references at the end of the chapter. Here, we will show only how a simple approximate formula for the polarizability can be derived from basic principles for the example of the loop-gap resonator of Figure 13.19(b). Write the impedance of the gap capacitance as $Z_g = 1/(j\omega C_g)$; in fact we could load the gap with any desired impedance and the derivation that follows remains valid. If the width of the conductor is small compared to its diameter, an applied magnetic field $\mathbf{B}^{\text{ext}} = \mathbf{u}_n B^{\text{ext}}$ perpendicular to the plane of the loop will result in a magnetic flux passing through the loop of

$$\psi_m = \int_{\text{loop}} \mathbf{B}^{\text{ext}} \cdot \mathbf{u}_n \, dS \simeq S B^{\text{ext}} \tag{13.139}$$

where S is the area contained within the loop and n is the direction normal to the surface enclosed by the loop. If the magnetic field produced by currents induced in the loop is negligible in comparison to B^{ext}, then Faraday's law tells us that the induced emf in the loop is

$$\oint_C \mathbf{E} \cdot d\mathbf{l} = -j\omega \psi_m \tag{13.140}$$

On the other hand, this emf must be the sum of the voltage drops across the impedance Z_g and that of the self-inductance L of the closed loop. Therefore,

$$\oint_C \mathbf{E} \cdot d\mathbf{l} = (j\omega L + Z_g)I \tag{13.141}$$

where I is the current induced in the loop. From the foregoing, we have

$$I = -\frac{j\omega S}{j\omega L + Z_g} B^{\text{ext}} \tag{13.142}$$

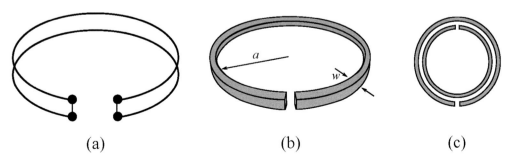

Figure 13.19 Variants of the conducting split-ring resonator: (a) wrapped half-wave transmission line; (b) loop-gap resonator; (c) split-ring resonator (top view).

The magnetic dipole (directed normal to the plane of the loop) corresponding to this current is

$$m_n = IS \tag{13.143}$$

so by the definition (10.23) of magnetic polarizability, we have

$$\alpha_m^{nn} = -\frac{j\omega\mu_0 S^2}{j\omega L + Z_g} \tag{13.144}$$

If Z_g is due to the gap capacitance, this becomes

$$\alpha_m^{nn} = \frac{\omega^2 \mu_0 C_g S^2}{1 - \omega^2 L C_g} \tag{13.145}$$

which exhibits resonant behavior at $\omega_0 = 1/\sqrt{LC_g}$. The loop-gap structure also possesses electric polarizability in directions parallel to the plane of the loop. If the radius of the loop is a and the width of the conductors from which the ring is made is w, then an approximate expression for the tangential electric polarizability is

$$\alpha_e^{tt} = \frac{2\pi^2 a^3}{\ln(8a/w) - 2} \tag{13.146}$$

This polarizability does not have a resonant behavior.

It should be noted that the asymmetry of the loop-gap resonator implies that there will be magnetoelectric coupling—i.e., an applied magnetic field can produce an electric dipole, and an electric field can generate a magnetic dipole. Treatments of this property and of other polarizabilities can be found in the references at the end of the chapter.

13.7   NOTES AND REFERENCES

The mode concept in RLC networks is explored in:

C. A. Desoer, "Modes in linear circuits," *IRE Trans. Circ. Theory*, vol. 7, pp. 211-223 (1960).

Zadeh and Desoer (1963), Chapter 5.

R. R. Parker, "Normal modes in RLC networks," *Proc. IEEE*, vol. 57, pp. 39-44 (1969).

Huang and Parker (1971), Chapters 5-7.

Modes in networks containing both lumped elements and classical transmission lines are investigated in:

W. H. Ingram, "Electrical oscillations in a non-uniform transmission line," *Univ. Washington Publ. Math.*, vol. 2, no. 1, pp. 17-38 (1930).

Adler, Chu and Fano (1960), Chapter 6.

Perturbation of a complex resonant frequency of a lumped-element resonant circuit is examined in

A. Papoulis, "Perturbations of the natural frequencies and eigenvectors of a network," *IEEE Trans. Circ. Theory*, vol. 13, pp. 188-195 (1966),

from which our treatment has been adapted.

The notion of the quality factor can be traced back to the early part of the twentieth century, and has undergone generalizations and modifications such that many different (and sometimes inconsistent) definitions can be found in the literature. For the history of Q, see:

E. I. Green, "The story of Q," *Amer. Scientist*, vol. 43, pp. 584-594 (1955).

P. B. Fellgett, "Origins of the usage of 'Q'," *J. IERE*, vol. 56, pp. 45-46 (1986).

K. L. Smith, " 'Q'," *Electron. Wireless World*, vol. 92, no. 1605, pp. 51-53 (1986).

It is surprising how difficult it has proven to give precise definitions to the notions of "Q" and resonant frequency. The "official" definition of Q given in the *IEEE Standard Dictionary of Electrical and Electronics Terms* is

$$Q = \frac{\omega W_{\max}}{P_L}$$

where ω is the frequency of the time-harmonic source(s), $W_{\max}$ is the *maximum* (it is not stated whether this is a maximum with respect to time or to frequency) stored energy and P_L is the time-average power dissipated in the resonator. The frequency is usually taken to be a resonance frequency, which is itself a somewhat ambiguous concept. A resonant frequency is defined "officially" to be the frequency of a source or sources driving a system which results in a locally maximum response (i.e., some induced voltage or current) from the system.

It is readily seen from consideration of simple examples that in general these maxima may occur at different frequencies if different responses are considered. Moreover, one can construct examples to show that the definition given above does not give a value of Q consistent with the bandwidth definition of this quantity. Consider for example the circuit shown in Figure 13.20. Note that this configuration is not allowed according to the specifications of Section 13.2. If we set $R = 1/G = \sqrt{L/C}$, it is an easy exercise to show that the input impedance $Z_{\text{in}} = R$ at all frequencies; i. e., this circuit is indistinguishable (to the outside world) from a single resistor R. Clearly, when current flows into this circuit, energy will be stored in the inductor and capacitor, while the single resistor stores no energy at all. Thus, for any resonant circuit that has a resistor, that resistor can be replaced by a constant-resistence circuit like Figure 13.20 and will contain more stored energy, in spite of the fact that its behavior at the terminals (including bandwidth) remains identical. It turns out that this additional stored energy is not recoverable to the outside world through the terminals of the network, and is always eventually dissipated internally. Another example of this occurs in systems containing transmission line sections terminated in matched loads, where there can be an energy stored in these sections merely until it is dissipated in the impedances which terminate these sections. By varying the length of these sections, the stored energy is varied, but the power dissipated is not changed proportionately, so different values of Q are obtainable for a network with the same terminal response. To get around this problem, the notion of observable energies in the definition of Q for systems of infinite extent was introduced by

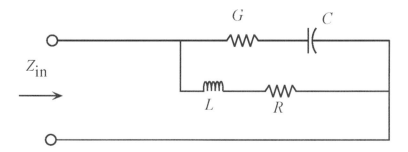

Figure 13.20 Constant-resistance circuit containing nonrecoverable stored energy.

D. R. Rhodes, "Observable stored energies of electromagnetic systems," *J. Franklin Inst.*, vol. 302, pp. 225-237 (1976).

A. D. Yaghjian and S. R. Best, "Impedance, bandwidth and Q of antennas," *IEEE Trans. Ant. Prop.*, vol. 53, pp. 1298-1324 (2005).

Studies of energy recoverable from the terminals of a network have been carried out in

W. E. Smith, "The energy storage of a prescribed impedance," *J. Proc. Roy. Soc. New South Wales*, vol. 102, pp. 203-218 (1969).

V. G. Polevoi, "Maximum energy extractable from an electromagnetic field," *Radiophys. Quant. Electron.*, vol. 33, pp. 603-609 (1990).

R. H. Direen, "Fundamental limitations on the terminal behavior of antennas and nonuniform transmission lines ," Ph. D. thesis, University of Colorado Boulder, 2010, Chapter 4.

The upshot of these considerations would seem to be that the bandwidth definition of Q is to be regarded as fundamental, and agrees with the energy definition only in certain cases or with modifications.

If, as we have done, one makes use of the mode concept in defining Q, the definitions become precise. Only when more than one mode is present does ambiguity set in, because then the Q's of the individual modes do not combine in any simple way to form an unambiguous Q for the network as a whole. However, since near resonant frequencies one mode is often excited with a much larger amplitude than all the others, use of the Q and resonant frequency for that mode will be a good approximation to those of the entire circuit obtained from the "official" definitions (the more so the larger Q is).

The transition from resonators made up purely from finite lumped-element circuits or circuits containing finite lengths of transmission line to ones of infinite extent can result in a much more complicated spectrum of modes which we do not investigate here. See, for example,

Gross and Braga (1961).

A. E. Karbowiak, "Solution of the excitation problem in resonant dissipative and inhomogeneous structures," *Alta Frequenza*, vol. 38 (supplement), pp. 109-115 (1969).

The Foster canonical equivalent networks were first given in:

R. M. Foster, "A reactance theorem," *Bell Syst. Tech. J.*, vol. 3, pp. 259-267 (1924).

for the case of a lossless (LC) circuit. The issue of whether the expansion of the input impedance or admittance of a resonator as a series of terms due to individual modes will correspond to a physically realizable equivalent circuit when losses (resistors) are present is addressed in

Schelkunoff (1943), Chapter 5.
Guillemin (1957), Chapter 9.
Montgomery, Dicke and Purcell (1965), Chapter 7.

as well as

S. A. Schelkunoff, "Representation of impedance functions in terms of resonant frequencies," *Proc. IRE*, vol. 32, pp. 83-90 (1944).

M. K. Zinn, "Network representation of transcendental impedance functions," *Bell Syst. Tech. J.*, vol. 31, pp. 378-404 (1952).

F. M. Reza, "*RLC* canonic forms," *J. Appl. Phys.*, vol. 25, pp. 297-301 (1954).

E. V. Zelyakh and A. V. Kisel', "Canonical configurations of two-terminal networks consisting of impedors of two types," *Telecommun. Radio Eng.*, vol. 19/20, no. 7, pp. 67-73 (1965).

E. V. Zelyakh and A. V. Kisel', "Equivalent-circuit theory," *Telecommun. Radio Eng.*, vol. 22/23, no. 1, pp. 108-113 (1968).

For various treatments of resonant elements in transmission line and waveguide circuits, see

Adler, Chu and Fano (1960), Chapter 6;
Ghose (1963), Chapter 8;
Altman (1964), Chapter 5;
Montgomery, Dicke and Purcell (1965);
Slater (1969), Chapter 4;
Kurokawa (1969), Chapter 4;
Collin (1991), Chapter 5;
Collin (2001), Chapter 7;
Pozar (2005), Chapter 6,

as well as many of the other references given at the end of Chapter 11. In their treatments of cavity resonators, most authors have assumed that the cavity is filled with a homogeneous material; an exception is:

Vainshtein (1988), Chapters 15 and 16.

whose treatment has inspired what is presented here. Mirror-type open resonators are treated in

Vainshtein (1969a);
Arnaud (1976), Chapters 1-2;
Siegman (1986), Chapters 14-23.

An equivalent circuit similar to those of Fig. 13.3 but based on the field theory presented in Section 13.4 is given in

C. Lange and M. Leone, "Broadband circuit model for electromagnetic-interference analysis in metallic enclosures," *IEEE Trans. Electromag. Compat.*, vol. 60, pp. 368-375 (2018).

Dielectric resonators are treated in

Il'chenko and Kudinov (1973);
Bahl and Bhartia (1988), Section 3.6;
Il'chenko (1989);
Kajfez and Guillon (1990);
Collin (1991), Section 6.6;
Collin (2001), Section 7.5;
Van Bladel (2007), Section 10.5.

Equations (13.136) and (13.137) are found, for example, in

C. L. Holloway, M. A. Mohamed, E. F. Kuester, and A. Dienstfrey, "Reflection and transmission properties of a metafilm: With an application to a controllable surface composed of resonant particles," *IEEE Trans. Electromag. Compat.*, vol. 47, pp. 853-865 (2005).

The split-ring resonator was originally proposed as a fundamental element in metamaterials:

J. B. Pendry, A. J. Holden, D. J. Robbins and W. J. Stewart, "Magnetism from conductors and enhanced nonlinear phenomena," *IEEE Trans. Micr. Theory Tech.*, vol. 47, pp. 2075-2084 (1999).

Equivalent circuit models for its polarizabilities are given there, as well as in the following for those of some structural variants of it:

M. V. Kostin and V. V. Shevchenko, "Artificial magnetic material based on ring currents," *Sov. J. Commun. Technol. Electron.*, vol. 33, no. 12, pp. 38-42 (1988).

R. Marqués, F. Medina and R. Rafii-El-Idrissi, "Role of bianisotropy in negative permeability and left-handed metamaterials," *Phys. Rev. B*, vol. 65, art. 144440 (2002).

Tretyakov (2003), Section 5.1.

M. Shamonin, E. Shamonina, V. Kalinin and L. Solymar, "Properties of a metamaterial element: Analytical solutions and numerical simulations for a singly split double ring," *J. Appl. Phys.*, vol. 95, pp. 3778-3784 (2004).

L. Jelinek, R. Marqués, F. Mesa and J. D. Baena, "Periodic arrangements of chiral scatterers providing negative refractive index bi-isotropic media," *Phys. Rev. B*, vol. 77, art. 205110 (2008).

C. É. Kriegler, M. S. Rill, S. Linden, and M. Wegener, "Bianisotropic photonic metamaterials," *IEEE J. Sel. Topics Quant. Electron.*, vol. 16, pp. 367-375 (2010).

13.8 PROBLEMS

p13-1 Suppose the LC resonator of Figure 13.1 is loaded with a small capacitor C_0 in series with a load resistor R as shown below.

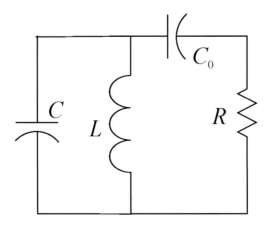

Use (13.37) to find an approximate expression for the complex resonant frequency Ω_1 of the mode of this circuit. Give an expression for the quality factor Q_1 of this mode.

p13-2 Consider a resonator formed by a length l of lossless classical transmission line whose characteristic impedance is Z_c and whose propagation coefficient is $\beta = \omega/v_p$. Both Z_c and v_p are real and independent of frequency. As the limit of lumped element circuits with indefinitely large numbers of components, this resonator in principle has an infinite number of tree branches and links, and therefore an infinite number of possible entry points. We will consider only one: the terminals at one end.

(a) What are the complex natural frequencies Ω_m for the modes of this resonator if the ends of the line are open-circuited? What are the (unloaded) Q's of these modes? What are the natural distributions of voltage and current along the line for each mode (the analogs of $[V^{(m)}]$ and $[I^{(m)}]$ in the lumped-element case)?

(b) Find the impedance Z_{in} of the resonator as seen at one end of the transmission line resonator (the other end being left open-circuited). Near an unloaded resonant frequency of the resonator, find an approximate expression for Z_{in} in the form of a pair of terms

$$\frac{A_m}{\omega - \Omega_m} + \frac{A_{-m}}{\omega - \Omega_{-m}}$$

where $A_{\pm m}$ are complex constants, and express this impedance in the form of a parallel LC module representing the special case of a module from Figure 13.3(a) with R_m and G_m equal to zero. Hence, sketch an equivalent circuit for the resonator in the form of Figure 13.3(a).

(c) A current generator consisting of an ideal current source I_G in parallel with a generator admittance of Y_G is connected to one end of the transmission line resonator as a soldering-iron entry point. Suppose that $|Y_G Z_c| \ll 1$, so that the resonator is only slightly perturbed by the connection of the generator. Near the resonant frequency of one of the modes, the circuit seen by the ideal current source is that of Y_G in parallel with the LC module for this mode only (why?). Find the perturbed natural frequency for this mode caused by the loading of the resonator by Y_G, and find the *loaded* Q for the mode. Express your answers first in terms of the equivalent parameters in the LC module, and then directly in terms of the parameters of the transmission line section.

(d) Instead of the current generator, suppose that a feed transmission line (with the same characteristic impedance Z_c as the resonator) is connected to the resonator through a small ($\omega C_0 Z_c \ll 1$) series capacitance C_0 as shown.

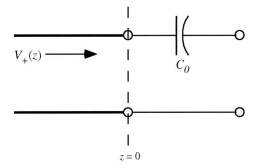

If an incident voltage wave V_+ is present on the feed line, what is the Norton equivalent circuit for the feed line/C_0 combination? Use your result to find an expression for the reflection coefficient ρ on the feed line, in the vicinity of one of the resonant modes. Sketch the a plot of $|\rho|$ versus frequency in this range. You may use any numerical values for Z_c, v_p, etc., that you want, provided that they meet the criteria stated above.

p13-3 Consider the transmission cavity shown below:

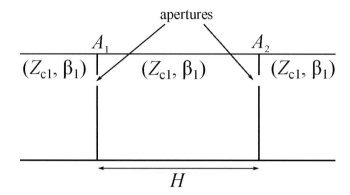

In contrast with the cavity shown in Figure 13.13, this is a two-port device. Find the natural modes of this cavity by the same method as used in Section 13.3.3, and comment on any differences from that case.

p13-4 Using perturbation theory, determine an approximate expression for the decay constant s_m of a mode of a closed cavity resonator with perfectly conducting walls, if the permittivity of the medium filling the cavity is complex: $\epsilon = \epsilon' - j\epsilon''$, and may be a function of position.

p13-5 Find the excitation amplitudes A_m and B_m in (13.121), where m denotes the TE$_{101}$ mode of a rectangular cavity occupying $0 < x < a$, $0 < y < d$ and $0 < z < l$, and excitation is due to an electric dipole of dipole moment $\mathbf{p} = p\mathbf{u}_y$ located at a point (x_0, y_0, z_0) inside the cavity.

p13-6 Using (13.136) and (13.137), obtain expressions for the reflection and transmission coefficients S_{11} and S_{21} of a spherical dielectric resonator centered at the point $(x_0, y_0, 0)$ in an air-filled rectangular metallic waveguide (see Section 10.6.2). Adjust your answer so that conservation of energy is obeyed by your result. Choose some suitable values for the parameters of the resonator and waveguide and plot the magnitudes of these scattering parameters as functions of frequency at least through the first resonant frequency.

p13-7 Repeat problem p13-6 for the case of a loop-gap resonator, using the polarizabilities given in (13.145) and (13.146).

Appendices

A PROPERTIES OF SOLUTIONS TO NETWORK AND TRANSMISSION LINE EQUATIONS

Since the classical transmission line is a limiting case of a lumped-element circuit, we may prove any result for the case of an arbitrary circuit, and carry it over to the transmission-line case by taking that limit. We will treat the properties of the network and classical transmission-line equations by emphasizing their structural analogy with Maxwell's equations. This allows us to see how theorems such as reciprocity, power balance, equivalence and so on bear a close mathematical resemblance in all three situations, and to recognize their essential similarity.

A.1 A STANDARD FORM FOR NETWORK EQUATIONS

Consider a linear lumped-element network with $n+1$ nodes, including a *reference* (or *datum* or *ground*) node labeled 0. In a somewhat abstract way, we can represent this network as a graph, in which each branch represents a circuit element such as a resistor, capacitor, inductor or voltage or current source, and the connection between elements occurs at the nodes. An example of such a network graph is shown in Fig. A.1. Let the network contain b branches connecting the various nodes. On each branch, reference directions for current in the branch and voltage across the branch are chosen as shown to obey the reference convention of circuit analysis.

We next define some terms from graph theory commonly used in circuit analysis. A *tree* of a graph is a set of branches from the graph chosen so that (i) all nodes of the network are met by at least one of the branches in the tree, and (ii) removal of any branch from the tree leaves at least one node not connected to the tree. A graph of $n+1$ nodes has trees with n tree branches. The remaining $l = b - n$ branches of the network are called *links*. A *loop* is a connected set of branches such that each node is connected by exactly two of these branches. The *fundamental loop* of a link is the unique loop which is formed from certain tree branches along with that link only. Examples of a tree and a fundamental loop are shown in Fig. A.1. A *cut-set* is a set of branches chosen so that (i) removal of all cut-set branches from the graph leaves two disconnected subgraphs remaining, and (ii) removal of all but any one of the cut-set branches leaves a connected graph remaining. The *fundamental cut-set* of a tree branch is the unique cut-set formed from that tree branch together with certain links only and no other tree branches. In Fig. A.1, the fundamental cut-set of branch $r = 3$ is that branch together with link $k = 2$ (observe that removal of these two branches leaves nodes 0, 1 and 2 connected in a subgraph, and a second disconnected subgraph consisting of only node 3).

A.1.1 TOPOLOGICAL LAWS (KVL AND KCL)

Now choose any tree of the network, labeling the voltages across the tree branches $V_{\text{int},r}$; $r = 1, \ldots, n$, and the currents through the links $I_{\text{int},k}$; $k = 1, \ldots, l$. The subscript $_{\text{int}}$ indicates that these voltages and currents are usually thought of as being internal to a multiport network, and are not to be confused with the port voltages and currents. Denote

483

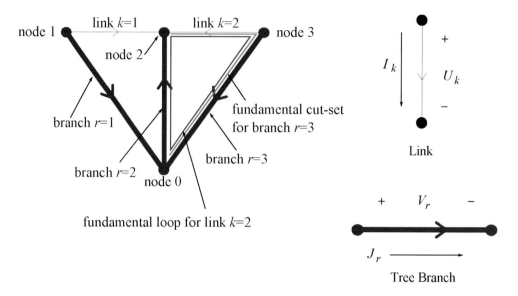

Figure A.1 Graph of example linear network graph with $n = 3$ and $b = 5$. A possible choice of tree is shown in blue, the fundamental loop for link $k = 2$ is shown in red, and the fundamental cut-set for branch $r = 3$ is shown in green.

the link voltages by U_k; $k = 1, \ldots, l$ and the tree branch currents by J_r; $r = 1, \ldots, n$. Then Kirchhoff's voltage law enforced around the fundamental loop of each of the links in the circuit can be expressed in matrix form as

$$[U] = -[D]^T [V_{\text{int}}] \qquad (A.1)$$

Kirchhoff's current law can be enforced on the fundamental cut-set for each of the tree branches: the sum of the currents through all the cut-set branches (measured from one of the resulting disconnected subgraphs to the other) must be zero. This is expressed in matrix form as

$$[J] = [D][I_{\text{int}}] \qquad (A.2)$$

where $[U]$, $[V_{\text{int}}]$, $[J]$ and $[I_{\text{int}}]$ are column vectors constructed from the tree branch and link currents and voltages described above, and the superscript ^{T} denotes the matrix transpose.

The matrix $[D]$ is a submatrix of the fundamental cut-set matrix of the graph of the network, whose elements are given by:

$$D_{rk} = +1$$

if tree branch r is in the fundamental loop of link k, with the same reference direction as the loop is traversed;

$$D_{rk} = -1$$

if tree branch r is in the fundamental loop of link k, with the opposite reference direction as the loop is traversed; and

$$D_{rk} = 0$$

if tree branch r is not in the fundamental loop of link k.[1]

---

[1]It can also be shown that $D_{rk} = -1$ if link k is in the fundamental cut-set of tree branch r, and passes between the resulting pair of disconnected subgraphs with the same reference direction; $D_{rk} = +1$ if link

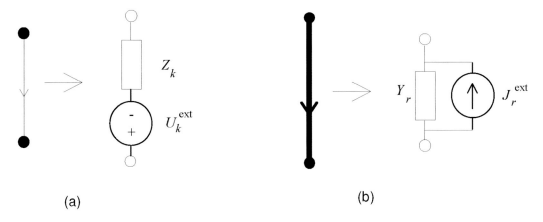

Figure A.2 (a) Typical link and (b) typical tree branch, in the absence of mutual coupling.

A.1.2 CONSTITUTIVE (ELEMENT) LAWS

To account for the constitutive properties of the elements making up the circuit, let us first assume that the tree branches consist of admittances Y_r in parallel with externally impressed current sources (that is, ideal current sources) J_r^{ext} (Fig. A.2(a)). The links are assumed to consist of impedances Z_k in series with ideal voltage sources U_k^{ext} (Fig. A.2(b)). If no mutual coupling between elements is present, this can always be done, by virtue of Thévenin's and Norton's equivalence theorems. Then Ohm's law for each element leads to the equations

$$[J] = [Y_{\text{int}}][V_{\text{int}}] + [J^{\text{ext}}]$$

$$[U] = [Z_{\text{int}}][I_{\text{int}}] + [U^{\text{ext}}]$$

where $[Y_{\text{int}}]$ and $[Z_{\text{int}}]$ are diagonal matrices formed by the element admittances and impedances indicated above. Then, eliminating the vectors $[U]$ and $[J]$ from the foregoing, we have

$$[D][I_{\text{int}}] = [Y_{\text{int}}][V_{\text{int}}] - [J^{\text{ext}}] \qquad (A.3)$$

$$-[D]^T[V_{\text{int}}] = [Z_{\text{int}}][I_{\text{int}}] - [U^{\text{ext}}] \qquad (A.4)$$

which is a matrix analog of the telegraphers' equations and of Maxwell's equations, with $[D]$ playing the role of the differential operator $d(\cdot)/dz$ or $\nabla \times (\cdot)$.

When mutual couplings are present between branches of the network, we can account for them by using dependent voltage or current sources in addition to the impressed sources already present above. For the case of sources linearly dependent on other voltages or currents in the network, we can add a column vector of dependent current sources to the tree branches:

$$[J^{\text{dep}}] = [Y'][V_{\text{int}}] + [T][I_{\text{int}}]$$

and dependent voltage sources to the links:

$$[U^{\text{dep}}] = [Z'][I_{\text{int}}] + [W][V_{\text{int}}]$$

---

k is in the fundamental cut-set of tree branch r, and passes between the resulting pair of disconnected subgraphs with the opposite reference direction; and $D_{rk} = 0$ if link k is not in the fundamental cut-set of tree branch r.

so that

$$[D][I_{\text{int}}] = [Y_{\text{int}}][V_{\text{int}}] + [T][I_{\text{int}}] - [J^{\text{ext}}] \qquad (A.5)$$

$$-[D]^T[V_{\text{int}}] = [Z_{\text{int}}][I_{\text{int}}] + [W][V_{\text{int}}] - [U^{\text{ext}}] \qquad (A.6)$$

where $[Y_{\text{int}}]$ and $[Z_{\text{int}}]$ are now general (not necessarily diagonal) square matrices, and $[T]$ and $[W]$ are general rectangular matrices. The analogy of these equations to the Maxwell field equations is preserved if the medium is regarded as bianisotropic (e.g., a magnetoelectric or Tellegen medium).

A.2   RECIPROCITY FOR LUMPED CIRCUITS AND CLASSICAL TRANSMISSION LINES

A.2.1   LUMPED NETWORKS

Consider two possible *states* of the network considered above. That is, consider the same network, with two possible external excitations: $[J^a]$ and $[U^a]$ (state a), or $[J^b]$ and $[U^b]$ (state b).[2] The link voltages and tree branch currents will obey (A.5)-(A.6), with the appropriate source vectors present. We now form the matrix products

$$[V^a_{\text{int}}]^T[D][I^b_{\text{int}}] = [V^a_{\text{int}}]^T[Y_{\text{int}}][V^b_{\text{int}}] + [V^a_{\text{int}}]^T[T][I^b_{\text{int}}] - [V^a_{\text{int}}]^T[J^b]$$

$$-[I^b_{\text{int}}]^T[D]^T[V^a_{\text{int}}] = [I^b_{\text{int}}]^T[Z_{\text{int}}][I^a_{\text{int}}] + [I^b_{\text{int}}]^T[W][V^a_{\text{int}}] - [I^b_{\text{int}}]^T[U^a]$$

These quantities are scalars, and the left sides of these expressions are the negatives of each other by the rule for taking the transpose of a matrix product. Thus adding these equations gives

$$\begin{aligned} 0 = \ & [V^a_{\text{int}}]^T[Y_{\text{int}}][V^b_{\text{int}}] + [V^a_{\text{int}}]^T[T][I^b_{\text{int}}] - [V^a_{\text{int}}]^T[J^b] \\ & + [I^b_{\text{int}}]^T[Z_{\text{int}}][I^a_{\text{int}}] + [I^b_{\text{int}}]^T[W][V^a_{\text{int}}] - [I^b_{\text{int}}]^T[U^a] \end{aligned}$$

A similar relation is obtained by interchanging the states a and b. If we subtract this from the original one, we get

$$\begin{aligned} & [V^a_{\text{int}}]^T\left\{[Y_{\text{int}}] - [Y_{\text{int}}]^T\right\}[V^b_{\text{int}}] + [V^a_{\text{int}}]^T\left\{[T] + [W]^T\right\}[I^b_{\text{int}}] \\ & -[V^b_{\text{int}}]^T\left\{[T] + [W]\right\}[I^a_{\text{int}}] + [I^b_{\text{int}}]^T\left\{[Z_{\text{int}}] - [Z_{\text{int}}]^T\right\}[I^a_{\text{int}}] \\ & -[V^a_{\text{int}}]^T[J^b] + [V^b_{\text{int}}]^T[J^a] - [I^b_{\text{int}}]^T[U^a] + [I^a_{\text{int}}]^T[U^b] = 0 \end{aligned}$$

If the matrices arising from the constitutive equations of the elements and dependent sources obey the conditions that $[Y]$ and $[Z]$ are symmetric, and $[T] = -[W]^T$, then the first four terms in this equation drop out, and we are left with the simpler form

$$[V^a_{\text{int}}]^T[J^b] - [I^a_{\text{int}}]^T[U^b] = [V^b_{\text{int}}]^T[J^a] - [I^b_{\text{int}}]^T[U^a] \qquad (A.7)$$

In other words, under these conditions, the indicated combination of products of impressed voltages of one state with induced currents of another and impressed currents of one with induced voltages of the other does not change when the states are interchanged. This is a form of the *reciprocity* property for networks. One special case for which it is true is that of a lumped element network with no mutual coupling between branches of the network.

---

[2]We drop the superscript ^{ext} hereafter for brevity.

As an application, consider a linear source-free N-port network as in Fig. 2.17. Whatever network is internal to this multiport, choose a tree of that network so that each port of the network is spanned by a branch of the tree, with reference directions for the tree branch voltages chosen the same as for the port voltages. Let the network be driven by ideal current sources J_1, J_2, ..., J_N connected to each port; there are no voltage sources exciting this network. The current sources become the port currents I_1, I_2, ..., I_N (not to be confused with the internally defined link currents). The general reciprocity property above gives us:

$$[V_{\text{int}}^a]^T [I_{\text{int}}^b] = [V_{\text{int}}^b]^T [I_{\text{int}}^a] \tag{A.8}$$

where $[V]$ and $[I]$ are now the $N \times 1$ column vectors representing port voltages and currents respectively. If $[Z_{\text{int}}]$ is now taken to represent the $N \times N$ impedance matrix for the N-port, we find that this $[Z_{\text{int}}]$ matrix must be symmetric if the internal network of the multiport satisfies the criteria given above for reciprocity, and in particular if it is made up entirely of uncoupled lumped impedances or admittances.

A.2.2 TRANSMISSION LINES

Although as stated above we could simply make limiting arguments to derive the reciprocity theorem for classical transmission lines, it is instructive to see the derivation done directly. The classical transmission line equations (with nonuniform line parameters and distributed sources) are:

$$
\begin{aligned}
\frac{dV(z)}{dz} &= -z(z)I(z) + E_{\text{ext}}(z) \\
\frac{dI(z)}{dz} &= -y(z)V(z) + H_{\text{ext}}(z)
\end{aligned}
\tag{A.9}
$$

Considering these equations for two possible states, a and b, arising from two separate excitations of the line, we can obtain:

$$\frac{d}{dz}\left(V^a I^b - V^b I^a\right) = \left(V^a H_{\text{ext}}^b - I^a E_{\text{ext}}^b\right) - \left(V^b H_{\text{ext}}^a - I^b E_{\text{ext}}^a\right)$$

If integrated between two points z_1 and z_2 on the line, this yields

$$\left(V^a I^b - V^b I^a\right)_{z_1}^{z_2} = \int_{z_1}^{z_2} \left[\left(V^a H_{\text{ext}}^b - I^a E_{\text{ext}}^b\right) - \left(V^b H_{\text{ext}}^a - I^b E_{\text{ext}}^a\right)\right] dz \tag{A.10}$$

These are the analogs of the lumped-circuit reciprocity properties of the previous subsection, and of the Lorentz reciprocity properties for Maxwell's equations derived in Chapter 9.

A.3 POWER FLOW IN NETWORKS AND CLASSICAL TRANSMISSION LINES

A.3.1 LUMPED NETWORKS

The time-dependent power flow from all the independent sources into a lumped-element network is the sum of the power flows from the individual sources:

$$p(t) = \sum_s p_s(t) \equiv \sum_k u_k(t) i_{\text{int},k}(t) + \sum_r v_{\text{int},r}(t) j_r(t)$$

where $p_s(t) = u_k(t) i_{\text{int},k}(t)$ for the ideal voltage source u_k in link k, and $i_{\text{int},k}$ is the current flowing out of its positive terminal, or $p_s(t) = v_{\text{int},r}(t) j_r(t)$ for the ideal current source j_r in tree branch r, and $v_{\text{int},r}$ is the voltage across its terminals.

When all sources are time-harmonic with the same angular frequency ω, the product of any pair of resultant time-harmonic quantities (e. g., voltage and current, voltage squared, etc.) can be represented in terms of the corresponding phasors (cf. (1.3)) in the following way:

$$
\begin{aligned}
p(t) &= v(t)i(t) \\
&= \frac{1}{4}[Ve^{j\omega t} + V^*e^{-j\omega t}][Ie^{j\omega t} + I^*e^{-j\omega t}] \\
&= \frac{1}{2}\mathrm{Re}[VI^* + VIe^{2j\omega t}] \\
&= p_{\mathrm{av}} + p_{\mathrm{osc}}(t)
\end{aligned}
$$

Thus, the power consists of a constant part, p_{av}, which is the time-average power flow, and an oscillatory part p_{osc} which has time average value zero, but in this case indicates the temporary exchange of energy back and forth from the terminals where v and i are measured. It is customary for calculations of power flow to define the *complex power*[3]

$$
\hat{P} = \frac{1}{2}VI^* \equiv P + jQ \tag{A.11}
$$

where P is sometimes called the active power and Q the reactive power (not to be confused with the quality factor Q, a notation used elsewhere), and the complex oscillatory power

$$
\hat{P}_{\mathrm{osc}} = \frac{1}{2}VI \tag{A.12}
$$

so that the average and oscillatory powers can be written as

$$
p_{\mathrm{av}} = \mathrm{Re}(\hat{P}) = P \tag{A.13}
$$

and

$$
p_{\mathrm{osc}}(t) = \mathrm{Re}(\hat{P}_{\mathrm{osc}}e^{2j\omega t}) \tag{A.14}
$$

The two complex power quantities are independent of t and are given solely by phasor voltage and current quantities, so they can be used to completely characterize the power flow in a time-harmonic lumped-element network. Note that the reactive power Q does not directly appear in the expression for time-dependent power flow.

It should be noted that when the power is given in terms of a single voltage and current as in the previous paragraph, the magnitudes of $\hat{P}$ and $\hat{P}_{\mathrm{osc}}$ are equal; only their phases are different. However, when a number of such terms is added together to compute a total power flow (as in what follows below), we cannot expect $|\hat{P}| = |\hat{P}_{\mathrm{osc}}|$ any longer unless special conditions hold (such as the phases of all currents being the same, for example). In general, the time-average and oscillatory parts of the power flow are completely unrelated.

The total power flow is related to the energy storage in the network. The same techniques used to derive the reciprocity theorem can be used to obtain this relationship. Consider the total complex oscillatory power *delivered* by all the sources in a network:

$$
P_{\mathrm{osc}} = \sum_s \hat{P}_{s,\mathrm{osc}} = \frac{1}{2}\sum_k U_k I_{\mathrm{int},k} + \frac{1}{2}\sum_r V_{\mathrm{int},r} J_r = \frac{1}{2}\left\{[I_{\mathrm{int}}]^T[U] + [V_{\mathrm{int}}]^T[J]\right\} \tag{A.15}
$$

---

[3]It is conventional to define complex power using the complex conjugate of the current, rather than of the voltage (which would have been equally possible, as the real part of $\hat{P}$ is not affected).

Using the circuit equations (A.3) and (A.4) for a lumped element network with no mutual couplings (this restriction can be relaxed), we have

$$P_{\text{osc}} = \frac{1}{2} \left\{ [I_{\text{int}}]^T [Z_{\text{int}}][I_{\text{int}}] + [V_{\text{int}}]^T [Y_{\text{int}}][V_{\text{int}}] \right\} = P_{L,\text{osc}} + 2j\omega \left(W_{M,\text{osc}} + W_{E,\text{osc}} \right) \quad (A.16)$$

Let us split the matrices $[Z_{\text{int}}]$ and $[Y_{\text{int}}]$ into their real and imaginary parts:

$$[Z_{\text{int}}] = [R] + j\omega[L]; \qquad [Y_{\text{int}}] = [G] + j\omega[C] \qquad (A.17)$$

where $[R]$ and $[L]$ denote the matrices consisting of self and mutual link resistances and inductances, while $[G]$ and $[C]$ denote the matrices consisting of self and mutual branch conductances and capacitances. Then

$$P_{L,\text{osc}} = \frac{1}{2} \left\{ [I_{\text{int}}]^T [R][I_{\text{int}}] + [V_{\text{int}}]^T [G][V_{\text{int}}] \right\} \qquad (A.18)$$

is the complex oscillatory power dissipated in the circuit resistances and conductances, and

$$W_{E,\text{osc}} = \frac{1}{4}[V_{\text{int}}]^T [C][V_{\text{int}}]; \qquad W_{M,\text{osc}} = \frac{1}{4}[I_{\text{int}}]^T [L][I_{\text{int}}] \qquad (A.19)$$

are the complex oscillatory electric and magnetic stored energies in the circuit capacitances and inductances respectively.

Likewise, the complex power $\hat{P}$ delivered to the circuit can be expressed as[4]

$$\hat{P} = \frac{1}{2} \left\{ [V_{\text{int}}]^T [J]^* + [I_{\text{int}}]^\dagger [U] \right\} = p_{L,\text{av}} + 2j\omega \left(w_{M,\text{av}} - w_{E,\text{av}} \right) \qquad (A.20)$$

where

$$p_{L,\text{av}} = \frac{1}{2} \left\{ [I_{\text{int}}]^\dagger [R][I_{\text{int}}] + [V_{\text{int}}]^\dagger [G][V_{\text{int}}] \right\} \qquad (A.21)$$

is the time-average power lost to the circuit conductances and resistances, and

$$w_{E,\text{av}} = \frac{1}{4}[V_{\text{int}}]^\dagger [C][V_{\text{int}}]; \qquad w_{M,\text{av}} = \frac{1}{4}[I_{\text{int}}]^\dagger [L][I_{\text{int}}] \qquad (A.22)$$

are the time-average electric and magnetic stored energies in the circuit capacitances and inductances respectively. Note that the reactive power Q delivered to the network is twice the angular frequency ω times the *difference* between the average stored magnetic energy and average stored electric energy in the circuit. It is thus positive if more magnetic energy is stored than electric, and negative otherwise.

In the special case of a lossless LC-network with only frequency-independent inductors in the links and frequency-independent capacitors in the tree branches, excited by independent sources $[J]$ and $[U]$ that are also independent of ω, we can perform similar manipulations with the frequency derivatives $[V_{\text{int},\omega}] = d[V_{\text{int}}]/d\omega$ and $[I_{\text{int},\omega}] = d[I_{\text{int}}]/d\omega$ of $[V_{\text{int}}]$ and $[I_{\text{int}}]$ and the matrix equations they satisfy. Leaving the details to the reader, we arrive at the following identity:

$$w_{M,\text{av}} + w_{E,\text{av}} = -\frac{j}{4} \left([V_{\text{int},\omega}]^T [J]^* + [I_{\text{int},\omega}]^T [U]^* \right) \qquad (A.23)$$

Now if we set $[U] = 0$ and regard the current sources $[J]$ as applied to the ports of an N-port network whose $N \times N$ impedance matrix is $[Z]$ as in Chapter 2, we find that

$$w_{M,\text{av}} + w_{E,\text{av}} = -\frac{j}{4}[J]^\dagger [Z_\omega][J] = \frac{1}{4}[J]^\dagger [X_\omega][J] \qquad (A.24)$$

---

[4]The superscript $\dagger$ denotes the Hermitian conjugate, or complex conjugate of the transpose, of a matrix.

since the impedance matrix for a lossless network is purely reactive: $[Z] = j[X]$. Similarly, driving the ports of an N-port network with voltage sources $[U]$ only, we have

$$w_{M,\text{av}} + w_{E,\text{av}} = \frac{1}{4}[U]^\dagger[B_\omega][U] \qquad (A.25)$$

where the admittance matrix of the network is purely susceptive: $[Y] = j[B]$. In both cases, the subscript ω again denotes differentiation with respect to frequency. Thus the *total* time-average stored energy (in a lossless network only) can be related to the frequency derivatives of the reactance or susceptance matrices of a multiport. For a lossy network, no such general statement is possible.

A.3.2 TRANSMISSION LINES

As with the reciprocity property, we will derive the power flow results for classical transmission lines directly. The power carried toward positive z at the point z by the current and voltage on a classical transmission line is $p(z,t) = v(z,t)i(z,t)$. As with the lumped-element circuit, when v and i are time-harmonic and expressed in terms of the phasors $V(z)$ and $I(z)$, we have

$$
\begin{aligned}
p(z,t) &= \frac{1}{4}[V(z)e^{j\omega t} + V^*(z)e^{-j\omega t}][I(z)e^{j\omega t} + I^*(z)e^{-j\omega t}] \\
&= \frac{1}{2}\text{Re}[V(z)I^*(z) + V(z)I(z)e^{2j\omega t}] \qquad (A.26) \\
&= p_{\text{av}}(z) + p_{\text{osc}}(z,t) \qquad (A.27)
\end{aligned}
$$

Defining the complex power

$$\hat{P}(z) = \frac{1}{2}V(z)I^*(z) \qquad (A.28)$$

and the complex oscillatory power

$$\hat{P}_{\text{osc}}(z) = \frac{1}{2}V(z)I(z) \qquad (A.29)$$

at a position z along the line, the average and oscillatory powers can be written as

$$p_{\text{av}}(z) = \text{Re}[\hat{P}(z)] \qquad (A.30)$$

and

$$p_{\text{osc}}(z,t) = \text{Re}[\hat{P}_{\text{osc}}(z)e^{2j\omega t}] \qquad (A.31)$$

The two complex power quantities depend only on z and are given solely by phasor voltage and current quantities, and can be used to completely characterize the power flow in a classical transmission line.

The complex power quantities can be related to stored and dissipated energies on the line. Consider a line on which the line parameters can possibly be functions of the position z on the line. Then the telegrapher's equations are again as in (A.9). Using these, we can obtain the identities:

$$\frac{d}{dz}\hat{P}(z) = -\frac{1}{2}\left\{z|I|^2 + y^*|V|^2\right\} \qquad (A.32)$$

$$\frac{d}{dz}\hat{P}_{\text{osc}}(z) = -\frac{1}{2}\left\{zI^2 + yV^2\right\} \qquad (A.33)$$

The real part of (A.32) relates the change in power flow with respect to z to the time-average rate of energy dissipation $[r|I|^2 + g|V|^2]/2$ per unit length along the line. The imaginary part of (A.32) has no direct interpretation in terms of energy conservation (since it plays no part in the description of time dependent power flow), but does relate the rate of change of the imaginary part of $\hat{P}$ to the difference between stored time-average inductive energy $l|I|^2/4$ and capacitive energy $c|V|^2/4$. Equation (A.33) relates the oscillatory power flow to the oscillatory parts of the stored and dissipated energies.

A.3.3 DECIBELS (dB)

The power flow from a source to a passive load is a positive number. The unit for power is a Watt (W), but often a relative unit of power, the *decibel* (dB) is used to express the ratio of two powers in a convenient way:

$$P_{\mathrm{dB}} = 10 \log\left(\frac{P_1}{P_2}\right)$$

and we say that P_1 is P_{dB} decibels above (or larger than) P_2. A single power level can be expressed in the form

$$P_{\mathrm{dB}} = 10 \log\left(\frac{P}{P_{\mathrm{ref}}}\right), \tag{A.34}$$

where P is the power we are measuring or calculating, and P_{ref} is some given reference power level. At microwave frequencies, very often this reference power level is 1 mW, and in this case the unit is called a dBm:

$$P_{\mathrm{dBm}} = 10 \log\left(\frac{P}{1\mathrm{mW}}\right) \tag{A.35}$$

but many other reference power levels are in common use.

A positive number of decibels corresponds to a ratio greater than 1 (gain), while a negative number represents a ratio less than 1 (loss). Decibels are convenient for two reasons: (1) they are easier to write (for example, the range between +63 dB to -153 dB corresponds to $2 \cdot 10^6$ to a factor of 0.5×10^{-15}), and (2) adding decibels corresponds to multiplication, which is useful whenever there are several stages cascaded in some system (for example, in a multistage amplifier, the gain can be found simply by adding the individual gains in dB).

Since power is proportional to the square of voltage in a circuit or on a transmission line, we can also use voltage ratios to obtain decibel levels for gain or loss, *provided the same reference impedance or characteristic impedance is used for both voltages being compared.* The same is, incidentally, true of currents. Thus, a voltage V_1 is said to be

$$P_{\mathrm{dB}} = 20 \log\left(\frac{V_1}{V_2}\right)$$

decibels above the voltage V_2 on a transmission line of identical characteristic impedance. It is important to keep in mind that the decibel is always a measure of a *power* ratio; to say that voltages are so many dB apart is somewhat an abuse of the terminology, although a common one. In an obvious way, we can also define dB relative to some given voltage reference level, such as the dBμV: the number of decibels above 1 microvolt. Although it is done less frequently, we can express ratios of currents, or electric or magnetic fields, in dB as well.

A.4 NOTES AND REFERENCES

Good sources for the treatment of circuit analysis, including modern aspects such as cut-set analysis, are

Seshu and Balabanian (1959), Chapters 3 and 4.
Carlin and Giordano (1964), Chapter 1.
Balabanian and Bickart (1969), Chapters 2-4.
Desoer and Kuh (1969), Chapters 9-11.

and also

T. R. Bashkow, "The A matrix, new network description," *IRE Trans. Circ. Theory*, vol. 4, pp. 117-119 (1957).

C.-W. Ho, A. E. Ruehli and P. A. Brennan, "The modified nodal approach to network analysis," *IEEE Trans. Circ. Syst.*, vol. 22, pp. 504-509 (1975).

Some treatments emphasizing the algebraic-topological aspects of network analysis and the formal analogies between the network equations and Maxwell's equations (using differential forms) are given in

F. H. Branin, Jr., "The algebraic-topological basis for network analogies and the vector calculus," in *Proceedings of the Symposium on Generalized Networks* (J. Fox, ed.). Brooklyn, NY: Polytechnic Press (1966), pp. 453-491.

Strang (1986), Sect. 2.3.

G. Strang, "A framework for equilibrium equations," *SIAM Review*, vol. 30, pp. 283-297 (1988).

Bamberg and Sternberg (1990).

B FORMULAS FROM VECTOR, MATRIX AND DYADIC ANALYSIS

B.1 VECTORS

An ordered set of numbers is called a *vector*. Sometimes this is a set of similar physical quantities, such as the voltages at the numbered nodes of a circuit. Other times, a vector may consist of exactly three numbers which represent the components in some coordinate system of a directed physical quantity such as current density, velocity, force, electric field, etc. The latter are often called Gibbsian vectors, and are denoted by boldface letters in this book. Some formulas from the algebra and analysis of Gibbsian vectors are given below.

We denote the unit vector in a given coordinate direction (say, v) by $\mathbf{u}_v$. Hence in Cartesian coordinates we have $(\mathbf{u}_x, \mathbf{u}_y, \mathbf{u}_z)$ and in cylindrical coordinates we have $(\mathbf{u}_\rho, \mathbf{u}_\phi, \mathbf{u}_z)$. A vector is generally represented in boldface, as $\mathbf{A}$. When a vector has no z-component we call it a transverse vector, and denote it by $\mathbf{A}_T$, for example. The operator ∇ is defined as

$$\nabla = \mathbf{u}_x \frac{\partial}{\partial x} + \mathbf{u}_y \frac{\partial}{\partial y} + \mathbf{u}_z \frac{\partial}{\partial z} \tag{B.1}$$

in Cartesian coordinates, while

$$\nabla_T = \mathbf{u}_x \frac{\partial}{\partial x} + \mathbf{u}_y \frac{\partial}{\partial y} \tag{B.2}$$

We can define certain vector differentiations on vectors (in Cartesian coordinates only) by:

$$\nabla^2 \mathbf{A} \equiv \mathbf{u}_x \nabla^2 A_x + \mathbf{u}_y \nabla^2 A_y + \mathbf{u}_z \nabla^2 A_z \tag{B.3}$$

$$(\mathbf{A} \cdot \nabla) \mathbf{B} \equiv A_x \frac{\partial \mathbf{B}}{\partial x} + A_y \frac{\partial \mathbf{B}}{\partial y} + A_z \frac{\partial \mathbf{B}}{\partial z} \tag{B.4}$$

The vector expressed in Cartesian coordinates as

$$\mathbf{r} = x\mathbf{u}_x + y\mathbf{u}_y + z\mathbf{u}_z \tag{B.5}$$

is called the *position vector*. In cylindrical coordinates it is

$$\mathbf{r} = \rho \mathbf{u}_\rho + z \mathbf{u}_z \tag{B.6}$$

and in spherical coordinates it is

$$\mathbf{r} = r\mathbf{u}_r \tag{B.7}$$

The following vector identities are often found useful; each holds equally well if ∇ is replaced by ∇_T:

$$\mathbf{A} \cdot (\mathbf{B} \times \mathbf{C}) = \mathbf{B} \cdot (\mathbf{C} \times \mathbf{A}) = \mathbf{C} \cdot (\mathbf{A} \times \mathbf{B}) \tag{B.8}$$

$$\mathbf{A} \times (\mathbf{B} \times \mathbf{C}) = (\mathbf{A} \cdot \mathbf{C})\mathbf{B} - (\mathbf{A} \cdot \mathbf{B})\mathbf{C} \tag{B.9}$$

$$\nabla(fg) = f\nabla g + g\nabla f \tag{B.10}$$

$$\nabla \cdot (f\mathbf{A}) = f\nabla \cdot \mathbf{A} + \mathbf{A} \cdot \nabla f \tag{B.11}$$

$$\nabla \times (f\mathbf{A}) = f\nabla \times \mathbf{A} - \mathbf{A} \times \nabla f \tag{B.12}$$

$$\nabla \cdot (\mathbf{A} \times \mathbf{B}) = \mathbf{B} \cdot \nabla \times \mathbf{A} - \mathbf{A} \cdot \nabla \times \mathbf{B} \tag{B.13}$$

$$\nabla \times \nabla f = 0 \tag{B.14}$$

$$\nabla \cdot (\nabla \times \mathbf{A}) = 0 \tag{B.15}$$

$$\nabla \times (\nabla \times \mathbf{A}) = \nabla (\nabla \cdot \mathbf{A}) - \nabla^2 \mathbf{A} \tag{B.16}$$

$$\nabla (\mathbf{A} \cdot \mathbf{B}) = (\mathbf{A} \cdot \nabla)\mathbf{B} + (\mathbf{B} \cdot \nabla)\mathbf{A} + \mathbf{A} \times (\nabla \times \mathbf{B}) + \mathbf{B} \times (\nabla \times \mathbf{A}) \tag{B.17}$$

There also hold the following miscellaneous identities:

$$\mathbf{u}_z \times (\mathbf{u}_z \times \mathbf{A}_T) = -\mathbf{A}_T \tag{B.18}$$

$$\mathbf{u}_z \times (\nabla \times \mathbf{A}) = \nabla A_z - \frac{\partial \mathbf{A}}{\partial z} \tag{B.19}$$

$$\nabla_T \times (\mathbf{u}_z \times \mathbf{A}_T) = \mathbf{u}_z (\nabla_T \cdot \mathbf{A}_T) \tag{B.20}$$

If V is a volume bounded by a closed surface S and $\mathbf{u}_n$ is the outward unit normal to S, then we have the divergence theorem

$$\int_V (\nabla \cdot \mathbf{A})\, dV = \oint_S \mathbf{u}_n \cdot \mathbf{A}\, dS \tag{B.21}$$

and its variants

$$\int_V (\nabla f)\, dV = \oint_S f\mathbf{u}_n\, dS \tag{B.22}$$

$$\int_V (\nabla \times \mathbf{A})\, dV = \oint_S (\mathbf{u}_n \times \mathbf{A})\, dS \tag{B.23}$$

If S is an open surface whose unit normal is $\mathbf{u}_n$ and whose closed boundary is the contour C, then we have Stokes' theorem

$$\int_S (\nabla \times \mathbf{A}) \cdot \mathbf{u}_n\, dS = \oint_C \mathbf{A} \cdot d\boldsymbol{l} \tag{B.24}$$

and its variant

$$\int_S (\mathbf{u}_n \times \nabla f)\, dS = \oint_C f\, d\boldsymbol{l} \tag{B.25}$$

where the direction of the vector line element $d\boldsymbol{l}$ is related to $\mathbf{u}_n$ by the right-hand rule. If S is a surface lying in a plane $z = \text{constant}$, C is its closed boundary, and $\mathbf{u}_n$ is an outward unit normal to C lying in the plane of S, then

$$\int_S (\nabla_T \cdot \mathbf{A}_T)\, dS = \oint_C \mathbf{A}_T \cdot \mathbf{u}_n\, dl \tag{B.26}$$

$$\int_S (\nabla_T f)\, dS = \oint_C f\mathbf{u}_n\, dl \tag{B.27}$$

where dl is the scalar arc length element on C.

B.1.1   VECTOR DIFFERENTIATION IN RECTANGULAR (CARTESIAN) COORDINATES

The expressions for the standard vector differential operations in Cartesian coordinates are:

$$\nabla f = \mathbf{u}_x \frac{\partial f}{\partial x} + \mathbf{u}_y \frac{\partial f}{\partial y} + \mathbf{u}_z \frac{\partial f}{\partial z} \tag{B.28}$$

$$\nabla \cdot \mathbf{A} = \frac{\partial A_x}{\partial x} + \frac{\partial A_y}{\partial y} + \frac{\partial A_z}{\partial z} \tag{B.29}$$

$$\nabla \times \mathbf{A} = \mathbf{u}_x \left(\frac{\partial A_z}{\partial y} - \frac{\partial A_y}{\partial z} \right) + \mathbf{u}_y \left(\frac{\partial A_x}{\partial z} - \frac{\partial A_z}{\partial x} \right) + \mathbf{u}_z \left(\frac{\partial A_y}{\partial x} - \frac{\partial A_x}{\partial y} \right) \tag{B.30}$$

B.1.2   VECTOR DIFFERENTIATION IN CIRCULAR CYLINDRICAL COORDINATES

The expressions for the standard vector differential operations in circular cylindrical coordinates are:

$$\nabla f = \mathbf{u}_\rho \frac{\partial f}{\partial \rho} + \mathbf{u}_\phi \frac{1}{\rho} \frac{\partial f}{\partial \phi} + \mathbf{u}_z \frac{\partial f}{\partial z} \tag{B.31}$$

$$\nabla \cdot \mathbf{A} = \frac{1}{\rho} \frac{\partial (\rho A_\rho)}{\partial \rho} + \frac{1}{\rho} \frac{\partial A_\phi}{\partial \phi} + \frac{\partial A_z}{\partial z} \tag{B.32}$$

$$\nabla \times \mathbf{A} = \mathbf{u}_\rho \left(\frac{1}{\rho} \frac{\partial A_z}{\partial \phi} - \frac{\partial A_\phi}{\partial z} \right) + \mathbf{u}_\phi \left(\frac{\partial A_\rho}{\partial z} - \frac{\partial A_z}{\partial \rho} \right) + \mathbf{u}_z \left(\frac{1}{\rho} \frac{\partial (\rho A_\phi)}{\partial \rho} - \frac{1}{\rho} \frac{\partial A_\rho}{\partial \phi} \right) \tag{B.33}$$

B.1.3   VECTOR DIFFERENTIATION IN SPHERICAL COORDINATES

The expressions for the standard vector differential operations in spherical coordinates are:

$$\nabla f = \mathbf{u}_r \frac{\partial f}{\partial r} + \mathbf{u}_\theta \frac{1}{r} \frac{\partial f}{\partial \theta} + \mathbf{u}_\phi \frac{1}{r \sin \theta} \frac{\partial f}{\partial \phi} \tag{B.34}$$

$$\nabla \cdot \mathbf{A} = \frac{1}{r^2} \frac{\partial (r^2 A_r)}{\partial r} + \frac{1}{r \sin \theta} \frac{\partial (\sin \theta A_\theta)}{\partial \theta} + \frac{1}{r \sin \theta} \frac{\partial A_\phi}{\partial \phi} \tag{B.35}$$

$$\nabla \times \mathbf{A} = \mathbf{u}_r \frac{1}{r \sin \theta} \left(\frac{\partial (\sin \theta A_\phi)}{\partial \theta} - \frac{\partial A_\theta}{\partial \phi} \right) + \mathbf{u}_\theta \frac{1}{r} \left(\frac{1}{\sin \theta} \frac{\partial A_r}{\partial \phi} - \frac{\partial (r A_\phi)}{\partial r} \right) + \mathbf{u}_\phi \frac{1}{r} \left(\frac{\partial (r A_\theta)}{\partial r} - \frac{\partial A_r}{\partial \theta} \right)$$
$$\tag{B.36}$$

B.2   MATRICES

In linear algebra, one often has occasion to transform an ordered set of numbers into another; this kind of transformation is called a *matrix*. Some formulas from matrix algebra are given below.

If $[A]$ and $[B]$ are square, $N \times N$ matrices, with elements A_{ik} and B_{ik} respectively, then $[A]^T$ (the transpose of $[A]$ is the square matrix whose elements are A_{ki}, and the Hermitian conjugate $[A]^\dagger$ of $[A]$ is the square matrix whose elements are A_{ki}^*, where * denotes the complex conjugate. A *symmetric* matrix is one for which $[A]^T = [A]$. A *Hermitian* matrix is one for which $[A]^\dagger = [A]$. A matrix all of whose elements are zero is denoted $[\mathbf{0}]$, although sometimes we write it simply as 0. The identity matrix $[\mathbf{1}]$ is the identity matrix, defined by $\mathbf{1}_{ii} = 1$, and $\mathbf{1}_{ik} = 0$ for $i \neq k$. The inverse $[A]^{-1}$ of a square matrix is defined by

$$[A]^{-1}[A] = [A][A]^{-1} = [\mathbf{1}] \tag{B.37}$$

The inverse matrix does not always exist; if it does not, $[A]$ is said to be *singular*. If a nonsingular matrix is partitioned as follows:

$$[A] = \begin{bmatrix} [E] & [F] \\ [G] & [H] \end{bmatrix}$$

where $[E]$, $[F]$, $[G]$ and $[H]$ are suitable matrices of smaller size, then its inverse is expressed as:

$$[A]^{-1} = \begin{bmatrix} [E]^{-1} + [E]^{-1}[F][M]^{-1}[G][E]^{-1} & -[E]^{-1}[F][M]^{-1} \\ -[M]^{-1}[G][E]^{-1} & [M]^{-1} \end{bmatrix} \tag{B.38}$$

where

$$[M] = [H] - [G][E]^{-1}[F]$$

provided that $[E]$ and $[M]$ are nonsingular. Repeated application of (B.38) allows the inverse of any nonsingular square matrix to be computed. The following identities can be useful:

$$([A][B])^T = [B]^T[A]^T \tag{B.39}$$

$$([A][B])^{-1} = [B]^{-1}[A]^{-1} \tag{B.40}$$

$$([A]^T)^{-1} = ([A]^{-1})^T \tag{B.41}$$

A Hermitian matrix $[A]$ for which

$$[X]^\dagger[A][X] > 0 \qquad \text{for any } [X] \neq [\mathbf{0}]$$

is said to be *positive definite* (sometimes this is expressed as $[A] \succ [\mathbf{0}]$), while if merely

$$[X]^\dagger[A][X] \geq 0 \qquad \text{for any } [X]$$

we say that $[A]$ is *positive semidefinite* ($[A] \succeq [\mathbf{0}]$). We define a kind of matrix inequality $[A] \succ [B]$ of two Hermitian matrices $[A]$ and $[B]$ to mean that the matrix $[A] - [B]$ is positive definite, and $[A] \succeq [B]$ to mean that $[A] - [B]$ is positive semidefinite. It is important to recognize that $[A] \succ [B]$ does *not* necessarily mean that $A_{11} > B_{11}$, $A_{12} > B_{12}$, etc., especially since the off-diagonal elements of a Hermitian matrix are complex in general.

An N-element column vector $[X]$ is constructed from a set of elements $X_i; i = 1, 2, \ldots, N$. Then its transpose $[X]^T$ is an N-element row vector whose elements are X_i. Two N-element column vectors $[X]$ and $[Y]$ are said to be orthogonal if $[Y]^T[X] = 0$. If $[X_1], [X_2], \ldots, [X_M]$ (with $M \leq N$) are N-element column vectors such that $[X_m]^T[X_m] \neq 0$ for $m = 1, \ldots, M$, we can always form a set of M linear combinations of them that are all orthogonal to each other. One method for doing this is called Gram-Schmidt orthogonalization. The procedure is to construct a new set of vectors $[Y_1], [Y_2], \ldots, [Y_M]$ as follows. First, put

$$[Y_1] = [X_1] \tag{B.42}$$

Next, form the vector

$$[Y_2] = [X_2] + a[X_1] \tag{B.43}$$

where the constant a is chosen to make $[Y_2]^T[Y_1] = 0$, i.e.,

$$a = -\frac{[X_2]^T[X_1]}{[X_1]^T[X_1]} \tag{B.44}$$

A similar procedure is used to construct each successive $[Y_i]$. For example, form the vector

$$[Y_3] = [X_3] + b[X_2] + c[X_1] \tag{B.45}$$

and choose the constants b and c to make $[Y_3]^T[Y_1] = [Y_3]^T[Y_2] = 0$.

B.3 DYADICS OR TENSORS

When we deal with vectors in Cartesian 3-space such as electromagnetic fields, currents, etc., it is common practice to refer to linear transformations of them not as matrices, but as *dyadics* or second-rank *tensors*. Such tensors are defined to be sums of one or more terms called *dyads*. A dyad is an association of two vectors—not a scalar or vector product—which can operate on ordinary vectors according to the rules we lay out below. We write a dyad $\overset{\leftrightarrow}{D}$ of two vectors $\mathbf{A}$ and $\mathbf{B}$ symbolically as

$$\overset{\leftrightarrow}{D} = \mathbf{AB} \tag{B.46}$$

without the use of a dot or cross-product sign. We then define right and left dot products of $\overset{\leftrightarrow}{D}$ with a third vector $\mathbf{F}$ as

$$\begin{aligned}
\mathbf{F} \cdot \overset{\leftrightarrow}{D} &= (\mathbf{F} \cdot \mathbf{A})\mathbf{B} \\
\overset{\leftrightarrow}{D} \cdot \mathbf{F} &= \mathbf{A}(\mathbf{B} \cdot \mathbf{F})
\end{aligned} \tag{B.47}$$

In this sense, we can think of $\overset{\leftrightarrow}{D}$ as if it were the 3×3 matrix:

$$\mathbf{D} \rightarrow \begin{bmatrix} A_x B_x & A_x B_y & A_x B_z \\ A_y B_x & A_y B_y & A_y B_z \\ A_z B_x & A_z B_y & A_z B_z \end{bmatrix} \tag{B.48}$$

The transpose of the dyad $\overset{\leftrightarrow}{D}$ given by (B.46) is defined to be

$$\overset{\leftrightarrow}{D}^T = \mathbf{BA} \tag{B.49}$$

in analogy with the transpose of a matrix.

The identity dyadic or tensor is defined by

$$\overset{\leftrightarrow}{I} = \mathbf{u}_x \mathbf{u}_x + \mathbf{u}_y \mathbf{u}_y + \mathbf{u}_z \mathbf{u}_z \tag{B.50}$$

The inverse $\overset{\leftrightarrow}{D}^{-1}$ of a dyadic $\overset{\leftrightarrow}{D}$ is defined as the dyadic associated with the inverse of the matrix associated with $\overset{\leftrightarrow}{D}$. In other words, if $\overset{\leftrightarrow}{D} \cdot \mathbf{F} = \mathbf{G}$, then $\mathbf{F} = \overset{\leftrightarrow}{D}^{-1} \cdot \mathbf{G}$, or more simply,

$\overset{\leftrightarrow}{D}{}^{-1} \cdot \overset{\leftrightarrow}{D} = \overset{\leftrightarrow}{I}$. We may take the gradient of a vector to get a dyadic; the definition is related to (B.4):

$$\nabla \mathbf{A} \equiv \mathbf{u}_x \frac{\partial \mathbf{A}}{\partial x} + \mathbf{u}_y \frac{\partial \mathbf{A}}{\partial y} + \mathbf{u}_z \frac{\partial \mathbf{A}}{\partial z} \tag{B.51}$$

As an example, the gradient of the position vector is the identity dyadic:

$$\nabla \mathbf{r} = \overset{\leftrightarrow}{I} \tag{B.52}$$

An operation which is a little less commonly encountered with matrices is the cross product, but it can readily be defined for dyads as:

$$\begin{aligned} \mathbf{F} \times \overset{\leftrightarrow}{D} &= (\mathbf{F} \times \mathbf{A})\mathbf{B} \\ \overset{\leftrightarrow}{D} \times \mathbf{F} &= \mathbf{A}(\mathbf{B} \times \mathbf{F}) \end{aligned} \tag{B.53}$$

This definition is readily extended to include any tensor by superposition. Notice that the result of taking the cross product of a tensor with a vector is once again a tensor.

B.4   PROBLEMS

pB-1 Prove that $\mathbf{u}_z \times (\mathbf{u}_z \times \mathbf{A}_T) = -\mathbf{A}_T$ for any transverse vector $\mathbf{A}_T$.

pB-2 Show that:

$$\nabla r = \mathbf{u}_r$$

$$\nabla \left(\frac{1}{r} \right) = -\frac{\mathbf{u}_r}{r^2} = -\frac{\mathbf{r}}{r^3}$$

and

$$\nabla \left(\frac{1}{r^2} \right) = -\frac{2\mathbf{u}_r}{r^3} = -\frac{2\mathbf{r}}{r^4}$$

pB-3 Show that

$$\nabla^2 \left(\frac{1}{r} \right) = 0$$

if $r \neq 0$.

pB-4 Show that

$$\nabla^2 \left(\frac{e^{-jkr}}{r} \right) = -k^2 \frac{e^{-jkr}}{r}$$

if $r \neq 0$.

pB-5 Show that

$$\nabla \left(\frac{\mathbf{r}}{r^2} \right) = \frac{\overset{\leftrightarrow}{I}}{r^2} - 2\frac{\mathbf{u}_r \mathbf{u}_r}{r^2} = \frac{\overset{\leftrightarrow}{I}}{r^2} - 2\frac{\mathbf{r}\mathbf{r}}{r^4}$$

where $\overset{\leftrightarrow}{I}$ is the identity dyadic.

pB-6 Show that

$$\nabla \nabla \left(\frac{1}{r} \right) = 3\frac{\mathbf{u}_r \mathbf{u}_r}{r^3} - \frac{\overset{\leftrightarrow}{I}}{r^3} = 3\frac{\mathbf{r}\mathbf{r}}{r^5} - \frac{\overset{\leftrightarrow}{I}}{r^3}$$

where $\overset{\leftrightarrow}{I}$ is the identity dyadic.

pB-7 Show that

$$\nabla \nabla \left(\frac{e^{-jkr}}{r} \right) = e^{-jkr} \left[\left(3\frac{\mathbf{r}\mathbf{r}}{r^5} - \frac{\overset{\leftrightarrow}{I}}{r^3} \right) (1 + jkr) - k^2 \frac{\mathbf{r}\mathbf{r}}{r^3} \right]$$

where $\overset{\leftrightarrow}{I}$ is the identity dyadic.

C SPECIAL FUNCTIONS

C.1 BESSEL FUNCTIONS

Bessel's differential equation

$$x^2 \frac{d^2 f}{dx^2} + x \frac{df}{dx} + \left(x^2 - \nu^2\right) f = 0 \tag{C.1}$$

is a linear, second-order ordinary differential equation, and therefore will have two linearly independent solutions. One of these can be obtained by power series methods, and is given by

$$
\begin{aligned}
f(x) &= J_\nu(x) \\
&= \sum_{m=0}^{\infty} \frac{(-1)^m (x/2)^{2m+\nu}}{m! \, \Gamma(m + \nu + 1)}
\end{aligned}
\tag{C.2}
$$

where Γ is the gamma function or factorial function (see Section C.2). The function $J_\nu(x)$ is known as the Bessel function of the *first* kind of *index* ν (or *order* ν) and argument x.

Clearly $J_{-\nu}(x)$ is also a solution of (C.1). However, when $\nu = n$ is an integer, it turns out that $J_{-n}(x)$ is not an independent solution; in fact

$$J_{-n}(x) = (-1)^n J_n(x) \tag{C.3}$$

So instead of using $J_{-\nu}(x)$ as our second solution, we define the Bessel function of the second kind $Y_\nu(x)$ as:

$$Y_\nu(x) = \frac{J_\nu(x) \cos \nu\pi - J_{-\nu}(x)}{\sin \nu\pi} \tag{C.4}$$

Even though this expression becomes indeterminate $(0/0)$ when $\nu \to n$, the limit exists and is still independent of $J_n(x)$. As a consequence of (C.3) and (C.4), we have also

$$Y_{-n}(x) = (-1)^n Y_n(x) \tag{C.5}$$

The following *Wronskian* relations exist between two independent solutions of Bessel's equation:

$$J_\nu(z) Y'_\nu(z) - Y_\nu(z) J'_\nu(z) = \frac{2}{\pi z} \tag{C.6}$$

$$J_\nu(z) J'_{-\nu}(z) - J_{-\nu}(z) J'_\nu(z) = -\frac{2 \sin \nu\pi}{\pi z} \tag{C.7}$$

Plots of J_0, J_1, Y_0, and Y_1 are given in Figure C.1. Notice that these functions oscillate much like sinusoidal functions, but do not have precisely regular periods and decay slowly as x increases. Also note that Y_0 and Y_1 (and indeed all the Y_ν functions) have a singularity (i.e., "blow up") at the origin $x = 0$, while J_0, J_1, etc., are well-behaved there. In fact, by (C.2):

$$J_\nu(x) \simeq \frac{1}{\Gamma(\nu + 1)} \left(\frac{x}{2}\right)^\nu ; \qquad (\nu \neq -1, -2, \ldots) \tag{C.8}$$

as $x \to 0$, but

$$Y_\nu(x) \simeq -\frac{\Gamma(\nu)}{\pi} \left(\frac{2}{x}\right)^\nu ; \qquad (\nu > 0) \tag{C.9}$$

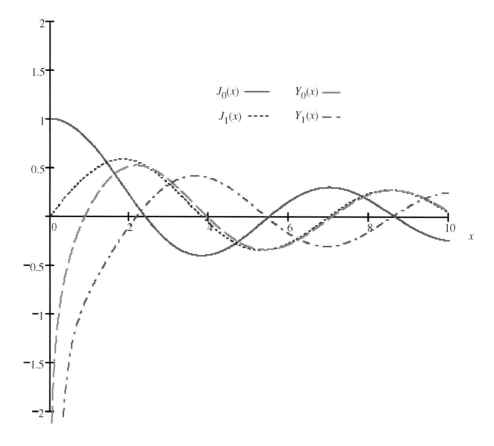

Figure C.1 Graphs of $J_0(x)$, $Y_0(x)$, $J_1(x)$ and $Y_1(x)$.

in the same limit. For the special cases of $\nu = 0$ or 1,

$$Y_0(x) \simeq \frac{2}{\pi}\left[\ln\left(\frac{x}{2}\right) + \gamma_E\right]$$

$$Y_1(x) \simeq -\frac{2}{\pi x} \tag{C.10}$$

where

$$\gamma_E = .57721\ldots \tag{C.11}$$

is known as *Euler's constant*, while

$$J_0(x) \simeq 1; \qquad J_1(x) \simeq \frac{x}{2} \tag{C.12}$$

as $x \to 0$.

The zeroes of J_0 and Y_0, J_1 and Y_1, etc., *interlace*, like those of the sine and cosine, but unlike those of the trigonometric functions, are *not* equally spaced. The positive zeroes of the Bessel function $J_n(x)$ are denoted by j_{nm}, n denoting the index of the Bessel function, and m denoting the ordering of the zero in increasing sequence. Equation (C.6) implies that the zeroes of $J_n(x)$ are simple; that is, that $J_n'(j_{nm}) \neq 0$; since $Y_n(x)$ is finite for $x > 0$. The first several such zeroes are listed in Table C.1 to five decimal places of accuracy. The first

Table C.1

The first three nonzero roots j_{nm} of the Bessel function $J_n(x)$ for $n = 0$ to 4, to five decimal places.

m \ n	0	1	2	3	4
1	2.40483	3.83171	5.13562	6.38016	7.58834
2	5.52008	7.01559	8.41724	9.76102	11.06471
3	8.65373	10.17347	11.61984	13.01520	14.37254

Table C.2

The first three nonzero roots j'_{nm} of $J'_n(x)$ for $n = 0$ to 4, to five decimal places.

m \ n	0	1	2	3	4
1	3.83171	1.84118	3.05424	4.20119	5.31755
2	7.01559	5.33144	6.70613	8.01524	9.28240
3	10.17347	8.53632	9.96947	11.34592	12.68191

three zeroes of $J_0(x)$ have also been indicated in Figure C.1. It is also sometimes necessary to know the zeroes of $J'_n(x)$. We denote these by j'_{nm}, so that $J'_n(j'_{nm}) = 0$. The first few values of j'_{nm} are given in Table C.2. Note that the roots of j'_{1m} are all *smaller* in magnitude than those corresponding to other values of n. If $|\nu| \geq 4$, $\mathrm{Re}(\nu) > 0$ and $m = 1$, 2 or 3, an approximate formula for the mth nonzero root of the Bessel function $J_\nu(x)$ is:

$$j_{\nu,m} \simeq \nu + r_m \left(\frac{\nu}{2}\right)^{1/3} + \frac{3r_m^2}{20}\left(\frac{\nu}{2}\right)^{-1/3} \tag{C.13}$$

Likewise, the mth nonzero root of the derivative $J'_\nu(x)$ is:

$$j'_{\nu,m} \simeq \nu + r'_m \left(\frac{\nu}{2}\right)^{1/3} + \left[\frac{3r_m'^2}{20} - \frac{1}{10r'_m}\right]\left(\frac{\nu}{2}\right)^{-1/3} \tag{C.14}$$

The error for these approximations is no more than 1.5%, the worst case being $\nu = 4$ and $m = 3$. The accuracy is often considerably better. Values for the quantities r_m and r'_m used in these expressions are given in Table C.3.

Bessel functions of half-integer order can be expressed in closed form using elementary functions. For example,

$$J_{1/2}(x) = \sqrt{\frac{2}{\pi x}} \sin x \tag{C.15}$$

Table C.3
Values of r_m and r'_m, to eight decimal places.

m	r_m	r'_m
1	2.33810741	1.01879297
2	4.08794944	3.24819758
3	5.52055983	4.82009921

and

$$Y_{1/2}(x) = -\sqrt{\frac{2}{\pi x}} \cos x \tag{C.16}$$

The zeroes of (C.15) are obviously integer multiples of π, while those of (C.16) are $(m+\frac{1}{2})\pi$. Zeroes of $J'_{1/2}(x)$ are not obtainable by elementary means; its lowest root is $j'_{1/2,1} = 1.16556$ to five decimal places.

When the argument of the Bessel functions is large ($|x| \gg |\nu|\pi \geq 0$), J_ν and Y_ν can be approximated by damped trigonometric functions:

$$J_\nu(x) \sim \sqrt{\frac{2}{\pi x}} \cos\left(x - \frac{\nu\pi}{2} - \frac{\pi}{4}\right) \tag{C.17}$$

$$Y_\nu(x) \sim \sqrt{\frac{2}{\pi x}} \sin\left(x - \frac{\nu\pi}{2} - \frac{\pi}{4}\right) \tag{C.18}$$

Evidently, the zeroes of J_ν for $m \gg \nu$ are approximately

$$j_{\nu m} \simeq \frac{\pi}{2}\left(2m + \nu - \frac{1}{2}\right) \tag{C.19}$$

In this sense, J_ν and Y_ν are analogous to the cosine and sine functions.

It is sometimes useful to define solutions of Bessel's equation which are analogous to complex exponential functions (i. e., $e^{\pm jx}$). These are the *Hankel functions* of the first and second kind:

$$H_\nu^{(1)}(x) \equiv J_\nu(x) + jY_\nu(x) \tag{C.20}$$

$$H_\nu^{(2)}(x) \equiv J_\nu(x) - jY_\nu(x) \tag{C.21}$$

The behavior of $H_\nu^{(1)}$ and $H_\nu^{(2)}$ for large or small x is easily deduced from (C.10), (C.12), (C.17) and (C.18). In particular,

$$H_\nu^{(1)}(x) \sim \sqrt{\frac{2}{\pi x}} e^{j(x - \frac{\nu\pi}{2} - \frac{\pi}{4})} \tag{C.22}$$

$$H_\nu^{(2)}(x) \sim \sqrt{\frac{2}{\pi x}} e^{-j(x - \frac{\nu\pi}{2} - \frac{\pi}{4})} \tag{C.23}$$

for $x \gg \nu\pi \geq 0$. From (C.4), (C.20) and (C.21), we have the relations

$$H_{-\nu}^{(1)}(x) = e^{j\nu\pi} H_\nu^{(1)}(x); \qquad H_{-\nu}^{(2)}(x) = e^{-j\nu\pi} H_\nu^{(2)}(x) \tag{C.24}$$

for expressing negative-index Hankel functions in terms of those with positive index.

C.1.1 MODIFIED BESSEL FUNCTIONS

Consider now the following variation of Bessel's equation:

$$y^2 \frac{d^2 g}{dy^2} + y \frac{dg}{dy} - (y^2 + \nu^2)g = 0 \tag{C.25}$$

The change of variable $x = jy$ transforms (C.25) into (C.1). Therefore, solutions of (C.25) can be chosen from linear combinations of $J_\nu(jy)$ and $Y_\nu(jy)$. It is always desirable (when possible) to deal with functions which are real when the argument is real. In this case, such functions are the so-called *modified Bessel functions*, I_ν and K_ν, defined by

$$I_\nu(y) \equiv e^{-j\frac{\nu\pi}{2}} J_\nu(jy) \tag{C.26}$$

$$K_\nu(y) \equiv \frac{\pi j}{2} e^{j\frac{\nu\pi}{2}} H_\nu^{(1)}(jy) \tag{C.27}$$

The modified Bessel functions are analogous to the pure exponentials $e^{\pm y}$. Plots of I_0, I_1, K_0 and K_1 are given in Figure C.2. From (C.26)-(C.27), and other properties of the Bessel functions, including the Wronskian relation (C.6), the Wronskian relation for modified Bessel functions is found to be

$$K_\nu(y) I_\nu'(y) - I_\nu(y) K_\nu'(y) = \frac{1}{y} \tag{C.28}$$

Evidently, the K_ν must become infinite as $y \to 0$. More precisely, we have

$$I_\nu(y) \simeq \frac{1}{\Gamma(1+\nu)} \left(\frac{y}{2}\right)^\nu; \qquad (\nu \neq -1, -2, \ldots) \tag{C.29}$$

$$K_\nu(y) \simeq \frac{\Gamma(\nu)}{2} \left(\frac{2}{y}\right)^\nu; \qquad (\nu > 0) \tag{C.30}$$

as $y \to 0$. We note the special cases for integer order l:

$$I_0(y) \simeq 1 \tag{C.31}$$

$$I_l(y) \simeq \frac{1}{l!} \left(\frac{y}{2}\right)^l; \qquad (l \geq 1) \tag{C.32}$$

and

$$K_0(y) \simeq - \left[\ln\left(\frac{y}{2}\right) + \gamma_E\right] \tag{C.33}$$

$$K_l(y) \simeq \frac{(l-1)!}{2} \left(\frac{2}{y}\right)^l; \qquad (l \geq 1) \tag{C.34}$$

Finally, we have

$$I_{-\nu}(y) = I_\nu(y) + \frac{2 \sin \nu\pi}{\pi} K_\nu(y) \tag{C.35}$$

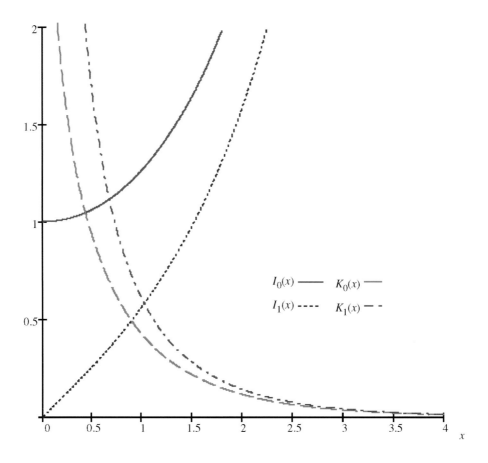

Figure C.2 Graphs of $I_0(y)$, $I_1(y)$, $K_0(y)$ and $K_1(y)$.

$$K_{-\nu}(y) = K_\nu(y) \tag{C.36}$$

for any value of ν. In particular,

$$I_{-n}(y) = I_n(y) \tag{C.37}$$

for any integer n.

If y is real and $y \gg 1$, the I_ν are essentially growing exponentials, while the K_ν are decaying ones:

$$I_\nu(y) \sim \frac{1}{\sqrt{2\pi y}} \left(e^y - \sin \nu\pi e^{-y}\right) \tag{C.38}$$

$$K_\nu(y) \sim \sqrt{\frac{\pi}{2y}} e^{-y} \tag{C.39}$$

If y is complex or negative, these expansions must be replaced by more complicated ones. Neither I_ν nor K_ν has any positive zeroes (in y) for $\nu \geq 0$. Note that $K_\nu(y) > 0$ and $K'_\nu(y) < 0$ for *any* real $y > 0$.

C.1.2   RECURRENCE RELATIONS

It is possible to express derivatives of Bessel functions or modified Bessel functions in terms of other orders of the same type of function. Thus, for example,

$$J_\nu'(x) = J_{\nu-1}(x) - \frac{\nu}{x} J_\nu(x) \tag{C.40}$$

$$J_\nu'(x) = -J_{\nu+1}(x) + \frac{\nu}{x} J_\nu(x) \tag{C.41}$$

$$I_\nu'(y) = I_{\nu-1}(y) - \frac{\nu}{y} I_\nu(y) \tag{C.42}$$

$$I_\nu'(y) = I_{\nu+1}(y) + \frac{\nu}{y} I_\nu(y) \tag{C.43}$$

$$K_\nu'(y) = -K_{\nu-1}(y) - \frac{\nu}{y} K_\nu(y) \tag{C.44}$$

$$K_\nu'(y) = -K_{\nu+1}(y) + \frac{\nu}{y} K_\nu(y) \tag{C.45}$$

among many other similar types of relations. These relations directly imply the connections

$$J_{\nu+1}(x) = -J_{\nu-1}(x) + \frac{2\nu}{x} J_\nu(x) \tag{C.46}$$

$$I_{\nu+1}(y) = I_{\nu-1}(y) - \frac{2\nu}{y} I_\nu(y) \tag{C.47}$$

$$K_{\nu+1}(y) = K_{\nu-1}(y) + \frac{2\nu}{y} K_\nu(y) \tag{C.48}$$

between Bessel functions of orders that differ by integers, without involving derivatives.

C.1.3   INTEGRAL FORMULAS

A wealth of formulas for evaluating integrals containing Bessel functions are available in the literature. We will quote only those which are needed in this book:

$$\int x Z_n^2(x)\,dx = \frac{x^2}{2} \left[Z_n^2(x) - Z_{n-1}(x) Z_{n+1}(x) \right] \tag{C.49}$$

$$\int y K_n^2(y)\,dy = \frac{y^2}{2} \left[K_n^2(y) - K_{n-1}(y) K_{n+1}(y) \right] \tag{C.50}$$

$$\int \frac{1}{x} Z_{n+1}^2(x)\,dx = \frac{n}{n+1} \int \frac{1}{x} Z_n^2(x)\,dx - \frac{1}{2(n+1)} [Z_n^2(x) + Z_{n+1}^2(x)] \tag{C.51}$$

In these formulas, $Z_n(x) = A J_n(x) + B Y_n(x)$ denotes an arbitrary linear combination of $J_n(x)$ and $Y_n(x)$.

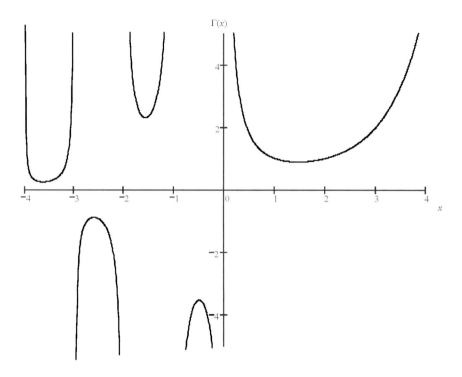

Figure C.3 The gamma function.

C.2   THE GAMMA FUNCTION

The *gamma function* (or factorial function) is defined by the integral formula

$$\Gamma(z) = \int_0^\infty e^{-t}t^{z-1}\, dt \tag{C.52}$$

if $z > 0$. It can be defined for all values of z except for 0 and the negative integers, where it becomes infinite. It is plotted in Figure C.3. For positive integers, it becomes an ordinary factorial:

$$\Gamma(n+1) = n!; \qquad n = 0, 1, 2, \ldots \tag{C.53}$$

For half-integer values of z, it is also possible to compute $\Gamma(z)$; we have in particular,

$$\Gamma(\tfrac{1}{2}) = 2\int_0^\infty e^{-u^2}\, du = \sqrt{\pi} \tag{C.54}$$

For other fractional values of the argument, closed form values do not exist, but we do have

$$\Gamma\left(\frac{1}{3}\right) = 2.6789\ldots; \qquad \Gamma\left(\frac{1}{4}\right) = 3.6256\ldots$$

For any z, we have the properties

$$\Gamma(z+1) = z\Gamma(z) \tag{C.55}$$

and

$$\Gamma(z)\Gamma(1-z) = \frac{\pi}{\sin \pi z} \tag{C.56}$$

C.3 THE ELLIPTIC INTEGRALS K AND E

The complete elliptic integral of the first kind is defined by

$$K(k) = \int_0^{\pi/2} \frac{d\theta}{\sqrt{1 - k^2 \sin^2 \theta}} \tag{C.57}$$

where k is called the *modulus* of K. The complete elliptic integral of the second kind is defined by

$$E(k) = \int_0^{\pi/2} \sqrt{1 - k^2 \sin^2 \theta} \, d\theta \tag{C.58}$$

Graphs of these two functions are shown in Figure C.4; they are clearly even functions of k. Both elliptic integrals can be computed using most standard mathematical software packages. However, it is important to be aware that some software and literature uses the notation $m = k^2$ as the argument of the elliptic integrals. Be sure to check carefully before using software or results found elsewhere.

The elliptic integrals obey a wide variety of identities. Of these we quote here only the following:

$$E(k)K(k') + E(k')K(k) = \frac{\pi}{2} + K(k)K(k') \tag{C.59}$$

$$\frac{dK(k)}{dk} = \frac{1}{k} \left[\frac{E(k)}{k'^2} - K(k) \right] \tag{C.60}$$

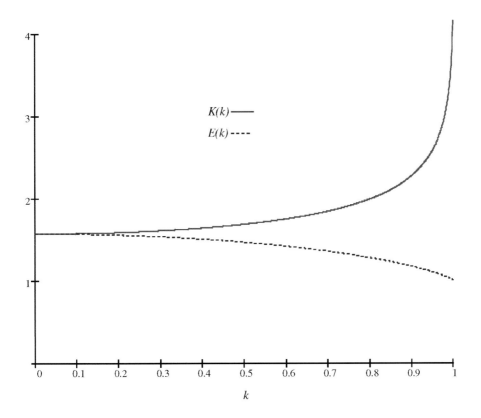

Figure C.4 The elliptic integrals K and E.

where k' is known as the *complementary modulus* $k' \equiv \sqrt{1 - k^2}$.

One rapid method to compute the elliptic integrals is the method of the arithmetic-geometric mean. Let

$$L_0 = 1; \quad M_0 = k' \tag{C.61}$$

Now generate the sequences L_n and M_n by

$$L_{n+1} = \frac{1}{2}(L_n + M_n); \quad M_{n+1} = \sqrt{L_n M_n} \tag{C.62}$$

These sequences converge to the same limit $m(L_0, M_0)$. Then

$$K(k) = \frac{\pi}{2m(1, k')} \tag{C.63}$$

For values of k not extremely close to 1, only a few steps of (C.62) are needed to obtain an extremely accurate value of K. The elliptic integral $E(k)$ can be computed with the help of an additional sequence N_n, determined by

$$N_0 = k; \quad N_{n+1} = \frac{1}{2}(L_n - M_n) \tag{C.64}$$

from which we then express $E(k)$ as

$$E(k) = K(k)\left[1 - \sum_{n=0}^{\infty} 2^{n-1} N_n^2\right] \tag{C.65}$$

The infinite series converges rapidly for most values of k, and can be truncated after a sufficiently large number of terms.

Often it is the ratio $K(k)/K(k')$ rather than individual elliptic integrals which is encountered in applications. There exist extremely accurate yet simple approximations for this ratio when k is real and between 0 and 1:

$$\frac{K(k)}{K(k')} \simeq \frac{\pi}{\ln\left(2\dfrac{1 + \sqrt{k'}}{1 - \sqrt{k'}}\right)} \quad \text{for } 0 \le k^2 \le \tfrac{1}{2}, \text{ i. e., } 0 \le \tfrac{K(k)}{K(k')} \le 1$$

$$\frac{K(k)}{K(k')} \simeq \frac{1}{\pi}\ln\left(2\dfrac{1 + \sqrt{k}}{1 - \sqrt{k}}\right) \quad \text{for } \tfrac{1}{2} \le k^2 \le 1, \text{ i. e., } 1 \le \tfrac{K(k)}{K(k')} \le \infty \tag{C.66}$$

The maximum relative error in (C.66) is 3×10^{-6}. The less accurate but simpler formulas

$$\frac{K(k)}{K(k')} \simeq \frac{\pi}{2\ln\left(\frac{4}{k}\right)} \quad \text{if } k \ll 1 \tag{C.67}$$

and

$$\frac{K(k)}{K(k')} \simeq \frac{2}{\pi}\ln\left(\frac{4}{k'}\right) \quad \text{if } k' \ll 1 \tag{C.68}$$

follow from (C.66).

Eqns. (C.66) are invertible to obtain k as an explicit function of the ratio $K(k)/K(k')$. This is commonly done using the so-called *nome*[1] q defined as

$$q = \exp[-\pi K(k')/K(k)] \tag{C.69}$$

---

[1]This should not be confused with the filling factor q defined in Section 9.4, just as the modulus k should not be confused with the wavenumber of a TEM mode or plane wave.

When $0 \le k^2 \le 1/2$, we have $0 \le K(k)/K(k') \le 1$, or $0 \le q \le 0.0432$. In this case,

$$k \simeq \frac{4\sqrt{q(1+4q^2)}}{(1+2q)^2}; \qquad 0 \le \frac{K(k)}{K(k')} \le 1 \qquad (C.70)$$

When $1/2 \le k^2 \le 1$, we have $1 \le K(k)/K(k') \le \infty$, or $0.0432 \le q \le 1$. In this case we define a complementary nome q' as

$$q' = \exp[-\pi K(k)/K(k')] \qquad (C.71)$$

so that $0 \le q' \le 0.0432$. Then,

$$k \simeq \left(\frac{1-2q'}{1+2q'}\right)^2; \qquad 1 \le \frac{K(k)}{K(k')} \le \infty \qquad (C.72)$$

C.4 THE LAMBERT *W*-FUNCTION AND RELATED FUNCTIONS

The Lambert W-function is a solution of the equation

$$W(x)e^{W(x)} = x \qquad (C.73)$$

It is actually a multivalued function like the arctangent function, consisting of an infinite number of branches. Most branches are complex when x is real, but two important ones are real for real x, as shown in Figure C.5. For $x > -e^{-1}$, the solution of (C.73) for which $W(x) > -1$ (the so-called principal branch) is denoted $W_0(x)$, while for $-e^{-1} < x < 0$ the solution for which $W(x) < -1$ is denoted $W_{-1}(x)$. For $|x| < e^{-1}$, there is the convergent

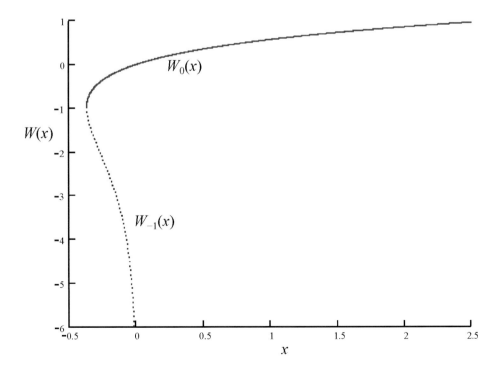

Figure C.5 The Lambert functions $W_0(x)$ and $W_{-1}(x)$.

series

$$W_0(x) = \sum_{n=1}^{\infty} \frac{(-n)^{n-1}x^n}{n!} \tag{C.74}$$

An empirical approximation for W_{-1} that has a relative error of less than 0.025% is

$$W_{-1}(x) \simeq \ln(-x) - \frac{\sqrt{2\sigma(x)}}{1 + M_1\sqrt{\frac{\sigma(x)}{2}} + M_2\sigma(x)e^{M_3\sqrt{\sigma(x)}}} \qquad \text{for } -e^{-1} < x < 0 \tag{C.75}$$

where $M_1 = 0.3361$, $M_2 = -0.0042$, $M_3 = -0.0201$ and

$$\sigma(x) = -1 - \ln(-x) \tag{C.76}$$

A wide class of related functions are called generalized Lambert functions, which are solutions of various equations analogous to (C.73). For example,

$$W_t(x)\tan W_t(x) = x \tag{C.77}$$

$$W_{ct}(x)\cot W_{ct}(x) = x \tag{C.78}$$

$$W_{se}(x)\sec W_{se}(x) = x \tag{C.79}$$

and so on. The theory of the generalized Lambert functions is much less well-developed than that of $W(x)$, but a few references can be found at the end of this appendix.

C.5   A FUNCTION ARISING IN MICROSTRIP ANALYSIS

A function that has no special name (it is related to the family of zeta functions) is

$$r(x) \equiv \exp[Q(x)] \tag{C.80}$$

where

$$Q(x) \equiv \sum_{m=1}^{\infty} x^m \ln\left(\frac{m+1}{m}\right) \tag{C.81}$$

When $|x| < 1$, which is the case if $x = (1 - \epsilon_r)/(1 + \epsilon_r)$ as above, a simple yet accurate approximation for $r(x)$ is

$$r(x) \simeq \sqrt{1 + 1.3471u + 0.3152u^2} \tag{C.82}$$

where $u = x/(1 - x)$. Graphs of $Q(x)$ and its approximation $\ln r(x)$ for $-1 < x < 1$ are shown in Figure C.6. This function arises in the analysis of microstrip structures.

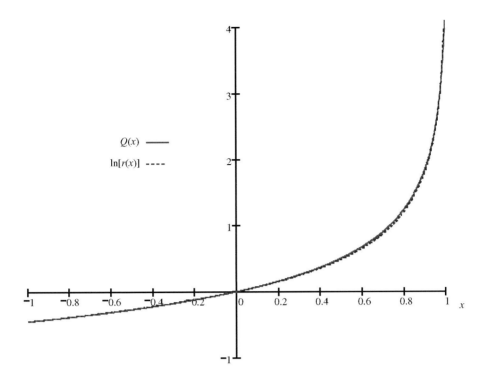

$Q(x)$ ——

$\ln[r(x)]$ ----

Figure C.6   The function $Q(x)$ and its approximation.

C.6   NOTES AND REFERENCES

For Bessel functions, see

> Abramowitz and Stegun (1964), Chapter 9;
> Watson (1966);
> N. N. Lebedev (1972), Chapters 5 and 6;
> Olver (1974), Chapters 7 and 10.

The expansions (C.13) and (C.14) are given in

> F. W. J. Olver, "A further method for the evaluation of zeros of Bessel functions and some new asymptotic expansions for zeros of functions of large order," *Proc. Camb. Phil. Soc.*, vol. 47, pp. 699-712 (1951).

along with further terms in the series which may be retained for more accuracy. More accurate (but also more elaborate) expansions can be found in

> F. W. J. Olver, "The asymptotic expansion of Bessel functions of large order," *Phil. Trans. Roy. Soc. London A*, vol. 247A, pp. 328-368 (1954).

The quantities $-r_m$ and $-r'_m$ are the roots of the Airy function and its derivative, respectively.

For gamma functions, see

> Abramowitz and Stegun (1964), Chapter 6;
> N. N. Lebedev (1972), Chapter 1.

Properties of elliptic integrals can be found in

Byrd and Friedman (1971).

The approximation (C.66), among others, can be found in

W. Hilberg, "From approximations to exact relations for characteristic impedances," *IEEE Trans. Micro. Theory Tech.*, vol. 17, pp. 259-265 (1969).

Simple programs for computing these functions on a programmable pocket calculator are given in:

Henrici (1977), pp. 214-221, 241-267.
Ball (1978), pp. 210-221.

A large amount of information about the Lambert W-functions is given in

R. M. Corless, G. H. Gonnet, D. E. G. Hare, D. J. Jeffrey and D. E. Knuth, "On the Lambert W function," *Adv. Comp. Math.*, vol. 5, pp. 329-359 (1996).

The approximation (C.75) was obtained by

D. A. Barry, J.-Y. Parlange , G .C. Sander and M. Sivaplan, "A class of exact solutions for Richards' equation," *J. Hydrol.*, vol. 142, pp. 29-46 (1993).

and other explicit approximations were given in

D. A. Barry, J.-Y. Parlange, L. Lia, H. Prommera, C. J. Cunningham and F. Stagnitti, "Analytical approximations for real values of the Lambert W-function," *Math. Computers Simul.*, vol. 53, pp. 95-103 (2000).

The generalized Lambert functions $W_t(x)$ and $W_{ct}(x)$ were discussed in

V. E. Markushin, R. Rosenfelder and A. W. Schreiber, "The W_t transcendental function and quantum-mechanical applications," *Nuov. Cim. B*, vol. 117, pp. 75-94 (2002).
Q. Luo, Z. Wang and J. Han, "A Padé approximant approach to two kinds of transcendental equations with applications in physics," *Eur. J. Phys.*, vol. 36, art. 035030 (2015).

and much other information can be found in

Dubinov *et al.* (2006).

The function treated in Section C.5 was studied in

E. F. Kuester, "Accurate approximations for a function appearing in the analysis of microstrip," *IEEE Trans. Micr. Theory Tech.*, vol. 32, pp. 131-133 (1984).

C.7 PROBLEMS

pC-1 Use (C.40)-(C.41) and (C.15) to find a closed-form expression for $J_{-1/2}(x)$.
pC-2 Use (C.66) to prove (C.67) and (C.68).

D PROPERTIES OF SOLUTIONS TO MAXWELL'S EQUATIONS

D.1 MAXWELL'S EQUATIONS AND BOUNDARY CONDITIONS

The electromagnetic field in any region of space must satisfy Maxwell's equations:

$$\nabla \times \mathbf{e} = -\mu \frac{\partial \mathbf{h}}{\partial t}$$

$$\nabla \times \mathbf{h} = \epsilon \frac{\partial \mathbf{e}}{\partial t} + \sigma \mathbf{e} + \mathbf{j}_{\text{ext}} \tag{D.1}$$

$$\nabla \cdot (\epsilon \mathbf{e}) = \rho_{\text{ext}}$$
$$\nabla \cdot (\mu \mathbf{h}) = 0$$

where ϵ, σ, and μ are the electrical parameters (possibly position-dependent) characterizing a region of space, $\mathbf{e}(x, y, z, t)$ is the time-dependent electric field, $\mathbf{h}(x, y, z, t)$ the magnetic field and $\mathbf{j}_{\text{ext}}(x, y, z, t)$ is the volume density of impressed electric current in the region.[1] The del operator ∇ is defined by

$$\nabla = \mathbf{u}_x \frac{\partial}{\partial x} + \mathbf{u}_y \frac{\partial}{\partial y} + \mathbf{u}_z \frac{\partial}{\partial z} \tag{D.2}$$

in Cartesian coordinates, where $\mathbf{u}_x$, $\mathbf{u}_y$, $\mathbf{u}_z$ are unit vectors in the x, y and z directions, respectively.

Phasor fields $\mathbf{E}$ and $\mathbf{H}$ corresponding to time-harmonic solutions $\mathbf{e}$ and $\mathbf{h}$ of (D.1) (or equally well, Fourier transforms of $\mathbf{e}$ and $\mathbf{h}$ with respect to t) are required to satisfy the time-harmonic Maxwell curl equations

$$\nabla \times \mathbf{E} = -j\omega\mu\mathbf{H}$$
$$\nabla \times \mathbf{H} = (j\omega\epsilon + \sigma)\mathbf{E} + \mathbf{J}_{\text{ext}} \tag{D.3}$$
$$= j\omega\hat{\epsilon}\mathbf{E} + \mathbf{J}_{\text{ext}}$$

which follow from (D.1); the divergence equations are redundant unless $\omega = 0$.

If fictitious magnetic currents and charges are allowed, the time-harmonic Maxwell equations have the form

$$\nabla \times \mathbf{E} = -j\omega\mu\mathbf{H} - \mathbf{M}_{\text{ext}}$$
$$\nabla \times \mathbf{H} = (j\omega\epsilon + \sigma)\mathbf{E} + \mathbf{J}_{\text{ext}} \tag{D.4}$$
$$= j\omega\hat{\epsilon}\mathbf{E} + \mathbf{J}_{\text{ext}}$$

Sometimes the Maxwell divergence equations (the Gauss laws) are needed; they take the forms:

$$\nabla \cdot (\epsilon \mathbf{E}) = \rho_{\text{ext}}$$
$$\nabla \cdot (\mu \mathbf{H}) = \rho_{\text{m,ext}} \tag{D.5}$$

---

[1]Remember that lowercase letters denote *real* functions (fields, currents, etc.) of (x, y, z) *and* time t.

where the electric and magnetic charge densities are found from the conservation relations

$$\rho_{\text{ext}} = -\frac{\nabla \cdot \mathbf{J}_{\text{ext}}}{j\omega}$$

$$\rho_{\text{m,ext}} = -\frac{\nabla \cdot \mathbf{M}_{\text{ext}}}{j\omega} \tag{D.6}$$

Standard arguments based on integrating Maxwell's equations over "pillbox" volumes or thin strip surfaces lead to boundary or jump conditions for the electromagnetic field across surfaces of discontinuity between different materials or surfaces containing surface-concentrated current distributions. The most general form of them applicable to time-harmonic fields with electric as well as magnetic sources is

$$\mathbf{J}_{S,\text{ext}} = \mathbf{u}_n \times (\mathbf{H}_1 - \mathbf{H}_2)_S$$

$$\mathbf{M}_{S,\text{ext}} = (\mathbf{E}_1 - \mathbf{E}_2)_S \times \mathbf{u}_n \tag{D.7}$$

where S is the surface $n = 0$ separating regions 1 (in $n > 0$) and 2 (in $n < 0$), and $\mathbf{u}_n$ is the unit vector normal to S pointing into region 1.

D.2 PLANE WAVES

The simplest solutions to the source-free ($\mathbf{J}_{\text{ext}} = 0$) time-harmonic Maxwell equations in an infinite homogeneous medium are *plane waves*. These have the form

$$\mathbf{E} = \mathbf{E}_0 e^{-j\mathbf{k}\cdot\mathbf{r}}; \qquad \mathbf{H} = \mathbf{H}_0 e^{-j\mathbf{k}\cdot\mathbf{r}} \tag{D.8}$$

where $\mathbf{r} = \mathbf{u}_x x + \mathbf{u}_y y + \mathbf{u}_z z$ is the position vector, and $\mathbf{k} = \mathbf{u}_x k_x + \mathbf{u}_y k_y + \mathbf{u}_z k_z$ is called the wave vector of the plane wave, which points in the direction of propagation of the wave if its components are real. By substitution of (D.8) into (D.3), we find that the fields are perpendicular to $\mathbf{k}$:

$$\mathbf{k} \cdot \mathbf{E}_0 = 0; \qquad \mathbf{k} \cdot \mathbf{H}_0 = 0 \tag{D.9}$$

that $\mathbf{k}$ is related to the *wavenumber* $k \equiv \omega\sqrt{\mu\epsilon}$ by:

$$\mathbf{k} \cdot \mathbf{k} = k^2 \tag{D.10}$$

and that $\mathbf{E}_0$ and $\mathbf{H}_0$ are mutually perpendicular (again if $\mathbf{k}$ is real) and related by:

$$\frac{\mathbf{k}}{k} \times \mathbf{E}_0 = \zeta \mathbf{H}_0 \tag{D.11}$$

where $\zeta \equiv (\mu/\epsilon)^{1/2}$ is the intrinsic wave impedance of the medium.

D.3 POYNTING'S THEOREM

Consider a region of space occupied by a possibly inhomogeneous but isotropic material ($\mu(\mathbf{r})$, $\epsilon(\mathbf{r})$, and $\sigma(\mathbf{r})$ may be functions of position but are scalars) with externally impressed volume electric current density $\mathbf{j}_{\text{ext}}(\mathbf{r}, t)$. Maxwell's equations (D.1) must be satisfied by the resultant fields $\mathbf{e}$ and $\mathbf{h}$. Using them, we can obtain the following identity:

$$\nabla \cdot (\mathbf{e} \times \mathbf{h}) = -\frac{\partial}{\partial t}\left[\frac{1}{2}\epsilon\mathbf{e}\cdot\mathbf{e} + \frac{1}{2}\mu\mathbf{h}\cdot\mathbf{h}\right] - \sigma\mathbf{e}\cdot\mathbf{e} - \mathbf{e}\cdot\mathbf{j}_{\text{ext}} \tag{D.12}$$

The term $\mathbf{e} \cdot \mathbf{j}_{\text{ext}}$ has the physical meaning of the rate of work being done (per unit volume) by the Lorentz force of the electric field on the charges (per unit volume) whose motion makes up the external impressed current density, or equivalently, the negative of the work done by the external impressed sources in delivering energy to the fields. Similarly, the term $\sigma \mathbf{e} \cdot \mathbf{e}$ represents the rate of energy dissipation per unit volume due to conduction losses in the medium. If we interpret the quantity

$$w = w_E + w_M = \frac{1}{2}\epsilon \mathbf{e} \cdot \mathbf{e} + \frac{1}{2}\mu \mathbf{h} \cdot \mathbf{h}$$

as the stored electric and magnetic energy in the electric and magnetic fields (as is rigorously true for static fields), then the term

$$\frac{\partial w}{\partial t} = \frac{\partial}{\partial t}\left[\frac{1}{2}\epsilon \mathbf{e} \cdot \mathbf{e} + \frac{1}{2}\mu \mathbf{h} \cdot \mathbf{h}\right]$$

represents the rate of increase of stored energy in the field, and finally, $\nabla \cdot (\mathbf{e} \times \mathbf{h})$ must be interpreted as the energy outflow rate (per unit volume) from the surface surrounding that unit volume. The vector function $\mathbf{e} \times \mathbf{h}$ is known as the Poynting vector.

The result is often stated as the integral form of Poynting's theorem:

$$p_{\text{ext}}(t) = -\int_V \mathbf{e} \cdot \mathbf{j}_{\text{ext}}\, dV = \oint_S (\mathbf{e} \times \mathbf{h}) \cdot \mathbf{u}_n\, dS + \frac{d}{dt}\int_V \left[\frac{1}{2}\epsilon \mathbf{e} \cdot \mathbf{e} + \frac{1}{2}\mu \mathbf{h} \cdot \mathbf{h}\right] dV + \int_V \sigma \mathbf{e} \cdot \mathbf{e}\, dV \tag{D.13}$$

where V is a volume of space bounded by the closed surface S. The left side of this equation represents the rate of energy deliverance (power flow) to the system by the sources within V, the first term on the right side is the rate of energy flow outward through the surface S (whose outward unit normal vector is $\mathbf{u}_n$), the second term on the right is the rate of increase of stored energy within V, and the third is the power dissipated as heat in the conductivity σ.

In the time-harmonic case, we can represent the fields and sources in terms of phasors, as we did with voltages and currents in Appendix A. Again, the time-dependent power and energy quantities are expressed as the sum of the time-average (constant) part and an oscillatory part. The time-average power flow through the surface S, for example, is given by

$$p_{\text{av}} = \text{Re}(\hat{P})$$

where the complex power $\hat{P}$ is

$$\hat{P} = \frac{1}{2}\oint_S (\mathbf{E} \times \mathbf{H}^*) \cdot \mathbf{u}_n\, dS = \frac{1}{2}\oint_S \hat{\mathbf{S}} \cdot \mathbf{u}_n\, dS$$

while the oscillatory part of this power is given by

$$p_{\text{osc}}(t) = \text{Re}(P_{\text{osc}}e^{j2\omega t})$$

where the complex oscillatory power is

$$P_{\text{osc}} = \frac{1}{2}\oint_S (\mathbf{E} \times \mathbf{H}) \cdot \mathbf{u}_n\, dS = \frac{1}{2}\oint_S \mathbf{S}_{\text{osc}} \cdot \mathbf{u}_n\, dS$$

We have defined the complex Poynting vector $\hat{\mathbf{S}} = \mathbf{E} \times \mathbf{H}^*$ and the complex oscillatory Poynting vector $\mathbf{S}_{\text{osc}} = \mathbf{E} \times \mathbf{H}$ in these equations. As was noted for the case of lumped-element circuits and transmission lines in Appendix A, we cannot expect that $|\hat{P}| = |\hat{P}_{\text{osc}}|$ in general.

The analogs of (D.12) for phasor quantities are:

$$\nabla \cdot \hat{\mathbf{S}} = -j\omega \left[\mu \mathbf{H} \cdot \mathbf{H}^* - \epsilon \mathbf{E} \cdot \mathbf{E}^* \right] - \sigma \mathbf{E} \cdot \mathbf{E}^* - \mathbf{E} \cdot \mathbf{J}_{\text{ext}}^* \tag{D.14}$$

$$\nabla \cdot \mathbf{S}_{\text{osc}} = -j\omega \left[\mu \mathbf{H} \cdot \mathbf{H} + \epsilon \mathbf{E} \cdot \mathbf{E} \right] - \sigma \mathbf{E} \cdot \mathbf{E} - \mathbf{E} \cdot \mathbf{J}_{\text{ext}} \tag{D.15}$$

Integrating these equations over the volume V and using the divergence theorem yields

$$\hat{P}_{\text{ext}} = \hat{P} + 2j\omega(w_{M,\text{av}} - w_{E,\text{av}}) + p_{L,\text{av}}$$

$$P_{\text{ext,osc}} = P_{\text{osc}} + 2j\omega(W_{M,\text{osc}} + W_{E,\text{osc}}) + P_{L,\text{osc}}$$

where

$$w_{M,\text{av}} = \frac{1}{4} \int_V \mu \mathbf{H} \cdot \mathbf{H}^* \, dV$$

$$w_{E,\text{av}} = \frac{1}{4} \int_V \epsilon \mathbf{E} \cdot \mathbf{E}^* \, dV$$

are the parts of the time-average stored magnetic and electric energies in V not due to material dispersion,

$$p_{L,\text{av}} = \frac{1}{2} \int_V \sigma \mathbf{E} \cdot \mathbf{E}^* \, dS$$

is the time-average ohmic power loss in V, and

$$\hat{P}_{\text{ext}} = -\frac{1}{2} \int_V \mathbf{E} \cdot \mathbf{J}_{\text{ext}}^* \, dV$$

is the complex power delivered by the sources to the system. The quantities

$$W_{M,\text{osc}} = \frac{1}{4} \int_V \mu \mathbf{H} \cdot \mathbf{H} \, dV$$

$$W_{E,\text{osc}} = \frac{1}{4} \int_V \epsilon \mathbf{E} \cdot \mathbf{E} \, dV$$

$$P_{L,\text{osc}} = \frac{1}{2} \int_V \sigma \mathbf{E} \cdot \mathbf{E} \, dS$$

$$P_{\text{ext,osc}} = -\frac{1}{2} \int_V \mathbf{E} \cdot \mathbf{J}_{\text{ext}} \, dV$$

are the corresponding complex oscillatory powers or energies.

The real part of (D.14) relates the change in power flow through the surface S to the difference between the time-average rate of energy generation and dissipation within V. The imaginary part of (D.14) has no direct interpretation in terms of energy conservation (since it plays no part in the description of time dependent power flow), but does relate the rate of change of the imaginary part of $\hat{P}$ to the difference between stored time-average magnetic energy and electric energy, and the imaginary part of the complex externally supplied power. Equation (D.15) relates the oscillatory power flow to the oscillatory parts of the stored and dissipated energies.

D.4 ELEMENTARY DIPOLES

Elementary (or Hertzian) dipole sources are idealizations of very small electric or magnetic current sources. The elementary electric dipole is described by a volume electric current density of

$$\mathbf{J} = j\omega\mathbf{p}\delta(x - x_0)\delta(y - y_0)\delta(z - z_0) \tag{D.16}$$

while the elementary magnetic dipole is described by the magnetic current density

$$\mathbf{M} = j\omega\mu\mathbf{m}\delta(x - x_0)\delta(y - y_0)\delta(z - z_0) \tag{D.17}$$

where the vector $\mathbf{p}$ is called the electric dipole moment and $\mathbf{m}$ the magnetic dipole moment. The strengths of these dipole moments are denoted by the scalars $p = |\mathbf{p}|$ and $m = |\mathbf{m}|$ respectively.

In (D.16) and (D.17), δ represents the so-called Dirac delta-function, more properly called a *generalized function*, which has the properties

$$\delta(x) = 0$$

if $x \neq 0$, and

$$\int_{x_1}^{x_2} \delta(x)\,dx = 1$$

if $x_1 < 0 < x_2$. Thus, the sources in (D.16) and (D.17) are concentrated at the single point in space (x_0, y_0, z_0).

The elementary electric dipole is a limiting case of a short electric current filament of length Δl carrying a current I, as $\Delta l \to 0$ and $I \to \infty$ in such a way that

$$I\Delta l \to j\omega p \tag{D.18}$$

with $\mathbf{p}$ pointing in the direction of positive I. The elementary magnetic dipole is a similar limiting case of a small loop of electric current I which spans an area $\Delta S \to 0$ such that $I\Delta S \to m$ and $\mathbf{m}$ is perpendicular to the surface spanned by the loop, and related to the direction of current flow by the right-hand rule (see problem p9-1).

The fields of such dipoles in a homogeneous medium of infinite extent (permittivity ϵ, permeability μ, wave impedance ζ and wavenumber k) are found by standard methods of antenna theory, for example. The fields of an electric dipole are given by:

$$\mathbf{H} = \frac{1}{\mu}\nabla \times \mathbf{A}; \quad \mathbf{E} = \frac{1}{j\omega\epsilon}\nabla \times \mathbf{H}$$

where the vector potential $\mathbf{A}$ is given by

$$\mathbf{A} = \mathbf{p}\frac{j\omega\mu}{4\pi R}e^{-jkR}$$

and $R = \sqrt{(x - x_0)^2 + (y - y_0)^2 + (z - z_0)^2}$ is the distance between the observation point (x, y, z) and the dipole location (x_0, y_0, z_0). Although these fields are often expressed in spherical coordinates when the dipole moment is in the z-direction, more universal and coordinate-free expressions for the fields of an arbitrarily oriented dipole are

$$\mathbf{E} = \frac{\overleftrightarrow{T}_{ee} \cdot \mathbf{P}}{4\pi\epsilon} \tag{D.19}$$

$$\mathbf{H} = \frac{j\omega}{4\pi}\overleftrightarrow{T}_{me} \cdot \mathbf{p} \tag{D.20}$$

where we have defined the dipole dyadics

$$\overset{\leftrightarrow}{T}_{ee} = e^{-jkR} \left[\left(3\frac{\mathbf{RR}}{R^5} - \frac{\overset{\leftrightarrow}{I}}{r^3} \right)(1 + jkR) - k^2\frac{\mathbf{RR}}{R^3} \right] \tag{D.21}$$

$$\overset{\leftrightarrow}{T}_{me} = -e^{-jkR}\frac{1 + jkR}{R^3} \mathbf{R} \times \overset{\leftrightarrow}{I} \tag{D.22}$$

and $\mathbf{R} = (x - x_0)\mathbf{u}_x + (y - y_0)\mathbf{u}_y + (z - z_0)\mathbf{u}_z$ (see problem pB-7). In a similar manner, the fields of a magnetic dipole with dipole moment $\mathbf{m}$ are

$$\mathbf{E} = -\frac{j\omega\mu}{4\pi}\overset{\leftrightarrow}{T}_{em} \cdot \mathbf{m} \tag{D.23}$$

$$\mathbf{H} = \frac{\overset{\leftrightarrow}{T}_{mm} \cdot \mathbf{m}}{4\pi} \tag{D.24}$$

where

$$\overset{\leftrightarrow}{T}_{mm} = \overset{\leftrightarrow}{T}_{ee}; \qquad \overset{\leftrightarrow}{T}_{em} = \overset{\leftrightarrow}{T}_{me} \tag{D.25}$$

D.5   IMAGE THEORY

Consider the electric and magnetic currents located above the plane $z = 0$ which is taken to be a perfect conductor, as shown in Figure D.1(a). We seek an equivalent auxiliary problem (the *image* problem) in which the ground plane is absent and additional sources are present in $z < 0$ such that the resulting fields in $z > 0$ are identical with those of case (a). These additional sources must be such that, together with the original sources in $z > 0$, they produce a zero value of $\mathbf{u}_z \times \mathbf{E}$ at $z = 0$ in the image problem (Figure D.1).

For example, if $\mathbf{M} = 0$ and $\mathbf{J} = \mathbf{u}_z J_z(x, y, z)$ in the original problem, and if we denote the fields this source distribution would produce in *infinite free space* (in the absence of the ground plane) as $\mathbf{E}_0(x, y, z)$ and $\mathbf{H}_0(x, y, z)$, then the image sources

$$\mathbf{J}^i(x, y, z) = \mathbf{u}_z J_z(x, y, -z)$$

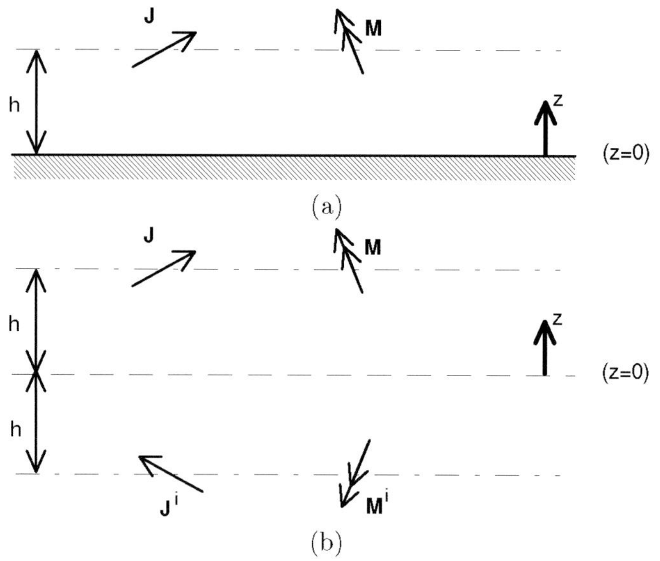

Figure D.1   (a) Sources above a perfectly conducting plane; (b) Image equivalent for $z > 0$.

$$\mathbf{M}^i(x, y, z) = 0 \tag{D.26}$$

will produce the fields (also in the absence of the ground plane)

$$\begin{aligned}
\mathbf{E}^i(x, y, z) &= \mathbf{u}_z E_{0z}(x, y, -z) - \mathbf{E}_{0t}(x, y, -z) \\
\mathbf{H}^i(x, y, z) &= -\mathbf{u}_z H_{0z}(x, y, -z) + \mathbf{H}_{0t}(x, y, -z)
\end{aligned} \tag{D.27}$$

Here "$_t$" denotes components of a vector transverse (perpendicular) to z in the xy-plane. By superposing these with the original fields to get the total field in case (b):

$$\begin{aligned}
\mathbf{E} &= \mathbf{E}_0 + \mathbf{E}^i \\
\mathbf{H} &= \mathbf{H}_0 + \mathbf{H}^i
\end{aligned} \tag{D.28}$$

we find that $\mathbf{u}_z \times \mathbf{E}|_{z=0}$ is indeed zero, so that (D.28) does give the proper total field in the region $z > 0$.

In similar fashion, we can show that the image currents for horizontal electric currents $\mathbf{J}_t$ are:

$$\mathbf{J}^i = -\mathbf{J}_t(x, y, -z) \tag{D.29}$$

while for vertical magnetic currents M_z we have

$$\mathbf{M}^i = -\mathbf{u}_z M_z(x, y, -z) \tag{D.30}$$

and for horizontal magnetic currents $\mathbf{M}_t$,

$$\mathbf{M}^i = \mathbf{M}_t(x, y, -z) \tag{D.31}$$

These results are summarized schematically in Figure D.1.

D.6   NOTES AND REFERENCES

For the fields of elementary dipoles, see, e.g.,

Balanis (1982), Chapters 3 and 4.

E ELECTROMAGNETIC MATERIAL PROPERTIES

The quantity

$$\hat{\epsilon} = \epsilon - j\frac{\sigma}{\omega} = \epsilon_0 \epsilon_r \left(1 - j \tan \delta\right)$$

is called the complex permittivity; we will often simply write ϵ, with the understanding that it can be made complex to account for dielectric or conduction losses if need be. The *loss tangent* $\tan \delta$ is a measure of the degree of losses in the material. Magnetic losses could similarly be accounted for by letting μ be complex, but these are less often encountered in waveguide applications. We have restricted the material parameters to be scalar quantities; this simplifies much of the analysis done in the book. Nevertheless, *anisotropic* and *gyrotropic* materials, for which ϵ and/or μ are *dyadics* or *tensors*, are of considerable interest in waveguide technology. Such materials include plasmas, ferrites, electro-optic materials and more, and make possible the implementation of isolators, circulators and other important devices. Detailed treatment of anisotropic materials can be found in the references given at the end of this appendix.

Some materials are *chiral* or *bianisotropic*. This means that **D** can depend not only on **E** but also on **B**, and likewise **H** depends not only on **B** but also on **E**. Typically, chiral materials are made up of complex molecules that have a spiral shape or other structural feature that gives it a "right-handed" or "left-handed" sense.

Many materials have electromagnetic properties that vary significantly with frequency. Such materials are said to be *dispersive*. This often occurs because of atomic or molecular resonances at one or more frequencies in the vicinity of the operating frequency. Very close to such resonances, it is common for a material to have increased losses, so that the loss tangent becomes larger. Losses may also be increased by the absorption of water in a humid environment. Some materials should be avoided for certain applications because of this effect.

If a material with some desired set of electromagnetic constitutive parameters cannot be found in nature, or is unsuitable because of its other physical properties (toxicity, mechanical strength, etc.), it may be possible to create engineered materials that have the desired behavior. In the early literature, these were called artificial dielectrics, but more recently the idea has been extended to a wider class of materials called *metamaterials*. These have made possible values of ϵ and μ that are simultaneously negative in some frequency ranges, leading to a negative index of refraction. This has become a very active area of research.

E.1 PERMITTIVITY AND CONDUCTIVITY

A brief collection of the permittivities and conductivities of common materials is given in Table E.1. Much information about other materials under various conditions of temperature, humidity, operating frequency and other parameters can be found in the references at the end of this appendix.

Table E.1

Electromagnetic properties of common materials.

Material	ϵ_r	σ (S/m)	$\tan \delta$
Aluminum Oxide (Sapphire; Al_2O_3) at $f = 1$ MHz	9.1		7×10^{-4}
Amber at $f = 3$ GHz	2.59		9×10^{-3}
Barium Strontium Titanate at $f = 10$ GHz	260		8×10^{-2}
Barium Titanate at $f = 1$ MHz	1200		10^{-2}
Concrete (varies by type, humidity and frequency)	$4 \sim 10$		$.01 \sim .4$
Copper	1	5.8×10^7	
FR-4 glass-reinforced epoxy laminate at $f = 1$ GHz	4.34		1.7×10^{-2}
Fused Quartz at $f = 3$ GHz	3.8		10^{-4}
Glass (typical) at $f = 3$ GHz	4.5		3×10^{-3}
Ice ($-12°$C) at $f = 3$ GHz	3.2		9×10^{-4}
Lucite at $f = 3$ GHz	2.57		5×10^{-3}
Paraffin Wax at $f = 3$ GHz	2.25		2×10^{-4}
Plexiglas at $f = 3$ GHz	2.6		5.7×10^{-3}
Polyethylene at $f = 3$ GHz	2.26		4×10^{-4}
Polystyrene at $f = 3$ GHz	2.5		10^{-3}
Polystyrene foam at $f = 3$ GHz	1.05		$< 3 \times 10^{-5}$
Ruby Mica at $f = 3$ GHz	5.4		3×10^{-4}
Teflon at $f = 100$ MHz	2.1		5×10^{-4}
Titanium Dioxide (TiO_2) at $f = 1$ MHz	85		5×10^{-4}
Water (distilled) at $f = 1$ MHz	78.2	2×10^{-4}	
Water (sea) at $f = 1$ MHz	77.7	2×10^4	
Wood (varies by type, humidity and frequency)	$1.2 \sim 5$		$.004 \sim .4$

E.2 NOTES AND REFERENCES

Good collections of data on the permittivity, permeability and conductivity of various materials are given in

Moreno (1958), pp. 201-209;
ITT Staff (1975), Chapter 4.
Komarov (2012)

and can also be found on the internet.

Properties and applications of anisotropic and gyrotropic materials are discussed in:

Johnson (1965), Chapters 11 and 12;
Kurokawa (1969), Sections 5.6-5.8;
I. Lebedev (1973), Sections 8.5 and 8.11;
Helszajn (1975);
Atwater (1981), Chapter 8;
Sazonov *et al.* (1982), Chapter 9;
Baden Fuller (1987);
Linkhart (1989).

For information on chiral materials, see:

Lakhtakia *et al.* (1989).

Treatments of artificial dielectrics and metamaterials can be found in:

J. Brown, "Artificial dielectrics," in *Progress in Dielectrics*, vol. 2, pp. 195- 225 (1960);
Eleftheriades and Balmain (2005);
Engheta and Ziolkowski (2006);

F EXPONENTIAL LINES, BESSEL LINES AND TURNING POINTS

In this appendix, we will consider the exact solutions for waves on nonuniform transmission lines of two special forms—those whose characteristic impedance varies exponentially with distance, and those whose parameters vary as powers of the distance along the line. The latter solutions will also allow us to examine the behavior of waves that encounter turning points: locations where either x or b becomes zero or infinite. We assume throughout this appendix that the line is lossless, i. e., $z = jx$ and $y = jb$. Our goal in this appendix is to get a chain parameter description for each line in the form (1.61):

$$
\begin{aligned}
V(z) &= \mathcal{A}(z,0)V(0) + \mathcal{B}(z,0)I(0) \\
I(z) &= \mathcal{C}(z,0)V(0) + \mathcal{D}(z,0)I(0)
\end{aligned}
\tag{F.1}
$$

Using (1.64) or (2.14), we can readily obtain the "reverse" chain parameters as well:

$$
\begin{aligned}
\mathcal{A}(0,z) &= \mathcal{D}(z,0) \\
\mathcal{B}(0,z) &= -\mathcal{B}(z,0) \\
\mathcal{C}(0,z) &= -\mathcal{C}(z,0) \\
\mathcal{D}(0,z) &= \mathcal{A}(z,0)
\end{aligned}
\tag{F.2}
$$

for which

$$
\begin{aligned}
V(0) &= \mathcal{A}(0,z)V(z) + \mathcal{B}(0,z)I(z) \\
I(0) &= \mathcal{C}(0,z)V(z) + \mathcal{D}(0,z)I(z)
\end{aligned}
\tag{F.3}
$$

F.1 THE EXPONENTIAL LINE

Let the line parameters vary with position as

$$
x(z) = X_0 e^{Kz}; \qquad b(z) = B_0 e^{-Kz}
\tag{F.4}
$$

where K is a real constant, and X_0 and B_0 are positive real constants. For this line, the propagation constant

$$
\beta = \sqrt{X_0 B_0}
\tag{F.5}
$$

is constant along the line, while the characteristic impedance

$$
Z_c(z) = Z_{c0} e^{Kz}
\tag{F.6}
$$

varies exponentially with distance, where

$$
Z_{c0} = \sqrt{\frac{X_0}{B_0}}
\tag{F.7}
$$

This parameter variation describes an *exponential line.*

Equations (1.4) yield the following equation for the voltage $V(z)$:

$$
\frac{d^2 V}{dz^2} - K\frac{dV}{dz} + \beta^2 V = 0
\tag{F.8}
$$

The general solution of (F.8) is

$$V(z) = e^{Kz/2} \left(V_1 e^{-wz} + V_2 e^{+wz} \right) \tag{F.9}$$

where V_1 and V_2 are some constants, and

$$w = \sqrt{\left(\frac{K}{2} \right)^2 - \beta^2} \tag{F.10}$$

The quantity w may be either real or imaginary, but it is important to note that the solution (F.9) does *not* represent a pair of waves, either traveling or attenuating, as is the case for a uniform transmission line. When line parameters vary with position, a unique separation of the solution into forward and backward traveling waves is not possible. An expression for $I(z)$ can be obtained by substituting (F.9) into the first of (1.4) and solving for I.

The constants V_1 and V_2 can be expressed in terms of $V(0)$ and $I(0)$, and the results used to obtain the chain parameters in (F.1). Leaving out the tedious but straightforward algebra, we obtain

$$
\begin{aligned}
\mathcal{A}(z,0) &= e^{Kz/2} \left(\cosh wz - \frac{K}{2w} \sinh wz \right) \\
\mathcal{B}(z,0) &= -\frac{\beta Z_{c0}}{w} e^{Kz/2} \sinh wz \\
\mathcal{C}(z,0) &= -\frac{\beta}{Z_{c0} w} e^{-Kz/2} \sinh wz \\
\mathcal{D}(z,0) &= e^{-Kz/2} \left(\cosh wz + \frac{K}{2w} \sinh wz \right)
\end{aligned}
\tag{F.11}
$$

The chain parameters can be used as in Section 1.3 to obtain the reflection coefficient of an exponential line terminated in a load, or any other quantity of interest for the line. For example, if a length d of exponential line is terminated in a load impedance $Z_L = Z_c(d) = Z_{c0} e^{Kd}$, then the input impedance at $z = 0$ is found from (1.65), (F.2) and (F.11) to be:

$$Z(0) = Z_{c0} \frac{w + \left(\beta + \frac{K}{2} \right) \tanh wd}{w + \left(\beta - \frac{K}{2} \right) \tanh wd} \tag{F.12}$$

and thus the reflection coefficient at $z = 0$ on a line of characteristic impedance Z_{c0} is

$$\rho(0) = \frac{Z(0) - Z_{c0}}{Z(0) + Z_{c0}} = \frac{K}{2} \frac{\tanh wd}{w + \beta \tanh wd} \tag{F.13}$$

F.2   THE BESSEL LINE

Let the line parameters vary with position for $z \geq 0$ as real powers of z:

$$\mathsf{x}(z) = X_0 z^p \quad (p > -1); \qquad \mathsf{b}(z) = B_0 z^q \quad (q > -1) \tag{F.14}$$

for some real constants X_0 and B_0, one (but not both) of which may be negative. The local propagation constant and characteristic impedance are defined as in Section 1.2. If $X_0 B_0 < 0$, then

$$\gamma(z) = \alpha(z) = \sqrt{-\mathsf{x}(z)\mathsf{b}(z)} = \sqrt{-X_0 B_0} \, z^{(p+q)/2} > 0; \qquad Z_c(z) = \frac{j\mathsf{x}(z)}{\gamma(z)} = j \frac{X_0}{\sqrt{-X_0 B_0}} z^{(p-q)/2} \tag{F.15}$$

while if X_0 and B_0 are positive, then

$$Z_c(z) = \sqrt{\frac{X_0}{B_0}} z^{(p-q)/2} > 0; \qquad \gamma(z) = j\beta(z) = j\mathrm{b}(z)Z_c(z) = j\sqrt{B_0 X_0} z^{(p+q)/2} \quad \text{(F.16)}$$

These parameters exhibit a classical turning-point behavior ($\gamma \to 0$) at $z = 0$ if $p + q$ is positive; we will consider $z = 0$ to be a generalized turning point as long as $p > -1$ and $q > -1$. This latter condition guarantees that the integral (1.99)

$$Q(z) = Q_r(z) + jQ_i(z) = \int_0^z \gamma(z')\,dz' \quad \text{(F.17)}$$

will be finite, and that the voltage $V(0)$ and current $I(0)$ will remain finite at the turning point. For later transformation of the telegraphers' equations we will introduce the quantities

$$r = \frac{p+1}{2}; \qquad s = \frac{p+q+2}{2} \quad \text{(F.18)}$$

and

$$\nu = \frac{p+1}{p+q+2} = \frac{r}{s} \quad \text{(F.19)}$$

Note that $r > 0$, $s > 0$ and $0 < \nu < 1$ under the conditions on p and q. Furthermore, we denote

$$g_0 = \frac{\sqrt{B_0 X_0}}{s} > 0 \qquad \text{(if } X_0, B_0 > 0\text{)}, \quad \text{(F.20)}$$

or

$$g_0 = -jg_1, \quad \text{where} \quad g_1 = \frac{\sqrt{-X_0 B_0}}{s} > 0 \qquad \text{(if } X_0 B_0 < 0\text{)} \quad \text{(F.21)}$$

Then we can write

$$\gamma(z) = jsg_0 z^{(p+q)/2} \quad \text{(F.22)}$$

and

$$Z_c(z) = Z_{c0} z^{(p-q)/2} \quad \text{(F.23)}$$

regardless of the sign of $X_0 B_0$, where we have denoted

$$Z_{c0} = \frac{X_0}{sg_0} = \frac{sg_0}{B_0} \quad \text{(F.24)}$$

F.2.1 PROPAGATING WAVES ON A BESSEL LINE

Equations (1.4) yield the following equation for the voltage $V(z)$:

$$\frac{d^2V}{dz^2} - \frac{p}{z}\frac{dV}{dz} + s^2 g_0^2 z^{p+q} V = 0 \quad \text{(F.25)}$$

Introducing the new variables

$$V = z^r U; \qquad w = z^s \quad \text{(F.26)}$$

transforms (F.25) to

$$\frac{d^2U}{dw^2} + \frac{1}{w}\frac{dU}{dw} + \left(g_0^2 - \frac{\nu^2}{w^2}\right)U = 0 \quad \text{(F.27)}$$

Equation (F.27) is none other than Bessel's differential equation (C.1), so its solution can be written in the form $U = AJ_\nu(g_0 w) + BJ_{-\nu}(g_0 w)$, where A and B are constants, and $J_{\pm\nu}$ is the Bessel function of order $\pm\nu$. Thus,

$$V = z^r \left[AJ_\nu(g_0 z^s) + BJ_{-\nu}(g_0 z^s) \right] \tag{F.28}$$

The current can then be recovered from the second of (1.4), using the recurrence relations (C.40)-(C.41):

$$I = j\frac{z^{s-r}}{Z_{c0}} \left[AJ_{\nu-1}(g_0 z^s) - BJ_{1-\nu}(g_0 z^s) \right] \tag{F.29}$$

The constants A and B can be expressed in terms of the voltage and current at the turning point by using (C.8), recalling that Γ in this equation denotes the gamma function (C.52). Using the identity (C.55) for the gamma function and taking the limits of (F.28) and (F.29) as $z \to 0$, we obtain

$$V(0) = \frac{B}{\Gamma(1-\nu)} \left(\frac{2}{g_0} \right)^\nu ; \qquad I(0) = j\frac{1}{Z_{c0}}\frac{A}{\Gamma(\nu)} \left(\frac{2}{g_0} \right)^{1-\nu} \tag{F.30}$$

Solving (F.30) for A and B, we have

$$A = -jZ_{c0}\Gamma(\nu) \left(\frac{g_0}{2} \right)^{1-\nu} I(0); \qquad B = \Gamma(1-\nu) \left(\frac{g_0}{2} \right)^\nu V(0) \tag{F.31}$$

With these results in hand, we can substitute (F.31) into (F.28) and (F.29) to get V and I at an arbitrary position z in terms of the voltage and current at the turning point, using chain parameters as in (F.1)-(F.3). We obtain after some algebra

$$\mathcal{A}(z,0) = \Gamma(1-\nu) \left[\frac{Q_i(z)}{2} \right]^\nu J_{-\nu}[Q_i(z)]$$

$$\mathcal{B}(z,0) = -j\Gamma(\nu)Z_c(z) \left[\frac{Q_i(z)}{2} \right]^{1-\nu} J_\nu[Q_i(z)]$$

$$\mathcal{C}(z,0) = -j\Gamma(1-\nu)\frac{1}{Z_c(z)} \left[\frac{Q_i(z)}{2} \right]^\nu J_{1-\nu}[Q_i(z)]$$

$$\mathcal{D}(z,0) = \Gamma(\nu) \left[\frac{Q_i(z)}{2} \right]^{1-\nu} J_{\nu-1}[Q_i(z)] \tag{F.32}$$

where $Q_i(z)$ is the phase integral (1.102) for this line:

$$Q_i(z) = \int_0^z \beta(z')\,dz' = g_0 z^s \geq 0 \tag{F.33}$$

Noting that

$$Z_c(z) = Z_{c0} \left(\frac{g_0}{2} \right)^{1-2\nu} \left[\frac{Q_i(z)}{2} \right]^{2\nu-1} \tag{F.34}$$

and assuming that z is sufficiently far from 0 that $|Q_i(z)| \gg 1$, we can approximate the chain parameters (F.32) using (C.17):

$$\mathcal{A}(z,0) \sim \frac{\Gamma(1-\nu)}{\sqrt{\pi}} \left(\frac{g_0}{2} \right)^{\nu-\frac{1}{2}} \sqrt{\frac{Z_c(z)}{Z_{c0}}} \cos\left[Q_i(z) + \frac{\pi}{2}\left(\nu - \frac{1}{2} \right) \right]$$

$$\mathcal{B}(z,0) \sim -j\frac{\Gamma(\nu)}{\sqrt{\pi}} \left(\frac{g_0}{2} \right)^{\frac{1}{2}-\nu} \sqrt{Z_c(z)Z_{c0}} \sin\left[Q_i(z) + \frac{\pi}{2}\left(\frac{1}{2} - \nu \right) \right]$$

$$\mathcal{C}(z,0) \quad \sim \quad -j\frac{\Gamma(1-\nu)}{\sqrt{\pi}}\left(\frac{g_0}{2}\right)^{\nu-\frac{1}{2}}\frac{1}{\sqrt{Z_c(z)Z_{c0}}}\sin\left[Q_i(z)+\frac{\pi}{2}\left(\nu-\frac{1}{2}\right)\right]$$

$$\mathcal{D}(z,0) \quad \sim \quad \frac{\Gamma(\nu)}{\sqrt{\pi}}\left(\frac{g_0}{2}\right)^{\frac{1}{2}-\nu}\sqrt{\frac{Z_{c0}}{Z_c(z)}}\cos\left[Q_i(z)+\frac{\pi}{2}\left(\frac{1}{2}-\nu\right)\right] \qquad \text{(F.35)}$$

which can be regarded as a generalization of the WKB results (1.110).

A similar procedure can be used to obtain the chain parameters of the Bessel line between any two points z_1 and z_2, provided that $z_2 > z_1 > 0$, with the only restriction on p and q being $p + q \neq -2$. We present only the final results here, leaving the mathematical manipulations as an exercise:

$$\mathcal{A}(z_2,z_1) \quad = \quad \frac{\pi}{2}[Q_i(z_2)]^\nu[Q_i(z_1)]^{1-\nu}\{J_\nu[Q_i(z_2)]Y_{\nu-1}[Q_i(z_1)]-Y_\nu[Q_i(z_2)]J_{\nu-1}[Q_i(z_1)]\}$$

$$\mathcal{B}(z_2,z_1) \quad = \quad \frac{j\pi}{2}Z_c(z_2)[Q_i(z_2)]^{1-\nu}[Q_i(z_1)]^\nu\{J_\nu[Q_i(z_2)]Y_\nu[Q_i(z_1)]-Y_\nu[Q_i(z_2)]J_\nu[Q_i(z_1)]\}$$

$$\mathcal{C}(z_2,z_1) \quad = \quad \frac{j\pi}{2}\frac{1}{Z_c(z_2)}[Q_i(z_2)]^\nu[Q_i(z_1)]^{1-\nu}\{J_{\nu-1}[Q_i(z_2)]Y_{\nu-1}[Q_i(z_1)]-Y_{\nu-1}[Q_i(z_2)]J_{\nu-1}[Q_i(z_1)]\}$$

$$\mathcal{D}(z_2,z_1) \quad = \quad \frac{\pi}{2}[Q_i(z_2)]^{1-\nu}[Q_i(z_1)]^\nu\{Y_{\nu-1}[Q_i(z_2)]J_\nu[Q_i(z_1)]-J_{\nu-1}[Q_i(z_2)]Y_\nu[Q_i(z_1)]\} \qquad \text{(F.36)}$$

If $p > -1$ and $q > -1$. then (F.32) can be obtained from (F.36) by taking the limit as $z_1 \to 0$.

F.2.2   ATTENUATING WAVES ON A BESSEL LINE

The case when $g_0 \equiv -jg_1$ is imaginary could be treated simply by making this substitution into (F.25)-(F.34) above, but it is generally more useful to employ a representation that explicitly displays the evanescent parts of the voltages and currents. It will prove convenient to replace X_0 and B_0 by X_1 and B_1 respectively for this case, so that we now use the attenuation integral (1.101) for this line in the form

$$Q_r(z) = \int_0^z \alpha(z')\,dz' = g_1z^s \geq 0 \qquad \text{(F.37)}$$

where (F.21) in this notation becomes

$$g_1 = \frac{\sqrt{-X_1B_1}}{s} > 0 \qquad \text{(F.38)}$$

Furthermore, the characteristic impedance is purely imaginary:

$$Z_c(z) = jX_{c1}z^{(p-q)/2} \qquad \text{(F.39)}$$

where we have denoted

$$X_{c1} = \frac{X_1}{sg_1} = -\frac{sg_1}{B_1} \qquad \text{(F.40)}$$

since $X_1B_1 < 0$ in this case. If $B_1 < 0$, then $X_{c1} > 0$; if $B_1 > 0$, then $X_{c1} < 0$. Note that

$$Z_c(z) = jX_{c1}\left(\frac{g_1}{2}\right)^{1-2\nu}\left[\frac{Q_r(z)}{2}\right]^{2\nu-1} \qquad \text{(F.41)}$$

and the ratio $Z_c(z)/(jX_1)$ is always positive for $z > 0$.

We observe that (F.27) now becomes the modified Bessel differential equation (C.25), and its solutions can be written in the form $AI_\nu(g_1w) + BK_\nu(g_1w)$, where I_ν and K_ν are the modified Bessel functions of order ν, and once again A and B are constants. We proceed analogously to the case when g_0 was real. We find

$$V(z) = z^r \left[AI_\nu \left(g_1 z^s \right) + BK_\nu \left(g_1 z^s \right) \right] \tag{F.42}$$

$$I(z) = j \frac{z^{s-r}}{X_{c1}} \left[AI_{\nu-1} \left(g_1 z^s \right) - BK_{\nu-1} \left(g_1 z^s \right) \right] \tag{F.43}$$

From (C.29), (C.30) and (C.36) we find

$$V(0) = B \frac{\Gamma(\nu)}{2} \left(\frac{2}{g_1} \right)^\nu \tag{F.44}$$

$$I(0) = j \frac{1}{X_{c1}} \left(\frac{g_1}{2} \right)^{\nu-1} \left[\frac{A}{\Gamma(\nu)} - \frac{B}{2} \Gamma(1-\nu) \right] \tag{F.45}$$

whence

$$A = \left(\frac{g_1}{2} \right)^\nu \Gamma(1-\nu) V(0) - j X_{c1} \left(\frac{g_1}{2} \right)^{1-\nu} \Gamma(\nu) I(0) \tag{F.46}$$

$$B = \frac{2}{\Gamma(\nu)} \left(\frac{g_1}{2} \right)^\nu V(0) \tag{F.47}$$

We thus again have the relations (F.1)-(F.3), where now

$$\begin{aligned}
\mathcal{A}(z,0) &= \Gamma(1-\nu) \left[\frac{Q_r(z)}{2} \right]^\nu I_{-\nu}[Q_r(z)] \\
\mathcal{B}(z,0) &= -\Gamma(\nu) Z_c(z) \left[\frac{Q_r(z)}{2} \right]^{1-\nu} I_\nu[Q_r(z)] \\
\mathcal{C}(z,0) &= -\Gamma(1-\nu) \frac{1}{Z_c(z)} \left[\frac{Q_r(z)}{2} \right]^\nu I_{1-\nu}[Q_r(z)] \\
\mathcal{D}(z,0) &= \Gamma(\nu) \left[\frac{Q_r(z)}{2} \right]^{1-\nu} I_{\nu-1}[Q_r(z)]
\end{aligned} \tag{F.48}$$

and we have used (C.35), (C.36) and (C.56). If z is far enough from $|Q_r(z)| \gg 1$, the chain parameters become [with the help of (C.38) and (F.41)]:

$$\begin{aligned}
\mathcal{A}(z,0) &\sim \frac{\Gamma(1-\nu)}{2\sqrt{\pi}} \left(\frac{g_1}{2} \right)^{\nu-\frac{1}{2}} \sqrt{\frac{Z_c(z)}{jX_{c1}}} \left[e^{Q_r(z)} + \sin\nu\pi e^{-Q_r(z)} \right] \\
\mathcal{B}(z,0) &\sim -jX_{c1} \frac{\Gamma(\nu)}{2\sqrt{\pi}} \left(\frac{g_1}{2} \right)^{\frac{1}{2}-\nu} \sqrt{\frac{Z_c(z)}{jX_{c1}}} \left[e^{Q_r(z)} - \sin\nu\pi e^{-Q_r(z)} \right] \\
\mathcal{C}(z,0) &\sim -\frac{1}{jX_{c1}} \frac{\Gamma(1-\nu)}{2\sqrt{\pi}} \left(\frac{g_1}{2} \right)^{\nu-\frac{1}{2}} \sqrt{\frac{jX_{c1}}{Z_c(z)}} \left[e^{Q_r(z)} - \sin\nu\pi e^{-Q_r(z)} \right] \\
\mathcal{D}(z,0) &\sim \frac{\Gamma(\nu)}{2\sqrt{\pi}} \left(\frac{g_1}{2} \right)^{\frac{1}{2}-\nu} \sqrt{\frac{jX_{c1}}{Z_c(z)}} \left[e^{Q_r(z)} + \sin\nu\pi e^{-Q_r(z)} \right]
\end{aligned} \tag{F.49}$$

which can also be put into a form more closely resembling (F.35):

$$\mathcal{A}(z,0) \sim \frac{\Gamma(1-\nu)}{\sqrt{\pi}} \left(\frac{g_1}{2} \right)^{\nu-\frac{1}{2}} \sqrt{\frac{Z_c(z)}{jX_{c1}}} \sqrt{\sin\nu\pi} \cosh\left[Q_r(z) + \frac{\ln(\csc\nu\pi)}{2} \right]$$

$$\mathcal{B}(z,0) \sim -jX_{c1}\frac{\Gamma(\nu)}{\sqrt{\pi}}\left(\frac{g_1}{2}\right)^{\frac{1}{2}-\nu}\sqrt{\frac{Z_c(z)}{jX_{c1}}}\sqrt{\sin\nu\pi}\sinh\left[Q_r(z)+\frac{\ln(\csc\nu\pi)}{2}\right]$$

$$\mathcal{C}(z,0) \sim -\frac{1}{jX_{c1}}\frac{\Gamma(1-\nu)}{\sqrt{\pi}}\left(\frac{g_1}{2}\right)^{\nu-\frac{1}{2}}\sqrt{\frac{jX_{c1}}{Z_c(z)}}\sqrt{\sin\nu\pi}\sinh\left[Q_r(z)+\frac{\ln(\csc\nu\pi)}{2}\right]$$

$$\mathcal{D}(z,0) \sim \frac{\Gamma(\nu)}{\sqrt{\pi}}\left(\frac{g_1}{2}\right)^{\frac{1}{2}-\nu}\sqrt{\frac{jX_{c1}}{Z_c(z)}}\sqrt{\sin\nu\pi}\cosh\left[Q_r(z)+\frac{\ln(\csc\nu\pi)}{2}\right] \tag{F.50}$$

F.3 CHAIN PARAMETERS, REFLECTION AND TRANSMISSION AT A TURNING POINT

The formulas in the previous section can be used to determine the chain parameters of a section of nonuniform transmission line that contains a generalized turning point at $z = 0$. From these, we will be able to obtain the reflection and transmission coefficients of the turning point. Consider the situation shown in Figure F.1. In $z < 0$, the transmission line has $x \geq 0$ and $b \geq 0$:

$$x(z) = X_0(-z)^{p_0}; \qquad b(z) = B_0(-z)^{q_0} \tag{F.51}$$

where X_0 and B_0 are both real and positive, and so supports propagating waves. The associated quantities r_0, s_0, ν_0, g_0 and Z_{c0} are then determined from (F.18)-(F.20) and (F.24). For $z > 0$, the transmission line has $xb \leq 0$:

$$x(z) = X_1 z^{p_1}; \qquad b(z) = B_1 z^{q_1} \tag{F.52}$$

where X_1 or B_1 are real, with one negative and the other positive. The line supports attenuating waves on this side of the turning point. The associated quantities r_1, s_1, ν_1, g_1 and X_{c1} are then determined from (F.18)-(F.19), (F.21) and (F.40). In fact, we only need the

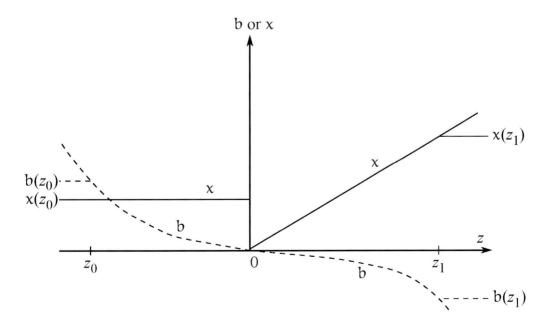

Figure F.1 Functions $x(z)$ and $b(z)$ near a generalized turning point.

line parameters to behave like those of a Bessel line sufficiently close to the turning point. Further away, the line can have arbitrary variations of its parameters with position, and the solutions for voltage and current can be represented by the ordinary WKB approximations of Section 1.5.

Designate a point $z_0 < 0$ to the left of the turning point, and another one $z_1 > 0$ to the right of the turning point. In order to make use of the results of Section F.2.1 for the region $z < 0$ [specifically (F.35)], we use (F.2) and the change of variable $u = -z$. It is convenient to introduce the notation

$$Q_i(0, z) \equiv -Q_i(z) = g_0(-z)^{s_o} \geq 0 \qquad (F.53)$$

for $z \leq 0$, whence

$$Z_c(z) = Z_{c0} \left(\frac{g_0}{2} \right)^{1-2\nu_0} \left[\frac{Q_i(0, z)}{2} \right]^{2\nu_0-1} \qquad (F.54)$$

For $|Q_i(0, z_0)| \gg 1$, the desired chain parameters for the interval between z_0 and the turning point are then approximately

$$\mathcal{A}_0(0, z_0) \sim \frac{\Gamma(\nu_0)}{\sqrt{\pi}} \left(\frac{g_0}{2} \right)^{\frac{1}{2}-\nu_0} \sqrt{\frac{Z_{c0}}{Z_c(z_0)}} \cos\left[Q_i(0, z_0) + \frac{\pi}{2}\left(\frac{1}{2} - \nu_0 \right) \right]$$

$$\mathcal{B}_0(0, z_0) \sim -j\frac{\Gamma(\nu_0)}{\sqrt{\pi}} \left(\frac{g_0}{2} \right)^{\frac{1}{2}-\nu_0} \sqrt{Z_c(z_0)Z_{c0}} \sin\left[Q_i(0, z_0) + \frac{\pi}{2}\left(\frac{1}{2} - \nu_0 \right) \right]$$

$$\mathcal{C}_0(0, z_0) \sim -j\frac{\Gamma(1-\nu_0)}{\sqrt{\pi}} \left(\frac{g_0}{2} \right)^{\nu_0-\frac{1}{2}} \frac{1}{\sqrt{Z_c(z_0)Z_{c0}}} \sin\left[Q_i(0, z_0) + \frac{\pi}{2}\left(\nu_0 - \frac{1}{2} \right) \right]$$

$$\mathcal{D}_0(0, z_0) \sim \frac{\Gamma(1-\nu_0)}{\sqrt{\pi}} \left(\frac{g_0}{2} \right)^{\nu_0-\frac{1}{2}} \sqrt{\frac{Z_c(z_0)}{Z_{c0}}} \cos\left[Q_i(0, z_0) + \frac{\pi}{2}\left(\nu_0 - \frac{1}{2} \right) \right] \qquad (F.55)$$

If $|Q_r(z_1)| \gg 1$, the chain parameters (F.49) for the interval between the turning point and z_1 are:

$$\mathcal{A}_1(z_1, 0) \sim \frac{\Gamma(1-\nu_1)}{2\sqrt{\pi}} \left(\frac{g_1}{2} \right)^{\nu_1-\frac{1}{2}} \sqrt{\frac{Z_c(z_1)}{jX_{c1}}} \left[e^{Q_r(z_1)} + \sin\nu_1\pi e^{-Q_r(z_1)} \right]$$

$$\mathcal{B}_1(z_1, 0) \sim -jX_{c1}\frac{\Gamma(\nu_1)}{2\sqrt{\pi}} \left(\frac{g_1}{2} \right)^{\frac{1}{2}-\nu_1} \sqrt{\frac{Z_c(z_1)}{jX_{c1}}} \left[e^{Q_r(z_1)} - \sin\nu_1\pi e^{-Q_r(z_1)} \right]$$

$$\mathcal{C}_1(z_1, 0) \sim -\frac{1}{jX_{c1}}\frac{\Gamma(1-\nu_1)}{2\sqrt{\pi}} \left(\frac{g_1}{2} \right)^{\nu_1-\frac{1}{2}} \sqrt{\frac{jX_{c1}}{Z_c(z_1)}} \left[e^{Q_r(z_1)} - \sin\nu_1\pi e^{-Q_r(z_1)} \right]$$

$$\mathcal{D}_1(z_1, 0) \sim \frac{\Gamma(\nu_1)}{2\sqrt{\pi}} \left(\frac{g_1}{2} \right)^{\frac{1}{2}-\nu_1} \sqrt{\frac{jX_{c1}}{Z_c(z_1)}} \left[e^{Q_r(z_1)} + \sin\nu_1\pi e^{-Q_r(z_1)} \right] \qquad (F.56)$$

where $Q_r(z)$ and $Z_c(z)$ are given by

$$Q_r(z) = \int_0^z \alpha(z')\, dz' = g_1 z_1^s \geq 0 \qquad (F.57)$$

and

$$Z_c(z) = jX_{c1} \left(\frac{g_1}{2} \right)^{1-2\nu_1} \left[\frac{Q_r(z)}{2} \right]^{2\nu_1-1} \qquad (F.58)$$

Since forward and backward traveling waves are not uniquely defined on nonuniform transmission lines, it is a delicate matter to define reflection and transmission coefficients on them. We will proceed as follows. Let a wave be incident toward the turning point from the region $z < 0$, in the sense that for $z < z_0$ we imagine that the line is continued uniformly toward $z \to -\infty$. The incident and reflected waves on this uniform section are (see Chapter 1):

$$V^{\mathrm{inc}}(z) = \frac{1}{2}\left[V(z) + Z_c(z)I(z)\right]; \qquad V^{\mathrm{refl}}(z) = \frac{1}{2}\left[V(z) - Z_c(z)I(z)\right] \tag{F.59}$$

and we will agree to use these definitions for any $z \leq 0$. The voltage reflection coefficient at z for this incident wave can then be defined, even on a nonuniform line, as

$$\begin{aligned}\rho(z) &= \frac{V^{\mathrm{refl}}(z)}{V^{\mathrm{inc}}(z)} = \frac{V(z) - Z_c(z)I(z)}{V(z) + Z_c(z)I(z)} \\[2mm] &= \frac{Z(0)\left[\mathcal{D}_0(0,z) + Z_c(z)\mathcal{C}_0(0,z)\right] - \mathcal{B}_0(0,z) - Z_c(z)\mathcal{A}_0(0,z)}{Z(0)\left[\mathcal{D}_0(0,z) - Z_c(z)\mathcal{C}_0(0,z)\right] - \mathcal{B}_0(0,z) + Z_c(z)\mathcal{A}_0(0,z)}\end{aligned} \tag{F.60}$$

where we have used (F.1) and defined

$$Z(0) = \frac{V(0)}{I(0)} \tag{F.61}$$

as the total impedance at the turning point, looking toward it from $z < 0$. When $z < z_0$, the reflection coefficient (F.60) is the usual definition on the uniformly extended line; when $z > z_0$, we take (F.60) to be the definition of the local reflection coefficient on the nonuniform line. We note that

$$V(z) = V^{\mathrm{inc}}(z)\left[1 + \rho(z)\right] \qquad \text{(for } z < 0) \tag{F.62}$$

From (F.55) we have in $z < 0$:

$$\mathcal{D}_0(0,z) \pm Z_c(z)\mathcal{C}_0(0,z) \sim \frac{\Gamma(1 - \nu_0)}{\sqrt{\pi}}\left(\frac{g_0}{2}\right)^{\nu_0 - \frac{1}{2}}\sqrt{\frac{Z_c(z)}{Z_{c0}}}\,e^{\mp j\left[Q_i(0,z_0) + \frac{\nu_0 \pi}{2} - \frac{\pi}{4}\right]} \tag{F.63}$$

and

$$Z_c(z)\mathcal{A}_0(0,z) \pm \mathcal{B}_0(0,z) \sim R_0 \frac{\Gamma(1 - \nu_0)}{\sqrt{\pi}}\left(\frac{g_0}{2}\right)^{\nu_0 - \frac{1}{2}}\sqrt{\frac{Z_c(z)}{Z_{c0}}}\,e^{\mp j\left[Q_i(0,z_0) - \frac{\nu_0 \pi}{2} + \frac{\pi}{4}\right]} \tag{F.64}$$

where by (F.53) and (F.54),

$$R_0 = \frac{\Gamma(\nu_0)}{\Gamma(1 - \nu_0)}Z_{c0}\left(\frac{g_0}{2}\right)^{1 - 2\nu_0} = \frac{\Gamma(\nu_0)}{\Gamma(1 - \nu_0)}(2s_0)^{2\nu_0 - 1}X_0\left(X_0 B_0\right)^{-\nu_0} \tag{F.65}$$

is a real constant. Thus, in this approximation the local reflection coefficient (F.60) is

$$\rho(z) \sim \rho(0)e^{-2jQ_i(0,z_0)} \tag{F.66}$$

where

$$\rho(0) = j\frac{Z(0)e^{-j\frac{\nu_0 \pi}{2}} + jR_0 e^{+j\frac{\nu_0 \pi}{2}}}{Z(0)e^{+j\frac{\nu_0 \pi}{2}} + jR_0 e^{-j\frac{\nu_0 \pi}{2}}} \tag{F.67}$$

is the reflection coefficient extrapolated to the turning point $z = 0$ in the WKB approximation. The value of $Z(0)$ must be obtained before we can calculate the value of $\rho(0)$.

To assure that no "incident" attenuating wave exists in $z > 0$ (i.e., there is no exponentially growing wave in either $V(z)$ or $I(z)$ as $z \to \infty$), we place a matched load at z_1:

$$V(z_1) = Z_c(z_1)I(z_1)$$

Expressing the voltage and current in terms of those at $z = 0$ by means of the chain parameters and solving for $Z(0)$ gives

$$Z(0) = \frac{Z_c(z_1)\mathcal{D}_1(z_1,0) - \mathcal{B}_1(z_1,0)}{\mathcal{A}_1(z_1,0) - Z_c(z_1)\mathcal{C}_1(z_1,0)} \simeq jR_1 \tag{F.68}$$

where

$$R_1 = \frac{\Gamma(\nu_1)}{\Gamma(1-\nu_1)}X_{c1}\left(\frac{g_1}{2}\right)^{1-2\nu_1} = \frac{\Gamma(\nu_1)}{\Gamma(1-\nu_1)}(2s_1)^{2\nu_1-1}X_1\left(-X_1B_1\right)^{-\nu_1} \tag{F.69}$$

is a real constant. Inserting (F.68) into (F.67), we find the reflection coefficient for this generalized turning point (in the WKB approximation):

$$\rho(0) = j\frac{R_1 e^{-j\frac{\nu_0\pi}{2}} + R_0 e^{+j\frac{\nu_0\pi}{2}}}{R_1 e^{+j\frac{\nu_0\pi}{2}} + R_0 e^{-j\frac{\nu_0\pi}{2}}} \tag{F.70}$$

It should be noted that $|\rho(0)| = 1$; a wave incident from the "propagating side" $z < 0$ of this turning point is always totally reflected if there is no further reflection of the attenuating wave in the region $z > 0$. Also, note that in spite of the rather complicated derivation that leads to it, expression (F.70) does not contain any Bessel functions, and its most complicated constituents are gamma functions.

The transmission coefficient of a wave incident from $z < 0$ can be defined in a similar way. Choose a point $z_1 > 0$ and imagine that the line is uniformly extended for $z > z_1$. We define the local voltage transmission coefficient as

$$\tau(z_1, z_0) = \frac{V(z_1)}{V(z_0)}[1 + \rho(z_0)] \tag{F.71}$$

if z_1 lies to the right of the turning point, and z_0 to the left of it. We have

$$V(z_1) = I(0)[Z(0)\mathcal{A}_1(z_1,0) + \mathcal{B}_1(z_1,0)] \tag{F.72}$$

and

$$V(z_0) = I(0)[Z(0)\mathcal{D}_0(0,z_0) - \mathcal{B}_0(0,z_0)] \tag{F.73}$$

while from (F.60) we have

$$1 + \rho(z) = \frac{2[Z(0)\mathcal{D}_0(0,z) - \mathcal{B}_0(0,z)]}{Z(0)[\mathcal{D}_0(0,z) - Z_c(z)\mathcal{C}_0(0,z)] - \mathcal{B}_0(0,z) + Z_c(z)\mathcal{A}_0(0,z)} \tag{F.74}$$

so that

$$\tau(z_1, z_0) = \frac{2[Z(0)\mathcal{A}_1(z_1,0) + \mathcal{B}_1(z_1,0)]}{Z(0)[\mathcal{D}_0(0,z) - Z_c(z)\mathcal{C}_0(0,z)] - \mathcal{B}_0(0,z) + Z_c(z)\mathcal{A}_0(0,z)} \tag{F.75}$$

Then from (F.55), (F.56) and (F.68), we get

$$\tau(z_1, z_0) \simeq T(0)e^{-jQ_i(0,z_0)-Q_r(z_1)}\sqrt{\frac{Z_c(z_1)}{jX_{c1}}}\sqrt{\frac{Z_{c0}}{Z_c(z_0)}} \tag{F.76}$$

where $T(0)$ is the (wave amplitude) transmission coefficient extrapolated to the turning point in the WKB approximation:

$$T(0) = \sin \nu_1 \pi \frac{\Gamma(1-\nu_1)}{\Gamma(1-\nu_0)} \left(\frac{g_0}{2}\right)^{\frac{1}{2}-\nu_0} \left(\frac{g_1}{2}\right)^{\nu_1-\frac{1}{2}} \frac{2R_1 e^{+j\pi/4}}{R_1 e^{+j\frac{\nu_0\pi}{2}} + R_0 e^{-j\frac{\nu_0\pi}{2}}} \qquad (F.77)$$

Formulas (F.66) with (F.70), and (F.76) with (F.77), are sometimes referred to as connection formulas, which allow us to link the expressions for waves on opposite sides of a turning point, even though the asymptotic (WKB) approximations for V and I break down at the turning point.

Even though we have derived these results for the specific power-law variations of x and b on a Bessel line, when we are far enough away from the turning point, we may still use these formulas if we replace expressions for Z_c and Q by those of any line. The behavior at the turning point is determined only by that of x and b as the turning point is crossed. If desired, we can obtain the chain parameters for the entire interval from z_0 to z_1 by multiplying the two chain matrices (F.55) and (F.56), resulting in

$$\begin{aligned} \mathcal{A}(z_1, z_0) &= \mathcal{A}_1(z_1, 0)\mathcal{A}_0(0, z_0) + \mathcal{B}_1(z_1, 0)\mathcal{C}_0(0, z_0) \\ \mathcal{B}(z_1, z_0) &= \mathcal{A}_1(z_1, 0)\mathcal{B}_0(0, z_0) + \mathcal{B}_1(z_1, 0)\mathcal{D}_0(0, z_0) \\ \mathcal{C}(z_1, z_0) &= \mathcal{C}_1(z_1, 0)\mathcal{A}_0(0, z_0) + \mathcal{D}_1(z_1, 0)\mathcal{C}_0(0, z_0) \\ \mathcal{D}(z_1, z_0) &= \mathcal{C}_1(z_1, 0)\mathcal{B}_0(0, z_0) + \mathcal{D}_1(z_1, 0)\mathcal{D}_0(0, z_0) \end{aligned} \qquad (F.78)$$

The foregoing results can be obtained in a more direct manner using these parameters. The reflection and transmission coefficients for an attenuating wave incident from $z > 0$ can also be obtained by the method of this section. Details are left as an exercise (problem pF-3).

F.3.1 EXAMPLE: LINEAR TURNING POINT

Two cases are of special interest. Consider first a linear capacitive turning point: $p_0 = p_1 = 0$, $X_0 = X_1 > 0$, $q_0 = q_1 = 1$ and $B_0 = -B_1 > 0$. Then $r_0 = r_1$, $s_0 = s_1$ and $\nu_0 = \nu_1 = 1/3$, implying $R_1 = R_0$ and $g_1 = g_0$, so that

$$\rho(0) = +j \qquad \text{(linear capacitive turning point)} \qquad (F.79)$$

On the other hand, a linear inductive turning point is defined by $p_0 = p_1 = 1$, $X_0 = -X_1 > 0$, $q_0 = q_1 = 0$ and $B_0 = B_1 > 0$. Then $\nu_0 = \nu_1 = 2/3$, $R_1 = -R_0$ and $g_1 = g_0$, and we find

$$\rho(0) = -j \qquad \text{(linear inductive turning point)} \qquad (F.80)$$

These results provide a more rigorous justification of (1.118) and (1.120) respectively. The corresponding transmission coefficients are

$$T(0) = e^{+j\pi/4} \qquad \text{(linear capacitive turning point)} \qquad (F.81)$$

and

$$T(0) = e^{-j\pi/4} \qquad \text{(linear inductive turning point)} \qquad (F.82)$$

This is a more rigorous way of obtaining the results of Section 1.5.3.

F.4 NOTES AND REFERENCES

This treatment of the exponential line is adapted from

Slater (1942), pp. 75-78,

but see also

Reich *et al.* (1953), Sect. 4-6.

This treatment of Bessel lines is adapted from

A. T. Starr, "The nonuniform transmission line," *Proc. IRE*, vol. 20, pp. 1052-1063 (1932).

and the briefer treatment in

Ghose (1963), Section 12.3.

The treatment of turning points is based on the ideas of Langer:

R. E. Langer, "On the asymptotic solutions of ordinary differential equations, with an application to the Bessel functions of large order," *Trans. Amer. Math. Soc.*, vol. 33, pp. 23-64 (1931).

R. E. Langer, "On the asymptotic solutions of differential equations, with an application to the Bessel functions of large complex order," *Trans. Amer. Math. Soc.*, vol. 34, pp. 447-480 (1932).

R. E. Langer, "On the asymptotic solutions of ordinary differential equations, with reference to the Stokes' phenomenon about a singular point," *Trans. Amer. Math. Soc.*, vol. 37, pp. 397-416 (1935).

F.5 PROBLEMS

pF-1 Show that the special case $p = q = 0$ of the Bessel line recovers that of a uniform transmission line; specifically, show that putting $p = q = 0$ in (F.32) yields (1.62).

pF-2 Provide the details of the derivation of (F.36).

pF-3 Obtain $\rho(0)$ and $\tau(0)$ for a wave incident on a generalized turning point from the attenuating side $z > 0$.

pF-4 Instead of interfacing to an attenuating region, suppose that a load impedance Z_L is connected at the turning point $z = 0$ with a wave incident from $z < 0$. Find an expression for the reflection coefficient $\rho(0)$ in this case. Show that for a short circuit load we have $\rho(0) = e^{-j\pi(\nu_0 + \frac{1}{2})}$ in this case, while for an open circuit load we have $\rho(0) = e^{j\pi(\nu_0 - \frac{1}{2})}$.

G FOURIER AND LAPLACE TRANSFORMS AND OTHER MATHEMATICAL DATA

In this appendix, we collect some useful properties of Fourier and Laplace transforms, as well as lists of some integrals and series that arise in certain applications. Further information can be found in the references at the end of this appendix. The notation $\theta(t)$ is used for the Heaviside unit step function:

$$\begin{aligned} \theta(t) &= 1 && (t > 0) \\ &= 0 && (t < 0) \end{aligned} \tag{G.1}$$

The gamma function $\Gamma(\nu)$ and the Bessel function $J_\nu(z)$ are defined in Appendix C. The Dirac delta function is denoted $\delta(t)$.

G.1 FOURIER TRANSFORMS

Given a (generally complex) function $f(t)$ with suitable properties, we define its Fourier transform $F(\omega)$ as

$$F(\omega) = \int_{-\infty}^{\infty} f(t) e^{-j\omega t}\, dt \tag{G.2}$$

The original function $f(t)$ can be recovered from a knowledge of its Fourier transform through the inverse transform relation

$$f(t) = \frac{1}{2\pi} \int_{-\infty}^{\infty} F(\omega) e^{j\omega t}\, d\omega \tag{G.3}$$

For two functions $f(t)$ and $g(t)$ and their corresponding Fourier transforms, we have the convolution theorem:

$$\int_{-\infty}^{\infty} f(t')g(t - t')\, dt' = \frac{1}{2\pi} \int_{-\infty}^{\infty} F(\omega)G(\omega) e^{j\omega t}\, d\omega \tag{G.4}$$

Since $G^*(\omega)$ is the Fourier transform of $g^*(-t)$, the convolution theorem implies the correlation theorem:

$$\int_{-\infty}^{\infty} f(t')g^*(t' - t)\, dt' = \frac{1}{2\pi} \int_{-\infty}^{\infty} F(\omega)G^*(\omega) e^{j\omega t}\, d\omega \tag{G.5}$$

of which a special case is the Parseval relation:

$$\int_{-\infty}^{\infty} f(t)g^*(t)\, dt = \frac{1}{2\pi} \int_{-\infty}^{\infty} F(\omega)G^*(\omega)\, d\omega \tag{G.6}$$

where * denotes the complex conjugate.

In Table G.1, we collect some general relations (beyond those given in Section G.1) which allow the Fourier transform of one function to be expressed in terms of the transform

Table G.1
Table of rules for Fourier transforms.

Formula	$f(t)$	$F(\omega) = \int_{-\infty}^{\infty} f(t)e^{-j\omega t}\, dt$
G.1.1	$f(t-b)$	$e^{-j\omega b} F(\omega)$
G.1.2	$e^{jat}g(t)$	$G(\omega - a)$
G.1.3	$g(at)$	$\frac{1}{a}G\left(\frac{\omega}{a}\right)$
G.1.4	$t^n g(t)$	$j^n \dfrac{d^n G(\omega)}{d\omega^n}$
G.1.5	$\dfrac{d^n g(t)}{dt^n}$	$(j\omega)^n G(\omega)$

Table G.2
Table of Fourier transforms.

Formula	$f(t)$	$F(\omega) = \int_{-\infty}^{\infty} f(t)e^{-j\omega t}\, dt$
G.2.1	$\delta(t)$	1
G.2.2	1	$2\pi\delta(\omega)$
G.2.3	$\dfrac{1}{t^2 + T^2}$	$\dfrac{\pi}{T}e^{-|\omega|T}$
G.2.4	e^{-t^2/T^2}	$T\sqrt{\pi}e^{-\omega^2 T^2/4}$
G.2.5	$\delta(t-b) - ab\dfrac{J_1(a\sqrt{t^2 - b^2})}{\sqrt{t^2 - b^2}}\theta(t-b)$	$e^{-jb\sqrt{\omega^2 - a^2}}$
G.2.6	$\dfrac{e^{-at}}{1 + e^{-bt}} \quad (0 < a < b)$	$\dfrac{-\pi j}{b\sinh\frac{\pi(\omega - ja)}{b}}$

of another related one. Here, the quantities a, T, etc., are real positive constants (unless otherwise noted), while n denotes a positive integer. The transform of a function denoted by a lowercase letter is represented by the corresponding upper-case letter ($g(t) \to G(\omega)$, for example). In all cases, it must be assumed that the various transform integrals are convergent for the formulas to be valid. In Table G.2, some Fourier transforms that are needed in this book are presented.

A word is in order about the derivation of the Fourier transform of the Gaussian function in entry G.2.4. It follows from the identity

$$\int_{-\infty}^{\infty} e^{-x^2}\, dx = \sqrt{\pi} \tag{G.7}$$

which can be proved as follows. Denote the integral on the left side of (G.7) by I. Then the square of I can be re-expressed using a change from rectangular to polar coordinates in the integrations:

$$I^2 = \int_{-\infty}^{\infty} e^{-x^2}\, dx \int_{-\infty}^{\infty} e^{-y^2}\, dy = \int_0^{2\pi} \int_0^{\infty} e^{-\rho^2} \rho\, d\rho\, d\phi = 2\pi \int_0^{\infty} e^{-\rho^2} \rho\, d\rho \tag{G.8}$$

Changing the integration variable to $u = \rho^2$ gives $I^2 = \pi$, or since I is obviously positive, (G.7). The desired Fourier transform then follows by some strategic changes of integration variables.

G.2 LAPLACE TRANSFORMS

The Laplace transform can be thought of as the Fourier transform of a function $f(t)$ that vanishes identically for $t < 0$ and for which $j\omega$ is replaced by the complex variable s. In Table G.3, we collect some general relations which allow the Laplace transform of one function to be expressed in terms of the transform of another related one. Here, the quantities a, T, etc., are real positive constants (unless otherwise noted), while n denotes a positive integer. As with the Fourier transform, the Laplace transform of a function denoted by

Table G.3
Table of rules for Laplace transforms.

Formula	$f(t)$	$F(s) = \int_0^{\infty} f(t)e^{-st}\, dt$
G.3.1	$g(t-T)\theta(t-T)$	$e^{-sT}G(s)$
G.3.2	$\mathrm{Re}\,[Ae^{-at}g(t)]$	$\mathrm{Re}\,[AG(s+a)]$
G.3.3	$g(at)$	$\dfrac{1}{a}G\left(\dfrac{s}{a}\right)$
G.3.4	$t^n g(t)$	$(-1)^n \dfrac{d^n G(s)}{ds^n}$
G.3.5	$\dfrac{d^n g(t)}{dt^n}$	$s^n G(s) - s^{n-1}f(0) - s^{n-2}f'(0) - \ldots - f^{[n-1]}(0)$
G.3.6	$\displaystyle\int_0^t g(u)\, du$	$\frac{1}{s}G(s)$
G.3.7	$\displaystyle\int_0^t g_1(u)g_2(t-u)\, du$	$G_1(s)G_2(s)$

Table G.4

Table of Laplace transforms.

Formula	$f(t)$	$F(s) = \int_0^\infty f(t)e^{-st}\, dt$
G.4.1	$\delta(t)$	1
G.4.2	$\theta(t)$	$\dfrac{1}{s}$
G.4.3	$t^\nu \theta(t) \quad (\nu > -1)$	$\dfrac{\Gamma(\nu+1)}{s^{\nu+1}}$
G.4.4	$\theta(t-b) \quad (b \geq 0)$	$\dfrac{1}{s}e^{-bs}$
G.4.5	$J_\nu(at)\theta(t) \quad (\nu > -1,\, a > 0)$	$\dfrac{1}{\sqrt{s^2+a^2}}\left(\dfrac{a}{s+\sqrt{s^2+a^2}}\right)^\nu$
G.4.6	$J_0(a\sqrt{t^2-b^2})\theta(t-b) \quad (a \geq 0,\, b \geq 0)$	$\dfrac{e^{-b\sqrt{s^2+a^2}}}{\sqrt{s^2+a^2}}$
G.4.7	$\dfrac{J_1(a\sqrt{t^2-b^2})}{\sqrt{t^2-b^2}}\theta(t-b) \quad (a \geq 0,\, b \geq 0)$	$\dfrac{e^{-bs}-e^{-b\sqrt{s^2+a^2}}}{ab}$
G.4.8	$ab\displaystyle\int_b^t \dfrac{J_1(a\sqrt{u^2-b^2})}{\sqrt{u^2-b^2}}\, du\, \theta(t-b) \quad (a \geq 0,\, b \geq 0)$	$\dfrac{e^{-bs}-e^{-b\sqrt{s^2+a^2}}}{s}$

a lowercase letter is represented by the corresponding upper-case letter ($g(t) \to G(s)$, for example). All functions of time are assumed implicitly to contain a factor $\theta(t)$. We present in Table G.4 a list of some Laplace transforms that are needed in this book.

As an example of the use of the rules in Table G.3, let $g_1(u) = g(u)$ and $g_2(u) = \theta(u)$ in Formula 7 of Table G.3. Then using Formula 2 of Table G.4, we obtain Formula 6 of Table G.3.

G.2.1   HEAVISIDE'S FIRST EXPANSION THEOREM

The behavior of a function of t for early times is related to the behavior of its Laplace transform for large $|s|$. This relationship is quantified by Heaviside's first expansion theorem, which is as follows. Suppose the Laplace transform of a function $f(t)$ can be expanded as a series in inverse powers of the complex frequency s:

$$F(s) = F_0 + \frac{F_1}{s} + \frac{F_2}{s^2} + \dots \tag{G.9}$$

as $s \to \infty$, where F_0, F_1, ..., are real constants. We identify the inverse transform of each term of this series from Table G.4, and add the results to form $f(t)$:

$$f(t) = F_0 \delta(t) + \theta(t)\sum_{i=0}^\infty F_{i+1}\frac{t^i}{i!} \tag{G.10}$$

G.3 INTEGRALS

For convenience, we list below some indefinite integrals that are needed in some applications. They, along with many others, can be found in standard mathematical tables.

$$\int \frac{dx}{\sqrt{x^2 + a^2}} = \ln\left(x + \sqrt{x^2 + a^2}\right) \tag{G.11}$$

$$\int \ln x \, dx = x \ln x - x \tag{G.12}$$

$$\int \ln\left(x^2 + a^2\right) dx = x \ln\left(x^2 + a^2\right) - 2x + 2a \tan^{-1}\left(\frac{x}{a}\right) \tag{G.13}$$

$$\int x \ln\left(\frac{a+x}{a-x}\right) dx = ax - \frac{a^2 - x^2}{2} \ln\left(\frac{a+x}{a-x}\right) \tag{G.14}$$

$$\int \tan^{-1} x \, dx = x \tan^{-1} x - \frac{1}{2} \ln\left(x^2 + 1\right) \tag{G.15}$$

$$\int x \ln\left(x^2 + a^2\right) dx = \frac{x^2 + a^2}{2} \ln\left(x^2 + a^2\right) - \frac{x^2}{2} \tag{G.16}$$

$$\int \sinh^{-1} x \, dx = x \sinh^{-1} x - \sqrt{1 + x^2} \tag{G.17}$$

$$\int x^3 \ln\left(\frac{a+x}{a-x}\right) dx = \frac{x^4 - a^4}{4} \ln\left(\frac{a+x}{a-x}\right) + \frac{ax}{2}\left(\frac{x^2}{3} + a^2\right) \tag{G.18}$$

G.4 SERIES

The following series can be useful on occasion.

$$\delta(x) = \frac{1}{a} + \frac{2}{a} \sum_{n=1}^{\infty} \cos\left(\frac{n\pi x}{a}\right) \tag{G.19}$$

$$\frac{\pi \coth \pi a}{2a} = \frac{1}{2} + \sum_{n=1}^{\infty} \frac{1}{n^2 + a^2} \tag{G.20}$$

$$-\frac{1}{2} \ln \tan \frac{|\theta|}{2} = \sum_{n=1,3,5,\dots}^{\infty} \frac{\cos m\theta}{m} \qquad (0 < |\theta| < \pi) \tag{G.21}$$

G.5 NOTES AND REFERENCES

More in-depth treatments of the Fourier transform are given in:

Davies (1978), Chapters 7-11;
Titchmarsh (1986).

Extensive tables of the Fourier transform are given in:

Erdélyi (1954), Chapters 1-3.

Good references for the properties and applications of the Laplace transform are:

Davies (1978), Chapters 1-6;
van der Pol and Bremmer (1987).

More extensive tables of the Laplace transform are given in:

Erdélyi (1954), Chapters 4 and 5;
Oberhettinger and Badii (1973).

Collections of integrals and series can be found in:

Jolley (1961);
Gradshteyn and Ryzhik (1980);
Collin (1991), Appendix A.6.

H MODES IN HOLLOW WAVEGUIDES WITH ZERO CUTOFF FREQUENCY

As mentioned in Chapter 5, the equations

$$(\nabla_T^2 + k_c^2)\mathcal{H}_z = 0; \quad \left.\frac{\partial \mathcal{H}_z}{\partial n}\right|_{C_0} = 0$$

describing TE modes in a hollow metallic waveguide filled with a homogeneous medium admit the special solution:

$$\mathcal{H}_z = K_0 = \text{constant}; \quad k_c = 0 \tag{H.1}$$

Because k_c vanishes, however, it is not possible to use (5.38) and (5.39) to recover the other field components, because they have the indeterminate form $0/0$. We must use (5.15)-(5.18) to study modes which have zero cutoff frequency.

Inserting $\mathcal{H}_z = K_0$ and $\mathcal{E}_z = 0$ into (5.17)-(5.20), we arrive at

$$\mathbf{u}_z \times \boldsymbol{\mathcal{E}}_T = \zeta \boldsymbol{\mathcal{H}}_T \tag{H.2}$$

$$\mathbf{u}_z \cdot \nabla_T \times \boldsymbol{\mathcal{H}}_T = 0 \tag{H.3}$$

$$\mathbf{u}_z \cdot \nabla_T \times \boldsymbol{\mathcal{E}}_T = -j\omega\mu K_0 \tag{H.4}$$

where $\zeta = (\mu/\epsilon)^{1/2}$ is the wave impedance of the medium filling the guide. We have used the fact that $\gamma = j\omega\sqrt{\mu\epsilon}$ (since $k_c = 0$), assuming the forward-propagating mode. The boundary conditions imply that

$$\mathcal{E}_l|_{C_0} = 0 \quad \text{and} \quad \mathcal{H}_n|_{C_0} = 0 \tag{H.5}$$

(cf. Figure 5.5).

Now, equations (H.2) and (H.4) imply that

$$\nabla_T \cdot \boldsymbol{\mathcal{H}}_T = j\omega\sqrt{\mu\epsilon}K_0 \tag{H.6}$$

If we apply the two-dimensional divergence theorem (cf. eqn. (B.26) in Appendix A),

$$-\oint_C \mathbf{A}_T \cdot \mathbf{u}_n \, dl = \int_S \nabla_T \cdot \mathbf{A}_T \, dS \tag{H.7}$$

(where S is a surface in the xy-plane bounded by a curve C whose *inward* unit normal vector is $\mathbf{u}_n$) to the vector

$$\mathbf{A}_T = \mathcal{H}_z \boldsymbol{\mathcal{H}}_T$$

and take C to be the conducting waveguide wall C_0, then

$$0 = -\oint_{C_0} \mathcal{H}_z \mathcal{H}_n \, dl$$

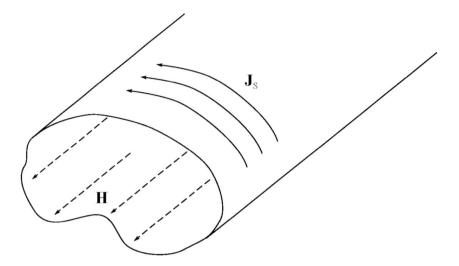

Figure H.1　Wall currents and magnetic fields for the "TE_0" mode.

$$= \int_{S_0} \nabla_T \cdot (\mathcal{H}_z \boldsymbol{\mathcal{H}}_T) \, dS$$

$$= \int_{S_0} [\nabla_T \mathcal{H}_z \cdot \boldsymbol{\mathcal{H}}_T + \mathcal{H}_z \nabla_T \cdot \boldsymbol{\mathcal{H}}_T] \, dS \qquad (\text{H.8})$$

$$= j\omega \sqrt{\mu\epsilon} K_0^2 S_0 \qquad (\text{H.9})$$

where we have used (H.5) and (H.6), and used the symbol S_0 also to denote the area of the waveguide cross-section.

Equation (H.9) implies that $\omega = 0$ or $K_0 = 0$, and so (H.4) becomes

$$\mathbf{u}_z \cdot \nabla_T \times \boldsymbol{\mathcal{E}}_T = 0 \qquad (\text{H.10})$$

and the equations (H.2), (H.3) and (H.10) which describe the transverse fields are completely decoupled from $\mathcal{H}_z$, and are, in fact, the equations for TEM mode fields in the waveguide. We study such modes in detail in Chapter 8, and show in particular that no TEM mode is possible in a hollow waveguide with a simply-connected boundary wall.

The nontrivial "TE_0" mode (one with $\mathcal{H}_z = $ constant $\neq 0$) which can exist therefore does so only at DC ($\omega = 0$), and consists of a constant, z-directed **H**-field maintained by transverse currents circulating in the waveguide wall (Figure H.1). Since this DC effect is usually not important at the operating frequency of a metallic waveguide, the contribution of this mode to the total field of the waveguide is generally ignored.

I FIELD LINE PLOTTING

Field-line plots of the transverse fields of waveguide modes are conventionally drawn with solid lines for the electric field and dashed lines for the magnetic field as shown in Figure I.1. The lines are drawn so as to be tangent to the appropriate field vector at each point, with the amplitude of the vector being proportional to the density of the lines in the vicinity of each point. An $\mathcal{E}$-field line in a two-dimensional plot (x, y or ρ, ϕ) should thus be everywhere tangent to $\mathcal{E}_T$; if the field line is described by the relation $y = y(x)$, we require that

$$\frac{dl_y}{dl_x} = \frac{dy}{dx} = \frac{\mathcal{E}_y}{\mathcal{E}_x} \tag{I.1}$$

be satisfied, whereas in cylindrical coordinates, a field line $\rho = \rho(\phi)$ should satisfy

$$\frac{dl_\rho}{dl_\phi} = \frac{d\rho}{\rho d\phi} = \frac{\mathcal{E}_\rho}{\mathcal{E}_\phi} \tag{I.2}$$

Inserting the specific expressions for the components of $\mathcal{E}_T$ into (I.1) or (I.2) allows us to solve for $y(x)$ or $\rho(\phi)$ and thus plot the field line (increasing the density of lines as the amplitude of $\mathcal{E}_T$ warrants). Clearly the ratio of the components of $\mathcal{E}_T$ must be real for (I.1) and (I.2) to make sense. Equations for the field lines of $\mathcal{H}_T$ can be treated in a similar manner.

In cases where $\mathcal{E}_T$ has been obtained by separation of variables, (I.1) and (I.2) can be solved more explicitly. For example, if $\mathcal{E}_\rho$ and $\mathcal{E}_\phi$ are the fields of a TE$_{lm}$ mode of a hollow metallic circular waveguide [equations (5.82) and (5.83) of Chapter 5], then we put $dl_\rho = d\rho$ and $dl_\phi = \rho \, d\phi$, and (I.2) gives

$$\frac{J_l'\left(\frac{j_{lm}'\rho}{a}\right)}{J_l\left(\frac{j_{lm}'\rho}{a}\right)} \frac{j_{lm}'}{a} d\rho = l \tan(l\phi) d\phi \tag{I.3}$$

or, integrating,

$$J_l\left(\frac{j_{lm}'\rho}{a}\right) \cos l\phi = K = \text{constant} \tag{I.4}$$

By choosing different values of the constant K, we can generate a whole family of curves corresponding to the mode fields. The plots of Figure 5.9 are special cases of the result (I.4).

A more general example is furnished by the two-dimensional magnetostatic field given in terms of the z-component of a vector potential:

$$\mathcal{B}_T = \nabla_T \times (\mathbf{u}_z A_z) = -\mathbf{u}_z \times \nabla_T A_z = \mathbf{u}_x \frac{\partial A_z}{\partial y} - \mathbf{u}_y \frac{\partial A_z}{\partial x} \tag{I.5}$$

Then the field line equations for $\mathcal{H}_T$ are given in rectangular coordinates as

$$\frac{dy}{dx} = \frac{B_y}{B_x} = -\frac{\partial A_z/\partial x}{\partial A_z/\partial y} \tag{I.6}$$

or

$$\frac{\partial A_z}{\partial x} dx + \frac{\partial A_z}{\partial y} dy = \nabla A_z \cdot d\mathbf{l} = 0 \tag{I.7}$$

where $d\mathbf{l} = \mathbf{u}_x \, dx + \mathbf{u}_y \, dy$. Hence, the $\mathcal{H}_T$ field lines must follow a path along which $\nabla A_z = 0$, or $A_z = \text{constant}$. The magnetostatic field of a line current shown in Figure I.2 illustrates this result.

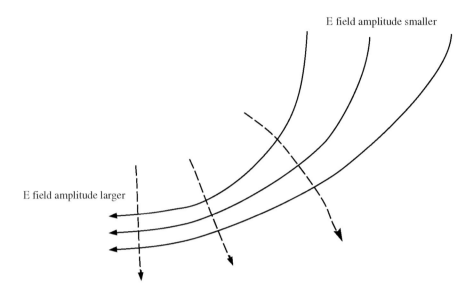

Figure I.1 Typical field plot: —————— E lines; – – – – H lines.

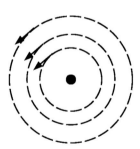

Figure I.2 Magnetostatic field of a line current.

J INTEGRAL IDENTITIES FOR FIELDS OF GUIDED MODES

It is possible, using the identities (9.5) and (9.6), or the Lorentz reciprocity theorem which follows from them, to obtain a number of useful relationships satisfied by the fields and propagation coefficient of a guided mode on an arbitrary waveguide structure. These can be used to gain useful physical insight into the properties of the mode, and even give approximate expressions for the propagation coefficient if suitable approximate knowledge of the fields is available. We assume throughout this appendix that $\omega \neq 0$, and that the boundary wall is a perfect conductor so that $\mathbf{u}_n \times \mathbf{E} = 0$. No assumption is made about the normalization of the mode fields, so that (5.27) is not assumed to hold.

J.1 IDENTITIES FOR GENERAL, LOSSY WAVEGUIDES

In the result of problem p9-4, let us take

$$\begin{aligned} \mathbf{E}^a &= \boldsymbol{\mathcal{E}}_m e^{-\gamma_m z} \\ \mathbf{H}^a &= \boldsymbol{\mathcal{H}}_m e^{-\gamma_m z} \end{aligned} \tag{J.1}$$

and

$$\begin{aligned} \mathbf{E}^b &= \boldsymbol{\mathcal{E}}_{-m} e^{\gamma_m z} \\ \mathbf{H}^b &= \boldsymbol{\mathcal{H}}_{-m} e^{\gamma_m z} \end{aligned} \tag{J.2}$$

where $(\boldsymbol{\mathcal{E}}_m, \boldsymbol{\mathcal{H}}_m)$ and $\gamma_m = \alpha_m + j\beta_m$ are the fields and propagation coefficient of a waveguide mode, while $(\boldsymbol{\mathcal{E}}_{-m}, \boldsymbol{\mathcal{H}}_{-m})$ and $-\gamma_m$ are those of the mode corresponding to it which propagates in the opposite direction [see (5.24)]. Since the fields are source-free, we obtain

$$\int_{S_0} \mu \boldsymbol{\mathcal{H}}_m \cdot \boldsymbol{\mathcal{H}}_{-m} dS = -\int_{S_0} \epsilon \boldsymbol{\mathcal{E}}_m \cdot \boldsymbol{\mathcal{E}}_{-m} dS \tag{J.3}$$

If, instead of (J.2), we had chosen the "b" fields to be the same as the "a" fields, we would have

$$\gamma_m = \frac{j\omega \int_{S_0} (\epsilon \boldsymbol{\mathcal{E}}_m^2 + \mu \boldsymbol{\mathcal{H}}_m^2) \, dS}{2 \int_{S_0} (\boldsymbol{\mathcal{E}}_m \times \boldsymbol{\mathcal{H}}_m) \cdot \mathbf{u}_z \, dS} = \frac{j\omega U_{m,\text{osc}}}{P_{m,\text{osc}}} \tag{J.4}$$

where by $\boldsymbol{\mathcal{E}}_m^2$, we mean $\boldsymbol{\mathcal{E}}_m \cdot \boldsymbol{\mathcal{E}}_m$,

$$P_{m,\text{osc}} = \frac{1}{2} \int_{S_0} (\boldsymbol{\mathcal{E}}_m \times \boldsymbol{\mathcal{H}}_m) \cdot \mathbf{u}_z \, dS \tag{J.5}$$

is the complex oscillatory power carried by the mode, and

$$U_{m,\text{osc}} = \frac{1}{4} \int_{S_0} (\epsilon \boldsymbol{\mathcal{E}}_m^2 + \mu \boldsymbol{\mathcal{H}}_m^2) \, dS \tag{J.6}$$

is the complex oscillatory energy per unit length stored by the mode. The latter, by virtue of (J.3), can also be expressed in the two alternative forms

$$U_{m,\text{osc}} = \frac{1}{2} \int_{S_0} (\epsilon \boldsymbol{\mathcal{E}}_{mT} \cdot \boldsymbol{\mathcal{E}}_{mT} + \mu \boldsymbol{\mathcal{H}}_{mz}^2) \, dS = \frac{1}{2} \int_{S_0} (\epsilon \boldsymbol{\mathcal{E}}_{mz}^2 + \mu \boldsymbol{\mathcal{H}}_{mT} \cdot \boldsymbol{\mathcal{H}}_{mT}) \, dS \tag{J.7}$$

Consider next the quantity $\mathbf{D}^a \times \mathbf{B}^b$. We can write

$$
\begin{aligned}
\mathbf{u}_z \cdot \mathbf{D}^a \times \mathbf{B}^b &= -\frac{1}{j\omega}\mathbf{u}_z \cdot [\mathbf{D}^a \times (\nabla \times \mathbf{E}^b)] - \frac{1}{j\omega}\mathbf{u}_z \cdot (\mathbf{D}^a \times \mathbf{M}^b) \\
&= \frac{1}{j\omega}\mathbf{D}^a \cdot [\mathbf{u}_z \times (\nabla \times \mathbf{E}^b)] - \frac{1}{j\omega}\mathbf{u}_z \cdot (\mathbf{D}^a \times \mathbf{M}^b) \qquad \text{(J.8)} \\
&= \frac{1}{j\omega}[\mathbf{D}^a \cdot \nabla E_z^b - \mathbf{D}^a \cdot \frac{\partial \mathbf{E}^b}{\partial z}] - \frac{1}{j\omega}\mathbf{u}_z \cdot (\mathbf{D}^a \times \mathbf{M}^b)
\end{aligned}
$$

where the last equality follows from the vector identity (B.19) in Appendix B. Then we have

$$
\begin{aligned}
\mathbf{u}_z \cdot \mathbf{D}^a \times \mathbf{B}^b &= \frac{1}{j\omega}\left[\mathbf{D}^a \cdot \nabla E_z^b + E_z^b \nabla \cdot \mathbf{D}^a - \mathbf{D}^a \cdot \frac{\partial \mathbf{E}^b}{\partial z}\right] \\
&\quad -\frac{1}{j\omega}\mathbf{u}_z \cdot (\mathbf{D}^a \times \mathbf{M}^b) - \frac{\nabla \cdot \mathbf{J}^a}{\omega^2} \qquad \text{(J.9)} \\
&= \frac{1}{j\omega}\left[\nabla \cdot (\mathbf{D}^a E_z^b) - \mathbf{D}^a \cdot \frac{\partial \mathbf{E}^b}{\partial z}\right] \\
&\quad -\frac{1}{j\omega}\mathbf{u}_z \cdot (\mathbf{D}^a \times \mathbf{M}^b) - \frac{\nabla \cdot \mathbf{J}^a}{\omega^2} E_z^b
\end{aligned}
$$

since

$$
-j\omega \nabla \cdot \mathbf{D}^a = \nabla \cdot \mathbf{J}^a
$$

and where the identity (B.11) has been used in the last step. Putting the "a" and "b" fields equal to those in (J.1) and (J.2), respectively, and taking into account that $\mathbf{J}^a$ and $\mathbf{M}^b$ are zero, we have

$$
\mu\epsilon\mathbf{u}_z \cdot (\boldsymbol{\mathcal{E}}_m \times \boldsymbol{\mathcal{H}}_{-m}) = \frac{1}{j\omega}\left\{\nabla \cdot [\epsilon\boldsymbol{\mathcal{E}}_m \mathcal{E}_{-mz}] - \gamma_m \epsilon\boldsymbol{\mathcal{E}}_m \cdot \boldsymbol{\mathcal{E}}_{-m}\right\} \qquad \text{(J.10)}
$$

Integrating over the same volume V and carrying out the z integrals as before leaves us with

$$
\int_{S_0} \mu\epsilon(\boldsymbol{\mathcal{E}}_m \times \boldsymbol{\mathcal{H}}_m) \cdot \mathbf{u}_z dS = \frac{\gamma_m}{j\omega}\int_{S_0} \epsilon\boldsymbol{\mathcal{E}}_m \cdot \boldsymbol{\mathcal{E}}_{-m} dS \qquad \text{(J.11)}
$$

Because of (J.3), there are several alternative ways to express (J.11):

$$
\begin{aligned}
\frac{1}{\gamma_m} &= \frac{\int_{S_0} \epsilon\boldsymbol{\mathcal{E}}_m \cdot \boldsymbol{\mathcal{E}}_{-m} dS}{j\omega \int_{S_0} \mu\epsilon(\boldsymbol{\mathcal{E}}_m \times \boldsymbol{\mathcal{H}}_m) \cdot \mathbf{u}_z dS} \\
&= -\frac{\int_{S_0} \mu\boldsymbol{\mathcal{H}}_m \cdot \boldsymbol{\mathcal{H}}_{-m} dS}{j\omega \int_{S_0} \mu\epsilon(\boldsymbol{\mathcal{E}}_m \times \boldsymbol{\mathcal{H}}_m) \cdot \mathbf{u}_z dS} \qquad \text{(J.12)} \\
&= \frac{\int_{S_0} (\epsilon\boldsymbol{\mathcal{E}}_m \cdot \boldsymbol{\mathcal{E}}_{-m} - \mu\boldsymbol{\mathcal{H}}_m \cdot \boldsymbol{\mathcal{H}}_{-m}) dS}{2j\omega \int_{S_0} \mu\epsilon(\boldsymbol{\mathcal{E}}_m \times \boldsymbol{\mathcal{H}}_m) \cdot \mathbf{u}_z dS}
\end{aligned}
$$

Another class of relationships can be obtained by taking derivatives with respect to ω. Let $\mathbf{E}, \mathbf{H}$ be any solution of source-free Maxwell's equations. Then their derivatives with respect to ω satisfy

$$
\nabla \times \frac{\partial \mathbf{E}}{\partial \omega} = -j\omega\mu\frac{\partial \mathbf{H}}{\partial \omega} - j\frac{\partial(\omega\mu)}{\partial \omega}\mathbf{H}
$$

$$\nabla \times \frac{\partial \mathbf{H}}{\partial \omega} = j\omega\epsilon\frac{\partial \mathbf{E}}{\partial \omega} + j\frac{\partial(\omega\epsilon)}{\partial \omega}\mathbf{E} \tag{J.13}$$

where we have allowed for the possibility that μ and ϵ may be dispersive (i.e., depend on frequency). Therefore, we may consider the fields

$$\mathbf{E}^b = \partial\mathbf{E}/\partial\omega \quad \text{and} \quad \mathbf{H}^b = \partial\mathbf{H}/\partial\omega \tag{J.14}$$

to be legitimate field solutions provided we regard the quantities

$$\mathbf{J}^b = j\frac{\partial(\omega\epsilon)}{\partial\omega}\mathbf{E} \quad \text{and} \quad \mathbf{M}^b = j\frac{\partial(\omega\mu)}{\partial\omega}\mathbf{H} \tag{J.15}$$

as the "sources" of $\mathbf{E}^b$ and $\mathbf{H}^b$.

Letting

$$\begin{aligned}
\mathbf{E} &= \boldsymbol{\mathcal{E}}_{-m}e^{\gamma_m z} \\
\mathbf{H} &= \boldsymbol{\mathcal{H}}_{-m}e^{\gamma_m z}
\end{aligned} \tag{J.16}$$

we have from (J.14) and (J.15) that

$$\begin{aligned}
\mathbf{E}^b &= \left[\frac{\partial\boldsymbol{\mathcal{E}}_{-m}}{\partial\omega} + z\frac{\partial\gamma_m}{\partial\omega}\boldsymbol{\mathcal{E}}_{-m}\right]e^{\gamma_m z} \\
\mathbf{H}^b &= \left[\frac{\partial\boldsymbol{\mathcal{H}}_{-m}}{\partial\omega} + z\frac{\partial\gamma_m}{\partial\omega}\boldsymbol{\mathcal{H}}_{-m}\right]e^{\gamma_m z}
\end{aligned} \tag{J.17}$$

and

$$\begin{aligned}
\mathbf{J}^b &= j\frac{\partial(\omega\epsilon)}{\partial\omega}\boldsymbol{\mathcal{E}}_{-m}e^{\gamma_m z} \\
\mathbf{M}^b &= j\frac{\partial(\omega\mu)}{\partial\omega}\boldsymbol{\mathcal{H}}_{-m}e^{\gamma_m z}
\end{aligned} \tag{J.18}$$

Using the reciprocity theorem (9.7) and integrating over V as before, we get:

$$\frac{\partial\gamma_m}{\partial\omega} = j\frac{\int_{S_0}\frac{\partial(\omega\epsilon)}{\partial\omega}\boldsymbol{\mathcal{E}}_m\cdot\boldsymbol{\mathcal{E}}_{-m}\,dS - \int_{S_0}\frac{\partial(\omega\mu)}{\partial\mu}\boldsymbol{\mathcal{H}}_m\cdot\boldsymbol{\mathcal{H}}_{-m}\,dS}{2\int_{S_0}\boldsymbol{\mathcal{E}}_m\times\boldsymbol{\mathcal{H}}_m\cdot\mathbf{u}_z\,dS} \tag{J.19}$$

Finally, the use of

$$\begin{aligned}
\mathbf{E} &= \boldsymbol{\mathcal{E}}_m e^{-\gamma_m z} \\
\mathbf{H} &= \boldsymbol{\mathcal{H}}_m e^{-\gamma_m z}
\end{aligned} \tag{J.20}$$

in (J.14) and (J.15), along with (J.1), followed by an application of Lorentz' theorem (9.7) and an integration over V yields

$$\gamma_m = j\frac{\int_{S_0}\left[\frac{\partial(\omega\epsilon)}{\partial\omega}\boldsymbol{\mathcal{E}}_m^2 - \frac{\partial(\omega\mu)}{\partial\omega}\boldsymbol{\mathcal{H}}_m^2\right]dS}{\int_{S_0}\left[\boldsymbol{\mathcal{E}}_m\times\frac{\partial\boldsymbol{\mathcal{H}}_m}{\partial\omega} - \frac{\partial\boldsymbol{\mathcal{E}}_m}{\partial\omega}\times\boldsymbol{\mathcal{H}}_m\right]\cdot\mathbf{u}_z\,dS} \tag{J.21}$$

which is interesting to compare with (J.4).

J.2 IDENTITIES FOR MODES IN LOSSLESS WAVEGUIDES

Of special interest is the case where ϵ and μ are real, making the waveguide a lossless one. Properties of the modes of lossless waveguides analogous to those of the foregoing section can be derived in a very similar fashion, and often have direct physical interpretations, which we will give below.

The derivation of these identities proceeds as before using (J.1) as a-state fields, but with a different choice for the b-state fields. An examination of the time-harmonic source-free Maxwell equations in a lossless medium (ϵ and μ are real) shows that, if $\mathbf{E}$, $\mathbf{H}$ is a solution, so is $\mathbf{E}^*$, $-\mathbf{H}^*$, where * denotes a complex conjugate. Thus, one possible choice for the b-state field would be

$$
\begin{aligned}
\mathbf{E}^b &= \boldsymbol{\mathcal{E}}_m^* e^{-\gamma_m^* z} \\
\mathbf{H}^b &= -\boldsymbol{\mathcal{H}}_m^* e^{-\gamma_m^* z}
\end{aligned}
\tag{J.22}
$$

Following the derivation that led to (J.3) leads to the result

$$
\alpha_m \hat{P} = j\omega(U_{M0} - U_{E0})
\tag{J.23}
$$

where

$$
U_{E0} = \frac{1}{4} \int_{S_0} \epsilon \boldsymbol{\mathcal{E}}_m \cdot \boldsymbol{\mathcal{E}}_m^* \, dS
\tag{J.24}
$$

$$
U_{M0} = \frac{1}{4} \int_{S_0} \mu \boldsymbol{\mathcal{H}}_m \cdot \boldsymbol{\mathcal{H}}_m^* \, dS
\tag{J.25}
$$

are respectively the nondispersive parts of the time-average electric and magnetic stored energy per unit length in the waveguide, and

$$
\hat{P} = \frac{1}{2} \int_{S_0} \boldsymbol{\mathcal{E}}_m \times \boldsymbol{\mathcal{H}}_m^* \cdot \mathbf{u}_z \, dS
\tag{J.26}
$$

is the complex power carried through the cross section S_0 of the waveguide (assuming it valid to make this interpretation even though S_0 is not a closed surface).

On the other hand, if we take as the b-state field

$$
\begin{aligned}
\mathbf{E}^b &= \boldsymbol{\mathcal{E}}_{-m}^* e^{\gamma_m^* z} \\
\mathbf{H}^b &= -\boldsymbol{\mathcal{H}}_{-m}^* e^{\gamma_m^* z}
\end{aligned}
\tag{J.27}
$$

then we are led to the result

$$
\beta_m \hat{P} = \omega T
\tag{J.28}
$$

where

$$
T = \frac{1}{4} \int_{S_0} \left[\epsilon \left(\boldsymbol{\mathcal{E}}_{mT} \cdot \boldsymbol{\mathcal{E}}_{mT}^* - |\mathcal{E}_{mz}|^2 \right) + \mu \left(\boldsymbol{\mathcal{H}}_{mT} \cdot \boldsymbol{\mathcal{H}}_{mT}^* - |\mathcal{H}_{mz}|^2 \right) \right] dS
\tag{J.29}
$$

is a quantity similar to but not quite identical with the stored energy. Since T, U_{E0} and U_{M0} are all real, (J.23) and (J.28) together imply that if $\hat{P}$ is not equal to zero, then either $\alpha_m = 0$ (the mode propagates, in which case $\hat{P}$ is real) or $\beta_m = 0$ (the mode is cutoff, in which case $\hat{P}$ is imaginary). Only if $\hat{P} = 0$ can both α_m and β_m be different from zero for a lossless waveguide, in which case the mode is called a *complex mode*.

For many purposes, the most important modes of lossless waveguides are the propagating ones. For these we set $\hat{P} \equiv P$, which is real, and by (J.23) we will have

$$
U_{E0} = U_{M0} = \frac{1}{2} U_0
\tag{J.30}
$$

where $U_0 = U_{E0} + U_{M0}$ is the part of the total stored energy per unit length of the waveguide not associated with material dispersion (i. e., the dependence of μ and ϵ on frequency). Other identities for the propagating modes of lossless waveguides can be found using (J.27) as the b-state fields. For example, a derivation parallel to that leading to (J.12) gives

$$v_p = \frac{\omega}{\beta} = \frac{U_0}{G_0} \tag{J.31}$$

where v_p is the phase velocity of the mode, while

$$G_0 = \frac{1}{2} \int_{S_0} \boldsymbol{\mathcal{D}}_m \times \boldsymbol{\mathcal{B}}_m^* \cdot \mathbf{u}_z \, dS \tag{J.32}$$

is the nondispersive part of the time average pseudomomentum of the fields of the mode per unit length of waveguide: $\boldsymbol{\mathcal{D}}_m = \epsilon \boldsymbol{\mathcal{E}}_m$ and $\boldsymbol{\mathcal{B}}_m = \mu \boldsymbol{\mathcal{H}}_m$.[1]

Analogous to (J.19), we find

$$v_g = \frac{1}{d\beta/d\omega} = \frac{P}{U} \tag{J.33}$$

where v_g is the group velocity of the mode, and U is the total stored energy per unit length of the waveguide:

$$
\begin{aligned}
U &= \frac{1}{4} \int_{S_0} \left[\frac{\partial(\omega\epsilon)}{\partial\omega} \boldsymbol{\mathcal{E}}_m \cdot \boldsymbol{\mathcal{E}}_m^* + \frac{\partial(\omega\mu)}{\partial\omega} \boldsymbol{\mathcal{H}}_m \cdot \boldsymbol{\mathcal{H}}_m^* \right] dS \\
&= U_0 + \frac{\omega}{4} \int_{S_0} \left[\frac{\partial\epsilon}{\partial\omega} \boldsymbol{\mathcal{E}}_m \cdot \boldsymbol{\mathcal{E}}_m^* + \frac{\partial\mu}{\partial\omega} \boldsymbol{\mathcal{H}}_m \cdot \boldsymbol{\mathcal{H}}_m^* \right] dS
\end{aligned} \tag{J.34}
$$

From (J.28), we have

$$v_p = \frac{P}{T} \tag{J.35}$$

We note from (J.31), (J.33) and (J.35) that T satisfies the relation

$$\frac{U_0}{G_0} = \frac{P}{T} = \frac{v_g U}{T} \tag{J.36}$$

or,

$$v_g = \frac{T}{G_0} \frac{U_0}{U} = \frac{T}{G} \tag{J.37}$$

where G is the total time average pseudomomentum carried by the mode:

$$
\begin{aligned}
G &= G_0 \left(\frac{U}{U_0} \right) \\
&= G_0 + \frac{G_0}{U_0} \frac{\omega}{4} \int_{S_0} \left[\frac{\partial\epsilon}{\partial\omega} \boldsymbol{\mathcal{E}}_m \cdot \boldsymbol{\mathcal{E}}_m^* + \frac{\partial\mu}{\partial\omega} \boldsymbol{\mathcal{H}}_m \cdot \boldsymbol{\mathcal{H}}_m^* \right] dS \\
&= G_0 + \frac{\beta}{4} \int_{S_0} \left[\frac{\partial\epsilon}{\partial\omega} \boldsymbol{\mathcal{E}}_m \cdot \boldsymbol{\mathcal{E}}_m^* + \frac{\partial\mu}{\partial\omega} \boldsymbol{\mathcal{H}}_m \cdot \boldsymbol{\mathcal{H}}_m^* \right] dS
\end{aligned} \tag{J.38}
$$

as follows from (J.31) and (J.34). Hence we interpret T as the time average of pseudo-momentum flow in the waveguide mode passing the cross section S_0, again assuming this interpretation to be valid for an unclosed surface like S_0.

---

[1]The momentum density associated with a field in free space is $\mathbf{d} \times \mathbf{b}$; the corresponding time-average momentum density is $\frac{1}{2}\text{Re}(\mathbf{D} \times \mathbf{B}^*)$. In material media, there is some cloudiness in the use of this expression, because some of it is associated with mechanical momentum of the atoms and molecules of the material itself, and only a portion can truly be claimed to be "electromagnetic" momentum. We thus use the term *pseudomomentum* for the integral of $\frac{1}{2}\mathbf{D} \times \mathbf{B}^*$ in an arbitrary medium.

J.3　THE PARTIAL POWER LAW

When ϵ and μ are nondispersive ($\partial\epsilon/\partial\omega = 0$ and $\partial\mu/\partial\omega = 0$), equations (J.31) and (J.33) can be combined to give the so-called *partial power law* of N. A. Semenov:

$$\frac{1}{v_p v_g} = \frac{G_0}{P} = \frac{\int_{S_0} \mu\epsilon \boldsymbol{\mathcal{E}}_m \times \boldsymbol{\mathcal{H}}_m^* \cdot \mathbf{u}_z \, dS}{\int_{S_0} \boldsymbol{\mathcal{E}}_m \times \boldsymbol{\mathcal{H}}_m^* \cdot \mathbf{u}_z \, dS} \tag{J.39}$$

This states that the inverse product of v_p and v_g is equal to an average of $\mu\epsilon$ over the guide cross-section, weighted by the z-component of the time-average Poynting vector. For a homogeneously-filled guide where ϵ and μ are independent of position, (J.39) simplifies to

$$v_p v_g = \frac{1}{\mu\epsilon} \tag{J.40}$$

which is easily verified for the hollow guides of Chapter 5 and the TEM transmission lines of Chapter 8. For more general modes on lossy waveguides, the relation corresponding to (J.39) follows from (J.19) and (J.12):

$$\frac{\gamma_m}{\omega}\frac{d\gamma_m}{d\omega} = \frac{d(\gamma_m^2)}{d(\omega^2)} = -\frac{\int_{S_0} \mu\epsilon \boldsymbol{\mathcal{E}}_m \times \boldsymbol{\mathcal{H}}_m \cdot \mathbf{u}_z \, dS}{\int_{S_0} \boldsymbol{\mathcal{E}}_m \times \boldsymbol{\mathcal{H}}_m \cdot \mathbf{u}_z \, dS} \tag{J.41}$$

J.4　EXPRESSIONS FOR FILLING FACTORS

We can obtain some interesting insights into the dielectric filling factors considered in Section 9.4 based upon a generalization of the identity (J.19). Suppose now that the frequency $\omega(s)$, the permittivity $\epsilon(x, y; s)$ and the permeability $\mu(x, y; s)$ are all functions of some parameter s. For example, s might be the relative permittivity ϵ_r of the substrate of a planar transmission line, whose variation causes $\epsilon(x, y; s)$ to vary only in a portion of space:

$$\begin{aligned} \epsilon(x, y; \epsilon_r) &= \epsilon_0 \quad \text{in } S_2 \ (y > h) \\ &= \epsilon_0 \epsilon_r \quad \text{in } S_1 \ (0 < y < h) \end{aligned} \tag{J.42}$$

We can repeat the derivation leading from (J.13) to (J.19), only now we consider the fields of the "b" state to be

$$\mathbf{E}^b = \partial\mathbf{E}/\partial s \quad \text{and} \quad \mathbf{H}^b = \partial\mathbf{H}/\partial s \tag{J.43}$$

Our result is

$$\frac{\partial\gamma_m}{\partial s} = j\frac{\int_{S_0} \left[\frac{\partial(\omega\epsilon)}{\partial s}\boldsymbol{\mathcal{E}}_m \cdot \boldsymbol{\mathcal{E}}_{-m} - \frac{\partial(\omega\mu)}{\partial s}\boldsymbol{\mathcal{H}}_m \cdot \boldsymbol{\mathcal{H}}_{-m}\right] dS}{2\int_{S_0} \boldsymbol{\mathcal{E}}_m \times \boldsymbol{\mathcal{H}}_m \cdot \mathbf{u}_z \, dS} \tag{J.44}$$

As in Section J.2, we can also obtain a version of (J.44) suitable for propagating modes on lossless waveguides:

$$\frac{\partial\beta_m}{\partial s} = \frac{\int_{S_0} \left[\frac{\partial(\omega\epsilon)}{\partial s}\boldsymbol{\mathcal{E}}_m \cdot \boldsymbol{\mathcal{E}}_m^* + \frac{\partial(\omega\mu)}{\partial s}\boldsymbol{\mathcal{H}}_m \cdot \boldsymbol{\mathcal{H}}_m^*\right] dS}{2\int_{S_0} \boldsymbol{\mathcal{E}}_m \times \boldsymbol{\mathcal{H}}_m^* \cdot \mathbf{u}_z \, dS} \tag{J.45}$$

Now consider the filling factor of a microstrip or other quasi-TEM line as defined by (9.44). If we assume ϵ_r to be independent of ω, but to affect $\epsilon(x, y)$ as in (J.42), then

$$\begin{aligned} \frac{\partial(\omega\epsilon)}{\partial\epsilon_r} &= \omega\epsilon_0 \quad \text{in } S_1 \\ &= 0 \quad \text{in } S_2 \end{aligned} \tag{J.46}$$

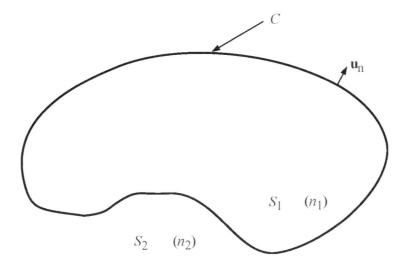

Figure J.1   Core and cladding geometry of a step-index dielectric waveguide of arbitrary geometry.

while $\partial(\omega\mu)/\partial\epsilon_r$ is identically zero. Hence (J.45) becomes

$$\frac{\partial\beta}{\partial\epsilon_r} = \frac{\omega\epsilon_0 \int_{S_1} \boldsymbol{\mathcal{E}}_m \cdot \boldsymbol{\mathcal{E}}_m^* \, dS}{2 \int_{S_0} \boldsymbol{\mathcal{E}}_m \times \boldsymbol{\mathcal{H}}_m^* \cdot \mathbf{u}_z \, dS} \tag{J.47}$$

But

$$q = \frac{\partial\epsilon_{\text{eff}}}{\partial\epsilon_r} = \frac{\partial(\beta^2/k_0^2)}{\partial\epsilon_r} = \frac{2\beta}{k_0^2}\frac{\partial\beta}{\partial\epsilon_r} \tag{J.48}$$

This in conjunction with (J.33) and (J.30) yields

$$q = \frac{c^2/\epsilon_r}{v_p v_g} \frac{\int_{S_1} \epsilon\boldsymbol{\mathcal{E}}_m \cdot \boldsymbol{\mathcal{E}}_m^* \, dS}{\int_{S_0} \epsilon\boldsymbol{\mathcal{E}}_m \cdot \boldsymbol{\mathcal{E}}_m^* \, dS} \tag{J.49}$$

We see that q is proportional to the fraction of stored energy carried in the substrate region of the microstrip, since S_1 represents the cross-section of that substrate. If further, the mode can be described quasistatically, where

$$\beta \simeq \omega\sqrt{l_s c_s}$$

and l_s, c_s are independent of ω, then

$$\frac{c}{v_p} \simeq \frac{c}{v_g} \simeq \sqrt{\epsilon_{\text{eff}}} \tag{J.50}$$

and hence

$$q = \frac{\partial\epsilon_{\text{eff}}}{\partial\epsilon_r} \simeq \frac{\epsilon_{\text{eff}}}{\epsilon_r} \frac{\int_{S_1} \epsilon\boldsymbol{\mathcal{E}}_m \cdot \boldsymbol{\mathcal{E}}_m^* \, dS}{\int_{S_0} \epsilon\boldsymbol{\mathcal{E}}_m \cdot \boldsymbol{\mathcal{E}}_m^* \, dS} \tag{J.51}$$

Equation (J.49) can clearly also be adapted to apply to the filling factors of a step-index dielectric waveguide as defined in Section 9.4. Let S_1 and S_2 represent the cross-sections of the core and cladding as shown in Figure J.1. The refractive indices of these regions are n_1 and n_2, respectively, and the boundary contour between them is denoted C. From (J.47) and (J.48) generalized to the case of the filling factors q_1 and q_2 defined by (9.40), we have

$$q_1 = \frac{cn_{\text{eff}} \int_{S_1} \epsilon\boldsymbol{\mathcal{E}}_m \cdot \boldsymbol{\mathcal{E}}_m^* \, dS}{n_1^2 \int_{S_0} \boldsymbol{\mathcal{E}} \times \boldsymbol{\mathcal{H}}_m^* \cdot \mathbf{u}_z \, dS} \tag{J.52}$$

$$q_2 = \frac{cn_{\text{eff}} \int_{S_2} \epsilon \boldsymbol{\mathcal{E}}_m \cdot \boldsymbol{\mathcal{E}}_m^* \, dS}{n_2^2 \int_{S_0} \boldsymbol{\mathcal{E}}_m \times \boldsymbol{\mathcal{H}}_m^* \cdot \mathbf{u}_z \, dS} \tag{J.53}$$

From (J.52) and (J.53), together with (J.30), (J.33) and the definition (9.55) for the group index n_g, we can show quite generally that

$$n_1^2 q_1 + n_2^2 q_2 = n_{\text{eff}} n_g \tag{J.54}$$

under no approximation.

An alternate representation for q_1 and q_2 is also possible, proceeding from the identity (J.10). If (J.27) is used in (J.10), we have for propagating mode fields on lossless guides:

$$\mu\epsilon \mathbf{u}_z \cdot (\boldsymbol{\mathcal{E}}_m \times \boldsymbol{\mathcal{H}}_m^*) = \frac{\beta_m}{\omega\epsilon} \boldsymbol{\mathcal{E}}_m \cdot \boldsymbol{\mathcal{E}}_m^* + \frac{j}{\omega} \nabla \cdot [\boldsymbol{\mathcal{D}}_m \boldsymbol{\mathcal{E}}_{mz}^*] \tag{J.55}$$

If this is integrated, first over S_1 and then over S_2, and we apply (J.52), (J.53) and the divergence theorem, then we obtain that

$$q_1 = \frac{\int_{S_1} \boldsymbol{\mathcal{E}}_m \times \boldsymbol{\mathcal{H}}_m^* \cdot \mathbf{u}_z \, dS}{\int_{S_0} \boldsymbol{\mathcal{E}}_m \times \boldsymbol{\mathcal{H}}_m^* \cdot \mathbf{u}_z \, dS} - \frac{jc^2}{\omega n_1^2} \frac{\oint_C \boldsymbol{\mathcal{E}}_{mz}^* \boldsymbol{\mathcal{D}}_m \cdot \mathbf{u}_n \, dl}{\int_{S_0} \boldsymbol{\mathcal{E}}_m \times \boldsymbol{\mathcal{H}}_m^* \cdot \mathbf{u}_z \, dS} \tag{J.56}$$

$$q_2 = \frac{\int_{S_2} \boldsymbol{\mathcal{E}}_m \times \boldsymbol{\mathcal{H}}_m^* \cdot \mathbf{u}_z \, dS}{\int_{S_0} \boldsymbol{\mathcal{E}}_m \times \boldsymbol{\mathcal{H}}_m^* \cdot \mathbf{u}_z \, dS} + \frac{jc^2}{\omega n_2^2} \frac{\oint_C \boldsymbol{\mathcal{E}}_{mz}^* \boldsymbol{\mathcal{D}}_m \cdot \mathbf{u}_n \, dl}{\int_{S_0} \boldsymbol{\mathcal{E}}_m \times \boldsymbol{\mathcal{H}}_m^* \cdot \mathbf{u}_z \, dS} \tag{J.57}$$

When the mode exists under weak-guidance conditions ($(n_1^2 - n_2^2)/n_2^2 \ll 1$), $\boldsymbol{\mathcal{E}}_{mz}^*$ can be set equal to zero, and we have (a) an independent proof of (9.54), and (b) that q_1 and q_2 in this case represent the percentages of *power* in the mode being carried in the core and cladding respectively.

J.5   BOUNDS ON PHASE AND GROUP VELOCITIES

We have seen examples (as for instance in Chapter 5) where $\beta/\omega \to 0$ and $d\beta/d\omega \to \infty$ for a propagating mode on a lossless guide; the situation occurs at cutoff and implies that $v_p \to \infty$ and $v_g \to 0$. There is thus no general upper bound on v_p nor lower bound on v_g. However, some partial bounds are obtainable from the results of this appendix, as we will show below.

In this section we will suppose that μ and ϵ are positive. Consider a forward mode with $\gamma = j\beta$. By the definition of a forward mode, $P > 0$.[2] Now for any fields $\boldsymbol{\mathcal{E}}$ and $\boldsymbol{\mathcal{H}}$, the inequality

$$|\boldsymbol{\mathcal{E}} \times \boldsymbol{\mathcal{H}}^* \cdot \mathbf{u}_z| \le \frac{1}{\sqrt{\mu\epsilon}} \left[\frac{1}{2}\epsilon \boldsymbol{\mathcal{E}} \cdot \boldsymbol{\mathcal{E}}^* + \frac{1}{2}\mu \boldsymbol{\mathcal{H}} \cdot \boldsymbol{\mathcal{H}}^* \right] \tag{J.58}$$

[2] It is possible, however, for β to be negative, implying that $v_p < 0$ as well. From (J.31), we see that G_0 is required to be negative for this to occur, and likewise from (J.35), we see that $T < 0$. In such a case, (J.33) implies that $v_g > 0$ since $U > 0$. Only if $\mu\epsilon$ is not a constant throughout the cross-section can $G_0 < 0$ at the same time that $P > 0$. A mode such as this, with $v_p v_g < 0$ is called a *backward-wave* mode, in contrast to the more usual *direct* or *normal-wave* mode for which $v_g v_p > 0$. Examples of backward-wave modes are well-known in the literature, often appearing in conjunction with complex modes. Properties analogous to those we are about to derive can also be found for backward-wave modes.

Note the distinction in terminology with *forward* and *reverse* modes, which are used merely to indicate in which direction ($\pm z$) energy flow occurs for a given mode of a lossless waveguide.

is true. We prove this by first noting that

$$|\mathcal{E} \times \mathcal{H}^* \cdot \mathbf{u}_z| \leq |\mathcal{E}||\mathcal{H}| \tag{J.59}$$

from the definition of the cross product. But also

$$\left[\left(\frac{\epsilon}{\mu}\right)^{1/4}|\mathcal{E}| - \left(\frac{\mu}{\epsilon}\right)^{1/4}|\mathcal{H}|\right]^2 \geq 0 \tag{J.60}$$

or,

$$|\mathcal{E}||\mathcal{H}| \leq \frac{1}{\sqrt{\mu\epsilon}}\left[\frac{1}{2}\epsilon\mathcal{E}\cdot\mathcal{E}^* + \frac{1}{2}\mu\mathcal{H}\cdot\mathcal{H}^*\right] \tag{J.61}$$

Combining (J.59) with (J.61) yields (J.58). Now (J.58) can be used to show that

$$
\begin{aligned}
|\mu\epsilon\mathcal{E} \times \mathcal{H}^* \cdot \mathbf{u}_z| &\leq \sqrt{\mu\epsilon}\left[\frac{1}{2}\epsilon\mathcal{E}\cdot\mathcal{E}^* + \frac{1}{2}\mu\mathcal{H}\cdot\mathcal{H}^*\right] \\
&\leq \sqrt{(\mu\epsilon)_{\max}}\left[\frac{1}{2}\epsilon\mathcal{E}\cdot\mathcal{E}^* + \frac{1}{2}\mu\mathcal{H}\cdot\mathcal{H}^*\right]
\end{aligned} \tag{J.62}
$$

where $(\mu\epsilon)_{\max}$ is the maximum value of the product $\mu\epsilon$ that occurs over the cross-section of a waveguide. Also,

$$
\begin{aligned}
\frac{1}{2}\epsilon\mathcal{E}\cdot\mathcal{E}^* + \frac{1}{2}\mu\mathcal{H}\cdot\mathcal{H}^* &\geq \sqrt{\mu\epsilon}|\mathcal{E} \times \mathcal{H}^* \cdot \mathbf{u}_z| \\
&\geq \sqrt{(\mu\epsilon)_{\min}}|\mathcal{E} \times \mathcal{H}^* \cdot \mathbf{u}_z|
\end{aligned} \tag{J.63}
$$

From (J.62) and (J.31), we have then

$$|v_p| \geq \frac{1}{\sqrt{(\mu\epsilon)_{\max}}} \equiv v_{p,\min} \tag{J.64}$$

or

$$|n_{\text{eff}}| = \frac{c}{|v_p|} \leq \frac{\sqrt{(\mu\epsilon)_{\max}}}{\sqrt{\mu_0\epsilon_0}} \equiv n_{\max} \tag{J.65}$$

From (J.63) and (J.33), if ϵ and μ can be reckoned independent of ω,

$$v_g \leq \frac{1}{\sqrt{(\mu\epsilon)_{\min}}} \equiv v_{g,\max} \geq v_{p,\min} \tag{J.66}$$

or

$$n_g = \frac{c}{v_g} \geq \frac{\sqrt{(\mu\epsilon)_{\min}}}{\sqrt{\mu_0\epsilon_0}} \equiv n_{\min} \tag{J.67}$$

Finally, note that from (J.29), (J.30), (J.24) and (J.25),

$$|T| \leq U_0 \tag{J.68}$$

Then by (J.35) and (J.33), again for ϵ and μ independent of ω, we have

$$|v_p| \geq v_g \tag{J.69}$$

This also implies that $n_g \geq n_{\text{eff}}$, and that

$$\frac{\omega}{\beta}\frac{d\beta}{d\omega} \geq 1$$

and thus that

$$\frac{dv_p}{d\omega} = \frac{1}{\beta}\left[1 - \frac{\omega}{\beta}\frac{d\beta}{d\omega}\right] \leq 0 \tag{J.70}$$

i.e., the phase velocity is a nonincreasing function of ω at any frequency, while the effective index $n_{\text{eff}} = \beta/k_0$ is a nondecreasing function of ω. We see many examples of this property in Chapters 5-8.

J.6   NOTES AND REFERENCES

The reader unfamiliar with the concept of momentum associated with an electromagnetic field should consult

Javid and Brown (1963), pp. 129-133.

A complete discussion of the energy-power-momentum relations for waveguides is given in

H. A. Haus and H. Kogelnik, "Electromagnetic momentum and momentum flow in dielectric waveguides," *J. Opt. Soc. Amer.*, vol. 66, pp. 320-327 (1976).

The various results are of differing ages, and can be traced in the following sources:

P. Chorney, "Power and energy relations in bidirectional waveguides," in *Proc. Symp. Electromagnetics and Fluid Dynamics of Gaseous Plasma.* New York: Polytechnic Press, 1962, pp. 195-210.

N. A. Semenov, "The partial-power law," *Radio Eng. Electron. Phys.*, vol. 8, pp. 1419-1420 (1963).

N. A. Semenov, "Proof of the partial-power law," *ibid.*, vol. 10, pp. 1320-1321 (1965).

J. Brown, "Electromagnetic momentum associated with waveguide modes," *Proc. IEE (London)*, vol. 113, pp. 27-34 (1966).

P. Chorney and P. Penfield, "Waveguide power-mode theorems for nonconservative systems," *IEEE Trans. Micr. Theory Tech.*, vol. 19, pp. 767-772 (1971).

S. Kawakami, "Relation between dispersion and power-flow distribution in a dielectric waveguide," *J. Opt. Soc. Amer.*, vol. 65, pp. 41-45 (1975).

S. Nemoto and T. Makimoto, "A relationship between phase and group indices of guided modes in dielectric waveguides," *Int. J. Electron.*, vol. 40, pp. 187-190 (1976).

E. F. Kuester, "Generalization of the partial-power law (Brown's identity) to waveguides with lossy media," *Electron. Lett.*, vol. 20, pp. 456-457 (1984).

Results similar to those of Section J.4 relating loss (and filling factor) to group velocity and stored energy fractions can be found in:

E. O. Schulz-DuBois, "Energy transport velocity of electromagnetic propagation in dispersive media," *Proc. IEEE*, vol. 57, pp. 1748-1757 (1969).

Some of the results of Sections J.4 and J.5 can also be found in:

R. B. Adler, "Properties of Guided Waves on Inhomogeneous Cylindrical Structures," *Tech. Rept. No. 102*, Research Laboratory of Electronics, Massachusetts Institute of Technology, 1949.

R. B. Adler, "Waves on inhomogeneous cylindrical structures," *Proc. IRE*, vol. 40, pp. 339-348 (1952).

More results of this type (though in somewhat abstract form applicable to very general—not just electromagnetic—waveguides) can be found in

Silbergleit and Kopilevich (1996), Sects. 20 and 28.

Equations (J.44) and (J.45) are electromagnetic vector analogs of the Hellmann-Feynman theorem; see, e.g.,

Arnaud (1976), pp. 216-218.

Equations (J.64)-(J.67) are given in

Mrozowski (1997), Section 11.3.4.

For more about complex modes, see:

T. Tamir and A. A. Oliner, "The spectrum of electromagnetic waves guided by a plasma layer," *Proc. IEEE*, vol. 51, pp. 317-332 (1963).

T. Tamir and A. A. Oliner, "Guided comples waves: Part 1. Fields at an interface," *Proc. IEE (London)*, vol. 110, pp. 310-324 (1963).

A. M. Belyantsev and A. V. Gaponov, "Waves with complex propagation coefficients in coupled transmission lines without energy dissipation," *Radio Eng. Electron. Phys.*, vol. 9, pp. 980-988 (1964).

A. S. Omar and K. F. Schünemann, "Complex and backward-wave modes in inhomogeneously and anisotropically filled waveguides," *IEEE Trans. Micr. Theory Tech.*, vol. 35, pp. 268-275 (1987).

A. Magos, "Singular frequencies in wave guides with inhomogeneous dielectric," *Int. J. Appl. Electromagnetics Mater.*, vol. 4, pp. 189-196 (1994).

T. F. Jabłoński, "Complex modes in open lossless dielectric waveguides," *J. Opt. Soc. Amer. A*, vol. 11, pp. 1272-1282 (1994)

and

Mrozowski (1997).

The terminology we have used (backward-wave versus direct or normal-wave modes) to indicate the relative directions of phase and group velocity for a mode is not universal, as can be seen from a review of the literature, but would seem to be most natural for our purposes. The reader should be aware of this when consulting other published work.

K NUMERICAL SOLUTION OF IMPLICIT EQUATIONS

Transcendental equations such as we have obtained for determining the propagation constants or other line parameters of dielectric waveguides or planar transmission lines can be solved numerically on a computer or programmable calculator by a variety of methods. Many commercial software packages offer the capability of solving these kinds of equations. It can, however, be critical to arrange the transcendental equation in a form which is stable for numerical solution procedures. Often, additional insight into the solution can be gained by writing an algorithm especially to solve a particular problem. We will briefly describe a few methods for finding roots (zeroes) of a real function $F(x)$ of a real variable x. We will denote the exact value of a root of F by x_e.

Even though a user may not have programmed the root-finding algorithm, it is important that properties of the algorithm be known so that the reliability of the results obtained can be judged. It should especially be noted that no algorithm can guarantee to find *all* roots of a given function unless more information about that function is available. Moreover, functions possessing singularities or varying rapidly may make it difficult to find their roots, so rearranging the function into a different form may sometimes be necessary, as will be illustrated for the iterative method below. Methods for finding complex roots of complex functions are in general more complicated, and will not be treated here (see the references for more information).

K.1 THE BISECTION METHOD

Perhaps the simplest numerical technique is the *bisection method.* Suppose we know that a root x_e of $F(x)$ lies between the lower limit x_l and the upper limit x_u. We might know this if the values of $F(x_l)$ and $F(x_u)$ have opposite signs, but this might not be the case if the function approaches infinity within the interval (x_l, x_u) (as for the function $\tan x$ at $x = \pi/2$, for example). We therefore restrict consideration to functions that are bounded and continuous on the interval (x_l, x_u). Now we consider the new point $x_n = (x_l + x_u)/2$ at the center of the interval. If $F(x_n)$ has the same sign as $F(x_l)$, then a root of F must lie within the smaller interval (x_n, x_u); if $F(x_n)$ has the same sign as $F(x_u)$, then a root of F must lie within (x_l, x_n). In either case, the region containing a root has been cut in half, and we can repeat this process until the length of the interval containing a root has become small enough to determine the root to a prespecified accuracy.

K.2 NEWTON'S METHOD

The bisection process is simple and reliable, but not particularly rapidly convergent. A more rapidly convergent (but sometimes less reliable) method is *Newton's method.* Here, we do not need an interval known to contain a root, but only an initial estimate x_0 of the location of a root. Suppose that we can compute not only the value of the function F at x_0, but also its derivative $F'(x_0)$. This can be done either analytically if the form of the function is available, or by the finite difference approximation

$$F'(x) \simeq \frac{F(x + \delta) - F(x)}{\delta} \tag{K.1}$$

for some sufficiently small quantity δ (though not *too* small, so that we avoid problems with computer roundoff error). Then the function F is approximated near x_0 by a linear function obtained from the first two terms of its Taylor series. This approximation is set equal to zero and solved for a new estimate x_1 of the root:

$$x_1 = x_0 - \frac{F(x_0)}{F'(x_0)} \tag{K.2}$$

This process is repeated until the approximation x_n to the root is as accurate as required:

$$x_{n+1} = x_n - \frac{F(x_n)}{F'(x_n)}; \qquad n = 1, 2, \ldots \tag{K.3}$$

Note that this will not work for multiple roots (where $F(x) = F'(x) = 0$), but when the process does converge, it does so much faster than does the bisection method.

K.3   ITERATIVE METHODS

One method that is well-suited to the equations we encounter in guided-wave theory is the iterative method, also known as the method of successive substitution or fixed-point method. The idea of this method is as follows. Suppose we have a transcendental equation $F(x) = 0$ for an unknown quantity x, and that this equation can be written in the form:

$$x = f(x) \tag{K.4}$$

where f is some given function of x. If the rearrangement into the form (K.4) is such that a given variation in x causes much less variation in $f(x)$ (logarithms, arctangents, etc., are examples of such functions), then we embark on the following succession of substitution. Choose a guess x_0 for the solution. Then compute the sequence

$$\begin{aligned}
x_1 &= f(x_0) \\
x_2 &= f(x_1) \\
&\vdots \\
x_{n+1} &= f(x_n)
\end{aligned} \tag{K.5}$$

of improved approximate solutions until further iteration results in negligible change in x_{N+1} over x_N. Then x_{N+1} is the solution of (K.4) to desired accuracy. Note that it is crucial that $f(x)$ varies less rapidly in x than x itself. Otherwise the iteration procedure can run away rather than converge to the desired answer. In fact, convergence of (K.5) will occur if $|f'(x_e)| < 1$ and x_0 is close enough to x_e. It should also be noted that Newton's method is a special case of the iteration algorithm (K.5), with

$$f(x) = x - \frac{F(x)}{F'(x)}$$

This method can be used to generate tables, plots or individual values of the solutions of the transcendental equations encountered in this text. As an example, consider two forms of the eigenvalue equation for the even TE modes of a dielectric slab waveguide. The form (6.48) that was used for the graphical solution can be written in the form:

$$h_1 a = \sqrt{V^2 - [h_1 a \tan(h_1 a/2)]^2} \tag{K.6}$$

This form is unsuitable for iterative solution, because the right side contains a tangent function of the variable $h_1 a$, which varies much more rapidly than $h_1 a$ itself—in fact, it becomes infinite when $h_1 a$ is an odd multiple of π. This form of the equation is also unstable for numerical use because for some values of $h_1 a$ the right side can become complex, whereas such solutions are not allowed from a physical standpoint.

On the other hand, the form (6.53) of this eigenvalue equation (which is in fact applicable to both even and odd modes) can be written as:

$$b = 1 - \left[\frac{m\pi}{V} + \frac{2}{V} \sin^{-1} \sqrt{b} \right]^2 \tag{K.7}$$

Here, the right side contains only an inverse trigonometric function of the unknown b, which varies more slowly than b itself, especially when V is not small. This equation has the added advantage that one can specify which mode m is desired, while (K.6) has more than one solution if V is large enough, and it may be difficult to make a numerical algorithm select the solution for a particular mode. The reader is encouraged to experiment with the iterative algorithm for different versions of the eigenvalue equation, to see how efficient and accurate it can be.

This example contains an inverse trigonometric function; these and logarithms must be treated with care in any root-finding technique. Both these kinds of function involve an arbitrary integer multiple of π, and mathematical software is designed to choose a specific value for this integer during calculations. This choice for these functions is usually called the principal branch of the function, and the program's documentation should be consulted for information on exactly what that choice is.

K.4   NOTES AND REFERENCES

For information on finding complex roots of complex functions, see

Stiefel (1963), Chapter 4;
Press *et al.* (1992), Chapter 9;
Hamming (1986), Chapter 5.

as well as

C. A. Barlow and E. L. Jones, "A method for the solution of roots of a nonlinear equation and for solution of the general eigenvalue problem," *J. Assoc. Comput. Mach.*, vol. 13, pp. 135-142 (1966).
B. Davies, "Locating the zeros of an analytic function," *J. Comp. Phys.*, vol. 66, pp. 36-49 (1986).

Sommerfeld was perhaps the first to investigate the use of iterative methods for solving guided wave problems (specifically, the surface wave guided by a finitely conducting cylindrical wire):

A. Sommerfeld, "Ueber die numerische Auflösung transcendenter Gleichungen durch successive Approximationen," *Nachr. Gess. Wiss. Gottingen, Math.-Phys. Klasse*, pp. 360-369 (1898).

More examples of iterative techniques can be found in

> K. S. Krane, "Iterative solutions of transcendental equations of mathematical physics with the programmable pocket calculator," *Amer. J. Phys.*, vol. 50, pp. 521-527 (1982).

A comprehensive treatment of methods (both analytical and numerical) for solving transcendental equations is

> Boyd (2014).

L DERIVATION OF THE SURFACE IMPEDANCE FOR A GOOD CONDUCTOR

In this appendix, we derive the expression for the surface impedance Z_S of a good conductor. Referring to Figure 9.7, we set up a coordinate system (x_1, x_2, n) corresponding to the unit vectors $\mathbf{u}_1, \mathbf{u}_2$ tangential to the interface and $\mathbf{u}_n$ normal to the interface. From what we know about the behavior of plane waves refracting into a dense or highly-conducting medium, we assume that, in the conducting medium

$$\mathbf{E} = \mathbf{E}_c(x_1, x_2, n)e^{\psi(k_c n)} \tag{L.1}$$

$$\mathbf{H} = \mathbf{H}_c(x_1, x_2, n)e^{\psi(k_c n)} \tag{L.2}$$

where

$$
\begin{aligned}
k_c &= \omega\sqrt{\mu_c(\epsilon_c - j\frac{\sigma_c}{\omega})} \\
&\simeq \frac{1-j}{\delta_c}
\end{aligned}
\tag{L.3}
$$

is the wavenumber in the conducting medium, and δ_c is the skin depth (9.58), under the condition $\sigma_c/\omega\epsilon_c \gg 1$. The function ψ must be determined, but it represents the expectation that $\mathbf{E}$ and $\mathbf{H}$ vary much more rapidly normal to the interface than in other directions, so that $\mathbf{E}_c$ and $\mathbf{H}_c$ are assumed to be slowly varying within distances on the order of $|k_c|^{-1}$.

Plugging (L.1) and (L.2) into Maxwell's equations, we get

$$
\begin{aligned}
\nabla \times \mathbf{E}_c + k_c\psi'(k_c n)\mathbf{u}_n \times \mathbf{E}_c &= -j\omega\mu_0\mathbf{H}_c \\
\nabla \times \mathbf{H}_c + k_c\psi'(k_c n)\mathbf{u}_n \times \mathbf{H}_c &= j\omega(\epsilon_c - j\frac{\sigma_c}{\omega})\mathbf{E}_c
\end{aligned}
\tag{L.4}
$$

Because $\mathbf{E}_c$ and $\mathbf{H}_c$ are slowly-varying on the scale of $|k_c|^{-1}$, we neglect the curl terms in (L.4) compared to the second terms on the left sides. From the resulting equations, we have that

$$E_{cn} \simeq 0; \qquad H_{cn} \simeq 0 \tag{L.5}$$

while the components tangential to the interface can be solved for to give

$$\psi' = \pm j \tag{L.6}$$

or

$$\psi(k_c n) = \pm jk_c n + \text{const} \tag{L.7}$$

We choose the $+$ sign in (L.7) in order that the solutions (L.1) and (L.2) decay away from the interface (as n becomes negative), so that we can obtain the relation

$$\mathbf{E}_{c(tan)} \simeq \hat{\zeta}_c \mathbf{u}_n \times \mathbf{H}_{c(tan)} \tag{L.8}$$

where $\hat{\zeta}_c$ is the wave impedance

$$\hat{\zeta}_c = \sqrt{\frac{\mu_c}{\epsilon_c - j\frac{\sigma_c}{\omega}}} \tag{L.9}$$

of the conducting medium. Since (L.8) holds in particular at the interface $(n = 0)$, and tangential field components are continuous across the interface, (L.8) implies the boundary condition (9.59) with

$$Z_S = \hat{\zeta}_c$$

REFERENCES

The analysis here is somewhat simplified from

S. M. Rytov, "Calcul du skin-effect par la mèthode des perturbations," *J. Phys. (USSR)*, vol. 2, pp. 233-242 (1940).

See also

T. B. A. Senior, "Impedance boundary conditions for imperfectly conducting surfaces," *Appl. Sci. Res.*, vol. B8, pp. 418-436 (1960).

In both papers, more details of how accurate the impedance boundary conditions are, are provided. See also

O. I. Panych, "Approximate boundary conditions in diffraction problems [Russian]," *Dokl. Akad. Nauk SSSR*, vol. 70, pp. 589-592 (1950).

O. I. Panych, "Asymptotic expansion of the solution of a boundary problem [Russian]," *Matem. Sbornik*, vol. 32(74), pp. 385-406 (1953).

K. M. Mitzner, "An integral equation approach to scattering from a body of finite conductivity," *Radio Science*, vol. 2, pp. 1459-1470 (1967).

Senior and Volakis (1995).

M CHANGE IN INDUCTANCE DUE TO DEFORMATION OF BOUNDARY

In this appendix, we will evaluate the change in the inductance l per unit length of a transmission line operating under quasistatic conditions when its boundary undergoes a small deformation. Specifically, consider the transmission line of cross-section S_0 bounded by two perfect conductors C_1 and C_g as shown in Figure M.1. Let the new boundaries C_1' and C_g' enclosing a new cross section S_0' result from displacing C_1 inward and C_g outward (that is, into the conductor in each case) by an amount δn normal to each curve. For the original line, there will be magnetic fields $\mathbf{H}$ and $\mathbf{B} = \nabla \times \mathbf{A} = \nabla \times (\mathbf{u}_z A_z)$ satisfying the static field equation

$$\nabla \times \mathbf{H} = 0 \tag{M.1}$$

inside S, and giving rise to an inductance l per unit length which can be defined by (see (8.43)):

$$\frac{1}{2}l I_1^2 = U_M \tag{M.2}$$

where I_1 is the total current on C_1,

$$I_1 = \oint_{C_1} \mathbf{H} \cdot d\boldsymbol{l} \tag{M.3}$$

and U_M is the static magnetic energy per unit length of the line,

$$
\begin{aligned}
U_M &= \frac{1}{2}\int_{S_0} \mathbf{H} \cdot \mathbf{B}\, dS = \frac{1}{2}\int_{S_0} \mathbf{H} \cdot \nabla \times \mathbf{A}\, dS \\
&= \frac{1}{2}\int_{S_0} [\nabla \cdot (\mathbf{A} \times \mathbf{H}) + \mathbf{A} \cdot \nabla \times \mathbf{H}]\, dS \\
&= -\frac{1}{2}\oint_{C_1} (\mathbf{A} \times \mathbf{H}) \cdot \mathbf{u}_n\, dl - \frac{1}{2}\oint_{C_g} (\mathbf{A} \times \mathbf{H}) \cdot \mathbf{u}_n\, dl
\end{aligned}
\tag{M.4}
$$

Since $\mathbf{u}_n \cdot \mathbf{B} = 0$ on C_1 and C_g, we have that

$$0 = \mathbf{u}_n \cdot \nabla \times \mathbf{A}\big|_{C_{1,g}} = \mathbf{u}_n \cdot \nabla \times (\mathbf{u}_z A_z)\big|_{C_{1,g}} = \frac{\partial A_z}{\partial l}\bigg|_{C_{1,g}} \tag{M.5}$$

or A_z is a constant (A_{z1}, say) on C_1 and also constant (A_{zg}, say) on C_g. Hence,

$$
\begin{aligned}
U_M &= \frac{1}{2}A_{z1}\oint_{C_1} \mathbf{H} \cdot d\boldsymbol{l} + \frac{1}{2}A_{zg}\oint_{C_g} \mathbf{H} \cdot d\boldsymbol{l} \\
&= \frac{1}{2}(A_{z1} - A_{zg})I_1
\end{aligned}
\tag{M.6}
$$

The deformed line will have slightly different fields $\mathbf{H}'$ and $\mathbf{B}' = \nabla \times (\mathbf{u}_z A_z')$ where

$$\nabla \times \mathbf{H}' = 0 \tag{M.7}$$

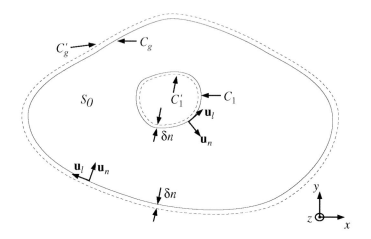

Figure M.1 Cross-section of TEM or quasi-TEM transmission line; original boundaries are C_1 and C_g; deformed boundaries are C_1' and C_g'.

in S_0', and will be characterized by a slightly different inductance l' such that

$$\frac{1}{2}l'(I_1')^2 = U_M' \tag{M.8}$$

where

$$I_1' = \oint_{C_1'} \mathbf{H}' \cdot d\mathbf{l} \tag{M.9}$$

and

$$U_M' = \frac{1}{2}(A_{z1}' - A_{zg}')I_1' \tag{M.10}$$

To obtain a formula for the change in l, let us consider the quantity

$$\nabla \cdot (\mathbf{A}' \times \mathbf{H}) = \mathbf{B}' \cdot \mathbf{H} \tag{M.11}$$

Integrating this identity over the cross section S_0 and using the two-dimensional divergence theorem, we get

$$\int_{S_0} (\mathbf{B}' \cdot \mathbf{H}) \, dS = -\oint_{C_1} (\mathbf{A}' \times \mathbf{H}) \cdot \mathbf{u}_n \, dl - \oint_{C_g} (\mathbf{A}' \times \mathbf{H}) \cdot \mathbf{u}_n \, dl$$

$$= (A_{z1}' - A_{zg}')I_1 + \oint_{C_1} (A_z' - A_{z1}')\mathbf{H} \cdot d\mathbf{l} + \oint_{C_g} (A_z' - A_{zg}')\mathbf{H} \cdot d\mathbf{l} \tag{M.12}$$

Now on C_1,

$$A_z' - A_{z1}' \simeq \delta n \frac{\partial A_z}{\partial n} = -\delta n B_l$$

and similarly on C_g. Since the first term on the right side of (M.12) is equal to $l'I_1 I_1'$, we can write

$$\int_{S_0} (\mathbf{B}' \cdot \mathbf{H}) \, dS \simeq l'I_1 I_1' - \oint_{C_1} \delta n \mathbf{B} \cdot \mathbf{H} \, dl - \oint_{C_g} \delta n \mathbf{B} \cdot \mathbf{H} \, dl \tag{M.13}$$

On the other hand, let us integrate the quantity

$$\nabla \cdot (\mathbf{A} \times \mathbf{H}') = \mathbf{B} \cdot \mathbf{H}' \tag{M.14}$$

over S_0 and proceed as above. We have

$$
\begin{aligned}
\int_{S_0} (\mathbf{B} \cdot \mathbf{H}') \, dS &= -\oint_{C_1} (\mathbf{A} \times \mathbf{H}') \cdot \mathbf{u}_n \, dl + \oint_{C_g} (\mathbf{A} \times \mathbf{H}') \cdot \mathbf{u}_n \, dl \\
&= (A_{z1} - A_{zg}) I_1'
\end{aligned}
\tag{M.15}
$$

From (M.15), (M.6) and (M.2), we have

$$
\int_{S_0} (\mathbf{B} \cdot \mathbf{H}') \, dS = \mathsf{L} I_1 I_1'
\tag{M.16}
$$

Now for an isotropic medium, $\mathbf{B}' \cdot \mathbf{H} = \mathbf{B} \cdot \mathbf{H}'$, so subtracting (M.16) from (M.13) and recognizing that we can replace $I_1 I_1'$ by I_1^2 to leading order, we get

$$
\mathsf{l}' - \mathsf{l} \simeq \frac{\oint_{C_1} \delta n (\mathbf{B} \cdot \mathbf{H}) \, dl + \oint_{C_g} \delta n (\mathbf{B} \cdot \mathbf{H}) \, dl}{I_1^2}
\tag{M.17}
$$

or, using (M.2) and (M.5):

$$
\begin{aligned}
\frac{\delta \mathsf{l}}{\mathsf{l}} \equiv \frac{\mathsf{l}' - \mathsf{l}}{\mathsf{l}} &\simeq \frac{\oint_{C_1} \delta n (\mathbf{B} \cdot \mathbf{H}) \, dl + \oint_{C_g} \delta n (\mathbf{B} \cdot \mathbf{H}) \, dl}{\int_{S_0} (\mathbf{B} \cdot \mathbf{H}) \, dS} \\
&= \frac{\delta U_M}{U_M}
\end{aligned}
\tag{M.18}
$$

That is, the relative change in l is proportional to the relative change in U_M. In fact, we can regard the ratio

$$
q_c = \frac{\oint_{C_1} \delta n (\mathbf{B} \cdot \mathbf{H}) \, dl + \oint_{C_g} \delta n (\mathbf{B} \cdot \mathbf{H}) \, dl}{\int_{S_0} (\mathbf{B} \cdot \mathbf{H}) \, dS}
\tag{M.19}
$$

as a kind of filling factor for the deformed region near the conducting boundary. Thus,

$$
\frac{\delta \mathsf{l}}{\mathsf{l}} \simeq q_c
\tag{M.20}
$$

N CORRECTION OF SMALL OBSTACLE AND SMALL HOLE THEORY FOR BOUNDARY EFFECTS

The definitions of the polarizabilities of a small scatterer or aperture as given in Sections 10.2.5 and 10.2.6 are made based on the dipole moments that would arise if the scatterer were located in an infinite homogeneous region, or the aperture were located in an infinite conducting screen in an otherwise infinite homogeneous region. If the scatterer or aperture is located in a more complicated environment such as a waveguide or resonator, the equivalent dipole moments create a field that will induce additional currents in the walls or medium of that environment.

For definiteness, let us consider a hollow metallic waveguide with perfectly conducting walls, containing an electrically small scatterer. Let a mode (say, the mode $m = 1$) be incident on this scatterer, centered at the location $(x_0, y_0, 0)$. To find the equivalent dipole moments, we must modify (10.20) and (10.21) to read:

$$\mathbf{p} = \epsilon_0 \overset{\leftrightarrow}{\alpha}_e \cdot \mathbf{E}^{\text{loc}} \tag{N.1}$$

$$\mathbf{m} = \frac{1}{\mu_0} \overset{\leftrightarrow}{\alpha}_m \cdot \mathbf{B}^{\text{loc}} \tag{N.2}$$

where the *local* or *acting* field is the sum of the incident waveguide mode field and a *reaction* field, which is what would be produced by the currents flowing in the waveguide wall if they acted in infinite space in the absence of the wall:

$$\mathbf{E}^{\text{loc}} = \mathbf{E}^{\text{inc}} + \mathbf{E}^{\text{reac}}; \qquad \mathbf{B}^{\text{loc}} = \mathbf{B}^{\text{inc}} + \mathbf{B}^{\text{reac}} \tag{N.3}$$

The reaction field will be proportional to the dipole moments $\mathbf{p}$ and $\mathbf{m}$, which are of course themselves determined by the local field. By (10.46), (10.47), (10.84), (10.85), (10.88) and (10.89), the total field excited by the dipoles can be written as:

$$\mathbf{E}^{\text{tot}}(x, y, z) = \frac{1}{4\pi\epsilon} \overset{\leftrightarrow}{T}_{ee}^{\text{wg}}(x_0, y_0, x, y, z) \cdot \mathbf{p} - \frac{j\omega\mu}{4\pi} \overset{\leftrightarrow}{T}_{em}^{\text{wg}}(x_0, y_0, x, y, z) \cdot \mathbf{m} \tag{N.4}$$

where the dyadics $\overset{\leftrightarrow}{T}_{ee}^{\text{wg}}$ and $\overset{\leftrightarrow}{T}_{em}^{\text{wg}}$ are given by

$$\overset{\leftrightarrow}{T}_{ee}^{\text{wg}}(x_0, y_0, x, y, z) = -j\omega\epsilon\pi \sum_{m>0} \boldsymbol{\mathcal{E}}_{\pm m}(x, y)\boldsymbol{\mathcal{E}}_{\mp m}(x_0, y_0)e^{\mp\gamma_m z} \qquad (z \gtrless 0) \tag{N.5}$$

$$\overset{\leftrightarrow}{T}_{em}^{\text{wg}}(x_0, y_0, x, y, z) = -\pi \sum_{m>0} \boldsymbol{\mathcal{E}}_{\pm m}(x, y)\boldsymbol{\mathcal{H}}_{\mp m}(x_0, y_0)e^{\mp\gamma_m z} \qquad (z \gtrless 0) \tag{N.6}$$

Likewise, the total magnetic field excited by the dipoles will be

$$\mathbf{H}^{\text{tot}}(x, y, z) = \frac{j\omega}{4\pi} \overset{\leftrightarrow}{T}_{me}^{\text{wg}}(x_0, y_0, x, y, z) \cdot \mathbf{p} + \frac{1}{4\pi} \overset{\leftrightarrow}{T}_{mm}^{\text{wg}}(x_0, y_0, x, y, z) \cdot \mathbf{m} \tag{N.7}$$

where

$$\overset{\leftrightarrow}{T}{}^{\text{wg}}_{me}(x_0, y_0, x, y, z) = -\pi \sum_{m>0} \boldsymbol{\mathcal{H}}_{\pm m}(x, y) \boldsymbol{\mathcal{E}}_{\mp m}(x_0, y_0) e^{\mp \gamma_m z} \qquad (z \gtrless 0) \qquad (\text{N.8})$$

$$\overset{\leftrightarrow}{T}{}^{\text{wg}}_{mm}(x_0, y_0, x, y, z) = j\omega\mu\pi \sum_{m>0} \boldsymbol{\mathcal{H}}_{\pm m}(x, y) \boldsymbol{\mathcal{H}}_{\mp m}(x_0, y_0) e^{\mp \gamma_m z} \qquad (z \gtrless 0) \qquad (\text{N.9})$$

To get the reaction field, we subtract from the total field that which would be produced by the dipoles in infinite homogeneous space, as given in Section D.4. We thus have

$$\mathbf{E}^{\text{reac}}(x_0, y_0, 0) = \frac{1}{4\pi\epsilon} \overset{\leftrightarrow}{T}{}^{\text{reac}}_{ee}(x_0, y_0) \cdot \mathbf{p} - \frac{j\omega\mu}{4\pi} \overset{\leftrightarrow}{T}{}^{\text{reac}}_{em}(x_0, y_0) \cdot \mathbf{m} \qquad (\text{N.10})$$

$$\mathbf{H}^{\text{reac}}(x_0, y_0, 0) = \frac{j\omega}{4\pi} \overset{\leftrightarrow}{T}{}^{\text{reac}}_{me}(x_0, y_0) \cdot \mathbf{p} + \frac{1}{4\pi} \overset{\leftrightarrow}{T}{}^{\text{reac}}_{mm}(x_0, y_0) \cdot \mathbf{m} \qquad (\text{N.11})$$

where

$$\overset{\leftrightarrow}{T}{}^{\text{reac}}_{uv}(x_0, y_0) = \lim_{(x,y,z)\to(x_0,y_0,0)} \left[\overset{\leftrightarrow}{T}{}^{\text{wg}}_{uv}(x_0, y_0, x, y, z) - \overset{\leftrightarrow}{T}_{uv}(x_0, y_0, x, y, z) \right] \qquad (\text{N.12})$$

for $uv = ee$, em, me or mm, and the infinite-region dipole dyadics are given by (D.21), (D.22) and (D.25).

In general, it is difficult to obtain simple analytical expressions for the reaction field. In an arbitrary waveguide, we may not even have analytical expressions for the mode fields and propagation constants, and the convergence of the infinite series (N.5), (N.6), (N.8) and (N.9) will in general be slow. Indeed, as $(x, y, z) \to (x_0, y_0, 0)$ these series will diverge, corresponding to the singularities in (D.21), (D.22) and (D.25) as $R \to 0$. These singularities must cancel in the computation of (N.12), but in practice the numerical evaluation is difficult. In the special case of a rectangular waveguide we can evaluate (N.12) in relatively simple form by making repeated use of image theory (Section D.5). Consider an electric dipole with moment $\mathbf{p}$ as shown in Figure N.1. In each of the four walls, an image dipole is created. In turn, these images are themselves imaged, and so on, resulting in an infinite array of dipoles as shown in the figure, which radiate in infinite space (without the conducting walls present). The reaction field is then the sum of the fields of all the imaged dipoles, not including the field of the original dipole inside the waveguide. If the original dipole is not too close to any of the walls, this infinite sum will converge reasonably rapidly.

For other kinds of waveguide, however, the image approach is not applicable. In those cases, it is common practice to approximate $\overset{\leftrightarrow}{T}{}^{\text{reac}}_{uv}(x_0, y_0)$ by keeping only the propagating mode(s) in (N.5), (N.6), (N.8) and (N.9), and discarding the infinite set of cutoff modes, assuming that they will contribute the singular near field that will essentially cancel the singular term $\overset{\leftrightarrow}{T}_{uv}$, leaving us with

$$\overset{\leftrightarrow}{T}{}^{\text{reac}}_{ee}(x_0, y_0) \simeq -j\omega\epsilon\pi \sum_{m>0, \gamma_m=j\beta_m} \boldsymbol{\mathcal{E}}_{\pm m}(x_0, y_0) \boldsymbol{\mathcal{E}}_{\mp m}(x_0, y_0) \qquad (z = 0^\pm) \qquad (\text{N.13})$$

$$\overset{\leftrightarrow}{T}{}^{\text{reac}}_{em}(x_0, y_0) \simeq -\pi \sum_{m>0, \gamma_m=j\beta_m} \boldsymbol{\mathcal{E}}_{\pm m}(x_0, y_0) \boldsymbol{\mathcal{H}}_{\mp m}(x_0, y_0) \qquad (z = 0^\pm) \qquad (\text{N.14})$$

$$\overset{\leftrightarrow}{T}{}^{\text{reac}}_{me}(x_0, y_0) \simeq -\pi \sum_{m>0, \gamma_m=j\beta_m} \boldsymbol{\mathcal{H}}_{\pm m}(x_0, y_0) \boldsymbol{\mathcal{E}}_{\mp m}(x_0, y_0) \qquad (z = 0^\pm) \qquad (\text{N.15})$$

$$\overset{\leftrightarrow}{T}{}^{\text{reac}}_{mm}(x_0, y_0) \simeq j\omega\mu\pi \sum_{m>0, \gamma_m=j\beta_m} \boldsymbol{\mathcal{H}}_{\pm m}(x_0, y_0) \boldsymbol{\mathcal{H}}_{\mp m}(x_0, y_0) \qquad (z = 0^\pm) \qquad (\text{N.16})$$

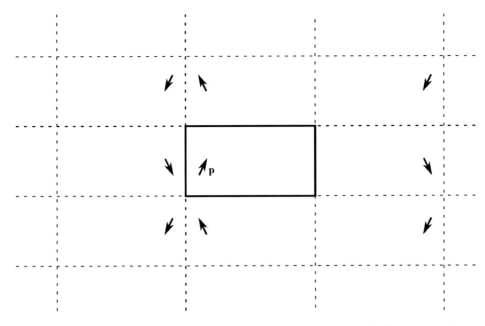

Figure N.1 Images of an electric dipole due to the boundary walls of a hollow rectangular waveguide.

The limit point can be taken at either $z = 0^+$ or 0^- so long as it is done consistently.

However the dyadics $\overset{\leftrightarrow}{T}{}_{uv}^{\text{reac}}(x_0, y_0)$ are computed, we can now use (N.10) and (N.11) with (N.1)-(N.3) to obtain the system of equations

$$\mathbf{p} = \overset{\leftrightarrow}{\alpha}_e \cdot \left[\epsilon_0 \mathbf{E}^{\text{inc}} + \frac{1}{4\pi}\overset{\leftrightarrow}{T}{}_{ee}^{\text{reac}}(x_0, y_0) \cdot \mathbf{p} - \frac{j\omega\mu_0\epsilon_0}{4\pi}\overset{\leftrightarrow}{T}{}_{em}^{\text{reac}}(x_0, y_0) \cdot \mathbf{m}\right]$$

$$\mathbf{m} = \overset{\leftrightarrow}{\alpha}_m \cdot \left[\mathbf{H}^{\text{inc}} + \frac{j\omega}{4\pi}\overset{\leftrightarrow}{T}{}_{me}^{\text{reac}}(x_0, y_0) \cdot \mathbf{p} - \frac{1}{4\pi}\overset{\leftrightarrow}{T}{}_{mm}^{\text{reac}}(x_0, y_0) \cdot \mathbf{m}\right] \qquad (N.17)$$

Solving these equations for $\mathbf{p}$ and $\mathbf{m}$ is straightforward but algebraically cumbersome in the general case. If there is sufficient symmetry that $\overset{\leftrightarrow}{T}{}_{me}^{\text{reac}}(x_0, y_0) = \overset{\leftrightarrow}{T}{}_{em}^{\text{reac}}(x_0, y_0) = 0$, we have

$$\mathbf{p} = \epsilon_0 \left[\overset{\leftrightarrow}{I} - \frac{1}{4\pi}\overset{\leftrightarrow}{\alpha}_e \cdot \overset{\leftrightarrow}{T}{}_{ee}^{\text{reac}}(x_0, y_0)\right]^{-1} \cdot \overset{\leftrightarrow}{\alpha}_e \cdot \mathbf{E}^{\text{inc}} \qquad (N.18)$$

$$\mathbf{m} = \left[\overset{\leftrightarrow}{I} + \frac{1}{4\pi}\overset{\leftrightarrow}{\alpha}_m \cdot \overset{\leftrightarrow}{T}{}_{mm}^{\text{reac}}(x_0, y_0)\right]^{-1} \cdot \overset{\leftrightarrow}{\alpha}_m \cdot \mathbf{H}^{\text{inc}} \qquad (N.19)$$

If the approximations (N.13)-(N.16) are used, the scattering parameters that are obtained as a result of (N.18) and (N.19) are the same as we obtain using the more empirical method of correction applied in Section 10.6.2.

The case of a small aperture, or a scatterer in a different environment such as a cavity resonator can be treated in a similar way, and details are left to the reader.

REFERENCES

The basic ideas used in this appendix are implicit in:

Slater (1942), Chapters 6 and 7.

Most treatments of corrections to small-obstacle and small-hole aperture theory have sought only to enforce energy conservation; see

I. Kaufman, "Calculation of radiation fields in waveguides by a principle of power balance," *IEEE Trans. Micr. Theory Tech.*, vol. 18, pp. 418-425 (1970).

C.-L. Ren, "Implementation of conservation-of-energy condition in small aperture and small obstacle theory," *IEEE Trans. Micr. Theory Tech.*, vol. 20, pp. 488-490 (1972).

L. V. Alekseychik, V. M. Gevorkyan and Yu. A. Kazantsev, "Excitation of an open dielectric resonator in a transmission line," *Radio Eng. Electron. Phys.*, vol. 17, pp. 1814-1821 (1972).

C. D. Taylor, "Electromagnetic pulse penetration through small apertures," *IEEE Trans. Electrom. Compat.*, vol. 15, pp. 17-26 (1973).

R. E. Collin, "Rayleigh scattering and power conservation," *IEEE Trans. Ant. Prop.*, vol. 29, pp. 795-798 (1981).

R. E. Collin, "Small aperture coupling between dissimilar regions," *Electromagnetics*, vol. 2, pp. 1-24 (1982).

D. K. Cheng and C.-H. Liang, "Generalised hybrid-network parameters for electromagnetic coupling between dissimilar regions through a small aperture," *IEE Proc. pt. H*, vol. 129, pp. 325-332 (1982).

J. R. Mautz and R. F. Harrington, "An admittance solution for electromagnetic coupling through a small aperture," *Appl. Sci. Res.*, vol. 40, pp. 39-69 (1983).

J. Reinert and A. F. Jacob, "Theoretical and experimental waveguide characterization of small wire scatterers," *IEEE Trans. Micr. Theory Tech.*, vol. 49, pp. 1266-1269 (2001).

as well as

Collin (1991), Chapter 7.

However, it was also recognized by a few authors that the reactive near field could be significant, for example in the establishment of resonant frequencies. See for instance

L. B. Felsen and W. K. Kahn, "Network properties of discontinuities in multi-mode circular waveguide, Part 1," *Proc. IEE (London) pt. C*, vol. 109, pp. 388-400 (1962).

A. G. Gurevich, "Ferrite ellipsoid in a waveguide," *Radio Eng. Electron. Phys.*, vol. 8, pp. 799-808 (1963).

L. B. Felsen and C. L. Ren, "Scattering by obstacles in a multimode waveguide," *Proc. IEE (London)*, vol. 113, pp. 16-26 (1966).

D. A. James, "Reflexion coefficient of a magnetized ferrite sphere in a rectangular waveguide," *Radio Electron. Engineer*, vol. 44, pp. 481-490 (1974).

The local field is analogous to that used in the determination of the bulk permittivity and permeability of a natural or artificial medium according to the Clausius-Mossotti formulas. See, for example,

Lorentz (1952), Chapter 4;
Collin (1991), Chapter 12.

O OVERLAP INTEGRALS FOR COUPLED-MODE THEORY

The overlap integrals between fields of normalized modes on different dielectric waveguides that are encountered in Chapter 12 can be transformed using the result of problem p9-4 and some of the identities found in Appendix J. We remind the reader that all regions are assumed to be nonmagnetic, so that $\mu = \mu_0$ everywhere. We assume that the permittivity of both waveguides is the same (ϵ_c) in the common cross-sectional region S_c.

For the fields of the "a" and "b" states, we choose

$$
\begin{aligned}
\mathbf{E}^a &= \boldsymbol{\mathcal{E}}_1 e^{-j\beta_1 z}; & \mathbf{H}^a &= \boldsymbol{\mathcal{H}}_1 e^{-j\beta_1 z} \\
\mathbf{E}^b &= \boldsymbol{\mathcal{E}}_{\pm 2} e^{\mp j\beta_2 z}; & \mathbf{H}^b &= \boldsymbol{\mathcal{H}}_{\pm 2} e^{\mp j\beta_2 z}
\end{aligned}
\tag{O.1}
$$

in the case when the entire cross section S is filled with the background medium for which the permittivity is ϵ_c. We require the equivalent sources

$$
\begin{aligned}
\mathbf{J}^a_{\text{eq}} &= j\omega(\epsilon_1 - \epsilon_c)\mathbf{E}^a & \text{in } S_1 \\
&= 0 & \text{in } S_c \text{ or } S_2
\end{aligned}
\tag{O.2}
$$

$$
\begin{aligned}
\mathbf{J}^b_{\text{eq}} &= j\omega(\epsilon_2 - \epsilon_c)\mathbf{E}^b & \text{in } S_2 \\
&= 0 & \text{in } S_c \text{ or } S_1
\end{aligned}
\tag{O.3}
$$

with respect to the background medium. On C_g (if present) we have $\mathbf{u}_n \times \boldsymbol{\mathcal{E}}_{1,2} = 0$.

We apply the result of problem p9-4 to the entire cross section S shown in Figure 12.5; the result is

$$
-j\left(\beta_1 \pm \beta_2\right) \int_S \boldsymbol{\mathcal{E}}_1 \times \boldsymbol{\mathcal{H}}_{\pm 2} \cdot \mathbf{u}_z \, dS = -j\omega \int_S \left(\mu_0 \boldsymbol{\mathcal{H}}_1 \cdot \boldsymbol{\mathcal{H}}_{\pm 2} + \epsilon_c \boldsymbol{\mathcal{E}}_1 \cdot \boldsymbol{\mathcal{E}}_{\pm 2}\right) dS - j\omega \int_{S_2} (\epsilon_2 - \epsilon_c)\boldsymbol{\mathcal{E}}_1 \cdot \boldsymbol{\mathcal{E}}_{\pm 2} \, dS
\tag{O.4}
$$

$$
\beta_2 C_{21} = \omega\left(U_{e12} + T_{e12}\right)
\tag{O.5}
$$

and

$$
\beta_1 C_{21} = \omega\left(U_{m12} + T_{m12}\right)
\tag{O.6}
$$

where

$$
\begin{aligned}
U_{e12} &= \frac{1}{2}\int_S \left(\epsilon_c \boldsymbol{\mathcal{E}}_{1T} \cdot \boldsymbol{\mathcal{E}}_{2T} + \mu_0 \mathcal{H}_{1z}\mathcal{H}_{2z}\right) dS \\
U_{m12} &= \frac{1}{2}\int_S \left(\mu_0 \boldsymbol{\mathcal{H}}_{1T} \cdot \boldsymbol{\mathcal{H}}_{2T} + \epsilon_c \mathcal{E}_{1z}\mathcal{E}_{2z}\right) dS
\end{aligned}
\tag{O.7}
$$

are energy overlap integrals. Repeating this procedure with the subscripts 1 and 2 interchanged gives

$$
\beta_1 C_{12} = \omega\left(U_{e12} + T_{e21}\right)
\tag{O.8}
$$

and

$$
\beta_2 C_{12} = \omega\left(U_{m12} + T_{m21}\right)
\tag{O.9}
$$

where C_{12} is defined in (12.57). In the above relations, the quantities T_{e12} etc. are defined in (12.60).

The quantities U_{e12} and U_{m12} can be eliminated from (O.5), (O.6), (O.8) and (O.9) to give

$$\beta_1 C_{12} - \beta_2 C_{21} = \omega \left(T_{e21} - T_{e12} \right); \qquad \beta_2 C_{12} - \beta_1 C_{21} = \omega \left(T_{m21} - T_{m12} \right) \qquad (O.10)$$

This means that the power overlap integrals can be expressed in terms of the polarization overlap integrals:

$$
\begin{aligned}
C_{21} &= \frac{\omega}{\beta_1^2 - \beta_2^2} \left[\beta_2 \left(T_{e21} - T_{e12} \right) - \beta_1 \left(T_{m21} - T_{m12} \right) \right] \\
C_{12} &= \frac{\omega}{\beta_2^2 - \beta_1^2} \left[\beta_1 \left(T_{e12} - T_{e21} \right) - \beta_2 \left(T_{m12} - T_{m21} \right) \right]
\end{aligned}
\qquad (O.11)
$$

and as a consequence

$$N_{12} = \frac{C_{12} + C_{21}}{2} = \frac{\omega}{\beta_1 - \beta_2} \left(T_{e21} - T_{e12} - T_{m21} + T_{m12} \right) \qquad (O.12)$$

In the special case when $\beta_1 = \beta_2 = \beta$, we have

$$C_{12} - C_{21} = \frac{\omega}{\beta} \left(T_{e21} - T_{e12} \right) = \frac{\omega}{\beta} \left(T_{m21} - T_{m12} \right) \qquad (O.13)$$

REFERENCES

See the references for Chapter 12 for similar derivations to those given here.

Bibliography

This listing contains books to which the reader can refer for a deeper insight or alternative treatment of the topics covered in this text. Although by no means exhaustive, this list has grown over the years to a point where it rivals Jorge Luis Borges' *The Library of Babel*. Readers will no doubt add their own favorites to these, letting the size of the collection tend to infinity as Borges preferred to dream about his library.

These books are cited in the references at the end of each chapter in an abbreviated form, e.g.:

Abramowitz and Stegun (1964), Chapter 6.

In a few cases, important books in languages other than English are cited when no translation is available, for the edification of the fluent or the brave.

M. Abramowitz and I. A. Stegun (eds.), *Handbook of Mathematical Functions*. Washington, DC: US Government Printing Office, 1964.

S. F. Adam, *Microwave Theory and Applications*. Englewood Cliffs, NJ: Prentice-Hall, 1969.

M. J. Adams, *An Introduction to Optical Waveguides*. New York: Wiley, 1981.

R. B. Adler, L. J. Chu and R. M. Fano, *Electromagnetic Energy Transmission and Radiation*. New York: Wiley, 1960.

G. P. Agrawal, *Nonlinear Fiber Optics*. San Diego: Academic Press, 2001.

J. L. Altman, *Microwave Circuits*. Princeton, NJ: Van Nostrand, 1964.

J. A. Arnaud, *Beam and Fiber Optics*. New York: Academic Press, 1976.

H. A. Atwater, *Introduction to Microwave Theory*. Malabar, FL: Krieger, 1981.

A. J. Baden Fuller, *Ferrites at Microwave Frequencies*. London: Peter Peregrinus, 1989.

E. J. Baghdady (ed.), *Lectures on Communication System Theory*. New York: McGraw-Hill, 1961.

I. Bahl and P. Bhartia, *Microwave Solid State Circuit Design*. New York: Wiley, 1988.

N. Balabanian and T. A. Bickart, *Electrical Network Theory*. New York: Wiley, 1969.

C. A. Balanis, *Antenna Theory: Analysis and Design*. New York: Harper and Row, 1982.

J. A. Ball, *Algorithms for RPN Calculators*. New York: Wiley, 1978.

P. Bamberg and S. Sternberg, *A Course in Mathematics for Students of Physics*, vol. 2. Cambridge, UK: Cambridge University Press, 1990.

A. A. Barybin and V. A. Dmitriev, *Modern Electrodynamics and Coupled-Mode Theory: Application to Guided-Wave Optics*. Princeton, NJ: Rinton Press, 2002.

L. V. Bewley, *Traveling Waves on Transmission Systems*. New York: Dover, 1963.

B. Bhat and S. K. Koul, *Analysis, Design and Applications of Fin Lines*. Norwood, MA: Artech House, 1987.

M. Born and E. Wolf, *Principles of Optics*. Oxford: Pergamon Press, 1975.

J. P. Boyd, *Solving Transcendental Equations*. Philadelphia: SIAM, 2014.

L. M. Brekhovskikh, *Waves in Layered Media*. New York: Academic Press, 1960.

H. Bremmer, *Terrestrial Radio Waves*. Amsterdam: Elsevier, 1949.

L. Brillouin, *Wave Propagation and Group Velocity*. New York: Academic Press, 1960.

W. M. Brown, *Analysis of Linear Time-Invariant Systems*. New York: McGraw-Hill, 1963.

K. G. Budden, *The Wave-Guide Mode Theory of Wave Propagation*. London: Logos Press, 1961.

P. F. Byrd and M. D. Friedman, *Handbook of Elliptic Integrals for Engineers and Scientists*. Berlin: Springer-Verlag, 1971.

C. Caloz and T. Itoh, *Electromagnetic Metamaterials: Transmission Line Theory and Microwave Applications*. Hoboken, NJ: Wiley, 2006.

H. J. Carlin and A. B. Giordano, *Network Theory*. Englewood Cliffs, NJ: Prentice-Hall, 1964.

H. S. Carslaw and J. C. Jaeger, *Operational Methods in Applied Mathematics*. New York: Dover, 1963.

E. Chiprout and M. S. Nakhla, *Asymptotic Waveform Evaluation and Moment Matching for Interconnect Analysis*. Boston: Kluwer, 1994.

L. Cohen, *Time-Frequency Analysis*. Englewood Cliffs, NJ: Prentice-Hall, 1995.

R. E. Collin, *Field Theory of Guided Waves*, second edition. New York: IEEE Press, 1991.

R. E. Collin, *Foundations for Microwave Engineering*, second edition. New York: IEEE Press, 2001.

G. Conciauro, M. Guglielmi and R. Sorrentino, *Advanced Modal Analysis*. Chichester, UK: Wiley, 2000.

R. Courant and D. Hilbert, *Methods of Mathematical Physics*, vol. 1. New York: Interscience, 1953.

B. Davies, *Integral Transforms and Their Applications*. New York: Springer, 1978.

P. Delogne, *Leaky Feeders and Subsurface Radio Communication*. Stevenage, UK: Peter Peregrinus, 1982.

C. A. Desoer and E. S. Kuh, *Basic Circuit Theory*. New York: McGraw-Hill, 1969.

A. E. Dubinov, I. D. Dubinova and S. K. Saikov, *W-Funktsiya Lamberta i ee Primenenie v Matematicheskikh Zadachakh Fiziki*. Sarov, Russia: Rossiiskii Federal'nyi Yadernyi Tsentr VNIIEF, 2006.

L. N. Dworsky, *Modern Transmission Line Theory and Applications*. New York: Wiley, 1979.

T. C. Edwards, *Foundations for Microstrip Circuit Design*. New York: Wiley, 1981.

G. V. Eleftheriades and K. G. Balmain, *Negative-Refraction Metamaterials*. Hoboken, NJ: Wiley, 2005.

R. S. Elliott, *An Introduction to Guided Waves and Microwave Circuits*. Englewood Cliffs, NJ: Prentice-Hall, 1993.

N. Engheta and R. W. Ziolkowski, *Metamaterials: Physics and Engineering Explorations*. Hoboken, NJ: Wiley, 2006.

A. Erdélyi (ed.), *Tables of Integral Transforms*, vol. I. New York: McGraw-Hill, 1954.

N. Faché, F. Olyslager and D. De Zutter, *Electromagnetic and Circuit Modelling of Multiconductor Transmission Lines*. Oxford: Clarendon Press, 1993.

L. B. Felsen and N. Marcuvitz, *Radiation and Scattering of Waves*. Englewood Cliffs, NJ: Prentice-Hall, 1973.

N. H. Fletcher and T. D. Rossing, *The Physics of Musical Instruments*. New York: Springer-Verlag, 1991.

S. Frankel, *Multiconductor Transmission Line Analysis*. Dedham, MA: Artech House, 1977.

L. Franks, *Signal Theory*. Englewood Cliffs, NJ: Prentice-Hall, 1969.

P. R. Garabedian, *Partial Differential Equations*. New York: Wiley, 1964.

F. E. Gardiol, *Lossy Transmission Lines*. Norwood, MA: Artech House, 1987.

R. Garg, I. Bahl and M. Bozzi, *Microstrip Lines and Slotlines* (third edition). Boston: Artech House, 2013.

R. N. Ghose, *Microwave Circuit Theory and Analysis*. New York: McGraw-Hill, 1963.

D. Gloge (ed.), *Optical Fiber Technology*. New York: IEEE Press, 1976.

I. S. Gradshteyn and I. M. Ryzhik, *Table of Integrals, Series, and Products*. New York: Academic Press, 1980.

B. Gross and E. P. Braga, *Singularities of Linear System Functions*. Amsterdam: Elsevier, 1961.

E. A. Guillemin, *Synthesis of Passive Networks*. New York: Wiley, 1957.

M. A. R. Gunston, *Microwave Transmission-Line Impedance Data*. London: Van Nostrand Reinhold, 1972.

K. C. Gupta, R. Garg and R. Chadha, *Computer-Aided Design of Microwave Circuits*. Dedham, MA: Artech House, 1981.

K. C. Gupta and A. Singh, *Microwave Integrated Circuits*. New York: Wiley, 1974.

R. W. Hamming, *Numerical Methods for Scientists and Engineers*. New York: Dover, 1986.

R. F. Harrington, *Time-Harmonic Electromagnetic Fields*. New York: McGraw-Hill, 1961.

A. F. Harvey, *Microwave Engineering*. London: Academic Press, 1963.

H. A. Haus, *Waves and Fields in Optoelectronics*. Englewood Cliffs, NJ: Prentice-Hall, 1984.

H. A. Haus and R. B. Adler, *Circuit Theory of Noisy Networks*. New York: Wiley, 1959.

J. Helszajn, *Nonreciprocal Microwave Junctions and Circulators*. New York: Wiley, 1975.

J. Helszajn, *Microwave Planar Passive Circuits and Filters*. Chichester, UK: Wiley, 1994.

P. Henrici, *Computational Analysis with the HP-25 Pocket Calculator*. New York: Wiley, 1977.

W. Hilberg, *Electrical Characteristics of Transmission Lines*. Dedham, MA: Artech House, 1979.

R. K. Hoffmann, *Handbook of Microwave Integrated Circuits*. Norwood, MA: Artech House, 1987.

Huang Hung-Chia, *Coupled-Mode Theory*. Utrecht, The Netherlands: VNU Science Press, 1984.

T. S. Huang and R. R. Parker, *Network Theory: An Introductory Course*. Reading, MA: Addison-Wesley, 1971.

M. E. Il'chenko (ed.), *Dielektricheskie Rezonatory*. Moscow: Radio i Svyaz', 1989.

M. E. Il'chenko and E. V. Kudinov, *Ferritovye i Dielektricheskie Rezonatory*. Kiev: Izdat. Kievsk. Univ., 1973.

A. S. Ilyinsky, G. Ya. Slepyan and A. Ya. Slepyan, *Propagation, Scattering and Dissipation of Electromagnetic Waves*. Stevenage, UK: Peter Peregrinus, 1993.

ITT Staff, *Reference Data for Radio Engineers*. Indianapolis: Howard W. Sams, 1975.

J. D. Jackson, *Classical Electrodynamics*. New York: Wiley, 1975, second edition.

M. Javid and P. M. Brown, *Field Analysis and Electromagnetics*. New York: McGraw-Hill, 1963.

C. T. A. Johnk, *Engineering Electromagnetic Fields and Waves*. New York: Wiley, 1988.

C. C. Johnson, *Field and Wave Electrodynamics*. New York: McGraw-Hill, 1965.

L. B. W. Jolley, *Summation of Series*. New York: Dover, 1961.

D. S. Jones, *The Theory of Electromagnetism*. Oxford: Pergamon Press, 1964.

D. Kajfez and P. Guillon, *Dielectric Resonators*. Oxford, MS: Vector Fields, 1990.

C. Kao (ed.), *Optical Fiber Technology II*. New York: IEEE Press, 1981.

N. S. Kapany and J. J. Burke, *Optical Waveguides*. New York: Academic Press, 1974.

J. J. Karakash, *Transmission Lines and Filter Networks*. New York: Macmillan, 1950.

B. Z. Katsenelenbaum, L. Mercader del Rio, M. Pereyaslavets, M. Sorolla Ayza and M. Thumm, *Theory of Nonuniform Waveguides*. London: The Institution of Electrical Engineers, 1998.

D. M. Kerns and R. W. Beatty, *Basic Theory of Waveguide Junctions and Introductory Microwave Network Analysis*. Oxford: Pergamon Press, 1967.

R. W. P. King, *Transmission-Line Theory*. New York: Dover, 1965.

R. W. P. King, H. R. Mimno and A. H. Wing, *Transmission Lines, Antennas and Wave Guides*. New York: Dover, 1965.

J. Klapper, *Selected Papers on Frequency Modulation*. New York: Dover, 1970.

L. Koenigsberger, *Hermann von Helmholtz*. New York: Dover, 1965.

V. V. Komarov, *Handbook of Dielectric and Thermal Properties of Materials at Microwave Frequencies*. Boston: Artech House, 2012.

M. Koshiba, *Optical Waveguide Analysis*. New York: McGraw-Hill, 1992.

K. Kurokawa, *An Introduction to the Theory of Microwave Circuits*. New York: Academic Press, 1969.

P. I. Kuznetsov and R. L. Stratonovich, *Waves in Multiconductor Transmission Lines*. New York: Macmillan, 1964.

A. Lakhtakia, V. K. Varadan and V. V. Varadan, *Time-Harmonic Electromagnetic Fields in Chiral Media*. Berlin: Springer-Verlag, 1989.

I. Lebedev, *Microwave Engineering*. Moscow: Mir, 1973.

N. N. Lebedev, *Special Functions and Their Applications*. New York: Dover, 1972.

D. L. Lee, *Electromagnetic Principles of Integrated Optics*. New York: Wiley, 1986.

L. Lewin, *Theory of Waveguides*. London: Newnes-Butterworths, 1975.

L. Lewin, D. C. Chang and E. F. Kuester *Electromagnetic Waves and Curved Structures*. Stevenage, UK: Peter Peregrinus, 1977.

D. K. Linkhart, *Microwave Circulator Design*. Norwood, MA: Artech House, 1989.

J. T. Londergan, J. P. Carini and D. P. Murdock, *Binding and Scattering in Two-Dimensional Systems*. Berlin: Springer-Verlag, 1999.

H. A. Lorentz, *The Theory of Electrons*. New York: Dover, 1952.

W. H. Louisell, *Coupled Mode and Parametric Electronics*. New York: Wiley, 1960.

S. A. Maas, *The RF and Microwave Circuit Design Cookbook*. Boston: Artech House, 1998.

L. M. Magid, *Electromagnetic Fields, Energy, and Waves*. New York: Wiley, 1972.

D. Marcuse (ed.), *Integrated Optics*. IEEE Press, 1973.

D. Marcuse, *Theory of Dielectric Optical Waveguides*. New York: Academic Press, 1974.

D. Marcuse, *Light Transmission Optics*. Princeton, NJ: Van Nostrand Reinhold, 1982.

N. Marcuvitz (ed.), *Waveguide Handbook*. New York: Dover, 1965.

J. Martin, *Telecommunications and the Computer*. Englewood Cliffs, NJ: Prentice-Hall, 1969.

J. Martin, *Future Developments in Telecommunications*. Englewood Cliffs, NJ: Prentice-Hall, 1971.

B. Mashkovtsev, K. N. Tsibizov and B. F. Emelin, *Teoriya Volnovodov*. Moscow: Nauka, 1966.

R. E. Matick, *Transmission Lines for Digital and Communication Networks*. New York: McGraw-Hill, 1969.

A. Matsumoto (ed.), *Microwave Filters and Circuits*. New York: Academic Press, 1970.

G. L. Matthaei, L. Young and E. M. T. Jones (eds.), *Microwave Filters, Impedance-Matching Networks, and Coupling Structures*. New York: McGraw-Hill, 1964.

A. R. Mickelson, *Guided Wave Optics*. New York: Van Nostrand Reinhold, 1993.

K. A. Milton and J. Schwinger, *Electromagnetic Radiation: Variational Methods, Waveguides and Accelerators*. Berlin: Springer-Verlag, 2006.

R. Mittra and S. W. Lee, *Analytical Techniques in the Theory of Guided Waves*. New York: Macmillan, 1971.

C. G. Montgomery, *Technique of Microwave Measurements*. New York: Dover, 1966.

C. G. Montgomery, R. H. Dicke and E. M. Purcell (eds.), *Principles of Microwave Circuits*. New York: Dover, 1965.

T. Moreno, *Microwave Transmission Design Data*. New York: Dover, 1958.

P. M. Morse and K. U. Ingard, *Theoretical Acoustics*. Princeton, NJ: Princeton University Press, 1968.

M. Mrozowski, *Guided Electromagnetic Waves: Properties and Analysis*. Taunton, UK: Research Studies Press, 1997.

E. I. Nefedov and A. T. Fialkovskii, *Poloskovye Linii Peredachi*. Moscow: Nauka, 1980.

R. W. Newcomb, *Linear Multiport Synthesis*. New York: McGraw-Hill, 1966.

F. Oberhettinger and L. Badii, *Tables of Laplace Transforms*. New York: Springer-Verlag, 1973.

F. W. J. Olver, *Asymptotics and Special Functions*. New York: Academic Press, 1974.

F. Olyslager, *Electromagnetic Waveguides and Transmission Lines*. Oxford: Clarendon Press, 1999.

K. E. Oughstun and G. C. Sherman, *Electromagnetic Pulse Propagation in Causal Dielectrics*. Berlin: Springer-Verlag, 1994.

C. R. Paul, *Introduction to Electromagnetic Compatibility*. New York: Wiley, 1992.

C. R. Paul, *Analysis of Multiconductor Lines*. New York: Wiley, 1994.

L. T. Pillage, R. A. Rohrer and C. Visweswariah, *Electronic Circuit and System Simulation Methods*. New York: McGraw-Hill, 1995.

G. Pólya and G. Szegö, *Isoperimetric Inequalities in Mathematical Physics*. Princeton, NJ: Princeton University Press, 1951.

A. M. Portis, *Electromagnetic Fields: Sources and Media*. New York: Wiley, 1978.

D. M. Pozar, *Microwave Engineering*. Hoboken, NJ: Wiley, 2005.

W. H. Press, S. A. Teukolsky, W. T. Vettering and B. P. Flannery, *Numerical Recipes in C*. Cambridge, UK: Cambridge University Press, 1992.

S. Ramo, J. R. Whinnery and T. van Duzer, *Fields and Waves in Communication Electronics*. New York: Wiley, 1994.

H. J. Reich, P. F. Ordnung, H. L. Krauss and J. G. Skalnik, *Microwave Theory and Techniques*. Princeton, NJ: Van Nostrand, 1953.

M. Remoissenet, *Waves Called Solitons*. Berlin: Springer-Verlag, 1996.

P. A. Rizzi, *Microwave Engineering: Passive Circuits*. Englewood Cliffs, NJ: Prentice-Hall, 1988.

H. E. Rowe, *Signals and Noise in Communication Systems*. Princeton, NJ: Van Nostrand, 1965.

T. Rozzi and M. Farina, *Advanced Electromagnetic Analysis of Passive and Active Planar Structures*. London: Institution of Electrical Engineers, 1999.

T. Rozzi and M. Mongiardo, *Open Electromagnetic Waveguides*. London: Institution of Electrical Engineers, 1997.

R. Rüdenberg, *Electrical Shock Waves in Power Systems.* Cambridge, MA: Harvard University Press, 1968.

T. S. Saad (ed.), *Microwave Engineer's Handbook.* Dedham, MA: Artech House, 1971.

D. M. Sazonov, A. N. Gridin and B. A. Mishustin, *Microwave Circuits.* Moscow: Mir, 1982.

S. A. Schelkunoff, *Electromagnetic Waves.* New York: Van Nostrand, 1943.

J. Schwinger and D. S. Saxon, *Discontinuities in Waveguides.* New York: Gordon and Breach, 1968.

A. Scott, *Active and Nonlinear Wave Propagation in Electronics.* New York: Wiley, 1970.

T. B. A. Senior and J. L. Volakis, *Approximate Boundary Conditions in Electromagnetics.* London: Institution of Electrical Engineers, 1995.

S. Seshu and N. Balabanian, *Linear Network Analysis.* New York: Wiley, 1959.

J. Sevick, *Transmission Line Transformers.* Newington, CT: American Radio Relay League, 1990.

V. P. Shestopalov, *Physical Foundations of the Millimeter and Submillimeter Waves Technique. Vol. 1: Open Structures.* Utrecht, The Netherlands: VSP, 1997.

V. V. Shevchenko, *Continuous Transitions in Open Waveguides.* Boulder, CO: Golem Press, 1971.

A. E. Siegman, *Lasers.* Sausalito, CA: University Science Books, 1986.

A. S. Silbergleit and Yu. I. Kopilevich, *Spectral Theory of Guided Waves.* Bristol, UK: Institute of Physics Publishing, 1996.

R. N. Simons, *Coplanar Waveguide Circuits, Components, and Systems.* New York: Wiley, 2001.

J. C. Slater, *Microwave Transmission.* New York: McGraw-Hill, 1942.

J. C. Slater, *Microwave Electronics.* New York: Van Nostrand, 1950.

A. A. Smith, *Coupling of External Electromagnetic Fields to Transmission Lines.* New York: Wiley, 1977.

A. W. Snyder and J. D. Love, *Optical Waveguide Theory.* London: Chapman and Hall, 1983.

M. S. Sodha and A. K. Ghatak, *Inhomogeneous Optical Waveguides.* New York: Plenum Press, 1977.

S. Solimeno, B. Crosignani and P. DiPorto, *Guiding, Diffraction and Confinement of Optical Radiation.* Orlando, FL: Academic Press, 1986.

F. Sporleder and H.-G. Unger, *Waveguide Tapers, Transitions and Couplers.* Stevenage, England: Peter Peregrinus, 1979.

E. L. Stiefel, *An Introduction to Numerical Mathematics.* New York: Academic Press, 1963.

G. Strang, *Introduction to Applied Mathematics.* Wellesley, MA: Wellesley-Cambridge Press, 1986.

F. G. Stremler, *Introduction to Communication Systems.* Reading, MA: Addison-Wesley, 1990.

E. D. Sunde, *Earth Conduction Effects in Transmission Systems.* New York: Dover, 1968.

J. B. Thomas, *An Introduction to Statistical Communication Theory.* New York: Wiley, 1969.

E. C. Titchmarsh, *Introduction to the Theory of Fourier Integrals.* New York: Chelsea, 1986.

S. A. Tretyakov, *Analytical Modeling in Applied Electromagentics.* Norwood, MA: Artech House, 2003.

H. C. Tuckwell, *Introduction to Theoretical Neurobiology. Volume 1: Linear Cable Theory and Dendritic Structure.* New York: Cambridge University Press, 1988.

H. Uchida, *Fundamentals of Coupled Lines and Multiwire Antennas.* Sendai, Japan: Sasaki Printing and Publishing Co., 1967.

J. Uher, J. Bornemann and U. Rosenberg, *Waveguide Components for Antenna Feed Systems: Theory and CAD.* Boston: Artech House, 1993.

H.-G. Unger, *Planar Optical Waveguides and Fibers.* Oxford: Clarendon Press, 1977.

L. A. Vainshtein (Weinstein), *Open Resonators and Open Waveguides.* Boulder, CO: Golem Press, 1969a.

L. A. Vainshtein (Weinstein), *The Theory of Diffraction and the Factorization Method.* Boulder, CO: Golem Press, 1969b.

L. A. Vainshtein (Weinstein), *Elektromagnitnye Volny.* Moscow: Radio i Svyaz', 1988.

L. A. Vainshtein (Weinstein) and D. E. Vakman, *Razdelenie Chastot v Teorii Kolebanii i Voln.* Moscow: Nauka, 1983.

D. E. Vakman, *Signals, Oscillations and Waves.* Boston: Artech House, 1998.

J. Van Bladel, *Electromagnetic Fields,* second ed. Hoboken, NJ: Wiley, 2007.

B. van der Pol and H. Bremmer, *Operational Calculus.* New York: Chelsea, 1987.

E. F. Vance, *Coupling to Shielded Cables.* New York: Wiley, 1978.

C. Vassallo, *Théorie des Guides d'Ondes Électromagnétiques.* Paris: Eyrolles, 1985, tomes 1 et 2.

C. Vassallo, *Optical Waveguide Concepts.* Amsterdam: Elsevier, 1991.

V. F. Vzyatyshev, *Dielektricheskie Volnovody.* Moscow: Sovetskoe Radio, 1970.

R. A. Waldron, *Theory of Guided Electromagnetic Waves.* London: Van Nostrand Reinhold, 1969.

D. A. Watkins, *Topics in Electromagnetic Theory.* New York: Wiley, 1958.

G. N. Watson, *A Treatise on the Theory of Bessel Functions.* Cambridge, UK: Cambridge University Press, 1966.

W. L. Weeks, *Antenna Engineering.* New York: McGraw-Hill, 1968.

L. A. Weinstein - see L. A. Vainshtein.

I. Wolff, *Coplanar Microwave Integrated Circuits.* Hoboken, NJ: Wiley, 2006.

T. T. Y. Wong, *Fundamentals of Distributed Amplification.* Boston: Artech House, 1993.

L. Young (ed.), *Parallel Coupled Lines and Directional Couplers.* Dedham, MA: Artech House, 1972.

L. A. Zadeh and C. A. Desoer, *Linear System Theory.* New York: McGraw-Hill, 1963.

Index